Aspects of Multivariate Statistical Theory

Aspects of Multivariate Statistical Theory

ROBB J. MUIRHEAD

Senior Statistical Scientist
Pfizer Global Research and Development
New London, Connecticut

WILEY-INTERSCIENCE

A JOHN WILEY & SONS, INC., PUBLICATION

Published by John Wiley & Sons, Inc., Hoboken, New Jersey.
Published simultaneously in Canada.

For general information on our other products and services or for technical support, please contact our Customer Care Department within the U.S. at (800) 762-2974, outside the U.S. at (317) 572-3993 or fax (317) 572-4002.

Wiley also publishes its books in a variety of electronic formats. Some content that appears in print may not be available in electronic format. For information about Wiley products, visit our web site at www.wiley.com.

Library of Congress Cataloging-in-Publication is available.

ISBN-13 978-0-471-76985-9
ISBN-10 0-471-76985-1

10 9 8 7 6 5 4 3

To
Nan and Bob
and
Maria and Mick

Preface

This book has grown out of lectures given in first- and second-year graduate courses at Yale University and the University of Michigan. It is designed as a text for graduate level courses in multivariate statistical analysis, and I hope that it may also prove to be useful as a reference book for research workers interested in this area.

Any person writing a book in multivariate analysis owes a great debt to T. W. Anderson for his 1958 text, *An Introduction to Multivariate Statistical Analysis*, which has become a classic in the field. This book synthesized various subareas for the first time in a broad overview of the subject and has influenced the direction of recent and current research in theoretical multivariate analysis. It is also largely responsible for the popularity of many of the multivariate techniques and procedures in common use today.

The current work builds on the foundation laid by Anderson in 1958 and in large part is intended to describe some of the developments that have taken place since then. One of the major developments has been the introduction of zonal polynomials and hypergeometric functions of matrix argument by A. T. James and A. G. Constantine. To a very large extent these have made possible a unified study of the noncentral distributions that arise in multivariate analysis under the standard assumptions of normal sampling. This work is intended to provide an introduction to some of this theory.

Most books of this nature reflect the author's tastes and interests, and this is no exception. The main focus of this work is on distribution theory, both exact and asymptotic. Multivariate techniques depend heavily on latent roots of random matrices; all of the important latent root distributions are introduced and approximations to them are discussed. In testing problems the primary emphasis here is on likelihood ratio tests and the distributions of likelihood ratio test statistics. The noncentral distributions are needed to evaluate power functions. Of course, in the absence of "best" tests simply computing power functions is of little interest; what is needed is a comparison of powers of competing tests over a wide range of alternatives. Wherever possible the results of such power studies in the literature are discussed. It should be mentioned, however, that although the emphasis is on likelihood ratio statistics, many of the techniques introduced here for studying and approximating their distributions can be applied to other test statistics as well.

A few words should be said about the material covered in the text. Matrix theory is used extensively, and matrix factorizations are extremely important. Most of the relevant material is reviewed in the Appendix, but some results also appear in the text and as exercises. Chapter 1 introduces the multivariate normal distribution and studies its properties, and also provides an introduction to spherical and elliptical distributions. These form an important class of non-normal distributions which have found increasing use in robustness studies where the aim is to determine how sensitive existing multivariate techniques are to multivariate normality assumptions. In Chapter 2 many of the Jacobians of transformations used in the text are derived, and a brief introduction to invariant measures via exterior differential forms is given. A review of matrix Kronecker or direct products is also included here. The reason this is given at this point rather than in the Appendix is that very few of the students that I have had in multivariate analysis courses have been familiar with this product, which is widely used in later work. Chapter 3 deals with the Wishart and multivariate beta distributions and their properties. Chapter 4, on decision-theoretic estimation of the parameters of a multivariate normal distribution, is rather an anomaly. I would have preferred to incorporate this topic in one of the other chapters, but there seemed to be no natural place for it. The material here is intended only as an introduction and certainly not as a review of the current state of the art. Indeed, only admissibility (or rather, inadmissibility) results are presented, and no mention is even made of Bayes procedures. Chapter 5 deals with ordinary, multiple, and partial correlation coefficients. An introduction to invariance theory and invariant tests is given in Chapter 6. It may be wondered why this topic is included here in view of the coverage of the relevant basic material in the books by E. L. Lehmann, *Testing Statistical Hypotheses*, and T. S. Ferguson, *Mathematical Statistics: A Decision Theoretic Approach*. The answer is that most of the students that have taken my multivariate analysis courses have been unfamiliar with invariance arguments, although they usually meet them in subsequent courses. For this reason I have long felt that an introduction to invariant tests in a multivariate text would certainly not be out of place.

Chapter 7 is where this book departs most significantly from others on multivariate statistical theory. Here the groundwork is laid for studying the noncentral distribution theory needed in subsequent chapters, where the emphasis is on testing problems in standard multivariate procedures. Zonal polynomials and hypergeometric functions of matrix argument are introduced, and many of their properties needed in later work are derived. Chapter 8 examines properties, and central and noncentral distributions, of likelihood ratio statistics used for testing standard hypotheses about covariance matrices and mean vectors. An attempt is also made here to explain what happens if these tests are used and the underlying distribution is non-normal. Chapter 9 deals with the procedure known as principal components, where much attention is focused on the latent roots of the sample covariance matrix. Asymptotic distributions of these roots are obtained and are used in various inference problems. Chapter 10 studies the multivariate general linear model and the distribution of latent roots and functions of them used for testing the general linear hypothesis. An introduction to discriminant analysis is also included here, although the coverage is rather brief. Finally, Chapter 11 deals with the problem of testing independence between a number of sets of variables and also with canonical correlation analysis.

The choice of the material covered is, of course, extremely subjective and limited by space requirements. There are areas that have not been mentioned and not everyone will agree with my choices; I do believe, however, that the topics included form the core of a reasonable course in classical multivariate analysis. Areas which are not covered in the text include factor analysis, multiple time series, multidimensional scaling, clustering, and discrete multivariate analysis. These topics have grown so large that there are now separate books devoted to each. The coverage of classification and discriminant analysis also is not very extensive, and no mention is made of Bayesian approaches; these topics have been treated in depth by Anderson and by Kshirsagar, *Multivariate Analysis*, and Srivastava and Khatri, *An Introduction to Multivariate Statistics*, and a person using the current work as a text may wish to supplement it with material from these references.

This book has been planned as a text for a two-semester course in multivariate statistical analysis. By an appropriate choice of topics it can also be used in a one-semester course. One possibility is to cover Chapters 1, 2, 3, 5, and possibly 6, and those sections of Chapters 8, 9, 10 and 11 which do not involve noncentral distributions and consequently do not utilize the theory developed in Chapter 7. The book is designed so that for the most part these sections can be easily identified and omitted if desired. Exercises are provided at the end of each chapter. Many of these deal with points which are alluded to in the text but left unproved. A few words are also in order concerning the Bibliography. I have not felt it necessary to cite the source of every result included here. Many of the original results due to such people as Wilks, Hotelling, Fisher, Bartlett, Wishart, and Roy have become so well known that they are now regarded as part of the folklore of multivariate analysis. T. W. Anderson's book provides an extensive bibliography of work prior to 1958, and my references to early work are indiscriminate at best. I have tried to be much more careful concerning references to the more recent work presented in this book, particularly in the area of distribution theory. No doubt some references have been missed, but I hope that the number of these is small. Problems which have been taken from the literature are for the most part not referenced unless the problem is especially complex or the reference itself develops interesting extensions and applications that the problem does not cover.

This book owes much to many people. My teachers, A. T. James and A. G. Constantine, have had a distinctive influence on me and their ideas are in evidence throughout, and especially in Chapters 2, 3, 7, 8, 9, 10, and 11. I am indebted to them both. Many colleagues and students have read, criticized, and corrected various versions of the manuscript. J. A. Hartigan read the first four chapters, and Paul Sampson used parts of the first nine chapters for a course at the University of Chicago; I am grateful to both for their extensive comments, corrections, and suggestions. Numerous others have also helped to weed out errors and have influenced the final version; especially deserving of thanks are D. Bancroft, W. J. Glynn, J. Kim, M. Kramer, R. Kuick, D. Marker, and J. Wagner. It goes without saying that the responsibility for all remaining errors is mine alone. I would greatly appreciate being informed about any that are found, large and small.

A number of people tackled the unenviable task of typing various parts and revisions of the manuscript. For their excellent work and their patience with my handwriting I would like to thank Carol Hotton, Terri Lomax Hunter, Kelly Kane, and Deborah Swartz.

ROBB J. MUIRHEAD

Ann Arbor, Michigan
February 1982

Contents

Tables

Commonly Used Notation

R^m	Euclidean space of dimension m consisting of $m\times 1$ real vectors
det	determinant
S_m	unit sphere in R^m centered at the origin
$\wedge$	exterior or wedge product
Re	real part
tr	trace
etr	exp tr
$A>0$	A is positive definite
$V_{m,n}$	Stiefel manifold of $n\times m$ matrices with orthonormal columns
$O(m)$	Group of orthogonal $m\times m$ matrices
$\otimes$	direct or Kronecker product
$\lVert\mathbf{X}\rVert$	norm of $\mathbf{X}$ i.e. $\left(\sum_{i=1}^{m} X_i^2\right)^{\frac{1}{2}}$
$\mathcal{S}_m$	set of all $m\times m$ positive definite matrices
$\mathcal{G}\ell(m,R)$	general linear group of $m\times m$ nonsingular real matrices
$\mathcal{A}\ell(m,R)$	affine group $\{(B,\mathbf{c});\ B\in\mathcal{G}\ell(m,R);\mathbf{c}\in R^m\}$
$C_\kappa(\cdot)$	zonal polynomial
$M_\kappa(\cdot)$	monomial symmetric function
$(\alpha)_\kappa$	generalized hypergeometric coefficient
$\lVert X\rVert$	max of absolute values of latent roots of the matrix X
$\binom{\kappa}{\sigma}$	generalized binomial coefficient
$L_\kappa^\gamma(\cdot)$	generalized Laguerre polynomial
$N_m(\boldsymbol{\mu},\Sigma)$	m-variate normal distribution with mean $\boldsymbol{\mu}$ and covariance matrix Σ
$W_m(n,\Sigma)$	$m\times m$ matrix variate Wishart distribution with n degrees of freedom and covariance matrix Σ
$\text{Beta}_m(\alpha,\beta)$	$m\times m$ matrix variate beta distribution with parameters α,β

Aspects of Multivariate Statistical Theory

CHAPTER 1

The Multivariate Normal and Related Distributions

1.1. INTRODUCTION

The basic, central distribution and building block in classical multivariate analysis is the multivariate normal distribution. There are two main reasons why this is so. First, it is often the case that multivariate observations are, at least approximately, normally distributed. This is especially true of sample means and covariance matrices used in formal inferential procedures, due to a *central limit theorem* effect. This effect is also felt, of course, when the observations themselves can be regarded as sums of independent random vectors or effects, a realistic model in many situations. Secondly, the multivariate normal distribution and the sampling distributions it gives rise to are, in the main, tractable. This is not generally the case with other multivariate distributions, even for ones which appear to be close to the normal.

We will be concerned primarily with classical multivariate analysis, that is, techniques, distributions, and inferences based on the multivariate normal distribution. This distribution is defined in Section 1.2 and various properties are also derived there. This is followed by a review of the noncentral χ^2 and F distributions in Section 1.3 and some results about quadratic forms in normal variables in Section 1.4.

A natural question is to ask what happens to the inferences we make under the assumption of normality if the observations are not normal. This is an important question, leading into the area that has come to be known generally as robustness. In Section 1.5 we introduce the class of elliptical distributions; these distributions have been commonly used as alternative models in robustness studies. Section 1.6 reviews some results about multivariate cumulants. For our purposes, these are important in asymptotic distributions of test statistics which are functions of a sample covariance matrix.

It is expected that the reader is familiar with basic distributions such as the normal, gamma, beta, t, and F and with the concepts of jointly distributed random variables, marginal distributions, moments, conditional distributions, independence, and related topics covered in such standard probability and statistics texts as Bickel and Doksum (1977) and Roussas (1973).

Characteristic functions and basic limit theorems are also important and useful references are Cramér (1946), Feller (1971), and Rao (1973). Matrix notation and theory is used extensively; some of this theory appears in the text and some is reviewed in the Appendix.

1.2. THE MULTIVARIATE NORMAL DISTRIBUTION

1.2.1. *Definition and Properties*

Before proceeding to the multivariate normal distribution we need to define some moments of a *random* vector, i.e., a vector whose components are jointly distributed. The *mean* or *expectation* of a random $m\times 1$ vector $\mathbf{X}=(X_1,\ldots,X_m)'$ is defined to be the vector of expectations:

$$E(\mathbf{X})=\begin{pmatrix} E(X_1) \\ \vdots \\ E(X_m) \end{pmatrix}.$$

More generally, if $Z=(z_{ij})$ is a $p\times q$ random matrix then $E(Z)$, the expectation of Z, is the matrix whose $i-j$th element is $E(z_{ij})$. It is a simple matter to check that if B, C and D are $m\times p$, $q\times n$ and $m\times n$ matrices of constants, then

$$E(BZC+D)=BE(Z)C+D. \tag{1}$$

If $\mathbf{X}$ has mean $\boldsymbol{\mu}$ the covariance matrix of $\mathbf{X}$ is defined to be the $m\times m$ matrix

$$\Sigma\equiv\text{Cov}(\mathbf{X})=E[(\mathbf{X}-\boldsymbol{\mu})(\mathbf{X}-\boldsymbol{\mu})'].$$

The $i-j$th element of Σ is

$$\sigma_{ij}=E[(X_i-\mu_i)(X_j-\mu_j)],$$

the covariance between X_i and X_j, and the $i-i$th element is

$$\sigma_{ii} = E\left[(X_i - \mu_i)^2\right],$$

the variance of X_i, so that the diagonal elements of Σ must be nonnegative. Obviously Σ is symmetric, i.e., $\Sigma = \Sigma'$. Indeed, the class of covariance matrices coincides with the class of *non-negative definite* matrices. Recall that an $m \times m$ symmetric matrix A is called non-negative definite if

$$\alpha' A \alpha \geq 0 \qquad \text{for all} \quad \alpha \in R^m$$

and positive definite if

$$\alpha' A \alpha > 0 \qquad \text{for all} \quad \alpha \in R^m, \quad \alpha \neq \mathbf{0}$$

(Here, and throughout the book, R^m denotes Euclidean space of m dimensions consisting of $m \times 1$ vectors with real components.)

LEMMA 1.2.1. The $m \times m$ matrix Σ is a covariance matrix if and only if it is non-negative definite.

Proof. Suppose Σ is the covariance matrix of a random vector $\mathbf{X}$, where $\mathbf{X}$ has mean $\boldsymbol{\mu}$, Then for all $\alpha \in R^m$,

$$
\begin{aligned}
(2) \qquad \operatorname{Var}(\alpha' \mathbf{X}) &= E\left[(\alpha' \mathbf{X} - \alpha' \boldsymbol{\mu})^2\right] \\
&= E\left[(\alpha'(\mathbf{X} - \boldsymbol{\mu}))^2\right] \\
&= E\left[\alpha'(\mathbf{X} - \boldsymbol{\mu})(\mathbf{X} - \boldsymbol{\mu})'\alpha\right] \\
&= \alpha' \Sigma \alpha \geq 0
\end{aligned}
$$

so that Σ is non-negative definite. Now suppose Σ is a non-negative definite matrix of rank r, say ($r \leq m$). Write $\Sigma = CC'$, where C is an $m \times r$ matrix of rank r (see Theorem A9.4). Let $\mathbf{Y}$ be an $r \times 1$ vector of independent random variables with mean $\mathbf{0}$ and $\operatorname{Cov}(\mathbf{Y}) = I$ and put $\mathbf{X} = C\mathbf{Y}$. Then $E(\mathbf{X}) = \mathbf{0}$ and

$$
\begin{aligned}
\operatorname{Cov}(\mathbf{X}) &= E[\mathbf{X}\mathbf{X}'] = E[C\mathbf{Y}\mathbf{Y}'C'] \\
&= CE(\mathbf{Y}\mathbf{Y}')C' \\
&= CC' = \Sigma,
\end{aligned}
$$

so that Σ is a covariance matrix.

As a direct consequence of the inequality (2) we see that if the covariance matrix Σ of a random vector $\mathbf{X}$ is not positive definite then, with probability 1, the components X_i of $\mathbf{X}$ are linearly related. For then there exists $\boldsymbol{\alpha}\in R^m$, $\boldsymbol{\alpha}\neq\mathbf{0}$, such that

$$\operatorname{Var}(\boldsymbol{\alpha}'\mathbf{X})=\boldsymbol{\alpha}'\Sigma\boldsymbol{\alpha}=0$$

so that, with probability 1, $\boldsymbol{\alpha}'\mathbf{X}=k$, where $k=\boldsymbol{\alpha}'E(\mathbf{X})$—which means that $\mathbf{X}$ lies in a hyperplane.

We will commonly make linear transformations of random vectors and will need to know how covariance matrices are transformed. Suppose $\mathbf{X}$ is an $m\times 1$ random vector with mean $\boldsymbol{\mu}_x$ and covariance matrix Σ_x and let $\mathbf{Y}=B\mathbf{X}+\mathbf{b}$, where B is $k\times m$ and $\mathbf{b}$ is $k\times 1$. The mean of $\mathbf{Y}$ is, by (1), $\boldsymbol{\mu}_y=B\boldsymbol{\mu}_x+\mathbf{b}$, and the covariance matrix of $\mathbf{Y}$ is

$$(3)\qquad \begin{aligned}\Sigma_y&=E\left[(Y-\boldsymbol{\mu}_y)(Y-\boldsymbol{\mu}_y)'\right]\\&=E\left[(B\mathbf{X}+\mathbf{b}-(B\boldsymbol{\mu}_x+\mathbf{b}))(B\mathbf{X}+\mathbf{b}-(B\boldsymbol{\mu}_x+\mathbf{b}))'\right]\\&=BE\left[(\mathbf{X}-\boldsymbol{\mu}_x)(\mathbf{X}-\boldsymbol{\mu}_x)'\right]B'\\&=B\Sigma_xB'.\end{aligned}$$

In order to define the multivariate normal distribution we will use the following result.

THEOREM 1.2.2. If $\mathbf{X}$ is an $m\times 1$ random vector then its distribution is uniquely determined by the distributions of linear functions $\boldsymbol{\alpha}'\mathbf{X}$, for every $\boldsymbol{\alpha}\in R^m$.

Proof. The characteristic function of $\boldsymbol{\alpha}'\mathbf{X}$ is

$$\phi(t,\boldsymbol{\alpha})=E\left[e^{it\boldsymbol{\alpha}'\mathbf{X}}\right]$$

so that

$$\phi(1,\boldsymbol{\alpha})=E\left[e^{i\boldsymbol{\alpha}'\mathbf{X}}\right],$$

which, considered as a function of $\boldsymbol{\alpha}$, is the characteristic function of $\mathbf{X}$ (i.e., the joint characteristic function of the components of $\mathbf{X}$). The required result then follows by invoking the fact that a distribution in R^m is uniquely determined by its characteristic function [see, e.g., Cramér (1946), Section 10.6, or Feller (1971), Section XV.7].

DEFINITION 1.2.3. The $m\times 1$ random vector $\mathbf{X}$ is said to have an m-variate normal distribution if, for every $\boldsymbol{\alpha}\in R^m$, the distribution of $\boldsymbol{\alpha}'\mathbf{X}$ is univariate normal.

Proceeding from this definition we will now establish some properties of the multivariate normal distribution.

THEOREM 1.2.4. If $\mathbf{X}$ has an m-variate normal distribution then both $\boldsymbol{\mu}\equiv E(\mathbf{X})$ and $\Sigma\equiv\text{Cov}(\mathbf{X})$ exist and the distribution of $\mathbf{X}$ is determined by $\boldsymbol{\mu}$ and Σ.

Proof. If $\mathbf{X}=(X_1,\ldots,X_m)'$ then, for each $i=1,\ldots,m$, X_i is univariate normal (using Definition 1.2.3) so that $E(X_i)$ and $\text{Var}(X_i)$ exist and are finite. Thus $\text{Cov}(X_i, X_j)$ exists. (Why?) Putting $\boldsymbol{\mu}=E(\mathbf{X})$ and $\Sigma=\text{Cov}(\mathbf{X})$, we have, from (1) and (3),

$$E(\boldsymbol{\alpha}'\mathbf{X})=\boldsymbol{\alpha}'\boldsymbol{\mu}$$

and

$$\text{Var}(\boldsymbol{\alpha}'\mathbf{X})=\boldsymbol{\alpha}'\Sigma\boldsymbol{\alpha}$$

so that the distribution of $\boldsymbol{\alpha}'\mathbf{X}$ is $N(\boldsymbol{\alpha}'\boldsymbol{\mu}, \boldsymbol{\alpha}'\Sigma\boldsymbol{\alpha})$ for each $\boldsymbol{\alpha}\in R^m$. Since these univariate distributions are determined by $\boldsymbol{\mu}$ and Σ so is the distribution of $\mathbf{X}$ by Theorem 1.2.2.

The m-variate normal distribution of the random vector $\mathbf{X}$ of Theorem 1.2.4 will be denoted by $N_m(\boldsymbol{\mu},\Sigma)$ and we will write that $\mathbf{X}$ is $N_m(\boldsymbol{\mu}, \Sigma)$.

THEOREM 1.2.5. If $\mathbf{X}$ is $N_m(\boldsymbol{\mu}, \Sigma)$ then the characteristic function of $\mathbf{X}$ is

$$\phi_{\mathbf{X}}(\mathbf{t})=\exp(i\boldsymbol{\mu}'\mathbf{t}-\tfrac{1}{2}\mathbf{t}'\Sigma\mathbf{t}). \tag{4}$$

Proof. Here

$$\phi_{\mathbf{X}}(\mathbf{t})=E[e^{i\mathbf{t}'\mathbf{X}}]=\phi_{\mathbf{t}'\mathbf{X}}(1),$$

where the right side denotes the characteristic function of the random variable $\mathbf{t}'\mathbf{X}$ evaluated at 1. Since $\mathbf{X}$ is $N_m(\boldsymbol{\mu}, \Sigma)$ then $\mathbf{t}'\mathbf{X}$ is $N(\mathbf{t}'\boldsymbol{\mu},\mathbf{t}'\Sigma\mathbf{t})$ so that

$$\phi_{\mathbf{t}'\mathbf{X}}(1)=\exp(i\mathbf{t}'\boldsymbol{\mu}-\tfrac{1}{2}\mathbf{t}'\Sigma\mathbf{t}),$$

completing the proof.

The alert reader may have noticed that we have not yet established the existence of the multivariate normal distribution. It could be that Definition 1.2.3 is vacuous! To sew things up we will show that the function given by (4) is indeed the characteristic function of a random vector. Let Σ be an $m \times m$ covariance matrix (i.e., a non-negative definite matrix) of rank r and let $U_1, \dots, U_r$ be independent standard normal random variables. The vector $\mathbf{U} = (U_1, \dots, U_r)'$ has characteristic function

$$\begin{aligned}
\phi_{\mathbf{u}}(\mathbf{t}) &= E[\exp(i\mathbf{t}'\mathbf{U})] \\
&= \prod_{j=1}^{r} E[\exp(it_j U_j)] \qquad \text{(by independence)} \\
&= \prod_{j=1}^{r} \exp(-\tfrac{1}{2}t_j^2) \qquad \text{(by normality)} \\
&= \exp(-\tfrac{1}{2}\mathbf{t}'\mathbf{t}).
\end{aligned}$$

Now put

$$\mathbf{X} = C\mathbf{U} + \boldsymbol{\mu} \tag{5}$$

where C is an $m \times r$ matrix of rank r such that $\Sigma = CC'$, and $\boldsymbol{\mu} \in R^m$. Then $\mathbf{X}$ has characteristic function (4), for

$$\begin{aligned}
E[\exp(i\mathbf{t}'\mathbf{X})] &= E[\exp(i\mathbf{t}'C\mathbf{U})]\exp(i\mathbf{t}'\boldsymbol{\mu}) \\
&= \phi_{\mathbf{u}}(C'\mathbf{t})\exp(i\mathbf{t}'\boldsymbol{\mu}) \\
&= \exp(-\tfrac{1}{2}\mathbf{t}'CC'\mathbf{t})\exp(i\boldsymbol{\mu}'\mathbf{t}) \\
&= \exp(i\boldsymbol{\mu}'\mathbf{t} - \tfrac{1}{2}\mathbf{t}'\Sigma\mathbf{t}).
\end{aligned}$$

It is worth remarking that we could have *defined* the multivariate normal distribution $N_m(\boldsymbol{\mu}, \Sigma)$ by means of the linear transformation (5) on independent standard normal variables. Such a representation is often useful; see, for example, the proof of Theorem 1.2.9.

Getting back to the properties of the multivariate normal distribution our next result shows that any linear transformation of a normal vector has a normal distribution.

THEOREM 1.2.6. If $\mathbf{X}$ is $N_m(\boldsymbol{\mu}, \Sigma)$ and B is $k \times m$, $\mathbf{b}$ is $k \times 1$ then

$$\mathbf{Y} = B\mathbf{X} + \mathbf{b} \quad \text{is} \quad N_k(B\boldsymbol{\mu} + \mathbf{b}, B\Sigma B').$$

Proof. The fact that $\mathbf{Y}$ is k-variate normal is a direct consequence of Definition 1.2.3, since all linear functions of the components of $\mathbf{Y}$ are linear functions of the components of $\mathbf{X}$ and these are all normal. The mean and covariance matrix of $\mathbf{Y}$ are clearly those stated.

A very important property of the multivariate normal distribution is that *all* marginal distributions are normal.

THEOREM 1.2.7. If $\mathbf{X}$ is $N_m(\boldsymbol{\mu}, \Sigma)$ then the marginal distribution of any subset of $k(<m)$ components of $\mathbf{X}$ is k-variate normal.

Proof. This follows directly from the definition, or from Theorem 1.2.6. For example, partition $\mathbf{X}$, $\boldsymbol{\mu}$, and Σ as

$$\mathbf{X}=\begin{pmatrix}\mathbf{X}_1\\ \mathbf{X}_2\end{pmatrix}, \qquad \boldsymbol{\mu}=\begin{pmatrix}\boldsymbol{\mu}_1\\ \boldsymbol{\mu}_2\end{pmatrix}, \qquad \Sigma=\begin{bmatrix}\Sigma_{11} & \Sigma_{12}\\ \Sigma_{21} & \Sigma_{22}\end{bmatrix}$$

where $\mathbf{X}_1$ and $\boldsymbol{\mu}_1$ are $k\times 1$ and Σ_{11} is $k\times k$. Putting

$$B=[I_k:0] \qquad (k\times m), \qquad \mathbf{b}=\mathbf{0}$$

in Theorem 1.2.6 shows immediately that $\mathbf{X}_1$ is $N_k(\boldsymbol{\mu}_1, \Sigma_{11})$. Similarly, the marginal distribution of any subvector of k components of $\mathbf{X}$ is normal, where the mean and covariance matrix are obtained from $\boldsymbol{\mu}$ and Σ by picking out the corresponding subvector and submatrix in an obvious way.

One consequence of this theorem (or of Definition 1.2.3) is that the marginal distribution of each component of $\mathbf{X}$ is univariate normal. The converse is not true in general; that is, the fact that each component of a random vector is (marginally) normal does not imply that the vector has a multivariate normal distribution. [This is one reason why the problem of *testing* multivariate normality is such a thorny one in practice. See Gnanadesikan (1977), Chapter 5.] As a counterexample, suppose U_1, U_2, U_3 are independent $N(0, 1)$ random variables and Z is an arbitrary random variable, independent of U_1, U_2 and U_3. Define X_1 and X_2 by

$$X_1=\frac{U_1+ZU_3}{\sqrt{1+Z^2}}$$

$$X_2=\frac{U_2+ZU_3}{\sqrt{1+Z^2}}.$$

Conditional on Z, X_1 is $N(0,1)$, and since this distribution does not depend on Z it is the unconditional distribution of X_1. Similarly X_2 is $N(0, 1)$. Again, conditional on Z, the joint distribution of X_1 and X_2 is bivariate normal but the unconditional distribution clearly need not be. Other examples are given in Problems 1.7, 1.8, and 1.9. Obviously the converse *is* true if the components of $\mathbf{X}$ are all independent and normal, or if $\mathbf{X}$ consists of independent subvectors, each of which is normally distributed. For then linear functions of the components of $\mathbf{X}$ are linear functions of independent normal random variables and hence are normal. This fact will be used in the proof of the next theorem.

The reader will recall that independence of two random variables implies that the covariance between them, if it exists, is zero, but that the converse is not true in general. It is, however, for the multivariate normal distribution, as the following result shows.

THEOREM 1.2.8. If $\mathbf{X}$ is $N_m(\boldsymbol{\mu}, \Sigma)$ and $\mathbf{X}$, $\boldsymbol{\mu}$, and Σ are partitioned as

$$\mathbf{X}=\begin{pmatrix}\mathbf{X}_1\\ \mathbf{X}_2\end{pmatrix},\qquad \boldsymbol{\mu}=\begin{pmatrix}\boldsymbol{\mu}_1\\ \boldsymbol{\mu}_2\end{pmatrix},\qquad \Sigma=\begin{pmatrix}\Sigma_{11} & \Sigma_{12}\\ \Sigma_{21} & \Sigma_{22}\end{pmatrix},$$

where $\mathbf{X}_1$ and $\boldsymbol{\mu}_1$ are $k\times 1$ and Σ_{11} is $k\times k$, then the subvectors $\mathbf{X}_1$ and $\mathbf{X}_2$ are independent if and only if $\Sigma_{12}=0$.

Proof. Σ_{12} is the matrix of covariances between the components of $\mathbf{X}_1$ and the components of $\mathbf{X}_2$, so independence of $\mathbf{X}_1$ and $\mathbf{X}_2$ implies that $\Sigma_{12}=0$. Now suppose that $\Sigma_{12}=0$. Let $\mathbf{Y}_1, \mathbf{Y}_2$ be independent random vectors where $\mathbf{Y}_1$ is $N_k(\boldsymbol{\mu}_1, \Sigma_{11})$ and $\mathbf{Y}_2$ is $N_{m-k}(\boldsymbol{\mu}_2, \Sigma_{22})$ and put $\mathbf{Y}=(\mathbf{Y}_1', \mathbf{Y}_2')'$. Then both $\mathbf{X}$ and $\mathbf{Y}$ are $N_m(\boldsymbol{\mu}, \Sigma)$, where

$$\Sigma=\begin{pmatrix}\Sigma_{11} & 0\\ 0 & \Sigma_{22}\end{pmatrix},$$

so that they are identically distributed. Hence $\mathbf{X}_1$ and $\mathbf{X}_2$ are independent. Alternatively this result is also easily established using the fact that the characteristic function (4) of $\mathbf{X}$ factors into the product of the characteristic functions of $\mathbf{X}_1$ and $\mathbf{X}_2$ when $\Sigma_{12}=0$ (see Problem 1.1)

Theorem 1.2.8 can be extended easily and in an obvious way to the case where $\mathbf{X}$ is partitioned into a number of subvectors (see Problem 1.2). The important message here is that in order to determine whether two subvectors of a normally distributed vector are independent it suffices to check that the matrix of covariances between the two subvectors is zero.

Let us now address the problem of finding the density function of a random vector $\mathbf{X}$ having the $N_m(\boldsymbol{\mu}, \Sigma)$ distribution. We have already noted that if Σ is not positive definite, and hence is singular, then $\mathbf{X}$ lies in some hyperplane with probability 1 so that a density function for $\mathbf{X}$ (with respect to Lebesgue measure on R^m) can not exist. In this case $\mathbf{X}$ is said to have a *singular* normal distribution. If Σ is positive definite, and hence nonsingular, the density function of $\mathbf{X}$ does exist and is easily found using the representation (5) of $\mathbf{X}$ in terms of independent standard normal random variables.

THEOREM 1.2.9. If $\mathbf{X}$ is $N_m(\boldsymbol{\mu},\Sigma)$ and Σ is positive definite then the density function of $\mathbf{X}$ is

$$(6)\qquad f_{\mathbf{X}}(\mathbf{x})=(2\pi)^{-m/2}(\det\Sigma)^{-1/2}\exp\left[-\tfrac{1}{2}(\mathbf{x}-\boldsymbol{\mu})'\Sigma^{-1}(\mathbf{x}-\boldsymbol{\mu})\right].$$

(Here, and throughout the book, det denotes determinant.)

Proof. Write $\Sigma=CC'$ where C is a nonsingular $m\times m$ matrix and put

$$\mathbf{X}=C\mathbf{U}+\boldsymbol{\mu},$$

where $\mathbf{U}$ is an $m\times 1$ vector of independent $N(0,1)$ random variables, i.e., $\mathbf{U}$ is $N_m(\mathbf{0},I_m)$. The joint density function of $U_1,\dots,U_m$ is

$$\begin{aligned} f_{\mathbf{U}}(\mathbf{u})&=\prod_{i=1}^{m}(2\pi)^{-1/2}\exp\left(-\tfrac{1}{2}u_i^2\right)\\ &=(2\pi)^{-m/2}\exp\left(-\tfrac{1}{2}\mathbf{u}'\mathbf{u}\right). \end{aligned}$$

The inverse transformation is $\mathbf{U}=B(\mathbf{X}-\boldsymbol{\mu})$, with $B=C^{-1}$, and the Jacobian of this transformation is

$$\begin{aligned} \det\begin{bmatrix} \dfrac{\partial u_1}{\partial x_1} & \cdots & \dfrac{\partial u_1}{\partial x_m}\\ \vdots & & \\ \dfrac{\partial u_m}{\partial x_1} & \cdots & \dfrac{\partial u_m}{\partial x_m}\end{bmatrix} &=\det\begin{bmatrix} b_{11} & b_{12} & \cdots & b_{1m}\\ \vdots & & & \\ b_{m1} & b_{m2} & \cdots & b_{mm}\end{bmatrix}\\ &=\det B=\det C^{-1}=(\det C)^{-1}\\ &=\left[\det(CC')\right]^{-1/2}=(\det\Sigma)^{-1/2} \end{aligned}$$

so that the density function of X is

$$f_{\mathbf{X}}(\mathbf{x})=(2\pi)^{-m/2}(\det\Sigma)^{-1/2}\exp\left[-\tfrac{1}{2}(\mathbf{x}-\boldsymbol{\mu})'C^{-1\prime}C^{-1}(\mathbf{x}-\boldsymbol{\mu})\right];$$

and since $\Sigma^{-1}=C^{-1\prime}C^{-1}$, we are done.

The density function (6) is constant whenever the quadratic form in the exponent is, so that it is constant on the ellipsoid

$$(\mathbf{x}-\boldsymbol{\mu})'\Sigma^{-1}(\mathbf{x}-\boldsymbol{\mu})=k$$

in R^m, for every $k>0$. This ellipsoid has center $\boldsymbol{\mu}$, while Σ determines its shape and orientation.

It is worthwhile looking explicitly at the bivariate normal distribution ($m=2$). In this case

$$\mathbf{X}=\begin{pmatrix} X_1 \\ X_2 \end{pmatrix}, \qquad \boldsymbol{\mu}=\begin{pmatrix} \mu_1 \\ \mu_2 \end{pmatrix}$$

and

$$\Sigma=\begin{bmatrix} \sigma_{11} & \sigma_{12} \\ \sigma_{12} & \sigma_{22} \end{bmatrix}=\begin{bmatrix} \sigma_1^2 & \rho\sigma_1\sigma_2 \\ \rho\sigma_1\sigma_2 & \sigma_2^2 \end{bmatrix},$$

where $\text{Var}(X_1)=\sigma_1^2$, $\text{Var}(X_2)=\sigma_2^2$, and the correlation between X_1 and X_2 is ρ. For the distribution of $\mathbf{X}$ to be nonsingular normal we need $\sigma_1^2>0$, $\sigma_2^2>0$, and

$$\det\Sigma=\sigma_1^2\sigma_2^2(1-\rho^2)>0$$

so that $-1<\rho<1$. When this holds,

$$\Sigma^{-1}=\frac{1}{1-\rho^2}\begin{bmatrix} \dfrac{1}{\sigma_1^2} & -\dfrac{\rho}{\sigma_1\sigma_2} \\ -\dfrac{\rho}{\sigma_1\sigma_2} & \dfrac{1}{\sigma_2^2} \end{bmatrix},$$

and the joint density function of X_1 and X_2 is

(7)

$$f_{\mathbf{X}}(x_1,x_2)=\frac{1}{2\pi\sigma_1\sigma_2(1-\rho^2)^{1/2}}\exp\left\{-\frac{1}{2(1-\rho^2)}\left[\left(\frac{x_1-\mu_1}{\sigma_1}\right)^2+\left(\frac{x_2-\mu_2}{\sigma_2}\right)^2 -2\rho\frac{(x_1-\mu_1)(x_2-\mu_2)}{\sigma_1\sigma_2}\right]\right\}.$$

The "standard" bivariate normal density function is obtained from this by transforming to standardized variables. Putting $Z_i=(X_i-\mu_i)/\sigma_i$ $(i=1,2)$, the joint density function of Z_1 and Z_2 is

$$(8)\quad f_z(z_1,z_2)=\frac{1}{2\pi(1-\rho^2)^{1/2}}\exp\left[-\frac{1}{2(1-\rho^2)}(z_1^2+z_2^2-2\rho z_1z_2)\right].$$

This density is constant on the ellipse

$$(9)\qquad \frac{1}{1-\rho^2}(z_1^2+z_2^2-2\rho z_1z_2)=k$$

in R^2, for every $k>0$. (Some properties of this ellipse are explored in Problem 1.3.)

In order to prove the next theorem we will use the following lemma. In this lemma the notations $R(M)$ and $K(M)$ for an $n\times r$ matrix M denote the range (or column space) and kernel (or null space) respectively:

$$(10)\qquad R(M)=\{\mathbf{v}\in R^n;\quad \mathbf{v}=M\mathbf{u}\quad \text{for some}\quad \mathbf{u}\in R^r\}$$

$$(11)\qquad K(M)=\{\mathbf{u}\in R^r;\quad M\mathbf{u}=\mathbf{0}\}.$$

Clearly $R(M)$ is a subspace of R^n, and $K(M)$ is a subspace of R^r.

LEMMA 1.2.10. If the $m\times m$ matrix Σ is non-negative definite and is partitioned as

$$\Sigma=\begin{bmatrix}\Sigma_{11} & \Sigma_{12}\\ \Sigma_{21} & \Sigma_{22}\end{bmatrix}$$

where Σ_{11} is $k\times k$ and Σ_{22} is $(m-k)\times(m-k)$ then:

(a) $K(\Sigma_{22})\subset K(\Sigma_{12})$

(b) $R(\Sigma_{21})\subset R(\Sigma_{22})$

Proof. (a) Suppose $\mathbf{z}\in K(\Sigma_{22})$. Then, for all $\mathbf{y}\in R^k$ and $\alpha\in R^1$ we have

$$\begin{aligned}(\mathbf{y}'\alpha\mathbf{z}')\begin{bmatrix}\Sigma_{11} & \Sigma_{12}\\ \Sigma_{21} & \Sigma_{22}\end{bmatrix}\begin{pmatrix}\mathbf{y}\\ \alpha\mathbf{z}\end{pmatrix}&=\mathbf{y}'\Sigma_{11}\mathbf{y}+2\alpha\mathbf{y}'\Sigma_{12}\mathbf{z}+\alpha^2\mathbf{z}'\Sigma_{22}\mathbf{z}\\ &=\mathbf{y}'\Sigma_{11}\mathbf{y}+2\alpha\mathbf{y}'\Sigma_{12}\mathbf{z}\qquad(\text{because }\Sigma_{22}\mathbf{z}=\mathbf{0})\\ &\geq 0\qquad(\text{because }\Sigma\text{ is non-negative definite}).\end{aligned}$$

Taking $\mathbf{y}=\Sigma_{12}\mathbf{z}$ then gives

$$\mathbf{z}'\Sigma_{21}\Sigma_{11}\Sigma_{12}\mathbf{z}+2\alpha(\Sigma_{12}\mathbf{z})'(\Sigma_{12}\mathbf{z})\geq 0$$

for all α, which means that $\Sigma_{12}\mathbf{z}=\mathbf{0}$, i.e., $\mathbf{z}\in K(\Sigma_{12})$. Hence $K(\Sigma_{22})\subset K(\Sigma_{12})$. Then, part (b) follows immediately on noting that $K(\Sigma_{12})^{\perp}\subset K(\Sigma_{22})^{\perp}$, where $K(M)^{\perp}$ denotes the orthogonal complement of $K(M)$ [i.e., the set of vectors orthogonal to every vector in $K(M)$] and using the easily proved fact that

$$K(M)^{\perp}=R(M'). \tag{12}$$

Our next theorem shows that the conditional distribution of a subvector of a normally distributed vector given the remaining components is also normal.

THEOREM 1.2.11. Let $\mathbf{X}$ be $N_m(\boldsymbol{\mu},\Sigma)$ and partition $\mathbf{X}$, $\boldsymbol{\mu}$ and Σ as

$$\mathbf{X}=\begin{pmatrix}\mathbf{X}_1\\ \mathbf{X}_2\end{pmatrix},\qquad \boldsymbol{\mu}=\begin{pmatrix}\boldsymbol{\mu}_1\\ \boldsymbol{\mu}_2\end{pmatrix},\qquad \Sigma=\begin{bmatrix}\Sigma_{11} & \Sigma_{12}\\ \Sigma_{21} & \Sigma_{22}\end{bmatrix}$$

where $\mathbf{X}_1$ and $\boldsymbol{\mu}_1$ are $k\times 1$ and Σ_{11} is $k\times k$. Let Σ_{22}^{-} be a generalized inverse of Σ_{22}, i.e., a matrix satisfying

$$\Sigma_{22}\Sigma_{22}^{-}\Sigma_{22}=\Sigma_{22} \tag{13}$$

and let $\Sigma_{11\cdot 2}=\Sigma_{11}-\Sigma_{12}\Sigma_{22}^{-}\Sigma_{21}$. Then

(a) $\mathbf{X}_1-\Sigma_{12}\Sigma_{22}^{-}\mathbf{X}_2$ is $N_k(\boldsymbol{\mu}_1-\Sigma_{12}\Sigma_{22}^{-}\boldsymbol{\mu}_2,\Sigma_{11\cdot 2})$

and is independent of $\mathbf{X}_2$, and

(b) the conditional distribution of $\mathbf{X}_1$ given $\mathbf{X}_2$ is

$$N_k(\boldsymbol{\mu}_1+\Sigma_{12}\Sigma_{22}^{-}(\mathbf{X}_2-\boldsymbol{\mu}_2),\Sigma_{11\cdot 2}).$$

Proof. From Lemma 1.2.10 we have $R(\Sigma_{21})\subset R(\Sigma_{22})$ so that there exists a $k\times(m-k)$ matrix B satisfying

$$\Sigma_{12}=B\Sigma_{22}. \tag{14}$$

Now note that

$$\Sigma_{12}\Sigma_{22}^{-}\Sigma_{22}=B\Sigma_{22}\Sigma_{22}^{-}\Sigma_{22}=B\Sigma_{22}=\Sigma_{12}, \tag{15}$$

where we have used (13) and (14). Put

$$C=\begin{bmatrix} I_k & -\Sigma_{12}\Sigma_{22}^- \\ 0 & I_{m-k} \end{bmatrix}$$

then, by Theorem 1.2.6,

$$C\mathbf{X}=\begin{pmatrix} \mathbf{X}_1-\Sigma_{12}\Sigma_{22}^-\mathbf{X}_2 \\ \mathbf{X}_2 \end{pmatrix}$$

is m-variate normal with mean

$$\begin{pmatrix} \boldsymbol{\mu}_1-\Sigma_{12}\Sigma_{22}^-\boldsymbol{\mu}_2 \\ \boldsymbol{\mu}_2 \end{pmatrix}$$

and covariance matrix

$$\begin{aligned} C\Sigma C' &= \begin{bmatrix} I_k & -\Sigma_{12}\Sigma_{22}^- \\ 0 & I_{m-k} \end{bmatrix}\begin{bmatrix} \Sigma_{11} & \Sigma_{12} \\ \Sigma_{21} & \Sigma_{22} \end{bmatrix}\begin{bmatrix} I_k & 0 \\ -\Sigma_{22}^-\Sigma_{21} & I_{m-k} \end{bmatrix} \\ &= \begin{bmatrix} \Sigma_{11\cdot 2} & 0 \\ \Sigma_{21} & \Sigma_{22} \end{bmatrix}\begin{bmatrix} I_k & 0 \\ -\Sigma_{22}^-\Sigma_{21} & I_{m-k} \end{bmatrix} \quad \text{using (15)} \\ &= \begin{bmatrix} \Sigma_{11\cdot 2} & 0 \\ 0 & \Sigma_{22} \end{bmatrix}. \end{aligned}$$

The first assertion (a) is a direct consequence of Theorems 1.2.7 and 1.2.8 while the second (b) follows immediately from (a) by conditioning on $\mathbf{X}_2$.

When the matrix Σ_{22} is nonsingular, which happens, for example, when Σ is nonsingular, then $\Sigma_{22}^-=\Sigma_{22}^{-1}$ and $\Sigma_{11\cdot 2}=\Sigma_{11}-\Sigma_{12}\Sigma_{22}^{-1}\Sigma_{21}$. The theorem is somewhat easier to prove in this case.

The mean of the conditional distribution of $\mathbf{X}_1$ given $\mathbf{X}_2$, namely,

$$E(\mathbf{X}_1|\mathbf{X}_2)=\boldsymbol{\mu}_1+\Sigma_{12}\Sigma_{22}^-(\mathbf{X}_2-\boldsymbol{\mu}_2) \tag{16}$$

is called the *regression function* of $\mathbf{X}_1$ on $\mathbf{X}_2$ with matrix of regression coefficients $\Sigma_{12}\Sigma_{22}^-$. It is a *linear* regression function since it depends linearly on the variables $\mathbf{X}_2$ being held fixed. The covariance matrix $\Sigma_{11\cdot 2}$ of the conditional distribution of $\mathbf{X}_1$ given $\mathbf{X}_2$ does *not* depend on $\mathbf{X}_2$, the variables being held fixed.

There are many characterizations of the multivariate normal distribution. We will look at just one; others may be found in Rao (1973) and Kagan et al. (1972). We will need the following famous result due to Cramér (1937), which characterizes the univariate normal distribution.

LEMMA 1.2.12. If X and Y are independent random variables whose sum $X+Y$ is normally distributed, then both X and Y are normally distributed.

A proof of this lemma is given by Feller (1971), Section XV.8.

THEOREM 1.2.13. If the $m\times 1$ random vectors $\mathbf{X}$ and $\mathbf{Y}$ are independent and $\mathbf{X}+\mathbf{Y}$ has an m-variate normal distribution, then both $\mathbf{X}$ and $\mathbf{Y}$ are normal.

Proof. For each $\boldsymbol{\alpha}\in R^m$, $\boldsymbol{\alpha}'(\mathbf{X}+\mathbf{Y})=\boldsymbol{\alpha}'\mathbf{X}+\boldsymbol{\alpha}'\mathbf{Y}$ is normal (by Definition 1.2.3, since $\mathbf{X}+\mathbf{Y}$ is normal). Since $\boldsymbol{\alpha}'\mathbf{X}$ and $\boldsymbol{\alpha}'\mathbf{Y}$ are independent, Lemma 1.2.12 implies that they are both normal, and hence $\mathbf{X}$ and $\mathbf{Y}$ are both m-variate normal.

This proof looks easy and uses the obvious trick of reducing the problem to a univariate one by using our definition of multivariate normality. We have, however, glossed over the hard part, namely, the proof of Lemma 1.2.12.

A well-known property of the univariate normal distribution is that linear combinations of independent normal random variables are normal. This generalizes to the multivariate situation in an obvious way.

THEOREM 1.2.14. If $\mathbf{X}_1,\dots,\mathbf{X}_N$ are all independent, and $\mathbf{X}_i$ is $N_m(\boldsymbol{\mu}_i,\Sigma_i)$ for $i=1,\dots,N$, then for any fixed constants $\alpha_1,\dots,\alpha_N$,

$$\sum_{i=1}^N \alpha_i\mathbf{X}_i \quad\text{is}\quad N_m\left(\sum_{i=1}^N \alpha_i\boldsymbol{\mu}_i,\ \sum_{i=1}^N \alpha_i^2\Sigma_i\right).$$

The proof is immediate from Definition 1.2.3, or by inspection of the characteristic function of $\sum_{i=1}^N\alpha_i\mathbf{X}_i$. It is left to the reader to fill in the details (Problem 1.5).

COROLLARY 1.2.15. If $\mathbf{X}_1,\dots,\mathbf{X}_N$ are independent, each having the $N_m(\boldsymbol{\mu},\Sigma)$ distribution, then the distribution of the *sample mean vector*

$$\overline{\mathbf{X}}=\frac{1}{N}\sum_{i=1}^N \mathbf{X}_i$$

is $N_m(\boldsymbol{\mu},(1/N)\Sigma)$.

1.2.2. Asymptotic Distributions of Sample Means and Covariance Matrices

Corollary 1.2.15 says that the distribution of

$$N^{1/2}(\bar{\mathbf{X}}-\boldsymbol{\mu})=N^{-1/2}\sum_{i=1}^{N}(\mathbf{X}_i-\boldsymbol{\mu})$$

is $N_m(\mathbf{0},\Sigma)$: When the vectors $\mathbf{X}_1,\ldots,\mathbf{X}_N$ are not normal we still have

$$E(\bar{\mathbf{X}})=\boldsymbol{\mu} \quad \text{and} \quad \operatorname{Cov}(\bar{\mathbf{X}})=\frac{1}{N}\Sigma,$$

but it is the asymptotic distribution which is normal, as the following version of the *multivariate central limit theorem* due to Cramér (1946), Sections 21.11 and 24.7, and Anderson (1958), page 74, shows.

THEOREM 1.2.16. Let $\mathbf{X}_1,\mathbf{X}_2,\ldots$ be a sequence of independent and identically distributed random vectors with mean $\boldsymbol{\mu}$ and covariance matrix Σ and let

$$\bar{\mathbf{X}}_N=\frac{1}{N}\sum_{i=1}^{N}\mathbf{X}_i, \qquad N\geq 1.$$

Then, as $N\rightarrow\infty$, the asymptotic distribution of

$$N^{1/2}(\bar{\mathbf{X}}_N-\boldsymbol{\mu})=N^{-1/2}\sum_{i=1}^{N}(\mathbf{X}_i-\boldsymbol{\mu})$$

is $N_m(\mathbf{0},\Sigma)$.

Proof. Put $\mathbf{Y}_N=N^{-1/2}\Sigma_{i=1}^{N}(\mathbf{X}_i-\boldsymbol{\mu})$. By the continuity theorem for characteristic functions [see Cramér (1946), Section 10.7], it suffices to show that $\phi_N(\mathbf{t})$, the characteristic function of $\mathbf{Y}_N$, converges to $\exp(-\frac{1}{2}\mathbf{t}'\Sigma\mathbf{t})$, the characteristic function of the $N_m(\mathbf{0},\Sigma)$ distribution. Now, the characteristic function of $\mathbf{t}'\mathbf{Y}_N$, where $\mathbf{t}\in R^m$, is

$$f_N(\alpha,\mathbf{t})=E\left[\exp(i\alpha\mathbf{t}'\mathbf{Y}_N)\right],$$

considered as a function of $\alpha\in R^1$. Also

$$\mathbf{t}'\mathbf{Y}_N=N^{-1/2}\sum_{j=1}^{N}(\mathbf{t}'\mathbf{X}_j-\mathbf{t}'\boldsymbol{\mu})$$

and since $\mathbf{t}'\mathbf{X}_1-\mathbf{t}'\boldsymbol{\mu},\mathbf{t}'\mathbf{X}_2-\mathbf{t}'\boldsymbol{\mu},\ldots$ is a sequence of independent and identically distributed random variables with zero mean and variance $\mathbf{t}'\Sigma\mathbf{t}$, it follows by the univariate central limit theorem that, as $N\to\infty$, the asymptotic distribution of $\mathbf{t}'Y_N$ is $N(\mathbf{0},\mathbf{t}'\Sigma\mathbf{t})$ and, hence, as $N\to\infty$

$$f_N(\alpha,\mathbf{t})\to\exp(-\tfrac{1}{2}\alpha^2\mathbf{t}'\Sigma\mathbf{t})$$

for all $\mathbf{t}$ and α, where the right side is the characteristic function of the $N(0,\mathbf{t}'\Sigma\mathbf{t})$ distribution. Putting $\alpha=1$ shows that

$$\phi_N(\mathbf{t})=f_N(1,\mathbf{t})=E\left[\exp(i\mathbf{t}'\mathbf{Y}_N)\right]\to\exp(-\tfrac{1}{2}\mathbf{t}'\Sigma\mathbf{t})$$

as $N\to\infty$, which completes the proof.

To introduce an application of this theorem, let $\mathbf{X}_1,\ldots,\mathbf{X}_N$ be a random sample of size N from any m-variate distribution and suppose

$$E(\mathbf{X}_i)=\boldsymbol{\mu},\qquad \mathrm{Cov}(\mathbf{X}_i)=\Sigma\qquad (i=1,\ldots,N).$$

Put

$$S=\frac{1}{n}A \tag{17}$$

where

$$A=\sum_{i=1}^{N}(\mathbf{X}_i-\overline{\mathbf{X}})(\mathbf{X}_i-\overline{\mathbf{X}})', \tag{18}$$

$n=N-1$, and $\overline{\mathbf{X}}=N^{-1}\Sigma_{i=1}^{N}\mathbf{X}_i$. The $m\times m$ matrix S is called the *sample covariance matrix* and is an unbiased estimate of Σ, that is, $E(S)=\Sigma$. To see this, write A as

$$A=\sum_{i=1}^{N}\left[(\mathbf{X}_i-\boldsymbol{\mu})-(\overline{\mathbf{X}}-\boldsymbol{\mu})\right]\left[(\mathbf{X}_i-\boldsymbol{\mu})-(\overline{\mathbf{X}}-\boldsymbol{\mu})\right]' \tag{19}$$

$$=\sum_{i=1}^{N}(\mathbf{X}_i-\boldsymbol{\mu})(\mathbf{X}_i-\boldsymbol{\mu})'-N(\overline{\mathbf{X}}-\boldsymbol{\mu})(\overline{\mathbf{X}}-\boldsymbol{\mu})'.$$

Then,

$$(20)\qquad E(A)=\sum_{i=1}^{N}E[(\mathbf{X}_i-\boldsymbol{\mu})(\mathbf{X}_i-\boldsymbol{\mu})']-NE[(\bar{\mathbf{X}}-\boldsymbol{\mu})(\bar{\mathbf{X}}-\boldsymbol{\mu})']$$

$$=N\Sigma-N\frac{1}{N}\Sigma$$

$$=(N-1)\Sigma$$

$$=n\Sigma$$

so that $E(n^{-1}A)=E(S)=\Sigma$.

In a moment we will show that the asymptotic joint distribution of the elements of A is normal.

First, some notation and terminology. If T is a $p\times q$ matrix then by vec(T) we mean the $pq\times 1$ vector formed by stacking the columns of T under each other; that is, if

$$T=[\mathbf{t}_1\mathbf{t}_2\ldots\mathbf{t}_q],$$

where $\mathbf{t}_i$ is $p\times 1$ for $i=1,\ldots,q$, then

$$(21)\qquad \operatorname{vec}(T)=\begin{pmatrix}\mathbf{t}_1\\ \mathbf{t}_2\\ \vdots\\ \mathbf{t}_q\end{pmatrix}.$$

When we talk about the asymptotic normality of a *random matrix* T (as in Theorem 1.2.17 below) we will mean the asymptotic normality of vec(T).

Now, from (19), we have

$$A(N)=\sum_{i=1}^{N}Z_i-NB(N),$$

where $Z_i=(\mathbf{X}_i-\boldsymbol{\mu})(\mathbf{X}_i-\boldsymbol{\mu})'$, $B(N)=(\bar{\mathbf{X}}_N-\boldsymbol{\mu})(\bar{\mathbf{X}}_N-\boldsymbol{\mu})'$, and we are indexing A, $\bar{\mathbf{X}}$ and B by N to reflect the fact that they are formed from the first N random vectors $\mathbf{X}_1,\ldots,\mathbf{X}_N$. Hence

$$(22)\qquad \operatorname{vec}(A(N))=\sum_{i=1}^{N}\operatorname{vec}(Z_i)-N\operatorname{vec}(B(N)),$$

where these vectors are all $m^2 \times 1$. Let

$$V=\text{Cov}(\text{vec}(Z_i))$$

(assuming this exists), then by Theorem 1.2.16

$$\text{(23)} \qquad \frac{1}{N^{1/2}}\sum_{i=1}^{N}\left[\text{vec}(Z_i)-\text{vec}(\Sigma)\right]\to N_{m^2}(\mathbf{0},V)$$

in distribution, as $N\to\infty$. Again, by Theorem 1.2.16

$$N^{1/2}(\overline{\mathbf{X}}_N-\boldsymbol{\mu})\to N_m(\mathbf{0},\Sigma)$$

in distribution as $N\to\infty$, and

$$\frac{1}{N^{1/4}}N^{1/2}(\overline{\mathbf{X}}_N-\boldsymbol{\mu})\to\mathbf{0}$$

in probability. [This means that each component $V_i(N)$, say, of the vector on the left converges to zero in probability, i.e., for each $\varepsilon>0$, $\lim_{N\to\infty}P(|V_i(N)|>\varepsilon)=0$. A similar definition holds for random matrices also.] Thus

$$\frac{1}{N^{1/2}}N(\overline{\mathbf{X}}_N-\boldsymbol{\mu})(\overline{\mathbf{X}}_N-\boldsymbol{\mu})'\to 0$$

in probability and hence

$$\text{(24)} \qquad \frac{1}{N^{1/2}}N\,\text{vec}(B(N))=N^{1/2}\text{vec}(B(N))\to\mathbf{0}$$

in probability as $N\to\infty$.

As a consequence of (23) and (24) we then have

$$\frac{1}{N^{1/2}}(\text{vec}(A)-N\,\text{vec}(\Sigma))=\frac{1}{N^{1/2}}\sum_{i=1}^{N}\left[\text{vec}(Z_i)-\text{vec}(\Sigma)\right]-N^{1/2}\text{vec}(B)$$
$$\to N_{m^2}(\mathbf{0},V)$$

in distribution as $N\to\infty$. [Here we have used the fact that if $\mathbf{Y}_N\to\mathbf{Y}$ in distribution and $\mathbf{Z}_N\to\mathbf{0}$ in probability, then $\mathbf{Y}_N+\mathbf{Z}_N\to\mathbf{Y}$ in distribution; see, e.g., Rao (1973), Section 2c.]

We can summarize our results in the following:

THEOREM 1.2.17. Let $\mathbf{X}_1, \mathbf{X}_2, \ldots$ be a sequence of independent and identically distributed $m \times 1$ random vectors with finite fourth moments and mean $\boldsymbol{\mu}$ and covariance matrix Σ and let

$$A(N) = \sum_{i=1}^{N} (\mathbf{X}_i - \overline{\mathbf{X}}_N)(\mathbf{X}_i - \overline{\mathbf{X}}_N)'$$

where

$$\overline{\mathbf{X}}_N = N^{-1} \sum_{i=1}^{N} \mathbf{X}_i.$$

Then the asymptotic distribution of

$$T(N) = N^{-1/2}[A(N) - N\Sigma]$$

is normal with mean 0 and covariance matrix

$$V = \mathrm{Cov}\big[\mathrm{vec}((\mathbf{X}_1 - \boldsymbol{\mu})(\mathbf{X}_1 - \boldsymbol{\mu})')\big].$$

The following corollary expresses this asymptotic result in terms of the sample covariance matrix.

COROLLARY 1.2.18. Let $n = N - 1$ and put $S(n) = n^{-1}A(N)$. Under the conditions of Theorem 1.2.17 the asymptotic distribution of $U(n) = n^{1/2}[S(n) - \Sigma]$ is normal with mean 0 and covariance matrix V.

This follows directly from Theorem 1.2.17 by putting $A(N) = nS(n)$ and replacing N by n, a modification which clearly has no effect on the limiting distribution.

Note that this asymptotic normal distribution is singular, because V is singular. This is due to the fact V is the $m^2 \times m^2$ covariance matrix in the asymptotic distribution of $\mathrm{vec}(T(N))$ or $\mathrm{vec}(U(n))$ and, because $T(N)$ and $U(n)$ are symmetric, these vectors have repeated elements.

In general, given an underlying distribution for the $\mathbf{X}_i$, it is rather tedious (in terms of the algebraic manipulation involved) to find the elements of the asymptotic covariance matrix, since this involves finding all the fourth order mixed moments of the distribution. However, the calculations are fairly straightforward when sampling from a $N_m(\boldsymbol{\mu}, \Sigma)$ distribution. In this case the elements of the asymptotic covariance matrix are given by

$$\text{Cov}(u_{ij}(n), u_{kl}(n)) = \sigma_{ik}\sigma_{jl} + \sigma_{il}\sigma_{jk} \tag{25}$$

(see Problem 1.6). For general distributions the asymptotic covariances have been expressed in terms of the cumulants by Cook (1951) and others; this work will be reviewed in Section 1.6.

1.3. THE NONCENTRAL χ^2 AND F DISTRIBUTIONS

Many statistics of interest in multivariate analysis and elsewhere have noncentral χ^2 and F distributions. Usually these distributions occur when a null hypothesis of interest is not true, hence the terms "non-null" and "noncentral." Here we will review these two distributions, and this will afford us an opportunity to introduce some definitions and notation that will be used later.

DEFINITION 1.3.1. The generalized hypergeometric function (or series) is

$$ {}_pF_q(a_1,\dots,a_p; b_1,\dots,b_q; z) = \sum_{k=0}^{\infty} \frac{(a_1)_k \cdots (a_p)_k}{(b_1)_k \cdots (b_q)_k} \frac{z^k}{k!}, \tag{1}$$

where $(a)_k = a(a+1)\cdots(a+k-1)$.

Here $a_1,\dots,a_p, b_1,\dots,b_q$ are (possibly complex) *parameters* and z, the *argument* of the function, is a complex variable. No denominator parameter b_j is allowed to be zero or a negative integer (otherwise one of the denominators in the series is zero), and, if any numerator parameter is zero or a negative integer, the series terminates to give a polynomial in z. It is easy to show using the ratio test that the series converges for all finite z if $p \le q$, it converges for $|z|<1$ and diverges for $|z|>1$ if $p=q+1$, and it diverges for all $z \ne 0$ if $p>q+1$. The term "generalized hypergeometric function" refers to the fact that ${}_pF_q$ is a generalization of the classical (or Gaussian) hypergeometric function ${}_2F_1$. For a detailed discussion of these functions and their properties the reader is referred to Erdélyi et al. (1953a). For our purposes we will make use of the results in the following two lemmas. The first gives a special integral for ${}_0F_1$ (which is related to a Bessel function) and the second shows that ${}_{p+1}F_q$ is essentially a Laplace transform of ${}_pF_q$.

LEMMA 1.3.2.

$$(2)\qquad \frac{\Gamma(\frac{1}{2}n)}{\Gamma(\frac{1}{2})\Gamma[\frac{1}{2}(n-1)]}\int_0^{\pi} e^{z\cos\theta}\sin^{n-2}\theta\,d\theta = {}_0F_1(\tfrac{1}{2}n;\tfrac{1}{4}z^2)$$

$$= \sum_{k=0}^{\infty}\frac{(\frac{1}{4}z^2)^k}{(\frac{1}{2}n)_k k!}$$

Proof. Let $I(n,z)$ denote the left side of (2). Expand the exponential term in the integrand and integrate term by term. (Why is this permissible?) Noting that terms corresponding to odd powers of z are integrals of odd functions and hence vanish, we get

$$I(n,z) = \frac{\Gamma(\frac{1}{2}n)}{\Gamma(\frac{1}{2})\Gamma[\frac{1}{2}(n-1)]}\sum_{k=0}^{\infty}\frac{z^{2k}}{(2k)!}2\int_0^{\pi/2}\cos^{2k}\theta\sin^{n-2}\theta\,d\theta.$$

To evaluate this last integral, make the change of variables $x=\sin^2\theta$ to give

$$2\int_0^{\pi/2}\cos^{2k}\theta\sin^{n-2}\theta\,d\theta = \int_0^1 x^{(n-3)/2}(1-x)^{k-1/2}\,dx$$

$$= \frac{\Gamma[\frac{1}{2}(n-1)]\Gamma(k+\frac{1}{2})}{\Gamma(k+\frac{1}{2}n)}$$

so that

$$I(n,z) = \sum_{k=0}^{\infty} z^{2k}\frac{\Gamma(\frac{1}{2}n)}{\Gamma(k+\frac{1}{2}n)}\,\frac{\Gamma(k+\frac{1}{2})}{\Gamma(\frac{1}{2})(2k)!}.$$

The desired result now follows, since

$$\frac{\Gamma(k+\frac{1}{2}n)}{\Gamma(\frac{1}{2}n)} = (\tfrac{1}{2}n)_k$$

and

$$\frac{\Gamma(k+\frac{1}{2})}{\Gamma(\frac{1}{2})(2k)!} = \frac{1}{4^k k!}.$$

LEMMA 1.3.3.

$$\int_0^\infty e^{-zt}t^{a-1}{}_pF_q(a_1,\dots,a_p;b_1,\dots,b_q;kt)\,dt$$
$$=\Gamma(a)z^{-a}{}_{p+1}F_q(a_1,\dots,a_p,a;b_1,\dots,b_q;kz^{-1})$$

for

$$p<q,\qquad \mathrm{Re}(a)>0,\qquad \mathrm{Re}(z)>0$$

or

$$p=q,\qquad \mathrm{Re}(a)>0,\qquad \mathrm{Re}(z)>\mathrm{Re}(k)$$

[Here $\mathrm{Re}(\cdot)$ denotes the real part of the argument.]

To prove this lemma, integrate the ${}_pF_q$ series term by term. The details are left as an exercise (Problem 1.15).

We will now derive an expression for the density function of the noncentral χ^2 distribution. Recall that the usual or central χ^2 distribution is the distribution of the sum of squares of independent standard normal random variables. The noncentral χ^2 distribution is the distribution of the sum of squares where the means need not be zero.

THEOREM 1.3.4. If $\mathbf{X}$ is $N_n(\boldsymbol{\mu}, I_n)$ then the random variable $Z=\mathbf{X}'\mathbf{X}$ has the density function

$$e^{-\delta/2}{}_0F_1\left(\tfrac{1}{2}n;\tfrac{1}{4}\delta z\right)\frac{1}{2^{n/2}\Gamma\left(\frac{1}{2}n\right)}e^{-z/2}z^{n/2-1}\qquad (z>0),$$

where $\delta=\boldsymbol{\mu}'\boldsymbol{\mu}$. Z is said to have the noncentral χ^2 distribution with n degrees of freedom and noncentrality parameter δ, to be written as $\chi_n^2(\delta)$.

Proof. Put $\mathbf{Y}=H\mathbf{X}$, where H is an $n\times n$ orthogonal matrix whose elements in the first row are

$$\frac{\mu_i}{(\boldsymbol{\mu}'\boldsymbol{\mu})^{1/2}}\qquad (i=1,\dots,n).$$

Then $\mathbf{Y}$ is $N_n(\boldsymbol{\nu}, I_n)$ (using Theorem 1.2.6) with $\boldsymbol{\nu}=(\delta^{1/2},0,\dots,0)'$, so that

$$Z=\mathbf{X}'\mathbf{X}=\mathbf{Y}'\mathbf{Y}=Y_1^2+U,$$

where $U=\sum_{i=2}^{n} Y_i^2$ is χ_{n-1}^2 and is independent of Y_1, which is $N(\delta^{1/2},1)$. Consequently the joint density function of Y_1 and U is

$$\frac{1}{(2\pi)^{1/2}}\exp\left[-\tfrac{1}{2}\left(y_1-\delta^{1/2}\right)^2\right]\frac{1}{2^{(n-1)/2}\Gamma\left[\frac{1}{2}(n-1)\right]}\exp\left(-\tfrac{1}{2}u\right)u^{(n-3)/2}.$$

Now make the change of variables

$$y_1=z^{1/2}\cos\theta \qquad (0<z<\infty, 0<\theta<\pi)$$

$$u=z\sin^2\theta.$$

The Jacobian is easily calculated to be $z^{1/2}\sin\theta$ so that the joint density function of Z and θ is

$$\exp\left(-\tfrac{1}{2}\delta\right)\frac{\Gamma\left(\frac{1}{2}n\right)}{\Gamma\left(\frac{1}{2}\right)\Gamma\left[\frac{1}{2}(n-1)\right]}\exp\left(z^{1/2}\delta^{1/2}\cos\theta\right)\sin^{n-2}\theta$$

$$\cdot\frac{1}{2^{n/2}\Gamma\left(\frac{1}{2}n\right)}\exp\left(-\tfrac{1}{2}z\right)z^{n/2-1}.$$

Now integrating with respect to θ over $0<\theta<\pi$ using Lemma 1.3.2 gives the desired marginal density function of Z.

An alternative way of expressing the noncentral χ^2 distribution is as a mixture of central χ^2 density functions where the weights are Poisson probabilities. The result, of independent interest and often useful in investigating distributions of functions of random variables where one or more of them is noncentral χ^2, is given in the following corollary, which is an immediate consequence of Theorem 1.3.4.

COROLLARY 1.3.5. If Z is $\chi_n^2(\delta)$ then its density function can be expressed as

$$\sum_{k=0}^{\infty} P(K=k)g_{n+2k}(z) \qquad (z>0),$$

where K is a Poisson random variable with mean $\delta/2$ so that

$$P(K=k)=\frac{e^{-\delta/2}\left(\frac{1}{2}\delta\right)^k}{k!} \qquad (k=0,1,\dots)$$

and

$$g_r(z)=\frac{1}{2^{r/2}\Gamma(\frac{1}{2}r)}e^{-z/2}z^{r/2-1},$$

the density function of the χ_r^2 distribution.

The characteristic function of Z is easily obtained, either from the definition of Z in terms of a sum of squares of independent normal variables or from Corollary 1.3.5, as

$$\phi_z(t)=(1-2it)^{-n/2}e^{it\delta/(1-2it)},$$

from which it follows that

$$E(Z)=n+\delta \qquad \text{and} \qquad \text{Var}(Z)=2n+4\delta.$$

It is also apparent from the characteristic function that if Z_1 is $\chi_{n_1}^2(\delta_1)$, Z_2 is $\chi_{n_2}^2(\delta_2)$, and Z_1 and Z_2 are independent, then Z_1+Z_2 is $\chi_{n_1+n_2}^2(\delta_1+\delta_2)$.

We now turn to the noncentral F distribution. Recall that the usual or central F distribution is obtained by taking the ratio of two independent χ^2 variables divided by their degrees of freedom. The noncentral F distribution is obtained by allowing the numerator variable to be noncentral χ^2.

THEOREM 1.3.6. If Z_1 is $\chi_{n_1}^2(\delta)$, Z_2 is $\chi_{n_2}^2$, and Z_1 and Z_2 are independent, then

$$F=\frac{Z_1/n_1}{Z_2/n_2}$$

has the density function

$$e^{-\delta/2}{}_1F_1\left(\frac{1}{2}(n_1+n_2);\frac{1}{2}n_1;\frac{\frac{1}{2}\frac{n_1}{n_2}\delta f}{1+\frac{n_1}{n_2}f}\right)$$

$$\times\frac{\Gamma[\frac{1}{2}(n_1+n_2)]}{\Gamma(\frac{1}{2}n_1)\Gamma(\frac{1}{2}n_2)}\frac{f^{n_1/2-1}\left(\frac{n_1}{n_2}\right)^{n_1/2}}{\left(1+\frac{n_1}{n_2}f\right)^{(n_1+n_2)/2}} \qquad (f>0);$$

F is said to have the noncentral F distribution with n_1 and n_2 degrees of freedom and noncentrality parameter δ, to be written as $F_{n_1,n_2}(\delta)$.

Proof. Using Theorem 1.3.4, the joint density function of Z_1 and Z_2 is

$$e^{-\delta/2}{}_0F_1\left(\tfrac{1}{2}n_1;\tfrac{1}{4}\delta z_1\right)\cdot\frac{e^{-(z_1+z_2)/2}z_1^{n_1/2-1}z_2^{n_2/2-1}}{2^{(n_1+n_2)/2}\Gamma\left(\frac{1}{2}n_1\right)\Gamma\left(\frac{1}{2}n_2\right)}.$$

Making the change of variables

$$f=\frac{n_2z_1}{n_1z_2},\qquad t=z_2\qquad (f>0,\quad t>0),$$

with Jacobian n_1t/n_2, the joint density function of F and T is

$$e^{-\delta/2}{}_0F_1\left(\frac{1}{2}n_1;\frac{1}{4}\delta\frac{n_1}{n_2}ft\right)\frac{\exp\left[-\frac{1}{2}t\left(1+\frac{n_1}{n_2}f\right)\right]t^{(n_1+n_2)/2-1}\left(\frac{n_1}{n_2}\right)^{n_1/2}f^{n_1/2-1}}{2^{(n_1+n_2)/2}\Gamma\left(\frac{1}{2}n_1\right)\Gamma\left(\frac{1}{2}n_2\right)}.$$

Now integrating with respect to t over $0<t<\infty$ using Lemma 1.3.3 gives the desired marginal density function of F.

We could also have derived an expression for the distribution of F rather more directly using the mixture representation for the noncentral χ^2 distribution of Z_1 given in Corollary 1.3.5. Since Z_2 is independent of Z_1 this means that the distribution of F can be expressed as a mixture of ratios of independent χ^2 variables. Specifically, the distribution function of F can be expressed in the form

$$P(F\le x)=\sum_{k=0}^{\infty}\frac{e^{-\delta/2}\left(\frac{1}{2}\delta\right)^k}{k!}P\left(F_{n_1+2k,n_2}\le\frac{n_1x}{n_1+2k}\right),$$

where here $F_{r,s}$ denotes a random variable with an F distribution on r and s degrees of freedom. It is not difficult to show that the mean and variance of the $F_{n_1,n_2}(\delta)$ distribution are

$$E(F)=\frac{n_2(n_1+\delta)}{n_1(n_2-2)}\qquad (n_2>2)$$

and

$$\mathrm{Var}(F)=2\left(\frac{n_2}{n_1}\right)^2\left[\frac{(n_1+\delta)^2+(n_1+2\delta)(n_2-2)}{(n_2-2)^2(n_2-4)}\right] \qquad (n_2>4).$$

For further properties of the noncentral χ^2 and F distributions and for information about tables of their distribution functions, a useful reference is Johnson and Kotz (1970), Chapters 28 and 30.

1.4. SOME RESULTS ON QUADRATIC FORMS

There is a vast literature on the distributions of quadratic forms in normal variables. Here we will prove some standard theorems which give just a flavor of results in this area. Our first theorem, although implied by more general ones, will be used often and is worth stating by itself. Note that part (a) says that the exponent in the $N_m(\boldsymbol{\mu},\Sigma)$ density function has a χ^2_m distribution.

THEOREM 1.4.1. If $\mathbf{X}$ is $N_m(\boldsymbol{\mu},\Sigma)$, where Σ is nonsingular, then

(a) $(\mathbf{X}-\boldsymbol{\mu})'\Sigma^{-1}(\mathbf{X}-\boldsymbol{\mu})$ is χ^2_m,

and

(b) $\mathbf{X}'\Sigma^{-1}\mathbf{X}$ is $\chi^2_m(\delta)$,

where $\delta=\boldsymbol{\mu}'\Sigma^{-1}\boldsymbol{\mu}$.

Proof. The central idea in the proofs of both parts (and in many proofs like these) is to transform the components of $\mathbf{X}$ to a set of *independent* normal variables. Write $\Sigma=CC'$, where C is nonsingular. To prove (a), put $\mathbf{U}=C^{-1}(\mathbf{X}-\boldsymbol{\mu})$ so that $\mathbf{U}$ is $N_m(\mathbf{0},I_m)$ and

$$(\mathbf{X}-\boldsymbol{\mu})'\Sigma^{-1}(\mathbf{X}-\boldsymbol{\mu})=\mathbf{U}'\mathbf{U},$$

which is a sum of squares of m independent $N(0,1)$ variables, hence is χ^2_m. To prove (b), put $\mathbf{V}=C^{-1}\mathbf{X}$ so that $\mathbf{V}$ is $N(C^{-1}\boldsymbol{\mu},I_m)$ and

$$\mathbf{X}'\Sigma^{-1}\mathbf{X}=\mathbf{V}'\mathbf{V}$$

which is $\chi^2_m(\delta)$, with $\delta=(C^{-1}\boldsymbol{\mu})'(C^{-1}\boldsymbol{\mu})=\boldsymbol{\mu}'\Sigma^{-1}\boldsymbol{\mu}$.

This theorem is of obvious use in testing hypotheses about the mean of a multivariate normal distribution when the covariance matrix is known. For suppose that $\mathbf{X}_1,\ldots,\mathbf{X}_N$ is a random sample from the $N_m(\boldsymbol{\mu}, \Sigma)$ distribution, then $\overline{\mathbf{X}} = N^{-1}\Sigma_{i=1}^N \mathbf{X}_i$ is $N_m(\boldsymbol{\mu}, (1/N)\Sigma)$ by Corollary 1.2.13 so that from (a) of Theorem 1.4.1,

$$N(\overline{\mathbf{X}}-\boldsymbol{\mu})'\Sigma^{-1}(\overline{\mathbf{X}}-\boldsymbol{\mu}) \quad \text{is} \quad \chi_m^2.$$

To test the *null hypothesis* H_0: $\boldsymbol{\mu}=\boldsymbol{\mu}_0$ against general alternatives H: $\boldsymbol{\mu} \neq \boldsymbol{\mu}_0$, where $\boldsymbol{\mu}_0$ is a specified $m \times 1$ vector, a test of size α is to reject H_0 if

$$W \equiv N(\overline{\mathbf{X}}-\boldsymbol{\mu}_0)'\Sigma^{-1}(\overline{\mathbf{X}}-\boldsymbol{\mu}_0) > c_m(\alpha),$$

where $c_m(\alpha)$ denotes the upper $100\alpha\%$ point of the χ_m^2 distribution. When H_0 is not true, $\overline{\mathbf{X}}-\boldsymbol{\mu}_0$ is $N_m(\boldsymbol{\mu}-\boldsymbol{\mu}_0, (1/N)\Sigma)$ so that from (b) of Theorem 1.4.1, W is $\chi_m^2(\delta)$ with $\delta = N(\boldsymbol{\mu}-\boldsymbol{\mu}_0)'\Sigma^{-1}(\boldsymbol{\mu}-\boldsymbol{\mu}_0)$. Hence the power of the test is a function of δ, namely,

$$\beta(\delta) = P_\delta[W > c_m(\alpha)] = P[\chi_m^2(\delta) > c_m(\alpha)]$$

which can be found from tables of the $\chi_m^2(\delta)$ distribution function.

THEOREM 1.4.2. If $\mathbf{X}$ is $N_m(\boldsymbol{\mu}, I_m)$ and B is an $m \times m$ symmetric matrix then $\mathbf{X}'B\mathbf{X}$ has a noncentral χ^2 distribution if and only if B is idempotent ($B^2 = B$), in which case the degrees of freedom and the noncentrality parameter are respectively $k = \operatorname{rank}(B) = \operatorname{tr} B$ (where $\operatorname{tr} B$ denotes the trace of B) and $\delta = \boldsymbol{\mu}'B\boldsymbol{\mu}$.

Proof. Suppose B is idempotent of rank k and let H be an $m \times m$ orthogonal matrix such that

$$H'BH = \begin{bmatrix} I_k & 0 \\ 0 & 0 \end{bmatrix}.$$

Put $\mathbf{V} = H'\mathbf{X}$ then $\mathbf{V}$ is $N_m(H'\boldsymbol{\mu}, I_m)$ and

$$\mathbf{X}'B\mathbf{X} = \mathbf{V}'\begin{bmatrix} I_k & 0 \\ 0 & 0 \end{bmatrix}\mathbf{V} = \sum_{i=1}^k V_i^2,$$

which, being the sum of squares of k independent normal variables, is noncentral χ^2 with k degrees of freedom. To calculate the noncentrality parameter δ note that

$$\begin{aligned} k+\delta = E\left[\chi_k^2(\delta)\right] &= E(\mathbf{X}'B\mathbf{X}) \\ &= E(\operatorname{tr}\mathbf{X}'B\mathbf{X}) \\ &= E(\operatorname{tr}\mathbf{X}\mathbf{X}'B) \\ &= \operatorname{tr}E(\mathbf{X}\mathbf{X}'B) \\ &= \operatorname{tr}E(\mathbf{X}\mathbf{X}')B \\ &= \operatorname{tr}(I+\boldsymbol{\mu}\boldsymbol{\mu}')B \\ &= k+\boldsymbol{\mu}'B\boldsymbol{\mu} \end{aligned}$$

and hence $\delta=\boldsymbol{\mu}'B\boldsymbol{\mu}$.

Now suppose $\mathbf{X}'B\mathbf{X}$ is $\chi_k^2(\delta)$. If B has rank p, say, let H be an $m\times m$ orthogonal matrix such that

$$H'BH=\begin{bmatrix} \lambda_1 & & & & & 0 \\ & \ddots & & & & \\ & & \lambda_p & & & \\ & & & 0 & & \\ & & & & \ddots & \\ 0 & & & & & 0 \end{bmatrix}$$

where $\lambda_1,\ldots,\lambda_p$ are the nonzero latent roots of B. Put $\mathbf{V}=H'\mathbf{X}$, then

$$\mathbf{X}'B\mathbf{X}=\mathbf{V}'H'BH\mathbf{V}=\sum_{j=1}^{p}\lambda_j V_j^2.$$

Now, $\mathbf{V}$ is $N_m(\boldsymbol{\nu}, I_m)$ with $\boldsymbol{\nu}=H'\boldsymbol{\mu}$, so that V_j^2 is $\chi_1^2(\nu_j^2)$ and has characteristic function

$$(1-2it)^{-1/2}\exp\left(\frac{it\nu_j^2}{1-2it}\right).$$

Hence the characteristic function of $\sum_{j=1}^{p}\lambda_j V_j^2$ is

$$(1)\quad \prod_{j=1}^{p}\left[\exp\left(\frac{it\lambda_j\nu_j^2}{1-2it\lambda_j}\right)(1-2it\lambda_j)^{-1/2}\right]$$

$$=\exp\left(it\sum_{j=1}^{p}\frac{\lambda_j\nu_j^2}{1-2it\lambda_j}\right)\cdot\prod_{j=1}^{p}(1-2it\lambda_j)^{-1/2}$$

where we have used the fact that $V_1,\dots,V_p$ are independent. But since $\mathbf{X}'B\mathbf{X}$ is $\chi_k^2(\delta)$ its characteristic function is

$$(2)\qquad \exp\left(\frac{it\delta}{(1-2it)}\right)(1-2it)^{-k/2}.$$

Equating (1) and (2) it is seen that we must have $p=k$, $\lambda_j=1$ $(j=1,\dots,p)$ and $\delta=\sum_{j=1}^{p}\nu_j^2$. Consequently $H'BH$ has the form

$$H'BH=\begin{bmatrix} I_k & 0\\ 0 & 0\end{bmatrix},$$

which is idempotent. Hence

$$H'BH=(H'BH)(H'BH)=H'B^2H,$$

giving $B=B^2$.

As an application of Theorem 1.4.2 we will prove the following result, which is of interest in the theory of linear models.

THEOREM 1.4.3. If $\mathbf{X}$ is $N_m(\boldsymbol{\mu},\Sigma)$ where Σ is nonsingular and $\mathbf{X}$, $\boldsymbol{\mu}$ and Σ are partitioned as

$$\mathbf{X}=\begin{pmatrix}\mathbf{X}_1\\ \mathbf{X}_2\end{pmatrix},\qquad \boldsymbol{\mu}=\begin{pmatrix}\boldsymbol{\mu}_1\\ \boldsymbol{\mu}_2\end{pmatrix},\qquad \Sigma=\begin{bmatrix}\Sigma_{11} & \Sigma_{12}\\ \Sigma_{21} & \Sigma_{22}\end{bmatrix},$$

where $\mathbf{X}_1$ and $\boldsymbol{\mu}_1$ are $k\times 1$ and Σ_{11} is $k\times k$, then

$$Q\equiv(\mathbf{X}-\boldsymbol{\mu})'\Sigma^{-1}(\mathbf{X}-\boldsymbol{\mu})-(\mathbf{X}_1-\boldsymbol{\mu}_1)'\Sigma_{11}^{-1}(\mathbf{X}_1-\boldsymbol{\mu}_1)$$

is χ_{m-k}^2.

Proof. Write $\Sigma = CC'$ where C is nonsingular and partition C as

$$C = \begin{bmatrix} C_1 \\ C_2 \end{bmatrix},$$

where C_1 is $k \times m$, so that $\Sigma_{11} = C_1 C_1'$. Put $\mathbf{U} = C^{-1}(\mathbf{X} - \boldsymbol{\mu})$, then $\mathbf{U}$ is $N_m(\mathbf{0}, I_m)$ and $\mathbf{X}_1 - \boldsymbol{\mu}_1 = C_1\mathbf{U}$. Then

$$\begin{aligned} Q &= \mathbf{U}'\mathbf{U} - \mathbf{U}'C_1'(C_1C_1')^{-1}C_1\mathbf{U} \\ &= \mathbf{U}'\left[I_m - C_1'(C_1C_1')^{-1}C_1\right]\mathbf{U} \\ &= \mathbf{U}'(I_m - P)\mathbf{U} \end{aligned}$$

where $P = C_1'(C_1C_1')^{-1}C_1$. Now P is symmetric and idempotent, hence so is $B = I_m - P$. Applying Theorem 1.4.2 it follows that Q is χ^2_f, where $f = \text{rank}(B) = m - k$.

Theorem 1.4.2 can be generalized to the case where $\mathbf{X}$ has an arbitrary covariance matrix Σ. Suppose $\mathbf{X}$ is $N_m(\mathbf{0}, \Sigma)$. A necessary and sufficient condition then for $\mathbf{X}'B\mathbf{X}$ to have a χ^2 distribution is

$$\Sigma B \Sigma B \Sigma = \Sigma B \Sigma,$$

in which case the degrees of freedom are $k = \text{rank}(B\Sigma)$. This result is due to Ogasawara and Takahashi (1951). If Σ is nonsingular this condition becomes $B\Sigma B = B$. The above condition is implied by the assumptions of the following theorem which we will prove without recourse to the general result.

THEOREM 1.4.4. If $\mathbf{X}$ is $N_m(\mathbf{0}, \Sigma)$, where Σ has rank r $(\leq m)$, and if B is a generalized inverse of Σ (so that $\Sigma B \Sigma = \Sigma$), then $\mathbf{X}'B\mathbf{X}$ is χ^2_r.

Proof. Put $\mathbf{Y} = C\mathbf{X}$ where C is a nonsingular $m \times m$ matrix such that

$$C\Sigma C' = \begin{bmatrix} I_r & 0 \\ 0 & 0 \end{bmatrix}.$$

Partitioning $\mathbf{Y}$ as $(\mathbf{Y}_1'\mathbf{Y}_2')'$, where $\mathbf{Y}_1$ is $r \times 1$, we have that $\mathbf{Y}_1$ is $N_r(\mathbf{0}, I_r)$ and $\mathbf{Y}_2 = \mathbf{0}$ with probability 1. Hence $\mathbf{Y} = (\mathbf{Y}_1'\mathbf{0}')'$ with probability 1. Now note that

$$(3)\qquad \begin{bmatrix} I_r & 0 \\ 0 & 0 \end{bmatrix} = C\Sigma C'$$

$$= C\Sigma B\Sigma C'$$

$$= C\Sigma C'C'^{-1}BC^{-1}C\Sigma C'$$

$$= \begin{bmatrix} I_r & 0 \\ 0 & 0 \end{bmatrix}(C'^{-1}BC^{-1})\begin{bmatrix} I_r & 0 \\ 0 & 0 \end{bmatrix}.$$

So, with probability 1,

$$\mathbf{X}'B\mathbf{X} = \mathbf{Y}'C^{-1\prime}BC^{-1}\mathbf{Y}$$

$$= (\mathbf{Y}_1'\mathbf{0}')C^{-1\prime}BC^{-1}\begin{pmatrix} \mathbf{Y}_1 \\ \mathbf{0} \end{pmatrix}$$

$$= (\mathbf{Y}_1'\mathbf{0}')\begin{bmatrix} I_r & 0 \\ 0 & 0 \end{bmatrix}C^{-1\prime}BC^{-1}\begin{bmatrix} I_r & 0 \\ 0 & 0 \end{bmatrix}\begin{pmatrix} \mathbf{Y}_1 \\ \mathbf{0} \end{pmatrix}$$

$$= (\mathbf{Y}_1'\mathbf{0}')\begin{bmatrix} I_r & 0 \\ 0 & 0 \end{bmatrix}\begin{pmatrix} \mathbf{Y}_1 \\ \mathbf{0} \end{pmatrix} \qquad \text{using (3)}$$

$$= \mathbf{Y}_1'\mathbf{Y}_1$$

which is χ_r^2.

Our final theorem assumes that the covariance matrix Σ is nonsingular but is otherwise quite general and incorporates Theorems 1.4.1 and 1.4.2 as special cases.

THEOREM 1.4.5. If $\mathbf{X}$ is $N_m(\boldsymbol{\mu}, \Sigma)$, where Σ is nonsingular, and B is an $m \times m$ symmetric matrix then $\mathbf{X}'B\mathbf{X}$ is $\chi_k^2(\delta)$, where $k = \text{rank}(B)$, $\delta = \boldsymbol{\mu}'B\boldsymbol{\mu}$, if and only if $B\Sigma$ is idempotent ($B\Sigma B\Sigma = B\Sigma$, i.e., $B\Sigma B = B$).

Proof. Put $\mathbf{Y} = C\mathbf{X}$, where C is a nonsingular $m \times m$ matrix such that $C\Sigma C' = I_m$. Then

$$\mathbf{X}'B\mathbf{X} = \mathbf{Y}'C^{-1\prime}BC^{-1}\mathbf{Y},$$

where $\mathbf{Y}$ is $N_m(C\boldsymbol{\mu}, I_m)$. From Theorem 1.4.2 it follows that $\mathbf{X}'B\mathbf{X}$ is noncentral χ^2 if and only if $C^{-1\prime}BC^{-1}$ is idempotent. Hence it suffices to show that this is so if and only if $B\Sigma$ is idempotent. If $B\Sigma$ is idempotent we have

$$B = B\Sigma B = BC^{-1}C^{-1\prime}B;$$

hence

$$C^{-1\prime}BC^{-1} = (C^{-1\prime}BC^{-1})(C^{-1\prime}BC^{-1})$$

so that $C^{-1\prime}BC^{-1}$ is idempotent. If $C^{-1\prime}BC^{-1}$ is idempotent then

$$C^{-1\prime}BC^{-1} = C^{-1\prime}BC^{-1}C^{-1\prime}BC^{-1} = C^{-1\prime}B\Sigma BC^{-1}$$

so that $B = B\Sigma B$ and hence $B\Sigma$ is idempotent.

Later we will look at some more results about quadratic forms, where the matrices of the quadratic forms are random, but for the moment enough is enough. It should be emphasized that we have barely scratched the surface. For the reader with a taste for more results relating to both the distributions and independence of quadratic forms, useful references are Johnson and Kotz (1970), Chapter 29; Rao (1973), Section 3b.4; Graybill (1961), Chapter 4; Styan (1970); and Srivastava and Khatri (1979), Chapter 2.

1.5. SPHERICAL AND ELLIPTICAL DISTRIBUTIONS

Although most of classical multivariate analysis has been concerned with the multivariate normal distribution, an increasing amount of attention is being given to alternative distributional models. This is particularly true in *robustness* studies where it is of interest to know how sensitive certain procedures are to the assumption of multivariate normality. As a starting point in such investigations it makes sense to consider a class of density functions whose contours of equal density have the same elliptical shape as the normal, and which contains long-tailed and short-tailed distributions (relative to the normal). Many properties of such distributions have been obtained by Kelker (1970); we will begin by looking at some of these. Our first definition characterizes a *spherical* distribution in an intuitively appealing way.

DEFINITION 1.5.1. A $m\times 1$ random vector $\mathbf{X}$ is said to have a spherical distribution if $\mathbf{X}$ and $H\mathbf{X}$ have the same distribution for all $m\times m$ orthogonal matrices H.

If $\mathbf{X}$ has a spherical distribution with a density function (with respect to Lebesgue measure on R^m) then it is clear that the density function depends on x only through the value of $\mathbf{x}'\mathbf{x}$. Some examples follow:

(i) the $N_m(\mathbf{0}, \sigma^2 I_m)$ distribution with density function

$$(1) \qquad \frac{1}{(2\pi\sigma^2)^{m/2}} \cdot \exp\left(-\frac{1}{2\sigma^2}\mathbf{x}'\mathbf{x}\right);$$

(ii) the "ε-contaminated" normal distribution with density function

$$(2) \quad (1-\varepsilon)\frac{1}{(2\pi)^{m/2}} \cdot \exp\left(-\frac{1}{2}\mathbf{x}'\mathbf{x}\right) + \varepsilon\frac{1}{(2\pi\sigma^2)^{m/2}} \cdot \exp\left(-\frac{1}{2\sigma^2}\mathbf{x}'\mathbf{x}\right)$$

$$(0 \le \varepsilon \le 1);$$

(iii) The multivariate t distribution with n degrees of freedom and density function

$$(3) \qquad \frac{\Gamma\left[\frac{1}{2}(n+m)\right]}{\Gamma\left(\frac{1}{2}n\right)(n\pi)^{m/2}} \frac{1}{\left(1+\frac{1}{n}\mathbf{x}'\mathbf{x}\right)^{(n+m)/2}}.$$

When $n = 1$ this is called the multivariate Cauchy distribution.

A convenient way of *generating* some spherical distributions is as follows. Let $X_1, \ldots, X_m$, Z $(Z > 0)$ be random variables such that given Z, $X_1, \ldots, X_m$ have independent $N(0, Z)$ distributions. If Z has distribution function G then the joint (marginal) density function of $X_1, \ldots, X_m$ is

$$f(x_1, \ldots, x_m) = \int_0^\infty (2\pi Z)^{-m/2} \exp\left(-\frac{1}{2Z}\sum_{i=1}^m x_i^2\right) dG(Z)$$

which is, of course, spherical and is called a *scale mixture* of normal distributions. The class of such spherical distributions formed by varying G is called the class of *compound normal* distributions. It follows that $\mathbf{X} = Z^{1/2}\mathbf{Y}$ where $\mathbf{Y}$ is $N_m(\mathbf{0}, I_m)$ and $Z, \mathbf{Y}$ are independent, so that values of $\mathbf{X}$ can be generated by generating values of independent $N(0,1)$ variables and multiplying them by values of an independent variable Z. Note that if Z takes the values 1 and σ^2 with probabilities $1-\varepsilon$ and ε, respectively, $\mathbf{X}$ has the ε-contaminated normal distribution given by (2). Also, if n/Z is χ^2_n, $\mathbf{X}$ has the m-variate t distribution with n degrees of freedom given by (3) (see Problem 1.30).

The class of elliptical distributions can be defined in a number of ways. We will assume the existence of a density function.

DEFINITION 1.5.2. The $m\times 1$ random vector $\mathbf{X}$ is said to have an elliptical distribution with parameters $\boldsymbol{\mu}(m\times 1)$ and $V(m\times m)$ if its density function is of the form

$$c_m(\det V)^{-1/2}h\big((\mathbf{x}-\boldsymbol{\mu})'V^{-1}(\mathbf{x}-\boldsymbol{\mu})\big) \tag{4}$$

for some function h, where V is positive definite.

Clearly the normalizing constant c_m could be absorbed into the function h, but with this notation h can be independent of m. If $\mathbf{X}$ has an elliptical distribution we will write that $\mathbf{X}$ is $E_m(\boldsymbol{\mu},V)$. Note that this does not mean that $\mathbf{X}$ has a particular elliptical distribution but only that its distribution belongs to the class of elliptical distributions. If $\mathbf{X}$ is $E_m(\mathbf{0},I_m)$ then obviously $\mathbf{X}$ has a spherical distribution. Also, if $\mathbf{Y}$ has an m-variate spherical distribution with a density function and $X=C\mathbf{Y}+\boldsymbol{\mu}$, where C is a nonsingular $m\times m$ matrix, then $\mathbf{X}$ is $E_m(\boldsymbol{\mu},V)$ with $V=CC'$.

The following two assertions are fairly easily proved and are left to the reader (Problem 1.27). If $\mathbf{X}$ is $E_m(\boldsymbol{\mu},V)$ then:

(a) The characteristic function $\phi(\mathbf{t})=E(e^{i\mathbf{t}'\mathbf{X}})$ has the form

$$\phi(\mathbf{t})=e^{i\mathbf{t}'\boldsymbol{\mu}}\psi(\mathbf{t}'V\mathbf{t}) \qquad \text{for some function } \psi. \tag{5}$$

(b) Provided they exist, $E(\mathbf{X})=\boldsymbol{\mu}$ and $\mathrm{Cov}(\mathbf{X})=\alpha V$ for some constant α. In terms of the characteristic function this constant is $\alpha=-2\psi'(0)$.

It follows from (b) that *all* distributions in the class $E_m(\boldsymbol{\mu},V)$ have the same mean $\boldsymbol{\mu}$ and the same correlation matrix $P=(\rho_{ij})$, where

$$\rho_{ij}=\frac{\mathrm{Cov}(X_i,X_j)}{\left[\mathrm{Var}(X_i)\,\mathrm{Var}(X_j)\right]^{1/2}}=\frac{\alpha v_{ij}}{(\alpha v_{ii}\alpha v_{jj})^{1/2}}=\frac{v_{ij}}{(v_{ii}v_{jj})^{1/2}}.$$

From (a) it follows that *all marginal distributions are elliptical* and all marginal density functions of dimension $j<m$ have the same functional form. For example, partitioning $\mathbf{X}$, $\boldsymbol{\mu}$, and V as

$$\mathbf{X}=\begin{pmatrix}\mathbf{X}_1\\ \mathbf{X}_2\end{pmatrix},\qquad \boldsymbol{\mu}=\begin{pmatrix}\boldsymbol{\mu}_1\\ \boldsymbol{\mu}_2\end{pmatrix},\qquad V=\begin{bmatrix}V_{11} & V_{12}\\ V_{21} & V_{22}\end{bmatrix},$$

where $\mathbf{X}_1$ and $\boldsymbol{\mu}_1$ are $k\times 1$ and V_{11} is $k\times k$, the characteristic function of $\mathbf{X}_1$, obtained from (a) by putting $\mathbf{t}=(\mathbf{t}_1'\mathbf{0}')'$, where $\mathbf{t}_1$ is $k\times 1$, is

$$\exp(i\mathbf{t}_1'\boldsymbol{\mu}_1)\psi(\mathbf{t}_1'V_{11}\mathbf{t}_1)$$

which is the characteristic function of a random vector with an $E_k(\boldsymbol{\mu}_1, V_{11})$ distribution. It is worth noting that if *any* marginal distribution is normal then $\mathbf{X}$ is normal, for the characteristic function of $\mathbf{X}$ has the same functional form as the characteristic function of the marginal distribution, i.e, normal.

We know from Theorem 1.2.8 that if $\mathbf{X}$ is $N_m(\boldsymbol{\mu},\Sigma)$ and Σ is diagonal then the components $X_1,\ldots,X_m$ of $\mathbf{X}$ are all independent. Within the class of elliptical distributions independence when Σ is diagonal characterizes the normal distribution, as the following theorem shows.

THEOREM 1.5.3. Let $\mathbf{X}$ be $E_m(\boldsymbol{\mu},V)$, where V is diagonal. If $X_1,\ldots,X_m$ are all independent then $\mathbf{X}$ is normal.

Proof. Without loss of generality we can assume $\boldsymbol{\mu}=\mathbf{0}$. Then the characteristic function of $\mathbf{X}$ has the form

$$\phi(\mathbf{t})=\psi(\mathbf{t}'V\mathbf{t})=\psi\left(\sum_{i=1}^m t_i^2 v_{ii}\right) \tag{6}$$

for some function ψ, because $V=\operatorname{diag}(v_{11},\ldots,v_{mm})$. Since $X_1,\ldots,X_m$ are independent we have

$$\psi(\mathbf{t}'V\mathbf{t})=\prod_{i=1}^m \psi(t_i^2 v_{ii}). \tag{7}$$

Equating (6) and (7) and putting $u_i=t_i v_{ii}^{1/2}$ gives

$$\psi\left(\sum_{i=1}^m u_i^2\right)=\prod_{i=1}^m \psi(u_i^2).$$

This equation is known as Hamel's equation and its only continuous solution is

$$\psi(z)=e^{kz}$$

for some constant k [see, e.g., Feller (1971), page 305]. Hence the characteristic function of $\mathbf{X}$ has the form

$$\phi(\mathbf{t})=e^{k\mathbf{t}'V\mathbf{t}}$$

and, because it is a characteristic function, we must have $k \le 0$ (why?) which implies that $\mathbf{X}$ has a normal distribution.

We now turn to an examination of some conditional distribution properties. If $\mathbf{X}$ is $N_m(\boldsymbol{\mu}, \Sigma)$ then by Theorem 1.2.11 the conditional expectation of a subvector of $\mathbf{X}$ given the remaining components is linear in the fixed variables, and the conditional covariance matrix does not depend on the fixed variables. The first of these properties carries over to the class of elliptical distributions.

THEOREM 1.5.4. If $\mathbf{X}$ is $E_m(\boldsymbol{\mu}, V)$ and $\mathbf{X}$, $\boldsymbol{\mu}$ and V are partitioned as

$$\mathbf{X}=\begin{pmatrix}\mathbf{X}_1\\ \mathbf{X}_2\end{pmatrix},\qquad \boldsymbol{\mu}=\begin{pmatrix}\boldsymbol{\mu}_1\\ \boldsymbol{\mu}_2\end{pmatrix},\qquad V=\begin{bmatrix}V_{11} & V_{12}\\ V_{21} & V_{22}\end{bmatrix}$$

where $\mathbf{X}_1$ and $\boldsymbol{\mu}_1$ are $k\times 1$ and V_{11} is $k\times k$, then, provided they exist,

$$E(\mathbf{X}_1|\mathbf{X}_2)=\boldsymbol{\mu}_1+V_{12}V_{22}^{-1}(\mathbf{X}_2-\boldsymbol{\mu}_2) \tag{8}$$

$$\operatorname{Cov}(\mathbf{X}_1|\mathbf{X}_2)=g(\mathbf{X}_2)(V_{11}-V_{12}V_{22}^{-1}V_{21}) \tag{9}$$

for some function g. Moreover the conditional distribution of $\mathbf{X}_1$ given $\mathbf{X}_2$ is k-variate elliptical.

A proof can be constructed which is similar to that of Theorem 1.2.11 and is left as an exercise (Problem 1.28). It can also be shown that if the conditional covariance matrix of $\mathbf{X}_1$ given $\mathbf{X}_2$ does not depend on $\mathbf{X}_2$ then $\mathbf{X}$ *must* be normal, i.e., this property characterizes the multivariate normal distribution in the class of elliptical distributions [see Kelker (1970)].

An interesting property of a spherically distributed vector $\mathbf{X}$ is that a transformation to polar coordinates yields angles and a radius which are all independently distributed, with the angles having the same distributions for all $\mathbf{X}$.

THEOREM 1.5.5. If $\mathbf{X}$ is $E_m(\mathbf{0}, I_m)$ with density function $c_m h(\mathbf{x}'\mathbf{x})$ and

$$\begin{aligned}
X_1&=r\sin\theta_1\sin\theta_2\ldots\sin\theta_{m-2}\sin\theta_{m-1}\\
X_2&=r\sin\theta_1\sin\theta_2\ldots\sin\theta_{m-2}\cos\theta_{m-1}\\
X_3&=r\sin\theta_1\sin\theta_2\ldots\cos\theta_{m-2}\\
&\;\vdots\\
X_{m-1}&=r\sin\theta_1\cos\theta_2\\
X_m&=r\cos\theta_1
\end{aligned} \tag{10}$$

$(r>0, 0<\theta_i\le\pi, i=1,\ldots,m-2, 0<\theta_{m-1}\le 2\pi)$ then $r,\theta_1,\ldots,\theta_{m-1}$ are independent, the distributions of $\theta_1,\ldots,\theta_{m-1}$ are the same for all $\mathbf{X}$, with θ_k having density function proportional to $\sin^{m-1-k}\theta_k$ (so that θ_{m-1} is uniformly distributed on $(0,2\pi)$), and $r^2=\mathbf{X}'\mathbf{X}$ has density function

$$(11)\qquad f_{r^2}(y)=\frac{c_m\pi^{m/2}}{\Gamma(\frac{1}{2}m)}y^{m/2-1}h(y)\qquad (y>0).$$

Proof. The Jacobian of the transformation from $X_1,\ldots,X_m$ to $r,\theta_1,\ldots,\theta_{m-1}$ given by (10) is $r^{m-1}\sin^{m-2}\theta_1\sin^{m-3}\theta_2\ldots\sin\theta_{m-2}$. (For the reader who is unfamiliar with this result it will be derived in Theorem 2.1.3). It follows then that the joint density function of $r^2,\theta_1,\ldots,\theta_{m-1}$ is

$$(12)\qquad \tfrac{1}{2}c_m(r^2)^{m/2-1}\sin^{m-2}\theta_1\sin^{m-3}\theta_2\ldots\sin\theta_{m-2}h(r^2)$$

from which it is apparent that $r,\theta_1,\ldots,\theta_{m-1}$ are all independent and θ_k has density function proportional to $\sin^{m-1-k}\theta_k$. Integrating (12) with respect to $\theta_1,\ldots,\theta_{m-1}$ yields the factor $2\pi^{m/2}/\Gamma(\frac{1}{2}m)$ which is, of course, the surface area of a sphere of unit radius in R^m. It then follows that r^2 has the density function given by (11).

As an example, if $\mathbf{X}$ is $N_m(\mathbf{0}, I_m)$ then $c_m=(2\pi)^{-m/2}$ and $h(u)=e^{-u/2}$ so that $r^2=\mathbf{X}'\mathbf{X}$ has density function

$$f_{r^2}(y)=\frac{1}{2^{m/2}\Gamma(\frac{1}{2}m)}e^{-y/2}y^{m/2-1}\qquad (y>0),$$

the familiar χ^2_m density.

It follows readily from Theorem 1.5.5 that if $\mathbf{X}$ is spherically distributed with a density function then $\mathbf{X}$ may be expressed as $\mathbf{X}=r\mathbf{T}$ where $r^2=\mathbf{X}'\mathbf{X}$ and $\mathbf{T}$ is a function of the angular variables $\theta_1,\ldots,\theta_{m-1}$. The variables r and $\mathbf{T}$ are independent and the distribution of $\mathbf{X}$ is characterized by the distribution of r, and it is easily shown that $\mathbf{T}$, for all $\mathbf{X}$, is uniformly distributed on

$$S_m=\{\mathbf{x}\in R^m;\mathbf{x}'\mathbf{x}=1\},$$

the unit sphere in R^m. The assumption that $\mathbf{X}$ has a density function is unnecessary, as Theorem 1.5.6 will show. In the proof, due to Kariya and Eaton (1977) and Eaton (1977) we will use the fact that the uniform distribution on S_m is the *unique* distribution on S_m which is invariant under orthogonal transformations. That it is invariant is clear; the uniqueness is a somewhat more subtle matter [see, for example, Dempster (1969), Section 12.2 and the discussion later in Section 2.1.4].

THEOREM 1.5.6. If $\mathbf{X}$ has an m-variate spherical distribution with $P(\mathbf{X}=0)=0$ and

$$r=\|\mathbf{X}\|=(\mathbf{X}'\mathbf{X})^{1/2},\ \mathbf{T}(\mathbf{X})=\|\mathbf{X}\|^{-1}\mathbf{X},$$

then $\mathbf{T}(\mathbf{X})$ is uniformly distributed on S_m and $\mathbf{T}(\mathbf{X})$ and r are independent.

Proof. For any $m\times m$ orthogonal matrix H

$$\mathbf{T}(H\mathbf{X})=\|H\mathbf{X}\|^{-1}H\mathbf{X}=\|\mathbf{X}\|^{-1}H\mathbf{X}=H\mathbf{T}(\mathbf{X}),$$

so that $\mathbf{T}(H\mathbf{X})$ and $H\mathbf{T}(\mathbf{X})$ have the same distribution. Since $\mathbf{X}$ has a spherical distribution both $\mathbf{X}$ and $H\mathbf{X}$ have the same distribution (by Definition 1.5.1), hence so do $\mathbf{T}(\mathbf{X})$ and $\mathbf{T}(H\mathbf{X})$. Consequently both $\mathbf{T}(\mathbf{X})$ and $H\mathbf{T}(\mathbf{X})$ have the same distribution. Since the uniform distribution on S_m is the unique distribution invariant under orthogonal transformations it follows that $\mathbf{T}(\mathbf{X})$ is uniformly distributed on S_m. For the independence part define a measure μ on S_m by

(13) $$\mu(B)=P(\mathbf{T}(\mathbf{X})\in B\,|\,r\in C)$$

for a fixed Borel set C with $P(r\in C)\neq 0$, where B is a Borel set in S_m. It is easily shown that μ is a probability measure on S_m which is invariant under orthogonal transformations so that μ is the probability measure of the uniform distribution on S_m; that is, the distribution of $\mathbf{T}(\mathbf{X})$. Hence

$$\mu(B)=P(\mathbf{T}(\mathbf{X})\in B),$$

and this, together with (13), shows that $\mathbf{T}(\mathbf{X})$ and r are independently distributed.

This theorem is used to generalize well-known results for normal random variables.

THEOREM 1.5.7. Let $\mathbf{X}$ have an m-variate spherical distribution with $P(\mathbf{X}=0)=0$.

(i) If $W=\dfrac{\boldsymbol{\alpha}'\mathbf{X}}{\|\mathbf{X}\|}$, where $\boldsymbol{\alpha}\in R^m$, $\boldsymbol{\alpha}'\boldsymbol{\alpha}=1$, then

$$Y=\frac{(m-1)^{1/2}W}{(1-W^2)^{1/2}}$$

has the t_{m-1} distribution.

(ii) If B is an $m \times m$ symmetric idempotent matrix of rank k then

$$Z = \frac{\mathbf{X}'B\mathbf{X}}{\|\mathbf{X}\|^2}$$

has the beta distribution with parameters $\frac{1}{2}k$ and $\frac{1}{2}(m-k)$.

Proof. Both parts are proved using Theorem 1.5.6 by noting that Y and Z are functions of a random vector $\mathbf{T}(\mathbf{X}) = \mathbf{X}/\|\mathbf{X}\|$ uniformly distributed on S_m. To prove (i), note that $W = \boldsymbol{\alpha}'\mathbf{T}(\mathbf{X})$ so, without loss of generality, we can assume that $\mathbf{X}$ is $N_m(\mathbf{0}, I_m)$ and take $\boldsymbol{\alpha} = (1,0,\ldots,0)'$. Then

$$Y = \frac{(m-1)^{1/2} X_1}{\left(\sum_{i=2}^{m} X_i^2 \right)^{1/2}}$$

and clearly has the t_{m-1} distribution. To prove (ii), note that

$$Z = \mathbf{T}(\mathbf{X})' B \mathbf{T}(\mathbf{X})$$

and so we can again assume that $\mathbf{X}$ is $N_m(\mathbf{0}, I_m)$. Putting $\mathbf{U} = H\mathbf{X}$ where H is an orthogonal $m \times m$ matrix such that

$$HBH' = \begin{bmatrix} I_k & 0 \\ 0 & 0 \end{bmatrix}$$

we then have

$$Z = \frac{\sum_{i=1}^{k} U_i^2}{\sum_{i=1}^{m} U_i^2} = \frac{V_1}{V_1 + V_2},$$

where $V_1 = \sum_{i=1}^{k} U_i^2$ is χ_k^2, $V_2 = \sum_{i=k+1}^{m} U_i^2$ is χ_{m-k}^2 and V_1 and V_2 are independent. It then follows easily that Z has the beta $(\frac{1}{2}k, \frac{1}{2}(m-k))$ distribution.

This theorem will be used in Chapter 5 to weaken normality assumptions usually made in order to derive the distributions of correlation coefficients. Another simple example of statistical interest where a normality assumption can be dropped is the following one, noted by Efron (1969). If $X_1, \ldots, X_N$

are independent $N(\mu, \sigma^2)$ random variables and the null hypothesis $\mu = 0$ is true, it is well known that the statistic

$$t = \frac{N^{1/2}\bar{X}}{S} = \frac{\dfrac{1}{N^{1/2}} \sum_{i=1}^{N} X_i}{\left[\dfrac{1}{N-1} \sum_{i=1}^{N} (X_i - \bar{X})^2\right]^{1/2}}$$

has the t_{N-1} distribution. For this result to hold it is enough that the vector $\mathbf{X} = (X_1, \ldots, X_N)'$ has a spherical distribution with $P(\mathbf{X} = \mathbf{0}) = 0$ as (i) of Theorem 1.5.7, with $\boldsymbol{\alpha} = (N^{-1/2}, \ldots, N^{-1/2})' \in R^N$, shows.

There is a growing literature on spherical and elliptical distributions; as well as the papers by Kelker (1970) and Kariya and Eaton (1977) already mentioned, a useful review paper by Devlin et al. (1976) gives many additional references, as does another by Chmielewski (1981).

1.6. MULTIVARIATE CUMULANTS

We now turn to a discussion of cumulants of multivariate distributions in general, and elliptical distributions in particular. Let $\mathbf{X}$ be an $m \times 1$ random vector with characteristic function $\phi(\mathbf{t})$ and suppose for simplicity that all the moments exist. The characteristic function of X_j is $\phi_j(t_j) = \phi(\mathbf{t})$, where

$$\mathbf{t} = (0, \ldots, 0, \; t_j, \; 0, \ldots, 0)',$$

and the cumulants of X_j are the coefficients κ_k^j in

$$\log \phi_j(t_j) = \sum_{k=0}^{\infty} \kappa_k^j \frac{(it_j)^k}{k!}.$$

(The superscript on κ refers to the variable, the subscript to the *order* of the cumulant.) The first four cumulants in terms of the moments $\mu_k^j = E(X_j^k)$ of X_j are [see, for example, Cramér (1946), Section 15.10]

$$\kappa_1^j = \mu_1^j,$$

$$\kappa_2^j = \mu_2^j - (\mu_1^j)^2 = \mathrm{Var}(X_j),$$

$$\kappa_3^j = \mu_3^j - 3\mu_1^j\mu_2^j + 2(\mu_1^j)^3,$$

$$\kappa_4^j = \mu_4^j - 4\mu_1^j\mu_3^j - 3(\mu_2^j)^2 + 12\mu_2^j(\mu_1^j)^2 - 6(\mu_1^j)^4.$$

The *skewness* γ_1^j and *kurtosis* γ_2^j of the (marginal) distribution of X_j are

$$\gamma_1^j=\frac{\kappa_3^j}{(\kappa_2^j)^{3/2}}\quad\text{and}\quad \gamma_2^j=\frac{\kappa_4^j}{(\kappa_2^j)^2}.$$

If X_j has a normal distribution, all cumulants κ_k^j of order $k>2$ are zero.

The mixed cumulants or cumulants of a joint distribution are defined in a similar way. For example, denoting the joint characteristic function of X_j and X_k by $\phi_{jk}(t_j,t_k)$, the cumulants of their joint distribution are the coefficients $\kappa_{r_1r_2}^{jk}$ in

$$\log\phi_{jk}(t_j,t_k)=\sum_{r_1,r_2=0}^{\infty}\kappa_{r_1r_2}^{jk}\frac{(it_j)^{r_1}(it_k)^{r_2}}{r_1!r_2!}$$

[where $\kappa_{11}^{jk}=\text{Cov}(X_j,X_k)$], and this can be extended in an obvious way to define the cumulants of the joint distribution of any number of the variables $X_1,\ldots,X_m$. The cumulants of the joint distribution of $X_1,\ldots,X_m$ then, are the coefficients $\kappa_{r_1r_2\ldots r_m}^{12\ldots m}$ in

$$\log\phi(\mathbf{t})=\sum_{r_1,\ldots,r_m=0}^{\infty}\kappa_{r_1r_2\ldots r_m}^{12\ldots m}\frac{(it_1)^{r_1}(it_2)^{r_2}\ldots(it_m)^{r_m}}{r_1!r_2!\ldots r_m!}.$$

If $\mathbf{X}$ is normal, all cumulants for which $\sum r_i>2$ are zero.

If $\mathbf{X}$ has an m-variate elliptical distribution $E_m(\boldsymbol{\mu},V)$ (see Section 1.5) with characteristic function

$$\phi(\mathbf{t})=e^{i\boldsymbol{\mu}'\mathbf{t}}\psi(\mathbf{t}'V\mathbf{t}),$$

covariance matrix $\Sigma=-2\psi'(0)V=(\sigma_{ij})$, and finite fourth moments, then it is easy to show (by differentiating $\log\phi(\mathbf{t})$) that:

(a) The marginal distributions of X_j $(j=1,\ldots,m)$ all have zero skewness and the *same* kurtosis

$$\text{(1)}\qquad \gamma_2^j=\frac{\kappa_4^j}{(\kappa_2^j)^2}=\frac{3[\psi''(0)-\psi'(0)^2]}{\psi'(0)^2}=3\kappa,\quad\text{say}\quad(j=1,\ldots,m)$$

(b) All fourth-order cumulants are determined by this *kurtosis parameter* κ as

$$\text{(2)}\qquad \kappa_{1111}^{ijkl}=\kappa(\sigma_{ij}\sigma_{kl}+\sigma_{ik}\sigma_{jl}+\sigma_{il}\sigma_{jk})$$

(see problem 1.33). Why is this of interest? We have already noted in Corollary 1.2.16 that the asymptotic distribution of a sample covariance matrix is normal, with a covariance matrix depending on the fourth-order moments of the underlying distribution. It follows that statistics which are "smooth" functions of the elements of the sample covariance matrix will also have limiting distributions depending on these fourth-order moments. The result (b) above shows that if we are sampling from an elliptical distribution these moments have reasonably simple forms. We will examine some specific limiting distributions later.

Finally, and for the sake of completeness, we will give the elements of the covariance matrix in the asymptotic normal distribution of a sample covariance matrix. From Corollary 1.2.18 we know that if

$$U(n)=n^{1/2}[S(n)-\Sigma],$$

where $S(n)$ is the sample covariance matrix constructed from a sample of $N=n+1$ independent and identically distributed $m\times 1$ vectors $\mathbf{X}_1,\dots,\mathbf{X}_N$ with finite fourth moments then the asymptotic distribution of $U(n)$ is normal with mean 0. The covariances, expressed in terms of the cumulants of the distribution of $\mathbf{X}_1$, are

$$\text{(3)} \qquad \operatorname{Cov}(u_{ij}(n),u_{kl}(n))=\kappa_{1111}^{ijkl}+\kappa_{11}^{ik}\kappa_{11}^{jl}+\kappa_{11}^{il}\kappa_{11}^{jk}.$$

In this formula the convention is that if any of the variables are identical the subscripts are amalgamated; for example,

$$\kappa_{1111}^{iijk}=\kappa_{211}^{ijk}, \qquad \kappa_{11}^{ii}=\kappa_2^i,$$

and so on. These covariances (3) have been given by Cook (1951); for related work the reader is referred to Kendall and Stuart (1969), Chapters 3, 12 and 13, and Waternaux (1976).

PROBLEMS

1.1. Prove Theorem 1.2.8 using the characteristic function of $\mathbf{X}$.

1.2. State and prove an extension of Theorem 1.2.8 when the $m\times 1$ vector $\mathbf{X}$ is partitioned into r subvectors $\mathbf{X}_1,\dots,\mathbf{X}_r$ of dimensions $m_1,\dots,m_r$, respectively ($\sum_1^r m_i=m$).

1.3. Consider the ellipse

$$\frac{1}{1-\rho^2}(z_1^2+z_2^2-2\rho z_1 z_2)=k$$

for $k>0$. For $\rho>0$ show that the principal axes are along the lines $z_1=z_2$, $z_1=-z_2$ with lengths $2\sqrt{k(1+\rho)}$ and $2\sqrt{k(1-\rho)}$, respectively. What happens if $\rho<0$?

1.4. If M is an $n\times r$ matrix and $R(M)$, $K(M)$ are defined by (10) and (11) of Section 1.2 prove (12), i.e., that $K(M)^{\perp}=R(M')$.

1.5. Prove Theorem 1.2.14.

1.6. If $U(n)=n^{1/2}[S(n)-\Sigma]$, where $S(n)$ is the sample covariance matrix formed from a random sample of size $N=n+1$ from a $N_m(\mu,\Sigma)$ distribution, show that the elements of the covariance matrix in the asymptotic normal distribution for $U(n)$ are given by

$$\text{Cov}(u_{ij}(n),u_{kl}(n))=\sigma_{ik}\sigma_{jl}+\sigma_{il}\sigma_{jk}$$

where $\Sigma=(\sigma_{ij})$.

1.7. Suppose that the random variables X and Y have joint distribution function

$$F(x,y)=\Phi(x)\Phi(y)[1+\alpha(1-\Phi(x))(1-\Phi(y))],$$

where $|\alpha|\le 1$ and $\Phi(x)$ denotes the standard normal distribution function. Show that the marginal distributions of X and Y are standard normal.

1.8. Let $\phi_1(x_1,x_2)$ and $\phi_2(x_1,x_2)$ be two bivariate normal density functions with zero means, unit variances and different correlation coefficients ρ_1 and ρ_2 respectively (i.e., ϕ_1 and ϕ_2 have the form (8) of Section 1.2). Show that the density function $\frac{1}{2}[\phi_1(x_1,x_2)+\phi_2(x_1,x_2)]$ is not normal but that its two marginal density functions are normal.

1.9. Let $h(x)$ be an odd continuous function such that $|h(x)|<(2\pi e)^{-1/2}$ for all x and $h(x)=0$ for $x\notin(-1,1)$, and let $\phi(x)$ be the standard normal density function. Show that the function

$$f(x,y)=\phi(x)\phi(y)+h(x)h(y)$$

is a non-normal bivariate density function with normal marginal density functions.

1.10. Suppose that $\mathbf{X}=\begin{pmatrix} X_1 \\ X_2 \end{pmatrix}$ has the $N_2(\mathbf{0},\Sigma)$ distribution, with

$$\Sigma=\begin{bmatrix} 1 & \rho \\ \rho & 1 \end{bmatrix}, \qquad |\rho|<1.$$

Changing to polar coordinates, put $X_1=r\cos\theta$, $X_2=r\sin\theta$ $(r>0, 0<\theta<2\pi)$.

(a) Show that the marginal density function of θ is

$$\frac{\sqrt{1-\rho^2}}{2\pi(1-2\rho\sin\theta\cos\theta)} \qquad 0<\theta<2\pi.$$

(b) Show that

$$P(X_1>0, X_2>0)=\frac{1}{2}-\frac{1}{2\pi}\cos^{-1}\rho.$$

(c) Show that

$$P(X_1X_2>0)=\frac{1}{2}+\frac{1}{\pi}\sin^{-1}\rho.$$

(d) Show that

$$P(X_1X_2<0)=\frac{1}{\pi}\cos^{-1}\rho.$$

1.11. Let $\mathbf{X}_1,\mathbf{X}_2,\ldots$ be independent $N_m(\boldsymbol{\mu},\Sigma)$ random vectors and let

$$\mathbf{S}_N=\sum_{i=1}^{N}\mathbf{X}_i.$$

For $N_1<N_2$:

(a) Find the distribution of $(\mathbf{S}'_{N_1},\mathbf{S}'_{N_2})'$.
(b) Find the conditional distribution of $\mathbf{S}_{N_1}$ given $\mathbf{S}_{N_2}$.

1.12. Suppose that $\mathbf{X}$ is $N_3(\mathbf{0},\Sigma)$, where

$$\Sigma=\begin{bmatrix} 1 & \sigma_{12} & \sigma_{13} \\ \sigma_{12} & 1 & \sigma_{23} \\ \sigma_{13} & \sigma_{23} & 1 \end{bmatrix}.$$

Show that

$$P(X_1>0, X_2>0, X_3>0)=\frac{1}{8}+\frac{1}{4\pi}(\sin^{-1}\sigma_{12}+\sin^{-1}\sigma_{13}+\sin^{-1}\sigma_{23}).$$

1.13. Suppose that $\mathbf{X}$ is $N_3(\mathbf{0}, \Sigma)$, where

$$\Sigma=\begin{bmatrix} 1 & \rho & 0 \\ \rho & 1 & \rho \\ 0 & \rho & 1 \end{bmatrix}.$$

Is there a value of ρ for which $X_1+X_2+X_3$ and $X_1-X_2-X_3$ are independent?

1.14. Suppose that the vector

$$\begin{pmatrix} Y \\ \mathbf{X} \end{pmatrix},$$

where $\mathbf{X}$ is $(m-1)\times 1$ and Y is 1×1, has mean vector $\mu\mathbf{1}$, $\mathbf{1}=(1,1,\ldots,1)'$ and covariance matrix

$$\begin{bmatrix} \sigma_{11} & \boldsymbol{\sigma}_{12}' \\ \boldsymbol{\sigma}_{12} & \Sigma_{22} \end{bmatrix},$$

where $\sigma_{11}=\text{Var}(Y)$, $\Sigma_{22}=\text{Cov}(\mathbf{X})$. Find the coefficient vector $\boldsymbol{\alpha}$ of a linear function $\boldsymbol{\alpha}'\mathbf{X}$ which minimizes $\text{Var}(Y-\boldsymbol{\alpha}'\mathbf{X})$ subject to the condition $E(\boldsymbol{\alpha}'\mathbf{X})=E(Y)$.

1.15. Prove Lemma 1.3.3.

1.16. If Z is $\chi_n^2(\delta)$ where n is an even integer, prove that

$$P(Z\le x)=P(X_1-X_2\ge\tfrac{1}{2}n),$$

where X_1 and X_2 are independent Poisson random variables with means $\frac{1}{2}x$ and $\frac{1}{2}\delta$, respectively.

1.17. If Z is $\chi_n^2(\delta)$ show that its characteristic function is

$$\Phi_Z(t)=\exp\left(\frac{it\delta}{1-2it}\right)(1-2it)^{-n/2}.$$

Hence, show that $E(Z)=n+\delta$, $\text{Var}(\delta)=2n+4\delta$ and that the skewness γ_1

and kurtosis γ_2 of Z (see Section 1.6) are

$$\gamma_1=\frac{\sqrt{8}\,(n+3\delta)}{(n+2\delta)^{3/2}},\qquad \gamma_2=\frac{12(n+4\delta)}{(n+2\delta)^2}.$$

1.18. If Z is $\chi_n^2(\delta)$ prove that the asymptotic distribution of

$$\frac{Z-(n+\delta)}{(2n+4\delta)^{1/2}}$$

is $N(0,1)$ as either $n\to\infty$ with δ fixed or $\delta\to\infty$ with n fixed.

1.19. Let $f(z;n,\delta)$ denote the density function of the $\chi_n^2(\delta)$ distribution (see Theorem 1.3.4 and Corollary 1.3.5). Show that

$$f(z;n,\delta)=\tfrac{1}{2}\left[P\left(\chi_n^2(\delta)>x\right)-P\left(\chi_{n-2}^2(\delta)>x\right)\right].$$

1.20. If F is $F_{n_1,n_2}(\delta)$ show that

$$E(F)=\frac{n_2(n_1+\delta)}{n_1(n_2-2)}\qquad (n_2>2)$$

and

$$\mathrm{Var}(F)=2\left(\frac{n_2}{n_1}\right)^2\left[\frac{(n_1+\delta)^2+(n_1+2\delta)(n_2-2)}{(n_2-2)^2(n_2-4)}\right]\qquad (n_2>4).$$

1.21. If F is $F_{n_1,n_2}(\delta)$, where n_1 is even, prove that

$$P\left(F<\frac{x}{n_1}\right)=P\left(X_1-X_2\geq\frac{1}{2}n_1\right),$$

where X_1 and X_2 are independent with X_1 having a Poisson distribution with mean $\frac{1}{2}\delta$ and X_2 having a negative binomial distribution, i.e.,

$$P(X_2=k)=\binom{\frac{1}{2}n_2+k-1}{\frac{1}{2}n_2-1}\left(\frac{x}{n_2+x}\right)^k\left(\frac{n_2}{n_2+x}\right)^{n_2/2}\qquad (k=0,1,2,\ldots).$$

1.22. If $\mathbf{X}$ is $N_m(\boldsymbol{\mu}, \Sigma)$, where Σ is positive definite, A is an $m \times m$ symmetric matrix, and B is an $r \times m$ matrix, prove that $\mathbf{X}'A\mathbf{X}$ and $B\mathbf{X}$ are independent if and only if $B\Sigma A = 0$.

1.23. If $\mathbf{X}$ is $N_m(\boldsymbol{\mu}, \Sigma)$, where Σ is positive definite and A and B are $m \times m$ symmetric matrices, prove that $\mathbf{X}'A\mathbf{X}$ and $\mathbf{X}'B\mathbf{X}$ are independent if and only if $A\Sigma B = 0$.

1.24. If $\mathbf{X}$ is $N_m(\boldsymbol{\mu}, \Sigma)$ prove that:

(a) $E(\mathbf{X}'A\mathbf{X}) = \text{tr}(A\Sigma) + \boldsymbol{\mu}'A\boldsymbol{\mu}$;

(b) $\text{Var}(\mathbf{X}'A\mathbf{X}) = 2[\text{tr}(A\Sigma A\Sigma) + 2\boldsymbol{\mu}'A\Sigma A\boldsymbol{\mu}]$.

1.25. Let $X_1, \ldots, X_m$ be independent random variables with means $\theta_1, \ldots, \theta_m$, common variance σ^2, and common third and fourth moments about their means μ_3, μ_4, respectively; i.e.,

$$\mu_k = E\left[(X_i - \theta_i)^k\right]; \qquad k = 3, 4; \quad i = 1, \ldots, m.$$

If A is an $m \times m$ symmetric matrix prove that

$$\text{Var}(\mathbf{X}'A\mathbf{X}) = (\mu_4 - 3\sigma^4)\mathbf{a}'\mathbf{a} + 2\sigma^4 \text{tr}(A^2) + 4\sigma^2\boldsymbol{\theta}'A^2\boldsymbol{\theta} + 4\mu_3\boldsymbol{\theta}'A\mathbf{a},$$

where $\mathbf{a}$ is the $m \times 1$ vector of diagonal elements of A.

1.26. Let $\mathbf{X}$ be $N_m(\mu\mathbf{1}, \Sigma)$, where $\mathbf{1} = (1, 1, \ldots, 1)' \in R^m$ and $\Sigma = (\sigma_{ij})$ with $\sigma_{ii} = \sigma^2$, $\sigma_{ij} = \sigma^2(1 - \rho^2)$, $i \neq j$. Show that

$$\bar{X} = m^{-1} \sum_{i=1}^{m} X_i \quad \text{and} \quad \sum_{i=1}^{m} (X_i - \bar{X})^2$$

are independent. (*Hint*: Use Problem 1.22.)

1.27. If $\mathbf{X}$ is $E_m(\boldsymbol{\mu}, V)$ (i.e. m-variate elliptical with parameters $\boldsymbol{\mu}$ and V) prove that:

(a) The characteristic function of $\mathbf{X}$ has the form

$$\phi(\mathbf{t}) = e^{i\mathbf{t}'\boldsymbol{\mu}}\psi(\mathbf{t}'V\mathbf{t}) \qquad \text{for some function } \psi.$$

(b) Provided they exist, $E(\mathbf{X}) = \boldsymbol{\mu}$ and $\text{Cov}(\mathbf{X}) = \alpha V$, where $\alpha = -2\psi'(0)$.

1.28. If $\mathbf{X}$ is $E_m(\boldsymbol{\mu}, V)$ and $\mathbf{X}$, $\boldsymbol{\mu}$ and V are partitioned as

$$\mathbf{X} = \begin{pmatrix} \mathbf{X}_1 \\ \mathbf{X}_2 \end{pmatrix}, \qquad \boldsymbol{\mu} = \begin{pmatrix} \boldsymbol{\mu}_1 \\ \boldsymbol{\mu}_2 \end{pmatrix}, \qquad V = \begin{bmatrix} V_{11} & V_{12} \\ V_{21} & V_{22} \end{bmatrix},$$

where $\mathbf{X}_1$ and $\boldsymbol{\mu}_1$ are $k\times 1$ and V_{11} is $k\times k$, show that the conditional distribution of $\mathbf{X}_1$ given $\mathbf{X}_2$ is k-variate elliptical. Show also that, if they exist, the conditional mean and covariance matrix are given by

$$E(\mathbf{X}_1|\mathbf{X}_2)=\boldsymbol{\mu}_1+V_{12}V_{22}^{-1}(\mathbf{X}_2-\boldsymbol{\mu}_2)$$

and

$$\mathrm{Cov}(\mathbf{X}_1|\mathbf{X}_2)=g(\mathbf{X}_2)(V_{11}-V_{12}V_{22}^{-1}V_{21})$$

for some function g.

1.29. Let $\mathbf{X}$ have the m-variate elliptical t-distribution on n degrees of freedom and parameters $\boldsymbol{\mu}$ and V, i.e., $\mathbf{X}$ has density function

$$\frac{\Gamma[\frac{1}{2}(n+m)]}{\Gamma(\frac{1}{2}n)(n\pi)^{m/2}}(\det V)^{-1/2}\frac{1}{\left[1+\frac{1}{n}(\mathbf{x}-\boldsymbol{\mu})'V^{-1}(\mathbf{x}-\boldsymbol{\mu})\right]^{(n+m)/2}}.$$

(a) Show that $E(\mathbf{X})=\boldsymbol{\mu}$ and $\mathrm{Cov}(\mathbf{X})=\dfrac{n}{n-2}V$ $(n>2)$.

(b) If $\mathbf{X}=(\mathbf{X}_1'\mathbf{X}_2')'$, where $\mathbf{X}_1$ is $k\times 1$, show that the marginal distribution of $\mathbf{X}_1$ is k-variate elliptical t.

(c) If $\mathbf{X}$ is partitioned as in (b), find the conditional distribution of $\mathbf{X}_1$ given $\mathbf{X}_2$. Give $E(\mathbf{X}_1|\mathbf{X}_2)$ and $\mathrm{Cov}(\mathbf{X}_1|\mathbf{X}_2)$, assuming these exist.

1.30. Suppose that $\mathbf{Y}$ is $N_m(\mathbf{0}, I_m)$ and Z is χ^2_n, and that $\mathbf{Y}$ and Z are independent. Let V be a positive definite matrix and $V^{1/2}$ be a symmetric square root of V. If

$$\mathbf{X}=\boldsymbol{\mu}+Z^{-1/2}(nV)^{1/2}\mathbf{Y}$$

show that $\mathbf{X}$ has the m-variate elliptical t-distribution of Problem 1.29. Use this to show that

$$\frac{1}{m}(\mathbf{X}-\boldsymbol{\mu})'V^{-1}(\mathbf{X}-\boldsymbol{\mu}) \quad \text{is} \quad F_{m,n}.$$

1.31. Suppose that $\mathbf{X}$ is $E_m(\boldsymbol{\mu},V)$ with density function

$$(\det V)^{-1/2}\frac{\Gamma(\frac{1}{2}m+1)}{[(m+2)\pi]^{m/2}}I_A(\mathbf{x})$$

where I_A is the indicator function of the set

$$A=\{\mathbf{x};(\mathbf{x}-\boldsymbol{\mu})'V^{-1}(\mathbf{x}-\boldsymbol{\mu})\le m+2\}.$$

Show that $E(\mathbf{X})=\boldsymbol{\mu}$ and $\mathrm{Cov}(\mathbf{X})=V$.

1.32. Let $\mathbf{T}$ be uniformly distributed on S_m and partition $\mathbf{T}$ as $\mathbf{T}'=(\mathbf{T}_1':\mathbf{T}_2')$ where $\mathbf{T}_1$ is $k\times 1$ and $\mathbf{T}_2$ is $(m-k)\times 1$.

(a) Prove that $\mathbf{T}_1$ has density function

$$f_{\mathbf{T}_1}(\mathbf{u})=\frac{\Gamma(\frac{1}{2}m)}{\pi^{k/2}\Gamma\left[\frac{1}{2}(m-k)\right]}(1-\mathbf{u}'\mathbf{u})^{(m-k)/2-1}\qquad 0<\mathbf{u}'\mathbf{u}<1$$

(b) Prove that $\mathbf{T}_1'\mathbf{T}_1$ has the beta $(\frac{1}{2}k,\frac{1}{2}(m-k))$ distribution. Eaton (1981).

1.33. If $\mathbf{X}$ is $E_m(\boldsymbol{\mu},V)$ with characteristic function $\phi(\mathbf{t})=e^{i\mathbf{t}'\boldsymbol{\mu}}\psi(\mathbf{t}'V\mathbf{t})$, covariance matrix $\Sigma=-2\psi'(0)V=(\sigma_{ij})$ and finite fourth moments prove that:

(a) The marginal distributions of X_j $(j=1,\dots,m)$ all have zero skewness and the same kurtosis $\gamma_2^j=3\kappa$ $(j=1,\dots,m)$, where

$$\kappa=\frac{\psi''(0)}{\psi'(0)^2}-1.$$

(b) In terms of the kurtosis parameter κ, all fourth-order cumulants can be expressed as

$$\kappa_{1111}^{ijkl}=\kappa(\sigma_{ij}\sigma_{kl}+\sigma_{ik}\sigma_{jl}+\sigma_{il}\sigma_{jk}).$$

1.34. Show that the kurtosis parameter for the ε-contaminated m-variate elliptical normal distribution with density function

$$\frac{(1-\varepsilon)(\det V)^{-1/2}}{(2\pi)^{m/2}}\exp\left(-\frac{1}{2}\mathbf{x}'V^{-1}\mathbf{x}\right)+\frac{\varepsilon(\det V)^{-1/2}}{(2\pi\sigma^2)^{m/2}}\exp\left(-\frac{1}{2\sigma^2}\mathbf{x}'V^{-1}\mathbf{x}\right)$$

is

$$\kappa=\frac{1+\varepsilon(\sigma^4-1)}{\left[1+\varepsilon(\sigma^2-1)\right]^2}-1.$$

1.35. Show that for the elliptical t-distribution of Problem 1.29 the kurtosis parameter is $\kappa=2/(n-4)$.

CHAPTER 2

Jacobians, Exterior Products, Kronecker Products, and Related Topics

2.1. JACOBIANS, EXTERIOR PRODUCTS, AND RELATED TOPICS

2.1.1. *Jacobians and Exterior Products*

In subsequent distribution theory, functions of random vectors and matrices will be of interest and we will need to know how density functions are transformed. This involves computing the Jacobians of these transformations. To review the relevant theory, let $\mathbf{X}$ be an $m\times 1$ random vector having a density function $f(\mathbf{x})$ which is positive on a set $S\subset R^m$. Suppose that the transformation $\mathbf{y}=\mathbf{y}(\mathbf{x})=(y_1(\mathbf{x}),\ldots,y_m(\mathbf{x}))'$ is 1-1 of S onto T, where T denotes the image of S under $\mathbf{y}$, so that the inverse transformation $\mathbf{x}=\mathbf{x}(\mathbf{y})$ exists for $\mathbf{y}\in T$. Assuming that the partial derivatives $\partial x_i/\partial y_j$ $(i, j=1,\ldots,m)$ exist and are continuous on T, it is well-known that the density function of the random vector $\mathbf{Y}=\mathbf{y}(\mathbf{X})$ is

$$g(\mathbf{y})=f(\mathbf{x}(\mathbf{y}))|J(\mathbf{x}\rightarrow\mathbf{y})| \qquad (\mathbf{y}\in T)$$

where $J(\mathbf{x}\rightarrow\mathbf{y})$, the Jacobian of the transformation from $\mathbf{x}$ to $\mathbf{y}$, is

$$(1) \qquad J(\mathbf{x}\rightarrow\mathbf{y})=\det\begin{bmatrix} \dfrac{\partial x_1}{\partial y_1} & \cdots & \dfrac{\partial x_1}{\partial y_m} \\ \vdots & & \\ \dfrac{\partial x_m}{\partial y_1} & \cdots & \dfrac{\partial x_m}{\partial y_m}\end{bmatrix}=\det\left(\frac{\partial x_i}{\partial y_j}\right).$$

Often when dealing with many variables it is tedious to explicitly write out the determinant (1). We will now sketch an equivalent approach which is often simpler and is based on an anticommutative or skew-symmetric multiplication of differentials. The treatment here follows that of James (1954).

Consider the multiple integral

$$I=\int_A f(x_1,\ldots,x_m)\,dx_1\ldots dx_m \tag{2}$$

where $A\subset R^m$. This represents the probability that $\mathbf{X}$ takes values in the set A. On making the change of variables

$$\begin{aligned} x_1 &= x_1(y_1,\ldots,y_m)\\ &\vdots\\ x_m &= x_m(y_1,\ldots,y_m)\end{aligned}$$

(2) becomes

$$I=\int_{A'} f(\mathbf{x}(\mathbf{y}))\det\left(\frac{\partial x_i}{\partial y_j}\right) dy_1\ldots dy_m \tag{3}$$

where A' denotes the image of A. Instead of writing out the matrix of partial derivatives $(\partial x_i/\partial y_j)$ and then calculating its determinant we will now indicate another way in which this can be evaluated.

Recall that the *differential* of the function $x_i=x_i(y_1,\ldots,y_m)$ is

$$dx_i=\frac{\partial x_i}{\partial y_1}dy_1+\cdots+\frac{\partial x_i}{\partial y_m}dy_m \qquad (i=1,\ldots,m). \tag{4}$$

Now substitute these *linear differential forms* (4) (in $dy_1,\ldots,dy_m$) in (2). For simplicity and concreteness, consider the case $m=2$; the reader can readily generalize what follows. We then have

$$I=\int_{A'} f(\mathbf{x}(\mathbf{y}))\left(\frac{\partial x_1}{\partial y_1}dy_1+\frac{\partial x_1}{\partial y_2}dy_2\right)\left(\frac{\partial x_2}{\partial y_1}dy_1+\frac{\partial x_2}{\partial y_2}dy_2\right). \tag{5}$$

Now, we must answer the question: Can the two differential forms in (5) be multiplied together in such a way that the result is $\det(\partial x_i/\partial y_j)\,dy_1\,dy_2$, that is,

$$(6)\qquad \left(\frac{\partial x_1}{\partial y_1}\frac{\partial x_2}{\partial y_2}-\frac{\partial x_1}{\partial y_2}\frac{\partial x_2}{\partial y_1}\right)dy_1\,dy_2?$$

Well, let's see. Suppose we multiply them in a formal way using the associative and distributive laws. This gives

$$(7)\quad \left(\frac{\partial x_1}{\partial y_1}dy_1+\frac{\partial x_1}{\partial y_2}dy_2\right)\left(\frac{\partial x_2}{\partial y_1}dy_1+\frac{\partial x_2}{\partial y_2}dy_2\right)$$

$$=\frac{\partial x_1}{\partial y_1}\frac{\partial x_2}{\partial y_1}dy_1\,dy_1+\frac{\partial x_1}{\partial y_1}\frac{\partial x_2}{\partial y_2}dy_1\,dy_2+\frac{\partial x_1}{\partial y_2}\frac{\partial x_2}{\partial y_1}dy_2\,dy_1+\frac{\partial x_1}{\partial y_2}\frac{\partial x_2}{\partial y_2}dy_2\,dy_2.$$

Comparing (6) and (7), we clearly must have

$$dy_1\,dy_2=-\,dy_2\,dy_1.$$

Hence, when multiplying two differentials dy_i and dy_j we will use a skew-symmetric or alternating product instead of a commutative one; that is, we will put

$$(8)\qquad dy_i\,dy_j=-\,dy_j\,dy_i$$

so that, in particular, $dy_i\,dy_i=-\,dy_i\,dy_i=0$. Such a product is called the *exterior* product and will be denoted by the symbol $\wedge$ (usually read "wedge product"), so that (8) becomes

$$dy_i\wedge dy_j=-\,dy_j\wedge dy_i.$$

Using this product, the right side of (7) becomes

$$\left(\frac{\partial x_1}{\partial y_1}\frac{\partial x_2}{\partial y_2}-\frac{\partial x_1}{\partial y_2}\frac{\partial x_2}{\partial y_1}\right)dy_1\wedge dy_2$$

This formal procedure of multiplying differential forms is equivalent to calculating the Jacobian as the following theorem shows.

THEOREM 2.1.1. If $d\mathbf{y}$ is an $m\times 1$ vector of differentials and if $d\mathbf{x}=B\,d\mathbf{y}$, where B is an $m\times m$ nonsingular matrix (so that $d\mathbf{x}$ is a vector of linear differential forms), then

$$(9) \qquad \bigwedge_{i=1}^{m} dx_i = \det B \bigwedge_{i=1}^{m} dy_i .$$

Proof. It is clear that the left side of (9) can be written as

$$\bigwedge_{i=1}^{m} dx_i = p(B) \bigwedge_{i=1}^{m} dy_i ,$$

where $p(B)$ is a polynomial in the elements of B. For example, with $m=3$ and $B=(b_{ij})$ it can be readily checked that

$$dx_1 \wedge dx_2 \wedge dx_3 = (b_{11}b_{22}b_{33} - b_{12}b_{21}b_{33} - b_{11}b_{23}b_{32} + b_{13}b_{21}b_{32} + b_{12}b_{23}b_{31} - b_{13}b_{22}b_{31})\, dy_1 \wedge dy_2 \wedge dy_3 .$$

In general:

(i) $p(B)$ is *linear* in each row of B.

(ii) If the order of two factors dx_i, dx_j is reversed then the sign of $\bigwedge_{i=1}^{m} dx_i$ is reversed. But this is also equivalent to interchanging the ith and jth rows of B. Hence interchanging two rows of B reverses the sign of $p(B)$.

(iii) $p(I_m)=1$.

But (i), (ii), and (iii) characterize the determinant function; in fact, they form the Weierstrass definition of a determinant [see, for example, MacDuffee (1943), Chapter 3]. Hence $p(B)=\det B$.

Now, returning to our general discussion, we have

$$I = \int_A f(x_1,\ldots,x_m)\, dx_1 \wedge \cdots \wedge dx_m ,$$

where the exterior product sign $\wedge$ has been used but where this integral is to be understood as the integral (2). Putting

$$x_i = x_i(y_1,\ldots,y_m) \qquad (i=1,\ldots,m)$$

we have

$$dx_i = \sum_{j=1}^{m} \frac{\partial x_i}{\partial y_j} dy_j \qquad (i=1,\ldots,m)$$

so that, in matrix notation,

$$d\mathbf{x} = \begin{bmatrix} \frac{\partial x_1}{\partial y_1} & \cdots & \frac{\partial x_1}{\partial y_m} \\ \vdots & & \vdots \\ \frac{\partial x_m}{\partial y_1} & \cdots & \frac{\partial x_m}{\partial y_m} \end{bmatrix} d\mathbf{y}.$$

Hence, by Theorem 2.1.1

$$\bigwedge_{i=1}^{m} dx_i = \det\left(\frac{\partial x_i}{\partial y_j}\right) \bigwedge_{i=1}^{m} dy_i$$

and the Jacobian is the absolute value of the determinant on the right.

DEFINITION 2.1.2. An *exterior differential form* of degree r in R^m is an expression of the type

$$(10) \qquad \sum_{i_1 < i_2 < \cdots < i_r} h_{i_1 \ldots i_r}(\mathbf{x})\, dx_{i_1} \wedge \cdots \wedge dx_{i_r}$$

where the $h_{i_1 \ldots i_r}(\mathbf{x})$ are analytic functions of $x_1, \ldots, x_m$.

A simple example of a form of degree 1 is the differential (4). We can regard (10) as the integrand of an r-dimensional surface integral. There are two things worth noting here about exterior differential forms:

(a) A form of degree m has only one term, namely, $h(\mathbf{x})\, dx_1 \wedge \cdots \wedge dx_m$.

(b) A form of degree greater than m is zero because at least one of the symbols dx_i is repeated in each term.

Exterior products and exterior differential forms were given a systematic treatment by Cartan (1922) in his theory of integral invariants. Since then they have found wide use in differential geometry and mathematical physics; see, for example, Sternberg (1964), Cartan (1967), and Flanders (1963).

Definition 2.1.2 can be extended to define exterior differential forms on differentiable and analytic manifolds and, under certain conditions, these in turn can be used to construct invariant measures on such manifolds. Details of this construction can be found in James (1954) for manifolds of particular interest in multivariate analysis. We will not go further into the formal theory here but will touch briefly on some aspects of it later (see Section 2.1.4).

We now turn to the calculation of some Jacobians of particular interest to us. The first result, chosen because the proof is particularly instructive, concerns the transformation to polar coordinates used in the proof of Theorem 1.5.5.

THEOREM 2.1.3. For the following transformation from rectangular coordinates $x_1,\dots,x_m$ to polar coordinates $r,\theta_1,\dots,\theta_{m-1}$:

$$\begin{aligned}
x_1 &= r\sin\theta_1\sin\theta_2\dots\sin\theta_{m-2}\sin\theta_{m-1}\\
x_2 &= r\sin\theta_1\sin\theta_2\dots\sin\theta_{m-2}\cos\theta_{m-1}\\
x_3 &= r\sin\theta_1\sin\theta_2\dots\cos\theta_{m-2}\\
&\vdots\\
x_{m-1} &= r\sin\theta_1\cos\theta_2\\
x_m &= r\cos\theta_1
\end{aligned}$$

$$[r>0,\qquad 0<\theta_i\le\pi\quad (i=1,\dots,m-2),\qquad 0<\theta_{m-1}\le 2\pi]$$

we have

$$\bigwedge_{i=1}^{m} dx_i = r^{m-1}\sin^{m-2}\theta_1\sin^{m-3}\theta_2\dots\sin\theta_{m-2}\left(\bigwedge_{i=1}^{m-1} d\theta_i\right)\wedge dr$$

(so that

$$J(\mathbf{x}\to r,\theta_1,\dots,\theta_{m-1}) = r^{m-1}\sin^{m-2}\theta_1\sin^{m-3}\theta_2\dots\sin\theta_{m-2}).$$

Proof. First note that

$$\begin{aligned}
x_1^2 &= r^2\sin^2\theta_1\sin^2\theta_2\dots\sin^2\theta_{m-2}\sin^2\theta_{m-1}\\
x_1^2+x_2^2 &= r^2\sin^2\theta_1\sin^2\theta_2\dots\sin^2\theta_{m-2}\\
&\vdots\\
x_1^2+\dots+x_m^2 &= r^2.
\end{aligned}$$

Differentiating the first of these gives

$$2x_1\,dx_1 = 2r^2\sin^2\theta_1\dots\sin^2\theta_{m-2}\sin\theta_{m-1}\cos\theta_{m-1}\,d\theta_{m-1}$$
$$+\text{terms involving } dr, d\theta_1,\dots,d\theta_{m-2}.$$

Differentiating the second gives

$$2x_1\,dx_1+2x_2\,dx_2=2r^2\sin^2\theta_1\dots\sin\theta_{m-2}\cos\theta_{m-2}\,d\theta_{m-2}$$
$$+\text{terms involving } dr,\, d\theta_1,\dots,d\theta_{m-3},$$

and so on, down to the last which gives

$$2x_1\,dx_1+\cdots+2x_m\,dx_m=2r\,dr.$$

Now take the exterior products of all the terms on the left and of all the terms on the right, remembering that repeated products of differentials are zero. The exterior product on the left side is

$$(11)\qquad 2^m x_1\dots x_m \bigwedge_{i=1}^{m} dx_i.$$

The exterior product of the right side is

$$2^m r^{2m-1}\sin^{2m-3}\theta_1\sin^{2m-5}\theta_2\dots\sin\theta_{m-1}\cos\theta_1\cos\theta_2\dots\cos\theta_{m-1}\bigwedge_{i=1}^{m-1}d\theta_i\wedge dr,$$

which equals

$$(12)\qquad 2^m x_1\dots x_m r^{m-1}\sin^{m-2}\theta_1\sin^{m-3}\theta_2\dots\sin\theta_{m-2}\bigwedge_{i=1}^{m-1}d\theta_i\wedge dr$$

since

$$x_1\dots x_m=r^m\sin^{m-1}\theta_1\sin^{m-2}\theta_2\dots\sin\theta_{m-1}\cos\theta_1\dots\cos\theta_{m-1}.$$

Equating (11) and (12) gives the derived result.

Before calculating more Jacobians we will make explicit a convention and introduce some more notation. First the convention. We will not concern ourselves with, or keep track of, the signs of exterior differential forms. Since we are, or will be, integrating exterior differential forms representing probability density functions we can avoid any difficulty with sign simply by defining only positive integrals. Now, the notation. For any matrix X, dX denotes the matrix of differentials (dx_{ij}). It is easy to check that if X is

$n \times m$ and Y is $m \times p$ then

$$d(XY) = X.dY + dX.Y$$

(see Problem 2.1). For an *arbitrary* $n \times m$ matrix X, the symbol (dX) will denote the exterior product of the mn elements of dX:

$$(dX) \equiv \bigwedge_{j=1}^{m} \bigwedge_{i=1}^{n} dx_{ij}.$$

If X is a *symmetric* $m \times m$ matrix, the symbol (dX) will denote the exterior product of the $\frac{1}{2}m(m+1)$ *distinct* elements of dX:

$$(dX) \equiv \bigwedge_{1 \le i \le j \le m} dx_{ij}.$$

Similarly, if X is a skew-symmetric matrix ($X = -X'$), then (dX) will denote the exterior product of the $\frac{1}{2}m(m-1)$ distinct elements of dX (either the sub-diagonal or super-diagonal elements), and if X is upper-triangular,

$$(dX) \equiv \bigwedge_{i \le j} dx_{ij}.$$

There will be occasions when the above notation will not be used. In these cases (dX) will be explicitly defined (as, for example, in Theorem 2.1.13).

The next few theorems give the Jacobians of some transformations which are commonly used in multivariate distribution theory.

THEOREM 2.1.4. If $X = BY$ where X and Y are $n \times m$ matrices and B is a (fixed) nonsingular $n \times n$ matrix then

$$(dX) = (\det B)^m (dY)$$

so that $J(X \to Y) = (\det B)^m$.

Proof. Since $X = BY$ it follows that $dX = B\,dY$. Putting $dX = [d\mathbf{x}_1 \ldots d\mathbf{x}_m]$ and $dY = [d\mathbf{y}_1 \ldots d\mathbf{y}_m]$, we then have $d\mathbf{x}_j = B\,d\mathbf{y}_j$ and hence, by Theorem 2.1.1,

$$\bigwedge_{i=1}^{n} dx_{ij} = (\det B) \bigwedge_{i=1}^{n} dy_{ij}.$$

From this it follows that

$$(dX)\equiv\bigwedge_{j=1}^{m}\bigwedge_{i=1}^{n}dx_{ij}=\bigwedge_{j=1}^{m}(\det B)\bigwedge_{i=1}^{n}dy_{ij}$$

$$=(\det B)^{m}(dY),$$

as desired.

THEOREM 2.1.5. If $X=BYC$, where X and Y are $n\times m$ matrices and B and C are $n\times n$ and $m\times m$ nonsingular matrices, then

$$(dX)=(\det B)^{m}(\det C)^{n}(dY)$$

so that $J(X\rightarrow Y)=(\det B)^{m}(\det C)^{n}$.

Proof. First put $Z=BY$, then $X=ZC$, so that $dX=dZ.C$. Using an argument similar to that used in the proof of Theorem 2.1.4, we get $(dX)=(\det C)^{n}(dZ)$, and, since $dZ=B\,dY$, Theorem 2.1.4 gives $(dZ)=(\det B)^{m}(dY)$, and the desired result follows.

THEOREM 2.1.6. If $X=BYB'$, where X and Y are $m\times m$ symmetric matrices and B is a nonsingular $m\times m$ matrix, then

$$(dX)=(\det B)^{m+1}(dY).$$

Proof. Since $X=BYB'$ we have $dX=B\,dY\,B'$ and it is clear that

(13) $$(dX)\equiv(B\,dY\,B')=p(B)(dY),$$

where $p(B)$ is a polynomial in the elements of B. This polynomial satisfies the equation

(14) $$p(B_1B_2)=p(B_1)p(B_2)$$

for all B_1 and B_2. To see this, first note that from (13),

(15) $$p(B_1B_2)(dY)=(B_1B_2dY(B_1B_2)').$$

But, by repeated application of (13),

(16) $$\begin{aligned}(B_1B_2dY(B_1B_2)')&=(B_1B_2dYB_2'B_1')\\&=p(B_1)(B_2dYB_2')\\&=p(B_1)p(B_2)(dY).\end{aligned}$$

Equating (15) and (16) gives (14). The only polynomials in the elements of a matrix satisfying (14) for all B_1 and B_2 are integer powers of $\det B$ [see MacDuffee (1943), Chapter 3], so that

$$p(B)=(\det B)^k \qquad \text{for some integer } k.$$

To calculate k we can take a special form for B. Taking $B=\mathrm{diag}(b,1,\dots,1)$, we compute

$$BYB'=\begin{bmatrix} b^2y_{11} & by_{12} & \cdots & by_{1m} \\ by_{12} & y_{22} & \cdots & y_{2m} \\ \vdots & & & \\ by_{1m} & y_{2m} & \cdots & y_{mm} \end{bmatrix},$$

so that the exterior product of the elements on and above the diagonal is

$$(B\,dYB')=b^{m+1}(dY).$$

Hence $p(B)=b^{m+1}=(\det B)^{m+1}$, so that $k=m+1$, and the proof is complete.

THEOREM 2.1.7. If $X=BYB'$ where X and Y are skew-symmetric $m\times m$ matrices and B is a nonsingular $m\times m$ matrix then

$$(dX)=(\det B)^{m-1}(dY).$$

The proof is almost identical to that of Theorem 2.1.6 and is left as an exercise (see Problem 2.2).

THEOREM 2.1.8. If $X=Y^{-1}$, where Y is a symmetric $m\times m$ matrix, then

$$(dX)=(\det Y)^{-(m+1)}(dY).$$

Proof. Since $YX=I_m$ we have $dY.X+Y.dX=0$, so that

$$\begin{aligned} dY&=-Y\,dX\,X^{-1} \\ &=-X^{-1}dX\,X^{-1}. \end{aligned}$$

Hence

$$(dY)=(X^{-1}dX\,X^{-1})=(\det X)^{-(m+1)}(dX)$$

by Theorem 2.1.6.

The next result is extremely useful and uses the fact (see Theorem A9.7) that any positive definite $m\times m$ matrix A has a unique decomposition as $A=T'T$, where T is an upper-triangular $m\times m$ matrix with positive diagonal elements.

THEOREM 2.1.9. If A is an $m\times m$ positive definite matrix and $A=T'T$, where T is upper-triangular with positive diagonal elements, then

$$(dA)=2^m\prod_{i=1}^{m} t_{ii}^{m+1-i}(dT).$$

Proof. We have

$$\begin{bmatrix} a_{11} & a_{12} & \cdots & a_{1m} \\ a_{12} & a_{22} & \cdots & a_{2m} \\ \vdots & & & \\ a_{1m} & a_{2m} & \cdots & a_{mm} \end{bmatrix} = \begin{bmatrix} t_{11} & 0 & \cdots & 0 \\ t_{12} & t_{22} & \cdots & 0 \\ \vdots & & & \\ t_{1m} & t_{2m} & \cdots & t_{mm} \end{bmatrix} \begin{bmatrix} t_{11} & t_{12} & \cdots & t_{1m} \\ 0 & t_{22} & \cdots & t_{2m} \\ \vdots & & & \\ 0 & 0 & \cdots & t_{mm} \end{bmatrix}.$$

Now express each of the elements of A on and above the diagonal in terms of each of the elements of T and take differentials. Remember that we are going to take the exterior product of these differentials and that products of repeated differentials are zero; hence there is no need to keep track of differentials in the elements of T which have previously occurred. We get

$$\begin{aligned}
a_{11}&=t_{11}^2, &\quad \text{so}\quad da_{11}&=2t_{11}\,dt_{11} \quad \text{and similarly}\\
a_{12}&=t_{11}t_{12}, & da_{12}&=t_{11}\,dt_{12}+\cdots\\
&\vdots & &\\
a_{1m}&=t_{11}t_{1m}, & da_{1m}&=t_{11}\,dt_{1m}+\cdots\\
a_{22}&=t_{12}^2+t_{22}^2, & da_{22}&=2t_{22}\,dt_{22}+\cdots\\
&\vdots & &\\
a_{2m}&=t_{12}t_{1m}+t_{22}t_{2m}, & da_{2m}&=t_{22}\,dt_{2m}+\cdots\\
&\vdots & &\\
a_{mm}&=t_{1m}^2+\cdots+t_{mm}^2, & da_{mm}&=2t_{mm}\,dt_{mm}+\cdots
\end{aligned}$$

Hence taking exterior products gives

$$(dA)\equiv\bigwedge_{i\le j}^{m} da_{ij}=2^m t_{11}^m t_{22}^{m-1}\cdots t_{mm}\bigwedge_{i\le j}^{m} dt_{ij},$$

as desired.

2.1.2. *The Multivariate Gamma Function*

We will use Theorem 2.1.9 in a moment to evaluate the multidimensional integral occurring in the following definition, which is of some importance and dates back to Wishart (1928), Ingham (1933), and Siegel (1935).

DEFINITION 2.1.10. The multivariate gamma function, denoted by $\Gamma_m(a)$, is defined to be

$$\Gamma_m(a)=\int_{A>0}\operatorname{etr}(-A)(\det A)^{a-(m+1)/2}(dA) \tag{17}$$

$[\operatorname{Re}(a)>\frac{1}{2}(m-1)]$, where $\operatorname{etr}(.)\equiv\exp\operatorname{tr}(.)$ and the integral is over the space of positive definite (and hence symmetric) $m\times m$ matrices. (Here, and subsequently, the notation $A>0$ means that A is positive definite.)

Note that when $m=1$, (17) just becomes the usual definition of a gamma function, so that $\Gamma_1(a)\equiv\Gamma(a)$. At first sight an integral like (17) may appear formidable, but let's look closer. A symmetric $m\times m$ matrix has $\frac{1}{2}m(m+1)$ elements and hence the set of all such matrices is a Euclidean space of distinct elements and hence the set of all such matrices is a Euclidean space of subset of this Euclidean space and in fact forms an open cone described by the following system of inequalities:

$$A>0\Leftrightarrow a_{11}>0,\quad \det\begin{bmatrix}a_{11}&a_{12}\\a_{12}&a_{22}\end{bmatrix}>0,\quad \ldots,\quad \det A>0.$$

It is a useful exercise to attempt to draw this cone in three dimensions when $m=2$ (see Problem 2.8). The integral (17) is simply an integral over this subset with respect to Lebesgue measure

$$(dA)\equiv da_{11}\wedge da_{12}\wedge\cdots\wedge da_{mm}\equiv da_{11}\,da_{12}\ldots da_{mm}.$$

Before evaluating $\Gamma_m(a)$ the following result is worth noting.

THEOREM 2.1.11. If $\operatorname{Re}(a)>\frac{1}{2}(m-1)$ and Σ is a symmetric $m\times m$ matrix with $\operatorname{Re}(\Sigma)>0$ then

$$\int_{A>0}\operatorname{etr}\left(-\tfrac{1}{2}\Sigma^{-1}A\right)(\det A)^{a-(m+1)/2}(dA)=\Gamma_m(a)(\det\Sigma)^a 2^{ma} \tag{17}$$

Proof. First suppose that $\Sigma>0$ is real. In the integral make the change of variables $A=2\Sigma^{1/2}V\Sigma^{1/2}$, where $\Sigma^{1/2}$ denotes the positive definite square root of Σ (see Theorem A9.3). By Theorem 2.1.6, $(dA)=$

$2^{m(m+1)/2}(\det\Sigma)^{(m+1)/2}(dV)$ so that the integral becomes

$$\int_{V>0}\operatorname{etr}(-V)(\det V)^{a-(m+1)/2}(dV)2^{ma}(\det\Sigma)^{a},$$

which, by Definition 2.1.10, is equal to the right side of (17). Hence, the theorem is true for real Σ and it follows for complex Σ by analytic continuation. Since $\operatorname{Re}(\Sigma)>0$, $\det\Sigma\neq 0$ and $(\det\Sigma)^{a}$ is well defined by continuation.

Put $a=\frac{1}{2}n$ in Theorem 2.1.11, where n ($>m-1$) is a real number, and suppose that $\Sigma>0$. It then follows that the function

(18)

$$f(A)=\frac{1}{2^{mn/2}\Gamma_m(\frac{1}{2}n)(\det\Sigma)^{n/2}}\operatorname{etr}\left(-\frac{1}{2}\Sigma^{-1}A\right)(\det A)^{(n-m-1)/2}\qquad (A>0)$$

is a density function, since it is nonnegative and integrates to 1. It is called the *Wishart* density function, and it plays an extremely important role in multivariate distribution theory since, as we will see in Chapter 3, when n is an integer ($>m-1$) it is the density function of nS, where S is a sample covariance matrix formed from a random sample of size $n+1$ from the $N_m(\mu,\Sigma)$ distribution.

The multivariate gamma function can be expressed as a product of ordinary gamma functions, as the following theorem shows.

THEOREM 2.1.12.

$$\Gamma_m(a)=\pi^{m(m-1)/4}\prod_{i=1}^{m}\Gamma\left[a-\tfrac{1}{2}(i-1)\right],\qquad \left[\operatorname{Re}(a)>\tfrac{1}{2}(m-1)\right].$$

Proof. By Definition 2.1.10

$$\Gamma_m(a)=\int_{A>0}\operatorname{etr}(-A)(\det A)^{a-(m+1)/2}(dA).$$

Put $A=T'T$ where T is upper-triangular with positive diagonal elements. Then

$$\operatorname{tr}A=\operatorname{tr}T'T=\sum_{i\le j}^{m}t_{ij}^2$$

$$\det A=\det T'T=(\det T)^2=\prod_{i=1}^{m}t_{ii}^2,$$

and from Theorem 2.1.9

$$(dA)=2^m \prod_{i=1}^{m} t_{ii}^{m+1-i} \bigwedge_{i\leq j}^{m} dt_{ij}.$$

Hence,

$$\Gamma_m(a)=2^m \int_{t_{ij}} \cdots \int \exp\left(-\sum_{i\leq j}^{m} t_{ij}^2\right) \prod_{i=1}^{m} t_{ii}^{2a-i} \bigwedge_{i\leq j}^{m} dt_{ij}$$

$$= \prod_{i<j}^{m} \left(\int_{-\infty}^{\infty} e^{-t_{ij}^2} dt_{ij}\right) \prod_{i=1}^{m} \left(\int_{0}^{\infty} e^{-t_{ii}^2} (t_{ii}^2)^{a-(i+1)/2} dt_{ii}^2\right)$$

The desired result now follows using

$$\int_{-\infty}^{\infty} e^{-t_{ij}^2} dt_{ij} = \pi^{1/2}$$

and

$$\int_{0}^{\infty} e^{-t_{ii}^2} (t_{ii}^2)^{a-(i+1)/2} dt_{ii}^2 = \Gamma[a-\tfrac{1}{2}(i-1)].$$

2.1.3. *More Jacobians*

Our next theorem uses the fact that any $n\times m$ $(n\geq m)$ real matrix Z with rank m can be uniquely decomposed as $Z=H_1T$, where T is an upper-triangular $m\times m$ matrix with positive diagonal elements and H_1 is an $n\times m$ matrix with orthonormal columns (see Theorem A9.8).

THEOREM 2.1.13. Let Z be an $n\times m$ $(n\geq m)$ matrix of rank m and write $Z=H_1T$, where H_1 is an $n\times m$ matrix with $H_1'H_1=I_m$ and T is an $m\times m$ upper-triangular matrix with positive diagonal elements. Let H_2 (a function of H_1) be an $n\times(n-m)$ matrix such that $H=[H_1 : H_2]$ is an orthogonal $n\times n$ matrix and write $H=[\mathbf{h}_1 \ldots \mathbf{h}_m : \mathbf{h}_{m+1} \ldots \mathbf{h}_n]$, where $\mathbf{h}_1,\ldots,\mathbf{h}_m$ are the columns of H_1 and $\mathbf{h}_{m+1},\ldots,\mathbf{h}_n$ are the columns of H_2. Then

$$(dZ)=\prod_{i=1}^{m} t_{ii}^{n-i}(dT)(H_1'dH_1) \tag{19}$$

where

$$(H_1'dH_1)\equiv \bigwedge_{i=1}^{m} \bigwedge_{j=i+1}^{n} \mathbf{h}_j' d\mathbf{h}_i. \tag{20}$$

Proof. Since $Z=H_1T$ we have $dZ=dH_1.T+H_1.dT$ and hence

$$(21)\qquad H'dZ=\begin{bmatrix}H_1'\\ \cdots\\ H_2'\end{bmatrix}dZ=\begin{bmatrix}H_1'dH_1T+H_1'H_1dT\\ H_2'dH_1T+H_2'H_1dT\end{bmatrix}$$

$$=\begin{bmatrix}H_1'dH_1T+dT\\ H_2'dH_1T\end{bmatrix}$$

since $H_1'H_1=I_m$, $H_2'H_1=0$. By Theorem 2.1.4 the exterior product of the elements on the left side of (21) is

$$(H'dZ)=(\det H')^m(dZ)=(dZ).$$

(ignoring sign). It remains to be shown that the exterior product of the elements on the right side of (21) is the right side of (19).

First consider the matrix $H_2'\,dH_1T$. The $(j-m)$th row of $H_2'\,dH_1T$ is

$$(\mathbf{h}_j'd\mathbf{h}_1,\ldots,\mathbf{h}_j'd\mathbf{h}_m)T\qquad(m+1\le j\le n).$$

Using Theorem 2.1.1, it follows that the exterior product of the elements in this row is

$$(\det T)\bigwedge_{i=1}^{m}\mathbf{h}_j'd\mathbf{h}_i.$$

Hence, the exterior product of all the elements in $H_2'\,dH_1T$ is

$$(22)\qquad \bigwedge_{j=m+1}^{n}\left[(\det T)\bigwedge_{i=1}^{m}\mathbf{h}_j'd\mathbf{h}_i\right]=(\det T)^{n-m}\bigwedge_{j=m+1}^{n}\bigwedge_{i=1}^{m}\mathbf{h}_j'd\mathbf{h}_i.$$

Now consider the upper matrix on the right side of (21), namely, $H_1'dH_1T+dT$. First note that since $H_1'H_1=I_m$ we have

$$H_1'dH_1+dH_1'.H_1=0$$

so that

$$H_1'dH_1=-dH_1'H_1=-(H_1'dH_1)',$$

and hence $H_1' dH_1$ is skew-symmetric:

$$H_1' dH_1 = \begin{bmatrix} 0 & -\mathbf{h}_2' d\mathbf{h}_1 & \cdots & -\mathbf{h}_m' d\mathbf{h}_1 \\ \mathbf{h}_2' d\mathbf{h}_1 & 0 & \cdots & -\mathbf{h}_m' d\mathbf{h}_2 \\ \mathbf{h}_3' d\mathbf{h}_1 & \mathbf{h}_3' d\mathbf{h}_2 & \cdots & -\mathbf{h}_m' d\mathbf{h}_3 \\ \vdots & \vdots & & \vdots \\ \mathbf{h}_m' d\mathbf{h}_1 & \mathbf{h}_m' d\mathbf{h}_2 & & 0 \end{bmatrix} \qquad (m \times m).$$

Postmultiplying this by the upper-triangular matrix T gives the following matrix, where only the subdiagonal elements are given, and where, in addition, terms of the form $\mathbf{h}_i' d\mathbf{h}_j$ are ignored if they have appeared already in a previous column:

$$H_1' dH_1 T = \begin{bmatrix} 0 & * & \cdots & * & * \\ \mathbf{h}_2' d\mathbf{h}_1 t_{11} & * & \cdots & * & * \\ \mathbf{h}_3' d\mathbf{h}_1 t_{11} & \mathbf{h}_3' d\mathbf{h}_2 t_{22} {}^{+*} & \cdots & * & * \\ \vdots & \vdots & & & \vdots \\ \mathbf{h}_m' d\mathbf{h}_1 t_{11} & \mathbf{h}_m' d\mathbf{h}_2 t_{22} {}^{+*} & & \mathbf{h}_m' d\mathbf{h}_{m-1} t_{m-1,m-1} {}^{+*} & * \end{bmatrix}$$

Column by column, the exterior product of the subdiagonal elements of $H_1' dH_1 T + dT$ is (remember that dT is upper-triangular)

$$(23) \quad t_{11}^{m-1} \bigwedge_{j=2}^{m} \mathbf{h}_j' d\mathbf{h}_1 \wedge t_{22}^{m-2} \bigwedge_{j=3}^{m} \mathbf{h}_j' d\mathbf{h}_2 \ldots \wedge t_{m-1,m-1} \mathbf{h}_m' d\mathbf{h}_{m-1}$$

$$= t_{11}^{m-1} t_{22}^{m-2} \ldots t_{m-1,m-1} \bigwedge_{i=1}^{m} \bigwedge_{j=i+1}^{m} \mathbf{h}_j' d\mathbf{h}_i.$$

It follows from (22) and (23) that the exterior product of the elements of $H_2' dH_1 T$ and the subdiagonal elements of $H_1' dH_1 T + dT$ is

$$(24) \quad \left(\prod_{i=1}^{m} t_{ii}^{n-m} \right) \left(\bigwedge_{i=1}^{m} \bigwedge_{j=m+1}^{n} \mathbf{h}_j' d\mathbf{h}_i \right) \left(\prod_{i=1}^{m} t_{ii}^{m-i} \right) \left(\bigwedge_{i=1}^{m} \bigwedge_{j=i+1}^{m} \mathbf{h}_j' d\mathbf{h}_i \right)$$

$$= \prod_{i=1}^{m} t_{ii}^{n-i} \bigwedge_{i=1}^{m} \bigwedge_{j=i+1}^{n} \mathbf{h}_j' d\mathbf{h}_i = \prod_{i=1}^{m} t_{ii}^{n-i} (H_1' dH_1),$$

using (20). The exterior product of the elements of $H_1' dH_1 T + dT$ on and above the diagonal is

$$\bigwedge_{i \leq j}^{m} dt_{ij} + \text{terms involving } dH_1. \tag{25}$$

We now multiply together (24) and (25) to get the exterior product of the elements of the right side of (21). The terms involving dH_1 in (25) will contribute nothing to this exterior product because (24) is already a differential form of maximum degree in H_1. Hence the exterior product of the elements of the right side of (21) is

$$\prod_{i=1}^{m} t_{ii}^{n-i}(dT)(H_1' dH_1).$$

and the proof is complete.

The following theorem is a consequence of Theorems 2.1.9 and 2.1.13 and plays a key role in the derivation of the Wishart distribution in Chapter 3.

THEOREM 2.1.14. With the assumption of Theorem 2.1.13,

$$(dZ) = 2^{-m}(\det A)^{(n-m-1)/2}(dA)(H_1' dH_1)$$

where $A = Z'Z$.

Proof. From Theorem 2.1.13

$$(dZ) = \prod_{i=1}^{m} t_{ii}^{n-i}(dT)(H_1' dH_1). \tag{26}$$

Also, $A = Z'Z = T'T$. Hence from Theorem 2.1.9,

$$(dA) = 2^m \prod_{i=1}^{m} t_{ii}^{m+1-i}(dT)$$

so that

$$(dT) = 2^{-m} \prod_{i=1}^{m} t_{ii}^{-m-1+i}(dA).$$

Substituting this for (dT) in (26) gives

$$(dZ)=2^{-m}\prod_{i=1}^{m} t_{ii}^{n-m-1}(dA)(H_1' dH_1)$$

$$=2^{-m}(\det A)^{(n-m-1)/2}(dA)(H_1' dH_1)$$

since $\prod_{i=1}^{m} t_{ii}=\det T=(\det T'T)^{1/2}=\det A^{1/2}$.

2.1.4. *Invariant Measures*

It is time to look a little more closely at the differential form (20), namely,

$$(H_1' dH_1)=\bigwedge_{i=1}^{m}\bigwedge_{j=i+1}^{n}\mathbf{h}_j' d\mathbf{h}_i,$$

which occurs in the previous two theorems. Recall that H_1 is an $n\times m$ matrix $(n\geq m)$ with orthonormal columns, so that $H_1'H_1=I_m$. The set (or space) of all such matrices H_1 is called the *Stiefel manifold*, denoted by $V_{m,n}$. Thus

$$V_{m,n}=\{H_1(n\times m);\, H_1'H_1=I_m\}.$$

The reader can check that there are $\frac{1}{2}m(m+1)$ functionally independent conditions on the mn elements of $H_1\in V_{m,n}$ implied by the equation $H_1'H_1=I_m$. Hence the elements of H_1 can be regarded as the coordinates of a point on a $mn-\frac{1}{2}m(m+1)$-dimensional surface in mn-dimensional Euclidean space. If $H_1=(h_{ij})$ $(i=1,\dots,n;\ j=1,\dots,m)$ then since $\sum_{i=1}^{n}\sum_{j=1}^{m}h_{ij}^2=m$ this surface is a subset of the sphere of radius $m^{1/2}$ in mn-dimensional space. Two special cases are the following:

(a) $m=n$.
Then

$$V_{m,m}\equiv O(m)=\{H(m\times m);\, H'H=I_m\},$$

the set of orthogonal $m\times m$ matrices. This is a group, called the orthogonal group, with the group operation being matrix multiplication. Here the elements of $H\in O(m)$ can be regarded as the coordinates of a point on a $\frac{1}{2}m(m-1)$-dimensional surface in Euclidean

m^2-space and the surface is a subset of the sphere of radius $m^{1/2}$ in m^2-space.

(b) $m=1$.
Then

$$V_{1,n}\equiv S_n=\{\mathbf{h}(n\times 1);\mathbf{h}'\mathbf{h}=1\},$$

the unit sphere in R^n. This is, of course, an $n-1$ dimensional surface in R^n.

Now let us look at the differential form (20). Consider first the special case $n=m$, corresponding to the orthogonal group $O(m)$; then, for $H\in O(m)$,

$$(H'dH)\equiv\bigwedge_{i<j}^{m}\mathbf{h}_j'd\mathbf{h}_i.$$

This differential form is just the exterior product of the subdiagonal elements of the skew-symmetric matrix $H'dH$. First note that it is invariant under *left* translation $H\rightarrow QH$ for $Q\in O(m)$, for then $H'dH\rightarrow H'Q'Q\,dH=H'dH$ and hence $(H'dH)\rightarrow(H'dH)$. It is also invariant under right translation $H\rightarrow HQ'$ for $Q\in O(m)$, for $H'dH\rightarrow QH'dHQ'$ and hence, by Theorem 2.1.7, $(H'\,dH)\rightarrow(QH'\,dHQ')=(\det Q)^{m-1}(H'\,dH)=(H'\,dH)$, ignoring the sign. This *invariant differential form* defines a measure μ on $O(m)$ given by

$$\mu(\mathfrak{D})=\int_{\mathfrak{D}}(H'dH)\qquad[\mathfrak{D}\subset O(m)]$$

where $\mu(\mathfrak{D})$ represents the surface area (usually referred to as the *volume*) of the region $\mathfrak{D}$ on the orthogonal manifold. Since the differential form $(H'dH)$ is invariant, it is easy to check that the measure μ is also. What this means in this instance is

$$\mu(Q\mathfrak{D})=\mu(\mathfrak{D}Q)=\mu(\mathfrak{D})\qquad\forall\, Q\in O(m)\tag{27}$$

(see Problem 2.9). The measure μ is called the *invariant* measure on $O(m)$. It is also often called the Haar measure on $O(m)$ in honor of Haar (1933), who proved the existence of an invariant measure on any locally compact topological group [see, for example, Halmos (1950) and Nachbin (1965)]. It can be shown that it is *unique* in the sense that any other invariant measure on $O(m)$ is a finite multiple of μ. The surface area (or, as it is more often called, the volume) of $O(m)$ is

$$\operatorname{Vol}[O(m)] = \mu[O(m)] = \int_{O(m)} (H'\,dH).$$

We will evaluate this explicitly in a moment. As a simple example consider the invariant measure on the proper orthogonal group $O^+(2)$ when $m=2$; that is, the subgroup of $O(2)$, or part of the orthogonal manifold or surface, of 2×2 orthogonal matrices H with $\det H = 1$. Such a matrix can be parameterized as

$$H = \begin{bmatrix} \cos\theta & -\sin\theta \\ \sin\theta & \cos\theta \end{bmatrix} \qquad (0<\theta\le 2\pi)$$

$$= [\mathbf{h}_1 \quad \mathbf{h}_2].$$

The invariant differential form $(H'\,dH)$ is

$$(H'\,dH) = \mathbf{h}_2'\,d\mathbf{h}_1 = (-\sin\theta \quad \cos\theta)\begin{pmatrix} -\sin\theta\,d\theta \\ \cos\theta\,d\theta \end{pmatrix} = d\theta$$

and

(28) $$\operatorname{Vol}[O^+(2)] = \int_{O^+(2)} (H'\,dH) = \int_0^{2\pi} d\theta = 2\pi.$$

Now consider the differential form (20) in general, so that $H_1 \in V_{m,n}$. Here we have (see the statement of Theorem 2.1.13)

$$(H_1'\,dH_1) \equiv \bigwedge_{i=1}^{m} \bigwedge_{j=i+1}^{n} \mathbf{h}_j'\,d\mathbf{h}_i$$

where $[H_1 : H_2] = [\mathbf{h}_1 \ldots \mathbf{h}_m : \mathbf{h}_{m+1} \ldots \mathbf{h}_n] \in O(n)$ is a function of H_1. It can be shown that this differential form does not depend on the choice of the matrix H_2 and that it is invariant under the transformations

$$H_1 \to QH_1 \qquad [Q \in O(n)]$$

and

$$H_1 \to H_1P \qquad [P \in O(m)]$$

and defines an invariant measure on the Stiefel manifold $V_{m,n}$. For proofs of these assertions, and much more besides, the interested reader is referred to James (1954). The surface area or volume of the Stiefel manifold $V_{m,n}$ is

$$\operatorname{Vol}(V_{m,n})=\int_{V_{m,n}}(H_1'dH_1).$$

We now evaluate this integral.

THEOREM 2.1.15.

$$\int_{V_{m,n}}(H_1'dH_1)=\frac{2^m\pi^{mn/2}}{\Gamma_m(\frac{1}{2}n)}$$

Proof. Let Z be an $n\times m$ $(n\geq m)$ random matrix whose elements are all independent $N(0,1)$ random variables. The density function of Z (that is, the joint density function of the mn elements of Z) is

$$(2\pi)^{-mn/2}\exp\left(-\frac{1}{2}\sum_{i=1}^{n}\sum_{j=1}^{m}z_{ij}^2\right)$$

which, in matrix notation, is the same as

$$(2\pi)^{-mn/2}\operatorname{etr}(-\tfrac{1}{2}Z'Z).$$

Since this is a density function, it integrates to 1, so

$$\int_{-\infty<z_{ij}<\infty}\cdots\int\operatorname{etr}(-\tfrac{1}{2}Z'Z)(dZ)=(2\pi)^{mn/2}. \tag{29}$$

Put $Z=H_1T$, where $H_1\in V_{m,n}$ and T is upper-triangular with positive diagonal elements, then

$$\operatorname{tr}Z'Z=\operatorname{tr}T'T=\sum_{i\leq j}^{m}t_{ij}^2,$$

$$(dZ)=\prod_{i=1}^{m}t_{ii}^{n-i}(dT)(H_1'dH_1)$$

(from Theorem 2.1.13) and (29) becomes

$$(30)\quad \int_{t_{ij}}\cdots\int \exp\left(-\frac{1}{2}\sum_{i\le j}^{m} t_{ij}^2\right)\prod_{i=1}^{m} t_{ii}^{n-i}\bigwedge_{i\le j}^{m} dt_{ij}\int_{V_{m,n}}(H_1'dH_1)$$

$$=(2\pi)^{mn/2}.$$

The integral involving the t_{ij} on the left side of (30) can be written as

$$\prod_{i<j}^{m}\left[\int_{-\infty}^{\infty}\exp\left(-\tfrac{1}{2}t_{ij}^2\right)dt_{ij}\right]\prod_{i=1}^{m}\left[\int_0^{\infty}\exp\left(-\tfrac{1}{2}t_{ii}^2\right)t_{ii}^{n-i}\,dt_{ii}\right]$$

$$=\prod_{i<j}^{m}\left[(2\pi)^{1/2}\right]\prod_{i=1}^{m}\left\{2^{(n-i-1)/2}\Gamma\left[\tfrac{1}{2}(n-i+1)\right]\right\}$$

$$=\pi^{m(m-1)/4}\prod_{i=1}^{m}\Gamma\left[\tfrac{1}{2}(n-i+1)\right]\cdot 2^{mn/2-m}$$

$$=\Gamma_m\left(\tfrac{1}{2}n\right)2^{mn/2-m},$$

using Theorem 2.1.12. Substituting back in (30) it then follows that

$$\int_{V_{m,n}}(H_1'dH_1)=\frac{2^m\pi^{mn/2}}{\Gamma_m\left(\frac{1}{2}n\right)}$$

and the proof is complete.

A special case of this theorem is when $m=n$, in which case it gives the volume of the orthogonal group $O(m)$. This is given in the following corollary.

COROLLARY 2.1.16.

$$\operatorname{Vol}[O(m)]=\frac{2^m\pi^{m^2/2}}{\Gamma_m\left(\frac{1}{2}m\right)}$$

Note that $\operatorname{Vol}[O(2)]=2^2\pi^2/\Gamma_2(1)=4\pi$, which is twice the volume of $O^+(2)$ found in (28), as is to be expected.

Another special case is when $m=1$ in which case Theorem 2.1.15 gives the surface area of the unit sphere S_n in R^n as $2\pi^{1/2}/\Gamma(\frac{1}{2}n)$, a result which has already previously been noted in the proof of Theorem 1.5.5.

The measures defined above via the differential form (20) on $V_{m,n}$ and $O(m)$ are "unnormalized" measures, equivalent to ordinary Lebesgue measure, regarding these spaces as point sets in Euclidean spaces of appropriate dimensions. Often it is more convenient to *normalize* the measures so that they are *probability measures*. For example, in the case of the orthogonal group, if we denote by (dH) the differential form

$$(dH)\equiv\frac{1}{\text{Vol}[O(m)]}(H'dH)=\frac{\Gamma_m(\frac{1}{2}m)}{2^m\pi^{m^2/2}}\bigwedge_{i<j}^{m}\mathbf{h}_j'd\mathbf{h}_i$$

then

$$\int_{O(m)}(dH)=1$$

and the measure μ^* on $O(m)$ defined by

$$\mu^*(\mathfrak{D})\equiv\int_{\mathfrak{D}}(dH)=\frac{1}{\text{Vol}[O(m)]}\int_{\mathfrak{D}}(H'dH)\qquad[\mathfrak{D}\subset O(m)] \tag{29}$$

is a probability measure representing what is often called the "Haar invariant" distribution [on $O(m)$]; see for example, Anderson (1958), page 321. In a similar way the differential form $(H_1'dH_1)$ representing the invariant measure on $V_{m,n}$ can be normalized by dividing by $\text{Vol}(V_{m,n})$, to give a probability distribution on $V_{m,n}$. In the special case $m=1$ this distribution, the uniform distribution on the unit sphere S_n in R^n, is the unique distribution invariant under orthogonal transformations, a fact alluded to in Section 1.5.

We have derived most of the results we need concerning Jacobians and invariant measures. Some other results about Jacobians appear in the problems and, in addition, others will be derived in the text as the need arises. For the interested reader useful reference papers on Jacobians in multivariate analysis are those by Deemer and Olkin (1951) and Olkin (1953).

2.2. KRONECKER PRODUCTS

Many of the results derived later can be expressed neatly and succinctly in terms of the Kronecker product of matrices. Rather than cover this in the Appendix the definition and some of the properties of this product will be reviewed in this section.

DEFINITION 2.2.1. Let $A=(a_{ij})$ be a $p\times q$ matrix and $B=(b_{ij})$ be an $r\times s$ matrix. The Kronecker product of A and B, denoted by $A\otimes B$, is the $pr\times qs$ matrix

$$A\otimes B=\begin{bmatrix} a_{11}B & a_{12}B & \dots & a_{1q}B \\ a_{21}B & a_{22}B & \dots & a_{2q}B \\ \vdots & & & \\ a_{p1}B & a_{p2}B & \dots & a_{pq}B \end{bmatrix}.$$

The Kronecker product is also often called the *direct product*; actually the connection between this product and the German mathematician Kronecker (1823–1891) seems rather obscure.

An important special Kronecker product, and one which occurs often is the following: If B is an $r\times s$ matrix then the $pr\times ps$ block-diagonal matrix with B occurring p times on the diagonal is $I_p\otimes B$; that is

$$I_p\otimes B=\begin{bmatrix} B & 0 & \dots & 0 \\ 0 & B & \dots & 0 \\ \vdots & & & \\ 0 & 0 & & B \end{bmatrix}.$$

Some of the important properties of the Kronecker product are now summarized.

(a) $(\alpha A)\otimes(\beta B)=\alpha\beta(A\otimes B)$ for any scalars α, β.

(b) If A and B are both $p\times q$ and C is $r\times s$, then

$$(A+B)\otimes C=A\otimes C+B\otimes C.$$

(c) $(A\otimes B)\otimes C=A\otimes(B\otimes C)$.

(d) $(A\otimes B)'=A'\otimes B'$.

(e) If A and B are both $m \times m$ then

$$\operatorname{tr}(A \otimes B) = (\operatorname{tr} A)(\operatorname{tr} B).$$

(f) If A is $m \times n$, B is $p \times q$, C is $n \times r$, and D is $q \times s$ then

$$(A \otimes B)(C \otimes D) = AC \otimes BD.$$

(g) If A and B are nonsingular then

$$(A \otimes B)^{-1} = A^{-1} \otimes B^{-1}.$$

(h) If H and Q are both orthogonal matrices, so is $H \otimes Q$.

(i) If A is $m \times m$, B is $n \times n$ then

$$\det(A \otimes B) = (\det A)^n (\det B)^m.$$

(j) If A is $m \times m$ with latent roots $a_1, \ldots, a_m$ and B is $n \times n$ with latent roots $b_1, \ldots, b_n$ then $A \otimes B$ has latent roots $a_i b_j$ ($i = 1, \ldots, m$; $j = 1, \ldots, n$).

(k) If $A > 0$, $B > 0$ (i.e., A and B are both positive definite) then $A \otimes B > 0$.

These results are readily proved from the definition and are left to the reader to verify. A useful reference is Graybill (1969), Chapter 8.

Now recall the vec notation introduced in (21) of Section 1.2; that is, if $T = [\mathbf{t}_1 \mathbf{t}_2 \ldots \mathbf{t}_q]$ is a $p \times q$ matrix then

$$\operatorname{vec}(T) = \begin{pmatrix} \mathbf{t}_1 \\ \mathbf{t}_2 \\ \vdots \\ \mathbf{t}_q \end{pmatrix} \qquad (pq \times 1).$$

The connection between direct products and the vec of a matrix specified in the following lemma is often useful. The proof is straightforward (see Problem 2.12).

LEMMA 2.2.2. If B is $r \times m$, X is $m \times n$, and C is $n \times s$ then

$$\operatorname{vec}(BXC) = (C' \otimes B)\operatorname{vec}(X).$$

As an application of this lemma, suppose that X is an $m\times n$ random matrix whose columns are independent $m\times 1$ random vectors, each with the same covariance matrix Σ. That is,

$$X=[\mathbf{X}_1\ldots\mathbf{X}_n]$$

where $\text{Cov}(\mathbf{X}_i)=\Sigma$, $i=1,\ldots,n$. We then have

$$\text{vec}(X)=\begin{pmatrix}\mathbf{X}_1\\ \vdots\\ \mathbf{X}_n\end{pmatrix}$$

and since the $\mathbf{X}_i$ are all independent with the same covariance matrix it follows that

$$\text{(1)}\qquad \text{Cov}[\text{vec}(X)]=\begin{bmatrix}\Sigma & 0 & \cdots & 0\\ 0 & \Sigma & \cdots & 0\\ \vdots & & & \\ 0 & 0 & & \Sigma\end{bmatrix}\qquad (mn\times nm)$$
$$=I_n\otimes\Sigma.$$

Now suppose we transform to a new random matrix Y given by $Y=BXC$, where B and C are $r\times m$ and $n\times s$ matrices of constants. Then $E(Y)=BE(X)C$ and, from Lemma 2.2.2,

$$\text{vec}(Y)=(C'\otimes B)\,\text{vec}(X)$$

so that

$$E[\text{vec}(Y)]=(C'\otimes B)E[\text{vec}(X)].$$

Also, using (3) of Section 1.2,

$$\begin{aligned}\text{Cov}(\text{vec}(Y))&=(C'\otimes B)\,\text{Cov}[\text{vec}(X)](C'\otimes B)'\\ &=(C'\otimes B)(I_n\otimes\Sigma)(C\otimes B')\\ &=C'C\otimes B\Sigma B',\end{aligned}$$

where we have used (1) and properties (d) and (f), above.

Some other connections between direct products and vec are summarized in the following lemma due to Neudecker (1969), where it is assumed that the sizes of the matrices are such that the statements all make sense.

LEMMA 2.2.3.

(i) $\text{vec}(BC)=(I\otimes B)\text{vec}(C)=(C'\otimes I)\text{vec}(B)=(C'\otimes B)\text{vec}(I)$

(ii) $\text{tr}(BCD)=(\text{vec}(B'))'(I\otimes C)\text{vec}(D)$

(iii) $\text{tr}(BX'CXD)=(\text{vec}(X))'(B'D'\otimes C)\text{vec}(X)$
$\qquad\qquad\qquad\quad =(\text{vec}(X))'(DB\otimes C')\text{vec}(X)$

Proof. Statement (i) is a direct consequence of Lemma 2.2.2. Statement (ii) is left as an exercise (Problem 2.13). To prove the first line of statement (iii), write

$$
\begin{aligned}
\text{tr}(BX'CXD)&=\text{tr}(BX')C(XD)\\
&=(\text{vec}(XB'))'(I\otimes C)\text{vec}(XD) \quad \text{using (ii)}\\
&=[(B\otimes I)\text{vec}(X)]'(I\otimes C)(D'\otimes I)\text{vec}(X) \quad \text{using (i)}\\
&=(\text{vec}(X))'(B'\otimes I)(I\otimes C)(D'\otimes I)\text{vec}(X) \quad \text{using property (d)}\\
&=\text{vec}(X)'(B'D'\otimes C)\text{vec}(X) \quad \text{using property (f).}
\end{aligned}
$$

The second line of statement (iii) is simply the transpose of the first.

PROBLEMS

2.1. If X is $n\times m$ and Y is $m\times p$ prove that

$$d(XY)=X.dY+dX.Y.$$

2.2. Prove Theorem 2.1.7.

2.3. Prove that if X, Y and B are $m\times m$ lower-triangular matrices with $X=YB$ where B is fixed then

$$(dX)=\prod_{i=1}^{m} b_{ii}^{m+1-i}(dY).$$

2.4. Show that if $X=Y+Y'$, where Y is $m\times m$ lower-triangular, then

$$(dX)=2^m(dY).$$

2.5. Prove that if $X = YB + BY'$ where Y and B are $m \times m$ lower-triangular then

$$(dX) = 2^m \prod_{i=1}^{m} b_{ii}^{m+1-i}(dY).$$

2.6. Prove that if $X = YB + BY'$, where Y and B are $m \times m$ upper-triangular, then

$$(dX) = 2^m \prod_{i=1}^{m} b_{ii}^{i}(dY).$$

2.7. Prove that if X is $m \times m$ nonsingular and $X = Y^{-1}$ then

$$(dX) = (\det Y)^{-2m}(dY).$$

2.8. The space of positive definite 2×2 matrices is a subset of R^3 defined by the inequalities

$$a_{11} > 0, \qquad \det\begin{bmatrix} a_{11} & a_{12} \\ a_{12} & a_{22} \end{bmatrix} = a_{11}a_{22} - a_{12}^2 > 0.$$

Sketch the region in R^3 described by these inequalities.

2.9. Verify equation (27) of Section 2.1:

$$\mu(Q\mathfrak{D}) = \mu(\mathfrak{D}Q) = \mu(\mathfrak{D}) \qquad \forall Q \in O(m)$$

where

$$\mu(\mathfrak{D}) = \int_{\mathfrak{D}} (H'dH) \qquad \mathfrak{D} \subset O(m).$$

2.10. Show that the measure μ on $O(m)$ defined by

$$\mu(\mathfrak{D}) = \int_{\mathfrak{D}} (H'dH) \qquad [\mathfrak{D} \subset O(m)]$$

is invariant under the transformation $H \to H'$.
[*Hint:* Define a new measure ν on $O(m)$ by

$$\nu(\mathfrak{D}) = \mu(\mathfrak{D}^{-1}),$$

where $\mathfrak{D}^{-1} = \{H \in O(m);\ H' \in \mathfrak{D}\}$. Show that ν is invariant under left translations, i.e., $\nu(Q\mathfrak{D}) = \nu(\mathfrak{D})$ for all $Q \in O(m)$. From the uniqueness of invariant measures $\nu = k\mu$ for some constant k. Show that $k = 1$.]

2.11. If

$$H=[\mathbf{h}_1\,\mathbf{h}_2\mathbf{h}_2]=\begin{bmatrix} \sin\theta_1\sin\theta_2 & \cos\theta_2 & \cos\theta_1\sin\theta_2 \\ \sin\theta_1\cos\theta_2 & -\sin\theta_2 & \cos\theta_1\cos\theta_2 \\ \cos\theta_1 & 0 & -\sin\theta_1 \end{bmatrix}$$

(where $0\leq\theta_1<\pi$, $0\leq\theta_2<2\pi$) show that

$$(\mathbf{h}_2'\,d\mathbf{h}_1)\wedge(\mathbf{h}_3'\,d\mathbf{h}_1)=\sin\theta_1\,d\theta_2\wedge d\theta_1.$$

Show also that its integral agrees with the result of Theorem 2.1.15.

2.12. Prove Lemma 2.2.2.

2.13. If B is $r\times m$, C is $m\times n$, and D is $n\times r$, prove that

$$\operatorname{tr}(BCD)=(\operatorname{vec}(B'))'(I\otimes C)\operatorname{vec}(D).$$

CHAPTER 3

Samples from a Multivariate Normal Distribution, and the Wishart and Multivariate Beta Distributions

3.1. SAMPLES FROM A MULTIVARIATE NORMAL DISTRIBUTION AND MAXIMUM LIKELIHOOD ESTIMATION OF THE PARAMETERS

In this section we will derive the distributions of the mean and covariance matrix formed from a sample from a multivariate normal distribution. First, a convention to simplify notation. When we write that an $r\times s$ random *matrix* Y is normally distributed, say, Y is $N(M, C\otimes D)$, where M is $r\times s$ and C and D are $r\times r$ and $s\times s$ positive definite matrices, we will simply mean that $E(Y)=M$ and that $C\otimes D$ is the covariance matrix of the vector $\mathbf{y}=\operatorname{vec}(Y')$ (see Section 2.2). That is, the statement "Y is $N(M, C\otimes D)$" is equivalent to the statement that "$\mathbf{y}$ is $N_{rs}(\mathbf{m}, C\otimes D)$," with $\mathbf{m}=\operatorname{vec}(M')$. The following result gives the joint density function of the elements of Y.

THEOREM 3.1.1. If the $r\times s$ matrix Y is $N(M, C\otimes D)$, where $C(r\times r)$ and $D(s\times s)$ are positive definite, then the density function of Y is

$$(1)\quad (2\pi)^{-rs/2}(\det C)^{-s/2}(\det D)^{-r/2}\operatorname{etr}\left[-\tfrac{1}{2}C^{-1}(Y-M)D^{-1}(Y-M)'\right]$$

Proof. Since $\mathbf{y}\equiv\operatorname{vec}(Y')$ is $N_{rs}(\mathbf{m}, C\otimes D)$, with $\mathbf{m}=\operatorname{vec}(M')$, the joint density function of the elements of $\mathbf{y}$ is

$$(2\pi)^{-rs/2}\det(C\otimes D)^{-1/2}\exp\left[-\tfrac{1}{2}(\mathbf{y}-\mathbf{m})'(C\otimes D)^{-1}(\mathbf{y}-\mathbf{m})\right].$$

That this is the same as (1) follows from Lemma 2.2.3 and the fact that $\det(C\otimes D)=(\det C)^s(\det D)^r$.

Now, let $\mathbf{X}_1,\ldots,\mathbf{X}_N$ be independent $N_m(\mu,\Sigma)$ random vectors. We will assume throughout this chapter that Σ is positive definite ($\Sigma>0$). Let X be the $N\times m$ matrix

$$X=\begin{bmatrix}\mathbf{X}_1'\\ \vdots\\ \mathbf{X}_N'\end{bmatrix}$$

then

$$E(X)=\begin{bmatrix}\mu'\\ \vdots\\ \mu'\end{bmatrix}=\mathbf{1}\mu',\qquad \text{where}\quad \mathbf{1}=(1,\ldots,1)'\in R^N,$$

and $\mathrm{Cov}[\mathrm{vec}(X')]=I_N\otimes\Sigma$, so that by our convention, X is $N(\mathbf{1}\mu',I_N\otimes\Sigma)$.

We have already noted in Section 1.2 that the sample mean vector $\overline{\mathbf{X}}$ and covariance matrix S, defined by

$$\text{(2)}\qquad \overline{\mathbf{X}}=\frac{1}{N}\sum_{i=1}^N\mathbf{X}_i=\frac{1}{N}X'\mathbf{1},\qquad S=\frac{1}{n}A,$$

where

$$\text{(3)}\qquad A=\sum_{i=1}^N(\mathbf{X}_i-\overline{\mathbf{X}})(\mathbf{X}_i-\overline{\mathbf{X}})'=(X-\mathbf{1}\overline{\mathbf{X}}')'(X-\mathbf{1}\overline{\mathbf{X}}')$$

and $n=N-1$, are unbiased estimates of μ and Σ, respectively. The following theorem shows that they are independently distributed and gives their distributions.

THEOREM 3.1.2. If the $N\times m$ matrix $\mathbf{X}$ is $N(\mathbf{1}\mu',I_N\otimes\Sigma)$ then $\overline{\mathbf{X}}$ and A, defined by (2) and (3), are independently distributed; $\overline{\mathbf{X}}$ is $N_m(\mu,(1/N)\Sigma)$ and A has the same distribution as $Z'Z$, where the $n\times m$ ($n=N-1$) matrix Z is $N(0,I_n\otimes\Sigma)$ (i.e., the n rows of Z are independent $N_m(\mathbf{0}',\Sigma)$ random vectors).

Proof. Note that we know the distribution of $\overline{\mathbf{X}}$ from Corollary 1.2.15. Using (1), the density function of X is

$$\text{(4)}\qquad (2\pi)^{-mN/2}(\det\Sigma)^{-N/2}\mathrm{etr}\left[-\tfrac{1}{2}\Sigma^{-1}(X-\mathbf{1}\mu')'(X-\mathbf{1}\mu')\right].$$

Now put $V=HX$, where H is an orthogonal $N\times N$ matrix [i.e., $H\in O(N)$], with elements in the last row all equal to $N^{-1/2}$. The Jacobian of this transformation is, from Theorem 2.1.4, $|\det H|^m=1$. Partition V as

$$V=\begin{bmatrix} Z \\ \cdot\cdot \\ \mathbf{v}' \end{bmatrix},$$

where Z is $n\times m$ $(n=N-1)$, and $\mathbf{v}$ is $m\times 1$. Then

$$X'X=V'V=Z'Z+\mathbf{v}\mathbf{v}'.$$

The term $(X-\mathbf{1}\boldsymbol{\mu}')'(X-\mathbf{1}\boldsymbol{\mu}')$ which appears in the exponent of (4) can be expanded as

$$\begin{aligned}(5)\qquad (X-\mathbf{1}\boldsymbol{\mu}')'(X-\mathbf{1}\boldsymbol{\mu}')&=X'X-X'\mathbf{1}\boldsymbol{\mu}'-\boldsymbol{\mu}\mathbf{1}'X+N\boldsymbol{\mu}\boldsymbol{\mu}'\\ &=Z'Z+\mathbf{v}\mathbf{v}'-X'\mathbf{1}\boldsymbol{\mu}'-(X'\mathbf{1}\boldsymbol{\mu}')'+N\boldsymbol{\mu}\boldsymbol{\mu}'.\end{aligned}$$

Now note that $H\mathbf{1}=(0,\ldots,0,N^{1/2})'$, since the first $n=N-1$ rows of H are orthogonal to $\mathbf{1}\in R^N$, and so

$$X'\mathbf{1}\boldsymbol{\mu}'=V'H\mathbf{1}\boldsymbol{\mu}'=[Z':\mathbf{v}]\begin{pmatrix}0\\ \vdots\\ 0\\ N^{1/2}\end{pmatrix}\boldsymbol{\mu}'=N^{1/2}\mathbf{v}\boldsymbol{\mu}'.$$

Substituting back in (5) then gives

$$\begin{aligned}(6)\qquad (X-\mathbf{1}\boldsymbol{\mu}')'(X-\mathbf{1}\boldsymbol{\mu}')&=Z'Z+\mathbf{v}\mathbf{v}'-N^{1/2}\boldsymbol{\mu}\mathbf{v}'-N^{1/2}\mathbf{v}\boldsymbol{\mu}'+N\boldsymbol{\mu}\boldsymbol{\mu}'\\ &=Z'Z+(\mathbf{v}-N^{1/2}\boldsymbol{\mu})(\mathbf{v}-N^{1/2}\boldsymbol{\mu})'.\end{aligned}$$

Hence the joint density function of Z and $\mathbf{v}$ can be written as

$$\begin{aligned}&(2\pi)^{-mn/2}(\det\Sigma)^{-n/2}\operatorname{etr}(-\tfrac{1}{2}\Sigma^{-1}Z'Z)\cdot(2\pi)^{-m/2}(\det\Sigma)^{-1/2}\\ &\qquad\exp\left[-\tfrac{1}{2}(\mathbf{v}-N^{1/2}\boldsymbol{\mu})'\Sigma^{-1}(\mathbf{v}-N^{1/2}\boldsymbol{\mu})\right],\end{aligned}$$

which implies that Z is $N(0,I_n\otimes\Sigma)$ (see Theorem 3.1.1) and is independent of $\mathbf{v}$, which is $N_m(N^{1/2}\boldsymbol{\mu},\Sigma)$. This shows immediately that $\overline{\mathbf{X}}$ is $N_m(\boldsymbol{\mu},(1/N)\Sigma)$ and is independent of Z since

$$(7)\qquad \mathbf{v}=N^{-1/2}X'\mathbf{1}=N^{1/2}\overline{\mathbf{X}}.$$

The only thing left to show is that $A=Z'Z$; this follows by replacing μ by $\overline{\mathbf{X}}$ in the identity (6) and using (7).

DEFINITION 3.1.3. If $A=Z'Z$, where the $n\times m$ matrix Z is $N(0, I_n\otimes\Sigma)$, then A is said to have the *Wishart distribution* with n degrees of freedom and covariance matrix Σ. We will write that A is $W_m(n,\Sigma)$, the subscript on W denoting the size of the matrix A.

The Wishart distribution is extremely important to us and some of its properties will be studied in the next section. Note that since $A=Z'Z$ from Theorem 3.1.2 then the sample covariance matrix S is

$$S=\frac{1}{n}A=\frac{1}{n}Z'Z=\tilde{Z}'\tilde{Z},$$

where $\tilde{Z}=n^{-1/2}Z$ is $N(0, I_n\otimes(1/n)\Sigma)$, so that S is $W_m(n,(1/n)\Sigma)$. Since S is an unbiased estimate for Σ it is of interest to know whether it, like Σ, is positive definite. The answer to this is given in the following theorem, whose proof is due to Dykstra (1970).

THEOREM 3.1.4. The matrix A given by (3) (and hence the sample covariance matrix $S=n^{-1}A$) is positive definite with probability 1 if and only if $n\geq m$ (i.e., $N>m$).

Proof. From Theorem 3.1.2, $A=Z'Z$ where the $n\times m$ matrix Z is $N(0, I_n\otimes\Sigma)$. Since $Z'Z$ is nonnegative definite it suffices to show that $Z'Z$ is nonsingular with probability 1 if and only if $n\geq m$. First, suppose that $n=m$; then the columns $\mathbf{z}_1,\ldots,\mathbf{z}_m$ of Z' are independent $N_m(0,\Sigma)$ random vectors. Now

$$\begin{aligned}
&P(\mathbf{z}_1,\ldots,\mathbf{z}_m \text{ are linearly dependent})\\
&\quad\leq\sum_{i=1}^{m}P(\mathbf{z}_i \text{ is a linear combination of } \mathbf{z}_1,\ldots,\mathbf{z}_{i-1},\mathbf{z}_{i+1},\ldots,\mathbf{z}_m)\\
&\quad=mP(\mathbf{z}_1 \text{ is a linear combination of } \mathbf{z}_2,\ldots,\mathbf{z}_m)\\
&\quad=mE\left[P(\mathbf{z}_1 \text{ is a linear combination of } \mathbf{z}_2,\ldots,\mathbf{z}_m|\mathbf{z}_2,\ldots,\mathbf{z}_m)\right]\\
&\quad=mE(0)=0,
\end{aligned}$$

where we have used the fact that $\mathbf{z}_1$ lies in a space of dimension less than m with probability 0 because $\Sigma>0$. We have then proved that, in the case $n=m$, Z has rank m with probability 1. Now, when $n>m$, the rank of Z is m with probability 1 because adding more rows to Z cannot decrease its rank, and when $n<m$ the rank of Z must be less than m. We conclude that Z has rank m with probability one if and only if $n \geq m$ and hence $A=Z'Z$ has rank m with probability 1 if and only if $n \geq m$.

Normality plays a key part in the above proof, but the interesting part of the theorem holds under much more general assumptions. Eaton and Perlman (1973) have shown that if S is the sample covariance matrix formed from N independent and identically distributed (not necessarily normal) $m\times 1$ random vectors $\mathbf{X}_1,\ldots,\mathbf{X}_N$ with $N>m$ then S is positive definite with probability 1 if and only if $P(\mathbf{X}_1 \in F_s)=0$ for *all* s-flats F_s in $R^m(0 \leq s<m)$, a condition which is implied by normality. [An s-flat is the translate $F_s=\{\mathbf{x}\}+F_{s(0)}$ of an s-dimensional linear subspace or s-subspace $F_{s(0)}$ in R^m.] A similar result has also been obtained by Das Gupta (1971).

The density function (4) considered as a function of the parameters $\boldsymbol{\mu}$ and Σ (for fixed observed X) is the *likelihood function*. Since

$$(X-\mathbf{1}\boldsymbol{\mu}')'(X-\mathbf{1}\boldsymbol{\mu}')=A+N(\overline{\mathbf{X}}-\boldsymbol{\mu})(\overline{\mathbf{X}}-\boldsymbol{\mu})'$$

[from (6) with $A=Z'Z$ and $\mathbf{v}=N^{1/2}\overline{\mathbf{X}}$], (4) can be written in the form

$$(8)\qquad (2\pi)^{-mN/2}(\det\Sigma)^{-N/2}\operatorname{etr}\left\{-\tfrac{1}{2}\Sigma^{-1}\left[A+N(\overline{\mathbf{X}}-\boldsymbol{\mu})(\overline{\mathbf{X}}-\boldsymbol{\mu})'\right]\right\}$$

so that in order to determine the likelihood function the only functions of the sample needed are $\overline{\mathbf{X}}$ and A. From this we conclude that $(\overline{\mathbf{X}}, A)$ [or $(\overline{\mathbf{X}}, S)$] is *sufficient* for $\boldsymbol{\mu}$ and Σ (or for the normal family of distributions (4) for $\boldsymbol{\mu}\in R^m, \Sigma>0$).

We conclude this section by finding the *maximum likelihood estimates* of $\boldsymbol{\mu}$ and Σ, that is, those values of $\boldsymbol{\mu}$ and Σ which maximize the likelihood function (8).

THEOREM 3.1.5. If $\mathbf{X}_1,\ldots,\mathbf{X}_N$ are independent $N_m(\boldsymbol{\mu},\Sigma)$ random vectors and $N>m$ then the maximum likelihood estimates of $\boldsymbol{\mu}$ and Σ are $\hat{\boldsymbol{\mu}}=\overline{\mathbf{X}}$ and $\hat{\Sigma}=(1/N)A=(n/N)S$, where $n=N-1$ and $\overline{\mathbf{X}}$, A, and S are given by (2) and (3).

Proof. Ignoring the constant in (8), which is of no consequence, the likelihood function is

$$L(\boldsymbol{\mu},\Sigma)=(\det\Sigma)^{-N/2}\operatorname{etr}\left(-\tfrac{1}{2}\Sigma^{-1}A\right)\exp\left[-\tfrac{1}{2}N(\overline{\mathbf{X}}-\boldsymbol{\mu})'\Sigma^{-1}(\overline{\mathbf{X}}-\boldsymbol{\mu})\right].$$

Now

$$L(\boldsymbol{\mu},\Sigma)\leq(\det\Sigma)^{-N/2}\operatorname{etr}\left(-\tfrac{1}{2}\Sigma^{-1}A\right),$$

with equality if and only if $\boldsymbol{\mu}=\overline{\mathbf{X}}$, where we have used the fact that

$$(\overline{\mathbf{X}}-\boldsymbol{\mu})'\Sigma^{-1}(\overline{\mathbf{X}}-\boldsymbol{\mu})=0$$

if and only if $\boldsymbol{\mu}=\overline{\mathbf{X}}$, because Σ^{-1} is positive definite. This shows that $\overline{\mathbf{X}}$ is the maximum likelihood estimate of $\boldsymbol{\mu}$ for all Σ. It remains to maximize the function (of Σ)

$$L(\overline{\mathbf{X}},\Sigma)=(\det\Sigma)^{-N/2}\operatorname{etr}(-\tfrac{1}{2}\Sigma^{-1}A)$$

or, equivalently, the function

$$\begin{aligned}
g(\Sigma)=\log L(\overline{\mathbf{X}},\Sigma)&=-\tfrac{1}{2}N\log\det\Sigma-\tfrac{1}{2}\operatorname{tr}(\Sigma^{-1}A)\\
&=\tfrac{1}{2}N\log\det(\Sigma^{-1}A)-\tfrac{1}{2}\operatorname{tr}(\Sigma^{-1}A)-\tfrac{1}{2}N\log\det A\\
&=\tfrac{1}{2}N\log\det(A^{1/2}\Sigma^{-1}A^{1/2})-\tfrac{1}{2}\operatorname{tr}(A^{1/2}\Sigma^{-1}A^{1/2})-\tfrac{1}{2}N\log\det A\\
&=\tfrac{1}{2}\sum_{i=1}^{m}(N\log\lambda_i-\lambda_i)-\tfrac{1}{2}N\log\det A
\end{aligned}$$

where $\lambda_1,\ldots,\lambda_m$ are the latent roots of $A^{1/2}\Sigma^{-1}A^{1/2}$, i.e., of $\Sigma^{-1}A$. Since the function

$$f(\lambda)=N\log\lambda-\lambda$$

has a unique maximum at $\lambda=N$ of $N\log N-N$ it follows that

$$g(\Sigma)\le\tfrac{1}{2}Nm\log N-\tfrac{1}{2}mN-\tfrac{1}{2}N\log\det A,$$

or

$$L(\overline{\mathbf{X}},\Sigma)\le N^{mN/2}e^{-mN/2}(\det A)^{-N/2},$$

with equality if and only if $\lambda_i=N$ $(i=1,\ldots,m)$. This last condition is equivalent to $A^{1/2}\Sigma^{-1}A^{1/2}=NI_m$ and hence to $\Sigma=(1/N)A$. Therefore we conclude that

$$L(\boldsymbol{\mu},\Sigma)\le N^{mN/2}e^{-mN/2}(\det A)^{-N/2},$$

with equality if and only if $\boldsymbol{\mu}=\overline{\mathbf{X}}$ and $\Sigma=(1/N)A$, and the proof is complete.

The above proof, which avoids any differentiation of the likelihood function, is due to Watson (1964). It is left to the reader to determine why the condition $N>m$ is imposed, where it is used, and what happens if it does not hold. Finally, note that the maximum likelihood estimate $\hat{\Sigma}$ has expectation

$$E(\hat{\Sigma})=E\left(\frac{1}{N}A\right)=\frac{n}{N}\Sigma$$

so that it is not unbiased for Σ. It is, however, asymptotically unbiased since $n/N\rightarrow 1$ as $N\rightarrow\infty$.

3.2. THE WISHART DISTRIBUTION

3.2.1. *The Wishart Density Function*

We have defined the Wishart $W_m(n,\Sigma)$ distribution in Definition 3.1.3 as the distribution of the $m\times m$ random matrix $A=Z'Z$, where $Z(n\times m)$ is $N(0,I_n\otimes\Sigma)$. When $n<m$, A is singular (Theorem 3.1.4) and the $W_m(n,\Sigma)$ distribution does not have a density function. The following theorem gives the density function of A when $n\geq m$; most of the work involved in the derivation has already been done in Section 2.1 and it is only a matter of putting things together.

THEOREM 3.2.1. If A is $W_m(n,\Sigma)$ with $n\geq m$ then the density function of A is

$$(1)\qquad \frac{1}{2^{mn/2}\Gamma_m\left(\frac{1}{2}n\right)(\det\Sigma)^{n/2}}\operatorname{etr}\left(-\tfrac{1}{2}\Sigma^{-1}A\right)(\det A)^{(n-m-1)/2}\qquad (A>0)$$

where $\Gamma_m(\cdot)$ denotes the multivariate gamma function given in Definition 2.1.10.

Proof. Write $A=Z'Z$, where $Z(n\times m)$ is $N(0,I_n\otimes\Sigma)$. The density of Z is

$$(2\pi)^{-mn/2}(\det\Sigma)^{-n/2}\operatorname{etr}\left(-\tfrac{1}{2}\Sigma^{-1}Z'Z\right)(dZ)$$

where the volume element $(dZ)\equiv\bigwedge_{i=1}^{n}\bigwedge_{j=1}^{m}dz_{ij}$ has been included to facilitate the calculation of Jacobians when we make transformations on Z. Since $n\geq m$, Z has rank m with probability 1 (see the proof of Theorem

3.1.4). Put $Z=H_1T$ as in Theorems 2.1.13 and 2.1.14, where H_1 is $n\times m$ with $H_1'H_1=I_m$ (i.e., $H_1\in V_{m,n}$, the Stiefel manifold consisting of $n\times m$ matrices with orthonormal columns) and T is $m\times m$ upper-triangular. Then $A=Z'Z=T'T$, and from Theorem 2.1.14 the volume element (dZ) becomes

$$(dZ)=2^{-m}(\det A)^{(n-m-1)/2}(dA)(H_1'dH_1),$$

so that the joint density of A and H_1 is

$$(2\pi)^{-mn/2}(\det\Sigma)^{-n/2}\operatorname{etr}(-\tfrac{1}{2}\Sigma^{-1}A)2^{-m}(\det A)^{(n-m-1)/2}(dA)(H_1'\,dH_1)$$

The marginal density function of A given by (1) then follows from this by integrating with respect to H_1 over the Stiefel manifold $V_{m,n}$ using

$$\int_{V_{m,n}}(H_1'dH_1)=\frac{2^m\pi^{mn/2}}{\Gamma_m(\frac{1}{2}n)},$$

the result of Theorem 2.1.15.

The density function of the sample covariance matrix S follows immediately and is worth stating explicitly.

COROLLARY 3.2.2. If $\mathbf{X}_1,\dots,\mathbf{X}_N$ are independent $\mathrm{N}_m(\boldsymbol{\mu},\Sigma)$ random vectors and $N>m$ the density function of the sample covariance matrix

$$S=\frac{1}{n}\sum_{i=1}^{N}(\mathbf{X}_i-\overline{\mathbf{X}})(\mathbf{X}_i-\overline{\mathbf{X}})'\qquad(n=N-1)$$

is

(2)

$$\frac{1}{\Gamma_m(\frac{1}{2}n)(\det\Sigma)^{n/2}}(\tfrac{1}{2}n)^{mn/2}\operatorname{etr}(-\tfrac{1}{2}n\Sigma^{-1}S)(\det S)^{(n-m-1)/2}\qquad(S>0)$$

Proof. The proof follows either by recalling that S is $W_m(n,(1/n)\Sigma)$ (see the discussion following Definition 3.1.3) or by making the transformation $A=nS$ in (1).

In the univariate case $m=1$, these results reduce to familiar ones. In this case let us write

$$\Sigma\equiv\sigma^2,\qquad S\equiv s^2=\frac{1}{n}\sum_{i=1}^{N}(X_i-\overline{X})^2;$$

then the density function of s^2 is, from (2),

$$\left(\frac{n}{2\sigma^2}\right)^{n/2}\frac{1}{\Gamma(\frac{1}{2}n)}\exp\left(-\frac{ns^2/\sigma^2}{2}\right)(s^2)^{n/2-1} \qquad (s^2>0).$$

Putting $v=ns^2/\sigma^2$, we then obtain the density function of v as

$$\frac{1}{2^{n/2}\Gamma(\frac{1}{2}n)}e^{-v/2}v^{n/2-1} \qquad (v>0),$$

the χ_n^2 density function. This shows that if A is $W_1(n,\sigma^2)$ (so that A is 1×1) then A/σ^2 is χ_n^2, a result which we will use quite often.

It is worth remarking here that although n is an integer ($\geq m$) in the derivation of the Wishart density function of Theorem 3.2.1, the function (1) is still a density function when n is any real number greater than $m-1$ (not necessarily an integer), a fact which was noted in the discussion following Theorem 2.1.11. We can, therefore, extend our definition of the Wishart distribution to cover noninteger degrees of freedom n for $n>m-1$; for most *practical* purposes, however, Definition 3.1.3, which defines it for all positive integers n, suffices.

The density function (1) was first obtained by Fisher (1915) when $m=2$, and for general m by Wishart (1928) using a geometrical argument. Since that time a number of derivations have appeared. The derivation given in this section is due to James (1954) and Olkin and Roy (1954).

3.2.2. *Characteristic Function, Moments, and Asymptotic Distribution*

The reader will recall that if the random variable A is $W_1(n,\sigma^2)$ then A/σ^2 is χ_n^2 so that the characteristic function of A is $(1-2it\sigma^2)^{-n/2}$. The following theorem generalizes this result.

THEOREM 3.2.3. If A is $W_m(n,\Sigma)$ then the characteristic function of A [that is, the joint characteristic function of the $\frac{1}{2}m(m+1)$ variables a_{ij}, $1\leq i\leq j\leq m$] is

$$\phi(\Theta)\equiv E\left[\exp\left(i\sum_{j\leq k}^{m}\theta_{jk}a_{jk}\right)\right]=\det(I_m-i\Gamma\Sigma)^{-n/2};$$

where $\Gamma=(\gamma_{ij})$, where $i,j=1,\ldots,m$, with $\gamma_{ij}=(1+\delta_{ij})\theta_{ij}$, $\theta_{ji}=\theta_{ji}$, and δ_{ij} is the Kronecker delta,

$$\delta_{ij}=\begin{cases}1 & \text{if} \quad i=j\\ 0 & \text{if} \quad i\neq j\end{cases}.$$

Proof. The characteristic function $\phi(\Theta)$ can be written as

$$\phi(\Theta)=E\left\{\exp\left[\frac{i}{2}\sum_{j,k=1}^{m}(1+\delta_{jk})\theta_{jk}a_{jk}\right]\right\} \tag{3}$$

$$=E\left[\operatorname{etr}\left(\frac{i}{2}A\Gamma\right)\right].$$

There are two cases to consider:

(i) First, suppose that n is a positive integer. Then we can write $A=Z'Z$, where Z is $N(0, I_n\otimes\Sigma)$. Let $\mathbf{z}_1,\dots,\mathbf{z}_n$ be the columns of Z'; then $\mathbf{z}_1,\dots,\mathbf{z}_n$ are independent $N_m(\mathbf{0},\Sigma)$ random vectors and $A=Z'Z=\sum_{j=1}^{n}\mathbf{z}_j\mathbf{z}_j'$. Hence

$$\phi(\Theta)=E\left[\operatorname{etr}\left(\frac{i}{2}Z'Z\Gamma\right)\right]=E\left[\operatorname{etr}\left(\frac{i}{2}\sum_{j=1}^{n}\mathbf{z}_j\mathbf{z}_j'\Gamma\right)\right]$$

$$=E\left[\exp\frac{i}{2}\sum_{j=1}^{n}\operatorname{tr}(\mathbf{z}_j\mathbf{z}_j'\Gamma)\right]$$

$$=E\left[\exp\frac{i}{2}\sum_{j=1}^{n}\mathbf{z}_j'\Gamma\mathbf{z}_j\right]$$

$$=\prod_{j=1}^{n}E\left[\exp\frac{i}{2}\mathbf{z}_j'\Gamma\mathbf{z}_j\right] \qquad \text{(by independence)}$$

$$=\left(E\left[\exp\frac{i}{2}\mathbf{z}_1'\Gamma\mathbf{z}_1\right]\right)^n$$

Put $\mathbf{y}=\Sigma^{-1/2}\mathbf{z}_1$; then $\mathbf{y}$ is $N_m(\mathbf{0}, I_m)$ and

$$\phi(\Theta)=\left(E\left[\exp\frac{i}{2}\mathbf{y}'\Sigma^{1/2}\Gamma\Sigma^{1/2}\mathbf{y}\right]\right)^n.$$

Since $\Sigma^{1/2}\Gamma\Sigma^{1/2}$ is real symmetric there exists an orthogonal $m\times m$ matrix H such that

$$H\Sigma^{1/2}\Gamma\Sigma^{1/2}H'=\Lambda=\operatorname{diag}(\lambda_1,\dots,\lambda_m),$$

where $\lambda_1,\dots,\lambda_m$ are the latent roots of $\Sigma^{1/2}\Gamma\Sigma^{1/2}$. Put $\mathbf{u}=H\mathbf{y}$, then $\mathbf{u}$ is

$N_m(\mathbf{0}, I_m)$ and

$$\phi(\Theta)=\left(E\left[\exp\frac{i}{2}\mathbf{u}'\Lambda\mathbf{u}\right]\right)^n$$

$$=\left(E\left[\exp\frac{i}{2}\sum_{j=1}^m \lambda_j u_j^2\right]\right)^n$$

$$=\prod_{j=1}^m E\left(\exp\frac{i}{2}\lambda_j u_j^2\right)^n$$

$$=\prod_{j=1}^m (1-i\lambda_j)^{-n/2},$$

where we have used the fact that the u_j^2, $j=1,\dots,m$ are independent χ_1^2 random variables. The desired result now follows by noting that

$$\prod_{j=1}^m (1-i\lambda_j)=\det(I_m-i\Lambda)$$

$$=\det\left(I_m-i\Sigma^{1/2}\Gamma\Sigma^{1/2}\right)$$

$$=\det(I_m-i\Gamma\Sigma).$$

(ii) Now suppose that n is any real number with $n>m-1$. Then A has the density function (1) (see the discussion following Corollary 3.2.2) so that

$$\phi(\Theta)=E\left[\operatorname{etr}\left(\frac{i}{2}A\Gamma\right)\right]$$

$$=\frac{1}{2^{mn/2}\Gamma_m(\frac{1}{2}n)(\det\Sigma)^{n/2}}$$

$$\int_{A>0}\operatorname{etr}\left[-\tfrac{1}{2}A(\Sigma^{-1}-i\Gamma)\right](\det A)^{(n-m-1)/2}(dA).$$

Now apply Theorem 2.1.11 to give

$$\phi(\Theta)=\frac{\det(\Sigma^{-1}-i\Gamma)^{-n/2}}{(\det\Sigma)^{n/2}}=\det(I-i\Gamma\Sigma)^{-n/2},$$

as desired.

The moments of the elements of the Wishart matrix A of Theorem 3.2.3 can be found from the characteristic function in the usual way. We know already that

$$E(A)=n\Sigma,$$

and it is a straightforward matter to show that

$$\text{(4)} \qquad \operatorname{Cov}(a_{ij}, a_{kl})=n(\sigma_{ik}\sigma_{jl}+\sigma_{il}\sigma_{jk})$$

for $i, j, k, l=1,\ldots,m$ (see Problem 3.1). The matrix of covariances between the elements of A can be expressed in terms of a Kronecker product. Let H_{ij} denote the $m\times m$ matrix with $h_{ij}=1$ and all other elements zero and put

$$K=\sum_{i,j=1}^{m}(H_{ij}\otimes H'_{ij}),$$

so that K is $m^2\times m^2$. For example, with $m=2$ the reader can readily verify that

$$K=\begin{bmatrix} 1 & 0 & 0 & 0\\ 0 & 0 & 1 & 0\\ 0 & 1 & 0 & 0\\ 0 & 0 & 0 & 1\end{bmatrix}.$$

For any $m\times m$ matrix C, the matrix K has the property that it transforms $\operatorname{vec}(C)$ into $\operatorname{vec}(C')$,

$$K\operatorname{vec}(C)=\operatorname{vec}(C'),$$

and for this reason is sometimes called the "commutation matrix." If A is $W_m(n,\Sigma)$ the covariance matrix of $\operatorname{vec}(A)$ can be readily expressed in terms of the matrix K as

$$\text{(5)} \qquad \operatorname{Cov}[\operatorname{vec}(A)]=n(I_{m^2}+K)(\Sigma\otimes\Sigma)$$

(see Problem 3.2) a fact noted by Magnus and Neudecker (1979). Finally, we saw in Corollary 1.2.18 that under general conditions the sample covariance matrix $S(n)$ formed from a sample of size $n+1$ is asymptotically normal as $n\rightarrow\infty$. In the case of normal sampling $S(n)$ is $W_m(n,(1/n)\Sigma)$ so that the asymptotic distribution as $n\rightarrow\infty$ of

$$n^{1/2}[\operatorname{vec}(S(n))-\operatorname{vec}(\Sigma)]$$

is

$$N_{m^2}[\mathbf{0},(I_{m^2}+K)(\Sigma\otimes\Sigma)].$$

3.2.3. *Some Properties of the Wishart Distribution*

In this section some properties of the Wishart distribution are derived. Our first result says that the sum of independent Wishart matrices with the same covariance matrix is also Wishart.

THEOREM 3.2.4. If the $m\times m$ random matrices $A_1,\dots,A_r$ are all independent and A_i is $W_m(n_i,\Sigma)$, $i=1,\dots,r$, then $\sum_{i=1}^r A_i$ is $W_m(n,\Sigma)$, where $n=\sum_{i=1}^r n_i$.

Proof. The characteristic function of $\sum_{i=1}^r A_i$ is the product of the characteristic functions of $A_1,\dots,A_r$ and hence, with the notation of Theorem 3.2.3, is

$$\prod_{j=1}^{r}\det(I_m-i\Gamma\Sigma)^{-n_j/2}=\det(I_m-i\Gamma\Sigma)^{-n/2},$$

which is the characteristic function of the $W_m(n,\Sigma)$ distribution.

The above theorem is valid regardless of whether the n_i are positive integers or real numbers bigger than $m-1$. When the n_i are restricted to being positive integers one can, of course, give a proof in terms of the normal decomposition. Write $A_i=Z_i'Z_i$, where Z_i is $N(0,I_{n_i}\otimes\Sigma)$ $(i=1,\dots,r)$ and $Z_1,\dots,Z_r$ are independent, and put

$$Z=\begin{bmatrix}Z_1\\ \vdots\\ Z_2\\ \vdots\\ \vdots\\ Z_r\end{bmatrix}$$

so that Z is $N(0,I_n\otimes\Sigma)$. Then

$$\sum_{i=1}^{r}A_i=\sum_{i=1}^{r}Z_i'Z_i=Z'Z,$$

which is $W_m(n,\Sigma)$.

The next theorem, which will be used often, shows that the family of Wishart distributions is closed under certain linear transformations.

THEOREM 3.2.5. If A is $W_m(n,\Sigma)$ and M is $k\times m$ of rank k then MAM' is $W_k(n,M\Sigma M')$.

Proof. The characteristic function of MAM' is [see (3)]

$$E\left[\operatorname{etr}\left(\frac{i}{2}MAM'\Gamma\right)\right]=E\left[\operatorname{etr}\left(\frac{i}{2}AM'\Gamma M\right)\right] \tag{6}$$
$$=\det(I_m-iM'\Gamma M\Sigma)^{-n/2}$$
$$=\det(I_k-i\Gamma M\Sigma M')^{-n/2},$$

where we have used Theorem 3.2.3 and the fact that M is $k\times m$. The result follows immediately, since the right side of (6) is the characteristic function of the $W_k(n,M\Sigma M')$ distribution.

Again, this theorem is valid whenever the Wishart distribution is defined. If n is a positive integer a proof can be constructed in terms of the normal decomposition of A; it is left to the reader to fill in the details (see Problem 3.4). As a special case of this theorem we have:

COROLLARY 3.2.6. If A is $W_m(n,\Sigma)$ and A and Σ are partitioned as

$$A=\begin{bmatrix}A_{11} & A_{12}\\ A_{21} & A_{22}\end{bmatrix},\qquad \Sigma=\begin{bmatrix}\Sigma_{11} & \Sigma_{12}\\ \Sigma_{21} & \Sigma_{22}\end{bmatrix}, \tag{7}$$

where A_{11} and Σ_{11} are $k\times k$, then A_{11} is $W_k(n,\Sigma_{11})$.

Proof. Put $M=[I_k:0]$ $(k\times m)$ in Theorem 3.2.5, then $MAM'=A_{11}$, $M\Sigma M'=\Sigma_{11}$, and the result is immediate.

Corollary 3.2.6 tells us that the marginal distribution of any square submatrix of A located on the diagonal of A (so that the diagonal elements of the submatrix are diagonal elements of A) is Wishart. In particular, of course, A_{22} is $W_{m-k}(n,\Sigma_{22})$. The next result says that if $\Sigma_{12}=0$ then A_{11} and A_{22} are independent.

THEOREM 3.2.7. If A is $W_m(n,\Sigma)$, where A and Σ are partitioned as in (7) and $\Sigma_{12}=0$, then A_{11} and A_{22} are independent and their distributions are, respectively, $W_k(n,\Sigma_{11})$ and $W_{m-k}(n,\Sigma_{22})$.

A proof of this theorem can be constructed by observing that when $\Sigma_{12}=0$ the joint characteristic function of A_{11} and A_{22} is the product of the characteristic functions of A_{11} and A_{22}. The details are left to the reader (see Problem 3.5). As usual, when n is a positive integer a direct proof involving the normal decomposition of A is also available. Note that in the special case when n is an integer and Σ is diagonal, $\Sigma=\text{diag}(\sigma_{11},\dots,\sigma_{mm})$, an obvious extension of Theorem 3.2.7 states that the diagonal elements $a_{11},\dots,a_{mm}$ of A are all independent, and a_{ii} is $W_1(n,\sigma_{ii})$; that is, a_{ii}/σ_{ii} is χ^2_n, for $i=1,\dots,m$.

Our next result is also a direct consequence of Theorem 3.2.5.

THEOREM 3.2.8. If A is $W_m(n,\Sigma)$, where n is a positive integer and $\mathbf{Y}$ is any $m\times 1$ random vector which is independent of A with $P(\mathbf{Y}=\mathbf{0})=0$ then $\mathbf{Y}'A\mathbf{Y}/\mathbf{Y}'\Sigma\mathbf{Y}$ is χ^2_n, and is independent of $\mathbf{Y}$.

Proof. In Theorem 3.2.5 put $M=\mathbf{Y}'$ $(1\times m)$ then, conditional on $\mathbf{Y}$, $\mathbf{Y}'A\mathbf{Y}$ is $W_1(n,\mathbf{Y}'\Sigma\mathbf{Y})$; that is $\mathbf{Y}'A\mathbf{Y}/\mathbf{Y}'\Sigma\mathbf{Y}$ is χ^2_n. Since this distribution does not depend on $\mathbf{Y}$ it is also the unconditional distribution of $\mathbf{Y}'A\mathbf{Y}/\mathbf{Y}'\Sigma\mathbf{Y}$ and the theorem is proved.

The following corollary is an interesting consequence of this theorem.

COROLLARY 3.2.9. If $\overline{\mathbf{X}}$ and S are the mean and covariance matrix formed from a sample of size $N=n+1$ from the $N_m(\boldsymbol{\mu},\Sigma)$ distribution then

$$n\frac{\overline{\mathbf{X}}'S\overline{\mathbf{X}}}{\overline{\mathbf{X}}'\Sigma\overline{\mathbf{X}}}$$

is χ^2_n and is independent of $\overline{\mathbf{X}}$.

Proof. From Theorem 3.1.2 we know that $\overline{\mathbf{X}}$ and S are independent, and S is $W_m(n,(1/n)\Sigma)$. A direct application of Theorem 3.2.8 completes the proof.

Our next result is of some importance and will be very useful in a variety of situations.

THEOREM 3.2.10. Suppose that A is $W_m(n,\Sigma)$, where A and Σ are partitioned as in (7), and put $A_{11\cdot 2}=A_{11}-A_{12}A_{22}^{-1}A_{21}$ and $\Sigma_{11\cdot 2}=\Sigma_{11}-\Sigma_{12}\Sigma_{22}^{-1}\Sigma_{21}$. Then

(i) $A_{11\cdot 2}$ is $W_k(n-m+k,\Sigma_{11\cdot 2})$ and is independent of A_{12} and A_{22};

(ii) the conditional distribution of A_{12} given A_{22} is $N(\Sigma_{12}\Sigma_{22}^{-1}A_{22}, \Sigma_{11\cdot 2}\otimes A_{22})$; and

(iii) A_{22} is $W_{m-k}(n,\Sigma_{22})$.

Proof. The Wishart density function has not yet been used explicitly, so we will give a proof which utilizes it. This involves assuming that $n>m-1$. The density of A is, from Theorem 3.2.1,

$$\text{(8)}\qquad \frac{1}{2^{mn/2}\Gamma_m\left(\frac{1}{2}n\right)(\det\Sigma)^{n/2}}\operatorname{etr}\left(-\tfrac{1}{2}\Sigma^{-1}A\right)(\det A)^{(n-m-1)/2}$$

Make the change of variables $A_{11\cdot2}=A_{11}-A_{12}A_{22}^{-1}A_{21}$, $B_{12}=A_{12}$, $B_{22}=A_{22}$ so that

$$(dA)=(dA_{11})\wedge(dA_{12})\wedge(dA_{22})$$
$$=(dA_{11\cdot2})\wedge(dB_{12})\wedge(dB_{22}).$$

Note that

$$\text{(9)}\qquad \det A=\det A_{22}\det(A_{11}-A_{12}A_{22}^{-1}A_{21})$$
$$=\det B_{22}\det A_{11\cdot2}$$

and

$$\det\Sigma=\det\Sigma_{22}\det\Sigma_{11\cdot2}.$$

Now put

$$C=\Sigma^{-1}=\begin{bmatrix} C_{11} & C_{12}\\ C_{21} & C_{22}\end{bmatrix},$$

where C_{11} is $k\times k$. Then

$$\operatorname{tr}(\Sigma^{-1}A)=\operatorname{tr}\begin{bmatrix} C_{11} & C_{12}\\ C_{21} & C_{22}\end{bmatrix}\begin{bmatrix} A_{11\cdot2}+B_{12}B_{22}^{-1}B_{21} & B_{12}\\ B_{21} & B_{22}\end{bmatrix}$$
$$=\operatorname{tr}(C_{11}A_{11\cdot2})+\operatorname{tr}(C_{11}B_{12}B_{22}^{-1}B_{21})+2\operatorname{tr}(C_{12}B_{21})$$
$$+\operatorname{tr}(C_{22}B_{22}),$$

and it can be readily verified that this can be written as

$$\text{(10)}\quad \operatorname{tr}(\Sigma^{-1}A)=\operatorname{tr}\left[C_{11}(B_{12}+C_{11}^{-1}C_{12}B_{22})B_{22}^{-1}(B_{12}+C_{11}^{-1}C_{12}B_{22})'\right]$$
$$+\operatorname{tr}\left[B_{22}(C_{22}-C_{21}C_{11}^{-1}C_{12})\right]+\operatorname{tr}(C_{11}A_{11\cdot2})$$
$$=\operatorname{tr}\left[\Sigma_{11\cdot2}^{-1}(B_{12}-\Sigma_{12}\Sigma_{22}^{-1}B_{22})B_{22}^{-1}(B_{12}-\Sigma_{12}\Sigma_{22}^{-1}B_{22})'\right]$$
$$+\operatorname{tr}(B_{22}\Sigma_{22}^{-1})+\operatorname{tr}(\Sigma_{11\cdot2}^{-1}A_{11\cdot2}),$$

where we have used the relations $C_{11}=\Sigma_{11\cdot2}^{-1}$, $C_{22}-C_{21}C_{11}^{-1}C_{12}=\Sigma_{22}^{-1}$, and

$C_{11}^{-1}C_{12} = -\Sigma_{12}\Sigma_{22}^{-1}$, which are implied by the equation $\Sigma C = I$ (see Theorem A5.2). Substituting back in (8) using (9) and (10), the joint density of $A_{11\cdot 2}$, B_{12}, and B_{22} can then be written in the form

$$(11)\quad \frac{\operatorname{etr}\left(-\frac{1}{2}\Sigma_{11\cdot 2}^{-1}A_{11\cdot 2}\right)(\det A_{11\cdot 2})^{(n-m+k-k-1)/2}}{2^{k(n-m+k)/2}\Gamma_k\left[\frac{1}{2}(n-m+k)\right](\det \Sigma_{11\cdot 2})^{(n-m+k)/2}}$$

$$\cdot \frac{\operatorname{etr}\left(-\frac{1}{2}\Sigma_{22}^{-1}B_{22}\right)(\det B_{22})^{(n-m+k-1)/2}}{2^{(m-k)n/2}\Gamma_{m-k}\left(\frac{1}{2}n\right)(\det \Sigma_{22})^{n/2}}$$

$$\cdot \frac{\operatorname{etr}\left[-\frac{1}{2}\Sigma_{11\cdot 2}^{-1}\left(B_{12}-\Sigma_{12}\Sigma_{22}^{-1}B_{22}\right)B_{22}^{-1}\left(B_{12}-\Sigma_{12}\Sigma_{22}^{-1}B_{22}\right)'\right]}{(2\pi)^{k(m-k)/2}(\det \Sigma_{11\cdot 2})^{(m-k)/2}(\det B_{22})^{k/2}}$$

where we have used the fact that

$$\begin{aligned}\Gamma_m\left(\frac{1}{2}n\right) &= \pi^{m(m-1)/4}\prod_{i=1}^{m}\Gamma\left[\tfrac{1}{2}(n-i+1)\right]\\ &= \pi^{k(k-1)/4}\prod_{i=1}^{k}\Gamma\left[\tfrac{1}{2}(n-m+k-i+1)\right]\\ &\quad\cdot \pi^{(m-k)(m-k-1)/4}\prod_{i=1}^{m-k}\Gamma\left[\tfrac{1}{2}(n-i+1)\right]\\ &\quad\cdot \pi^{k(m-k)/2}\\ &= \Gamma_k\left[\tfrac{1}{2}(n-m+k)\right]\Gamma_{m-k}\left(\tfrac{1}{2}n\right)\pi^{k(m-k)/2}\end{aligned}$$

From (11) we see that $A_{11\cdot 2}$ is independent of B_{12} and B_{22}, i.e. of A_{12} and A_{22}, because the density function factors. The first line is the $W_k(n-m+k, \Sigma_{11\cdot 2})$ density function for $A_{11\cdot 2}$. The last two lines in (11) give the joint density function of B_{12} and B_{22}, i.e., of A_{12} and A_{22}. From Corollary 3.2.6, the distribution of A_{22} is $W_{m-k}(n, \Sigma_{22})$ with density function given by the second line in (11). The third line thus represents the conditional density function of B_{12} given B_{22}, i.e., of A_{12} given A_{22}. Using Theorem 3.1.1, it is seen that this is $N(\Sigma_{12}\Sigma_{22}^{-1}A_{22}, \Sigma_{11\cdot 2}\otimes A_{22})$, and the proof is complete.

The next result can be proved with the help of Theorem 3.2.10.

THEOREM 3.2.11. If A is $W_m(n,\Sigma)$ and M is $k\times m$ of rank k, then $(MA^{-1}M')^{-1}$ is $W_k(n-m+k,(M\Sigma^{-1}M')^{-1})$.

Proof. Put $B=\Sigma^{-1/2}A\Sigma^{-1/2}$, where $\Sigma^{1/2}$ is the positive definite square root of Σ. Then, from Theorem 3.2.5, B is $W_m(n, I_m)$. Putting $R=M\Sigma^{-1/2}$ we have

$$(MA^{-1}M')^{-1}=(R\Sigma^{1/2}\Sigma^{-1/2}B^{-1}\Sigma^{-1/2}\Sigma^{1/2}R')^{-1}=(RB^{-1}R')^{-1}$$

and $(M\Sigma^{-1}M')^{-1}=(RR')^{-1}$, so that we need to prove that $(RB^{-1}R')^{-1}$ is $W_k(n-m+k,(RR')^{-1})$. Put $R=L[I_k:0]H$, where L is $k\times k$ and nonsingular and H is $m\times m$ and orthogonal, then

$$\begin{aligned}(RB^{-1}R')^{-1}&=\left(L[I_k:0]HB^{-1}H'\begin{bmatrix}I_k\\ \cdots\\ 0\end{bmatrix}L'\right)^{-1}\\ &=L'^{-1}\left([I_k:0](HBH')^{-1}\begin{bmatrix}I_k\\ \cdots\\ 0\end{bmatrix}\right)^{-1}L^{-1}\\ &=L'^{-1}\left([I_k:0]C^{-1}\begin{bmatrix}I_k\\ \cdots\\ 0\end{bmatrix}\right)^{-1}L^{-1},\end{aligned}$$

where $C=HBH'$ is $W_m(n, I_m)$, using Theorem 3.2.5 again. Now, put

$$D=C^{-1}=\begin{bmatrix}D_{11} & D_{12}\\ D_{21} & D_{22}\end{bmatrix},\qquad C=\begin{bmatrix}C_{11} & C_{12}\\ C_{21} & C_{22}\end{bmatrix},$$

where D_{11} and C_{11} are $k\times k$, then $(RB^{-1}R')^{-1}=L'^{-1}D_{11}^{-1}L^{-1}$ and, since $D_{11}^{-1}=C_{11}-C_{12}C_{22}^{-1}C_{21}$, it follows from (i) of Theorem 3.2.10 that D_{11}^{-1} is $W_k(n-m+k, I_k)$. Hence, $L'^{-1}D_{11}^{-1}L^{-1}$ is $W_k(n-m+k,(LL')^{-1})$ and, since $(LL')^{-1}=(RR')^{-1}$, the proof is complete.

One consequence of Theorem 3.2.11 is the following result, which should be compared with Theorem 3.2.8.

THEOREM 3.2.12. If A is $W_m(n,\Sigma)$, where n is a positive integer, $n>m-1$, and $\mathbf{Y}$ is any $m\times 1$ random vector distributed independently of A with $P(\mathbf{Y}=\mathbf{0})=0$ then $\mathbf{Y}'\Sigma^{-1}\mathbf{Y}/\mathbf{Y}'A^{-1}\mathbf{Y}$ is χ^2_{n-m+1}, and is independent of $\mathbf{Y}$.

Proof. In Theorem 3.2.11 put $M=\mathbf{Y}'$ $(1\times m)$ then, conditional on $\mathbf{Y}$, $(\mathbf{Y}'A^{-1}\mathbf{Y})^{-1}$ is $W_1(n-m+1,(\mathbf{Y}'\Sigma^{-1}\mathbf{Y})^{-1})$; that is, $\mathbf{Y}'\Sigma^{-1}\mathbf{Y}/\mathbf{Y}'A^{-1}\mathbf{Y}$ is χ^2_{n-m+1}.

Since this distribution does not depend on $\mathbf{Y}$ it is also the unconditional distribution, and the proof is complete.

There are a number of interesting applications of this result. We will outline two of them here. First, if A is $W_m(n,\Sigma)$ then the distribution of A^{-1} is called the inverted Wishart distribution. Some of its properties are studied in Problem 3.6. The expectation of A^{-1} is easy to obtain using Theorem 3.2.12. For any fixed $\boldsymbol{\alpha}\in R^m$, $\boldsymbol{\alpha}\neq\mathbf{0}$, we know that $\boldsymbol{\alpha}'\Sigma^{-1}\boldsymbol{\alpha}/\boldsymbol{\alpha}'A^{-1}\boldsymbol{\alpha}$ is χ^2_{n-m+1}, so that

$$E(\boldsymbol{\alpha}'A^{-1}\boldsymbol{\alpha})=\boldsymbol{\alpha}'\Sigma^{-1}\boldsymbol{\alpha}E\left(\frac{1}{\chi^2_{n-m+1}}\right)$$

$$=\frac{1}{n-m-1}\boldsymbol{\alpha}'\Sigma^{-1}\boldsymbol{\alpha}\qquad (n-m-1>0).$$

Hence

$$\boldsymbol{\alpha}'E(A^{-1})\boldsymbol{\alpha}=\frac{\boldsymbol{\alpha}'\Sigma^{-1}\boldsymbol{\alpha}}{n-m-1}\qquad\text{for all }\boldsymbol{\alpha},$$

which implies that

$$(12)\qquad E(A^{-1})=\frac{1}{n-m-1}\Sigma^{-1}\qquad\text{for } n-m-1>0.$$

The second application is of great practical importance in testing hypotheses about the mean of a multivariate normal distribution when the covariance matrix is unknown. Suppose that $\mathbf{X}_1,\ldots,\mathbf{X}_N$ are independent $N_m(\boldsymbol{\mu},\Sigma)$ random vectors giving rise to a sample mean vector $\overline{\mathbf{X}}$ and sample covariance matrix S; Hotelling's T^2 statistic (Hotelling, 1931) is defined as

$$T^2=N\overline{\mathbf{X}}'S^{-1}\overline{\mathbf{X}}.$$

Note that when $m=1$, T^2 is the square of the usual t statistic used for testing whether $\mu=0$. In general, it is clear that $T^2\geq 0$, and if $\boldsymbol{\mu}=\mathbf{0}$ then $\overline{\mathbf{X}}$ should be close to $\mathbf{0}$, hence so should T^2. It therefore seems reasonable to reject the *null hypothesis* that $\boldsymbol{\mu}=\mathbf{0}$ if the observed value of T^2 is large enough. This test has certain optimal properties which will be studied later in Section 6.3. At this point however, we can easily derive the distribution of T^2 with the help of Theorem 3.2.12.

THEOREM 3.2.13. Let $\bar{\mathbf{X}}$ and S be the mean and covariance matrix formed from a random sample of size $N=n+1$ from the $N_m(\boldsymbol{\mu},\Sigma)$ distribution $(n\geq m)$, and let $T^2=N\bar{\mathbf{X}}'S^{-1}\bar{\mathbf{X}}$. Then

$$\frac{T^2}{n}\cdot\frac{n-m+1}{m}$$

is $F_{m,n-m+1}(\delta)$, $\delta=N\boldsymbol{\mu}'\Sigma^{-1}\boldsymbol{\mu}$ (i.e., noncentral F with m and $n-m+1$ degrees of freedom and noncentrality parameter δ).

Proof. From Theorem 3.1.2 $\bar{\mathbf{X}}$ and S are independent; $\bar{\mathbf{X}}$ is $N_m(\boldsymbol{\mu},(1/N)\Sigma)$ and S is $W_m(n,(1/n)\Sigma)$. Write T^2/n as

$$\begin{aligned}\frac{T^2}{n}&=\frac{N\bar{\mathbf{X}}'S^{-1}\bar{\mathbf{X}}}{n\bar{\mathbf{X}}'\Sigma^{-1}\bar{\mathbf{X}}}\cdot\bar{\mathbf{X}}'\Sigma^{-1}\bar{\mathbf{X}}\\&=\frac{N\bar{\mathbf{X}}'\Sigma^{-1}\bar{\mathbf{X}}}{\dfrac{n\bar{\mathbf{X}}'\Sigma^{-1}\bar{\mathbf{X}}}{\bar{\mathbf{X}}'S^{-1}\bar{\mathbf{X}}}}.\end{aligned}$$

Theorem 3.2.12 shows that

$$n\frac{\bar{\mathbf{X}}'\Sigma^{-1}\bar{\mathbf{X}}}{\bar{\mathbf{X}}'S^{-1}\bar{\mathbf{X}}}\quad\text{is}\quad\chi^2_{n-m+1}$$

and is independent of $\bar{\mathbf{X}}$. Moreover, since $\bar{\mathbf{X}}$ is $N_m(\boldsymbol{\mu},(1/N)\Sigma)$, Theorem 1.4.1 shows that

$$N\bar{\mathbf{X}}'\Sigma^{-1}\bar{\mathbf{X}}\quad\text{is}\quad\chi^2_m(\delta),\qquad\delta=N\boldsymbol{\mu}'\Sigma^{-1}\boldsymbol{\mu}.$$

Hence

$$\frac{T^2}{n}=\frac{\chi^2_m(\delta)}{\chi^2_{n-m+1}}$$

where the denominator and numerator are independent. Dividing them each by their respective degrees of freedom and using the definition of the noncentral F distribution (see Section 1.3) shows that

$$\frac{T^2}{n}\cdot\frac{n-m+1}{m}\quad\text{is}\quad F_{m,n-m+1}(\delta),$$

as required.

This derivation of the distribution of T^2 is due to Wijsman (1957). Note that when $\boldsymbol{\mu}=\mathbf{0}$, the distribution of $T^2(n-m+1)/nm$ is (central) $F_{m,n-m+1}$ and hence a test of size α of the null hypothesis H_0: $\boldsymbol{\mu}=\mathbf{0}$ against the alternative H: $\boldsymbol{\mu}\neq\mathbf{0}$ is to reject H_0 if

$$\frac{(n-m+1)T^2}{nm}>F^*_{m,n-m+1}(\alpha),$$

where $F^*_{m,n-m+1}(\alpha)$ denotes the upper $100\alpha\%$ point of the $F_{m,n-m+1}$ distribution. The power function of this test is a function of the noncentrality parameter δ, namely,

$$\begin{aligned}\beta(\delta)&=P_\delta\left[\frac{n-m+1}{nm}T^2>F^*_{m,n-m+1}(\alpha)\right]\\&=P\left[F_{m,n-m+1}(\delta)>F^*_{m,n-m+1}(\alpha)\right].\end{aligned}$$

3.2.4. Bartlett's Decomposition and the Generalized Variance

Our next result is concerned with the transformation of a Wishart matrix A to $T'T$, where T is upper-triangular. The following theorem, due to Bartlett (1933), is essentially contained in the proofs of Theorems 2.1.11 and 2.1.12 but is often useful and is worth repeating.

THEOREM 3.2.14. Let A be $W_m(n,I_m)$, where $n\geq m$ is an integer, and put $A=T'T$, where T is an upper-triangular $m\times m$ matrix with positive diagonal elements. Then the elements t_{ij} $(1\leq i\leq j\leq m)$ of T are all independent, t_{ii}^2 is χ^2_{n-i+1} $(i=1,\dots,m)$, and t_{ij} is $N(0,1)$ $(1\leq i<j\leq m)$.

Proof. The density of A is

$$(13)\qquad \frac{1}{2^{mn/2}\Gamma_m(\frac{1}{2}n)}\operatorname{etr}(-\tfrac{1}{2}A)(\det A)^{(n-m-1)/2}(dA).$$

Since $A=T'T$ we have

$$\operatorname{tr}A=\operatorname{tr}T'T=\sum_{i\leq j}^{m}t_{ij}^2,$$

$$\det A=\det(T'T)=(\det T)^2=\prod_{i=1}^{m}t_{ii}^2$$

and, from Theorem 2.1.9,

$$(dA)=2^m\prod_{i=1}^{m} t_{ii}^{m+1-i}\bigwedge_{i\le j}^{m} dt_{ij}.$$

Substituting these expressions in (13) and using

$$\Gamma_m(\tfrac{1}{2}n)=\pi^{m(m-1)/4}\prod_{i=1}^{m}\Gamma\left[\tfrac{1}{2}(n-i+1)\right]$$

we find that the joint density of the t_{ij} $(1\le i\le j\le m)$ can be written in the form

$$\prod_{i<j}^{m}\left[\frac{1}{(2\pi)^{1/2}}\exp\left(-\tfrac{1}{2}t_{ij}^2\right)dt_{ij}\right]$$

$$\cdot\prod_{i=1}^{m}\left[\frac{1}{2^{(n-i+1)/2}\Gamma\left[\frac{1}{2}(n-i+1)\right]}\exp\left(-\tfrac{1}{2}t_{ii}^2\right)\left(t_{ii}^2\right)^{(n-i-1)/2}dt_{ii}^2\right],$$

which is the product of the marginal density functions for the elements of T stated in the theorem.

If a multivariate distribution has a covariance matrix Σ then one overall measure of spread of the distribution is the scalar quantity $\det\Sigma$, called the *generalized variance* by Wilks (1932). In rather imprecise terms, if the elements of Σ are large one might expect that $\det\Sigma$ is also large. This often happens although it is easy to construct counter-examples. For example, if Σ is diagonal, $\det\Sigma$ will be close to zero if any diagonal element (variance) is close to zero, even if some of the other variances are large. The generalized variance is usually estimated by the *sample generalized variance*, $\det S$, where S is the sample covariance matrix. The following theorem gives the distribution of $\det S$ when S is formed from a sample of size $N=n+1$ from the $N_m(\mu,\Sigma)$ distribution. In this case $A=nS$ is $W_m(n,\Sigma)$.

THEOREM 3.2.15. If A is $W_m(n,\Sigma)$, where $n\ge m$ is an integer then $\det A/\det\Sigma$ has the same distribution as $\prod_{i=1}^{m}\chi_{n-i+1}^2$, where the χ_{n-i+1}^2 for $i=1,\dots,m$, denote independent χ^2 random variables.

Proof. Since A is $W_m(n,\Sigma)$ then $B=\Sigma^{-1/2}A\Sigma^{-1/2}$ is $W_m(n,I_m)$ by Theorem 3.2.5. Put $B=T'T$, where T is upper-triangular, then from

Theorem 3.2.14

$$\det B = \prod_{i=1}^{m} t_{ii}^2 = \prod_{i=1}^{m} \chi^2_{n-i+1},$$

where the χ^2_{n-i+1} are independent χ^2 variables. Noting that $\det B = \det A/\det\Sigma$ completes the proof.

Although Theorem 3.2.15 gives a tidy representation for the distribution of $\det A/\det\Sigma$, it is not an easy matter to obtain the density function of a product of independent χ^2 random variables; see Anderson (1958), page 172, for special cases. It is, however, easy to obtain an expression for the moments of the distribution and from this an asymptotic distribution. The rth moment of $\det A$ is, from Theorem 3.2.15,

$$(14) \qquad E[(\det A)^r] = (\det\Sigma)^r \prod_{i=1}^{m} E[(\chi^2_{n-i+1})^r]$$

$$= (\det\Sigma)^r \prod_{i=1}^{m} \left(\frac{2^r \Gamma[\frac{1}{2}(n-i+1)+r]}{\Gamma[\frac{1}{2}(n-i+1)]} \right),$$

where we have used the fact that

$$E[(\chi^2_k)^r] = \frac{2^r \Gamma(\frac{1}{2}k+r)}{\Gamma(\frac{1}{2}k)}.$$

In terms of the multivariate gamma function (14) becomes

$$(15) \qquad E[(\det A)^r] = (\det\Sigma)^r 2^{mr} \frac{\Gamma_m(\frac{1}{2}n+r)}{\Gamma_m(\frac{1}{2}n)}.$$

In particular, the mean and the variance of the sample generalized variance $\det S$ are

$$E(\det S) = n^{-m} E(\det A)$$

$$= (\det\Sigma) \prod_{i=1}^{m} \left[1 - \frac{1}{n}(i-1)\right]$$

and

$$\begin{aligned}\operatorname{Var}(\det S) &= n^{-2m}\operatorname{Var}(\det A)\\ &= n^{-2m}\left\{E\left[(\det A)^2\right] - E(\det A)^2\right\}\\ &= (\det\Sigma)^2\prod_{i=1}^{m}\left[1-\frac{1}{n}(i-1)\right]\\ &\quad\cdot\left\{\prod_{j=1}^{m}\left[1-\frac{1}{n}(j-3)\right]-\prod_{j=1}^{m}\left[1-\frac{1}{n}(j-1)\right]\right\}.\end{aligned}$$

Note that $E(\det S) < \det\Sigma$ for $m>1$ so that $\det S$ underestimates $\det\Sigma$. The following theorem gives the asymptotic distribution of $\log\det S$.

THEOREM 3.2.16. If S is $W_m(n,(1/n)\Sigma)$ then the asymptotic distribution as $n\to\infty$ of

$$v \equiv \sqrt{\frac{n}{2m}}\log\frac{\det S}{\det\Sigma}$$

is standard normal $N(0,1)$

Proof. The characteristic function of v is

$$\begin{aligned}\phi(t) = E\left[e^{itv}\right] &= E\left[\left(\frac{\det S}{\det\Sigma}\right)^{it\sqrt{n/2m}}\right]\\ &= \left(\tfrac{1}{2}n\right)^{-mit\sqrt{n/2m}}\prod_{j=1}^{m}\frac{\Gamma\left[\frac{1}{2}n+it\sqrt{n/2m}+\frac{1}{2}(1-j)\right]}{\Gamma\left[\frac{1}{2}n+\frac{1}{2}(1-j)\right]},\end{aligned}$$

using (14) with $A=nS$ and $r=it\sqrt{n/2m}$. Hence

$$(16)\quad \log\phi(t) = -mit\sqrt{\frac{n}{2m}}\log\frac{n}{2}+\sum_{j=1}^{m}\log\Gamma\left[\tfrac{1}{2}n+it\sqrt{\frac{n}{2m}}+\tfrac{1}{2}(1-j)\right] - \sum_{j=1}^{m}\log\Gamma\left[\tfrac{1}{2}n+\tfrac{1}{2}(1-j)\right]$$

Using the following asymptotic formula for $\log\Gamma(z+a)$,

$$(17)\qquad \log\Gamma(z+a) = \left(z+a-\tfrac{1}{2}\right)\log z - z + \tfrac{1}{2}\log 2\pi + O(z^{-1})$$

(see, for example, Erdélyi et al. (1953a), page 47), it is a simple matter to show that

$$\lim_{n\to\infty} \phi(t) = \exp(-\tfrac{1}{2}t^2). \tag{18}$$

For a more direct proof start with

$$n^m \frac{\det S}{\det \Sigma} = \sum_{i=1}^{m} \chi^2_{n-i+1},$$

where the χ^2_{n-i+1}, for $i=1,\dots,m$, denote independent χ^2 random variables. Taking logs then gives

$$\log \frac{\det S}{\det \Sigma} = \prod_{i=1}^{m} \left[\log \chi^2_{n-i+1} - \log n\right]$$

Using the easily proved fact that the asymptotic distribution as $n\to\infty$ of $(n/2)^{1/2}[\log \chi^2_{n-i+1} - \log n]$ is $N(0,1)$, it follows that the asymptotic distribution of $(n/2)^{1/2}\log(\det S/\det \Sigma)$ is $N(0,m)$, completing the proof.

Since v is asymptotically $N(0,1)$ a standard argument shows that the asymptotic distribution of $(n/2m)^{1/2}(\det S/\det \Sigma - 1)$ is also $N(0,1)$, a result established by Anderson (1958), page 173.

3.2.5. *The Latent Roots of a Wishart Matrix*

The latent roots of a sample covariance matrix play a very important part in principal component analysis, a multivariate technique which will be looked at in Chapter 9. Here a general result is given, useful in a variety of situations, which enables us to transform the density function of a positive definite matrix to the density function of its latent roots.

First we recall some of the notation and results of Section 2.1.4. Let $H=[\mathbf{h}_1 \dots \mathbf{h}_m]$ be an orthogonal $m\times m$ matrix [i.e., $H\in O(m)$], and let $(H'dH)$ denote the exterior product of the subdiagonal elements of the skew-symmetric matrix $H'dH$, that is,

$$(H'dH) = \bigwedge_{i<j}^{m} \mathbf{h}_j' \, d\mathbf{h}_i. \tag{19}$$

This differential form represents the invariant (Haar) measure on the orthogonal group $O(m)$; see (27) of Section 2.1.4. The differential form

$$(20) \qquad (dH) \equiv \frac{1}{\text{Vol}[O(m)]}(H'dH) = \frac{\Gamma_m(\frac{1}{2}m)}{2^m \pi^{m^2/2}}(H'dH)$$

has the property that

$$\int_{O(m)} (dH) = 1,$$

and it represents the "Haar invariant" probability measure on $O(m)$; see (29) of Section 2.1.4. In what follows (and in particular in the next theorem), (dH) will always represent the invariant measure on $O(m)$, normalized so that the volume of $O(m)$ is unity.

THEOREM 3.2.17. If A is an $m \times m$ positive definite random matrix with density function $f(A)$ then the joint density function of the latent roots $l_1, \dots, l_m$ of A is

$$\frac{\pi^{m^2/2}}{\Gamma_m(\frac{1}{2}m)} \prod_{i<j}^{m} (l_i - l_j) \int_{O(m)} f(HLH')(dH) \qquad (l_1 > l_2 > \cdots > l_m > 0),$$

where $L = \text{diag}(l_1, \dots, l_m)$.

Proof. Since the probability that any latent roots of A are equal is 0 we can let $l_1 > l_2 > \cdots > l_m > 0$ be the ordered latent roots. Make a transformation from A to its latent roots and vectors, i.e., put

$$A = HLH',$$

where $H \in O(m)$ and $L = \text{diag}(l_1, \dots, l_m)$. The ith column of H is a normalized latent vector of A corresponding to the latent root l_i. This transformation is not $1-1$ since A determines 2^m matrices $H = [\pm \mathbf{h}_1 \cdots \pm \mathbf{h}_m]$ such that $A = HLH'$. The transformation can be made $1-1$ by requiring, for example, that the first element in each column of H be nonnegative. This restricts the range of H (as A varies) to a 2^{-m}th part of the orthogonal group $O(m)$. When we make the transformation $A = HLH'$ and integrate with respect to (dH) over $O(m)$ the result must be divided by 2^m.

We now find the Jacobian of this transformation. First note that

$$dA = dH\,LH' + H\,dL\,H' + HL\,dH'$$

so that

$$H'\,dA\,H = H'\,dH\,L + dL + L\,dH'\,H \tag{21}$$
$$= H'\,dH\,L - LH'\,dH + dL$$

since $H'\,dH = -\,dH'H$, i.e., $H'\,dH$ is skew-symmetric. By Theorem 2.1.6 the exterior product of the distinct elements in the symmetric matrix on the left side of (21) is

$$(\det H)^{m+1}(dA) = (dA),$$

(ignoring sign). The exterior product of the diagonal elements on the right side of (21) is

$$\bigwedge_{i=1}^{m} dl_i$$

and for $i<j$ the $i-j$th element on the right side of (21) is $\mathbf{h}_j'\,d\mathbf{h}_i(l_i - l_j)$. Hence the exterior product of the distinct elements of the symmetric matrix on the right side of (21) is

$$\bigwedge_{i<j}^{m} \mathbf{h}_j'\,d\mathbf{h}_i \prod_{i<j}^{m} (l_i - l_j) \bigwedge_{i=1}^{m} dl_i.$$

Equating exterior products on both sides then gives

$$(dA) = \bigwedge_{i<j}^{m} \mathbf{h}_j'\,d\mathbf{h}_i \prod_{i<j}^{m} (l_i - l_j) \bigwedge_{i=1}^{m} dl_i \tag{22}$$
$$= (H'\,dH) \prod_{i<j}^{m} (l_i - l_j) \bigwedge_{i=1}^{m} dl_i$$
$$= \frac{2^m \pi^{m^2/2}}{\Gamma_m(\frac{1}{2}m)} (dH) \prod_{i<j}^{m} (l_i - l_j) \bigwedge_{i=1}^{m} dl_i,$$

using (19) and (20). Substituting $A = HLH'$ and (dA) from (22) in $f(A)(dA)$, integrating with respect to (dH) over $O(m)$, and dividing the result by 2^m gives the density function of $l_1, \dots, l_m$ as

$$\frac{\pi^{m^2/2}}{\Gamma_m(\frac{1}{2}m)} \prod_{i<j}^{m} (l_i - l_j) \int_{O(m)} f(HLH')(dH),$$

as required.

As an application of this theorem we consider the distribution of the latent roots of a Wishart matrix.

THEOREM 3.2.18. If A is $W_m(n,\Sigma)$ with $n>m-1$ the joint density function of the latent roots $l_1,\ldots,l_m$ of A is

$$(23)\qquad \frac{\pi^{m^2/2}2^{-mn/2}(\det\Sigma)^{-n/2}}{\Gamma_m(\frac{1}{2}m)\Gamma_m(\frac{1}{2}n)}\prod_{i=1}^{m} l_i^{(n-m-1)/2}\prod_{i<j}^{m}(l_i-l_j)$$

$$\cdot\int_{O(m)}\operatorname{etr}(-\tfrac{1}{2}\Sigma^{-1}HLH')(dH)\qquad (l_1>l_2>\cdots>l_m>0).$$

Proof. The proof follows immediately by applying Theorem 3.2.17 to the $W_m(n,\Sigma)$ density function for A, namely,

$$f(A)=\frac{2^{-mn/2}(\det\Sigma)^{-n/2}}{\Gamma_m(\frac{1}{2}n)}\operatorname{etr}(-\tfrac{1}{2}\Sigma^{-1}A)(\det A)^{(n-m-1)/2},$$

and noting that $\det A=\det HLH'=\prod_{i=1}^{m}l_i$.

The integral in (23) is, in general, not easy to evaluate. In Chapter 9 we will obtain an infinite series representation for this integral in terms of *zonal polynomials*. For the moment, however, two observations are worth making. The first is that the density function (23) depends on the population covariance matrix Σ only through its latent roots. To see this, write $\Sigma=Q\Lambda Q'$, where $Q\in O(m)$ and $\Lambda=\operatorname{diag}(\lambda_1,\ldots,\lambda_m)$, with $\lambda_1,\ldots,\lambda_m$ being the latent roots of Σ. Then $\det\Sigma=\prod_{i=1}^{m}\lambda_i$ and the integral in (23) is

$$I=\int_{O(m)}\operatorname{etr}(-\tfrac{1}{2}Q\Lambda^{-1}Q'HLH')(dH)$$

$$=\int_{O(m)}\operatorname{etr}(-\tfrac{1}{2}\Lambda^{-1}Q'HLH'Q)(dH).$$

Now put $\tilde{H}=Q'H$ then $\tilde{H}\in O(m)$ and $(d\tilde{H})=(dH)$ so that

$$I=\int_{O(m)}\operatorname{etr}(-\tfrac{1}{2}\Lambda^{-1}\tilde{H}L\tilde{H}')(d\tilde{H}),$$

which depends only on $\lambda_1,\ldots,\lambda_m$. The second observation is that when

$\Sigma=\lambda I_m$ the joint density function of $l_1,\dots,l_m$ is particularly simple and is given in the following corollary.

COROLLARY 3.2.19. If A is $W_m(n,\lambda I_m)$, with $n>m-1$, the joint density function of the latent roots $l_1,\dots,l_m$ of A is

$$\frac{\pi^{m^2/2}}{(2\lambda)^{mn/2}\Gamma_m(\frac{1}{2}m)\Gamma_m(\frac{1}{2}n)}\exp\left(-\frac{1}{2\lambda}\sum_{i=1}^m l_i\right)\prod_{i=1}^m l_i^{(n-m-1)/2}\prod_{i<j}^m (l_i-l_j)$$

$$(l_1>l_2>\cdots>l_m>0).$$

Proof. Putting $\Sigma=\lambda I_m$ in Theorem 3.2.18 and noting that

$$\int_{O(m)}\operatorname{etr}\left(-\frac{1}{2\lambda}HLH'\right)(dH)=\operatorname{etr}\left(-\frac{1}{2\lambda}L\right)\int_{O(m)}(dH)$$

$$=\exp\left(-\frac{1}{2\lambda}\sum_{i=1}^m l_i\right)$$

completes the proof.

It is interesting to note that when $\Sigma=\lambda I_m$ and $A=HLH'$ as in the proof of Theorem 3.2.17, where $H=[\mathbf{h}_1\dots\mathbf{h}_m]\in O(m)$ with the first element in each column being nonnegative, then H is *independent* of the latent roots $l_1,\dots,l_m$ because the joint density of H and L factors. The columns of H are the latent vectors of A. The distribution of H has been called the conditional Haar invariant distribution by Anderson (1958), page 322; it is the conditional distribution of an orthogonal $m\times m$ matrix whose distribution is the invariant distribution on $O(m)$, given that the first element in each column is nonnegative.

Our next result can be proved in a number of ways; we will establish it using Corollary 3.2.19.

THEOREM 3.2.20. If A is $W_m(n,\lambda I_m)$ where $n(\geq m)$ is an integer, then $u=(\det A)/[(1/m)\operatorname{tr}A]^m$ and $\operatorname{tr}A$ are independent, and $(1/\lambda)\operatorname{tr}A$ is χ^2_{mn}.

Proof. First note that $(1/\lambda)A$ is $W_m(n,I_m)$ so that by Corollary 3.2.6 the diagonal elements a_{ii}/λ $(i=1,\dots,m)$ are independent χ^2_n random variables. Hence

$$\frac{1}{\lambda}\operatorname{tr}A=\frac{1}{\lambda}\sum_{i=1}^m a_{ii}$$

is χ^2_{mn}. To show that $\operatorname{tr}A$ and u are independent we will show that their joint density factors. The joint density function of the latent roots $l_1,\ldots,l_m$ of A is, from Corollary 3.2.19,

$$\frac{\pi^{m^2/2}}{(2\lambda)^{mn/2}\Gamma_m(\frac{1}{2}m)\Gamma_m(\frac{1}{2}n)}\exp\left(-\frac{1}{2\lambda}\sum_{i=1}^m l_i\right)\prod_{i=1}^m l_i^{(n-m-1)/2}\prod_{i<j}^m (l_i-l_j).$$

Make the change of variables from $l_1,\ldots,l_m$ to $\bar{l}, y_1,\ldots,y_{m-1}$ given by

$$\bar{l}=\frac{1}{m}\sum_{i=1}^m l_i=\frac{1}{m}\operatorname{tr}A$$

$$y_i=\frac{l_i}{\bar{l}}\qquad i=1,\ldots,m$$

(Note that $y_1+\cdots+y_m=m$.) Then

$$u=\frac{\det A}{\left(\frac{1}{m}\operatorname{tr}A\right)^m}=\prod_{i=1}^m\frac{l_i}{\bar{l}}=\prod_{i=1}^m y_i$$

and the reader can readily check that the joint density function of $\bar{l}, y_1,\ldots,y_{m-1}$ is

$$\frac{\pi^{m^2/2}}{(2\lambda)^{mn/2}\Gamma_m(\frac{1}{2}m)\Gamma_m(\frac{1}{2}n)}\bar{l}^{mn/2-1}\exp\left(-\frac{m}{2\lambda}\bar{l}\right)\prod_{i=1}^m y_i^{(n-m-1)/2}\prod_{i<j}^m (y_i-y_j)$$

This shows that $\bar{l}$ is independent of $y_1,\ldots,y_{m-1}$ and hence is independent of u, completing the proof.

The statistic u defined in Theorem 3.2.20 is used to test the null hypothesis that $\Sigma=\lambda I_m$ and will be studied further in Chapter 8. For arbitrary Σ the distribution of $\operatorname{tr}A$ is rather complicated and will be derived in Chapter 8. The distribution in the case $m=2$ is reasonably tractable and is left as an exercise (see Problem 3.12).

3.3. THE MULTIVARIATE BETA DISTRIBUTION

Closely related to the Wishart distribution is the multivariate Beta distribution. This will be introduced via the following theorem, due to Hsu (1939), Khatri (1959), and Olkin and Rubin (1964).

THEOREM 3.3.1. Let A and B be independent, where A is $W_m(n_1, \Sigma)$ and B is $W_m(n_2, \Sigma)$, with $n_1 > m-1$, $n_2 > m-1$. Put $A+B = T'T$ where T is an upper-triangular $m \times m$ matrix with positive diagonal elements. Let U be the $m \times m$ symmetric matrix defined by $A = T'UT$. Then $A+B$ and U are independent; $A+B$ is $W_m(n_1+n_2, \Sigma)$ and the density function of U is

(1)

$$\frac{\Gamma_m[\frac{1}{2}(n_1+n_2)]}{\Gamma_m(\frac{1}{2}n_1)\Gamma_m(\frac{1}{2}n_2)}(\det U)^{(n_1-m-1)/2}\det(I_m-U)^{(n_2-m-1)/2} \qquad (0<U<I_m),$$

where $0<U<I_m$ means that $U>0$ (i.e., U is positive definite) and $I_m - U > 0$.

Proof. The joint density of A and B is

$$\frac{2^{-m(n_1+n_2)/2}(\det\Sigma)^{-(n_1+n_2)/2}}{\Gamma_m(\frac{1}{2}n_1)\Gamma_m(\frac{1}{2}n_2)}\operatorname{etr}\left[-\tfrac{1}{2}\Sigma^{-1}(A+B)\right](\det A)^{(n_1-m-1)/2}$$

$$\cdot(\det B)^{(n_2-m-1)/2}(dA)(dB).$$

First transform to the joint density of $C = A+B$ and A. Noting that $(dA)\wedge(dB) = (dA)\wedge(dC)$ (i.e., the Jacobian is 1), the joint density of C and A is

$$(2) \qquad \frac{2^{-m(n_1+n_2)/2}(\det\Sigma)^{-(n_1+n_2)/2}}{\Gamma_m(\frac{1}{2}n_1)\Gamma_m(\frac{1}{2}n_2)}\operatorname{etr}(-\tfrac{1}{2}\Sigma^{-1}C)(\det A)^{(n_1-m-1)/2}$$

$$\cdot\det(C-A)^{(n_2-m-1)/2}(dA)(dC)$$

Now put $C = T'T$, where T is upper-triangular, and $A = T'UT$. Remembering that T is a function of C alone we have

$$\begin{aligned}(dA)\wedge(dC) &= (T'\,dUT)\wedge(d(T'T))\\ &= (\det T)^{m+1}(dU)\wedge(d(T'T))\\ &= \det(T'T)^{(m+1)/2}(dU)\wedge(d(T'T)),\end{aligned}$$

where Theorem 2.1.6 has been used. Now substitute for C, A, and $(dA)(dC)$ in (2) using $\det A=\det(T'T)\det U$ and $\det(C-A)=\det(T'T)\det(I-U)$. Then the joint density function of $T'T$ and U is

$$\frac{2^{-m(n_1+n_2)/2}(\det\Sigma)^{-(n_1+n_2)/2}}{\Gamma_m[\frac{1}{2}(n_1+n_2)]}\operatorname{etr}(-\tfrac{1}{2}\Sigma^{-1}T'T)\det(T'T)^{(n_1+n_2-m-1)/2}$$

$$\cdot\frac{\Gamma_m[\frac{1}{2}(n_1+n_2)]}{\Gamma_m(\frac{1}{2}n_1)\Gamma_m(\frac{1}{2}n_2)}(\det U)^{(n_1-m-1)/2}\det(I-U)^{(n_2-m-1)/2},$$

which shows that $T'T=C=A+B$ is $W_m(n_1+n_2,\Sigma)$ and is independent of U, where U has the density function (1).

DEFINITION 3.3.2. A matrix U with density function (1) is said to have the multivariate beta distribution with parameters $\frac{1}{2}n_1$ and $\frac{1}{2}n_2$, and we will write that U is $\text{Beta}_m(\frac{1}{2}n_1,\frac{1}{2}n_2)$. It is obvious that if U is $\text{Beta}_m(\frac{1}{2}n_1,\frac{1}{2}n_2)$ then I_m-U is $\text{Beta}_m(\frac{1}{2}n_2,\frac{1}{2}n_1)$.

The multivariate beta distribution generalizes the usual beta distribution in much the same way that the Wishart distribution generalizes the χ^2 distribution. Some of its properties are similar to those of the Wishart distribution. As an example it was shown in Theorem 3.2.14 that if A is $W_m(n,I_m)$ and is written as $A=T'T$, where T is upper-triangular, then $t_{11},t_{22},\dots,t_{mm}$ are all independent and t_{ii}^2 is χ^2_{n-i+1}. A similar type of result holds for the multivariate beta distribution as the following theorem, due to Kshirsagar (1961, 1972), shows.

THEOREM 3.3.3. If U is $\text{Beta}_m(\frac{1}{2}n_1,\frac{1}{2}n_2)$ and $U=T'T$, where T is upper-triangular then $t_{11},\dots,t_{mm}$ are all independent and t_{ii}^2 is $\text{beta}[\frac{1}{2}(n_1-i+1),\frac{1}{2}n_2]$; $i=1,\dots,m$.

Proof. In the density function (1) for U, make the change of variables $U=T'T$; then

$$\det U=\det T'T=\prod_{i=1}^{m}t_{ii}^2$$

and, from Theorem 2.1.9,

$$(dU)=2^m\prod_{i=1}^{m}t_{ii}^{m+1-i}\bigwedge_{i\le j}^{m}dt_{ij}$$

so that the density of T is $f(T;m,n_1,n_2)$, where

(3)

$$f(T;m,n_1,n_2)=\frac{\Gamma_m\left[\frac{1}{2}(n_1+n_2)\right]}{\Gamma_m\left(\frac{1}{2}n_1\right)\Gamma_m\left(\frac{1}{2}n_2\right)}2^m\prod_{i=1}^{m}t_{ii}^{n_1-i}\det(I-T'T)^{(n_2-m-1)/2}$$

Now partition T as

$$T=\begin{bmatrix} t_{11} & \mathbf{t}' \\ 0 & T_{22}\end{bmatrix},$$

where $\mathbf{t}$ is $(m-1)\times 1$ and T_{22} is $(m-1)\times(m-1)$ and upper-triangular; note that

(4) $$\det(I-T'T)=\det\begin{bmatrix} 1-t_{11}^2 & -t_{11}\mathbf{t}' \\ -t_{11}\mathbf{t} & I-\mathbf{t}\mathbf{t}'-T_{22}'T_{22}\end{bmatrix}$$

$$=(1-t_{11}^2)\det(I-T_{22}'T_{22})\left[1-\frac{1}{1-t_{11}^2}\mathbf{t}'(I-T_{22}'T_{22})^{-1}\mathbf{t}\right]$$

(see Problem 3.20). Now make a change of variables from $t_{11},T_{22},\mathbf{t}$ to $t_{11},T_{22},\mathbf{v}$, where

$$\mathbf{v}=\frac{1}{(1-t_{11}^2)^{1/2}}(I-T_{22}'T_{22})^{-1/2}\mathbf{t},$$

then

$$\bigwedge_{i\le j}^{m}dt_{ij}=dt_{11}\wedge(dT_{22})\wedge(d\mathbf{t})$$

$$=(1-t_{11}^2)^{(m-1)/2}\det(I-T_{22}'T_{22})^{1/2}dt_{11}\wedge(dT_{22})\wedge(d\mathbf{v})$$

by Theorem 2.1.1, and hence the joint density of t_{11}, T_{22} and $\mathbf{v}$ is

$$\frac{2^m\Gamma_m\left[\frac{1}{2}(n_1+n_2)\right]}{\Gamma_m\left(\frac{1}{2}n_1\right)\Gamma_m\left(\frac{1}{2}n_2\right)}t_{11}^{n_1-1}(1-t_{11}^2)^{n_2/2-1}$$

$$\cdot\prod_{i=2}^{m}t_{ii}^{n_1-i}\det(I-T_{22}'T_{22})^{(n_2-m)/2}(1-\mathbf{v}'\mathbf{v})^{(n_2-m-1)/2}.$$

This shows immediately that t_{11}, T_{22}, and $\mathbf{v}$ are all independent and that t_{11}^2 has the beta$(\frac{1}{2}n_1, \frac{1}{2}n_2)$ distribution. The density function of T_{22} is proportional to

$$\prod_{i=2}^{m} t_{ii}^{n_1-i}\det(I-T_{22}'T_{22})^{(n_2-m)/2}$$

which has the same form as the density function (3) for T, with m replaced by $m-1$ and n_1 replaced by n_1-1. Hence the density function of T_{22} is $f(T_{22}; m-1, n_1-1, n_2)$. Repeating the argument above on this density function then shows that t_{22}^2 is beta$(\frac{1}{2}(n_1-1), \frac{1}{2}n_2)$, and is independent of $t_{33},\dots,t_{mm}$. The proof is completed in an obvious way by repetition of this argument.

The distribution of the latent roots of a multivariate beta matrix will occur extensively in later chapters; for future reference it is given here.

THEOREM 3.3.4. If U is Beta$_m(\frac{1}{2}n_1, \frac{1}{2}n_2)$ the joint density function of the latent roots $u_1,\dots,u_m$ of U is

$$\frac{\pi^{m^2/2}}{\Gamma_m(\frac{1}{2}m)} \frac{\Gamma_m[\frac{1}{2}(n_1+n_2)]}{\Gamma_m(\frac{1}{2}n_1)\Gamma_m(\frac{1}{2}n_2)} \prod_{i=1}^{m} \left[u_i^{(n_1-m-1)/2}(1-u_i)^{(n_2-m-1)/2}\right]$$

$$\cdot \prod_{i<j}^{m} (u_i-u_j) \qquad (1>u_1>\cdots>u_m>0).$$

The proof follows immediately by applying the latent roots theorem (Theorem 3.2.17) to the Beta$_m(\frac{1}{2}n_1, \frac{1}{2}n_2)$ density function (1). Note that the latent roots of U are, from Theorem 3.3.1, the latent roots of $A(A+B)^{-1}$, where A is $W_m(n_1, \Sigma)$, B is $W_m(n_2, \Sigma)$ (here $n_1>m-1$; $n_2>m-1$) and A and B are independent. The distribution of these roots was obtained independently by Fisher, Girshick, Hsu, Roy, and Mood, all in 1939, although Mood's derivation was not published until 1951.

PROBLEMS

3.1. If $A=(a_{ij})$ is $W_m(n,\Sigma)$, where $\Sigma=(\sigma_{ij})$, show that

$$\mathrm{Cov}(a_{ij}, a_{kl}) = n(\sigma_{ik}\sigma_{jl}+\sigma_{il}\sigma_{jk}).$$

3.2. Let $K=\Sigma_{i,j=1}^{m}(H_{ij}\otimes H_{ij}')$, where H_{ij} denotes the $m\times m$ matrix with $h_{ij}=1$ and all other elements zero. Show that if A is $W_m(n,\Sigma)$ then

$$\operatorname{Cov}(\operatorname{vec}(A))=n(I_{m^2}+K)(\Sigma\otimes\Sigma).$$

3.3. If $S(n)$ denotes the sample covariance matrix formed from a sample of size $n+1$ from an elliptical distribution with covariance matrix Σ and kurtosis parameter κ then the asymptotic distribution, as $n\to\infty$, of $U(n)=n^{1/2}[S(n)-\Sigma]$ is normal with mean zero (see Corollary 1.2.18). The elements of the covariance matrix in this asymptotic normal distribution are, from (2) and (3) of Section 1.6,

$$\operatorname{Cov}\left[u_{ij}(n),u_{kl}(n)\right]=\kappa(\sigma_{ij}\sigma_{kl}+\sigma_{ik}\sigma_{jl}+\sigma_{il}\sigma_{jk})+\sigma_{ik}\sigma_{jl}+\sigma_{il}\sigma_{jk}.$$

Show that $\operatorname{vec}(U(n))$ has asymptotic covariance matrix

$$\operatorname{Cov}\left[\operatorname{vec}(U(n))\right]=(1+\kappa)(I_{m^2}+K)(\Sigma\otimes\Sigma)+\kappa\operatorname{vec}(\Sigma)\left[\operatorname{vec}(\Sigma)\right]',$$

where K is the commutation matrix defined in Problem 3.2.

3.4. Prove Theorem 3.2.5 when n is a positive integer by expressing A in terms of normal variables.

3.5. Prove Theorem 3.2.7.

3.6. A random $m\times m$ positive definite matrix B is said to have the inverted Wishart distribution with n degrees of freedom and positive definite $m\times m$ parameter matrix V if its density function is

$$\frac{2^{-m(n-m-1)/2}}{\Gamma_m\left[\frac{1}{2}(n-m-1)\right]}\frac{(\det V)^{(n-m-1)/2}}{(\det B)^{n/2}}\operatorname{etr}\left(-\tfrac{1}{2}B^{-1}V\right)\qquad (B>0),$$

where $n>2m$. We will write that B is $W_m^{-1}(n,V)$.

(a) Show that if A is $W_m(n,\Sigma)$ then A^{-1} is $W_m^{-1}(n+m+1,\Sigma^{-1})$.

(b) If B is $W_m^{-1}(n,V)$ show that

$$E(B)=\frac{1}{n-2m-2}V^{-1}.$$

(c) Suppose that A is $W_m(n,\Sigma)$ and that Σ has a $W_m^{-1}(\nu,V)$ prior distribution, $\nu>2m$. Show that given A the posterior distribution of Σ is $W_m^{-1}(n+\nu,A+V)$.

(d) Suppose that B is $W_m^{-1}(n,V)$ and partition B and V as

$$B=\begin{bmatrix} B_{11} & B_{12} \\ B_{21} & B_{22} \end{bmatrix}, \qquad V=\begin{bmatrix} V_{11} & V_{12} \\ V_{21} & V_{22} \end{bmatrix},$$

where B_{11} and V_{11} are $k\times k$ and B_{22} and V_{22} are $(m-k)\times(m-k)$. Show that B_{11} is $W_k^{-1}(n-2m+2k,V_{11})$.

3.7. If A is a positive definite random matrix such that $E(A)$, $E(A^{-1})$ exist, prove that the matrix $E(A^{-1})-E(A)^{-1}$ is non-negative definite. [*Hint:* Put $E(A)=\Sigma$ and $\Delta=A-\Sigma$ and show that $\Sigma^{-1}E(\Delta A^{-1}\Delta)\Sigma^{-1}=E(A^{-1})-\Sigma^{-1}$.]

3.8. If A is $W_m(n,\Sigma)$, where $n>m-1$ and $\Sigma>0$, show that the maximum likelihood estimate of Σ is $(1/n)A$.

3.9. Suppose that A is $W_m(n,\Sigma)$, $n>m-1$, where Σ has the form

$$\Sigma=\sigma^2[(1-\rho)I_m+\rho\mathbf{1}\mathbf{1}']=\sigma^2\begin{bmatrix} 1 & \rho & \cdots & \rho \\ \rho & 1 & \cdots & \rho \\ \vdots & & & \\ \rho & \rho & \cdots & 1 \end{bmatrix},$$

where $\mathbf{1}$ is an $m\times 1$ vector of ones.

(a) Show that

$$\Sigma^{-1}=\frac{1}{\sigma^2(1-\rho)}I_m-\frac{\rho}{\sigma^2(1-\rho)[1+(m-1)\rho]}\mathbf{1}\mathbf{1}'$$

and that

$$\det\Sigma=(\sigma^2)^m(1-\rho)^{m-1}[1+(m-1)\rho].$$

(b) Show that the maximum likelihood estimates of σ^2 and ρ are

$$\hat\sigma^2=\frac{\operatorname{tr}A}{mn} \quad\text{and}\quad \hat\rho=\frac{\mathbf{1}'A\mathbf{1}-\operatorname{tr}A}{(m-1)\operatorname{tr}A}=\frac{2\Sigma_{i<j}^{m}a_{ij}}{(m-1)\operatorname{tr}A}.$$

3.10. Let X be an $n\times m$ random matrix and P be an $n\times n$ symmetric idempotent matrix of rank $k\geq m$

(a) If X is $N(0,P\otimes\Sigma)$ prove that $X'X$ is $W_m(k,\Sigma)$.
(b) If X is $N(0,I_n\otimes\Sigma)$ prove that $X'PX$ is $W_m(k,\Sigma)$.

3.11. If A is $W_m(n, \Sigma)$, $n > m-1$, show, using the Wishart density function, that

$$E\left[(\det A)^r\right] = (\det \Sigma)^r 2^{mr} \frac{\Gamma_m(\frac{1}{2}n + r)}{\Gamma_m(\frac{1}{2}n)}.$$

3.12. If A is $W_m(n, \Sigma)$ show that the characteristic function of $\operatorname{tr} A$ is

$$\phi(t) = E\left[\operatorname{etr}(itA)\right] = \det(I - 2it\Sigma)^{-n/2}.$$

Using this, show that when $m = 2$ the distribution function of $\operatorname{tr} A$ can be expressed in the form

$$P(\operatorname{tr} A \le x) = \sum_{k=0}^{\infty} c_k P\left(\chi^2_{2n+4k} \le \frac{\lambda_1 + \lambda_2}{2\lambda_1\lambda_2} x\right),$$

where λ_1 and λ_2 are the latent roots of Σ and c_k is the negative binomial probability

$$c_k = (-1)^k \binom{-\frac{1}{2}n}{k} p^{n/2}(1-p)^k$$

with $p = 4\lambda_1\lambda_2/(\lambda_1 + \lambda_2)^2$.
[*Hint:* Find the density function corresponding to this distribution function and then show that its characteristic function agrees with $\phi(t)$ when $m = 2$.]

3.13. Let A be $W_2(n, \Sigma)$ and let l_1, l_2 ($l_1 > l_2 > 0$) denote the latent roots of the sample covariance matrix $S = n^{-1}A$.

(a) Show that the joint density function of l_1 and l_2 can be expressed as

$$\left(\frac{n}{2}\right)^n \frac{\pi}{\Gamma_2(\frac{1}{2}n)} (\alpha_1\alpha_2)^{n/2} (l_1 l_2)^{(n-3)/2} (l_1 - l_2)$$

$$\cdot \int_{O(2)} \operatorname{etr}(-\tfrac{1}{2}n\Sigma^{-1}HLH')(dH),$$

where α_1 and α_2 are the latent roots of Σ^{-1}.

(b) Without loss of generality (see the discussion following Theorem 3.2.18) Σ^{-1} can be assumed diagonal, $\Sigma^{-1} = \operatorname{diag}(\alpha_1, \alpha_2)$, $0 < \alpha_1 \le \alpha_2$. Let $I(n; \Sigma^{-1}, L)$ denote the integral in (a). Show that

$$I(n;\Sigma^{-1},L)=\exp\left[-\tfrac{1}{4}n(l_1+l_2)(\alpha_1+\alpha_2)\right]$$
$$\cdot {}_0F_1\left(1;\frac{n^2}{64}(\alpha_1-\alpha_2)^2(l_1-l_2)^2\right)$$

where the function ${}_0F_1$ is defined in Definition 1.3.1. [*Hint:* Argue that

$$I(n;\Sigma^{-1},L)=2\int_{O^+(2)}\operatorname{etr}\left(-\tfrac{1}{2}n\Sigma^{-1}HLH'\right)(dH),$$

where $O^+(2)=\{H\in O(2);\det H=1\}$. Put

$$H=\begin{bmatrix}\cos\theta & -\sin\theta\\ \sin\theta & \cos\theta\end{bmatrix}\qquad(0<\theta<2\pi)$$

and then use Lemma 1.3.2.]

(c) Show that $I(n;\Sigma^{-1},L)$ can also be expressed in the form

$$I(n;\Sigma^{-1},L)=\frac{1}{\pi}\exp\left[-\tfrac{1}{2}n(\alpha_1 l_1+\alpha_2 l_2)\right]$$
$$\cdot\int_{-\pi/2}^{\pi/2}\exp\left[-\frac{nc}{4}(1-\cos 2\theta)\right]d\theta,$$

where $c=(l_1-l_2)(\alpha_2-\alpha_1)$.

(d) Laplace's method says that if a function $f(x)$ has a unique maximum at an interior point ξ of $[a,b]$ then, under suitable regularity conditions, as $n\rightarrow\infty$,

$$\int_a^b[f(x)]^n\,dx\sim\left(\frac{2\pi}{n}\right)^{1/2}\frac{f(\xi)}{[h''(\xi)]^{1/2}}$$

where $h(x)=-\log f(x)$ and $a\sim b$ means that $a/b\rightarrow 1$ as $n\rightarrow\infty$. (The regularity conditions in a multivariate generalization are given in Theorem 9.5.1). Assuming that $\alpha_1<\alpha_2$ use (c) to show that as $n\rightarrow\infty$

$$I(n;\Sigma^{-1},L)\sim\frac{1}{\pi}\exp\left[-\tfrac{1}{2}n(\alpha_1 l_1+\alpha_2 l_2)\right]\left(\frac{2\pi}{nc}\right)^{1/2}.$$

3.14. Suppose that A is $W_m(n,\Sigma)$ and partition A and Σ as

$$A=\begin{bmatrix} A_{11} & A_{12} \\ A_{21} & A_{22} \end{bmatrix}, \qquad \Sigma=\begin{bmatrix} \Sigma_{11} & 0 \\ 0 & \Sigma_{22} \end{bmatrix},$$

where A_{11} and Σ_{11} are $k\times k$ and A_{22} and Σ_{22} are $(m-k)\times(m-k)$, $m\geq 2k$. Note that $\Sigma_{12}=0$. Show that the matrices $A_{11\cdot 2}=A_{11}-A_{12}A_{22}^{-1}A_{21}$, A_{22}, and $A_{12}A_{22}^{-1}A_{21}$ are independently distributed and that $A_{12}A_{22}^{-1}A_{21}$ is $W_k(m-k,\Sigma_{11})$.

3.15. Suppose that $\mathbf{X}_1,\dots,\mathbf{X}_N$ are independent $N_m(\mathbf{0},\Sigma)$ random vectors, $N>m$.

(a) Write down the joint density function of $\mathbf{X}_1$ and $B=\Sigma_{i=2}^N \mathbf{X}_i\mathbf{X}_i'$.

(b) Put $A=B+\mathbf{X}_1\mathbf{X}_1'=\Sigma_{i=1}^N \mathbf{X}_i\mathbf{X}_i'$ and $\mathbf{Y}=A^{-1/2}\mathbf{X}_1$ and note that

$$\det B=(\det A)(1-\mathbf{Y}'\mathbf{Y}).$$

Find the Jacobian of the transformation from B and $\mathbf{X}_1$ to A and $\mathbf{Y}$, and show that the joint density function of A and $\mathbf{Y}$ is

$$\frac{2^{-mN/2}(\det\Sigma)^{-N/2}}{\pi^{m/2}\Gamma_m\left[\frac{1}{2}(N-1)\right]}\operatorname{etr}\left(-\tfrac{1}{2}\Sigma^{-1}A\right)(\det A)^{(N-m-1)/2} \cdot (1-\mathbf{Y}'\mathbf{Y})^{(N-m-2)/2}.$$

(c) Show that the marginal density function of $\mathbf{Y}$ is

$$\frac{\Gamma\left(\frac{1}{2}N\right)}{\pi^{m/2}\Gamma\left[\frac{1}{2}(N-m)\right]}(1-\mathbf{Y}'\mathbf{Y})^{(N-m-2)/2} \qquad (\mathbf{Y}'\mathbf{Y}<1).$$

(d) Using the fact that $\mathbf{Y}$ has a spherical distribution, show that the random variable $z=(\boldsymbol{\alpha}'\mathbf{Y})^2$ has a beta distribution with parameters $\frac{1}{2},\frac{1}{2}(N-1)$, where $\boldsymbol{\alpha}\neq\mathbf{0}$ is any fixed vector.

3.16. Suppose that A is $W_m(n,I_m)$, and partition A as

$$A=\begin{bmatrix} A_{11} & A_{12} \\ A_{21} & A_{22} \end{bmatrix},$$

where A_{11} is $k\times k$ and A_{22} is $(m-k)\times(m-k)$, with $m\geq 2k$.

(a) Show that the matrices A_{11}, A_{22}, and $B_{12} \equiv A_{11}^{-1/2} A_{12} A_{22}^{-1/2}$ are independently distributed and that B_{12} has density function

$$\frac{\Gamma_k(\frac{1}{2}n)\Gamma_{m-k}(\frac{1}{2}n)}{\Gamma_m(\frac{1}{2}n)} \det(I - B_{12}B_{12}')^{(n-m-1)/2}.$$

(b) Show that the matrix $U = B_{12}B_{12}' = A_{11}^{-1/2} A_{12} A_{22}^{-1} A_{21} A_{11}^{-1/2}$ is independent of A_{11} and A_{22} and has the $\text{Beta}_k[\frac{1}{2}(m-k), \frac{1}{2}(n-m+k)]$ distribution.

3.17. Suppose that A is $W_m(\nu, \Sigma)$, X is $N(0, I_N \otimes \Sigma)$ $(\nu \geq m)$ and that A and X are independent.

(a) Put $B = A + X'X$. Find the joint density function of B and X.

(b) Put $B = T'T$ and $Y = XT^{-1}$, where T is upper-triangular. Show that B and Y are independent and find their distributions.

3.18. Suppose that A is $W_m(n, \sigma^2 P)$, $\nu S^2/\sigma^2$ is χ^2_ν, and A and S^2 are independent. Here P is an $m \times m$ matrix with diagonal elements equal to 1. Show that the matrix $B = S^{-2}A$ has density function

$$\frac{\Gamma[\frac{1}{2}(mn+\nu)](\nu)^{-mn/2}}{\Gamma_m(\frac{1}{2}n)\Gamma(\frac{1}{2}\nu)} (\det P)^{-n/2} (\det B)^{(n-m-1)/2}$$

$$\cdot \left[1 + \frac{1}{\nu}\operatorname{tr}(P^{-1}B)\right]^{(nm+\nu)/2}$$

3.19. Suppose that A is $W_m(n, \Sigma)$, v is beta$[\frac{1}{2}\nu, \frac{1}{2}(n-\nu)]$, where $n > \nu$, and that A and v are independent. Put $B = vA$. If α is any $m \times 1$ fixed vector show that:

(a) $\alpha' B \alpha / \alpha' \Sigma \alpha$ is χ^2_ν provided $\alpha' \Sigma \alpha \neq 0$.

(b) $\alpha' B \alpha = 0$ with probability 1, if $\alpha' \Sigma \alpha = 0$.

(c) $E(B) = \nu \Sigma$.

Show that B does not have a Wishart distribution (cf. Theorem 3.2.8).

3.20. If T is an $m \times m$ upper-triangular matrix partitioned as

$$T = \begin{bmatrix} t_{11} & \mathbf{t}' \\ 0 & T_{22} \end{bmatrix},$$

where T_{22} is an $(m-1)\times(m-1)$ upper-triangular matrix, prove that

$$\det(I-T'T)=(1-t_{11}^2)\det(I-T_{22}'T_{22})\cdot\left[1-\frac{1}{1-t_{11}^2}\mathbf{t}'(I-T_{22}'T_{22})^{-1}\mathbf{t}\right].$$

3.21. Suppose that U has the $\text{Beta}_m(\frac{1}{2}n_1,\frac{1}{2}n_2)$ distribution and put $U=T'T$ where T is upper-triangular. Partition T as

$$T=\begin{bmatrix} t_{11} & \mathbf{t}' \\ \mathbf{0} & T_{22}\end{bmatrix},$$

where T_{22} is $(m-1)\times(m-1)$ upper-triangular, and put $\mathbf{v}_1=(1-t_{11}^2)^{-1/2}(I-T_{22}'T_{22})^{-1/2}\mathbf{t}$. In the proof of Theorem 3.3.3 it is shown that t_{11}, T_{22}, and $\mathbf{v}_1$ are independent, where t_{11}^2 has the beta$(\frac{1}{2}n_1,\frac{1}{2}n_2)$ distribution, T_{22} has the same density function as T with m replaced by $m-1$ and n_1 replaced by n_1-1, and $\mathbf{v}_1$ has the density function

$$\frac{\Gamma(\frac{1}{2}n_2)}{\pi^{(m-1)/2}\Gamma[\frac{1}{2}(n_2-m+1)]}(1-\mathbf{v}_1'\mathbf{v}_1)^{(n_2-m-1)/2}\qquad(\mathbf{v}_1'\mathbf{v}_1<1).$$

Now put $\mathbf{v}_1'=(v_{12},v_{13},\dots,v_{1m})$, and let

$$y_{12}=v_{12},\ y_{13}=\frac{v_{13}}{(1-v_{12}^2)^{1/2}},\dots,y_{1m}=\frac{v_{1m}}{(1-v_{12}^2)^{1/2}}.$$

Show that y_{12} is independent of $y_{13},\dots,y_{1m}$ and that y_{12}^2 has a beta distribution.

By repeating this argument for T_{22} and $(y_{13},\dots,y_{1m})$ and so on, show that the $\text{Beta}_m(\frac{1}{2}n_1,\frac{1}{2}n_2)$ density function for U can be decomposed into a product of density functions of independent univariate beta random variables.

3.22. Suppose that U has the $\text{Beta}_m(\frac{1}{2}n_1,\frac{1}{2}n_2)$ distribution, $n_1>m-1$, $n_2>m-1$.

(a) If $\boldsymbol{\alpha}\neq\mathbf{0}$ is a fixed $m\times 1$ vector show that $\boldsymbol{\alpha}'U\boldsymbol{\alpha}/\boldsymbol{\alpha}'\boldsymbol{\alpha}$ is beta$(\frac{1}{2}n_1,\frac{1}{2}n_2)$.

(b) If V has the $\text{Beta}_m[\frac{1}{2}(n_1+n_2),n_3]$ distribution and is independent of U show that $V^{1/2}UV^{1/2}$ is $\text{Beta}_m[\frac{1}{2}n_1,\frac{1}{2}(n_2+n_3)]$.

(c) Partition U as

$$U=\begin{bmatrix} U_{11} & U_{12} \\ U_{21} & U_{22} \end{bmatrix},$$

where U_{11} is $k\times k$ and U_{22} is $(m-k)\times(m-k)$, $n_1>k-1$, and put $U_{22\cdot 1}=U_{22}-U_{21}U_{11}^{-1}U_{12}$. Show that U_{11} is $\text{Beta}_k(\frac{1}{2}n_1,\frac{1}{2}n_2)$, $U_{22\cdot 1}$ is $\text{Beta}_{m-k}[\frac{1}{2}(n_1-k),\frac{1}{2}n_2]$, and U_{11} and $U_{22\cdot 1}$ are independent.

(d) If H is any $m\times m$ orthogonal matrix show that HUH' has the $\text{Beta}_m(\frac{1}{2}n_1,\frac{1}{2}n_2)$ distribution.

(e) If $\boldsymbol{\alpha}\neq\mathbf{0}$ is a fixed vector show that $\boldsymbol{\alpha}'\boldsymbol{\alpha}/\boldsymbol{\alpha}'U^{-1}\boldsymbol{\alpha}$ is $\text{beta}[\frac{1}{2}(n_1-m+1),\frac{1}{2}n_2]$.

3.23. Let A have the $W_m(n,\Sigma)$ distribution and let A_i and Σ_i be the matrices consisting of the first i rows and columns of A and Σ, respectively, with both $\det A_0$ and $\det\Sigma_0$ defined to be 1. Show that

$$v_i=\frac{\det A_i}{\det A_{i-1}}\cdot\frac{\det\Sigma_{i-1}}{\det\Sigma}$$

is χ^2_{n-i+1} and that $v_1,\dots,v_m$ are independent.

3.24. Let U have the $\text{Beta}_m(\frac{1}{2}n_1,\frac{1}{2}n_2)$ distribution and let U_i be the matrix consisting of the first i rows and columns of U, with $\det U_0\equiv 1$. Show that $v_i=\det U_i/\det U_{i-1}$ is $\text{Beta}[\frac{1}{2}(n_1-i+1),\frac{1}{2}n_2]$ and that $v_1,\dots,v_m$ are independent.

CHAPTER 4

Some Results Concerning Decision-Theoretic Estimation of the Parameters of a Multivariate Normal Distribution

4.1. INTRODUCTION

It was shown in Section 3.1 that, if $\mathbf{X}_1,\dots,\mathbf{X}_n$ are independent $N_m(\boldsymbol{\mu},\Sigma)$ random vectors, the maximum likelihood estimates of the mean $\boldsymbol{\mu}$ and covariance matrix Σ are, respectively,

$$\overline{\mathbf{X}}=\frac{1}{N}\sum_{i=1}^{N}\mathbf{X}_i \quad \text{and} \quad \hat{\Sigma}=\frac{1}{N}\sum_{i=1}^{N}(\mathbf{X}_i-\overline{\mathbf{X}})(\mathbf{X}_i-\overline{\mathbf{X}})'.$$

We saw also that $(\overline{\mathbf{X}},\hat{\Sigma})$ is sufficient, $\overline{\mathbf{X}}$ is unbiased for $\boldsymbol{\mu}$, and an unbiased estimate of Σ is the sample covariance matrix $S=(N/n)\hat{\Sigma}$ (where $n=N-1$). These estimates are easy to calculate and to work with, and their distributions are reasonably simple. However they are generally not optimal estimates from a decision theoretic viewpoint in the sense that they are *inadmissible*. In this chapter we will look at the estimation of $\boldsymbol{\mu}$, Σ, and Σ^{-1} from an admissibility standpoint and find estimates that are better than the usual ones (relative to particular loss functions).

First let us recall some of the terminology and definitions involved in decision-theoretic estimation. The discussion here will not attempt to be completely rigorous, and we will pick out the concepts needed; for more details an excellent reference is the book by Ferguson (1967).

Let X denote a random variable whose distribution depends on an unknown parameter θ. Here X can be a vector or matrix, as can θ. Let $d(X)$ denote an estimate of θ. A *loss function* $l(\theta, d(X))$ is a non-negative function of θ and $d(X)$ that represents the loss incurred (to the statistician) when θ is estimated by $d(X)$. The *risk function* corresponding to this loss function is

$$R(\theta, d) = E_\theta[l(\theta, d(X))],$$

namely, the average loss incurred when θ is estimated by $d(X)$. (This expectation is taken with respect to the distribution of X when θ represents the true value of the parameter.) In decision theory, how "good" an estimate is depends on its risk function. An estimate d_1 is said to be *as good as* an estimate d_2 if, for all θ, its risk function is no larger than the risk function for d_2; that is,

$$R(\theta, d_1) \leq R(\theta, d_2) \qquad \forall\theta.$$

An estimate d_1 is *better than*, or *beats* an estimate d_2 if

$$R(\theta, d_1) \leq R(\theta, d_2) \qquad \forall\theta$$

and

$$R(\theta, d_1) < R(\theta, d_2) \quad \text{for at least one } \theta.$$

An estimate is said to be *admissible* if there exists no estimate which beats it. If there is an estimate which beats it, it is called *inadmissible.*

The above definitions, of course, are all relative to a given loss function. If d_1 and d_2 are two estimates of θ it is possible for d_1 to beat d_2 using one loss function and for d_2 to beat d_1 using another. Hence the choice of a loss function can be a critical consideration. Having decided on a loss function, however, it certainly seems reasonable to rule out from further consideration an estimate which is inadmissible, since there exists one which beats it. It should be mentioned that, in many situations, this has the effect of eliminating estimates which are appealing on intuitive rather than on decision-theoretic grounds, such as maximum likelihood estimates and least-squares estimates, or estimates which are deeply rooted in our statistical psyches, such as uniformly minimum variance unbiased estimates.

4.2. ESTIMATION OF THE MEAN

Suppose that $\mathbf{Y}_1, \ldots, \mathbf{Y}_N$ are independent $N_m(\boldsymbol{\tau}, \Sigma)$ random vectors and that we are interested in estimating $\boldsymbol{\tau}$. We will assume that the covariance matrix

$\Sigma>0$ is *known*. Let $\mathbf{Z}_i=\Sigma^{-1/2}\mathbf{Y}_i\ (i=1,\dots,N)$, then $\mathbf{Z}_1,\dots,\mathbf{Z}_N$ are independent $N_m(\Sigma^{-1/2}\boldsymbol{\tau}, I_m)$ random vectors, so that $\overline{\mathbf{Z}}=N^{-1}\sum_{i=1}^{N}\mathbf{Z}_i$ is $N_m(\Sigma^{-1/2}\boldsymbol{\tau}, N^{-1}I_m)$, and $\overline{\mathbf{Z}}$ is sufficient. Putting $\mathbf{X}=N^{1/2}\overline{\mathbf{Z}}$ and $\boldsymbol{\mu}=N^{1/2}\Sigma^{-1/2}\boldsymbol{\tau}$, the problem can be restated as follows: Given a random vector $\mathbf{X}$ having the $N_m(\boldsymbol{\mu}, I_m)$ distribution [so that the components $X_1,\dots,X_m$ are independent and X_i is $N(\mu_i,1)$], estimate the mean vector $\boldsymbol{\mu}$.

The first consideration is the choice of a loss function. When estimating a single parameter a loss function which is appealing on both intuitive and technical grounds is squared-error loss (that is, the loss is the square of the difference between the parameter value and the value of the estimate), and a simple generalization of such a loss function to a multiparameter situation is the sum of squared errors. Our problem here, then, is to choose $\mathbf{d}(\mathbf{X})=[d_1(\mathbf{X}),\dots,d_m(\mathbf{X})]'$ to estimate $\boldsymbol{\mu}$ using as the loss function

$$
\begin{aligned}
l(\boldsymbol{\mu},\mathbf{d})&=(\mathbf{d}-\boldsymbol{\mu})'(\mathbf{d}-\boldsymbol{\mu})\\
&=\sum_{i=1}^{m}(d_i-\mu_i)^2\\
&=\|\mathbf{d}-\boldsymbol{\mu}\|^2.
\end{aligned}
$$

The maximum likelihood estimate of $\boldsymbol{\mu}$ is $\mathbf{d}_0(\mathbf{X})=\mathbf{X}$, which is unbiased for $\boldsymbol{\mu}$, and its risk function is

$$
\begin{aligned}
(1)\qquad R(\boldsymbol{\mu},\mathbf{d}_0)&=E_{\boldsymbol{\mu}}[l(\boldsymbol{\mu},\mathbf{d}_0(X))]\\
&=E_{\boldsymbol{\mu}}\left[\sum_{i=1}^{m}(X_i-\mu_i)^2\right]\\
&=m \qquad\qquad \forall\boldsymbol{\mu}\in R^m.
\end{aligned}
$$

For a long time this estimate was thought to be optimal in every sense, and certainly admissible. Stein (1956a) showed that it *is* admissible if $m\le 2$ but *inadmissible* if $m\ge 3$ and James and Stein (1961) exhibited a simple estimate which beats it in this latter case. These two remarkable papers have had a profound influence on current approaches to inference problems in multiparameter situations. Here we will indicate the argument used by James and Stein to derive a better estimate.

Consider the estimate

$$\mathbf{d}_\alpha(\mathbf{X})=\left(1-\frac{\alpha}{\mathbf{X}'\mathbf{X}}\right)\mathbf{X}$$

where $\alpha \geq 0$ is a constant. Note that this estimate pulls every component of the usual estimate $\mathbf{X}$ toward the origin and, in particular, the estimate of μ_i obtained by taking the ith component of $\mathbf{d}_\alpha$ will depend not only on X_i but, somewhat paradoxically, on all the other X_j's whose marginal distributions do not depend on μ_i. The risk function for the estimate $\mathbf{d}_\alpha$ is

$$(2) \qquad R(\boldsymbol{\mu},\mathbf{d}_\alpha)=E\left\{\left[\left(1-\frac{\alpha}{\mathbf{X}'\mathbf{X}}\right)\mathbf{X}-\boldsymbol{\mu}\right]'\left[\left(1-\frac{\alpha}{\mathbf{X}'\mathbf{X}}\right)\mathbf{X}-\boldsymbol{\mu}\right]\right\}$$

$$=m-2\alpha E\left[\frac{(\mathbf{X}-\boldsymbol{\mu})'\mathbf{X}}{\mathbf{X}'\mathbf{X}}\right]+\alpha^2 E\left[\frac{1}{\mathbf{X}'\mathbf{X}}\right],$$

where all expectations are taken with respect to the $N_m(\boldsymbol{\mu}, I_m)$ distribution of $\mathbf{X}$. From (1) and (2) it follows that

$$(3) \qquad R(\boldsymbol{\mu},\mathbf{d}_\alpha)-R(\boldsymbol{\mu},\mathbf{d}_0)=\alpha^2 E\left[\frac{1}{\mathbf{X}'\mathbf{X}}\right]-2\alpha E\left[\frac{(\mathbf{X}-\boldsymbol{\mu})'\mathbf{X}}{\mathbf{X}'\mathbf{X}}\right]$$

We need now to compute the expected values on the right side of (3). Expressions for these are given in the following lemma.

LEMMA 4.2.1. If $\mathbf{X}$ is $N_m(\boldsymbol{\mu}, I_m)$ then

$$\text{(i)} \quad E\left[\frac{1}{\mathbf{X}'\mathbf{X}}\right]=E\left[\frac{1}{m-2+2K}\right]$$

and

$$\text{(ii)} \quad E\left[\frac{(\mathbf{X}-\boldsymbol{\mu})'\mathbf{X}}{\mathbf{X}'\mathbf{X}}\right]=(m-2)E\left[\frac{1}{m-2+2K}\right],$$

where K is a random variable having a Poisson distribution with mean $\boldsymbol{\mu}'\boldsymbol{\mu}/2$.

Proof. Put $Z=\mathbf{X}'\mathbf{X}$, then Z is $\chi^2_m(\boldsymbol{\mu}'\boldsymbol{\mu})$, that is, noncentral χ^2 on m degrees of freedom and noncentrality parameter $\boldsymbol{\mu}'\boldsymbol{\mu}$. In Corollary 1.3.4 it was shown that the density function of Z can be written in the form

$$f(z)=\sum_{k=0}^{\infty} P(K=k)g_{m+2k}(z),$$

where K is a Poisson random variable with mean $\frac{1}{2}\boldsymbol{\mu}'\boldsymbol{\mu}$ and $g_r(\cdot)$ is the density function of the (central) χ^2_r distribution. This means that the distribution of Z can be obtained by taking a random variable K having a Poisson distribution with mean $\frac{1}{2}\boldsymbol{\mu}'\boldsymbol{\mu}$ and then taking the conditional distribution of Z given K to be (central) χ^2_{m+2K}. Now note that

$$E\left[\frac{1}{\mathbf{X}'\mathbf{X}}\right]=E\left[\frac{1}{Z}\right]=E\left[E\left[\frac{1}{Z}\middle|K\right]\right]$$
$$=E\left[E\left[\frac{1}{\chi^2_{m+2K}}\middle|K\right]\right]$$
$$=E\left[\frac{1}{m-2+2K}\right],$$

which proves (i).

To prove (ii) we first compute $E[\boldsymbol{\mu}'\mathbf{X}/\|\mathbf{X}\|^2]$. This can be evaluated, with the help of (i), as

$$\text{(4)}\quad E\left[\frac{\boldsymbol{\mu}'\mathbf{X}}{\|\mathbf{X}\|^2}\right]=\boldsymbol{\mu}'\boldsymbol{\mu}E\left[\frac{1}{\|\mathbf{X}\|^2}\right]+E\left[\frac{\boldsymbol{\mu}'(\mathbf{X}-\boldsymbol{\mu})}{\|\mathbf{X}\|^2}\right]$$
$$=\boldsymbol{\mu}'\boldsymbol{\mu}E\left[\frac{1}{m-2+2K}\right]+\sum_{i=1}^{m}\mu_i\frac{d}{d\mu_i}$$
$$\left\{\frac{1}{(2\pi)^{m/2}}\int\cdots\int\|\mathbf{x}\|^{-2}\cdot\exp\left[-\tfrac{1}{2}(\mathbf{x}-\boldsymbol{\mu})'(\mathbf{x}-\boldsymbol{\mu})\right]dx_1\ldots dx_m\right\}$$
$$=\boldsymbol{\mu}'\boldsymbol{\mu}E\left[\frac{1}{m-2+2K}\right]+\sum_{i=1}^{m}\mu_i\frac{d}{d\mu_i}E\left[\frac{1}{\|\mathbf{X}\|^2}\right]$$
$$=\boldsymbol{\mu}'\boldsymbol{\mu}E\left[\frac{1}{m-2+2K}\right]+\sum_{i=1}^{m}\mu_i\frac{d}{d\mu_i}E\left[\frac{1}{m-2+2K}\right]$$
$$=\boldsymbol{\mu}'\boldsymbol{\mu}E\left[\frac{1}{m-2+2K}\right]+\sum_{i=1}^{m}\mu_i\frac{d}{d\mu_i}\sum_{k=0}^{\infty}\frac{e^{-\|\boldsymbol{\mu}\|^2/2}\left(\frac{1}{2}\|\boldsymbol{\mu}\|^2\right)^k}{k!(m-2+2k)}$$
$$=\boldsymbol{\mu}'\boldsymbol{\mu}\left\{E\left[\frac{1}{m-2+2K}\right]\right.$$
$$\left.+\sum_{k=0}^{\infty}\frac{e^{-\|\boldsymbol{\mu}\|^2/2}\left[k\left(\frac{1}{2}\|\boldsymbol{\mu}\|^2\right)^{k-1}-\left(\frac{1}{2}\|\boldsymbol{\mu}\|^2\right)^k\right]}{k!(m-2+2k)}\right\}$$

$$=\mu'\mu\sum_{k=0}^{\infty}\frac{e^{-\|\mu\|^2/2}k\left(\frac{1}{2}\|\mu\|^2\right)^{k-1}}{k!(m-2+2k)}$$

$$=\sum_{k=0}^{\infty}\frac{e^{-\|\mu\|^2/2}2k\left(\frac{1}{2}\|\mu\|^2\right)^{k}}{k!(m-2+2k)}$$

$$=E\left[\frac{2K}{m-2+2K}\right].$$

Hence

$$E\left[\frac{(\mathbf{X}-\mu)'\mathbf{X}}{\mathbf{X}'\mathbf{X}}\right]=1-E\left[\frac{\mu'\mathbf{X}}{\|\mathbf{X}\|^2}\right] \tag{5}$$

$$=1-E\left[\frac{2K}{m-2+2K}\right]$$

$$=(m-2)E\left[\frac{1}{m-2+2K}\right]$$

which proves (ii).

Returning to our risk computations, it follows from Lemma 4.2.1 that (3) can be written as

$$R(\mu,\mathbf{d}_\alpha)-R(\mu,\mathbf{d}_0)=\left[\alpha^2-2\alpha(m-2)\right]E\left[\frac{1}{m-2+2K}\right], \tag{6}$$

where K is Poisson with mean $\frac{1}{2}\mu'\mu$. The right side of (6) is minimized, for all μ, when $\alpha=m-2$ and the minimum value is

$$-(m-2)^2E\left[\frac{1}{m-2+2K}\right].$$

Since this is less than zero for $m\geq 3$ it follows that, for $m\geq 3$, the estimate $\delta(\mathbf{X})$ given by

$$\delta(\mathbf{X})\equiv\mathbf{d}_{m-2}(\mathbf{X})=\left(1-\frac{m-2}{\mathbf{X}'\mathbf{X}}\right)\mathbf{X}, \tag{7}$$

with risk function

$$R(\mu,\delta)=m-(m-2)^2E\left[\frac{1}{m-2+2K}\right], \tag{8}$$

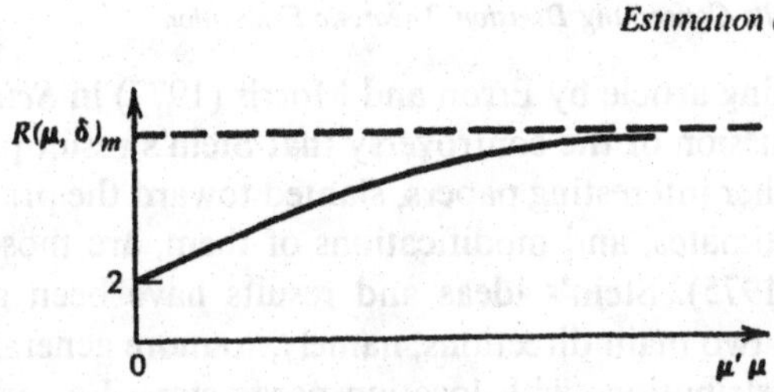

Figure 1.

beats the maximum likelihood estimate **X** which is therefore inadmissible. The risk (8) depends on μ only through $\mu'\mu$, and it is clear that if $\mu = \mathbf{0}$ the risk is 2; the risk approaches m (the risk for **X**) as $\mu'\mu \to \infty$, as shown in Figure 1. It is apparent that if m is large and μ is near $\mathbf{0}$, $\delta(\mathbf{X})$ represents a substantial improvement (in terms of risk) over the usual estimate **X**. It is also worth noticing that although **X** is inadmissible it can be shown that it is a *minimax* estimate of μ; that is, there is no other estimate of μ whose risk function has a smaller supremum. This being the case, it is clear that any estimate which beats **X**—for example, the James-Stein estimate $\delta(\mathbf{X})$ given by (7)—must also be minimax.

James and Stein (1961) also consider estimating μ when Σ is unknown and a sample of size N is drawn from the $N_m(\mu, \Sigma)$ distribution. Reducing the problem in an obvious way by sufficiency we can assume that we observe **X** and A, where **X** is $N_m(\mu, \Sigma)$, A is $W_m(n, \Sigma)$, **X** and A are independent, and $n = N - 1$. Using the loss function

$$l((\mu, \Sigma), \mathbf{d}) = (\mathbf{d} - \mu)'\Sigma^{-1}(\mathbf{d} - \mu)$$

it can be shown, using an argument similar to that above, that the estimate

$$\mathbf{d}(\mathbf{X}, A) = \left(1 - \frac{m-2}{n-m+3}\frac{1}{\mathbf{X}'A^{-1}\mathbf{X}}\right)\mathbf{X}$$

has risk function

$$R((\mu, \Sigma), \mathbf{d}) = m - \left(\frac{n-m+1}{n-m+3}\right)(m-2)^2 E\left[\frac{1}{m-2+2K}\right],$$

where K has a Poisson distribution with mean $\frac{1}{2}\mu'\Sigma^{-1}\mu$. The risk of the maximum likelihood and minimax estimate **X** is

$$R((\mu, \Sigma), \mathbf{X}) = m \qquad \forall \mu \in R^m, \quad \Sigma > 0$$

and hence, if $m \geq 3$, the estimate **d** beats **X** (see Problem 4.1).

An entertaining article by Efron and Morris (1977) in *Scientific American* provides a discussion of the controversy that Stein's result provoked among statisticians. Other interesting papers, slanted toward the practical use of the James–Stein estimates, and modifications of them, are those by Efron and Morris (1973, 1975). Stein's ideas and results have been generalized and expanded on in two main directions, namely, to more general loss functions, and to other distributions with location parameters. For examples of such extensions the reader is referred to Brown (1966), (1980), Berger et al. (1977), and Brandwein and Strawderman (1978, 1980), Berger (1980a,b) and to the references in these papers.

4.3. ESTIMATION OF THE COVARIANCE MATRIX

Let $\mathbf{X}_1,\dots,\mathbf{X}_N$ (where $N>m$) be independent $N_m(\boldsymbol{\mu},\Sigma)$ random vectors and put

$$A=\sum_{i=1}^{N}(\mathbf{X}_i-\overline{\mathbf{X}})(\mathbf{X}_i-\overline{\mathbf{X}})',$$

so that A is $W_m(n,\Sigma)$ with $n=N-1$. The maximum likelihood estimate of Σ is $\hat{\Sigma}=N^{-1}A$, and an unbiased estimate of Σ is the sample covariance matrix $S=n^{-1}A$. In this section we consider the problem of estimating Σ by an $m\times m$ positive definite matrix $\phi(A)$ whose elements are functions of the elements of A. Two loss functions which have been suggested and considered in the literature by James and Stein (1961), Olkin and Selliah (1977), and Haff (1980) are

$$(1)\qquad l_1(\Sigma,\phi)=\operatorname{tr}(\Sigma^{-1}\phi)-\log\det(\Sigma^{-1}\phi)-m$$

and

$$(2)\qquad l_2(\Sigma,\phi)=\operatorname{tr}(\Sigma^{-1}\phi-I_m)^2.$$

The respective risk functions will be similarly subscripted. Both loss functions are non-negative and are zero when $\phi\equiv\Sigma$. Certainly there are many other possible loss functions with these properties; the two above, however, have the attractive feature that they are relatively easy to work with. We will first consider the loss function $l_1(\Sigma,\phi)$. If we restrict attention to estimates of the form αA, where α is a constant, we can do no better than the sample covariance matrix S, as the following result shows.

THEOREM 4.3.1. Using the loss function $l_1(\Sigma, \phi)$, the best (smallest risk) estimate of Σ having the form αA is the unbiased estimate $S = n^{-1}A$.

Proof. The risk of the estimate αA is

$$(3) \qquad R_1(\Sigma, \alpha A) = E\left[\alpha \operatorname{tr}(\Sigma^{-1}A) - \log\det(\alpha\Sigma^{-1}A) - m\right]$$

$$= \alpha \operatorname{tr} \Sigma^{-1}E(A) - m\log\alpha - E\left[\log\frac{\det A}{\det \Sigma}\right] - m$$

$$= \alpha mn - m\log\alpha - E\left[\log \prod_{i=1}^{m} \chi^2_{n-i+1}\right] - m$$

$$= \alpha mn - m\log\alpha - \sum_{i=1}^{m} E\left[\log \chi^2_{n-i+1}\right] - m,$$

where we have used $E(A) = n\Sigma$ and the fact that $\det A/\det\Sigma$ has the same distribution as the product of independent χ^2 random variables $\prod_{i=1}^{m}\chi^2_{n-i+1}$, the result of Theorem 3.2.15. The proof is completed by noting that the value of α which minimizes the right side of (3) is $\alpha = 1/n$.

If we look outside the class of estimates of the form αA we can do better than the sample covariance matrix S, as James and Stein (1961) have shown using an *invariance argument*. It is reasonable to require that if $\phi(A)$ estimates Σ and L is a nonsingular $m \times m$ matrix then ϕ should satisfy

$$\phi(L'AL) = L'\phi(A)L,$$

for $L'AL$ is $W_m(n, L'\Sigma L)$, so that $\phi(L'AL)$ estimates $L'\Sigma L$, as does $L'\phi(A)L$. If this holds for all matrices L then $\phi(A) = \alpha A$. If the requirement is relaxed a little an estimate which beats any estimate of the form αA can be found. The approach taken by James and Stein is to find the best estimate ϕ out of all estimates satisfying

$$(4) \qquad \phi(L'AL) = L'\phi(A)L$$

for all *upper-triangular* matrices L. Note that all estimates of the form αA satisfy (4); the *best* estimate however turns out *not* to be of the form αA so that all such estimates, including S, are *inadmissible*. It also turns out that the best estimate is not particularly appealing; an estimate need not be attractive simply because it beats S.

Putting $A = I_m$ in (4) gives

$$(5) \qquad \phi(L'L) = L'\phi(I)L.$$

Now let

$$L=\begin{bmatrix} \pm 1 & & 0 \\ & \ddots & \\ 0 & & \pm 1 \end{bmatrix},$$

then $L'L = I_m$ and (5) becomes

$$\phi(I) = L'\phi(I)L$$

for all such matrices L, which implies that $\phi(I)$ is diagonal,

$$\phi(I) = \operatorname{diag}(\delta_1,\dots,\delta_m) = \Delta, \text{ say.} \tag{6}$$

Now write $A = T'T$, where T is upper-triangular with positive diagonal elements, then

$$\begin{aligned} \phi(A) &= \phi(T'T) \\ &= T'\phi(I)T \qquad \text{by (5)} \\ &= T'\Delta T \qquad \text{by (6).} \end{aligned} \tag{7}$$

What we have shown is that an estimate $\phi(A)$ is *invariant* under the group of *upper-triangular matrices* [that is, it satisfies (4)] if and only if it has the form (7) where $A = T'T$ with T upper-triangular and where Δ is an arbitrary diagonal matrix whose elements do not depend on A. We next note that the estimate $\phi(A)$ in (7) has *constant risk*; that is, the risk does not depend on Σ. To spell it out, the risk function is

$$\begin{aligned} R_1(\Sigma,\phi) &= E_\Sigma\left[\operatorname{tr}\Sigma^{-1}\phi(A) - \log\det\Sigma^{-1}\phi(A) - m\right] \\ &= \frac{2^{-mn/2}\det\Sigma^{-n/2}}{\Gamma_m(\frac{1}{2}n)}\int_{A>0}\left[\operatorname{tr}\Sigma^{-1}\phi(A) - \log(\det\Sigma)^{-1}\phi(A) - m\right] \\ &\quad \cdot \operatorname{etr}\left(-\tfrac{1}{2}\Sigma^{-1}A\right)(\det A)^{(n-m-1)/2}(dA). \end{aligned}$$

Now write Σ^{-1} as $\Sigma^{-1} = LL'$, where L is upper-triangular, and note that

$$\begin{aligned} \operatorname{tr}\Sigma^{-1}\phi(A) - \log\det\Sigma^{-1}\phi(A) - m &= \operatorname{tr}L'\phi(A)L - \log\det L'\phi(A)L - m \\ &= \operatorname{tr}\phi(L'AL) - \log\det\phi(L'AL) - m, \end{aligned}$$

using (4). Hence

$$R_1(\Sigma,\phi)=\frac{2^{-mn/2}\det\Sigma^{-n/2}}{\Gamma_m(\frac{1}{2}n)}\int_{A>0}\left[\operatorname{tr}\phi(L'AL)-\log\det\phi(L'AL)-m\right]$$

$$\cdot\operatorname{etr}(-\tfrac{1}{2}L'AL)(\det A)^{(n-m-1)/2}(dA).$$

Putting $U=L'AL$, this becomes

$$\begin{aligned}R_1(\Sigma,\phi)&=\frac{2^{-mn/2}}{\Gamma_m(\frac{1}{2}n)}\int_{U>0}\left[\operatorname{tr}\phi(U)-\log\det\phi(U)-m\right]\\&\qquad\cdot\operatorname{etr}(-\tfrac{1}{2}U)(\det U)^{(n-m-1)/2}(dU)\\&=R_1(I_m,\phi)\end{aligned}$$

and hence the risk does not depend on Σ. The next step is to compute the risk and to find the diagonal matrix Δ which minimizes this. We have

$$\begin{aligned}(8)\qquad R_1(I_m,\phi)&=E\left[\operatorname{tr}\phi(A)-\log\det\phi(A)-m\right]\\&=E\left[\operatorname{tr}T'\Delta T-\log\det T'\Delta T\right]-m\\&=E(\operatorname{tr}T'\Delta T)-\log\det\Delta-E\left[\log\det A\right]-m,\end{aligned}$$

where all expectations are computed with $\Sigma=I_m$.

Now, if $T=(t_{ij})$ then

$$\operatorname{tr}T'\Delta T=\sum_{i\le j}^{m}\delta_i t_{ij}^2$$

and, from Theorem 3.2.14, the elements t_{ij} of T are all independent, t_{ii}^2 is χ^2_{n-i+1} $(i=1,\dots,m)$ and t_{ij} is $N(0,1)$ for $i<j$. Hence

$$\begin{aligned}(9)\qquad E(\operatorname{tr}T'\Delta T)&=\sum_{i=1}^{m}\delta_i E\left[t_{ii}^2\right]+\sum_{i<j}\delta_i E\left[t_{ij}^2\right]\\&=\sum_{i=1}^{m}\delta_i(n-i+1)+\sum_{i<j}\delta_i\\&=\sum_{i=1}^{m}\delta_i(n+m-2i+1)\end{aligned}$$

Also, $\det A=\prod_{i=1}^{m}\chi^2_{n-i+1}$ by Theorem 3.2.15, so that

$$E[\log\det A]=\sum_{i=1}^{m}E\left[\log\chi^2_{n-i+1}\right]. \tag{10}$$

Substituting for (9) and (10) in (8) we then have

$$\begin{aligned}R_1(\Sigma,\phi)&=R_1(I_m,\phi)\\&=\sum_{i=1}^{m}\left[(n+m-2i+1)\delta_i-\log\delta_i\right]\\&\quad-\sum_{i=1}^{m}E\left[\log\chi^2_{n-i+1}\right]-m.\end{aligned}$$

This attains its minimum value when

$$\delta_i=\frac{1}{n+m-2i+1}\qquad(i=1,\dots,m).$$

We can summarize our results in the following theorem.

THEOREM 4.3.2. Using the loss function $l_1(\Sigma,\phi)$ given by (1), the best (smallest risk) estimate of Σ in the class of estimates satisfying

$$\phi(L'AL)=L'\phi(A)L$$

for all upper-triangular matrices L, is

$$\phi^*(A)=T'\begin{bmatrix}\frac{1}{n+m-1} & & & 0\\ & \frac{1}{n+m-3} & & \\ & & \ddots & \\ 0 & & & \frac{1}{n+m-2m+1}\end{bmatrix}T,$$

where $A=T'T$ with T upper-triangular. The minimum risk [i.e., the risk function for $\phi^*(A)$] is

$$R_1(\Sigma,\phi^*)=\sum_{i=1}^{m}\left\{\log(n+m-2i+1)-E\left[\log\chi^2_{n-i+1}\right]\right\}.$$

In particular, $\phi^*(A)$ beats all estimates of Σ of the form αA (which have $\delta_i = \alpha$; $i=1,\dots,m$), the best of which is $S = n^{-1}A$ with risk function

$$R_1(\Sigma, S) = m\log n - \sum_{i=1}^{m} E\left[\log \chi^2_{n-i+1}\right].$$

It can be shown that $\phi^*(A)$ is minimax (that is, there is no other estimate of Σ whose risk function has a smaller supremum), but that it is itself inadmissible. For details, the reader is referred to James and Stein (1961).

It is interesting to examine the estimate $\phi^*(A)$ when $m=2$. If $A=(a_{ij})$, the sample covarance matrix is

$$S = \frac{1}{n}A = \left(\frac{1}{n}a_{ij}\right)$$

whereas $\phi^*(A)$ is easily shown to be

$$\text{(11)} \qquad \phi^*(A) = \begin{bmatrix} \dfrac{a_{11}}{n+1} & \dfrac{a_{12}}{n+1} \\ \dfrac{a_{12}}{n+1} & \dfrac{a_{22}}{n-1} - \dfrac{2a_{12}^2}{(n^2-1)a_{11}} \end{bmatrix}$$

(see Problem 4.2). The expectations of these two estimates are $E(S) = \Sigma = (\sigma_{ij})$ and

$$\text{(12)} \qquad E[\phi^*(A)] = \begin{bmatrix} \dfrac{n\sigma_{11}}{n+1} & \dfrac{n\sigma_{12}}{n+1} \\ \dfrac{n\sigma_{12}}{n+1} & \left(\dfrac{n+2}{n+1}\right)\sigma_{22} - \dfrac{2}{n+1}\dfrac{\sigma_{12}^2}{\sigma_{11}} \end{bmatrix}$$

$$= \frac{n}{n+1}\Sigma + \frac{2}{n+1}\begin{bmatrix} 0 & 0 \\ 0 & \dfrac{\det\Sigma}{\sigma_{11}} \end{bmatrix}.$$

Note that although $\phi^*(A)$ beats S it has the rather unappealing feature of not being invariant under permutations of the variables.

A problem of considerable importance in principal components analysis (see Chapter 9) concerns the estimation of the latent roots of Σ. These are commonly estimated by the latent roots of the sample covariance matrix S. In view of the fact that $\phi^*(A)$ is a better estimate of Σ than S with respect to the rather special loss function l_1 one might also consider estimating the latent roots of Σ by the latent roots of $\phi^*(A)$. Relations between these two sets of estimates are investigated in Problem 4.3.

Finally, we will look briefly at the problem of estimating Σ using the loss function

$$l_2(\Sigma,\phi)=\operatorname{tr}(\Sigma^{-1}\phi-I_m)^2$$

considered by Olkin and Selliah (1977) and Haff (1980). We have seen that any estimate $\phi(A)$ satisfying

$$\phi(L'AL)=L'\phi(A)L$$

for all upper-triangular matrices L has the form

$$\phi(A)=T'\Delta T$$

where $A=T'T$ with T upper-triangular and where $\Delta=\operatorname{diag}(\delta_1,\dots,\delta_m)$ is an arbitrary diagonal matrix whose elements δ_i do not depend on A. (The argument used to show this had nothing to do with the loss function.) Again, one can easily show that the risk function for $\phi(A)$ does not depend on Σ, so that the risk need only be computed for $\Sigma=I_m$. Hence

$$\begin{aligned}R_2(\Sigma,\phi)&=R_2(I_m,\phi)\\&=E\left[l_2(I_m,\phi(A))\right]\\&=E\left[\operatorname{tr}(\phi(A)-I)^2\right]\\&=E\left[\operatorname{tr}(T'\Delta TT'\Delta T)\right]-2E\left[\operatorname{tr}(T'\Delta T)\right]+m.\end{aligned}$$

The Bartlett decomposition (Theorem 3.2.14) can be used to show that

(13)

$$\begin{aligned}R_2(\Sigma,\phi)=&\sum_{i=1}^{m}\delta_i^2(n+m-2i+1)(n+m-2i+3)\\&+2\sum_{i<j}^{m}\delta_i\delta_j(n+m-2i+1)-2\sum_{i=1}^{m}\delta_i(n+m-2i+1)+m.\end{aligned}$$

The δ_i's which minimize this are obtained by differentiating $R_2(\Sigma,\phi)$ with respect to the δ_i and equating the derivatives to zero. The minimizing δ's are given by the solutions of

$$(14) \qquad B\delta = \begin{pmatrix} n+m-1 \\ n+m-3 \\ \vdots \\ n+m-2m+1 \end{pmatrix},$$

where $\delta = (\delta_1, \ldots, \delta_m)'$ and $B = (b_{ij})$ is a symmetric $m \times m$ matrix with

$$b_{ii} = (n+m-2i+1)(n+m-2i+3), \qquad b_{ij} = n+m-2j+1 \qquad (i<j).$$

Summarizing we have the following theorem.

THEOREM 4.3.3. Using the loss function $l_2(\Sigma, \phi)$ given by (2) the best (smallest risk) estimate in the class of estimates satisfying

$$\phi(L'AL) = L'\phi(A)L$$

for all upper-triangular matrices L, is

$$\tilde{\phi}(A) = T'\Delta T,$$

where $A = T'T$ with T upper-triangular, and $\Delta = \text{diag}(\delta_1, \ldots, \delta_m)$, where the δ_i are given by the solution of (14). In particular, all estimates of the form αA are inadmissible.

It is difficult to obtain the δ_i's in $\tilde{\phi}(A)$ explicity. For $m=2$ they are given by

$$\delta_1 = \frac{(n+1)^2 - (n-1)}{(n+1)^2(n+2) - (n-1)}$$

and

$$\delta_2 = \frac{(n+1)(n+2)}{(n+1)^2(n+3) - (n-1)}.$$

Finally, the estimate $\tilde{\phi}(A)$ can be shown to be minimax (see Olkin and Selliah, 1977). And it is also worth noting that relative to the loss function $l_2(\Sigma, \phi)$ the estimate $\tilde{\phi}(A)$ beats the estimate $\phi^*(A)$ (given in Theorem 4.3.2) but that the opposite is true when the loss function $l_1(\Sigma, \phi)$ is used.

Estimates which are more appealing than $\phi^*(A)$ and $\tilde{\phi}(A)$ have been given by Haff (1980). With respect to the loss function $l_1(\Sigma,\phi)$ Haff has shown all estimates of the form

$$\phi_1(A)=\frac{1}{n}[A+ut(u)C]$$

beat S, where $u=1/\mathrm{tr}(A^{-1}C)$, C is an arbitrary positive definite matrix, and $t(u)$ is an absolutely continuous, nonincreasing function with $0\le t(u)\le 2(m-1)/n$. Similarly, using the loss function $l_2(\Sigma,\phi)$, the best estimate of the form αA is $(n+m+1)^{-1}A$ (see Problem 4.5), and this is beaten by all estimates of the form

$$\phi_2(A)=\frac{1}{n+m+1}[A+\gamma uC],$$

where $u=1/\mathrm{tr}(A^{-1}C)$, C is an arbitrary positive definite matrix, and γ is a constant with $0\le\gamma\le 2(m-1)/(n-m+3)$. For details and further references concerning the estimation of Σ the interested reader is referred to Haff (1980).

4.4. ESTIMATION OF THE PRECISION MATRIX

In this section we consider the problem of estimating the *precision matrix* Σ^{-1} by $\gamma(A)$, where A is $W_m(n,\Sigma)$, with $n>m+1$. Here we will concentrate primarily on the loss function

$$(1)\qquad l(\Sigma^{-1},\gamma)=\frac{\mathrm{tr}\left[(\gamma-\Sigma^{-1})^2A\right]}{n\,\mathrm{tr}\,\Sigma^{-1}}$$

introduced by Efron and Morris (1976) in an empirical Bayes estimation context. First recall from (12) of Section 3.2.3 that

$$(2)\qquad E[A^{-1}]=\frac{1}{n-m-1}\Sigma^{-1}$$

so that the estimate

$$(3)\qquad \gamma_0(A)\equiv(n-m-1)A^{-1}=\frac{n-m-1}{n}S^{-1}$$

is unbiased for Σ^{-1}. In the class of estimates of the form αA^{-1} this is the best estimate, as the following result demonstrates.

THEOREM 4.4.1. The best (smallest risk) estimate of Σ^{-1} having the form αA^{-1} is the unbiased estimate $\gamma_0(A)=(n-m-1)A^{-1}$.

Proof. The risk of the estimate αA^{-1} is

$$
\begin{aligned}
(4)\qquad R(\Sigma^{-1},\alpha A^{-1}) &= \frac{E\left[\operatorname{tr}(\alpha A^{-1}-\Sigma^{-1})^2 A\right]}{n\operatorname{tr}\Sigma^{-1}} \\
&= \frac{E\left[\operatorname{tr}(\alpha^2 A^{-1})+\operatorname{tr}(\Sigma^{-2}A)-2\alpha\operatorname{tr}(\Sigma^{-1})\right]}{n\operatorname{tr}\Sigma^{-1}} \\
&= \frac{\dfrac{\alpha^2}{n-m-1}\operatorname{tr}\Sigma^{-1}+n\operatorname{tr}\Sigma^{-1}-2\alpha\operatorname{tr}\Sigma^{-1}}{n\operatorname{tr}\Sigma^{-1}} \\
&= \frac{1}{n}\left[\frac{\alpha^2}{n-m-1}+n-2\alpha\right],
\end{aligned}
$$

where we have used

$$(5)\qquad E[\operatorname{tr}A^{-1}]=\operatorname{tr}E[A^{-1}]=\frac{1}{n-m-1}\operatorname{tr}\Sigma^{-1}$$

and

$$(6)\qquad E[\operatorname{tr}A]=\operatorname{tr}E[A]=n\operatorname{tr}\Sigma.$$

The proof is completed by noting that the value of α which minimizes the right side of (4) is $\alpha=n-m-1$ and the minimum risk is

$$(7)\qquad R(\Sigma^{-1},\gamma_0)=\frac{m+1}{n}.$$

Outside the class of estimates of the form αA Efron and Morris (1976) have shown that we can do better than the unbiased estimate $\gamma_0(A)$. To demonstrate this we will make use of the following lemma.

LEMMA 4.4.2. Suppose that A is $W_m(n,\Sigma)$ and put

$$\beta=\frac{1}{\omega}E\left[\frac{mn-2}{\operatorname{tr}A}\right],\qquad \text{where}\quad \omega\equiv\frac{1}{m}\operatorname{tr}\Sigma^{-1}.$$

Then

$$\text{(i)}\quad \beta=\frac{1}{\omega}\left[\frac{\operatorname{tr}\Sigma^{-1}A}{\operatorname{tr}A}\right]$$

and

(ii) $0<\beta\le 1$ for all $\Sigma>0$.

Proof. To prove (i) we have to show that

$$E\left[\frac{mn-2}{\operatorname{tr}A}\right]=E\left[\frac{\operatorname{tr}\Sigma^{-1}A}{\operatorname{tr}A}\right]. \tag{8}$$

Let H be an orthogonal $m\times m$ matrix such that

$$H\Sigma H'=\Lambda=\operatorname{diag}(\lambda_1,\dots,\lambda_m)$$

and put $B=HAH'$. Then B is $W_m(n,\Lambda)$ and the right side of (8) is

$$\begin{aligned}E\left[\frac{\operatorname{tr}\Sigma^{-1}A}{\operatorname{tr}A}\right]&=E\left[\frac{\operatorname{tr}H'\Lambda^{-1}HH'BH}{\operatorname{tr}H'BH}\right]\\&=E\left[\frac{\operatorname{tr}\Lambda^{-1}B}{\operatorname{tr}B}\right]\\&=E\left[\frac{\sum_{i=1}^m b_{ii}/\lambda_i}{\sum_{i=1}^m b_{ii}}\right].\end{aligned}$$

Let $u_i=b_{ii}/\lambda_i$; then from Theorem 3.2.7 it follows that $u_1,\dots,u_m$ are independent χ_n^2 random variables, and

$$\begin{aligned}E\left[\frac{\operatorname{tr}\Sigma^{-1}A}{\operatorname{tr}A}\right]&=E\left[\frac{\sum_{i=1}^m u_i}{\sum_{i=1}^m \lambda_i u_i}\right]\\&=E\left[\frac{1}{\sum_{i=1}^m \lambda_i v_i}\right],\end{aligned} \tag{9}$$

where $v_i=u_i/\sum_{j=1}^m u_j$. Now, it is well-known (and easily checked) that $\sum_{j=1}^m u_j$ is independent of $(v_1,\dots,v_m)$. The distribution of $\sum_{j=1}^m u_j$ *is* χ_{mn}^2, so that

$$E\left[\frac{1}{\sum_{j=1}^m u_j}\right]=E\left[\frac{1}{\chi_{mn}^2}\right]=\frac{1}{mn-2}.$$

Using this in (9) we then have

$$E\left[\frac{\operatorname{tr}\Sigma^{-1}A}{\operatorname{tr}A}\right]=E\left[\frac{1}{\sum_{i=1}^{m}\lambda_i v_i}\right]E\left[\frac{mn-2}{\sum_{j=1}^{m}u_j}\right]$$

$$=E\left[\frac{mn-2}{(\sum_{i=1}^{m}\lambda_i v_i)(\sum_{j=1}^{m}u_j)}\right]\qquad\text{(by independence)}$$

$$=E\left[\frac{mn-2}{\sum_{i=1}^{m}\lambda_i u_i}\right]$$

$$=E\left[\frac{mn-2}{\operatorname{tr}B}\right]$$

$$=E\left[\frac{mn-2}{\operatorname{tr}A}\right],$$

which proves (8) and hence establishes (i).

To prove (ii) note that

$$\beta=\frac{1}{\omega}E\left[\frac{\operatorname{tr}\Sigma^{-1}A}{\operatorname{tr}A}\right]$$

$$=\frac{1}{\omega}E\left[\frac{1}{\sum_{i=1}^{m}\lambda_i v_i}\right]\qquad\text{from (9).}$$

Clearly $\beta>0$, and since

$$\frac{1}{\sum_{i=1}^{m}\lambda_i v_i}\le\sum_{i=1}^{m}\frac{v_i}{\lambda_i}$$

it follows that

$$\beta\le\frac{1}{\omega}E\left[\sum_{i=1}^{m}\frac{v_i}{\lambda_i}\right]=\frac{1}{\omega m}\sum_{i=1}^{m}\frac{1}{\lambda_i}=\frac{1}{\omega m}\operatorname{tr}\Sigma^{-1}=1.$$

We are now ready to demonstrate the existence of an estimate which beats $\gamma_0(A)$.

THEOREM 4.4.3. The estimate

$$(10) \qquad \gamma_1(A) \equiv (n-m-1)A^{-1} + \frac{m^2+m-2}{\operatorname{tr} A} I_m$$

of Σ^{-1} beats $\gamma_0(A) = (n-m-1)A^{-1}$ if $m \geq 2$ [and hence $\gamma_0(A)$ is inadmissible if $m \geq 2$].

Proof. Define β and ω as in Lemma 4.4.2 and put $\delta = n-m-1$ and $\eta = m^2+m-2$ so that

$$\gamma_1(A) = \delta A^{-1} + \frac{\eta}{\operatorname{tr} A} I_m.$$

The risk function of γ_1 is

$$\begin{aligned} R(\Sigma^{-1}, \gamma_1) &= \frac{1}{mn\omega} E\left[\operatorname{tr}\left(\delta A^{-1} + \frac{\eta}{\operatorname{tr} A} I - \Sigma^{-1}\right)^2 A\right] \\ &= \frac{\delta^2}{mn\omega} E[\operatorname{tr} A^{-1}] + \frac{2\delta\eta}{n\omega} E\left[\frac{1}{\operatorname{tr} A}\right] - \frac{2\delta}{n} + \frac{\eta^2}{mn\omega} E\left[\frac{1}{\operatorname{tr} A}\right] \\ &\quad - \frac{2\eta}{mn\omega} E\left[\frac{\operatorname{tr} \Sigma^{-1} A}{\operatorname{tr} A}\right] + \frac{1}{mn\omega} E[\operatorname{tr} \Sigma^{-2} A] \\ &= \frac{\delta^2}{mn\omega} \cdot \frac{\operatorname{tr} \Sigma^{-1}}{n-m-1} + \frac{2\delta\eta\beta}{n(mn-2)} - \frac{2\delta}{n} + \frac{\eta^2\beta}{mn(mn-2)} - \frac{2\eta\beta}{mn} + 1, \end{aligned}$$

where we have used (5), (6), and (i) of Lemma 4.4.2. Substitution of $\delta = n-m-1$ and $\eta = m^2+m-2$ gives

$$R(\Sigma^{-1}, \gamma_1) = \frac{m+1}{n} - \frac{mn-2}{mn} c^2 \beta,$$

where

$$c = \frac{m^2+m-2}{mn-2}.$$

Note that $0 \leq c \leq 1$ and $0 < c < 1$ if $m > 1$ and $n > m+1$, and we know from Lemma 4.4.2 that $0 < \beta \leq 1$. It follows that for $m \geq 2$ and $n > m+1$

$$R(\Sigma^{-1}, \gamma_1) < \frac{m+1}{n} \qquad \text{for all } \Sigma.$$

The right side of this inequality is the risk function of the unbiased estimate $\gamma_0(A)$ of Σ^{-1} [see (7)], hence the estimate $\gamma_1(A)$ has uniformly smaller risk than $\gamma_0(A)$ and the proof is complete.

It is interesting to note that $\gamma_1(A)$ can be written as

$$\gamma_1(A)=\gamma_0(A)+\frac{m^2+m-2}{mn-2}\gamma^*(A),$$

where

$$\gamma^*(A)\equiv\frac{mn-2}{\operatorname{tr}A}I.$$

Here $\gamma^*(A)$ is the best unbiased estimate of Σ^{-1} when Σ is known to be proportional to I_m. The estimate $\gamma_1(A)$ increases the unbiased estimate $\gamma_0(A)$ by an amount proportional to $\gamma^*(A)$. It is also worth pointing out that $\gamma_0(A)$ is minimax and, as a consequence, so is $\gamma_1(A)$. For more details the reader is referred to Efron and Morris (1976).

Another loss function considered by Haff (1977, 1979) is

$$l(\Sigma^{-1},\gamma)=\operatorname{tr}(\gamma-\Sigma^{-1})^2Q, \tag{11}$$

where Q is an arbitrary positive definite matrix. We will not go into the details, but Haff has noted an interesting result. We saw that, using the loss function (1), the Efron–Morris estimate $\gamma_1(A)$ beats the unbiased estimate $\gamma_0(A)$. When the loss function (11) is used the reverse is true; that is, the unbiased estimate $\gamma_0(A)$ beats $\gamma_1(A)$. This is curious in view of the fact that the two loss functions (1) and (11) are expected to be close (up to a multiplicative factor) if $Q=\Sigma$ and n is large, for then $n^{-1}A\rightarrow\Sigma$ in probability.

PROBLEMS

4.1. Suppose that $\mathbf{Y}_1,\ldots,\mathbf{Y}_N$ are independent $N_m(\boldsymbol{\tau},\Sigma)$ random vectors where both $\boldsymbol{\tau}$ and Σ are unknown and $\boldsymbol{\tau}$ is to be estimated. Reducing the problem by sufficiency it can be assumed that $\mathbf{X}=N^{1/2}\overline{\mathbf{Y}}$ and $A=\Sigma_{i=1}^N(\mathbf{Y}_i-\overline{\mathbf{Y}})(\mathbf{Y}_i-\overline{\mathbf{Y}})'$ are observed; $\mathbf{X}$ is $N_m(\boldsymbol{\mu},\Sigma)$ with $\boldsymbol{\mu}=N^{1/2}\boldsymbol{\tau}$, A is $W_m(n,\Sigma)$ with $n=N-1$, and $\mathbf{X}$ and A are independent. Consider the problem of estimating $\boldsymbol{\mu}$ using the loss function

$$l((\boldsymbol{\mu},\Sigma),\mathbf{d})=(\mathbf{d}-\boldsymbol{\mu})'\Sigma^{-1}(\mathbf{d}-\boldsymbol{\mu}).$$

Let $\mathbf{d}_\alpha$ denote the estimate

$$\mathbf{d}_\alpha = \left(1 - \frac{\alpha}{\mathbf{X}'A\mathbf{X}}\right)\mathbf{X}$$

(a) Show that the risk function of $\mathbf{d}_\alpha$ can be written as

$$R((\boldsymbol{\mu},\Sigma),\mathbf{d}_\alpha) = E_{(\boldsymbol{\mu}^*, I_m)}[(\mathbf{d}_\alpha - \boldsymbol{\mu}^*)'(\mathbf{d}_\alpha - \boldsymbol{\mu}^*)]$$

where $\boldsymbol{\mu}^* = [(\boldsymbol{\mu}'\Sigma^{-1}\boldsymbol{\mu})^{1/2},0,\ldots,0]'$ and $E_{(\boldsymbol{\mu}^*, I_m)}$ denotes expectation taken with respect to the joint distribution of $\mathbf{X}$ and A; $\mathbf{X}$ is $N_m(\boldsymbol{\mu}^*, I_m)$, A is $W_m(n, I_m)$, and $\mathbf{X}$ and A are independent.

(b) From Theorem 3.2.12 it follows that conditional on $\mathbf{X}$, $\mathbf{X}'A^{-1}\mathbf{X} = \mathbf{X}'\mathbf{X}/U$, where U is χ^2_{n-m+1} and is independent of $\mathbf{X}$. Writing

$$\mathbf{d}_\alpha = \left(1 - \frac{\alpha U}{\mathbf{X}'\mathbf{X}}\right)\mathbf{X}$$

conditioning on $\mathbf{X}$, and using Lemma 4.2.1, show that

$$R((\boldsymbol{\mu},\Sigma),\mathbf{d}_\alpha) = m - 2\alpha(n-m+1)(m-2)E\left[\frac{1}{m-2+2K}\right]$$
$$+\alpha^2\left[2(n-m+1)+(n-m+1)^2\right]E\left[\frac{1}{m-2+2K}\right],$$

where K has a Poisson distribution with mean $\frac{1}{2}\boldsymbol{\mu}'\Sigma^{-1}\boldsymbol{\mu}$. Show that this risk is minimized, for all $\boldsymbol{\mu}$ and Σ, when $\alpha = (m-2)/(n-m+3)$, and show that with this value for α the estimate $\mathbf{d}_\alpha$ beats the maximum likelihood estimate $\mathbf{X}$ if $m \ge 3$.

4.2. Show that when $m=2$ the best estimate of Σ in Theorem 4.3.2 is (11) of Section 4.3 and that it has expectation given by (12) of Section 4.3.

4.3. Suppose that S is a sample covariance matrix and nS is $W_m(n,\Sigma)$ and consider the problem of estimating the latent roots $\lambda_1,\ldots,\lambda_m$ ($\lambda_1 \ge \cdots \ge \lambda_m > 0$) of Σ. A commonly used estimate of λ_i is l_i, where $l_1,\ldots,l_m$ ($l_1 \ge \cdots \ge l_m > 0$) are the latent roots of S. An estimate of λ_i obtained using the unbiased estimate $[(n-m-1)/n]S^{-1}$ of Σ^{-1} is $\tilde{\lambda}_i = nl_i/(n-m-1)$, $(i=1,\ldots,m)$. Let $\phi_1^*,\ldots,\phi_m^*$ ($\phi_1^* \ge \cdots \ge \phi_m^* > 0$) be the latent roots of the estimate $\phi^*(A)$ given in Theorem 4.3.2. Show that

$$\tilde{\lambda}_i \ge \phi_i^* \ge \phi_m^* \ge \frac{n-m-1}{n+m-1}\tilde{\lambda}_m \qquad (i=1,\ldots,m).$$

[*Hint*: The following two facts are useful [see Bellman, 1970, pp. 122 and 137]: (a) If F is a symmetric matrix whose latent roots all lie between 0 and 1 and E is positive definite then the latent roots of $E^{1/2}FE^{1/2}$ are all less than those of E. (b) For any two matrices E and F the square of the absolute value of any latent root of EF is at least as big as the product of the minimum latent root of EE' and the minimum latent root of FF'.]

4.4. If $\phi^*(A)$ is the best estimate of Σ in Theorem 4.3.2, show that

$$R_1(\Sigma,\phi^*)-R_1(\Sigma,S)=O(n^{-2}).$$

4.5. Suppose A is $W_m(n,\Sigma)$ and consider the problem of estimating Σ using the loss function $l_2(\Sigma,\phi)$ given by (2) of Section 4.3. Show that the best estimate having the form αA is $(n+m+1)^{-1}A$.

4.6. Suppose that $\phi^*(A)$ is the best estimate of Σ in Theorem 4.3.2 and put $\phi_L(A)=L'^{-1}\phi^*(L'AL)L^{-1}$ where L is an $m\times m$ nonsingular matrix. Show that

$$R_1(\Sigma,\phi_L)=R_1(I_m,\phi^*)=R_1(\Sigma,\phi^*).$$

4.7. When $m=2$, express the best estimate of Σ in Theorem 4.3.3 in terms of the elements of A and find its expectation.

4.8. Suppose A is $W_m(n,\Sigma)$ and consider the problem of estimating the generalized variance $\det\Sigma$ by $d(A)$ using the loss function

$$l(\det\Sigma,d(A))=\left[\frac{d(A)}{\det\Sigma}-1\right]^2$$

(a) Show that any estimate of $\det\Sigma$ which is invariant under the group of upper-triangular matrices, i.e., which satisfies

$$d(L'AL)=(\det L')\,d(A)\,(\det L)$$

for all upper-triangular nonsingular matrices L, has the form $d(A)=k\det A$.

(b) Show that the best estimate of $\det\Sigma$ which is invariant under the group of upper-triangular matrices is

$$d(A)=\prod_{i=1}^{m}(n-i+3)^{-1}\cdot\det A.$$

CHAPTER 5

Correlation Coefficients

5.1. ORDINARY CORRELATION COEFFICIENTS

5.1.1. Introduction

If the $m\times 1$ random vector $\mathbf{X}$ has covariance matrix $\Sigma=(\sigma_{ij})$ the correlation coefficient between two components of X, say, X_i and X_j, is

(1) $$\rho_{ij}=\frac{\sigma_{ij}}{\sqrt{\sigma_{ii}\sigma_{jj}}}=\frac{\operatorname{Cov}(X_i,X_j)}{\sqrt{\operatorname{Var}(X_i)\operatorname{Var}(X_j)}}.$$

The reader will recall that $|\rho_{ij}|\le 1$ and that $\rho_{ij}=\pm 1$ if and only if X_i and X_j are linearly related (with probability 1) so that ρ_{ij} is commonly regarded as a natural measure of *linear* dependence between X_i and X_j.

Now let $\mathbf{X}_1,\dots,\mathbf{X}_n$ be N independent observations on $\mathbf{X}$ and put

$$A=nS=\sum_{i=1}^{N}(\mathbf{X}_i-\overline{\mathbf{X}})(\mathbf{X}_i-\overline{\mathbf{X}})'$$

where $n=N-1$, so that S is the sample covariance matrix. The sample correlation coefficient between X_i and X_j is

(2) $$r_{ij}=\frac{a_{ij}}{\sqrt{a_{ii}a_{jj}}}=\frac{s_{ij}}{\sqrt{s_{ii}s_{jj}}}.$$

It is clear that if we are sampling from a multivariate normal distribution where all parameters are unknown then r_{ij} is the *maximum likelihood estimate* of ρ_{ij}. In this case $\rho_{ij}=0$ if and only if the variables X_i and X_j are *independent*. For other multivariate distributions $\rho_{ij}=0$ will not, in general, mean that X_i and X_j are independent although, of course, the converse is always true.

In the following subsections, exact and asymptotic distributions will be given for sample correlation coefficients, sometimes under fairly weak assumptions about the underlying distributions from which the sample is drawn. We will also indicate how these results can be used to test various hypotheses about population correlation coefficients.

5.1.2. *Joint and Marginal Distributions of Sample Correlation Coefficients in the Case of Independence*

In this section we will find the joint and marginal distributions of sample correlation coefficients formed from independent variables. First let us look at a *single* sample correlation coefficient; it is clear that in order to find its distribution we need only consider the distribution of those particular variables from which it is formed. Hence we consider N pairs of variables $(X_1,Y_1),\dots,(X_N,Y_N)$ and form the sample correlation coefficient

$$r=\frac{\Sigma_{i=1}^N(X_i-\overline{X})(Y_i-\overline{Y})}{\left[\Sigma_{i=1}^N(X_i-\overline{X})^2\Sigma_{i=1}^N(Y_i-\overline{Y})^2\right]^{1/2}} \tag{3}$$

where $\overline{X}=N^{-1}\Sigma_{i=1}^N X_i$ and $\overline{Y}=N^{-1}\Sigma_{i=1}^N Y_i$. The assumption that is commonly made is that the N 2×1 vectors

$$\begin{pmatrix} X_1 \\ Y_1 \end{pmatrix},\dots,\begin{pmatrix} X_N \\ Y_N \end{pmatrix}$$

are independent $N_2(\boldsymbol{\mu},\Sigma)$ random vectors, where

$$\Sigma=\begin{bmatrix} \sigma_{11} & \sigma_{12} \\ \sigma_{12} & \sigma_{22} \end{bmatrix}=\begin{bmatrix} \sigma_1^2 & \rho\sigma_1\sigma_2 \\ \rho\sigma_1\sigma_2 & \sigma_1^2 \end{bmatrix}$$

with $\rho=\sigma_{12}/(\sigma_{11}\sigma_{22})^{1/2}$. In this case the X's are independent of the Y's when $\rho=0$. If, *in general*, we assume that the X's are independent of the Y's, the normality assumption is not important as long as one set of these variables has a spherical distribution (see Section 1.5). This result, noted by Kariya and Eaton (1977), is given in the following theorem. In this theorem, $\mathbf{1}=(1,\dots,1)'\in R^N$ and $\{\mathbf{1}\}=\{k\mathbf{1};\,k\in R^1\}$, the *span* of $\mathbf{1}$.

THEOREM 5.1.1. Let $\mathbf{X}=(X_1,\ldots,X_N)'$ and $\mathbf{Y}=(Y_1,\ldots,Y_N)'$, with $N>2$, be two *independent* random vectors where $\mathbf{X}$ has an N-variate *spherical* distribution with $P(\mathbf{X}=\mathbf{0})=0$ and $\mathbf{Y}$ has *any* distribution with $P(\mathbf{Y}\in\{\mathbf{1}\})=0$. If r is the sample correlation coefficient given by (3) then

$$(N-2)^{1/2}\frac{r}{(1-r^2)^{1/2}}$$

has the t_{N-2} distribution.

Proof. Put $M=(1/N)\mathbf{1}\mathbf{1}'$; then r can be written as

$$r=\frac{\mathbf{X}'(I-M)\mathbf{Y}}{[\mathbf{X}'(I-M)\mathbf{X}\mathbf{Y}'(I-M)\mathbf{Y}]^{1/2}}.$$

Since $I-M$ is idempotent of rank $N-1$ there exists $H\in O(N)$ such that

$$H(I-M)H'=\begin{bmatrix} I_{N-1} & \mathbf{0} \\ \mathbf{0}' & 0 \end{bmatrix}.$$

Put $\mathbf{U}=H\mathbf{X}$ and $\mathbf{V}=H\mathbf{Y}$ and partition $\mathbf{U}$ and $\mathbf{V}$ as

$$\mathbf{U}=\begin{pmatrix} \mathbf{U}^* \\ \cdots \\ U_N \end{pmatrix}, \qquad \mathbf{V}=\begin{pmatrix} \mathbf{V}^* \\ \cdots \\ V_n \end{pmatrix},$$

where $\mathbf{U}^*$ and $\mathbf{V}^*$ are $(N-1)\times 1$. Then

$$(4) \qquad r=\frac{\mathbf{U}'H(I-M)H'\mathbf{V}}{[\mathbf{U}'H(I-M)H'\mathbf{U}\mathbf{V}'H(I-M)H'\mathbf{V}]^{1/2}}$$

$$=\frac{\mathbf{U}^{*\prime}\mathbf{V}^*}{[\mathbf{U}^{*\prime}\mathbf{U}^*\mathbf{V}^{*\prime}\mathbf{V}^*]^{1/2}}$$

$$=\frac{\mathbf{U}^{*\prime}\mathbf{V}^*}{\|\mathbf{U}^*\|\,\|\mathbf{V}^*\|}.$$

Note that $\mathbf{U}^*$ has an $(N-1)$-variate spherical distribution and is independent of $\mathbf{V}^*$. Conditioning on $\mathbf{V}^*$, part (i) of Theorem 1.5.7 with $\alpha=\|\mathbf{V}^*\|^{-1}\mathbf{V}^*$ then shows that $(N-2)^{1/2}r/(1-r^2)^{1/2}$ has the t_{N-2} distribution, and the proof is complete.

It is easy to see that $r=\cos\theta$, where θ is the angle between the two normalized vectors $\|\mathbf{U}^*\|^{-1}\mathbf{U}^*$ and $\|\mathbf{V}^*\|^{-1}\mathbf{V}^*$ in the proof of Theorem 5.1.1. Because $\mathbf{U}^*$ has a spherical distribution, $\|\mathbf{U}^*\|^{-1}\mathbf{U}^*$ has a uniform distribution over the unit sphere in R^{N-1} (see Theorem 1.5.6), and it is clear that in order to find the distribution of $\cos\theta$ we can regard $\|\mathbf{V}^*\|^{-1}\mathbf{V}^*$ as a fixed point on this sphere.

As noted previously, it is usually assumed that the X's and Y's are normal. This is a special case of Theorem 5.1.1, given explicitly in the following corollary.

COROLLARY 5.1.2. Let

$$\begin{pmatrix} X_1 \\ Y_1 \end{pmatrix},\dots,\begin{pmatrix} X_N \\ Y_N \end{pmatrix}$$

be independent $N_2(\boldsymbol{\mu},\Sigma)$ random vectors, where

$$\Sigma=\begin{bmatrix} \sigma_1^2 & \rho\sigma_1\sigma_2 \\ \rho\sigma_1\sigma_2 & \sigma_2^2 \end{bmatrix},$$

and let r be the sample correlation coefficient given by (3). Then, when $\rho=0$

$$(N-2)^{1/2}\frac{r}{(1-r^2)^{1/2}} \quad \text{is} \quad t_{N-2}.$$

Proof. Since the correlation between the standardized variables $(X_i-\mu_1)/\sigma_1$ and $(Y_i-\mu_2)/\sigma_2$ is the same as the correlation between X_i and Y_i we can assume without loss of generality that $\boldsymbol{\mu}=\mathbf{0}$ and $\Sigma=I_2$ (when $\rho=0$).

Then $X_1,\dots,X_N$ are independent $N(0,1)$ random variables and so $\mathbf{X}=(X_1,\dots,X_N)'$ certainly has a spherical distribution and is independent of $\mathbf{Y}=(Y_1,\dots,Y_N)'$ by assumption. The conditions of Theorem 5.1.1 are satisfied and the desired result follows immediately.

Suppose that the conditions of Theorem 5.1.1 are satisfied, so that $(n-1)^{1/2}r/(1-r^2)^{1/2}$ is t_{n-1}, where $n=N-1$. Starting with the density function of the t_{n-1} distribution the density function of r can be easily obtained as

$$(5) \qquad \frac{\Gamma(\frac{1}{2}n)}{\pi^{1/2}\Gamma[\frac{1}{2}(n-1)]}(1-r^2)^{(n-3)/2} \qquad (-1<r<1).$$

Equivalently, r^2 has the beta distribution with parameters $\frac{1}{2}$ and $\frac{1}{2}(n-1)$.

The density function (5) is symmetric about zero so that all odd moments are zero. The reader can easily check that the even moments are

$$E(r^{2k})=\frac{\Gamma(\tfrac{1}{2}n)\Gamma(k+\tfrac{1}{2})}{\pi^{1/2}\Gamma(\tfrac{1}{2}n+k)}, \tag{6}$$

so that $\text{Var}(r)=E(r^2)=n^{-1}$. In fact, if r is the sample correlation coefficient formed from two sets of independent variables then $E(r)=0$ and $\text{Var}(r)=n^{-1}$ under much more general conditions than those of Theorem 5.1.1, a result noted by Pitman (1937).

Let us now turn to the problem of finding the joint distribution of a *set* of correlation coefficients.

THEOREM 5.1.3. Let X be an $N\times m$ random matrix

$$X=(X_{ij})=\begin{bmatrix}\mathbf{X}_1'\\ \vdots\\ \mathbf{X}_N'\end{bmatrix}=[\mathbf{Y}_1 \quad \dots \quad \mathbf{Y}_m]$$

(so that the $\mathbf{X}_i'$ are the rows of X and the $\mathbf{Y}_i$ are the columns of X) and let $R=(r_{ij})$ be the $m\times m$ *sample correlation matrix* where

$$r_{ij}=\frac{\sum_{k=1}^N (X_{ik}-\bar{X}_i)(X_{jk}-\bar{X}_j)}{\left[\sum_{k=1}^N (X_{ik}-\bar{X}_i)^2\sum_{k=1}^N (X_{jk}-\bar{X}_j)^2\right]^{1/2}}$$

with $\bar{X}_i=N^{-1}\sum_{k=1}^N X_{ik}$. Suppose that $\mathbf{Y}_1,\dots,\mathbf{Y}_m$ are all *independent* random vectors where $\mathbf{Y}_i$ has an N-variate spherical distribution with $P(\mathbf{Y}_i=\mathbf{0})=0$ for $i=1,\dots,m$. (These spherical distributions need not be the same.) Then the density function of R (i.e., the joint density function of the $r_{ij}, i<j$) is

$$\frac{[\Gamma(\tfrac{1}{2}n)]^m}{\Gamma_m(\tfrac{1}{2}n)}(\det R)^{(n-m-1)/2} \qquad (-1<r_{ij}<1, i<j), \tag{7}$$

where $n=N-1$.

Proof. As in the proof of Theorem 5.1.1 we can write $r_{ij}=\hat{\mathbf{U}}_i'\hat{\mathbf{V}}_j$, where $\hat{\mathbf{U}}_i$ and $\hat{\mathbf{V}}_j$ are uniformly distributed over the unit sphere in R^n [see (4)]. This being the case we can assume that $\mathbf{Y}_1,\ldots,\mathbf{Y}_m$ are all independent $N_N(\mathbf{0},\mathbf{I}_N)$ random vectors since this leads to the same result. Thus $\mathbf{X}_1,\ldots,\mathbf{X}_N$ are independent $N_m(\mathbf{0},\mathbf{I}_m)$ random vectors so that the matrix

$$A=\sum_{i=1}^{N}(\mathbf{X}_i-\overline{\mathbf{X}})(\mathbf{X}_i-\overline{\mathbf{X}})'=(a_{ij}),$$

with $\overline{\mathbf{X}}=N^{-1}\sum_{i=1}^{N}\mathbf{X}_i$, is $W_m(n,I_m)$ and $r_{ij}=a_{ij}/(a_{ii}a_{jj})^{1/2}$. The density of A is then

$$\frac{\exp\left(-\frac{1}{2}\sum_{i=1}^{m}a_{ii}\right)\det A^{(n-m-1)/2}}{2^{mn/2}\Gamma_m\left(\frac{1}{2}n\right)}(dA)$$

Now make the change of variables

$$r_{ij}=\frac{a_{ij}}{(a_{ii}a_{jj})^{1/2}}\qquad (i<j),$$

$$t_i=a_{ii}\qquad (i=1,\ldots,m);$$

then $da_{ii}=dt_i$ and

$$da_{ij}=(t_it_j)^{1/2}\,dr_{ij}+\text{terms in } dt_i$$

so that

$$(dA)=\bigwedge_{i\le j}^{m}da_{ij}=\bigwedge_{i<j}^{m}(t_it_j)^{1/2}\,dr_{ij}\bigwedge_{i=1}^{m}dt_i$$

$$=\prod_{i=1}^{m}t_i^{(m-1)/2}\bigwedge_{i<j}^{m}dr_{ij}\bigwedge_{i=1}^{m}dt_i,$$

that is, the Jacobian is $\prod_{i=1}^{m}t_i^{(m-1)/2}$. The joint density function of the r_{ij} and t_i is, then,

$$(8)\qquad \frac{\exp\left(-\frac{1}{2}\sum_{i=1}^{m}t_i\right)\det\left(r_{ij}(t_it_j)^{1/2}\right)^{(n-m-1)/2}}{2^{mn/2}\Gamma_m\left(\frac{1}{2}n\right)}\prod_{i=1}^{m}t_i^{(m-1)/2}.$$

Now note that

$$\det\left(r_{ij}(t_it_j)^{1/2}\right)=\det\left(\begin{bmatrix} t_1^{1/2} & & 0 \\ & \ddots & \\ 0 & & t_m^{1/2}\end{bmatrix}\cdot R\begin{bmatrix} t_1^{1/2} & & 0 \\ & \ddots & \\ 0 & & t_m^{1/2}\end{bmatrix}\right)$$

$$=\left(\prod_{i=1}^{m} t_i\right)(\det R).$$

Substituting in (8) gives the joint density of the r_{ij} and the t_i as

$$\frac{\det R^{(n-m-1)/2}}{\Gamma_m(\frac{1}{2}n)}\prod_{i=1}^{m}\left[\frac{e^{-t_i/2}t_i^{n/2-1}}{2^{n/2}}\right].$$

Integrating with respect to $t_1,\dots,t_m$ using

$$2^{-n/2}\int_0^\infty e^{-t_i/2}t_i^{n/2-1}\,dt_i=\Gamma(\tfrac{1}{2}n)$$

gives the desired marginal density function of the sample correlation matrix, completing the proof.

The assumption commonly made is that the rows of the matrix X in Theorem 5.1.3 are independent $N_m(\mu,\Sigma)$ random vectors, where Σ is diagonal. This is a special case of Theorem 5.1.3 and follows in much the same way that Corollary 5.1.2 follows from Theorem 5.1.1.

Suppose that the conditions of Theorem 5.1.3 are satisfied, so that R has density function (7). From this we can easily find the moments of $\det R$, sometimes called the *scatter coefficient*. We have

$$(9)\qquad E\left[(\det R)^k\right]=\frac{\left[\Gamma(\frac{1}{2}n)\right]^m}{\Gamma_m(\frac{1}{2}n)}\int_{\substack{-1<r_{ij}<1\\ i<j}}\cdots\int \det R^{(n+2k-m-1)/2}(dR)$$

$$=\left[\frac{\Gamma(\frac{1}{2}n)}{\Gamma(\frac{1}{2}n+k)}\right]^m\frac{\Gamma_m(\frac{1}{2}n+k)}{\Gamma_m(\frac{1}{2}n)},$$

on adjusting the integrand so that it is the density function (7) with n

replaced by $n+2k$, and hence integrates to 1. In particular

$$E(\det R)=\prod_{i=1}^{m}\left(1-\frac{i-1}{n}\right)$$

and

$$\operatorname{Var}(\det R)=\prod_{i=1}^{m}\left(1-\frac{i-1}{n}\right)\left[\prod_{j=1}^{m}\left(1-\frac{j-1}{n+2}\right)-\prod_{j=1}^{m}\left(1-\frac{j-1}{n}\right)\right].$$

From the moments follows the characteristic function of $\log\det R$, and this can be used to show that the limiting distribution, as $n\to\infty$, of $-n\log\det R$ is $\chi^2_{m(m-1)/2}$ (see Problem 5.1).

5.1.3. The Non-null Distribution of a Sample Correlation Coefficient in the Case of Normality

In this section we will derive the distribution of the sample correlation coefficient r formed from a sample from a *bivariate normal* distribution with population correlation coefficient ρ. The distribution will be expressed in terms of a ${}_2F_1$ hypergeometric function (see Definition 1.3.1). We will make use of the following lemma, which gives an integral representation for this function.

LEMMA 5.1.4.

$$\frac{\Gamma(c)}{\Gamma(a)\Gamma(c-a)}\int_0^1 t^{a-1}(1-t)^{c-a-1}(1-tz)^{-b}\,dt={}_2F_1(a,b;c;z)$$

$$\equiv\sum_{k=0}^{\infty}\frac{(a)_k(b)_k}{(c)_k k!}z^k$$

for $\operatorname{Re}(c)>\operatorname{Re}(a)>0$ and $|z|<1$.

To prove this, expand $(1-tz)^{-b}$ in a binomial series and integrate term by term. The details are left as an exercise (see Problem 5.2).

The following theorem gives an expression for the non-null density function of r.

THEOREM 5.1.5. If r is the correlation coefficient formed from a sample of size $N=n+1$ from a bivariate normal distribution with correlation coefficient ρ then the density function of r is

$$(10)\quad \frac{\Gamma(n)(n-1)}{\Gamma(n+\frac{1}{2})(2\pi)^{1/2}}(1-\rho^2)^{n/2}(1-\rho r)^{-n+1/2}$$

$$\cdot(1-r^2)^{(n-3)/2}{}_2F_1\left(\tfrac{1}{2},\tfrac{1}{2};n+\tfrac{1}{2};\tfrac{1}{2}(1+\rho r)\right) \qquad (-1<r<1).$$

Proof. Let the sample be $\mathbf{X}_1,\dots,\mathbf{X}_N$ so that each of these vectors are independent and have the $N_2(\boldsymbol{\mu},\Sigma)$ distribution. Since we are only interested in the correlation between the components we can assume without loss of generality that

$$\Sigma=\begin{bmatrix}1 & \rho\\ \rho & 1\end{bmatrix}.$$

Put $A=\sum_{i=1}^N(\mathbf{X}_i-\overline{\mathbf{X}})(\mathbf{X}_i-\overline{\mathbf{X}})'$, then A is $W_2(n,\Sigma)$, where $n=N-1$, and the sample correlation coefficient is $r=a_{12}/(a_{11}a_{22})^{1/2}$. The density function of A (i.e., the joint density function of a_{11}, a_{12}, and a_{22}) is

$$\frac{\operatorname{etr}(-\frac{1}{2}\Sigma^{-1}A)(\det A)^{(n-3)/2}}{2^n(\det\Sigma)^{n/2}\Gamma_2(\frac{1}{2}n)}.$$

Now

$$\Sigma^{-1}=\frac{1}{1-\rho^2}\begin{bmatrix}1 & -\rho\\ -\rho & 1\end{bmatrix}$$

so that

$$\delta\equiv\tfrac{1}{2}\operatorname{tr}\Sigma^{-1}A=\frac{1}{2(1-\rho^2)}(a_{11}+a_{22}-2\rho a_{12}),$$

and hence the joint density function of a_{11}, a_{12}, and a_{22} is

$$(11)\quad \frac{e^{-\delta}(a_{11}a_{22}-a_{12}^2)^{(n-3)/2}}{2^n(1-\rho^2)^{n/2}\Gamma_2(\frac{1}{2}n)}.$$

Now make the change of variables

$$a_{11}=se^{-t},\qquad a_{22}=se^{t},\qquad a_{12}=rs$$

(so that $r=a_{12}/(a_{11}a_{22})^{1/2}$), then

$$da_{11}\wedge da_{12}\wedge da_{22}=2s^2ds\wedge dt\wedge dr$$

(i.e., the Jacobian is $2s^2$); the joint density function of r, s, and t is then

$$(12)\qquad \frac{e^{-\delta}s^{n-1}(1-r^2)^{(n-3)/2}}{2^{n-1}(1-\rho^2)^{n/2}\Gamma_2(\frac{1}{2}n)}\qquad (s>0,\quad |r|<1,\quad -\infty<t<\infty),$$

where now

$$\delta=\frac{s}{2(1-\rho^2)}(e^{-t}+e^{t}-2\rho r)$$
$$=\frac{s}{1-\rho^2}(\cosh t-\rho r).$$

Integrating (12) with respect to s from 0 to ∞ using

$$\int_0^\infty \exp\left[-\frac{s}{1-\rho^2}(\cosh t-\rho r)\right]s^{n-1}ds=\frac{\Gamma(n)(1-\rho^2)^n}{(\cosh t-\rho r)^n}$$

gives the joint density function of r and t as

$$\frac{\Gamma(n)(1-\rho^2)^{n/2}(1-r^2)^{(n-3)/2}}{2^{n-1}\Gamma_2(\frac{1}{2}n)(\cosh t-\rho r)^n}\qquad (|r|<1;\quad -\infty<t<\infty).$$

We must now integrate with respect to t from $-\infty$ to ∞ to get the marginal density function of r. Note that

$$I\equiv\int_{-\infty}^{\infty}(\cosh t-\rho r)^{-n}dt$$
$$=2\int_0^\infty(\cosh t-\rho r)^{-n}dt$$
$$=2^{1/2}(1-\rho r)^{-n+1/2}\int_0^1(1-u)^{n-1}u^{-1/2}\left[1-\frac{u}{2}(1+\rho r)\right]^{-1/2}du,$$

on making the change of variables

$$\cosh t = \frac{1-\rho r u}{1-u}.$$

Using Lemma 5.1.4 we then see that

$$I = 2^{1/2}(1-\rho r)^{-n+1/2}\frac{\Gamma(n)\Gamma(\frac{1}{2})}{\Gamma(n+\frac{1}{2})}\,{}_2F_1\left(\tfrac{1}{2},\tfrac{1}{2};n+\tfrac{1}{2};\tfrac{1}{2}(1+\rho r)\right)$$

Hence the density function of r is

$$(13)\qquad \frac{\Gamma(n)^2\pi^{1/2}}{2^{n-3/2}\Gamma_2(\frac{1}{2}n)\Gamma(n+\frac{1}{2})}(1-\rho^2)^{n/2}(1-\rho r)^{-n+1/2}(1-r^2)^{(n-3)/2}$$
$$\cdot{}_2F_1\left(\tfrac{1}{2},\tfrac{1}{2};n+\tfrac{1}{2};\tfrac{1}{2}(1+\rho r)\right)\qquad (|r|<1).$$

Using Legendre's duplication formula

$$(14)\qquad \Gamma(2z) = 2^{2z-1}\pi^{-1/2}\Gamma(z)\Gamma(z+\tfrac{1}{2})$$

[see, for example, Erdélyi et al. (1953a), Section 1.2] the constant in the density function (13) can be written

$$\frac{\Gamma(n)^2\pi^{1/2}}{2^{n-3/2}\Gamma_2(\frac{1}{2}n)\Gamma(n+\frac{1}{2})} = \frac{\Gamma(n)^2}{2^{n-3/2}\Gamma(\frac{1}{2}n)\Gamma[\frac{1}{2}(n-1)]\Gamma(n+\frac{1}{2})}$$
$$= \frac{\Gamma(n)^2}{2^{1/2}\pi^{1/2}\Gamma(n-1)\Gamma(n+\frac{1}{2})}$$
$$= \frac{(n-1)\Gamma(n)}{(2\pi)^{1/2}\Gamma(n+\frac{1}{2})},$$

and the proof is complete.

The density function of r can be expressed in many forms; the form (10), which converges rapidly even for small n, is due to Hotelling (1953). Other expressions had been found earlier by Fisher (1915). One of these is

$$(15)\qquad \frac{2^{n-2}(1-\rho^2)^{n/2}(1-r^2)^{(n-3)/2}}{\pi\Gamma(n-1)}\sum_{k=0}^{\infty}\left[\Gamma\left(\frac{n+k}{2}\right)\right]^2\frac{(2r\rho)^k}{k!},$$

which can be obtained from (11) by changing variables to $r = a_{12}/(a_{11}a_{22})^{1/2}$, $u = a_{11}$, $v = a_{22}$, expanding $\exp[\rho r(uv)^{1/2}/(1-\rho^2)]$ (which is part of the $\exp(-\delta)$ term) and integrating term by term with respect to u and v (see Problem 5.3). The form (15) for the density function of r is probably the easiest one to use in an attack on the moments of r. To derive these, it helps if one acquires a taste for carrying out manipulations with hypergeometric functions. For example, the mean of r is, using (15),

$$(16)\qquad E(r) = \frac{2^{n-2}(1-\rho^2)^{n/2}}{\pi\Gamma(n-1)} \sum_{k=0}^{\infty} \frac{\{\Gamma[(n+k)/2]\}^2 (2\rho)^k}{k!} \cdot \int_{-1}^{1} (1-r^2)^{(n-3)/2} r^{k+1}\, dr.$$

This last integral is zero unless k is odd so, putting $k = 2j+1$, we have

$$\int_{-1}^{1} (1-r^2)^{(n-3)/2} r^{2j+2}\, dr = \frac{\Gamma(3/2+j)\Gamma[\frac{1}{2}(n-1)]}{\Gamma(\frac{1}{2}n + j + 1)}.$$

Substituting back in (16) gives

$$E(r) = \frac{2^{n-2}(1-\rho^2)^{n/2}}{\pi\Gamma(n-1)} \Gamma[\tfrac{1}{2}(n-1)] \sum_{j=0}^{\infty} \frac{(2\rho)^{2j+1}}{(2j+1)!} \frac{[\Gamma(n/2 + j + \frac{1}{2})]^2}{\Gamma(n/2 + j + 1)} \cdot \Gamma\left(j + \frac{3}{2}\right).$$

On using

$$\frac{\Gamma(a+j)}{\Gamma(a)} = (a)_j \equiv a(a+1)\dots(a+j-1),$$

$$\frac{\Gamma(j+3/2)2^{2j+1}}{(2j+1)!} = \frac{\pi^{1/2}}{j!}$$

and the duplication formula (14) we get

$$E(r) = \frac{2}{n}\left\{\frac{\Gamma[(n+1)/2]}{\Gamma(\frac{1}{2}n)}\right\}^2 \rho(1-\rho^2)^{n/2} {}_2F_1(\tfrac{1}{2}(n+1), \tfrac{1}{2}(n+1); \tfrac{1}{2}n + 1; \rho^2)$$

This can be simplified a little more using the Euler relation

$$(17)\qquad {}_2F_1(a,b:c;z)=(1-z)^{c-a-b}{}_2F_1(c-a,c-b;c;z),$$

[see, for example, Erdélyi et al. (1953a), Section 2.9, or Rainville (1960), Section 38]; we then get

$$(18)\qquad E(r)=\frac{2}{n}\left[\frac{\Gamma[(n+1)/2]}{\Gamma(\frac{1}{2}n)}\right]^2\rho\,{}_2F_1(\tfrac{1}{2},\tfrac{1}{2};\tfrac{1}{2}n+1;\rho^2).$$

In a similar way the second moment can be obtained; we have

$$(19)\qquad E(1-r^2)=\frac{2^{n-2}(1-\rho^2)^{n/2}}{\pi\Gamma(n-1)}\sum_{k=0}^{\infty}\frac{\{\Gamma[(n+k)/2]\}^2(2\rho)^k}{k!}\cdot\int_{-1}^{1}r^k(1-r^2)^{(n-1)/2}\,dr.$$

This integral is zero unless k is even; putting $k=2j$ we have

$$\int_{-1}^{1}r^{2j}(1-r^2)^{(n-1)/2}\,dr=\frac{\Gamma(j+\frac{1}{2})\Gamma[\frac{1}{2}(n+1)]}{\Gamma(\frac{1}{2}n+j+1)}.$$

Substituting back in (19) and using

$$\frac{2^{2j}\Gamma(j+\frac{1}{2})}{\pi^{1/2}(2j)!}=\frac{1}{j!},$$

$$\frac{\Gamma(\frac{1}{2}n+j)}{\Gamma(\frac{1}{2}n+j+1)}=\frac{1}{\frac{1}{2}n+j}=\frac{2}{n}\frac{(\frac{1}{2}n)_j}{(\frac{1}{2}n+1)_j},$$

the duplication formula (14) and Euler's relation (17), we then find that

$$(20)\qquad E(r^2)=1-\frac{n-1}{n}(1-\rho^2)\,{}_2F_1(1,1;\tfrac{1}{2}n+1;\rho^2).$$

These moments, and others, have been given by Ghosh (1966). Expanding (18) and (20) in terms of powers of n^{-1} it is easily shown that

$$(21)\qquad E(r)=\rho-\frac{\rho(1-\rho^2)}{2n}+O(n^{-2})$$

and

$$\mathrm{Var}(r) = \frac{(1-\rho^2)^2}{n} + O(n^{-2}). \tag{22}$$

It is seen from (18) that r is a biased estimate of ρ. Olkin and Pratt (1958) have shown that an unbiased estimate of ρ is

$$T(r) = r\,{}_2F_1\left(\tfrac{1}{2}, \tfrac{1}{2}; \tfrac{1}{2}(n-1); 1-r^2\right), \tag{23}$$

which may be expanded as

$$T(r) = r + \frac{r(1-r^2)}{n-1} + O(n^{-2})$$

and hence differs from r only by terms of order n^{-1}. Since it is a function of a complete sufficient statistic, $T(r)$ is the unique minimum variance unbiased estimate of ρ.

5.1.4. *Asymptotic Distribution of a Sample Correlation Coefficient from an Elliptical Distribution*

Here we will derive the asymptotic distribution of a correlation coefficient as the sample size tends to infinity. Since it turns out to be not very different from the situation where the underlying distribution is normal, we will assume that we are sampling from a bivariate *elliptical* distribution. Thus, suppose that $S(n) = (s_{ij}(n))$ is the 2×2 covariance matrix formed from a sample of size $N = n+1$ from a bivariate elliptical distribution with *covariance matrix*

$$\Sigma = \begin{bmatrix} 1 & \rho \\ \rho & 1 \end{bmatrix}$$

and finite fourth moments. It has previously been noted that, as $n \to \infty$, the asymptotic joint distribution of the elements of $n^{1/2}[S(n) - \Sigma]$ is normal and that the asymptotic covariances are functions of the fourth order cumulants (see Corollary 1.2.18 and the discussion at the end of Section 1.6). We have also noted that, for elliptical distributions, all fourth-order cumulants are functions of the elements of Σ and a *kurtosis parameter* κ [see (1) and (2) of Section 1.6].

Put

$$U=\begin{bmatrix} u_{11} & u_{12} \\ u_{12} & u_{22} \end{bmatrix}=n^{1/2}[S(n)-\Sigma];$$

it then follows, using (2) and (3) of Section 1.6, that the asymptotic distribution of $\mathbf{u}\equiv(u_{11},u_{12},u_{22})'$ is *normal* with mean $\mathbf{0}$ and covariance matrix

$$V=\begin{bmatrix} 2+3\kappa & (2+3\kappa)\rho & 2\rho^2+\kappa(1+2\rho^2) \\ (2+3\kappa)\rho & \kappa(1+2\rho^2)+(1+\rho^2) & (2+3\kappa)\rho \\ 2\rho^2+\kappa(1+2\rho^2) & (2+3\kappa)\rho & 2+3\kappa \end{bmatrix}. \tag{24}$$

Now, in terms of the elements of U, the sample correlation coefficient $r(n)$ can be expanded as

$$\begin{aligned} r(n)&=s_{12}(n)(s_{11}(n)s_{22}(n))^{-1/2} \\ &=(\rho+n^{-1/2}u_{12})(1+n^{-1/2}u_{11})^{-1/2}(1+n^{-1/2}u_{22})^{-1/2} \\ &=(\rho+n^{-1/2}u_{12})(1-\tfrac{1}{2}n^{-1/2}u_{11}+O_p(n^{-1})) \\ &\quad\cdot(1-\tfrac{1}{2}n^{-1/2}u_{22}+O_p(n^{-1})) \\ &=\rho+n^{-1/2}(u_{12}-\tfrac{1}{2}\rho u_{11}-\tfrac{1}{2}\rho u_{22})+O_p(n^{-1}). \end{aligned}$$

[For the reader who is unfamiliar with the O_p notation, a useful reference is Bishop et al. (1975), Chapter 14.] It follows from this that

$$n^{1/2}[r(n)-\rho]=u_{12}-\tfrac{1}{2}\rho u_{11}-\tfrac{1}{2}\rho u_{22}+O_p(n^{-1/2})$$

and hence the asymptotic distribution of $n^{1/2}(r(n)-\rho)$ is the same as that of $u_{12}-\frac{1}{2}\rho u_{11}-\frac{1}{2}\rho u_{22}$. With $\alpha=(-\frac{1}{2}\rho\quad 1\quad -\frac{1}{2}\rho)'$, the asymptotic distribution of

$$\alpha'\mathbf{u}=u_{12}-\tfrac{1}{2}\rho u_{11}-\tfrac{1}{2}\rho u_{22}$$

is normal with mean zero and variance $\alpha' V\alpha$, which is easily verified to be equal to $(1+\kappa)(1-\rho^2)^2$.

Summarizing, we have the following theorem.

THEOREM 5.1.6. Let $r(n)$ be the correlation coefficient formed from a sample of size $n+1$ from a bivariate elliptical distribution with correlation coefficient ρ and kurtosis parameter κ. Then the asymptotic distribution, as $n\rightarrow\infty$, of

$$n^{1/2}\frac{r(n)-\rho}{1-\rho^2}$$

is $N(0,1+\kappa)$.

When the elliptical distribution in Theorem 5.1.6 is *normal*, the kurtosis parameter κ is zero and the limiting distribution of $n^{1/2}[r(n)-\rho]/(1-\rho^2)$ is $N(0,1)$. In this situation Fisher (1921) suggested the statistic

$$z=\tanh^{-1}r=\tfrac{1}{2}\log\frac{1+r}{1-r} \tag{25}$$

(known as Fisher's z transformation), since this approaches normality much faster than r, with an asymptotic variance which is independent of ρ. In this connection a useful reference is Hotelling (1953). For elliptical distributions a similar result holds and is given in the following theorem.

THEOREM 5.1.7. Let $r(n)$ be the correlation coefficient formed from a sample of size $n+1$ from a bivariate elliptical distribution with correlation coefficient ρ and kurtosis parameter κ and put

$$z(n)=\tanh^{-1}r(n)=\tfrac{1}{2}\log\frac{1+r(n)}{1-r(n)}$$

and

$$\xi=\tanh^{-1}\rho=\tfrac{1}{2}\log\frac{1+\rho}{1-\rho}. \tag{26}$$

Then, as $n\rightarrow\infty$, the asymptotic distribution of

$$n^{1/2}(z(n)-\xi)$$

is $N(0,1+\kappa)$.

This theorem follows directly from the asymptotic normality of $r(n)$ established in Theorem 5.1.6; the details are left as an exercise (see Problem 5.4).

Again, when the elliptical distribution here is normal we have $\kappa=0$, and the limiting distribution of $n^{1/2}[z(n)-\xi]$ is $N(0,1)$. In this particular case, z

is the *maximum likelihood estimate* of ξ. For general non-normal distributions Gayen (1951) has obtained expressions for the mean, variance, skewness, and kurtosis of z. These have been used by Devlin et al. (1976) to study Fisher's z transformation for some specific elliptical distributions. They state that "the main effect of the elliptically constrained departures from normality appears to be to increase the variabilty of z" and conclude that the distribution of z can be approximated quite well in many situations, even for small sample sizes, by taking z to be *normal* with mean $E(z)=\xi$ and variance

$$\text{Var}(z)=\frac{1}{n-2}+\frac{\kappa}{n+2} \tag{27}$$

($n=N-1$). (It should be noted that the kurtosis parameter ϕ^2 used by Devlin et al. is equal to $1+\kappa$ in our notation.)

5.1.5. *Testing Hypotheses about Population Correlation Coefficients*

The results of the preceding sections can be used in fairly obvious ways to test hypotheses about correlation coefficients and to construct confidence intervals. First, suppose that we have a sample of size $N=n+1$ from a bivariate *normal* distribution with correlation coefficient ρ and we wish to test the null hypothesis H_0: $\rho=0$ (that is, the two variables are uncorrelated and hence independent) against general alternatives H: $\rho\neq 0$. It is clear that this problem is equivalent to that of testing whether two specified variables are uncorrelated in an m-variate normal distribution. An *exact* test can be constructed using the results of Section 5.1.2. We know from Theorem 5.1.1 that, when H_0 is true, $(n-1)^{1/2}r/(1-r^2)^{1/2}$ has the t_{n-1} distribution so that a test of size α is to reject H_0 if

$$(n-1)^{1/2}\frac{|r|}{(1-r^2)^{1/2}}>t^*_{n-1}(\alpha), \tag{28}$$

where $t^*_{n-1}(\alpha)$ denotes the two-tailed $100\alpha\%$ point of the t_{n-1} distribution. This test is, in fact, the *likelihood ratio test* of the null hypothesis H_0 (Problem 5.5). The power function of this test is a function of ρ, namely,

$$\begin{aligned}\beta(\rho)&=P_\rho\left[\frac{(n-1)^{1/2}|r|}{(1-r^2)^{1/2}}>t^*_{n-1}(\alpha)\right]\\&=P_\rho(|r|>r^*),\end{aligned}$$

where

$$r^* = \frac{t^*_{n-1}(\alpha)}{\left[n-1+t^{*2}_{n-1}(\alpha)\right]^{1/2}}.$$

Expressions for the density function of r when $\rho \neq 0$ were given in Section 5.1.3. From these, expressions for the distribution function

$$F(x; n, \rho) = P_\rho(r \le x)$$

of r can be obtained. Tables of this function have been prepared by David (1938) for a wide range of values of x, ρ, and n. In terms of the distribution function the power is

$$\beta(\rho) = 1 - F(r^*; n, \rho) + F(-r^*; n, \rho).$$

Now consider testing the null hypothesis $H: \rho = \rho_0$ against *one-sided* alternatives $K: \rho > \rho_0$. A test of size α is to reject H if $r > k_\alpha$, where k_α is chosen so that

$$P_{\rho_0}(r > k_\alpha) = 1 - F(k_\alpha; n, \rho_0) = \alpha.$$

This test has the optimality property stated in the following theorem due to T. W. Anderson (1958).

THEOREM 5.1.8. In the class of tests of $H: \rho \le \rho_0$ against $K: \rho > \rho_0$ that are based on r, the test which rejects H if $r > k_\alpha$ is uniformly most powerful.

Proof. Because we are restricting attention to tests based on r we can assume that a value of r is observed from the distribution with density function specified in Theorem 5.1.5, namely,

(29)

$$f(r; n, \rho) = \frac{\Gamma(n)(n-1)}{\Gamma(n+\frac{1}{2})(2\pi)^{1/2}}(1-\rho^2)^{n/2}(1-\rho r)^{-n+1/2}(1-r^2)^{(n-3)/2}$$

$$\cdot {}_2F_1\left(\tfrac{1}{2}, \tfrac{1}{2}; n+\tfrac{1}{2}; \tfrac{1}{2}(1+\rho r)\right).$$

The desired conclusion will follow if we can show that the density function $f(r; n, \rho)$ has monotone likelihood ratio; that is, if $\rho > \rho'$ then $f(r; n, \rho)/f(r; n, \rho')$ is increasing in r [see, for example, Lehmann (1959),

Section 3.3, or Roussas (1973), Section 13.3]. To this end, it suffices to show

$$\frac{\partial^2 \log f(r; n, \rho)}{\partial\rho\,\partial r} \geq 0$$

for all ρ and r [see Lehmann (1959), page 111]. Writing the series expansion for the ${}_2F_1$ function in (29) as

$${}_2F_1\left(\tfrac{1}{2}, \tfrac{1}{2}; n+\tfrac{1}{2}; \tfrac{1}{2}(1+\rho r)\right) = \sum_{i=0}^{\infty} \delta_i z^i,$$

where

$$\delta_i = \left(\tfrac{1}{2}\right)_i \left(\tfrac{1}{2}\right)_i \big/ \left(n+\tfrac{1}{2}\right)_i i!\,2^i$$

and $z = 1 + \rho r$, it is reasonably straightforward to show that

$$\frac{\partial^2 \log f(r; n, \rho)}{\partial\rho\,\partial r} = \left(n-\tfrac{1}{2}\right)(1-\rho r)^{-2} + \frac{g(z)}{2\left(\sum_{i=0}^{\infty}\delta_i z^i\right)^2}$$

where

$$g(z) = \sum_{i,j=0}^{\infty} \delta_i \delta_j z^{i+j-2}\left[(j-i)^2(z-1)+i+j\right]$$

We now claim that $g(z) > 0$ for all $z > 0$. To see this note that

$$\begin{aligned} g(z) &= 2\sum_{i \leq j} \delta_i \delta_j z^{i+j-2}\left[(j-i)^2(z-1)+(i+j)\right] \\ &> 2\sum_{i<j} \delta_i \delta_j z^{i+j-2}\left[(j-i)^2(z-1)+(i+j)\right] \\ &= 2\sum_{i=0}^{\infty} \delta_i z^{i-2} \sum_{j=i+1}^{\infty} \delta_j z^j\left[(j-i)^2(z-1)+(i+j)\right]. \end{aligned}$$

Holding i fixed, the coefficient of z^j in the inner sum is

$$\delta_j\left[-(j-i)^2+(i+j)\right]+\delta_{j-1}(j-i-1)^2$$

for $j \geq i+1$. That this is non-negative now follows if we use the fact (easily proved) that

$$\delta_{j-1} > 2\delta_j,$$

and the proof is complete.

The test described by Theorem 5.1.8 is a uniformly most powerful *invariant* test; this means that if the sample is $(X_i, Y_i)'$, with $i = 1,\ldots,N$, then r is invariant under the transformations $\tilde{X}_i = aX_i + b$, $\tilde{Y}_i = cY_i + d$, where $a > 0$ and $c > 0$, and any function of the sufficient statistic which is invariant is a function of r. The invariant character of this test is discussed in Chapter 6 in Example 6.1.16.

The asymptotic results of Section 5.1.4 can also be used for testing hypotheses and, in fact, it is usually simpler to do this. Moreover, one can deal with a wider class of distributions. Suppose that we have a sample of size $N = n+1$ from an *ellipitical* distribution with correlation ρ and kurtosis parameter κ and we wish to test the null hypothesis $H_0\colon \rho = \rho_0$ against $H\colon \rho \neq \rho_0$. Putting $\xi_0 = \tanh^{-1}\rho_0$, we know that when H_0 is true the distribution of $z = \tanh^{-1} r$ is approximately

$$N\left(\xi_0, \frac{1}{n-2} + \frac{\kappa}{n+2}\right)$$

so that an approximate test of size α is to reject H_0 if

$$\frac{|z-\xi_0|}{\left[\dfrac{1}{n-2} + \dfrac{\kappa}{n+2}\right]^{1/2}} \geq d_\alpha,$$

where d_α is the two-tailed $100\alpha\%$ point of the $N(0,1)$ distribution. (If κ is not known it could be replaced by a consistent estimate $\hat{\kappa}$.) The asymptotic normality of z also enables us to easily construct confidence intervals for ξ, and hence for ρ. A confidence interval for ξ with confidence coefficient $1-\alpha$ (approximately) is

$$\left(z - d_\alpha\left[\frac{1}{n-2} + \frac{\kappa}{n+2}\right]^{1/2}, \quad z + d_\alpha\left[\frac{1}{n-2} + \frac{\kappa}{n+2}\right]^{1/2}\right)$$

and for ρ it is

$$\left(\tanh\left(z-d_\alpha\left[\frac{1}{n-2}+\frac{\kappa}{n+2}\right]^{1/2}\right),\quad \tanh\left(z+d_\alpha\left[\frac{1}{n-2}+\frac{\kappa}{n+2}\right]^{1/2}\right)\right).$$

It is also possible, for example, to test whether the correlation coefficients in two elliptical distributions are equal; the details are left as an exercise (see Problem 5.6).

A caveat is in order at this point; the procedure just described for testing a hypothesis about a correlation coefficient in an elliptical distribution may have poorer power properties than a test based on a statistic computed from a robust estimate of the covariance matrix, although if the kurtosis parameter κ is small there probably is not very much difference.

5.2. THE MULTIPLE CORRELATION COEFFICIENT

5.2.1. Introduction

Let $\mathbf{X}=(X_1,\dots,X_m)'$ be a random vector with covariance matrix $\Sigma>0$. Partition $\mathbf{X}$ and Σ as

$$\text{(1)} \qquad \mathbf{X}=\begin{pmatrix} X_1 \\ \mathbf{X}_2 \end{pmatrix}, \qquad \Sigma=\begin{bmatrix} \sigma_{11} & \boldsymbol{\sigma}_{12}' \\ \boldsymbol{\sigma}_{12} & \Sigma_{22} \end{bmatrix},$$

where $\mathbf{X}_2=(X_2,\dots,X_m)'$ and Σ_{22} is $(m-1)\times(m-1)$, so that $\text{Var}(X_1)=\sigma_{11}$, $\text{Cov}(\mathbf{X}_2)=\Sigma_{22}$, and $\boldsymbol{\sigma}_{12}$ is the $(m-1)\times 1$ vector of covariances between X_1 and each of the variables in $\mathbf{X}_2$. The multiple correlation coefficient can be characterized in various ways. We will use the following definition.

DEFINITION 5.2.1. The multiple correlation coefficient between X_1 and the variables $X_2,\dots,X_m$, denoted by $\bar{R}_{1\cdot 2\cdots m}$, is the maximum correlation between X_1 and any linear function $\boldsymbol{\alpha}'\mathbf{X}_2$ of $X_2,\dots,X_m$.

Using this definition, we have

$$\text{(2)} \qquad \bar{R}_{1\cdot 2\cdots m}=\max_{\boldsymbol{\alpha}}\frac{\text{Cov}(X_1,\boldsymbol{\alpha}'\mathbf{X}_2)}{[\text{Var}(X_1)\text{Var}(\boldsymbol{\alpha}'\mathbf{X}_2)]^{1/2}}$$

$$=\max_{\boldsymbol{\alpha}}\frac{\boldsymbol{\alpha}'\boldsymbol{\sigma}_{12}}{(\sigma_{11}\boldsymbol{\alpha}'\Sigma_{22}\boldsymbol{\alpha})^{1/2}}.$$

Now note that

$$\frac{\boldsymbol{\alpha}'\boldsymbol{\sigma}_{12}}{(\sigma_{11}\boldsymbol{\alpha}'\Sigma_{22}\boldsymbol{\alpha})^{1/2}}=\frac{\boldsymbol{\alpha}'\Sigma_{22}^{1/2}\Sigma_{22}^{-1/2}\boldsymbol{\sigma}_{12}}{(\alpha_{11}\boldsymbol{\alpha}'\Sigma_{22}\boldsymbol{\alpha})^{1/2}}$$

$$=\frac{\mathbf{u}'\mathbf{v}}{(\sigma_{11}\boldsymbol{\alpha}'\Sigma_{22}\boldsymbol{\alpha})^{1/2}},\qquad \text{where}\quad \mathbf{u}=\Sigma_{22}^{1/2}\boldsymbol{\alpha},\mathbf{v}=\Sigma_{22}^{-1/2}\boldsymbol{\sigma}_{12},$$

$$\leq\frac{(\mathbf{u}'\mathbf{u})^{1/2}(\mathbf{v}'\mathbf{v})^{1/2}}{(\sigma_{11}\boldsymbol{\alpha}'\Sigma_{22}\boldsymbol{\alpha})^{1/2}},\qquad \text{by the Cauchy–Schwarz inequality,}$$

$$=\frac{(\boldsymbol{\alpha}'\Sigma_{22}\boldsymbol{\alpha})^{1/2}(\boldsymbol{\sigma}_{12}'\Sigma_{22}^{-1}\boldsymbol{\sigma}_{12})^{1/2}}{(\sigma_{11}\boldsymbol{\alpha}'\Sigma_{22}\boldsymbol{\alpha})^{1/2}}$$

$$=\left(\frac{\boldsymbol{\sigma}_{12}'\Sigma_{22}^{-1}\boldsymbol{\sigma}_{12}}{\sigma_{11}}\right)^{1/2},$$

with equality if $\boldsymbol{\alpha}=\Sigma_{22}^{-1}\boldsymbol{\sigma}_{12}$. Using this in (2), we can show that

$$\text{(3)}\qquad \overline{R}_{1\cdot 2\cdots m}=\left(\frac{\boldsymbol{\sigma}_{12}'\Sigma_{22}^{-1}\boldsymbol{\sigma}_{12}}{\sigma_{11}}\right)^{1/2}.$$

Note that $0\leq\overline{R}_{1\cdot 2\cdots m}\leq 1$, unlike an ordinary correlation coefficient. We have now shown that $\overline{R}_{1\cdot 2\cdots m}$ is the correlation between X_1 and the linear function $\boldsymbol{\sigma}_{12}'\Sigma_{22}^{-1}\mathbf{X}_2$. Now recall that if $\mathbf{X}$ is $N_m(\boldsymbol{\mu},\Sigma)$ and $\boldsymbol{\mu}$ is partitioned similarly to $\mathbf{X}$ then the conditional distribution of X_1 given $\mathbf{X}_2$ is normal with mean

$$\text{(4)}\qquad E(X_1|\mathbf{X}_2)=\mu_1+\boldsymbol{\sigma}_{12}'\Sigma_{22}^{-1}(\mathbf{X}_2-\boldsymbol{\mu}_2)$$

and variance

$$\text{(5)}\qquad \operatorname{Var}(X_1|\mathbf{X}_2)\equiv\sigma_{11\cdot 2\cdots m}=\sigma_{11}-\boldsymbol{\sigma}_{12}'\Sigma_{22}^{-1}\boldsymbol{\sigma}_{12}$$

(see Theorem 1.2.11); hence we see in this case that the multiple correlation coefficient $\overline{R}_{1\cdot 2\cdots m}$ is the correlation between X_1 and the *regression function* $E(X_1|\mathbf{X}_2)$ of X_1 on $\mathbf{X}_2$ [see (16) of Section 1.2]. [In general, this will be true if $E(X_1|\mathbf{X}_2)$ is a linear function of $X_2,\ldots,X_m$.] Also, using (5) we have

$$\text{(6)}\qquad \overline{R}^2_{1\cdot 2\cdots m}=\frac{\sigma_{11}-\sigma_{11\cdot 2\cdots m}}{\sigma_{11}};$$

the numerator here is the amount that the variance of X_1 can be reduced by conditioning on $\mathbf{X}_2$ and hence $\bar{R}^2_{1\cdot 2\cdots m}$ measures the fraction of reduction in the variance of X_1 obtained by conditioning on $\mathbf{X}_2$.

It is worth noting that in the bivariate case where

$$\Sigma=\begin{bmatrix}\sigma_1^2 & \rho\sigma_1\sigma_2\\ \rho\sigma_1\sigma_2 & \sigma_2^2\end{bmatrix}$$

we have $\operatorname{Var}(X_1|X_2)\equiv\sigma_{11\cdot 2}=\sigma_1^2(1-\rho^2)$, so that

$$\bar{R}^2_{1\cdot 2}=\frac{\sigma_1^2-\sigma_1^2(1-\rho^2)}{\sigma_1^2}=\rho^2$$

and hence

$$\bar{R}_{1\cdot 2}=|\rho|,$$

the absolute value of the ordinary correlation coefficient.

We have defined the multiple correlation coefficient between X_1 and $\mathbf{X}_2$, where X_2 contains *all* the other variables, but we can obviously define a whole set of multiple correlation coefficients. Partition $\mathbf{X}$ and Σ as

$$\mathbf{X}=\begin{pmatrix}\mathbf{X}_1\\ \mathbf{X}_2\end{pmatrix},\qquad \Sigma=\begin{bmatrix}\Sigma_{11} & \Sigma_{12}\\ \Sigma_{21} & \Sigma_{22}\end{bmatrix},$$

where $\mathbf{X}_1$ is $k\times 1$, $\mathbf{X}_2$ is $(m-k)\times 1$, Σ_{11} is $k\times k$, and Σ_{22} is $(m-k)\times(m-k)$. Let X_i be a variable in the subvector $\mathbf{X}_1$ (with $i=1,\dots,k$). The multiple correlation coefficient between X_i and the variables $X_{k+1},\dots,X_m$ in $\mathbf{X}_2$, denoted by $\bar{R}_{i\cdot k+1,\dots,m}$, is the maximum correlation between X_i and any linear function $\boldsymbol{\alpha}'\mathbf{X}_2$ of $X_{k+1},\dots,X_m$. Arguing as before it follows that the maximizing value of $\boldsymbol{\alpha}$ is $\boldsymbol{\alpha}=\Sigma_{22}^{-1}\boldsymbol{\sigma}_i$, where $\boldsymbol{\sigma}_i'$ is the ith row of Σ_{12}, and hence that

$$\bar{R}_{i\cdot k+1,\dots,m}=\left(\frac{\boldsymbol{\sigma}_i'\Sigma_{22}^{-1}\boldsymbol{\sigma}_i}{\sigma_{ii}}\right)^{1/2}$$

Equivalently,

$$\bar{R}^2_{i\cdot k+1,\dots,m}=\frac{\sigma_{ii}-\sigma_{ii\cdot k+1,\dots,m}}{\sigma_{ii}},$$

where $\Sigma_{11\cdot2}\equiv\Sigma_{11}-\Sigma_{12}\Sigma_{22}^{-1}\Sigma_{21}=(\sigma_{ij\cdot k+1,\ldots,m})$. In the case where $\mathbf{X}$ is normal, $\Sigma_{11\cdot2}$ is the covariance matrix in the conditional distribution of $\mathbf{X}_1$ given $\mathbf{X}_2$.

For the remainder of the discussion we will restrict attention to the multiple correlation coefficient $\bar{R}_{1\cdot2\cdots m}$ between X_1 and the variables $X_2,\ldots,X_m$, and we shall drop the subscripts, so that $\bar{R}\equiv\bar{R}_{1\cdot2\cdots m}$. What follows will obviously apply to any other multiple correlation coefficient. We then have $\mathbf{X}$ and Σ partitioned as in (1). Now let $\mathbf{X}_1,\ldots,\mathbf{X}_N$ be N independent observations on $\mathbf{X}$ and put

$$A=nS=\sum_{i=1}^{N}(\mathbf{X}_i-\overline{\mathbf{X}})(\mathbf{X}_i-\overline{\mathbf{X}})'$$

where $n=N-1$, so that S is the sample covariance matrix. Partition A and S as

$$A=\begin{bmatrix} a_{11} & \mathbf{a}_{12}' \\ \mathbf{a}_{12} & A_{22}\end{bmatrix}, \qquad S=\begin{bmatrix} s_{11} & \mathbf{s}_{12}' \\ \mathbf{s}_{12} & S_{22}\end{bmatrix},$$

where A_{22} and S_{22} are $(m-1)\times(m-1)$. The *sample multiple correlation coefficient* between X_1 and $X_2,\ldots,X_m$ is defined as

$$R=\left[\frac{\mathbf{a}_{12}'A_{22}^{-1}\mathbf{a}_{12}}{a_{11}}\right]^{1/2}=\left[\frac{\mathbf{s}_{12}'S_{22}^{-1}\mathbf{s}_{12}}{s_{11}}\right]^{1/2}. \tag{7}$$

When the underlying distribution is normal, R is the *maximum likelihood estimate of* $\bar{R}$. Note that $\bar{R}=0$ implies that $\boldsymbol{\sigma}_{12}=\mathbf{0}$ [see (2)]; hence, in the case of normality, $\bar{R}=0$ if and only if X_1 is independent of $\mathbf{X}_2=(X_2,\ldots,X_m)'$.

In the following subsections exact and asymptotic distributions will be derived for the sample multiple correlation coefficient under various assumptions about the underlying distribution from which we are sampling. Some uses for these results in the area of hypothesis testing are also discussed.

5.2.2. *Distribution of the Sample Multiple Correlation Coefficient in the Case of Independence*

Here we will find the distribution of a multiple correlation coefficient formed from *independent* variables. We consider N random $m\times1$ vectors

$(N>m)$

$$\begin{pmatrix} Y_1 \\ \mathbf{X}_1 \end{pmatrix}, \quad \ldots, \quad \begin{pmatrix} Y_N \\ \mathbf{X}_N \end{pmatrix}$$

where each $\mathbf{X}_i$ is $(m-1)\times 1$ and form the $m\times N$ matrix

$$Z=\begin{bmatrix} Y_1 & \ldots & Y_N \\ \mathbf{X}_1 & \ldots & \mathbf{X}_N \end{bmatrix}=\begin{bmatrix} \mathbf{Y}' \\ X' \end{bmatrix}$$

where $\mathbf{Y}$ is $N\times 1$ and X is $N\times(m-1)$. The square of the sample multiple correlation coefficient is

$$R^2=\frac{\mathbf{a}_{12}'A_{22}^{-1}\mathbf{a}_{12}}{a_{11}}. \tag{8}$$

Here A is the usual matrix of sum of squares and sum of products

$$A=Z\left(I_N-\frac{1}{N}\mathbf{1}\mathbf{1}'\right)Z'=\begin{bmatrix} a_{11} & \mathbf{a}_{12}' \\ \mathbf{a}_{12} & A_{22} \end{bmatrix}, \tag{9}$$

where A_{22} is $(m-1)\times(m-1)$ and $\mathbf{1}=(1,1,\ldots,1)'\in R^N$. (For convenience the notation has been changed from that in Section 5.2.1. There we were looking at the multiple correlation coefficient between X_1 and $\mathbf{X}_2$; here X_1 has been replaced by Y and $\mathbf{X}_2$ by $\mathbf{X}$.) The assumption usually made is that the N vectors

$$\begin{pmatrix} Y_1 \\ \mathbf{X}_1 \end{pmatrix}, \quad \ldots, \quad \begin{pmatrix} Y_N \\ \mathbf{X}_N \end{pmatrix} \tag{10}$$

are independent $N_m(\boldsymbol{\mu},\Sigma)$ random vectors, where

$$\Sigma=\begin{bmatrix} \sigma_{11} & \boldsymbol{\sigma}_{12}' \\ \boldsymbol{\sigma}_{12} & \Sigma_{22} \end{bmatrix}$$

so that the population multiple correlation coefficient $\bar{R}$ is given by

$$\bar{R}^2=\frac{\boldsymbol{\sigma}_{12}'\Sigma_{22}^{-1}\boldsymbol{\sigma}_{12}}{\sigma_{11}}.$$

In this case the Y's are independent of the $\mathbf{X}$'s when $\bar{R}=0$. If, *in general*, we assume that the Y's are independent of the $\mathbf{X}$'s, the normality assumption is not important as long as the vector **Y** has a spherical distribution. This is noted in the following theorem.

THEOREM 5.2.2. Let **Y** be an $N\times 1$ random vector having a spherical distribution with $P(\mathbf{Y}=\mathbf{0})=0$, and let X be an $N\times(m-1)$ random matrix independent of **Y** and of rank $m-1$ with probability 1. If R is the sample multiple correlation coefficient given by (8) then R^2 has the beta distribution with parameters $\frac{1}{2}(m-1)$ and $\frac{1}{2}(N-m)$, or equivalently

$$\frac{N-m}{m-1}\cdot\frac{R^2}{1-R^2} \quad \text{is} \quad F_{m-1,N-m}.$$

Proof. Write the matrix A given by (9) as

$$A=Z(I-M)Z'$$

where

$$M=\frac{1}{N}\mathbf{1}\mathbf{1}' \quad \text{and} \quad Z=\begin{bmatrix}\mathbf{Y}'\\ X'\end{bmatrix}.$$

Then

$$a_{11}=\mathbf{Y}'(I-M)\mathbf{Y},$$

$$\mathbf{a}_{12}=X'(I-M)\mathbf{Y},$$

and

$$A_{22}=X'(I-M)X$$

so that

$$R^2=\frac{\mathbf{Y}'(I-M)X[X'(I-M)X]^{-1}X'(I-M)\mathbf{Y}}{\mathbf{Y}'(I-M)\mathbf{Y}}.$$

Since $I-M$ is idempotent of rank $N-1$ there exists $H\in O(N)$ such that

$$H(I-M)H'=\begin{bmatrix}I_{N-1} & \mathbf{0}\\ \mathbf{0}' & 0\end{bmatrix}.$$

Put $\mathbf{U}=H\mathbf{Y}$ and $V=HX$, and partition **U** and V as

$$\mathbf{U}=\begin{bmatrix}\mathbf{U}^*\\ U_N\end{bmatrix}, \qquad V=\begin{bmatrix}V^*\\ \mathbf{V}_N'\end{bmatrix},$$

where $\mathbf{U}^*$ is $(N-1)\times 1$ and V^* is $(N-1)\times(m-1)$. Then

$$(11)\quad R^2=\frac{\mathbf{U}'H(I-M)H'V[V'H(I-M)H'V]^{-1}V'H(I-M)H'\mathbf{U}}{\mathbf{U}'H(I-M)H'\mathbf{U}}$$

$$=\frac{\mathbf{U}^{*\prime}V^*(V^{*\prime}V^*)^{-1}V^{*\prime}\mathbf{U}^*}{\mathbf{U}^{*\prime}\mathbf{U}^*}.$$

Now, $V^*(V^{*\prime}V^*)^{-1}V^{*\prime}$ is idempotent of rank $m-1$ and is independent of $\mathbf{U}^*$, which has an $(N-1)$-variate spherical distribution. Conditioning on V^*, we can then use part (ii) of Theorem 1.5.7, with $B=V^*(V^{*\prime}V^*)^{-1}V^{*\prime}$, to show that R^2 has the beta distribution with parameters $\frac{1}{2}(m-1)$ and $\frac{1}{2}(N-m)$, and the proof is complete.

A geometrical interpretation of R is apparent from (11). Writing $\hat{\mathbf{U}}=\|\mathbf{U}^*\|^{-1}\mathbf{U}^*$ we have

$$R^2=\hat{\mathbf{U}}'V^*(V^{*\prime}V^*)^{-1}V^{*\prime}\hat{\mathbf{U}}$$

where $\hat{\mathbf{U}}$ has a uniform distribution over the unit sphere in R^{N-1}. Hence $R=\cos\theta$, where θ is the angle between $\hat{\mathbf{U}}$ and the orthogonal projection of $\hat{\mathbf{U}}$ onto the $m-1$ dimensional subspace of R^{N-1} spanned by the columns of V^*.

We noted previously that it is usually assumed that the vectors (10) are normal. This is a special case of Theorem 5.2.2, stated explicity in the following corollary.

COROLLARY 5.2.3. Let

$$\begin{pmatrix} Y_1 \\ \mathbf{X}_1 \end{pmatrix},\quad \dots,\quad \begin{pmatrix} Y_N \\ \mathbf{X}_N \end{pmatrix}$$

be independent $N_m(\boldsymbol{\mu},\Sigma)$ random vectors, where each $\mathbf{X}_i$ is $(m-1)\times 1$, and let R be the sample multiple correlation coefficient given by (8). Then, when the population multiple correlation coefficient $\bar{R}$ is,

$$\frac{N-m}{m-1}\,\frac{R^2}{1-R^2}$$

is $F_{m-1,N-m}$.

Proof. Partition $\boldsymbol{\mu}$ and Σ as

$$\boldsymbol{\mu}=\begin{pmatrix}\mu_1\\ \boldsymbol{\mu}_2\end{pmatrix} \text{ and } \Sigma=\begin{bmatrix}\sigma_{11} & \mathbf{0}'\\ \mathbf{0} & \Sigma_{22}\end{bmatrix}$$

where $\boldsymbol{\mu}_2$ is $(m-1)\times 1$ and Σ_{22} is $(m-1)\times(m-1)$. Note that $\boldsymbol{\sigma}_{12}=\mathbf{0}$ because $\bar{R}=0$. The reader can easily check that the multiple correlation between the standardized variables $(Y_i-\mu_1)/\sigma_{11}^{1/2}$ and $\Sigma_{22}^{-1/2}(\mathbf{X}_i-\boldsymbol{\mu}_2)$ is the same as that between Y_i and $\mathbf{X}_i$, so we can assume without loss of generality that $\boldsymbol{\mu}=\mathbf{0}$ and $\Sigma=I_m$. Then it is clear that the conditions of Theorem 5.2.2 are satisfied and the desired result follows immediately.

Suppose that the conditions of Theorem 5.2.2 are satisfied, so that R^2 has a beta distribution with parameters $\frac{1}{2}(m-1)$ and $\frac{1}{2}(N-m)$. Then the kth moment of R^2 is

$$E(R^{2k})=\frac{\Gamma\left[\frac{1}{2}(N-1)\right]}{\Gamma\left[\frac{1}{2}(N-1)+k\right]}\frac{\Gamma\left[\frac{1}{2}(m-1)+k\right]}{\Gamma\left[\frac{1}{2}(m-1)\right]}$$

$$=\frac{\left[\frac{1}{2}(m-1)\right]_k}{\left[\frac{1}{2}(N-1)\right]_k}.$$

In particular, the mean and variance of R^2 are

$$E(R^2)=\frac{m-1}{N-1}$$

and

$$\operatorname{Var}(R^2)=\frac{2(N-m)(m-1)}{(N^2-1)(N-1)}.$$

5.2.3. *The Non-null Distribution of a Sample Multiple Correlation Coefficient in the Case of Normality*

In this section we will derive the distribution of the sample multiple correlation coefficient R formed from a sample from a *normal* distribution, when the population multiple correlation coefficient $\bar{R}$ is non-zero.

THEOREM 5.2.4. Let $\begin{pmatrix} Y \\ \mathbf{X} \end{pmatrix}$ be $N_m(\boldsymbol{\mu}, \Sigma)$, where $\mathbf{X}$ is $(m-1)\times 1$ and partition Σ as

$$\Sigma = \begin{bmatrix} \sigma_{11} & \boldsymbol{\sigma}'_{12} \\ \boldsymbol{\sigma}_{12} & \Sigma_{22} \end{bmatrix},$$

where Σ_{22} is $(m-1)\times(m-1)$, so that the population multiple correlation coefficient between Y and $\mathbf{X}$ is $\bar{R} = (\boldsymbol{\sigma}'_{12}\Sigma_{22}^{-1}\boldsymbol{\sigma}_{12}/\sigma_{11})^{1/2}$. Let R be the sample multiple correlation coefficient between Y and $\mathbf{X}$ based on a sample of size $N(N>m)$; then the density function of R^2 is

$$(12)\qquad \frac{\Gamma(\frac{1}{2}n)}{\Gamma[\frac{1}{2}(m-1)]\Gamma[\frac{1}{2}(n-m+1)]}(R^2)^{(m-3)/2}(1-R^2)^{(n-m-1)/2}$$

$$\cdot(1-\bar{R}^2)^{n/2}{}_2F_1\left(\tfrac{1}{2}n, \tfrac{1}{2}n; \tfrac{1}{2}(m-1); \bar{R}^2R^2\right) \qquad (0<R^2<1),$$

where $n = N-1$.

Proof. Let $\hat{\Sigma}$ be the maximum likelihood estimate of Σ based on the N observations, and put $A = N\hat{\Sigma}$; then A is $W_m(n,\Sigma)$, $n = N-1$. If we partition A similarly to Σ as

$$A = \begin{bmatrix} a_{11} & \mathbf{a}'_{12} \\ \mathbf{a}_{12} & A_{22} \end{bmatrix},$$

the sample multiple correlation coefficient is given by

$$R^2 = \frac{\mathbf{a}'_{12}A_{22}^{-1}\mathbf{a}_{12}}{a_{11}},$$

so that

$$(13)\qquad \frac{R^2}{1-R^2} = \frac{\mathbf{a}'_{12}A_{22}^{-1}\mathbf{a}_{12}}{a_{11\cdot 2}},$$

where $a_{11\cdot 2} = a_{11} - \mathbf{a}'_{12}A_{22}^{-1}\mathbf{a}_{12}$. From Theorem 3.2.10 we know that the numerator and denominator on the right side of (13) are independent; $a_{11\cdot 2}/\sigma_{11\cdot 2}$ is χ^2_{n-m+1}, where $\sigma_{11\cdot 2} = \sigma_{11} - \boldsymbol{\sigma}'_{12}\Sigma_{22}^{-1}\boldsymbol{\sigma}_{12}$, and the conditional distribution of $\mathbf{a}_{12}$ given A_{22} is $N(A_{22}\Sigma_{22}^{-1}\boldsymbol{\sigma}_{12}, \sigma_{11\cdot2}A_{22})$. Hence, conditional on A_{22}, part (b) of Theorem 1.4.1 shows that

$$\frac{\mathbf{a}_{12}'A_{22}^{-1}\mathbf{a}_{12}}{\sigma_{11\cdot2}} \quad \text{is} \quad \chi_{m-1}^2(\delta),$$

where the noncentrality parameter δ is

$$(14) \qquad \delta = \frac{\boldsymbol{\sigma}_{12}'\Sigma_{22}^{-1}A_{22}\Sigma_{22}^{-1}\boldsymbol{\sigma}_{12}}{\sigma_{11\cdot2}}.$$

Hence, conditional on A_{22}, or equivalently on δ,

$$(15) \qquad Z \equiv \frac{n-m+1}{m-1}\frac{R^2}{1-R^2} = \frac{\chi_{m-1}^2(\delta)/(m-1)}{\chi_{n-m+1}^2/(n-m+1)}$$

is $F_{m-1,n-m+1}(\delta)$ (see Section 1.3 for the noncentral F distribution). At this point it is worth noting that if $\boldsymbol{\sigma}_{12}=\mathbf{0}$ (so that $\bar{R}=0$), then $\delta=0$ and the F distribution is central, the result given in Theorem 5.2.2. Now, using the noncentral F density function given in Theorem 1.3.6, the conditional density function of the random variable Z in (15) given δ is

$$e^{-\delta/2}\,{}_1F_1\left(\frac{1}{2}n; \frac{1}{2}(m-1); \frac{\frac{1}{2}\frac{m-1}{n-m+1}\delta z}{1+\frac{m-1}{n-m+1}z}\right)$$

$$\cdot\frac{\Gamma\left(\frac{1}{2}n\right)z^{(m-3)/2}}{\Gamma[\frac{1}{2}(m-1)]\Gamma[\frac{1}{2}(n-m+1)](1+[(m-1)/(n-m+1)]z)^{n/2}}$$

$$\cdot\left(\frac{m-1}{n-m+1}\right)^{(m-1)/2}, \qquad (z>0).$$

Changing variables from Z to R^2 via

$$Z = \frac{n-m+1}{m-1}\frac{R^2}{1-R^2}$$

the conditional density function of R^2 given δ is

$$(16)\qquad e^{-\delta/2}{}_1F_1\left(\tfrac{1}{2}n;\tfrac{1}{2}(m-1);\tfrac{1}{2}\delta R^2\right)\frac{\Gamma(\frac{1}{2}n)}{\Gamma[\frac{1}{2}(m-1)]\Gamma[\frac{1}{2}(n-m+1)]}$$

$$\cdot(R^2)^{(m-3)/2}(1-R^2)^{(n-m-1)/2}\qquad (0<R^2<1).$$

To get the (unconditional) density function of R^2 we first multiply this by the density function of δ to give the joint density function of R^2 and δ. Now, A_{22} is $W_{m-1}(n,\Sigma_{22})$ and hence $\sigma_{12}'\Sigma_{22}^{-1}A_{22}\Sigma_{22}^{-1}\sigma_{12}$ is $W_1(n,\sigma_{12}'\Sigma_{22}^{-1}\sigma_{12})$ (using Theorem 3.2.5); that is,

$$(17)\qquad v\equiv\frac{\sigma_{12}'\Sigma_{22}^{-1}A_{22}\Sigma_{22}^{-1}\sigma_{12}}{\sigma_{12}'\Sigma_{22}^{-1}\sigma_{12}}\quad\text{is}\quad\chi_n^2.$$

If we define the parameter θ as

$$(18)\qquad \theta=\frac{\bar{R}^2}{1-\bar{R}^2}=\frac{\sigma_{12}'\Sigma_{22}^{-1}\sigma_{12}}{\sigma_{11\cdot 2}},$$

it follows from (14) and (17) that $\delta=\theta v$. The joint density function of R^2 and v is obtained by multiplying (16) (with $\delta=\theta v$) by the χ_n^2 density for v and is

$$\frac{e^{-v/2}v^{n/2-1}}{2^{n/2}\Gamma(\frac{1}{2}n)}e^{-\theta v/2}{}_1F_1\left(\tfrac{1}{2}n;\tfrac{1}{2}(m-1);\tfrac{1}{2}\theta vR^2\right)$$

$$\cdot\frac{\Gamma(\frac{1}{2}n)}{\Gamma[\frac{1}{2}(m-1)]\Gamma[\frac{1}{2}(n-m+1)]}(R^2)^{(m-3)/2}(1-R^2)^{(n-m-1)/2}.$$

To get the marginal density function of R^2 we now integrate with respect to v from 0 to ∞. Now, since

$$\frac{1}{\Gamma(\frac{1}{2}n)}\int_0^\infty e^{-\theta v/2}e^{-v/2}\left(\frac{v}{2}\right)^{n/2-1}{}_1F_1\left(\tfrac{1}{2}n;\tfrac{1}{2}(m-1);\tfrac{1}{2}\theta vR^2\right)d\left(\frac{v}{2}\right)$$

$$=\frac{(1-\bar{R}^2)^{n/2}}{\Gamma(\frac{1}{2}n)}\int_0^\infty e^{-u}u^{n/2-1}{}_1F_1\left(\tfrac{1}{2}n;\tfrac{1}{2}(m-1);\bar{R}^2R^2u\right)du$$

$$\left[\text{on putting } \tfrac{1}{2}v = u\left(1-\bar{R}^2\right) \text{ and using (18)}\right]$$

$$=\left(1-\bar{R}^2\right)^{n/2} {}_2F_1\left(\tfrac{1}{2}n, \tfrac{1}{2}n; \tfrac{1}{2}(m-1); \bar{R}^2 R^2\right),$$

by Lemma 1.3.3 the result desired is obtained and the proof is complete.

The distribution of R^2 was first found by Fisher (1928) and can be expressed in many different forms. We will give one other due to Gurland (1968). First let $I_x(\alpha, \beta)$ denote the *incomplete beta function*:

(19)

$$I_x(\alpha,\beta)=\frac{\Gamma(\alpha+\beta)}{\Gamma(\alpha)\Gamma(\beta)}\int_0^x t^{\alpha-1}(1-t)^{\beta-1}\,dt, \qquad (0<x<1, \alpha>0, \beta>0).$$

It is well known (and easily verified), that the F_{n_1, n_2} distribution function and the incomplete beta function are related by the identity

$$P\left(F_{n_1,n_2}\le x\right)=I_z\left(\tfrac{1}{2}n_1, \tfrac{1}{2}n_2\right), \tag{20}$$

where $z = n_1x/(n_2+n_1x)$.

THEOREM 5.2.5. With the same assumptions as Theorem 5.2.4, the distribution function of R^2 can be expressed in the form

$$\begin{aligned} P(R^2\le x) &= \sum_{k=0}^{\infty} c_k I_x\left(\tfrac{1}{2}(m-1)+k, \tfrac{1}{2}(n-m+1)\right) \\ &= \sum_{k=0}^{\infty} c_k P\left(F_{m-1+2k,\, n-m+1} \le \frac{n-m+1}{m-1+2k}\,\frac{x}{1-x}\right), \end{aligned} \tag{21}$$

where c_k is the negative binomial probability

$$c_k = (-1)^k \binom{-\frac{1}{2}n}{k}\left(1-\bar{R}^2\right)^{n/2}\left(\bar{R}^2\right)^k. \tag{22}$$

Proof. Using the series expansion for the ${}_2F_1$ function, it follows from Theorem 5.2.4 that the density function of R^2 can be written as

(23)

$$\frac{\Gamma(\frac{1}{2}n)\left(1-\bar{R}^2\right)^{n/2}}{\Gamma\left[\frac{1}{2}(m-1)\right]\Gamma\left[\frac{1}{2}(n-m+1)\right]}\sum_{k=0}^{\infty}\frac{\left(\frac{1}{2}n\right)_k\left(\frac{1}{2}n\right)_k}{\left(\frac{1}{2}(m-1)\right)_k k!}\left(\bar{R}^2\right)^k\left(R^2\right)^{(m-3)/2+k}$$

$$\cdot\left(1-R^2\right)^{(n-m-1)/2}$$

Using

$$(a)_k\Gamma(a)=\Gamma(a+k)$$

[with a equal to $\frac{1}{2}n$ and $\frac{1}{2}(m-1)$], and

$$(\tfrac{1}{2}n)_k=\binom{-\frac{1}{2}n}{k}(-1)^k k!$$

in (23), and integrating with respect to R^2 from 0 to x gives the desired result.

Note that Theorem 5.2.5 expresses the distribution of R^2 as a mixture of beta distributions where the weights are negative binomial probabilities; that is, the distribution of R^2 can be obtained by taking a random variable K having a negative binomial distribution with $P(K=k)=c_k$, $k=0,1,\ldots$ and then taking the conditional distribution of R^2 given $K=k$ to be beta with parameters $\frac{1}{2}(m-1)+k$ and $\frac{1}{2}(n-m+1)$. An immediate consequence of Theorem 5.2.5 is given in the following corollary.

COROLLARY 5.2.6. If $U=R^2/(1-R^2)$ then

$$P(U\le x)=\sum_{k=0}^{\infty} c_k P\left(F_{m-1+2k,\,n-m+1}\le\frac{n-m+1}{m-1+2k}x\right), \tag{24}$$

where c_k is given by (22).

From this it follows that U can be expressed as $U=V_1/V_2$, where V_1 and V_2 are independent, V_2 has the χ^2_{n-m+1} distribution, and the distribution of V_1 is a mixture of χ^2 distributions, namely,

$$P(V_1\le x)=\sum_{k=0}^{\infty} c_k P(\chi^2_{m-1+2k}\le x). \tag{25}$$

A common way of *approximating* a mixture of χ^2 distributions is to fit a scaled central χ^2 distribution, $a\chi^2_b$ say, or, more correctly, since the degrees of freedom need not be an integer, a *gamma* distribution with parameters $\frac{1}{2}b$ and $2a$ and density function

$$\frac{1}{\Gamma(\frac{1}{2}b)(2a)^{b/2}}e^{-x/2a}x^{b/2-1} \qquad (x>0). \tag{26}$$

If the distribution of V_1 is approximated in this way by equating the first two moments of V_1 with those of this gamma distribution, one finds that the fitted values for a and b are

$$(27)\qquad a=\frac{n\theta(\theta+2)+m-1}{n\theta+m-1}, \qquad b=\frac{(n\theta+m-1)^2}{n\theta(\theta+2)+m-1},$$

where $\theta=\overline{R}^2/(1-\overline{R}^2)$ (see Problem 5.10). With these values of a and b we then get an approximation to the distribution function of R^2 as

(28)

$$P(R^2\le x)=P\left(U\le\frac{x}{1-x}\right)=P\left(\frac{V_1}{V_2}\le\frac{x}{1-x}\right)\approx P\left(\frac{a\chi_b^2}{\chi_{n-m+1}^2}\le\frac{x}{1-x}\right)$$

$$=I_z\left[\tfrac{1}{2}b,\tfrac{1}{2}(n-m+1)\right],$$

where $z=x/[a(1-x)+x]$. This approximation, due to Gurland (1968), appears to be quite accurate. Note that when $\overline{R}=0$ (so that $\theta=0$), the values of a and b are

$$a=1, \qquad b=m-1,$$

and the approximation (28) gives the *exact* null distribution for R^2 found in Theorem 5.2.2.

The moments of R^2 are easily obtained using the representation given in Theorem 5.2.5 for the distribution of R^2 as a mixture of beta distributions. Using the fact that the hth moment of a beta distribution with parameters α and β is

$$\frac{\Gamma(\alpha+h)\Gamma(\alpha+\beta)}{\Gamma(\alpha)\Gamma(\alpha+\beta+h)},$$

we find that

$$E\left[(1-R^2)^h\right]=\sum_{k=0}^{\infty}c_k\frac{\Gamma\left[\frac{1}{2}(n-m+1)+h\right]}{\Gamma\left[\frac{1}{2}(n-m+1)\right]}\frac{\Gamma(\frac{1}{2}n+k)}{\Gamma(\frac{1}{2}n+k+h)},$$

where c_k is

$$c_k=(-1)^k\binom{-\frac{1}{2}n}{k}(1-\overline{R}^2)^{n/2}(\overline{R}^2)^k=\frac{(\frac{1}{2}n)_k}{k!}(1-\overline{R}^2)^{n/2}(\overline{R}^2)^k.$$

Hence we can write

$$E\left[(1-R^2)^h\right]=\frac{\left[\frac{1}{2}(n-m+1)\right]_h}{\left(\frac{1}{2}n\right)_h}\left(1-\bar{R}^2\right)^{n/2}\sum_{k=0}^{\infty}\frac{(\frac{1}{2}n)_k(\frac{1}{2}n)_k}{(\frac{1}{2}n+h)_k}\frac{(\bar{R}^2)^k}{k!}$$

$$=\frac{\left[\frac{1}{2}(n-m+1)\right]_h}{\left(\frac{1}{2}n\right)_h}\left(1-\bar{R}^2\right)^{n/2}{}_2F_1\left(\tfrac{1}{2}n,\tfrac{1}{2}n;\tfrac{1}{2}n+h;\bar{R}^2\right).$$

If we use the Euler relation given by (17) of Section 5.1, this becomes

$$(29)\quad E\left[(1-R^2)^h\right]=\frac{\left[\frac{1}{2}(n-m+1)\right]_h}{\left(\frac{1}{2}n\right)_h}\left(1-\bar{R}^2\right)^h{}_2F_1\left(h,h;\tfrac{1}{2}n+h;\bar{R}^2\right).$$

In particular, the mean and the variance of R^2 are

$$(30)\quad E(R^2)=1-\left(\frac{n-m+1}{n}\right)\left(1-\bar{R}^2\right){}_2F_1\left(1,1;\tfrac{1}{2}n+1;\bar{R}^2\right)$$

$$=\bar{R}^2+\frac{m-1}{n}\left(1-\bar{R}^2\right)+\frac{2}{n+2}\bar{R}^2\left(1-\bar{R}^2\right)+O(n^{-2})$$

and

(31)

$$\mathrm{Var}(R^2)=E(R^4)-E(R^2)^2$$

$$=E\left[(1-R^2)^2\right]-E(1-R^2)^2$$

$$=\frac{\left[\frac{1}{2}(n-m+1)\right]_2}{\left(\frac{1}{2}n\right)_2}\left(1-\bar{R}^2\right)^2{}_2F_1\left(2,2;\tfrac{1}{2}n+2;\bar{R}^2\right)$$

$$-\left[\left(\frac{n-m+1}{n}\right)\left(1-\bar{R}^2\right){}_2F_1\left(1,1;\tfrac{1}{2}n+1;\bar{R}^2\right)\right]^2$$

$$=\frac{n-m+1}{n^2(n+2)}\left(1-\bar{R}^2\right)^2\left\{2(m-1)+4\bar{R}^2\left[\frac{4(m-1)+n(n-m+1)}{n+4}\right]\right.$$

$$\left.+O(n^{-2})\right\}.$$

Note the different orders of magnitude for $\operatorname{Var}(R^2)$ depending on whether $\bar{R}=0$ or $\bar{R}\neq 0$. For $\bar{R}\neq 0$, (31) gives

$$(32)\qquad \operatorname{Var}(R^2)=\frac{4\bar{R}^2(1-\bar{R}^2)^2(n-m+1)^2}{n(n+2)(n+4)}+O(n^{-2})$$

$$=\frac{4\bar{R}^2(1-\bar{R}^2)^2}{n}+O(n^{-2});$$

if $\bar{R}=0$, (31) gives

$$(33)\qquad \operatorname{Var}(R^2)=\frac{2(n-m+1)(m-1)}{n^2(n+2)},$$

which is the exact variance in the null case.

It is seen from (30) that R^2 is a biased estimate of $\bar{R}^2$ and that $E(R^2)>\bar{R}^2$; that is, R^2 overestimates $\bar{R}^2$. Olkin and Pratt (1958) have shown that an unbiased estimate of $\bar{R}^2$ is

$$(34)\quad T(R^2)=1-\left(\frac{n-2}{n-m+1}\right)(1-R^2)\,{}_2F_1\left(1,1;\tfrac{1}{2}(n-m+3);1-R^2\right)$$

(see Problem 5.11). This may be expanded as

$$(35)\quad T(R^2)=R^2-\frac{n-2}{n-m+1}(1-R^2)-\frac{2(n-2)}{(n-m+1)(n-m+3)}$$

$$\cdot(1-R^2)^2+O(n^{-2}),$$

from which it is clear that $T(R^2)<R^2$. $T(R^2)$ is in fact the unique minimum variance unbiased estimate of $\bar{R}^2$ since it is a function of a complete sufficient statistic. Obviously $T(1)=1$ and it can be shown that $T(0)=-(m-1)/(n-m+1)$. In fact it is clear from (35) that $T(R^2)<0$ for R^2 near zero, so that the unique unbiased estimate of $\bar{R}^2$ takes values outside the parameter space $[0,1]$.

5.2.4. *Asymptotic Distributions of a Sample Multiple Correlation Coefficient from an Elliptical Distribution*

In the case of sampling from a multivariate normal distribution, we noted in (32) and (33) the different orders of magnitude of $\operatorname{Var}(R^2)$, depending on whether $\bar{R}=0$ or $\bar{R}\neq 0$. This is true for more general populations and it reflects the fact that the limiting distributions of R^2 are different in these two situations. In this section we will derive these limiting distributions when the underlying distribution is *elliptical*; this is done mainly for the sake of concreteness and because the asymptotic distributions turn out to be very simple. The reader should note, however, that the only essential ingredient in the derivations is the asymptotic normality of the sample covariance matrix so that the arguments that follow will generalize with obvious modifications if the underlying distribution has finite fourth moments.

Thus, suppose that the $m\times 1$ random vector $(Y,\mathbf{X}')'$, where $\mathbf{X}$ is $(m-1)\times 1$, has an elliptical distribution with covariance matrix

$$\Sigma=\begin{bmatrix}\sigma_{11} & \sigma_{12}' \\ \sigma_{12} & \Sigma_{22}\end{bmatrix} \tag{36}$$

and *kurtosis parameter* κ [see (1) and (2) of Section 1.6]. The population multiple correlation coefficient $\bar{R}$ between Y and $\mathbf{X}$ is

$$\bar{R}=\left(\frac{\sigma_{12}'\Sigma_{22}^{-1}\sigma_{12}}{\sigma_{11}}\right)^{1/2}. \tag{37}$$

It helps at the outset to simplify the distribution theory by reducing the covariance structure. This is done in the following theorem.

THEOREM 5.2.7. If $\Sigma>0$ is partitioned as in (36) there exists a nonsingular $m\times m$ matrix

$$B=\begin{bmatrix}b & \mathbf{0}' \\ \mathbf{0} & C\end{bmatrix}$$

where C is $(m-1)\times(m-1)$, such that

$$B\Sigma B'=\begin{bmatrix}1 & \bar{R} & 0 & \dots & 0 \\ \bar{R} & 1 & 0 & \dots & 0 \\ \vdots & & & & \\ 0 & 0 & 0 & \dots & 1\end{bmatrix}$$

where $\bar{R}$ is given by (37).

Proof. Multiplying, we have

$$B\Sigma B' = \begin{bmatrix} b^2\sigma_{11} & b\boldsymbol{\sigma}'_{12}C' \\ bC\boldsymbol{\sigma}_{12} & C\Sigma_{22}C' \end{bmatrix}.$$

Put $b=\sigma_{11}^{-1/2}$ and $C=H\Sigma_{22}^{-1/2}$, where $H\in O(m-1)$; then $b^2\sigma_{11}=1$, $C\Sigma_{22}C'=I_{m-1}$, and

$$bC\boldsymbol{\sigma}_{12}=\sigma_{11}^{-1/2}H\Sigma_{22}^{-1/2}\boldsymbol{\sigma}_{12}.$$

Now let H be any orthogonal matrix whose first row is $\bar{R}^{-1}\sigma_{11}^{-1/2}\boldsymbol{\sigma}'_{12}\Sigma_{22}^{-1/2}$, then

$$bC\boldsymbol{\sigma}_{12}=\begin{pmatrix} \bar{R} \\ 0 \\ \vdots \\ 0 \end{pmatrix},$$

and the proof is complete.

Now, if we put

$$\begin{pmatrix} Y^* \\ \mathbf{X}^* \end{pmatrix} = B\begin{pmatrix} Y \\ \mathbf{X} \end{pmatrix} = \begin{pmatrix} bY \\ C\mathbf{X} \end{pmatrix},$$

it follows that $\text{Var}(Y^*)=1$, $\text{Cov}(\mathbf{X}^*)=I_{m-1}$ and the vector of covariances between Y^* and $\mathbf{X}^*$ is $(\bar{R},0,\dots,0)'$. Given a sample of size N, the reader can easily check that the sample multiple correlation coefficient between Y and $\mathbf{X}$ is the same as that between the transformed variables Y^* and $\mathbf{X}^*$, so there is no loss of generality in assuming that the covariance matrix in our elliptical distribution has the form

$$\Sigma = \begin{bmatrix} 1 & \mathbf{P}' \\ \mathbf{P} & I_{m-1} \end{bmatrix}, \tag{38}$$

where

$$\mathbf{P}=(\bar{R},0,\dots,0)'. \tag{39}$$

This is an example of an *invariance* argument commonly used in distribution theory as a means of reducing the number of parameters that need to be considered; we will look at the area of invariance in more detail in Chapter 6.

Now, let $S(n)=(s_{ij}(n))$ be the $m\times m$ sample covariance matrix formed from a sample of size $N=n+1$ from an m-variate elliptical distribution with covariance matrix Σ, given by (38), and kurtosis parameter κ. Partitioning $S(n)$ as

$$S(n)=\begin{bmatrix} s_{11} & \mathbf{s}_{12}' \\ \mathbf{s}_{12} & S_{22}\end{bmatrix}$$

(where we have supressed the dependence on n), the sample multiple correlation coefficient R is given by the positive square root of

$$R^2=\frac{\mathbf{s}_{12}'S_{22}^{-1}\mathbf{s}_{12}}{s_{11}}.$$

It is convenient to work in terms of the following variables constructed from $S(n)$,

(40)
$$u_{11}=\frac{n^{1/2}(s_{11}-1)}{1-\bar{R}^2}$$

$$\mathbf{u}_{12}=n^{1/2}(1-\bar{R}^2)^{-1/2}(I-\mathbf{P}\mathbf{P}')^{-1/2}(\mathbf{s}_{12}-\mathbf{P})$$

$$U_{22}=n^{1/2}(I-\mathbf{P}\mathbf{P}')^{-1/2}(S_{22}-I)(I-\mathbf{P}\mathbf{P}')^{-1/2},$$

where $\mathbf{P}$ is given by (39). Let $U=(u_{ij})$ be the $m\times m$ matrix

(41)
$$U=\begin{bmatrix} u_{11} & \mathbf{u}_{12}' \\ \mathbf{u}_{12} & U_{22}\end{bmatrix}.$$

The asymptotic normality of U follows from the asymptotic normality of $n^{1/2}(S-\Sigma)$ (see Corollary 1.2.18). In terms of the elements of U the sample multiple correlation coefficient R can be expanded as

(42)
$$\begin{aligned}R^2&=s_{11}^{-1}\mathbf{s}_{12}'S_{22}^{-1}\mathbf{s}_{12}\\&=\left[1+n^{-1/2}(1-\bar{R}^2)u_{11}\right]^{-1}\left[\mathbf{P}'+n^{-1/2}(1-\bar{R}^2)^{1/2}\mathbf{u}_{12}'(I-\mathbf{P}\mathbf{P}')^{1/2}\right]\\&\quad\cdot\left[I+n^{-1/2}(I-\mathbf{P}\mathbf{P}')^{1/2}U_{22}(I-\mathbf{P}\mathbf{P}')^{1/2}\right]^{-1}\end{aligned}$$

$$\cdot\left[\mathbf{P}+n^{-1/2}(1-\bar{R}^2)^{1/2}(I-\mathbf{PP}')^{1/2}\mathbf{u}_{12}\right]$$

$$=\left[1-n^{-1/2}(1-\bar{R}^2)u_{12}+O_p(n^{-1})\right]\left[\mathbf{P}'+n^{-1/2}(1-\bar{R}^2)^{1/2}\mathbf{u}'_{11}(I-\mathbf{PP}')^{1/2}\right]$$

$$\cdot\left[I-n^{-1/2}(I-\mathbf{PP}')^{1/2}U_{22}(I-\mathbf{PP}')^{1/2}+O_p(n^{-1})\right]$$

$$\cdot\left[\mathbf{P}+n^{-1/2}(1-\bar{R}^2)^{1/2}(I-\mathbf{PP}')^{1/2}\mathbf{u}_{12}\right]$$

$$=\bar{R}^2+n^{-1/2}\bar{R}(1-\bar{R}^2)(2u_{12}-\bar{R}u_{11}-\bar{R}u_{22})+O_p(n^{-1}),$$

so that, if $\bar{R}\neq 0,1$

$$\frac{n^{1/2}(R^2-\bar{R}^2)}{2\bar{R}(1-\bar{R}^2)}=u_{12}-\tfrac{1}{2}\bar{R}u_{11}-\tfrac{1}{2}\bar{R}u_{22}+O_p(n^{-1/2}),$$

and hence the asymptotic distribution of $n^{1/2}(R^2-\bar{R}^2)/2\bar{R}(1-\bar{R}^2)$ is the same as that of $u_{12}-\frac{1}{2}\bar{R}u_{11}-\frac{1}{2}\bar{R}u_{22}$. Now note that

$$\mathbf{u}\equiv\begin{pmatrix}u_{11}\\u_{12}\\u_{22}\end{pmatrix}=\frac{n^{1/2}}{(1-\bar{R}^2)}\left[\begin{pmatrix}s_{11}\\s_{12}\\s_{22}\end{pmatrix}-\begin{pmatrix}1\\\bar{R}\\1\end{pmatrix}\right],$$

and the asymptotic distribution of this vector, given in Section 5.1.4, is normal with mean $\mathbf{0}$ and covariance matrix $(1-\bar{R}^2)^{-2}V$, where V is given by (24) of Section 5.1.4, with ρ replaced by $\bar{R}$. Putting $\boldsymbol{\alpha}=(-\frac{1}{2}\bar{R}\ 1-\frac{1}{2}\bar{R})'$ it follows that the asymptotic distribution of

$$\boldsymbol{\alpha}'\mathbf{u}=u_{12}-\tfrac{1}{2}\bar{R}u_{11}-\tfrac{1}{2}\bar{R}u_{22}$$

is normal with mean 0 and variance

$$\frac{\boldsymbol{\alpha}'V\boldsymbol{\alpha}}{(1-\bar{R}^2)^2}=\frac{(1+\kappa)(1-\bar{R}^2)^2}{(1-\bar{R}^2)^2}=1+\kappa.$$

Summarizing, we have the following theorem.

THEOREM 5.2.8. Let $\bar{R}$ be the multiple correlation coefficient between Y and $\mathbf{X}$, where $(Y,\mathbf{X}')'$ has an m-variate elliptical distribution with kurtosis parameter κ, and let $R(n)$ be the sample multiple correlation coefficient between Y and $\mathbf{X}$ formed from a sample of size $n+1$ from this distribution.

If $\bar{R}\neq 0,1$, the asymptotic distribution as $n\to\infty$, of

$$\frac{n^{1/2}\left(R(n)^2-\bar{R}^2\right)}{2\bar{R}\left(1-\bar{R}^2\right)}$$

is $N(0,1+\kappa)$.

When the elliptical distribution in Theorem 5.2.8 is *normal*, the kurtosis parameter κ is zero and the limiting distribution of $n^{1/2}(R(n)^2-\bar{R}^2)/[2\bar{R}(1-\bar{R}^2)]$ is $N(0,1)$.

Let us now turn to the asymptotic distribution of R^2 in the null case when $\bar{R}=0$. In this situation it is clear that in the expansion (42) for R^2 we need the term of order n^{-1}. Defining the matrix U as in (41) as before, but with $\bar{R}=0$, we have

$$\begin{aligned}R^2&=s_{11}^{-1}\mathbf{s}_{12}'S_{22}^{-1}\mathbf{s}_{12}\\&=\left(1+n^{-1/2}u_{11}\right)^{-1}\left(n^{-1/2}\mathbf{u}_{12}'\right)\left(I+n^{-1/2}U_{22}\right)^{-1}\left(n^{-1/2}\mathbf{u}_{12}\right)\\&=\frac{1}{n}\left[1-n^{-1/2}u_{11}+O_p(n^{-1})\right]\mathbf{u}_{12}'\left[I-n^{-1/2}U_{22}+O_p(n^{-1})\right]\mathbf{u}_{12},\end{aligned}$$

so that

$$nR^2=\mathbf{u}_{12}'\mathbf{u}_{12}+O_p(n^{-1/2}).$$

Hence the asymptotic distribution of nR^2 (when $\bar{R}=0$) is the same as that of $\mathbf{u}_{12}'\mathbf{u}_{12}$. Using (2) and (3) of Section 1.6, we can show that the asymptotic distribution of $\mathbf{u}_{12}=n^{1/2}\mathbf{s}_{12}$ is $(m-1)$-variate normal, with mean $\mathbf{0}$ and covariance matrix $(1+\kappa)I_{m-1}$ and so the asymptotic distribution of $\mathbf{u}_{12}'\mathbf{u}_{12}/(1+\kappa)$ is χ^2_{m-1}. Summarizing, we have the following theorem.

THEOREM 5.2.9. With the assumptions of Theorem 5.2.8 but with $\bar{R}=0$, the asymptotic distribution of $nR^2/(1+\kappa)$ is χ^2_{m-1}.

Again, when the elliptical distribution is normal we have $\kappa=0$, and then the limiting distribution of nR^2 is χ^2_{m-1}. This is a special case of a result due to Fisher (1928), who established that if $n\to\infty$ and $\bar{R}^2\to 0$ in such a way that $n\bar{R}^2=\delta$ (fixed) then the asymptotic distribution of nR^2 is $\chi^2_{m-1}(\delta)$. A similar result holds also for elliptical distributions, as the following theorem shows.

THEOREM 5.2.10. With the assumptions of Theorem 5.2.8 but with $n\bar{R}^2=\delta$ (fixed), the asymptotic distribution of $nR^2/(1+\kappa)$ is $\chi^2_{m-1}(\delta^*)$, where the noncentrality parameter is $\delta^*=\delta/(1+\kappa)$.

The proof of this result is similar to that of Theorem 5.2.9 and is left as an exercise (see Problem 5.12).

It is natural to ask whether Fisher's variance-stabilizing transformation, which works so well in the case of an ordinary correlation coefficient, is useful in the context of multiple correlation. The answer is yes, as long as $\bar{R}>0$.

THEOREM 5.2.11. Assume the conditions of Theorem 5.2.8 hold, with $\bar{R}\neq 0$, and put

$$z=\tanh^{-1}R \quad \text{and} \quad \xi=\tanh^{-1}\bar{R}.$$

Then, as $n\to\infty$, the asymptotic distribution of

$$n^{1/2}(z-\xi)$$

is $N(0,1+\kappa)$.

This result follows readily from the asymptotic normality of R^2 (when $\bar{R}\neq 0$) established in Theorem 5.2.8; the details are left as an exercise (see Problem 5.13).

For further results on asymptotic distributions for R and approximations to the distribution of R, the reader is referred to Gajjar (1967) and Johnson and Kotz (1970), Chapter 32. Many of the results presented here appear also in Muirhead and Waternaux (1980).

5.2.5. Testing Hypotheses about a Population Multiple Correlation Coefficient

The results of the previous section can be used to test hypotheses about multiple correlation coefficients and to construct confidence intervals. Suppose $(Y,\mathbf{X}')'$ is $N_m(\boldsymbol{\mu},\Sigma)$, where

$$\Sigma=\begin{bmatrix}\sigma_{11} & \boldsymbol{\sigma}_{12}' \\ \boldsymbol{\sigma}_{12} & \Sigma_{22}\end{bmatrix},$$

and we wish to test the null hypothesis H_0: $\bar{R}=0$ against general alternatives H: $\bar{R}>0$, where $\bar{R}$ is the multiple correlation coefficient between Y and $\mathbf{X}$ given by

$$\bar{R}=\left(\frac{\boldsymbol{\sigma}_{12}'\Sigma_{22}^{-1}\boldsymbol{\sigma}_{12}}{\sigma_{11}}\right)^{1/2}.$$

Note that testing H_0 is equivalent to testing that Y and $\mathbf{X}$ are independent.

Given a sample of size $N=n+1$ from this distribution, an exact test can be constructed using the results of Section 5.2.2. If R^2 denotes the sample multiple correlation coefficient between Y and $\mathbf{X}$ we know from Corollary 5.2.3 that

$$\frac{n-m+1}{m-1}\frac{R^2}{1-R^2}$$

has the $F_{m-1,n-m+1}$ distribution when H_0 is true, so that a test of size α is to reject H_0 if

$$\frac{n-m+1}{m-1}\frac{R^2}{1-R^2}>F^*_{m-1,n-m+1}(\alpha),$$

where $F^*_{m-1,n-m+1}(\alpha)$ denotes the upper $100\alpha\%$ point of the $F_{m-1,n-m+1}$ distribution. This test is, in fact, the *likelihood ratio test* of the null hypothesis H_0 (see Problem 5.14). The power function of the test is a function of $\bar{R}$, namely,

$$\beta(\bar{R})=P_{\bar{R}}\left[\frac{n-m+1}{m-1}\frac{R^2}{1-R^2}>F^*_{m-1,n-m+1}(\alpha)\right].$$

An expression for the distribution function of $R^2/(1-R^2)$ was given in Corollary 5.2.6. Using this, it follows that the power function can be expressed as

$$\beta(\bar{R})=\sum_{k=0}^{\infty}c_kP\left[F_{m-1+2k,n-m+1}>\frac{m-1}{m-1+2k}F^*_{m-1,n-m+1}(\alpha)\right],$$

where c_k (with $k\geq 0$) denotes the negative binomial probability given by (22) of Section 5.2.3.

The test described above also has the property that it is a uniformly most powerful invariant test; this approach will be explored further in Section 6.2.

The asymptotic results of Section 5.2.4 can also be used for testing hypotheses. Suppose that we have a sample of size $N=n+1$ from an *elliptical* distribution for $(Y,\mathbf{X}')'$ with kurtosis parameter κ and that we wish to test H_0: $\bar{R}=0$ against H: $\bar{R}>0$. Bear in mind that $\kappa=0$ takes us back to the normal distribution. From Theorem 5.2.9, the asymptotic distribution of $nR^2/(1+\kappa)$ is χ^2_{m-1}, so that an approximate test of size α is to reject H_0 if

$$\frac{nR^2}{1+\kappa}>c_{m-1}(\alpha),$$

where $c_{m-1}(\alpha)$ denotes the upper $100\alpha\%$ point of the χ^2_{m-1} distribution. (If κ is not known, it can be replaced by a consistent estimate $\hat{\kappa}$.) The power function of this test may be calculated approximately using Theorem 5.2.10 for alternatives $\bar{R}$ which are close to zero and Theorem 5.2.8 for alternatives $\bar{R}$ further away from zero. Theorems 5.2.8 or 5.2.11 may also be used for testing the null hypothesis K_0: $\bar{R}=\bar{R}_0(>0)$ against general alternatives K: $\bar{R}\neq\bar{R}_0$. Putting $\xi_0=\tanh^{-1}\bar{R}_0$, we know from Theorem 5.2.11 that when K_0 is true the distribution of $z=\tanh^{-1}R$ is approximately

$$N\left(\xi_0,\frac{1+\kappa}{n}\right),$$

so that an approximate test of size α is to reject H_0 if

$$\frac{n^{1/2}|z-\xi_0|}{(1+\kappa)^{1/2}}\geq d_\alpha,$$

where d_α is the two-tailed $100\alpha\%$ point of the $N(0,1)$ distribution. It should be remembered that the asymptotic normality of R and hence of z holds only if $\bar{R}\neq 0$, and the normal approximation is not likely to be much good if $\bar{R}$ is close to zero. If $\bar{R}\neq 0$ the asymptotic normality of z also leads to confidence intervals for ξ and for $\bar{R}$. An interval for ξ with confidence coefficient $1-\alpha$ (approximately) is

$$z-d_\alpha\left(\frac{1+\kappa}{n}\right)^{1/2}\leq\xi\leq z+d_\alpha\left(\frac{1+\kappa}{n}\right)^{1/2},$$

and for $\bar{R}$ such an interval is

$$\tanh\left[z-d_\alpha\left(\frac{1+\kappa}{n}\right)^{1/2}\right]\leq\bar{R}\leq\tanh\left[z+d_\alpha\left(\frac{1+\kappa}{n}\right)^{1/2}\right].$$

The caveat mentioned at the end of Section 5.1.5 is also applicable here with regard to inferences concerning elliptical distributions.

5.3. PARTIAL CORRELATION COEFFICIENTS

Suppose that $\mathbf{X}$ is $N_m(\boldsymbol{\mu},\Sigma)$ and partition $\mathbf{X}$, $\boldsymbol{\mu}$, and Σ as

$$\mathbf{X}=\begin{pmatrix}\mathbf{X}_1\\ \mathbf{X}_2\end{pmatrix},\qquad \boldsymbol{\mu}=\begin{pmatrix}\boldsymbol{\mu}_1\\ \boldsymbol{\mu}_2\end{pmatrix},\qquad \Sigma=\begin{bmatrix}\Sigma_{11} & \Sigma_{12}\\ \Sigma_{21} & \Sigma_{22}\end{bmatrix},$$

where $\mathbf{X}_1$ and $\boldsymbol{\mu}_1$ are $k\times 1$, $\mathbf{X}_2$ and $\boldsymbol{\mu}_2$ are $(m-k)\times 1$, Σ_{11} is $k\times k$, and Σ_{22} is $(m-k)\times(m-k)$. From Theorem 1.2.11 the conditional distribution of $\mathbf{X}_1$ given $\mathbf{X}_2$ is $N_k(\boldsymbol{\mu}_1+\Sigma_{12}\Sigma_{22}^{-1}(\mathbf{X}_2-\boldsymbol{\mu}_2),\Sigma_{11\cdot 2})$, where

$$\Sigma_{11\cdot 2}=\Sigma_{11}-\Sigma_{12}\Sigma_{22}^{-1}\Sigma_{21}=(\sigma_{ij\cdot k+1,\ldots,m}),$$

i.e., $\sigma_{ij\cdot k+1,\ldots,m}$ denotes the $i-j$th element of the $k\times k$ matrix $\Sigma_{11\cdot 2}$. The partial correlation coefficient between two variables X_i and X_j, which are components of the subvector $\mathbf{X}_1$ when $\mathbf{X}_2$ is held fixed, is denoted by $\rho_{ij\cdot k+1,\ldots,m}$ and is defined as being the correlation between X_i and X_j in the conditional distribution of $\mathbf{X}_1$, given $\mathbf{X}_2$. Hence

$$\rho_{ij\cdot k+1,\ldots,m}=\frac{\sigma_{ij\cdot k+1,\ldots,m}}{(\sigma_{ii\cdot k+1,\ldots,m}\sigma_{jj\cdot k+1,\ldots,m})^{1/2}}.$$

Now suppose a sample of size N is drawn from this $N_m(\boldsymbol{\mu},\Sigma)$ distribution. Let $A=N\hat{\Sigma}$, where $\hat{\Sigma}$ is the maximum likelihood estimate of Σ, and partition A as

$$A=\begin{bmatrix} A_{11} & A_{12}\\ A_{21} & A_{22}\end{bmatrix}$$

where A_{11} is $k\times k$ and A_{22} is $(m-k)\times(m-k)$. The maximum likelihood estimate of $\Sigma_{11\cdot 2}$ is $\hat{\Sigma}_{11\cdot 2}=N^{-1}A_{11\cdot 2}$, where $A_{11\cdot 2}=A_{11}-A_{12}A_{22}^{-1}A_{21}=(a_{ij\cdot k+1,\ldots,m})$, and the maximum likelihood estimate of $\rho_{ij\cdot k+1,\ldots,m}$ is

$$r_{ij\cdot k+1,\ldots,m}=\frac{a_{ij\cdot k+1,\ldots,m}}{(a_{ii\cdot k+1,\ldots,m}a_{jj\cdot k+1,\ldots,m})^{1/2}}.$$

Now recall that we obtained the distribution of an ordinary correlation coefficient defined in terms of the matrix A having the $W_m(n,\Sigma)$ distribution, with $n=N-1$. Here we can obtain the distribution of a partial correlation coefficient starting with the distribution of the matrix $A_{11\cdot 2}$ which, from Theorem 3.2.10, is $W_k(n-m+k,\Sigma_{11\cdot 2})$. The derivation is exactly the same as that of Theorem 5.1.5, leading to the following result.

THEOREM 5.3.1. *If $r_{ij\cdot k+1,\ldots,m}$ is a sample partial correlation coefficient formed from a sample of size $N=n+1$ from a normal distribution then its density function is the same as that of an ordinary correlation coefficient given by (10) and (15) of Section 5.1 with n replaced by $n-m+k$.*

As a consequence of this theorem, the inference procedures discussed in Section 5.1.5 in the context of an ordinary correlation coefficient are all relevant to a partial correlation coefficient as well, as long as the underlying distribution is normal. The asymptotic normality of $r_{ij\cdot k+1,\ldots,m}$ and of $z=\tanh^{-1} r_{ij\cdot k+1,\ldots,m}$ follow directly as well, using Theorems 5.1.6 and 5.1.7 (with $\kappa=0$).

PROBLEMS

5.1. Let R be an $m\times m$ correlation matrix having the density function of Theorem 5.1.3 and moments given by (9) of Section 5.1. Find the characteristic function $\phi_n(t)$ of $-n\log\det R$. Using (17) of Section 3.2, show that

$$\lim_{n\to\infty}\log\phi_n(t)=-\tfrac{1}{4}m(m-1)\log(1-2it)$$

so that $-n\log\det R\to\chi^2_{m(m-1)/2}$ in distribution as $n\to\infty$.

5.2. Prove Lemma 5.1.4.

5.3. Show that the density function of a correlation coefficient r obtained from normal sampling can be expressed in the form (15) of Section 5.1.

5.4. Prove Theorem 5.1.7.
[*Hint*: A very useful result to know is that if $\{X_n\}$ is a sequence of random variables such that $n^{1/2}(X_n-\mu)\to N(0,\sigma^2)$ in distribution as $n\to\infty$, and if $f(x)$ is a function which is differentiable at $x=\mu$, then $n^{1/2}[f(X_n)-f(\mu)]\to N(0,f'(\mu)^2\sigma^2)$ in distribution as $n\to\infty$; see, e.g., Bickel and Doksum (1977), p. 461.]

5.5. Let $\mathbf{X}_1,\ldots,\mathbf{X}_n$ be independent $N_2(\boldsymbol{\mu},\Sigma)$ random vectors where

$$\Sigma=\begin{bmatrix}\sigma_1^2 & \rho\sigma_1\sigma_2\\ \rho\sigma_1\sigma_2 & \sigma_2^2\end{bmatrix}.$$

(a) Show that the likelihood ratio statistic for testing H_0: $\rho=\rho_0$ against H: $\rho\neq\rho_0$ is

$$\Lambda=\left[\frac{(1-\rho_0^2)(1-r^2)}{(1-\rho_0 r)^2}\right]^{N/2}$$

(b) Show that the likelihood ratio test of size α rejects H_0 if $r<r_1$ or $r>r_2$, where r_1 and r_2 are determined by the equations

$$P(r<r_1 \mid \rho=\rho_0)+P(r>r_2 \mid \rho=\rho_0)=\alpha$$

and

$$\frac{1-r_1^2}{(1-\rho_0 r_1)^2}=\frac{1-r_2^2}{(1-\rho_0 r_2)^2}.$$

(c) Show that when $\rho_0=0$ the likelihood ratio test of size α rejects H_0: $\rho=0$ if $(n-1)^{1/2}|r|/(1-r^2)^{1/2}>t^*_{n-1}(\alpha)$, where $t^*_{n-1}(\alpha)$ denotes the two-tailed $100\alpha\%$ point of the t_{n-1} distribution, with $n=N-1$.

5.6. Suppose r_i is the sample correlation coefficient from a sample of size $N_i=n_i+1$ from a bivariate elliptical distribution with correlation coefficient ρ_i and kurtosis parameter κ_i $(i=1,2)$, where κ_i is assumed known $(i=1,2)$. Explain how an approximate test of size α of H_0: $\rho_1=\rho_2$ against H: $\rho_1\neq\rho_2$ may be constructed.

5.7. Let r be the correlation coefficient formed from a sample of size $N=n+1$ from a bivariate normal distribution with correlation coefficient ρ, so that r has density function given by (15) of Section 5.1. Show that $E[\sin^{-1}r]=\sin^{-1}\rho$.

5.8. Let r be the sample correlation coefficient formed from a sample of size $N=n+1$ from a bivariate normal distribution with correlation coefficient ρ. Put $z=\tanh^{-1}r$ and $\xi=\tanh^{-1}\rho$ so that z is the maximum likelihood estimate of ξ. Show that

$$E(z)=\xi+\frac{\rho}{2n}+O(n^{-2})$$

and

$$\mathrm{Var}(z)=\frac{1}{n}+O(n^{-2}).$$

5.9. From Problem 5.8 the bias in z is of order n^{-1}. Often bias can be reduced by looking not at the maximum likelihood estimate but at an estimate which maximizes a "marginal" likelihood function depending only on the parameter of interest. The density function of r in Theorem 5.1.5 depends only on ρ; the part involving ρ can be regarded as a marginal likelihood function $L(\rho)$, where

$$L(\rho)=(1-\rho^2)^{n/2}(1-\rho r)^{-n+1/2}{}_2F_1\left(\tfrac{1}{2},\tfrac{1}{2};n+\tfrac{1}{2};\tfrac{1}{2}(1+\rho r)\right).$$

It is difficult to find the value of ρ which maximizes this but an approximation can be found. Since

$$_2F_1\left(\tfrac{1}{2},\tfrac{1}{2};n+\tfrac{1}{2};\tfrac{1}{2}(1+\rho r)\right)=1+O(n^{-1})$$

we have

$$L(\rho)=L_1(\rho)\left[1+O(n^{-1})\right]$$

where

$$L_1(\rho)=(1-\rho^2)^{n/2}(1-\rho r)^{-n+1/2}.$$

(a) Show that the value r^* of ρ which maximizes $L_1(\rho)$ may be written as

$$r^*=r-\frac{1}{2n}r(1-r^2)+O(n^{-2}).$$

(b) Let $z^*=\tanh^{-1}r^*$. Show that

$$z^*=\tanh^{-1}r-\frac{r}{2n}+O(n^{-2}).$$

(c) Show that

$$E(z^*)=\xi+O(n^{-2}).$$

and

$$\text{Var}(z^*)=\frac{1}{n}+O(n^{-2})$$

where $\xi=\tanh^{-1}\rho$. (This shows that the bias in z^* is of smaller order of magnitude than the bias in $z=\tanh^{-1}r$ given in Problem 5.8.)

5.10. Consider a random variable V_1 whose distribution is the mixture of χ^2 distributions given by (25) of Section 5.2. This says that the distribution of V_1 can be obtained by taking a random variable K having a negative binomial distribution with $P(K=k)=c_k$ $(k=0,1,\ldots)$, where c_k is given by (22) of Section 5.2, and then taking the conditional distribution of V_1 given $K=k$ to be χ^2_{m-1+2k}.

(a) By conditioning on K show that

$$E[V_1]=m-1+n\theta$$

and

$$\mathrm{Var}(V_1)=2m-2+4n\theta+2n\theta^2,$$

where $\theta=\bar{R}^2/(1-\bar{R}^2)$.

(b) Suppose that the distribution of V_1 is approximated by the gamma distribution, with parameters $\frac{1}{2}b$ and $2a$ given by (26) of Section 5.2, by equating the mean and variance of V_1 to the mean and variance of the gamma distribution. Show that the fitted values for a and b are

$$a=\frac{n\theta(\theta+2)+m-1}{n\theta+m-1}, \qquad b=\frac{(n\theta+m-1)^2}{n\theta(\theta+2)+m-1}.$$

5.11. Prove that:

(a) $$\sum_{k=0}^{\infty}\frac{(a)_k(b)_k}{(c)_k k!}(-z)^k {}_2F_1(a+k,b+k;c+k;1+z)$$
$$={}_2F_1(a,b;c;1).$$

(b) $${}_2F_1(a,b;c;1)=\frac{\Gamma(c)\Gamma(c-a-b)}{\Gamma(c-a)\Gamma(c-b)} \qquad [c\neq 0,-1,-2,\ldots;\ \mathrm{Re}(c)>\mathrm{Re}(a+b)]$$

(c) Let R be a sample multiple correlation coefficient obtained from normal sampling having the density function of Theorem 5.2.4, and consider the problem of estimating $\bar{R}^2$. Using parts (a) and (b) above and the moments of $1-R^2$ given by (29) of Section 5.2, show that the estimate

$$T(R^2)=1-\left(\frac{n-2}{n-m+1}\right)(1-R^2)\,{}_2F_1(1,1;\tfrac{1}{2}(n-m+3);1-R^2)$$

is an unbiased estimate of $\bar{R}^2$.

5.12. Prove Theorem 5.2.10.

5.13. Prove Theorem 5.2.11. (See the hint following Problem 5.4.)

5.14. Suppose that $(Y,\mathbf{X}')'$ is $N_m(\boldsymbol{\mu},\Sigma)$, where $\mathbf{X}$ is $(m-1)\times 1$ and

$$\Sigma=\begin{bmatrix}\sigma_{11} & \boldsymbol{\sigma}_{12}' \\ \boldsymbol{\sigma}_{12} & \Sigma_{22}\end{bmatrix}, \qquad \Sigma_{22} \text{ is } (m-1)\times(m-1),$$

and consider testing H_0: $\bar{R}=0$ against H: $\bar{R}>0$, where $\bar{R}=(\sigma_{12}'\Sigma_{22}^{-1}\sigma_{12}/\sigma_{11})^{1/2}$. Note that H_0 is true if and only if $\sigma_{12}=\mathbf{0}$. Suppose a sample of size $N=n+1$ is drawn.

(a) Show that the likelihood ratio statistic for testing H_0 against H is

$$\Lambda=(1-R^2)^{N/2},$$

where R is the sample multiple correlation coefficient between Y and $\mathbf{X}$.

(b) Show that the likelihood ratio test of size α rejects H_0 if

$$\frac{n-m+1}{m-1}\frac{R^2}{1-R^2}>F^*_{m-1,n-m+1}(\alpha)$$

where $F^*_{m-1,n-m+1}(\alpha)$ is the upper $100\alpha\%$ point of the $F_{m-1,n-m+1}$ distribution.

5.15. Let S be the sample covariance matrix formed from a sample of size $N=n+1$ on $\mathbf{X}=(X_1,X_2,\ldots,X_m)'$, which is a $N_m(\mu,\Sigma)$ random vector, so that $A=nS$ is $W_m(n,\Sigma)$. Suppose that $\Sigma=\text{diag}(\sigma_{11},\ldots,\sigma_{mm})$. Let $R_{j\cdot1,\ldots,j-1}$ denote the sample multiple correlation between X_j and $X_1,\ldots,X_{j-1}$ for $j=2,\ldots,m$. Put $A=T'T$, where T is an upper-triangular matrix with positive diagonal elements.

(a) Show that

$$R^2_{j\cdot1,\ldots,j-1}=\frac{\sum_{i=1}^{j-1}t_{ij}^2}{\sum_{i=1}^{j}t_{ij}^2}.$$

(b) Show that the joint density function of the t_{ij}'s is

$$\frac{2^{-m(n-2)/2}}{\Gamma_m(\frac{1}{2}n)}\prod_{i=1}^{m}\sigma_{ii}^{-n/2}\exp\left(-\frac{1}{2}\sum_{i\le j}\sigma_{jj}^{-1}t_{ij}^2\right)\prod_{i=1}^{m}t_{ii}^{n-i}.$$

(c) From part (b), above, find the joint density function of the t_{ij} for $i<j$ and the $R^2_{j\cdot1,\ldots,j-1}$ for $j=2,\ldots,m$. Hence show that the $R^2_{j\cdot1,\ldots,j-1}$ are independent and $R^2_{j\cdot1,\ldots,j-1}$ has the beta $[\frac{1}{2}(j-1),\frac{1}{2}(n-j+1)]$ distribution.

5.16. Show that:

(a) ${}_2F_1(a,b;c;x)=(1-x)^{-a}{}_2F_1\left(a,c-b;c;\dfrac{-x}{1-x}\right)$.

[*Hint*: The right side is equal to

$$\sum_{k=0}^{\infty} \frac{(a)_k(c-b)_k}{(c)_k k!} \frac{(-1)^x x^k}{(1-x)^{a+k}}.$$

Expand $(1-x)^{-a-k}$ using the binomial expansion and then interchange the order of summation. Use result (b) of Problem 5.11 to tidy up.]

(b) Suppose that a sample multiple correlation coefficient R^2 has the density function given in Theorem 5.2.4. Show that if $k=\frac{1}{2}(n+1-m)$ is a positive integer the distribution function of R^2 can be written in the form

$$P(R^2 \le x) = \sum_{j=0}^{k} b_j I_y\left(\tfrac{1}{2}(m-1)+j, k\right),$$

where I_y denotes the incomplete beta function given by (19) of Section 5.2, with $y=x(1-\bar{R}^2)/(1-x\bar{R}^2)$, and b_j denotes the binomial probability

$$b_j = \binom{k}{j}(\bar{R}^2)^j(1-\bar{R}^2)^{k-j}.$$

5.17. Prove that

$$1-\rho_{12\cdot 3,4,\ldots,m}^2 = \frac{\sigma_{11\cdot 2,3,\ldots,m}}{\sigma_{11\cdot 3,4,\ldots,m}}$$

and use this to show that

$$1-\bar{R}^2_{1\cdot 2,3,\ldots,m} = (1-\rho^2_{12\cdot 3,4,\ldots,m})(1-\rho^2_{13\cdot 4,5,\ldots,m})\cdots(1-\rho^2_{1,m-1\cdot m})(1-\rho^2_{1m}).$$

5.18. Show that

$$\rho_{12\cdot 3} = \frac{\rho_{12}-\rho_{13}\rho_{23}}{\left[(1-\rho_{23}^2)(1-\rho_{13}^2)\right]^{1/2}}.$$

5.19. If the random vector $\mathbf{X}=(X_1, X_2, X_3, X_4)'$ has covariance matrix

$$\Sigma=\begin{bmatrix} \sigma^2 & \sigma_{12} & \sigma_{13} & \sigma_{14} \\ \sigma_{12} & \sigma^2 & \sigma_{14} & \sigma_{13} \\ \sigma_{13} & \sigma_{14} & \sigma^2 & \sigma_{12} \\ \sigma_{14} & \sigma_{13} & \sigma_{12} & \sigma^2 \end{bmatrix}$$

show that the four multiple correlation coefficients between one variable and the other three are equal.

CHAPTER 6

Invariant Tests and Some Applications

6.1. INVARIANCE AND INVARIANT TESTS

Many inference problems in statistics have inherent properties of symmetry or invariance and thereby impose fairly natural restrictions on the possible procedures that should be used. As a simple example, suppose that $(X,Y)'$ has a bivariate normal distribution with correlation coefficient ρ and consider the problem of estimating ρ given a sample $(X_i, Y_i)'$, $i=1,\ldots,N$. The correlation coefficient ρ is unchanged by, or is invariant under, the transformations $\tilde{X}=b_1X+c_1$, $\tilde{Y}=b_2Y+c_2$ where $b_1>0$, $b_2>0$, so that it is natural to require that if the statistic $\phi(X_1,Y_1,\ldots,X_N,Y_N)$ is to be used as an estimate of ρ then ϕ should also be invariant; that is

$$\phi(X_1,Y_1,\ldots,X_N,Y_N)=\phi(\tilde{X}_1,\tilde{Y}_1,\ldots,\tilde{X}_N,\tilde{Y}_N),$$

since both sides are estimating the same parameter. The sample correlation coefficient r (see Section 5.1) is obviously an example of such an invariant estimate. The reader will recall that a similar type of invariance argument was used in Section 4.3 in connection with the estimation of a covariance matrix.

In many hypothesis-testing problems in multivariate analysis there is no uniformly most powerful or uniformly most powerful unbiased test. There is, however, often a natural group of transformations with respect to which a specific testing problem is invariant, and where it is sensible to restrict one's attention to the class of invariant tests; that is, to tests based on statistics that are invariant under this group of transformations. The likelihood ratio test under general conditions is such a test, but it need not be the "best" one. In some interesting situations it turns out that within this class there exists a test which is uniformly most powerful, and such a test is called, of course, a *uniformly most powerful invariant test*. Often such a test, if it exists, is the same as the likelihood ratio test, but this is not always the case. In what follows we will review briefly some of the relevant theory needed about invariance; much more detail can be found in Lehmann (1959), Chapter 6, and Ferguson (1967), Chapters 4 and 5. For further applications of invariance arguments to problems in multivariate analysis the reader is referred to T. W. Anderson (1958) and Eaton (1972).

Let G denote a *group* of transformations from a space $\mathcal{X}$ into itself; this means that, if $g_1 \in G$, $g_2 \in G$, then $g_1 g_2 \in G$ where $g_1 g_2$ is defined as the transformation $(g_1 g_2)x = g_1(g_2 x)$, and that if $g \in G$ then $g^{-1} \in G$, where g^{-1} satisfies $gg^{-1} = e$, with e the identity transformation in G. Obviously all transformations in G are 1–1 of $\mathcal{X}$ *onto* itself.

DEFINITION 6.1.1. Two points x_1, x_2 in $\mathcal{X}$ are said to be *equivalent* under G, written $x_1 \sim x_2 \pmod G$, if there exists a $g \in G$ such that $x_2 = gx_1$.

Clearly, this is an *equivalence* relation; that is, it has the properties that

(i) $x \sim x \pmod G$;

(ii) $x \sim y \pmod G \Rightarrow y \sim x \pmod G$; and

(iii) $x \sim y \pmod G$, $y \sim z \pmod G \Rightarrow x \sim z \pmod G$.

The equivalence classes are called the *orbits* of $\mathcal{X}$ under G; in particular, the set $\{gx;\, g \in G\}$ is called the orbit of x under G. Obviously two orbits are either identical or disjoint, and the orbits form a *partition* of $\mathcal{X}$. Two types of function defined on $\mathcal{X}$ are of fundamental importance.

DEFINITION 6.1.2. A function $\phi(x)$ on $\mathcal{X}$ is said to be *invariant* under G if

$$\phi(gx) = \phi(x) \qquad \text{for all} \quad x \in \mathcal{X} \quad \text{and} \quad g \in G.$$

Hence, ϕ is invariant if and only if it is constant on each orbit under G.

DEFINITION 6.1.3. A function $\phi(x)$ on $\mathcal{X}$ is said to be a *maximal invariant* under G if it is invariant under G and if

$$\phi(x_1) = \phi(x_2) \Rightarrow x_1 \sim x_2 \pmod G.$$

Hence ϕ is a maximal invariant if and only if it is constant on each orbit and assigns different values to each orbit. Any invariant function is a function of a maximal invariant, as the following theorem shows.

THEOREM 6.1.4. Let the function $\phi(x)$ on $\mathcal{X}$ be a maximal invariant under G. Then a function $\psi(x)$ on $\mathcal{X}$ is invariant under G if and only if ψ is a function of $\phi(x)$.

Proof. Suppose ψ is a function of $\phi(x)$; that is, there exists a function f such that

$$\psi(x) = f(\phi(x)) \qquad \text{for all} \quad x \in \mathcal{X}.$$

Then, for all $g \in G$, $x \in \mathcal{X}$

$$\psi(gx) = f(\phi(gx)) = f(\phi(x)) = \psi(x),$$

and hence ψ is invariant.

Now suppose that ψ is invariant. If $\phi(x_1) = \phi(x_2)$ then $x_1 \sim x_2 \pmod G$, because ϕ is a maximal invariant, and hence $x_2 = gx_1$ for some $g \in G$. Then

$$\psi(x_2) = \psi(gx_1) = \psi(x_1),$$

which establishes that $\psi(x)$ depends on x only through $\phi(x)$ and completes the proof.

DEFINITION 6.1.5. If $x_1 \sim x_2 \pmod G$ for all x_1, x_2 in $\mathcal{X}$ then the group G is said to act *transitively* on $\mathcal{X}$, and $\mathcal{X}$ is said to be *homogeneous* with respect to G.

Hence, G acts transitively on $\mathcal{X}$ if there is only one orbit, namely, $\mathcal{X}$ itself. In this case the only invariant functions are constant functions. Continuing, if x_0 is any point taken as origin in the homogeneous space $\mathcal{X}$, then the subgroup G_0 of G, consisting of all transformations which leave x_0 invariant, namely,

$$G_0 = \{g \in G;\, gx_0 = x_0\},$$

is called the *isotropy subgroup* of G at x_0. It is clear that if g is any group element transforming x_0 into x $(gx_0 = x)$ then the set of all group elements which transform x_0 into x is the *coset*

$$gG_0 = \{gg_0;\, g_0 \in G_0\}.$$

Hence the points $x \in \mathcal{X}$ are in 1–1 correspondence with the cosets gG_0 so that $\mathcal{X}$ may be *regarded* as the coset space G/G_0 consisting of the cosets gG_0.

We will now look at some examples which illustrate these concepts.

EXAMPLE 6.1.6. Suppose that $\mathfrak{X} = R^m$ and $G = O(m)$, the group of $m \times m$ orthogonal matrices. The action of $H \in O(m)$ on $\mathbf{x} \in R^m$ is

$$\mathbf{x} \to H\mathbf{x}$$

and the group operation is matrix multiplication. The orbit of $\mathbf{x}$ under $O(m)$ consists of all points of the form $\mathbf{y} = H\mathbf{x}$ for some $H \in O(m)$; this is the same as the set of all points in R^m which have the same distance from the origin as $\mathbf{x}$. For, if $\mathbf{y} = H\mathbf{x}$ then obviously $\mathbf{y}'\mathbf{y} = \mathbf{x}'\mathbf{x}$. Conversely, suppose that $\mathbf{y}'\mathbf{y} = \mathbf{x}'\mathbf{x}$. Choose H_x and H_Y in $O(m)$ such that

$$H_x\mathbf{x} = (\|\mathbf{x}\|,\ 0,\dots,0)' \qquad \text{and} \qquad H_Y\mathbf{y} = (\|\mathbf{y}\|,\ 0,\dots,0)'$$

then $H_x\mathbf{x} = H_Y\mathbf{y}$ so that $\mathbf{y} = H\mathbf{x}$, with $H = H_Y'H_x \in O(m)$. A maximal invariant under G is $\phi(\mathbf{x}) = \mathbf{x}'\mathbf{x}$, and any invariant function is a function of $\mathbf{x}'\mathbf{x}$.

EXAMPLE 6.1.7. Suppose that $\mathfrak{X} = R^2 \times \mathbb{S}_2$, where $\mathbb{S}_2$ is the space of positive definite 2×2 matrices $\Sigma = (\sigma_{ij})$, and G is the group of transformations

$$(1) \qquad G = \left\{ (B, \mathbf{c});\ B = \begin{bmatrix} b_1 & 0 \\ 0 & b_2 \end{bmatrix},\ \mathbf{c} = \begin{pmatrix} c_1 \\ c_2 \end{pmatrix},\ b_1 > 0,\ b_2 > 0 \right\}.$$

The group operation is defined by

$$(B_1, \mathbf{c}_1)(B_2, \mathbf{c}_2) = (B_1 B_2,\ B_1\mathbf{c}_2 + \mathbf{c}_1)$$

and the action of $(B, \mathbf{c}) \in G$ on $(\boldsymbol{\mu}, \Sigma) \in \mathfrak{X}$ is

$$(2) \qquad (\boldsymbol{\mu}, \Sigma) \to (B, \mathbf{c})(\boldsymbol{\mu}, \Sigma) = (B\boldsymbol{\mu} + \mathbf{c},\ B\Sigma B').$$

A maximal invariant under G is

$$\phi(\boldsymbol{\mu}, \Sigma) = \frac{\sigma_{12}}{(\sigma_{11}\sigma_{22})^{1/2}}.$$

To prove this, first note that if $(B, \mathbf{c}) \in G$, then

$$B\Sigma B' = \begin{bmatrix} b_1^2\sigma_{11} & b_1 b_2 \sigma_{12} \\ b_1 b_2 \sigma_{12} & b_2^2 \sigma_{22} \end{bmatrix}$$

so that

$$\phi(B\mu+\mathbf{c}, B\Sigma B')=\frac{b_1 b_2 \sigma_{12}}{(b_1^2\sigma_{11}b_2^2\sigma_{22})^{1/2}}=\frac{\sigma_{12}}{(\sigma_{11}\sigma_{22})^{1/2}}=\phi(\mu,\Sigma),$$

and hence ϕ is invariant. To show that it is maximal invariant, suppose that

$$\phi(\mu,\Sigma)=\phi(\tau,\Gamma),$$

that is

$$\frac{\sigma_{12}}{(\sigma_{11}\sigma_{22})^{1/2}}=\frac{\gamma_{12}}{(\gamma_{11}\gamma_{22})^{1/2}}.$$

Putting $b_1=(\gamma_{11}/\sigma_{11})^{1/2}$, $b_2=(\gamma_{22}/\sigma_{22})^{1/2}$ and $\mathbf{c}=-B\mu+\tau$, we have

$$(B,\mathbf{c})(\mu,\Sigma)=(\tau,\Gamma)$$

so that

$$(\mu,\Sigma)\sim(\tau,\Gamma) \quad (\operatorname{mod} G),$$

as required. Regarding $\mathfrak{X}=R^2\times\mathcal{S}_2$ as the set of all possible mean vectors μ and covariance matrices Σ of the random vector $\mathbf{X}=(X_1, X_2)'$, the transformation (2) is induced by the transformation $\mathbf{Y}=B\mathbf{X}+\mathbf{c}$ in the sense that the mean of $\mathbf{Y}$ is $B\mu+\mathbf{c}$ and the covariance matrix of $\mathbf{Y}$ is $B\Sigma B'$. We have shown that the correlation coefficient ρ between X_1 and X_2 is a maximal invariant under G, and so any invariant function is a function of ρ.

EXAMPLE 6.1.8. Suppose that $\mathfrak{X}=\mathcal{S}_m$, the space of positive definite $m\times m$ matrices, and $G=\mathcal{G}\ell(m,R)$, the general linear group of $m\times m$ nonsingular real matrices. The action of $L\in\mathcal{G}\ell(m,R)$ on $S\in\mathcal{S}_m$ is given by the congruence transformation

$$S\rightarrow LSL',$$

with the group operation being matrix multiplication. The group $\mathcal{G}\ell(m,R)$ acts *transitively* on $\mathcal{S}_m$ and the only invariant functions are constant functions. The *isotropy subgroup* of $\mathcal{G}\ell(m,R)$ at $I_m\in\mathcal{S}_m$ is clearly the orthogonal group $O(m)$. Given $S\in\mathcal{S}_m$ the coset corresponding to S is

$$LO(m)=\{LH;\ H\in O(m)\},$$

where L is any matrix in $\mathcal{GL}(m, R)$ such that $S = LL'$. Writing the homogeneous space $\mathcal{S}_m$ as a coset space of the isotropy subgroup, we have

$$\mathcal{S}_m = \mathcal{GL}(m, R)/O(m).$$

EXAMPLE 6.1.9. Suppose that $\mathcal{X} = V_{m,n}$, the *Stiefel manifold* of $n \times m$ matrices with orthonormal columns (see Section 2.1.4), and $G = O(n)$. The action of $H \in O(n)$ on $Q_1 \in V_{m,n}$ is given by

$$Q_1 \rightarrow HQ_1,$$

with the group operation being matrix multiplication. Then $O(n)$ acts transitively on $V_{m,n}$ (why?) so that the only invariant functions are constant functions. The isotropy subgroup of $O(n)$ at

$$\begin{bmatrix} I_m \\ \cdots \\ 0 \end{bmatrix} \in V_{m,n}$$

is clearly

$$G_0 = \left\{ \begin{bmatrix} I_m & 0 \\ 0 & H_1 \end{bmatrix} \in O(n); \quad H_1 \in O(n-m) \right\},$$

and the coset corresponding to $Q_1 \in V_{m,n}$ is $[Q_1 : Q_2]G_0$, where Q_2 is any $n \times (n-m)$ matrix such that $[Q_1 : Q_2] \in O(n)$. This coset consists of all orthogonal $n \times n$ matrices with Q_1 as the first m columns. Writing the homogeneous space $V_{m,n}$ as a coset space of the isotropy subgroup we have

$$V_{m,n} = O(n)/O(n-m).$$

Continuing, let X be a random variable with values in a space $\mathcal{X}$ and probability distribution P_θ, with $\theta \in \Omega$. (The distributions P_θ are, of course, defined over a σ-algebra $\mathcal{B}$ of subsets of $\mathcal{X}$, but measurability considerations will not be stressed in our discussion.) Let G be a group of transformations from $\mathcal{X}$ into itself. (These transformations are assumed measurable, so that for each $g \in G$, gX is also a random variable, taking the value gx when $X = x$.) The space $\mathcal{X}$ here is the sample space and Ω is the parameter space. In many important situations it turns out that the distributions P_θ are invariant, in the sense of the following definition.

DEFINITION 6.1.10. The family of distributions $\{P_\theta; \theta \in \Omega\}$ is said to be *invariant* under G if every $g \in G$, $\theta \in \Omega$ determine a unique element in Ω, denoted by $\bar{g}\theta$, such that when X has distribution P_θ, gX has distribution $P_{\bar{g}\theta}$.

This means that for every (measurable) set $B \subset \mathfrak{X}$

$$P_\theta(gX \in B) = P_{\bar{g}\theta}(X \in B),$$

which is equivalent to

$$P_\theta(g^{-1}B) = P_{\bar{g}\theta}(B)$$

and hence to

$$P_{\bar{g}\theta}(gB) = P_\theta(B).$$

Now, suppose that the family $\{P_\theta;\ \theta \in \Omega\}$ is invariant under G and let

$$\bar{G} = \{\bar{g};\ g \in G\}.$$

Then the elements of $\bar{G}$ are transformations of the parameter space into itself. In fact, as the following theorem shows, $\bar{G}$ is a group, called the group *induced* by G.

THEOREM 6.1.11. If the family of distributions $\{P_\theta;\ \theta \in \Omega\}$ is invariant under the group G then $\bar{G} = \{\bar{g};\ g \in G\}$ is a group of transformations from Ω into itself.

Proof. If the distribution of X is P_θ then $g_1 X$ has distribution $P_{\bar{g}_1\theta}$ and so $g_2(g_1 X)$ has distribution $P_{\bar{g}_2(\bar{g}_1\theta)}$. But $g_2(g_1 X) = (g_2 g_1)X$ also has distribution $P_{\overline{g_2 g_1}\theta}$. By uniqueness it follows that $\bar{g}_2\bar{g}_1 = \overline{g_2 g_1} \in \bar{G}$, so that $\bar{G}$ is closed under composition. To show that $\bar{G}$ is closed under inversion, put $g_2 = g_1^{-1}$, then $\overline{g_1^{-1}}\,\bar{g}_1 = \bar{e}$; now $\bar{e}$ is the identity element in $\bar{G}$, and so $\bar{g}_1^{-1} = \overline{g_1^{-1}} \in \bar{G}$.

Obviously, all transformations in $\bar{G}$ are 1–1 of Ω *onto* itself, and the mapping $G \to \bar{G}$ given by $g \to \bar{g}$ is a homomorphism.

The next result shows that if we have a family of distributions which is invariant under a group G then the distribution of any invariant function (under G) depends only on a maximal invariant parameter (under $\bar{G}$).

THEOREM 6.1.12. Suppose that the family of distributions $\{P_\theta;\ \theta \in \Omega\}$ is invariant under the group G. If $\phi(x)$ is invariant under G and $\psi(\theta)$ is a maximal invariant under the induced group $\bar{G}$ then the distribution of $\phi(X)$ depends only on $\psi(\theta)$.

Proof. It suffices to show that $P_\theta[\phi(X)\in B]$ is constant on the orbits of Ω under $\bar{G}$, for then it is invariant under $\bar{G}$ and by Theorem 6.1.4 must be a function of the maximal invariant $\psi(\theta)$. Thus, suppose that $\theta_2=\bar{g}\theta_1$ for some $\bar{g}\in\bar{G}$; then

$$\begin{aligned} P_{\theta_1}[\phi(X)\in B] &= P_{\theta_1}[\phi(gX)\in B]=P_{\bar{g}\theta_1}[\phi(X)\in B] \\ &= P_{\theta_2}[\phi(X)\in B], \end{aligned}$$

and the proof is complete.

This theorem is of great use in reducing the parameter space in complicated distribution problems. Two simple examples follow, and other applications will appear later.

EXAMPLE 6.1.13. Suppose that $\mathbf{X}$ is $N_m(\boldsymbol{\mu}, I_m)$. Here, both the sample space $\mathfrak{X}$ and the parameter space Ω are R^m. Take the group G to be $O(m)$ acting on $\mathfrak{X}=R^m$ as in Example 6.1.6. Since $H\mathbf{X}$ is $N_m(H\boldsymbol{\mu}, I_m)$ we see that the family of distributions is invariant and that the group $\bar{G}$ induced by G is $\bar{G}=O(m)$, where the action of $H\in O(m)$ on $\boldsymbol{\mu}\in\Omega$ is given by $\boldsymbol{\mu}\rightarrow H\boldsymbol{\mu}$. A maximal invariant parameter under $\bar{G}$ is $\psi(\boldsymbol{\mu})=\boldsymbol{\mu}'\boldsymbol{\mu}$ (see Example 6.1.6), so that by Theorem 6.1.12 any function $\phi(\mathbf{X})$ of $\mathbf{X}$ which is invariant under $O(m)$ has a distribution which depends only on $\boldsymbol{\mu}'\boldsymbol{\mu}$. In particular $\mathbf{X}'\mathbf{X}$, a maximal invariant under G, has a distribution which depends only on $\boldsymbol{\mu}'\boldsymbol{\mu}$ and is, of course, the $\chi^2_m(\boldsymbol{\mu}'\boldsymbol{\mu})$ distribution.

EXAMPLE 6.1.14. Suppose that A is $W_2(n,\Sigma)$, $n\geq 2$. Here both the sample space $\mathfrak{X}$ (consisting of the values of A) and the parameter space Ω (consisting of the values of Σ) are $\mathbb{S}_2$, the space of 2×2 positive definite matrices. Take G to be the group

$$G=\left\{B=\begin{bmatrix} b_1 & 0 \\ 0 & b_2 \end{bmatrix};\quad b_1>0,\quad b_2>0\right\},$$

where the action of $B\in G$ on $A\in\mathbb{S}_2$ is

$$A\rightarrow BAB'. \tag{3}$$

Since BAB' is $W_2(n, B\Sigma B')$ the family of distributions is invariant and the induced transformation on Ω corresponding to (3) is $\Sigma\rightarrow B\Sigma B'$, so that

$\bar{G}=G$. A maximal invariant parameter under $\bar{G}$ is the population correlation coefficient

$$\rho=\frac{\sigma_{12}}{(\sigma_{11}\sigma_{22})^{1/2}}$$

(see Example 6.1.7—actually a trivial modification of it); hence by Theorem 6.1.12 any function of A which is invariant under G has a distribution which depends only on the population correlation coefficient ρ. In particular the sample correlation coefficient

$$r=\frac{a_{12}}{(a_{11}a_{22})^{1/2}},$$

a maximal invariant under G, has a distribution which depends only on ρ. Hence, in order to find the distribution of r it can be assumed without loss of generality that

$$\Sigma=\begin{bmatrix} 1 & \rho \\ \rho & 1 \end{bmatrix};$$

this reduction was noted in the proof of Theorem 5.1.5. It is also worth noting that if Σ is restricted to being diagonal, so that the parameter space is

$$\Omega=\left\{\Sigma=\begin{bmatrix} \sigma_{11} & 0 \\ 0 & \sigma_{22} \end{bmatrix},\quad \sigma_{11}>0,\quad \sigma_{22}>0\right\},$$

then $\bar{G}$ acts *transitively* on Ω so that the only invariant functions are constant functions. Theorem 6.1.12 then tells us that the distribution of r, a maximal invariant under G, does not depend on any parameters. This, of course, corresponds to the case where $\rho=0$ and the distribution of r in this case is given in Theorem 5.1.1.

The next definition explains what is meant when one says that a *testing problem* is invariant.

DEFINITION 6.1.15. Let the family of distributions $\{P_\theta;\ \theta\in\Omega\}$ be invariant under G. The problem of testing H: $\theta\in\Omega_0$ against K: $\theta\in\Omega-\Omega_0$ is said to be invariant under G if $\bar{g}\Omega_0=\Omega_0$ for all $\bar{g}\in\bar{G}$.

If the testing problem is invariant under G then obviously we must also have $\bar{g}(\Omega-\Omega_0)=\Omega-\Omega_0$ for all $\bar{g}\in\bar{G}$. In an invariant testing problem (under G) an *invariant test* is one which is based on a statistic which is invariant under G. If $T(x)$ is a maximal invariant under G then all invariant test statistics are functions of T by Theorem 6.1.4, so that the class of all invariant test statistics is the same as the class of test statistics which are functions of the maximal invariant T.

There are some standard steps involved in the construction of invariant tests, and it may be worthwhile to list them here, at least informally.

(a) Reduce the problem by sufficiency. This means at the outset that all test statistics must be functions of a sufficient statistic; such a reduction usually has the effect of reducing the sample space.

(b) For the sample space $\mathcal{X}$ of the sufficient statistic find a group of transformations G on $\mathcal{X}$ under which the testing problem is invariant.

(c) Find a maximal invariant T under G; then any invariant test statistic is a function of T and by Theorem 6.1.12 its distribution depends only on a maximal invariant parameter under the induced group $\bar{G}$ acting on the parameter space Ω.

At this stage we are looking at test statistics which are functions of a maximal invariant T. Often there is no "best" test in this class, and the choice of a test now may be somewhat arbitrary. The likelihood ratio test is one possibility since, under fairly general conditions, this is invariant. In some cases, however, it is also possible to carry out one more step.

(d) In the class of invariant tests, find a uniformly most powerful test. If such a test exists it is called a *uniformly most powerful invariant test* under the group G. Often, but not always, it coincides with the likelihood ratio test. This, being an invariant test, can certainly be no better.

We will deal with some examples of uniformly most powerful invariant tests in the following sections. For now, by way of illustration, let us return to the example on the ordinary correlation coefficient (see Examples 6.1.7 and 6.1.14).

EXAMPLE 6.1.16. Let $\mathbf{X}_1,\ldots,\mathbf{X}_N$ be independent $N_2(\boldsymbol{\mu},\Sigma)$ random vectors and consider the problem of testing $H: \rho\le\rho_0$ against $K: \rho>\rho_0$, where ρ is the population correlation coefficient,

$$\rho=\frac{\sigma_{12}}{(\sigma_{11}\sigma_{22})^{1/2}}.$$

A sufficient statistic is the pair $(\bar{\mathbf{X}}, A)$, where

$$A=\sum_{i=1}^{N}(\mathbf{X}_i-\bar{\mathbf{X}})(\mathbf{X}_i-\bar{\mathbf{X}})', \qquad \bar{\mathbf{X}}=\frac{1}{N}\sum_{i=1}^{N}\mathbf{X}_i.$$

Here $\bar{\mathbf{X}}$ and A are independent; $\bar{\mathbf{X}}$ is $N_m(\boldsymbol{\mu},(1/N)\Sigma)$ and A is $W_m(n,\Sigma)$, with $n=N-1$. Reducing the problem by sufficiency, we consider only test statistics which are functions of $\bar{\mathbf{X}}$ and A. Consider the group of transformations G given by

$$\begin{aligned}\bar{\mathbf{X}}&\rightarrow B\bar{\mathbf{X}}+\mathbf{c}\\ A&\rightarrow BAB',\end{aligned} \tag{4}$$

where

$$B=\begin{bmatrix} b_1 & 0\\ 0 & b_2\end{bmatrix} \qquad (b_1>0,\quad b_2>0,\quad \text{and} \quad \mathbf{c}\in R^2).$$

(This is the group G of Example 6.1.7.) Obviously the family of distributions of $(\bar{\mathbf{X}}, A)$ is invariant, and the transformations induced on the parameter space by (4) are given by

$$\boldsymbol{\mu}\rightarrow B\boldsymbol{\mu}+\mathbf{c}$$

$$\Sigma\rightarrow B\Sigma B'.$$

Both H and K are invariant under these transformations, so the testing problem is invariant under G. A maximal invariant under G is the sample correlation coefficient $r=a_{12}/(a_{11}a_{22})^{1/2}$, and its distribution depends only on ρ. Thus any invariant test statistic is a function of r. Finally, we have already seen in Theorem 5.1.8 that of all tests based on r the one which rejects H if $r>k_\alpha$, with k_α being chosen so that the test has size α, is uniformly most powerful of size α for testing H against K. Hence, this test is a uniformly most powerful invariant test under the group G.

6.2. THE MULTIPLE CORRELATION COEFFICIENT AND INVARIANCE

We will now apply some of the invariance theory of Section 6.1 to the multiple correlation coefficient. Using the notation of Section 5.2.3, suppose that $(Y,\mathbf{X}')'$ is a $N_m(\boldsymbol{\mu},\Sigma)$ random vector, where $\mathbf{X}$ is $(m-1)\times 1$, and Σ is partitioned as

$$\Sigma=\begin{bmatrix}\sigma_{11} & \boldsymbol{\sigma}_{12}' \\ \boldsymbol{\sigma}_{12} & \Sigma_{22}\end{bmatrix},$$

where Σ_{22} is $(m-1)\times(m-1)$. The population multiple correlation coefficient between Y and $\mathbf{X}$ is

$$\bar{R}=\left(\frac{\boldsymbol{\sigma}_{12}'\Sigma_{22}^{-1}\boldsymbol{\sigma}_{12}}{\sigma_{11}}\right)^{1/2}.$$

Let $(Y_i,\mathbf{X}_i')'$, with $i=1,\ldots,N$, be a sample of size $N(>m)$; a sufficient statistic is the pair $((\bar{Y},\bar{\mathbf{X}}')', A)$, where $A=N\hat{\Sigma}$ is the usual matrix of sums of squares and sums of products. Under the transformations

$$(1)\qquad \begin{aligned} Y_i &\to b_1 Y_i + c_1 \\ \mathbf{X}_i &\to B_2\mathbf{X}_i + \mathbf{c}_2, \end{aligned}$$

where $b_1\neq 0$ and B_2 is $(m-1)\times(m-1)$ nonsingular [i.e., $B_2\in \mathcal{G}\ell(m-1, R)$] the sufficient statistic is transformed as

$$(2)\qquad \begin{aligned} \begin{pmatrix}\bar{Y}\\ \bar{\mathbf{X}}\end{pmatrix} &\to B\begin{pmatrix}\bar{Y}\\ \bar{\mathbf{X}}\end{pmatrix}+\mathbf{c} \\ A &\to BAB' \end{aligned}$$

where

$$B=\begin{bmatrix} b_1 & \mathbf{0}' \\ \mathbf{0} & B_2\end{bmatrix}\quad\text{and}\quad \mathbf{c}\in R^m.$$

The family of distributions of the sufficient statistic is invariant under this group of transformations, G say, and the group of transformations $\bar{G}$ induced on the parameter space is given by

$$(3)\qquad \begin{aligned} \boldsymbol{\mu} &\to B\boldsymbol{\mu}+\mathbf{c} \\ \Sigma &\to B\Sigma B'. \end{aligned}$$

The next result shows that the sample multiple correlation coefficient

$$R=\left(\frac{\mathbf{a}_{12}'A_{22}^{-1}\mathbf{a}_{12}}{a_{11}}\right)^{1/2}$$

is a maximal invariant under the group of transformations G given by (2) and that the population multiple correlation coefficient $\bar{R}$ is a maximal invariant under the group of transformations $\bar{G}$ given by (3). We will state the result for $\bar{R}$.

THEOREM 6.2.1. Under the group of transformations $\bar{G}$ a maximal invariant is

$$\bar{R}^2 = \frac{\sigma_{12}'\Sigma_{22}^{-1}\sigma_{12}}{\sigma_{11}}.$$

Proof. Let $\phi(\mu, \Sigma) = \sigma_{12}'\Sigma_{22}^{-1}\sigma_{12}/\sigma_{11} = \bar{R}^2$. First note that since

$$B\Sigma B' = \begin{bmatrix} b_1^2\sigma_{11} & b_1\sigma_{12}'B_2' \\ b_1B_2\sigma_{12} & B_2\Sigma_{22}B_2' \end{bmatrix}$$

we have

$$\begin{aligned}\phi(B\mu + c, B\Sigma B') &= \frac{b_1^2\sigma_{12}'B_2'(B_2\Sigma_{22}B_2')^{-1}B_2\sigma_{12}}{b_1^2\sigma_{11}} \\ &= \frac{\sigma_{12}'\Sigma_{22}^{-1}\sigma_{12}}{\sigma_{11}} \\ &= \phi(\mu, \Sigma),\end{aligned}$$

so that $\phi(\mu, \Sigma)$ is invariant. To show that it is maximal invariant, suppose that

$$\phi(\mu, \Sigma) = \phi(\tau, \Gamma),$$

i.e.,

$$\frac{\sigma_{12}'\Sigma_{22}^{-1}\sigma_{12}}{\sigma_{11}} = \frac{\gamma_{12}'\Gamma_{22}^{-1}\gamma_{12}}{\gamma_{11}}.$$

Then

$$\left(\frac{1}{\sigma_{11}^{1/2}}\sigma_{12}'\Sigma_{22}^{-1/2}\right)\left(\frac{1}{\sigma_{11}^{1/2}}\sigma_{12}'\Sigma_{22}^{-1/2}\right)' = \left(\frac{1}{\gamma_{11}^{1/2}}\gamma_{12}'\Gamma_{22}^{-1/2}\right)\left(\frac{1}{\gamma_{11}^{1/2}}\gamma_{12}'\Gamma_{22}^{-1/2}\right)'$$

By Vinograd's theorem (Theorem A9.5) there is an $(m-1)\times(m-1)$ orthogonal matrix H such that

$$\frac{1}{\sigma_{11}^{1/2}}H\Sigma_{22}^{-1/2}\sigma_{12}=\frac{1}{\gamma_{11}^{1/2}}\Gamma_{22}^{-1/2}\gamma_{12}.$$

Now, putting

$$b_1=(\gamma_{11}/\sigma_{11})^{1/2},$$

$$B_2=\Gamma_{22}^{1/2}H\Sigma_{22}^{-1/2},$$

$$B=\begin{bmatrix} b_1 & \mathbf{0}' \\ \mathbf{0} & B_2 \end{bmatrix},$$

and

$$\mathbf{c}=-B\boldsymbol{\mu}+\boldsymbol{\tau},$$

we have

$$B\boldsymbol{\mu}+\mathbf{c}=\boldsymbol{\tau}$$

and

$$B\Sigma B'=\begin{bmatrix} \gamma_{11} & \gamma_{12}' \\ \gamma_{12} & \Gamma_{22} \end{bmatrix}=\Gamma,$$

so that

$$(\boldsymbol{\mu},\Sigma)\sim(\boldsymbol{\tau},\Gamma)\ (\mathrm{mod}\,\overline{G}\,).$$

Hence ϕ is a maximal invariant, and the proof is complete.

It follows, using Theorems 6.1.4 and 6.1.12, that any function of the sufficient statistic which is invariant under G is a function of R^2 and has a distribution which depends only on the population multiple correlation coefficient $\overline{R}^2$, a maximal invariant under the induced group $\overline{G}$. In particular, R^2 has a distribution which depends only on $\overline{R}^2$, a result which is apparent from Theorem 5.2.4. In the proof of that theorem we could have assumed without any loss of generality that Σ has the form

$$\Sigma=\begin{bmatrix} 1 & \bar{R} & 0 & \dots & 0 \\ \bar{R} & 1 & 0 & \dots & 0 \\ \vdots & & & & \\ 0 & 0 & 0 & \dots & 1 \end{bmatrix} \tag{4}$$

(see Theorem 5.2.7), often called a *canonical form* for Σ under the group of transformations $\bar{G}$ since it depends only on the maximal invariant $\bar{R}$. The reader is encouraged to work through the proof of Theorem 5.2.4, replacing the arbitrary Σ there by (4).

Let us now consider testing the null hypothesis $H: \bar{R}=0$ (or, equivalently, $\boldsymbol{\sigma}_{12}=\mathbf{0}$, or Y and $\mathbf{X}$ are independent) against the alternative $K: \bar{R}>0$. We noted in Section 5.2.5 that a test of size α (in fact, the likelihood ratio test) is to reject H if

$$\frac{n-m+1}{m-1}\,\frac{R^2}{1-R^2}\geq F^*_{m-1,n-m+1}(\alpha),$$

where $n=N-1$ and $F^*_{m-1,n-m+1}(\alpha)$ denotes the upper $100\alpha\%$ point of the $F_{m-1,n-m+1}$ distribution. Equivalently, the test is to reject H if

$$R^2\geq\frac{(m-1)F^*_{m-1,n-m+1}(\alpha)}{n-m+1+(m-1)F^*_{m-1,n-m+1}(\alpha)}=c_\alpha. \tag{5}$$

This test is a uniformly most powerful invariant test, as the following theorem shows.

THEOREM 6.2.2. Under the group of transformations G given by (2) a uniformly most powerful invariant test of size α of $H: \bar{R}=0$ against $K: \bar{R}>0$ is to reject H if $R^2\geq c_\alpha$, where c_α is given by (5).

Proof. Clearly the testing problem is invariant under G, and we have already noted that R^2 is a maximal invariant under G. Restricting attention to invariant tests, we can assume that a value of R^2 is observed from the distribution with density function specified in Theorem 5.2.4, namely,

(6)

$$f(R^2;n,\bar{R}^2)=\frac{\Gamma(\frac{1}{2}n)}{\Gamma[\frac{1}{2}(m-1)]\Gamma[\frac{1}{2}(n-m+1)]}(R^2)^{(m-3)/2}(1-R^2)^{(n-m-1)/2}$$

$$\cdot(1-\bar{R}^2)^{n/2}{}_2F_1(\tfrac{1}{2}n,\tfrac{1}{2}n;\tfrac{1}{2}(m-1);\bar{R}^2R^2).$$

The Neyman–Pearson lemma says that in this class of tests the most powerful test of size α of H: $\bar{R}=0$ against a simple alternative K_1: $\bar{R}=\bar{R}_1$ (>0) is to reject H if

$$\text{(7)} \qquad \frac{f(R^2;n,0)}{f\left(R^2;n,\bar{R}_1^2\right)} \le k_\alpha$$

where k_α is chosen so that the size of the test is α. Substituting the density function (6) in (7) gives the inequality

$$\text{(8)} \qquad \left(1-\bar{R}_1^2\right)^{n/2} {}_2F_1\left(\tfrac{1}{2}n,\tfrac{1}{2}n;\tfrac{1}{2}(m-1);\bar{R}_1^2R^2\right) \ge \lambda_\alpha = \frac{1}{k_\alpha}.$$

Using the series expansion for the ${}_2F_1$ function it is easy to see that the left side of (8) is an increasing function of R^2. Hence this inequality is equivalent to $R^2 \ge c_\alpha$, where c_α is given by (5), so that the test has size α. Since this test is the same for all alternatives $\bar{R}_1$ it is a uniformly most powerful invariant test, and the proof is complete.

The test described by (5), as well as being the uniformly most powerful invariant and the likelihood ratio test, has a number of other optimal properties. Simaika (1941) has shown that it is uniformly most powerful in the class of all tests whose power function depends only on $\bar{R}^2$. Clearly this is a wider class than the class of invariant tests. The test is also admissable (see Kiefer and Schwartz, 1965); that is, there is no other test whose power function is at least as large and actually larger for some alternatives. For a discussion of other properties, the reader is referred to Giri (1977), Section 8.3, and the references therein.

6.3. HOTELLING'S T^2 STATISTIC AND INVARIANCE

The T^2 statistic proposed by Hotelling (1931) for testing hypotheses about mean vectors has already been introduced briefly in Section 3.2.3 (see Theorem 3.2.13). In this section we will indicate some testing problems for which a T^2 statistic is appropriate and look at some properties of such tests, concentrating primarily on those concerned with invariance. First, let us paraphrase Theorem 3.2.13.

THEOREM 6.3.1. Let $\mathbf{X}$ be $N_m(\boldsymbol{\mu},\Sigma)$ and $A=nS$ be $W_m(n,\Sigma)$ $(n\ge m)$, with $\mathbf{X}$ and S independent, and put $T^2=\mathbf{X}'S^{-1}\mathbf{X}$. Then

$$\frac{T^2}{n}\cdot\frac{n-m+1}{m}$$

is $F_{m,n-m+1}(\delta)$, where $\delta=\boldsymbol{\mu}'\Sigma^{-1}\boldsymbol{\mu}$.

This is merely the restatement of Theorem 3.2.13 obtained by replacing $N^{1/2}\bar{\mathbf{X}}$ by $\mathbf{X}$ and $N^{1/2}\boldsymbol{\mu}$ by $\boldsymbol{\mu}$.

Now, suppose that $\mathbf{X}_1,\ldots,\mathbf{X}_N$ are independent $N_m(\boldsymbol{\mu},\Sigma)$ random vectors where $\boldsymbol{\mu}$ and Σ are unknown and consider testing the null hypothesis that $\boldsymbol{\mu}$ is a specified vector. Obviously we can assume without loss of generality that the specified vector is $\mathbf{0}$ (otherwise subtract it from each $\mathbf{X}_i$ and it will be). Thus the problem is to test $H:\boldsymbol{\mu}=\mathbf{0}$ against the alternative $K:\boldsymbol{\mu}\neq\mathbf{0}$. Let $\bar{\mathbf{X}}$ and S be the sample mean and covariance matrix formed from $\mathbf{X}_1,\ldots,\mathbf{X}_N$ and put

$$T^2=N\bar{\mathbf{X}}'S^{-1}\bar{\mathbf{X}}. \tag{1}$$

The test of size α suggested in Section 3.2.3 consists of rejecting H if

$$T^2\geq\frac{mn}{n-m+1}F^*_{m,n-m+1}(\alpha)=c_\alpha, \tag{2}$$

where $n=N-1$ and $F^*_{m,n-m+1}(\alpha)$ denotes the upper $100\alpha\%$ point of the $F_{m,n-m+1}$ distribution. We will show in a moment that this test is a uniformly most powerful invariant test. That it is also the likelihood ratio test is established in the following theorem.

THEOREM 6.3.2. If $\mathbf{X}_1,\ldots,\mathbf{X}_N$ are independent $N_m(\boldsymbol{\mu},\Sigma)$ random vectors the likelihood ratio test of size α of $H:\boldsymbol{\mu}=\mathbf{0}$ against $K:\boldsymbol{\mu}\neq\mathbf{0}$ is given by (2).

Proof. Apart from a multiplicative constant which does not involve $\boldsymbol{\mu}$ or Σ, the likelihood function is

$$L(\boldsymbol{\mu},\Sigma)=(\det\Sigma)^{-N/2}\operatorname{etr}\left(-\tfrac{1}{2}\Sigma^{-1}A\right)\exp\left[-\tfrac{1}{2}N(\bar{\mathbf{X}}-\boldsymbol{\mu})'\Sigma^{-1}(\bar{\mathbf{X}}-\boldsymbol{\mu})\right],$$

where $A=nS$, $n=N-1$. [See, for example, (8) of Section 3.1.] The likelihood ratio statistic is

$$\Lambda=\frac{\sup_{\Sigma>0}L(\mathbf{0},\Sigma)}{\sup_{\boldsymbol{\mu}\in R^m,\Sigma>0}L(\boldsymbol{\mu},\Sigma)}. \tag{3}$$

The denominator in (3) is

$$\sup_{\boldsymbol{\mu},\Sigma}L(\boldsymbol{\mu},\Sigma)=L(\bar{\mathbf{X}},\hat{\Sigma})=N^{mN/2}e^{-mN/2}(\det A)^{-N/2} \tag{4}$$

where $\hat{\Sigma}=N^{-1}A$ (see the proof of Theorem 3.1.5), while the numerator in

(3) is

$$\sup_{\Sigma>0} L(\mathbf{0},\Sigma)=\sup_{\Sigma>0}(\det\Sigma)^{-N/2}\operatorname{etr}\left[-\tfrac{1}{2}\Sigma^{-1}(A+N\overline{\mathbf{X}}\overline{\mathbf{X}}')\right].$$

The same argument used in the proof of Theorem 3.1.5 shows that this supremum is attained when

$$\Sigma=\frac{1}{N}A+\overline{\mathbf{X}}\overline{\mathbf{X}}'$$

and is

$$\sup_{\Sigma>0} L(\mathbf{0},\Sigma)=\det\left(\frac{1}{N}A+\overline{\mathbf{X}}\overline{\mathbf{X}}'\right)^{-N/2}e^{-mN/2}. \tag{5}$$

Using (4) and (5) in (3), we get

$$\begin{aligned}\Lambda^{2/N}&=\frac{\det A}{\det(A+N\overline{\mathbf{X}}\overline{\mathbf{X}}')}\\&=\frac{1}{\det(I_m+NA^{-1}\overline{\mathbf{X}}\overline{\mathbf{X}}')}\\&=\frac{1}{1+N\overline{\mathbf{X}}'A^{-1}\overline{\mathbf{X}}}\\&=\frac{1}{1+T^2/n},\end{aligned}$$

where $T^2=N\overline{\mathbf{X}}'S^{-1}\overline{\mathbf{X}}$. The likelihood ratio test is to reject H if the likelihood ratio statistic Λ is small. Since Λ is a decreasing function of T^2 this is the same as rejecting H for large values of T^2, thus giving the test in (2) and completing the proof.

Now let us look at the problem of testing H: $\boldsymbol{\mu}=\mathbf{0}$ against K: $\boldsymbol{\mu}\neq\mathbf{0}$ from an invariance point of view. A sufficient statistic is $(\overline{\mathbf{X}},A)$, where $\overline{\mathbf{X}}$ is $N_m(\boldsymbol{\mu},(1/N)\Sigma)$, $A=nS$ is $W_m(n,\Sigma)$ with $n=N-1$, and $\overline{\mathbf{X}}$ and A are independent. Consider the general linear group $\mathcal{GL}(m,R)$ of $m\times m$ nonsingular real matrices acting on the space $R^m\times\mathbb{S}_m$ of pairs $(\overline{\mathbf{X}},A)$ by

$$\overline{\mathbf{X}}\rightarrow B\overline{\mathbf{X}},\qquad A\rightarrow BAB',\qquad B\in\mathcal{GL}(m,R). \tag{6}$$

($\mathbb{S}_m$ denotes, as always, the set of positive definite $m\times m$ matrices.) The corresponding induced group of transformations [also $\mathcal{G}\ell(m,R)$] on the parameter space (also $R^m \times \mathbb{S}_m$) of pairs (μ,Σ) is given by

$$\mu \to B\mu, \qquad \Sigma \to B\Sigma B', \qquad B \in \mathcal{G}\ell(m,R), \tag{7}$$

and it is clear that the problem of testing $H:\mu = \mathbf{0}$ against $K:\mu \neq \mathbf{0}$ is invariant under $\mathcal{G}\ell(m,R)$, for the family of distributions of $(\overline{\mathbf{X}}, A)$ is invariant and the null and alternative hypotheses are unchanged. Our next problem is to find a maximal invariant under the action of $\mathcal{G}\ell(m,R)$ on $R^m \times \mathbb{S}_m$ given by (6) or (7). This is done in the following theorem.

THEOREM 6.3.3. Under the group $\mathcal{G}\ell(m,R)$ of transformations (7) on $R^m \times \mathbb{S}_m$, a maximal invariant is

$$\phi(\mu,\Sigma) = \mu'\Sigma^{-1}\mu.$$

Proof. First note that for $B \in \mathcal{G}\ell(m,R)$,

$$\phi(B\mu, B\Sigma B') = \mu' B'(B\Sigma B')^{-1} B\mu = \mu'\Sigma^{-1}\mu = \phi(\mu,\Sigma),$$

so that $\phi(\mu,\Sigma)$ is invariant. To show that it is maximal invariant, suppose that

$$\phi(\mu,\Sigma) = \phi(\tau,\Gamma),$$

that is

$$\mu'\Sigma^{-1}\mu = \tau'\Gamma^{-1}\tau.$$

Then

$$(\mu'\Sigma^{-1/2})(\mu'\Sigma^{-1/2})' = (\tau'\Gamma^{-1/2})(\tau'\Gamma^{-1/2})'$$

so that, by Vinograd's theorem (Theorem A9.5) there exists an orthogonal $m \times m$ matrix H such that

$$H\Sigma^{-1/2}\mu = \Gamma^{-1/2}\tau.$$

Putting $B = \Gamma^{1/2} H \Sigma^{-1/2}$, we then have $B\mu = \tau$ and $B\Sigma B' = \Gamma$ so that $(\mu,\Sigma) \sim (\tau,\Gamma)$ [mod $\mathcal{G}\ell(m,R)$]. Hence ϕ is a maximal invariant, and the proof is complete.

As a consequence of this theorem, $T^2 = N\overline{\mathbf{X}}'S^{-1}\overline{\mathbf{X}}$ is a maximal invariant statistic under the group $\mathcal{G}\ell(m,R)$ acting on the sample space $R^m \times \mathbb{S}_m$ of the sufficient statistic. From Theorem 3.2.13 we know that the distribution of $(n-m+1)T^2/nm$ is $F_{m,n-m+1}(\delta)$, where $n=N-1$ and $\delta=N\boldsymbol{\mu}'\Sigma^{-1}\boldsymbol{\mu}$. Considering only invariant tests we can assume a value of T^2 is observed from this distribution. In terms of the noncentrality parameter δ we are now testing $H: \delta=0$ against $K: \delta>0$. The Neyman–Pearson lemma, applied to the noncentral F density function given in Theorem 1.3.6, says that in the class of tests based on T^2 the most powerful test of size α of $H: \delta=0$ against a simple alternative $K_1: \delta=\delta_1$ (>0) is to reject H if

$$(8)\qquad \exp(-\tfrac{1}{2}\delta_1)\,{}_1F_1\left(\tfrac{1}{2}(n+1);\tfrac{1}{2}m;\frac{\tfrac{1}{2}\delta_1 T^2/n}{1+T^2/n}\right)\geq\lambda_\alpha$$

where λ_α is chosen so that the test has size α. Using the series expansion for the ${}_1F_1$ function in (8), it is easy to see that it is an increasing function of $(T^2/n)(1+T^2/n)^{-1}$ and hence of T^2. Hence the inequality (8) is equivalent to $T^2\geq c_\alpha$, where c_α is given by (2) so that the size of the test is α. Since this test is the same for all alternatives δ_1 it is a uniformly most powerful invariant test. Summarizing, we have:

THEOREM 6.3.4. Under the group $\mathcal{G}\ell(m,R)$ of transformations given by (6) a uniformly most powerful invariant test of size α of $H: \boldsymbol{\mu}=\mathbf{0}$ against $K: \boldsymbol{\mu}\neq\mathbf{0}$ is to reject H if $T^2=N\overline{\mathbf{X}}'S^{-1}\overline{\mathbf{X}}\geq c_\alpha$, where c_α is given by (2).

Before looking at another testing problem note that the T^2 statistic can be used to construct *confidence regions* for the mean vector $\boldsymbol{\mu}$. Let $\mathbf{X}_1,\ldots,\mathbf{X}_N$ be independent $N_m(\boldsymbol{\mu},\Sigma)$ random vectors giving rise to the sample mean vector $\overline{\mathbf{X}}$ and sample covariance matrix S. These are independent; $N^{1/2}(\overline{\mathbf{X}}-\boldsymbol{\mu})$ is $N_m(\mathbf{0},\Sigma)$ and nS is $W_m(n,\Sigma)$ with $n=N-1$. From Theorem 6.3.1 [with $\mathbf{X}$ replaced by $N^{1/2}(\overline{\mathbf{X}}-\boldsymbol{\mu})$ and $\boldsymbol{\mu}$ replaced by $\mathbf{0}$] it follows that $(n-m+1)T^2/nm$ has the $F_{m,n-m+1}$ distribution, where $T^2=N(\overline{\mathbf{X}}-\boldsymbol{\mu})'S^{-1}(\overline{\mathbf{X}}-\boldsymbol{\mu})$. Thus, defining c_α by (2), we have

$$P\left[N(\overline{\mathbf{X}}-\boldsymbol{\mu})'S^{-1}(\overline{\mathbf{X}}-\boldsymbol{\mu})\leq c_\alpha\right]=1-\alpha,$$

from which it follows that the random ellipsoid defined by

$$N(\overline{\mathbf{X}}-\boldsymbol{\mu})'S^{-1}(\overline{\mathbf{X}}-\boldsymbol{\mu})\leq c_\alpha$$

has a probability of $1-\alpha$ of containing $\boldsymbol{\mu}$ and hence the region

$$(9)\qquad \left\{\boldsymbol{\mu}\in R^m;\, N(\overline{\mathbf{X}}-\boldsymbol{\mu})'S^{-1}(\overline{\mathbf{X}}-\boldsymbol{\mu})\leq c_\alpha\right\},$$

for observed $\bar{\mathbf{X}}$ and S, is a confidence region for $\boldsymbol{\mu}$ with confidence coefficient $1-\alpha$.

The T^2 statistic can also be used to test whether the mean vectors of two normal distributions with the same covariance matrix are equal. Let $\mathbf{X}_1,\ldots,\mathbf{X}_{N_1}$ be a random sample from the $N_m(\boldsymbol{\mu}_1,\Sigma)$ distribution, and let $\mathbf{Y}_1,\ldots,\mathbf{Y}_{N_2}$ be a random sample from the $N_m(\boldsymbol{\mu}_2,\Sigma)$ distribution. The sample mean and covariance matrix formed from the X's will be denoted by $\bar{\mathbf{X}}$ and S_x, and from the Y's by $\bar{\mathbf{Y}}$ and S_y. The problem here is to test that the two population mean vectors are equal, that is, to test $H: \boldsymbol{\mu}_1=\boldsymbol{\mu}_2$, against the alternative, $K: \boldsymbol{\mu}_1\neq\boldsymbol{\mu}_2$. It is a simple matter to construct a T^2 statistic appropriate for this task. First, let $A_x=n_1S_x$, $A_y=n_2S_y$, where $n_i=N_i-1$ $(i=1,2)$; then A_x is $W_m(n_1,\Sigma)$ and A_y is $W_m(n_2,\Sigma)$, and hence $A=A_x+A_y$ is $W_m(n_1+n_2,\Sigma)$. Now put $S=(n_1+n_2)^{-1}A$, the *pooled* sample covariance matrix, so that $(n_1+n_2)S$ is $W_m(n_1+n_2,\Sigma)$. This is independent of $\bar{\mathbf{X}}-\bar{\mathbf{Y}}$ and the distribution of

$$\left(\frac{N_1N_2}{N_1+N_2}\right)^{1/2}(\bar{\mathbf{X}}-\bar{\mathbf{Y}}) \quad\text{is}\quad N_m\left(\left(\frac{N_1N_2}{N_1+N_2}\right)^{1/2}(\boldsymbol{\mu}_1-\boldsymbol{\mu}_2),\Sigma\right).$$

From Theorem 6.3.1 (with $\mathbf{X}$ replaced by $[N_1N_2/(N_1+N_2)]^{1/2}(\bar{\mathbf{X}}-\bar{\mathbf{Y}})$, $\boldsymbol{\mu}$ by $[N_1N_2/(N_1+N_2)]^{1/2}(\boldsymbol{\mu}_1-\boldsymbol{\mu}_2)$ and n by n_1+n_2) it follows that if

$$(10)\qquad T^2=\frac{N_1N_2}{N_1+N_2}(\bar{\mathbf{X}}-\bar{\mathbf{Y}})'S^{-1}(\bar{\mathbf{X}}-\bar{\mathbf{Y}})$$

then

$$(11)\qquad \frac{T^2}{n_1+n_2}\cdot\frac{n_1+n_2-m+1}{m}\quad\text{is}\quad F_{m,n_1+n_2-m+1}(\delta),$$

where

$$(12)\qquad \delta=\frac{N_1N_2}{N_1+N_2}(\boldsymbol{\mu}_1-\boldsymbol{\mu}_2)'\Sigma^{-1}(\boldsymbol{\mu}_1-\boldsymbol{\mu}_2).$$

When the null hypothesis $H: \boldsymbol{\mu}_1=\boldsymbol{\mu}_2$ is true the noncentrality parameter δ is zero so that $(n_1+n_2-m+1)T^2/[m(n_1+n_2)]$ is F_{m,n_1+n_2-m+1}. Hence a test of size α of H against K is to reject H if

$$(13)\qquad T^2\geq\frac{m(n_1+n_2)}{n_1+n_2-m+1}F^*_{m,n_1+n_2-m+1}(\alpha)=c_\alpha,$$

where $F^*_{m,n_1+n_2-m+1}(\alpha)$ denotes the upper $100\alpha\%$ point of the F_{m,n_1+n_2-m+1} distribution. It should also be clear that a T^2 statistic can be used to construct confidence regions for $\mu_1-\mu_2$ (see Problem 6.3).

Now let us look at the test of equality of two mean vectors just described from the point of view of invariance. It is easy to check that a sufficient statistic is $(\overline{\mathbf{X}},\overline{\mathbf{Y}},A)$, where $\overline{\mathbf{X}}$ is $N_m(\mu_1,(1/N_1)\Sigma)$, $\overline{\mathbf{Y}}$ is $N_m(\mu_2,(1/N_2)\Sigma)$, A is $W_m(n_1+n_2,\Sigma)$, and these are all independent. Consider the *affine group* of transformations

$$\mathcal{A}\ell(m,R)=\{(B,\mathbf{c});\ B\in\mathcal{G}\ell(m,R),\mathbf{c}\in R^m\} \tag{14}$$

acting on the space $R^m\times R^m\times \mathbb{S}_m$ of triples $(\overline{\mathbf{X}},\overline{\mathbf{Y}},A)$ by

$$(B,\mathbf{c})(\overline{\mathbf{X}},\overline{\mathbf{Y}},A)=(B\overline{\mathbf{X}}+\mathbf{c},\ B\overline{\mathbf{Y}}+\mathbf{c},\ BAB'), \tag{15}$$

where the group operation is

$$(B_1,\mathbf{c}_1)(B_2,\mathbf{c}_2)=(B_1B_2,\ B_1\mathbf{c}_2+\mathbf{c}_1). \tag{16}$$

The corresponding induced group of transformations [also $\mathcal{A}\ell(m,R)$] on the parameter space (also $R^m\times R^m\times \mathbb{S}_m$) of triples (μ_1,μ_2,Σ) is given by

$$(B,\mathbf{c})(\mu_1,\mu_2,\Sigma)=(B\mu_1+\mathbf{c},\ B\mu_2+\mathbf{c},\ B\Sigma B'). \tag{17}$$

Clearly the problem of testing $H:\mu_1=\mu_2$ against $K:\mu_1\neq\mu_2$ is invariant under $\mathcal{A}\ell(m,R)$, for the family of distributions of $(\overline{\mathbf{X}},\overline{\mathbf{Y}},A)$ is invariant, as are the null and alternative hypotheses. A maximal invariant under the group $\mathcal{A}\ell(m,R)$ acting on the sample space of the sufficient statistic is

$$T^2=\frac{N_1N_2}{N_1+N_2}(\overline{\mathbf{X}}-\overline{\mathbf{Y}})'S^{-1}(\overline{\mathbf{X}}-\overline{\mathbf{Y}}); \tag{18}$$

the proof of this is similar to that of Theorem 6.3.3 and is left as an exercise (see Problem 6.2). We know from (11) that the distribution of $(n_1+n_2-m+1)T^2/[m(n_1+n_2)]$ is $F_{m,n_1+n_2-m+1}(\delta)$, where $\delta=[N_1N_2/(N_1+N_2)]\cdot(\mu_1-\mu_2)'\Sigma^{-1}(\mu_1-\mu_2)$. Considering only invariant tests we can assume that T^2 is observed from this distribution. In terms of the noncentrality parameter δ we are now testing $H:\delta=0$ against $K:\delta>0$. Exactly as in the proof of Theorem 6.3.4 there exists a uniformly most powerful invariant test which rejects H for large values of T^2. The result is summarized in the following theorem.

THEOREM 6.3.5. Under the group $\mathcal{Gl}(m, R)$ of transformations given by (15) a uniformly most powerful invariant test of size α of $H: \boldsymbol{\mu}_1 = \boldsymbol{\mu}_2$ against $K: \boldsymbol{\mu}_1 \neq \boldsymbol{\mu}_2$ is to reject H if

$$T^2 = \frac{N_1 N_2}{N_1 + N_2}(\overline{\mathbf{X}} - \overline{\mathbf{Y}})' S^{-1}(\overline{\mathbf{X}} - \overline{\mathbf{Y}}) \geq c_\alpha,$$

where c_α is given by (13).

There are many other situations for which a T^2 statistic is appropriate. We will indicate one more. A generalization of the first problem considered, that is, of testing $\boldsymbol{\mu} = \mathbf{0}$, is to test the null hypothesis $H: C\boldsymbol{\mu} = \mathbf{0}$, where C is a specified $p \times m$ matrix of rank p, given a sample of size $N = n + 1$ from the $N_m(\boldsymbol{\mu}, \Sigma)$ distribution. Let $\overline{\mathbf{X}}$ and S denote the sample mean vector and covariance matrix; then $N^{1/2} C\overline{\mathbf{X}}$ is $N_p(N^{1/2} C\boldsymbol{\mu}, C\Sigma C')$ and $nCSC'$ is $W_p(n, C\Sigma C')$, and these are independent. In Theorem 6.3.1 making the transformations $\overline{\mathbf{X}} \to N^{1/2} C\overline{\mathbf{X}}$, $\Sigma \to C\Sigma C'$, $\boldsymbol{\mu} \to N^{1/2} C\boldsymbol{\mu}$, $S \to CSC'$, $m \to p$ shows that

$$\frac{T^2}{n} \cdot \frac{n-p+1}{p} \quad \text{is} \quad F_{p,n-p+1}(\delta),$$

with $\delta = N\boldsymbol{\mu}' C'(C\Sigma C')^{-1} C\boldsymbol{\mu}$, where

$$T^2 = N\overline{\mathbf{X}}' C'(CSC')^{-1} C\overline{\mathbf{X}}.$$

When the null hypothesis $H: C\boldsymbol{\mu} = \mathbf{0}$ is true the noncentrality parameter δ is zero, and hence a test of size α is to reject H if

$$T^2 \geq \frac{np}{n-p+1} F^*_{p,n-p+1}(\alpha)$$

where $F^*_{p,n-p+1}(\alpha)$ is the upper $100\alpha\%$ point of the $F_{p,n-p+1}$ distribution. This test is a uniformly most powerful invariant test (see Problem 6.6) and the likelihood ratio test (see Problem 6.8).

There are many other situations for which a T^2 statistic is appropriate; some of these appear in the problems. For a discussion of applications of the T^2 statistic, useful references are Anderson (1958), Section 5.3, and Kshirsagar (1972), Section 5.4.

We have seen that the test described by (2) for testing $H: \boldsymbol{\mu} = \mathbf{0}$ against $K: \boldsymbol{\mu} \neq \mathbf{0}$ on the basis of $N = n + 1$ observations from the $N_m(\boldsymbol{\mu}, \Sigma)$ distribution is both the uniformly most powerful invariant and the likelihood ratio test. It has also a number of other optimal properties. It is uniformly most powerful in the class of tests whose power function depends only on $\boldsymbol{\mu}'\Sigma^{-1}\boldsymbol{\mu}$, a result due to Simaika (1941). The test is also admissible, a result established by Stein (1956b), and Kiefer and Schwartz (1965). Kariya (1981) has also demonstrated a robustness property of this T^2 test. Let X be an $N\times m$ random matrix with a density function h and having rows $\mathbf{x}'_1,\mathbf{x}'_2,\ldots,\mathbf{x}'_N$, let C_{Nm} be the class of all density functions on R^{Nm} [with respect to Lebesgue measure (dX)], and let Q be the set of nonincreasing convex functions from $[0,\infty)$ to $[0,\infty)$. For $\boldsymbol{\mu}\in R^m$ and $\Sigma\in \mathbb{S}_m$ define a class of density functions on R^{Nm} by

$$C_{Nm}(\boldsymbol{\mu},\Sigma)$$

$$=\left\{f\in C_{Nm};\ f(X;\boldsymbol{\mu},\Sigma)=\frac{q\left(\sum_{i=1}^{N}(\mathbf{x}_i-\boldsymbol{\mu})'\Sigma^{-1}(\mathbf{x}_i-\boldsymbol{\mu})\right)}{(\det\Sigma)^{N/2}},\ q\in Q\right\}.$$

Clearly, if X is $N(\mathbf{1}\boldsymbol{\mu}', I_N\otimes\Sigma)$, where $\mathbf{1}=(1,1,\ldots,1)'\in R^N$, then the density function h of X belongs to $C_{Nm}(\boldsymbol{\mu},\Sigma)$. If $f(X;\boldsymbol{\mu},\Sigma)\in C_{Nm}(\boldsymbol{\mu},\Sigma)$ then mixtures of the form

$$g(X;\boldsymbol{\mu},\Sigma)=\int_0^\infty f(X;\boldsymbol{\mu},a\Sigma)\,dG(a)$$

also belong to $C_{Nm}(\boldsymbol{\mu},\Sigma)$, where G is a distribution function on $(0,\infty)$. From this result it follows that $C_{Nm}(\boldsymbol{\mu},\Sigma)$ contains such elliptical distributions as the Nm-variate t distribution and contaminated normal distribution (see Section 1.5). Kariya (1981) considered the problem of testing H: $h\in C_{Nm}(\mathbf{0},\Sigma)$ against K: $h\in C_{Nm}(\boldsymbol{\mu},\Sigma)$, with $\boldsymbol{\mu}\neq 0$, and showed that the T^2 test is a uniformly most powerful invariant test, and that the null distribution of T^2 is the same as that under normality. For a discussion of other properties the reader is referred to Giri (1977), Section 7.2, and the references therein.

PROBLEMS

6.1. The Grassmann manifold $G_{k,r}$ is the set of all k dimensional subspaces in R^n (with $n=k+r$). When R^n is transformed by the orthogonal group $O(n)$ [$\mathbf{x}\rightarrow H\mathbf{x}$; $H\in O(n)$], a subspace p is transformed as $p\rightarrow Hp$, where Hp denotes the subspace spanned by the transforms $H\mathbf{x}$ of the vectors $\mathbf{x}\in p$.

(a) Show that $O(n)$ acts transitively on $G_{k,r}$.

(b) Let p_0 be the subspace of R^n spanned by the first k coordinate vectors. Show that the isotropy subgroup at p_0 is

$$G_0=\left\{H\in O(n);\quad H=\begin{bmatrix} H_1 & 0 \\ 0 & H_2 \end{bmatrix},\quad H_1\in O(k),\right.$$

$$\left. H_2\in O(n-k)\right\}.$$

(c) Find the coset corresponding to a point $p\in G_{k,r}$.

6.2. Let $\mathbf{X}_1,\dots,\mathbf{X}_{N_1}$ be independent $N_m(\boldsymbol{\mu}_1,\Sigma)$ random vectors and $\mathbf{Y}_1,\dots,\mathbf{Y}_{N_2}$ be independent $N_m(\boldsymbol{\mu}_2,\Sigma)$ random vectors. Let $\overline{\mathbf{X}}, S_X, \overline{\mathbf{Y}}, S_Y$ denote the respective sample mean vectors and sample covariance matrices, and put $S=(n_1+n_2)^{-1}(n_1S_X+n_2S_Y)$, where $n_i=N_i-1$, $i=1,2$. Consider the group $\mathcal{A}\ell(m,R)$ given by (14) of Section 6.3 acting on the space of the sufficient statistic $(\overline{\mathbf{X}},\overline{\mathbf{Y}},S)$ by

$$(\overline{\mathbf{X}},\overline{\mathbf{Y}},S)\rightarrow(B\overline{\mathbf{X}}+\mathbf{c},\ B\overline{\mathbf{Y}}+\mathbf{c},\ BSB')\qquad \left[B\in\mathcal{G}\ell(m,R);\quad \mathbf{c}\in R^m\right].$$

(a) Show that a maximal invariant under this group is

$$T^2=\frac{N_1N_2}{N_1+N_2}(\overline{\mathbf{X}}-\overline{\mathbf{Y}})'S^{-1}(\overline{\mathbf{X}}-\overline{\mathbf{Y}}).$$

(b) Consider the problem of testing $H:\boldsymbol{\mu}_1=\boldsymbol{\mu}_2$ against $K:\boldsymbol{\mu}_1\neq\boldsymbol{\mu}_2$. Show that the test which rejects H for large values of T^2 is a uniformly most powerful invariant test under the group $\mathcal{A}\ell(m,R)$.

6.3. Suppose that $\mathbf{X}_1,\dots,\mathbf{X}_{N_1}$ is a random sample from the $N_m(\boldsymbol{\mu}_1,\Sigma)$ distribution and that $\mathbf{Y}_1,\dots,\mathbf{Y}_{N_2}$ is a random sample from the $N_m(\boldsymbol{\mu}_2,\Sigma)$ distribution. Show how a T^2 statistic may be used to construct a confidence region for $\boldsymbol{\mu}_1-\boldsymbol{\mu}_2$ with confidence coefficient $1-\alpha$.

6.4. Let $\mathbf{X}_{i1},\dots,\mathbf{X}_{iN_i}$ be a random sample from the $N_m(\boldsymbol{\mu}_i,\Sigma)$ distribution, with $i=1,\dots,p$. Construct a T^2 statistic appropriate for testing the null hypothesis $H:\sum_{i=1}^{p}\alpha_i\boldsymbol{\mu}_i=\boldsymbol{\mu}$, where $\alpha_1,\dots,\alpha_p$ are specified numbers and $\boldsymbol{\mu}$ is a specified vector.

6.5. Let $\mathbf{X}_1,\dots,\mathbf{X}_{N_1}$ be a random sample from the $N_m(\boldsymbol{\mu}_1,\Sigma_1)$ distribution and $\mathbf{Y}_1,\dots,\mathbf{Y}_{N_2}$ be a random sample from the $N_m(\boldsymbol{\mu}_2,\Sigma_2)$ distribution. Here the covariance matrices Σ_1 and Σ_2 are unknown and unequal. The problem of testing $H: \boldsymbol{\mu}_1 = \boldsymbol{\mu}_2$ against $K: \boldsymbol{\mu}_1 \neq \boldsymbol{\mu}_2$ is called the multivariate Behrens–Fisher problem.

(a) Suppose $N_1 = N_2 = N$. Put $\mathbf{Z}_i = \mathbf{X}_i - \mathbf{Y}_i$ so that $\mathbf{Z}_1, \ldots, \mathbf{Z}_N$ are independent $N_m(\boldsymbol{\mu}_1 - \boldsymbol{\mu}_2, \Sigma_1 + \Sigma_2)$ random vectors. From $\mathbf{Z}_1, \ldots, \mathbf{Z}_N$ construct a T^2 statistic appropriate for testing H against K. What is the distribution of T^2? How does this differ from the distribution of T^2 when it is known that $\Sigma_1 = \Sigma_2$?

(b) Suppose $N_1 < N_2$. Put

$$\mathbf{Z}_i = \mathbf{X}_i - \left(\frac{N_1}{N_2}\right)^{1/2} \mathbf{Y}_i + (N_1 N_2)^{-1/2} \sum_{j=1}^{N_1} \mathbf{Y}_j - \frac{1}{N_2} \sum_{k=1}^{N_2} \mathbf{Y}_k$$

$$(i = 1, \ldots, N_1).$$

Show that

$$E(\mathbf{Z}_i) = \boldsymbol{\mu}_1 - \boldsymbol{\mu}_2 \quad \text{and} \quad \operatorname{Cov}(\mathbf{Z}_i) = \Sigma_1 + \frac{N_1}{N_2} \Sigma_2,$$

and that $\mathbf{Z}_1, \ldots, \mathbf{Z}_{N_1}$ are independently normally distributed. Using $\mathbf{Z}_1, \ldots, \mathbf{Z}_{N_1}$, construct a T^2 statistic appropriate for testing H against K. What is the distribution of T^2? How does this differ from the distribution of T^2 when it is known that $\Sigma_1 = \Sigma_2$?

6.6. Given a sample $\mathbf{X}_1, \ldots, \mathbf{X}_N$ from the $N_m(\boldsymbol{\mu}, \Sigma)$ distribution consider the problem of testing $H: C\boldsymbol{\mu} = 0$ against $K: C\boldsymbol{\mu} \neq 0$, where C is a specified $p \times m$ matrix of rank p. Put $C = B[I_p : 0]H$, where $B \in \mathcal{G}\ell(p, R)$ and $H \in O(m)$ and let $\mathbf{Y}_i = H\mathbf{X}_i$, $i = 1, \ldots, N$; then $\mathbf{Y}_1, \ldots, \mathbf{Y}_N$ are independent $N_m(\boldsymbol{\nu}, \Gamma)$, where $\boldsymbol{\nu} = H\boldsymbol{\mu}$ and $\Gamma = H\Sigma H'$. Put

$$A = \sum_{i=1}^{N} (\mathbf{Y}_i - \overline{\mathbf{Y}})(\mathbf{Y}_i - \overline{\mathbf{Y}})'$$

and partition $\overline{\mathbf{Y}}$, $\boldsymbol{\nu}$, A, and Γ as

$$\overline{\mathbf{Y}} = \begin{pmatrix} \overline{\mathbf{Y}}_1 \\ \overline{\mathbf{Y}}_2 \end{pmatrix}, \quad \boldsymbol{\nu} = \begin{pmatrix} \boldsymbol{\nu}_1 \\ \boldsymbol{\nu}_2 \end{pmatrix}, \quad A = \begin{bmatrix} A_{11} & A_{12} \\ A_{21} & A_{22} \end{bmatrix}, \quad \Gamma = \begin{bmatrix} \Gamma_{11} & \Gamma_{12} \\ \Gamma_{21} & \Gamma_{22} \end{bmatrix}$$

where $\overline{\mathbf{Y}}_1$ and $\boldsymbol{\nu}_1$ are $p \times 1$, $\mathbf{Y}_2$ and $\boldsymbol{\nu}_2$ are $(m-p) \times 1$, A_{11} and Γ_{11} are $p \times p$, and A_{22} and Γ_{22} are $(m-p) \times (m-p)$. Testing the null hypothesis $H: C\boldsymbol{\mu} = \mathbf{0}$ is equivalent to testing the null hypothesis $H: \boldsymbol{\nu}_1 = \mathbf{0}$. A sufficient statistic is

$(\bar{\mathbf{Y}}, A)$, where $\bar{\mathbf{Y}}$ is $N_m(\boldsymbol{\nu},(1/N)\Gamma)$, A is $W_m(n,\Gamma)$ with $n=N-1$, and $\bar{\mathbf{Y}}$ and A are independent. Consider the group of transformations

$$G=\left\{(B,\mathbf{c});\ B=\begin{bmatrix} B_{11} & 0 \\ B_{21} & B_{22}\end{bmatrix},\ B_{11}\in \mathcal{GL}(p,R),\ B_{22}\in \mathcal{GL}(m-p,R),\right.$$

$$\left.\mathbf{c}=\begin{pmatrix}\mathbf{0}\\ \mathbf{c}_2\end{pmatrix}\in R^m,\ \mathbf{c}_2\in R^{m-p}\right\},$$

acting on the space of the sufficient statistic by $(B,\mathbf{c})(\bar{\mathbf{Y}}, A)=(B\bar{\mathbf{Y}}+\mathbf{c}, BAB')$.

(a) Show that the problem of testing $H: \boldsymbol{\nu}_1=\mathbf{0}$ against $K: \boldsymbol{\nu}_1\neq\mathbf{0}$ is invariant under G.

(b) Prove that $\bar{\mathbf{Y}}_1'A_{11}^{-1}\bar{\mathbf{Y}}_1$ is a maximal invariant under G.

(c) Put $T^2=N\bar{\mathbf{Y}}_1'S_{11}^{-1}\bar{\mathbf{Y}}_1$, where $S_{11}=n^{-1}A_{11}$. Show that the test which rejects H for large values of T^2 is the uniformly most powerful invariant test under the group G. What is the distribution of T^2?

(d) Let

$$S_X=\frac{1}{n}\sum_{i=1}^{N}(\mathbf{X}_i-\bar{\mathbf{X}})(\mathbf{X}_i-\bar{\mathbf{X}})'.$$

Show that $CS_XC'=BS_{11}B'$ and hence that, in terms of the original sample $\mathbf{X}_1,\ldots,\mathbf{X}_N$,

$$T^2=N\bar{\mathbf{X}}'C'(CS_XC')^{-1}C\bar{\mathbf{X}}.$$

6.7. Let $\mathbf{X}_1,\ldots,\mathbf{X}_N$ be independent $N_m(\boldsymbol{\mu},\Sigma)$ random vectors, where $\boldsymbol{\mu}=(\mu_1,\ldots,\mu_m)'$, and consider testing the null hypothesis $H: \mu_1=\cdots=\mu_m$.

(a) Specify an $(m-1)\times m$ matrix C of rank $m-1$ such that the null hypothesis is equivalent to $C\boldsymbol{\mu}=\mathbf{0}$.

(b) Using the result of Problem 6.6 write down a T^2 statistic appropriate for testing H.

(c) The matrix C chosen in part (a), above, is clearly not unique. Show that any such matrix must satisfy $C\mathbf{1}=\mathbf{0}$, where $\mathbf{1}=(1,1,\ldots,1)'\in R^m$, and show that the T^2 statistic in part (b), above, does not depend upon the choice of C.

6.8. Show that the T^2 test of Problem 6.6 for testing $H: C\boldsymbol{\mu}=\mathbf{0}$ against $K: C\boldsymbol{\mu}\neq\mathbf{0}$ is the likelihood ratio test.

6.9. Let $\mathbf{X}_1,\ldots,\mathbf{X}_N$ be independent $N_m(\boldsymbol{\mu},\Sigma)$ random vectors and consider testing the null hypothesis H: $\boldsymbol{\mu}=k\boldsymbol{\varepsilon}$, where $\boldsymbol{\varepsilon}$ is a specified non-null vector and k is unknown, i.e., the null hypothesis says that $\boldsymbol{\mu}$ is proportional to $\boldsymbol{\varepsilon}$. (The case when $\boldsymbol{\varepsilon}=\mathbf{1}=(1,1,\ldots,1)'$ is treated in Problem 6.7.) Let C be an $(m-1)\times m$ matrix of rank $m-1$ such that H is equivalent to $C\boldsymbol{\mu}=\mathbf{0}$; clearly C must satisfy $C\boldsymbol{\varepsilon}=\mathbf{0}$. The T^2 statistic appropriate for testing H is

$$T^2=N\overline{\mathbf{X}}'C'(CSC')^{-1}C\overline{\mathbf{X}},$$

where $\overline{\mathbf{X}}$ and S are the sample mean vector and covariance matrix. Put $A=nS$, where $n=N-1$, and define

$$B=A^{-1}-\frac{1}{\boldsymbol{\varepsilon}'A^{-1}\boldsymbol{\varepsilon}}A^{-1}\boldsymbol{\varepsilon}\boldsymbol{\varepsilon}'A^{-1}.$$

Show that $B\boldsymbol{\varepsilon}=\mathbf{0}$ and that rank $(B)\le m-1$. Show that this implies that $B=DC$ for some $m\times(m-1)$ matrix D and use the fact that B is symmetric to conclude that $B=C'EC$ where E is a symmetric $(m-1)\times(m-1)$ matrix. Hence show that

$$B=C'EC=C'(CAC')^{-1}C.$$

Using this show that

$$T^2=N\left[\overline{\mathbf{X}}'A^{-1}\overline{\mathbf{X}}-\frac{(\overline{\mathbf{X}}'A^{-1}\boldsymbol{\varepsilon})^2}{\boldsymbol{\varepsilon}'A^{-1}\boldsymbol{\varepsilon}}\right].$$

This demonstrates that T^2 does not depend upon the choice of the matrix C and gives a form which may be calculated directly, once $\boldsymbol{\varepsilon}$ is specified.

6.10. Suppose that $\mathbf{X}_1,\ldots,\mathbf{X}_N$ are independent $N_m(\boldsymbol{\mu},\Sigma)$ random vectors. Partition $\boldsymbol{\mu}$ as $\boldsymbol{\mu}=(\boldsymbol{\mu}_1',\boldsymbol{\mu}_2',\boldsymbol{\mu}_3')'$, where $\boldsymbol{\mu}_1$ is $m_1\times 1$, $\boldsymbol{\mu}_2$ is $m_2\times 1$, and $\boldsymbol{\mu}_3$ is $m_3\times 1$, with $m_1+m_2+m_3=m$. It is known that $\boldsymbol{\mu}_3=0$.

(a) Derive the likelihood ratio statistic for testing H: $\boldsymbol{\mu}_2=\mathbf{0}$ against K: $\boldsymbol{\mu}_2\neq\mathbf{0}$ and find its distribution.

(b) Find a group of transformations which leaves the testing problem invariant and show that the likelihood ratio test is a uniformly most powerful invariant test.

6.11. Let F denote the class of spherically symmetric density functions (with respect to Lebesgue measure on R^m), i.e., $f\in F\Rightarrow f(\mathbf{x})=f(H\mathbf{x})$ for all $\mathbf{x}\in R^m$, $H\in O(m)$, and let $F(\Sigma)$ denote the class of elliptical density functions given by $f\in F(\Sigma)\Rightarrow f(\mathbf{x})=(\det\Sigma)^{-1/2}h(\mathbf{x}'\Sigma^{-1}\mathbf{x})$ for some funtion h

on $[0,\infty)$. Let $\mathbf{X}$ be an $m\times 1$ random vector with density function h and consider the problem of testing H_0: $h\in F$ against K: $h\in F(\Sigma)$, where $\Sigma\neq\sigma^2 I_m$ is a fixed positive definite $m\times m$ matrix.

(a) Show that this testing problem is invariant under the group of transformations

$$\mathbf{X}\rightarrow\alpha\mathbf{X}$$

for $\alpha>0$.

(b) Show that a maximal invariant is $\phi(\mathbf{x})=\|\mathbf{x}\|^{-1}\mathbf{x}$.

(c) Show that under H_0, $\phi(\mathbf{x})$ has the same distribution for all $h\in F$.

(d) Show that under K, $\mathbf{y}=\phi(\mathbf{x})$ has density function.

$$\frac{\Gamma(\frac{1}{2}m)}{2\pi^{m/2}}(\det\Sigma)^{-1/2}(\mathbf{y}'\Sigma^{-1}\mathbf{y})^{-m/2}$$

with respect to the uniform measure on $S_m=\{\mathbf{x}\in R^m;\mathbf{x}'\mathbf{x}=1\}$, so that under K, $\phi(\mathbf{x})$ has the same distribution for all $h\in F(\Sigma)$.

(e) Show that the test which rejects H_0 for small values of

$$\frac{\mathbf{x}'\Sigma^{-1}\mathbf{x}}{\mathbf{x}'\mathbf{x}}$$

is a uniformly most powerful invariant test (King, 1980).

CHAPTER 7

Zonal Polynomials and Some Functions of Matrix Argument

7.1. INTRODUCTION

Many noncentral distributions in classical multivariate analysis involve integrals, over orthogonal groups or Stiefel manifolds with respect to an invariant measure, which cannot be evaluated in closed form. We have already met such a distribution in Theorem 3.2.18, where it was shown that if the $m\times m$ random matrix A has the $W_m(n,\Sigma)$ distribution then the joint density function of the latent roots $l_1,\dots,l_m$ of A involves the integral

$$\int_{O(m)}\operatorname{etr}\left(-\tfrac{1}{2}\Sigma^{-1}HLH'\right)(dH)$$

where $L=\operatorname{diag}(l_1,\dots,l_m)$ and (dH) represents the invariant measure on the group $O(m)$ of orthogonal $m\times m$ matrices, normalized so that the volume of $O(m)$ is unity (see the discussion preceding Theorem 3.2.17). This integral depends on Σ only through its latent roots $\lambda_1,\dots,\lambda_m$ and it is easy to see that it is a symmetric function of $l_1,\dots,l_m$ and of $\lambda_1,\dots,\lambda_m$. To evaluate the integral an obvious approach is to expand the exponential in the integrand as an infinite series and attempt to integrate term by term. This is very difficult to carry out in general, unless one chooses the "right" symmetric functions to work with. It can be done, but first we need to develop some theory. We will return to this example in Chapter 9.

Let us see what types of results we might hope for by comparing a familiar univariate distribution with its multivariate counterpart. Suppose that $a=\mathbf{X}'\mathbf{X}$, where $\mathbf{X}$ is $N_n(\boldsymbol{\mu},I_n)$; then the random variable a has the noncentral $\chi_n^2(\delta)$ distribution with $\delta=\mu'\mu$, and density function (see Theorem 1.3.4)

$$(1)\qquad \begin{aligned}&\frac{1}{2^{n/2}\Gamma(\frac{1}{2}n)}e^{-a/2}a^{n/2-1}\\ &e^{-\delta/2}{}_0F_1\left(\tfrac{1}{2}n;\tfrac{1}{4}\delta a\right)\qquad\qquad (a>0).\end{aligned}$$

Now suppose that $A=Z'Z$, where Z is $N(M,I_n\otimes I_m)$; that is, $E(Z)=M$ and the elements of the $n\times m$ matrix Z are independent and normally distributed with unit variance. If $M=0$, A has the $W_m(n,I_m)$ distribution (recall Definition 3.1.3) with density function

$$\frac{1}{2^{mn/2}\Gamma_m(\frac{1}{2}n)}\operatorname{etr}\left(-\tfrac{1}{2}A\right)(\det A)^{(n-m-1)/2}\qquad (A>0),$$

which reduces to the first line of (1) when $m=1$. When $M\neq 0$ the distribution of A is called noncentral Wishart and it is clear (use invariance) that this depends on M only through a "noncentrality matrix" $\Delta=M'M$. Moreover, the noncentral Wishart density function must reduce to (1) when $m=1$. This being the case, we might hope that there is a "natural" generalization of the noncentral part

$$e^{-\delta/2}{}_0F_1\left(\tfrac{1}{2}n;\tfrac{1}{4}\delta a\right)$$

of the density function (1) when δ is replaced by Δ and a is replaced by A. It seems reasonable to anticipate that $e^{-\delta/2}$ would be generalized by $\operatorname{etr}(-\frac{1}{2}\Delta)$ and that the real problem will be to generalize the ${}_0F_1$ function, which has $\frac{1}{4}\delta a$ as its argument, to a function which has $\frac{1}{4}\Delta A$ as its argument. Recall that

$${}_0F_1(c;x)=\sum_{k=0}^{\infty}\frac{x^k}{(c)_k k!},$$

so that if the argument x is to be replaced by a matrix X (with the generalized function remaining real-valued), what is needed is a generalization of the powers x^k of x when x is replaced by a matrix X. This is the role played by *zonal polynomials*, which are symmetric polynomials in the latent roots of X. The general theory of zonal polynomials was developed in a series of papers by James (1960, 1961a, b, 1964, 1968, 1973, 1976) and Constantine (1963, 1966). Zonal polynomials are usually defined using the group representation theory of $\mathcal{G}\ell(m, R)$, the general linear group. The theory leading up to this definition is, however, quite difficult from a technical point of view, and for a detailed discussion of the group theoretic construction of zonal polynomials the reader is referred to Farrell (1976) and the papers of James and Constantine cited above, particularly James (1961b). Rather than outline a course in group representation theory, here we will start from another definition for the zonal polynomials which may appear somewhat arbitrary but probably has more pedagogic value. It should be emphasized that the treatment here is intended as an introduction to zonal polynomials and related topics. This is particularly true in Sections 7.2.1 and 7.2.2, where a rather informal approach is apparent. [For yet another approach, see an interesting paper by Saw (1977).]

7.2. ZONAL POLYNOMIALS

7.2.1. *Definition and Construction*

The zonal polynomials of a matrix are defined in terms of partitions of positive integers. Let k be a positive integer; a partition κ of k is written as $\kappa=(k_1, k_2,\dots)$, where $\Sigma_i k_i = k$, with the convention unless otherwise stated, that $k_1 \geq k_2 \geq \cdots$, where $k_1, k_2,\dots$ are non-negative integers. We will order the partitions of k *lexicographically*; that is, if $\kappa=(k_1, k_2,\dots)$ and $\lambda=(l_1, l_2,\dots)$ are two partitions of k we will write $\kappa>\lambda$ if $k_i>l_i$ for the first index i for which the parts are unequal. For example, if $k=6$,

$$\kappa=(2,2,2)>\lambda=(2,2,1,1).$$

Now suppose that $\kappa=(k_1,\dots,k_m)$ and $\lambda=(l_1,\dots,l_m)$ are two partitions of k (some of the parts may be zero) and let $y_1,\dots,y_m$ be m variables. If $\kappa>\lambda$ we will say that the monomial $y_1^{k_1}\dots y_m^{k_m}$ is of *higher weight* than the monomial $y_1^{l_1}\dots y_m^{l_m}$.

We are now ready to define a zonal polynomial. Before doing so, recall from the discussion in Section 7.1 that what we would like is a generalization of the function $f_k(x)=x^k$, which satisfies the differential equation $x^2 f_k''(x)= k(k-1)x^k$. Bearing this in mind may help to make the following definition seem a little less arbitrary. It is based on papers by James in 1968 and 1973.

DEFINITION 7.2.1. Let Y be an $m\times m$ symmetric matrix with latent roots $y_1,\dots,y_m$ and let $\kappa=(k_1,\dots,k_m)$ be a partition of k into not more than m parts. The zonal polynomial of Y corresponding to κ, denoted by $C_\kappa(Y)$, is a symmetric, homogeneous polynomial of degree k in the latent roots $y_1,\ldots,y_m$ such that:

(i) The term of highest weight in $C_\kappa(Y)$ is $y_1^{k_1}\ldots y_m^{k_m}$; that is,

$$C_\kappa(Y)=d_\kappa y_1^{k_1}\ldots y_m^{k_m}+\text{terms of lower weight}, \tag{1}$$

where d_κ is a constant.

(ii) $C_\kappa(Y)$ is an *eigenfunction* of the differential operator Δ_Y given by

$$\Delta_Y=\sum_{i=1}^{m} y_i^2\frac{\partial^2}{\partial y_i^2}+\sum_{i=1}^{m}\sum_{\substack{j=1\\ j\neq i}}^{m}\frac{y_i^2}{y_i-y_j}\frac{\partial}{\partial y_i}. \tag{2}$$

(iii) As κ varies over all partitions of k the zonal polynomials have unit coefficients in the expansion of $(\operatorname{tr} Y)^k$; that is,

$$(\operatorname{tr} Y)^k=(y_1+\cdots+y_m)^k=\sum_\kappa C_\kappa(Y). \tag{3}$$

We will now comment on various aspects of this definition.

Remark 1. By a symmetric, homogeneous polynomial of degree k in $y_1,\ldots,y_m$ we mean a polynomial which is unchanged by a permutation of the subscripts and such that every term in the polynomial has degree k.

For example, if $m=2$ and $k=3$,

$$y_1^3+y_2^3+10y_1^2y_2+10y_1y_2^2$$

is a symmetric, homogeneous polynomial of degree 3 in y_1 and y_2.

Remark 2. The zonal polynomial $C_\kappa(Y)$ is a function only of the latent roots $y_1,\ldots,y_m$ of Y and so could be written, for example, as $C_\kappa(y_1,\ldots,y_m)$. However, for many purposes it is more convenient to use the matrix notation of the definition; see, for example, Theorem 7.2.4 later.

Remark 3. By saying that $C_\kappa(Y)$ is an eigenfunction of the differential operator Δ_Y given by (2) we mean that

$$\Delta_Y C_\kappa(Y)=\alpha C_\kappa(Y),$$

where α is a constant which does not depend on $y_1,\ldots,y_m$ (but which can depend on κ) and which is called the eigenvalue of Δ_Y corresponding to $C_\kappa(Y)$. This constant will be found in Theorem 7.2.2.

Remark 4. It has yet to be established that Definition 7.2.1 is not vacuous and that indeed there exists a unique polynomial in $y_1,\ldots,y_m$ satisfying all the conditions of this definition. Basically what happens is that condition (i), along with the condition that $C_\kappa(Y)$ is a symmetric, homogeneous polynomial of degree k, establishes what types of terms appear in $C_\kappa(Y)$. The differential equation for $C_\kappa(Y)$ provided by (ii) and Theorem 7.2.2 below then gives recurrence relations between the coefficients of these terms which determine $C_\kappa(Y)$ uniquely up to some normalizing constant. The normalization is provided by condition (iii), and this is the only role this condition plays. At this point it should be stated that no *general* formula for zonal polynomials is known; however, the above description provides a general *algorithm* for their calculation. We will illustrate the steps involved with concrete examples later. Before doing so, let us find the eigenvalue implicit in condition (ii).

THEOREM 7.2.2. The zonal polynomial $C_\kappa(Y)$ corresponding to the partition $\kappa=(k_1,\ldots,k_m)$ of k satisfies the partial differential equation

$$\text{(4)}\qquad \Delta_Y C_\kappa(Y)=[\rho_\kappa+k(m-1)]C_\kappa(Y),$$

where Δ_Y is given by (2) and

$$\text{(5)}\qquad \rho_\kappa=\sum_{i=1}^{m}k_i(k_i-i).$$

[Hence the eigenvalue α in Remark 3 is $\alpha=\rho_\kappa+k(m-1)$.]

Proof. By conditions (i) and (ii) it suffices to show that

$$\Delta_Y y_1^{k_1}\cdots y_m^{k_m}=[\rho_\kappa+k(m-1)]y_1^{k_1}\cdots y_m^{k_m}+\text{terms of lower weight.}$$

By straightforward differentiation it is seen that

$$\Delta_Y y_1^{k_1}\cdots y_m^{k_m}=y_1^{k_1}\cdots y_m^{k_m}\left[\sum_{i=1}^{m}k_i(k_i-1)+\sum_{i=1}^{m}\sum_{\substack{j=1\\ j\neq i}}^{m}\frac{y_ik_i}{y_i-y_j}\right]$$

$$=y_1^{k_1}\cdots y_m^{k_m}\left[\sum_{i=1}^{m}k_i^2-k+\sum_{i=1}^{m-1}\sum_{j=i+1}^{m}\left(\frac{y_ik_i}{y_i-y_j}+\frac{y_jk_j}{y_j-y_i}\right)\right]$$

Since

$$\frac{y_ik_i}{y_i-y_j}=k_i+\frac{y_jk_i}{y_i-y_j}$$

it follows that

$$\Delta_Y y_1^{k_1}\cdots y_m^{k_m} = y_1^{k_1}\cdots y_m^{k_m}\left\{\sum_{i=1}^{m} k_i^2 - k + \sum_{i=1}^{m-1}\sum_{j=i+1}^{m}\left[k_i + \frac{y_j}{y_j - y_i}(k_j - k_i)\right]\right\}$$

$$= y_1^{k_1}\cdots y_m^{k_m}\left[\sum_{i=1}^{m} k_i^2 - k + \sum_{i=1}^{m-1} k_i(m-i)\right] + \text{terms of lower weight.}$$

Noting that

$$\sum_{i=1}^{m-1} k_i(m-i) = \sum_{i=1}^{m} k_i(m-i) = km - \sum_{i=1}^{m} ik_i$$

we then have

$$\Delta_Y y_1^{k_1}\cdots y_m^{k_m} = y_1^{k_1}\cdots y_m^{k_m}\left[\sum_{i=1}^{m} k_i(k_i - i) + k(m-1)\right] + \text{terms of lower weight,}$$

and the proof is complete.

Before proceeding further it is worth pointing out explicitly two consequences of Definition 7.2.1. The first is that if $m=1$, condition (iii) becomes $y^k = C_{(k)}(Y)$ so that the zonal polynomials of a matrix variable are analogous to powers of a single variable. The second consequence is that if β is a constant then the fact that $C_\kappa(Y)$ is homogeneous of degree k implies that $C_\kappa(\beta Y) = \beta^k C_\kappa(Y)$.

We will now illustrate how Definition 7.2.1 can be used to construct an algorithm for calculating zonal polynomials by using it to find explicit formulas corresponding to the values $k=1$, 2, and 3. We will express these zonal polynomials in terms of the *monomial symmetric functions*. If $\kappa = (k_1,\ldots,k_m)$, the monomial symmetric function of $y_1,\ldots,y_m$ corresponding to κ is defined as

$$M_\kappa(Y) = \sum\cdots\sum y_{i_1}^{k_1} y_{i_2}^{k_2}\cdots y_{i_p}^{k_p},$$

where p is the number of nonzero parts in the partition κ and the summation is over the distinct permutations $(i_1,\ldots,i_p)$ of p different integers from the integers $1,\ldots,m$. Hence

$$M_\kappa(Y) = y_1^{k_1}\cdots y_m^{k_m} + \text{symmetric terms.}$$

Thus, for example,

$$M_{(1)}(Y) = y_1 + \cdots + y_m,$$

$$M_{(2)}(Y) = y_1^2 + \cdots + y_m^2,$$

$$M_{(1,1)}(Y) = \sum_{i<j}^{m} y_i y_j,$$

and so on.

$\underline{k=1}$: When $k=1$ there is only one partition $\kappa=(1)$ so, by condition (iii), $C_{(1)}(Y) = \operatorname{tr} Y = y_1 + \cdots + y_m = M_{(1)}(Y)$.

$\underline{k=2}$: When $k=2$ there are two zonal polynomials corresponding to the partitions (2) and (1,1) of the integer 2. Using condition (i) and the fact that the zonal polynomials are symmetric and homogeneous of degree 2 we have

$$\begin{aligned} C_{(2)}(Y) &= d_{(2)} y_1^2 + \text{terms of lower weight} \\ &= d_{(2)} y_1^2 + \beta y_1 y_2 + \text{symmetric terms} \\ &= d_{(2)}(y_1^2 + \cdots + y_m^2) + \beta(y_1 y_2 + \cdots + y_{m-1} y_m) \\ &= d_{(2)} M_{(2)}(Y) + \beta M_{(1,1)}(Y) \end{aligned}$$

for some constant β, and

$$\begin{aligned} C_{(1,1)}(Y) &= d_{(1,1)} y_1 y_2 + \text{terms of lower weight} \\ &= d_{(1,1)}(y_1 y_2 + \cdots + y_{m-1} y_m) \\ &= d_{(1,1)} M_{(1,1)}(Y). \end{aligned}$$

By condition (iii) we have

$$\begin{aligned} (\operatorname{tr} Y)^2 \equiv M_{(2)}(Y) + 2M_{(1,1)}(Y) &= C_{(2)}(Y) + C_{(1,1)}(Y) \\ &= d_{(2)} M_{(2)}(Y) + (\beta + d_{(1,1)}) M_{(1,1)}(Y), \end{aligned}$$

and equating coefficients of $M_{(2)}(Y)$ and $M_{(1,1)}(Y)$ on both sides shows that

$$d_{(2)} = 1, \qquad d_{(1,1)} = 2 - \beta,$$

so that

$$C_{(2)}(Y)=M_{(2)}(Y)+\beta M_{(1,1)}(Y) \tag{6}$$

and

$$C_{(1,1)}(Y)=(2-\beta)M_{(1,1)}(Y).$$

The constant β is now found using the differential equation for $C_{(2)}(Y)$. Since $\rho_{(2)}=2(2-1)=2$, Theorem 7.2.2 shows that $C_{(2)}(Y)$ satisfies the partial differential equation

$$\Delta_Y C_{(2)}(Y)=2mC_{(2)}(Y), \tag{7}$$

where Δ_Y is the differential operator given by (2). It is easily verified that

$$\Delta_Y M_{(2)}(Y)=2mM_{(2)}(Y)+2M_{(1,1)}(Y)$$

and

$$\Delta_Y M_{(1,1)}(Y)=(2m-3)M_{(1,1)}(Y),$$

and hence substitution of (6) in (7) yields

$$2mM_{(2)}(Y)+2M_{(1,1)}(Y)+\beta(2m-3)M_{(1,1)}(Y)$$
$$=2mM_{(2)}(Y)+2m\beta M_{(1,1)}(Y).$$

Equating coefficients of $M_{(1,1)}(Y)$ on both sides then gives $\beta=2/3$. Hence the two zonal polynomials corresponding to $k=2$ are

$$C_{(2)}(Y)=M_{(2)}(Y)+\frac{2}{3}M_{(1,1)}(Y)$$

and

$$C_{(1,1)}(Y)=\frac{4}{3}M_{(1,1)}(Y).$$

<u>$k=3$</u>: When $k=3$ there are three zonal polynomials corresponding to the partitions (3), (2, 1), and (1, 1, 1); we will indicate how these can be evaluated, leaving the details as an exercise. Conditions (i) and (iii) of Definition 7.2.1, together with the symmetric homogeneous nature of the zonal polynomials,

are sufficient to show that

$$C_{(3)}(Y)=M_{(3)}(Y)+\beta M_{(2,1)}(Y)+\gamma M_{(1,1,1)}(Y), \tag{8}$$

$$C_{(2,1)}(Y)=(3-\beta)M_{(2,1)}(Y)+\delta M_{(1,1,1)}(Y),$$

and

$$C_{(1,1,1)}(Y)=(6-\gamma-\delta)M_{(1,1,1)}(Y),$$

for some constants β, γ, and δ. The partial differential equation for $C_{(3)}(Y)$ then determines β and γ. Once β is known, the partial differential equation for $C_{(2,1)}(Y)$ determines δ. To demonstrate this, we need the effect of the operator Δ_Y on the monomial symmetric functions $M_{(3)}(Y)$, $M_{(2,1)}(Y)$, and $M_{(1,1,1)}(Y)$. It can be readily verified that

$$\Delta_Y M_{(3)}(Y)=3(m+1)M_{(3)}(Y)+3M_{(2,1)}(Y), \tag{9}$$

$$\Delta_Y M_{(2,1)}(Y)=(3m-2)M_{(2,1)}(Y)+6M_{(1,1,1)}(Y),$$

and

$$\Delta_Y M_{(1,1,1)}(Y)=3(m-2)M_{(1,1,1)}(Y).$$

Since $\rho_{(3)}=3(3-1)=6$, Theorem 7.2.2 shows that $C_{(3)}(Y)$ satisfies the partial differential equation

$$\Delta_Y C_{(3)}(Y)=3(m+1)C_{(3)}(Y). \tag{10}$$

Substituting for $C_{(3)}(Y)$ from (8) in (10), using the differential relations (9), and equating coefficients of $M_{(2,1)}(Y)$ and $M_{(1,1,1)}(Y)$ on both sides then gives $\beta=3/5$ and $\gamma=2/5$. Since $\rho_{(2,1)}=2(2-1)+1(1-2)=1$, the partial differential equation given by Theorem 7.2.2 for $C_{(2,1)}(Y)$ is

$$\Delta_Y C_{(2,1)}(Y)=(3m-2)C_{(2,1)}(Y). \tag{11}$$

Substituting for $C_{(2,1)}(Y)$ from (8), with $\beta=3/5$, in (11), using the differential relations (9), and equating coefficients of $M_{(1,1,1)}(Y)$ on both sides then gives $\delta=18/5$. Hence the three zonal polynomials of degree 3 are

$$C_{(3)}(Y)=M_{(3)}(Y)+\frac{3}{5}M_{(2,1)}(Y)+\frac{2}{5}M_{(1,1,1)}(Y), \tag{12}$$

$$C_{(2,1)}(Y)=\frac{12}{5}M_{(2,1)}(Y)+\frac{18}{5}M_{(1,1,1)}(Y),$$

and

$$C_{(1,1,1)}(Y)=2M_{(1,1,1)}(Y).$$

In general, it should now be apparent that the differential equation for $C_\kappa(Y)$ gives rise to a recurrence relation between the coefficients of the monomial symmetric functions in $C_\kappa(Y)$; once the coefficient of the term of highest weight is given, the other coefficients are uniquely determined by the recurrence relation. We will *state* a general result, due to James (1968). Let κ be a partition of k; condition (i) of Definition 7.2.1 and the fact that the zonal polynomial $C_\kappa(Y)$ is symmetric and homogeneous of degree k show that $C_\kappa(Y)$ can be expressed in terms of the monomial symmetric functions as

$$C_\kappa(Y)=\sum_{\lambda\le\kappa} c_{\kappa,\lambda}M_\lambda(Y), \tag{13}$$

where the $c_{\kappa,\lambda}$ are constants and the summation is over all partitions λ of k with $\lambda\le\kappa$ (that is, λ is below or equal to κ in the lexicographical ordering). Substituting this expression (13) in the partial differential equation

$$\Delta_Y C_\kappa(Y)=[\rho_\kappa+k(m-1)]C_\kappa(Y)$$

and equating coefficients of like monomial symmetric functions on both sides leads to a recurrence relation for the coefficients, namely,

$$c_{\kappa,\lambda}=\sum_{\lambda<\mu\le\kappa}\frac{[(l_i+t)-(l_j-t)]}{\rho_\kappa-\rho_\lambda}c_{\kappa,\mu} \tag{14}$$

where $\lambda=(l_1,\dots,l_m)$ and $\mu=(l_1,\dots,l_i+t,\dots,l_j-t,\dots,l_m)$ for $t=1,\dots,l_j$ such that, when the parts of the partition μ are arranged in descending order, μ is above λ and below or equal to κ in the lexicographical ordering. The summation in (14) is over all such μ, including possibly, nondescending ones, and any empty sum is taken to be zero. This recurrence relation determines $C_\kappa(Y)$ uniquely once the coefficient of the term of highest weight is given. Using condition (iii) of Definition 7.2.1 it follows that for $\kappa=(k)$ the coefficient of the term of highest weight in $C_{(k)}(Y)$ is unity; that is, $c_{(k),(k)}=1$. This determines all the other coefficients $c_{(k),\lambda}$ in the expansion (13) of $C_{(k)}(Y)$ in terms of monomial symmetric functions. These determine, in turn, the coefficient of the term of highest weight in $C_{(k-1,1)}(Y)$, and once this is known, the recurrence relation gives all the other coefficients, and so on. The reader can readily verify that the general recurrence relation (14) gives the coefficients of the monomial symmetric functions found earlier in the expressions for the zonal polynomials of degree $k=1$, 2, and 3. We will look at one further example, namely, $k=4$. Here there are five zonal polynomials, corresponding to the partitions (4), (3,1), (2,2), (2,1,1), and (1,1,1,1). Consider the zonal polynomial $C_{(4)}(Y)$. Using (13) this can be written in terms of the monomial symmetric functions as

$$C_{(4)}(Y)=M_{(4)}(Y)+c_{(4),(3,1)}M_{(3,1)}(Y)+c_{(4),(2,2)}M_{(2,2)}(Y)$$
$$+c_{(4),(2,1,1)}M_{(2,1,1)}(Y)+c_{(4),(1,1,1,1)}M_{(1,1,1,1)}(Y),$$

where we have used the fact that $c_{(4),(4)}=1$. Consider the coefficient $c_{(4),(3,1)}$. Putting $\kappa=(4)$, $\lambda=(3,1)$ in (14) and using $\rho_{(4)}=12$, $\rho_{(3,1)}=5$ gives

$$c_{(4),(3,1)}=\frac{4-0}{12-5}=\frac{4}{7}.$$

The coefficient $c_{(4),(2,2)}$ comes from the partitions (3,1) and (4) and, since $\rho_{(2,2)}=2$, it is

$$c_{(4),(2,2)}=\frac{4-0}{12-2}c_{(4),(4)}+\frac{3-1}{12-2}c_{(4),(3,1)}$$
$$=\frac{18}{35}.$$

The coefficient $c_{(4),(2,1,1)}$ comes from the partitions (3,1,0), (3,0,1), and (2,2,0) and, since $\rho_{(2,1,1)}=-1$, it is

$$c_{(4),(2,1,1)}=2\cdot\frac{3-0}{12+1}c_{(4),(3,1)}+\frac{2-0}{12+1}c_{(4),(2,2)}$$
$$=\frac{12}{35}.$$

The coefficient $c_{(4),(1,1,1,1)}$ comes from the partitions (2,0,1,1), (2,1,0,1), (2,1,1,0), (1,2,0,1), (1,2,1,0), and (1,1,2,0) and, since $\rho_{(1,1,1,1)}=-6$, it is

$$c_{(4),(1,1,1,1)}=6\cdot\frac{2-0}{12+6}c_{(4),(2,1,1)}=\frac{8}{35}.$$

Hence the zonal polynomial $C_{(4)}(Y)$ is

$$(15)\qquad C_{(4)}(Y)=M_{(4)}(Y)+\frac{4}{7}M_{(3,1)}(Y)+\frac{18}{35}M_{(2,2)}(Y)+\frac{12}{35}M_{(2,1,1)}(Y)+\frac{8}{35}M_{(1,1,1,1)}(Y).$$

The next zonal polynomial $C_{(3,1)}(Y)$ can be written

$$(16)\quad C_{(3,1)}(Y)=c_{(3,1),(3,1)}M_{(3,1)}(Y)+c_{(3,1),(2,2)}M_{(2,2)}(Y)+c_{(3,1),(2,1,1)}M_{(2,1,1)}(Y)+c_{(3,1),(1,1,1,1)}M_{(1,1,1,1)}(Y),$$

and condition (iii) of Definition 7.2.1, in conjunction with the expression (15) for $C_{(4)}(Y)$, shows that $c_{(3,1),(3,1)}=24/7$. The recurrence relation (14) then determines the other coefficients in (16); the remaining computations for $k=4$ are left as an exercise (see Problem 7.1).

Without delving deeply into the details we will give two properties of zonal polynomials which can be proved using the recurrence relation (14). They are consequences of the following lemma.

LEMMA 7.2.3. Let the coefficients $c_{\kappa,\lambda}$ be given by (13) and suppose that κ is a partition of k into p nonzero parts. If the partition λ of k has less than p nonzero parts and $\lambda<\kappa$ then $c_{\kappa,\lambda}=0$.

Rather than giving a tedious algebraic proof, we will illustrate the lemma with an example. The partition $\kappa=(4,1,1,1)$ of $k=7$ is followed in the lexicographical ordering by two partitions with less than four parts, namely, $(3,3,1)$ and $(3,2,2)$. Considering first $\lambda=(3,3,1)$, the recurrence relation (14) immediately shows that $c_{\kappa,\lambda}=0$ because there are no partitions μ satisfying $\lambda<\mu\leq\kappa$ [see the discussion following (14)]. Now taking $\lambda=(3,2,2)$, the coefficient $c_{\kappa,\lambda}$ comes from the partition $(3,3,1)$ so that

$$c_{\kappa,\lambda}=2\cdot\frac{3-1}{\rho_\kappa-\rho_\lambda}c_{\kappa,\mu},$$

where $\mu=(3,3,1)$, and it has just been established that $c_{\kappa,\mu}=0$.

The two aforementioned properties of zonal polynomials are given in the following corollary.

COROLLARY 7.2.4.

(i) If the $m\times m$ symmetric matrix Y has rank r, so that $y_{r+1}=\cdots=y_m=0$, and if κ is a partition of k into more than r parts, then $C_\kappa(Y)=0$.

(ii) If Y is a positive definite matrix ($Y>0$) then $C_\kappa(Y)>0$.

Proof. To prove (i), write $C_\kappa(Y)$ as

$$C_\kappa(Y)=\sum_{\lambda\leq\kappa} c_{\kappa,\lambda}M_\lambda(Y).$$

Now note that $M_\lambda(Y)=0$ if the number of nonzero parts in λ is greater than or equal to the number of nonzero parts in κ, while if the reverse is true then $c_{\kappa,\lambda}=0$ by Lemma 7.2.3. Part (ii) is proved by noting that the monomial symmetric functions are positive when $Y>0$, and the coefficients $c_{\kappa,\lambda}$ generated by the recurrence relation (14) are non-negative.

Zonal polynomials have so far been defined only for symmetric matrices. The definition can be extended: if Y is symmetric and X is positive definite then the latent roots of XY are the same as the latent roots of $X^{1/2}YX^{1/2}$ and we define $C_\kappa(XY)$ as

$$C_\kappa(XY)=C_\kappa(X^{1/2}YX^{1/2}). \tag{17}$$

As stated earlier there is no known general formula for zonal polynomials. Expressions are known for some special cases (see James, 1964, 1968). One of these special cases is when $Y=I_m$. Although we will not derive the result here, it is worth stating. If the partition κ of k has p nonzero parts, the value of the zonal polynomial at I_m is given by

$$C_\kappa(I_m)=2^{2k}k!(\tfrac{1}{2}m)_\kappa\frac{\prod_{i<j}^{p}(2k_i-2k_j-i+j)}{\prod_{i=1}^{p}(2k_i+p-i)!} \tag{18}$$

where

$$(\tfrac{1}{2}m)_\kappa=\prod_{i=1}^{p}\left(\tfrac{1}{2}(m-i+1)\right)_{k_i},$$

with $(a)_k=a(a+1)\dots(a+k-1),(a)_0=1$. For a proof of this result the reader is referred to Constantine (1963). Although no general formula is known, the recurrence relation (14) enables the zonal polynomials to be computed quite readily. The coefficients $c_{\kappa,\lambda}$ of the monomial symmetric functions $M_\lambda(Y)$ in $C_\kappa(Y)$ obtained from (14) are given in Table 1 to $k=5$. They have been tabulated to $k=12$ by Parkhurst and James (1974) in terms of the sums of powers of the latent roots and in terms of the elementary

Table 1. Coefficients of monomial symmetric functions $M_\lambda(Y)$ in the zonal polynomial $C_\kappa(Y)$

$k=2$

κ \ λ	(2)	(1,1)
(2)	1	2/3
(1,1)	0	4/3

$k=3$

κ \ λ	(3)	(2,1)	(1,1,1)
(3)	1	3/5	2/5
(2,1)	0	12/5	18/5
(1,1,1)	0	0	2

$k=4$

κ \ λ	(4)	(3,1)	(2,2)	(2,1,1)	(1,1,1,1)
(4)	1	4/7	18/35	12/35	8/35
(3,1)	0	24/7	16/7	88/21	32/7
(2,2)	0	0	16/5	32/15	16/5
(2,1,1)	0	0	0	16/3	64/5
(1,1,1,1)	0	0	0	0	16/5

$k=5$

κ \ λ	(5)	(4,1)	(3,2)	(3,1,1)	(2,2,1)	(2,1,1,1)	(1,1,1,1,1)
(5)	1	5/9	10/21	20/63	2/7	4/21	8/63
(4,1)	0	40/9	8/3	46/9	4	14/3	40/9
(3,2)	0	0	48/7	32/7	176/21	64/7	80/7
(3,1,1)	0	0	0	10	20/3	130/7	200/7
(2,2,1)	0	0	0	0	32/3	16	32
(2,1,1,1)	0	0	0	0	0	80/7	800/21
(1,1,1,1,1)	0	0	0	0	0	0	16/3

symmetric functions of the roots. For larger values of k tabulation of zonal polynomials seems prohibitive in terms of space; indeed, for $k=12$, there are already 77 zonal polynomials corresponding to the 77 partitions of 12. However, the recurrence relation (14) has been used as the basis of a subroutine due to McLaren (1976) which calculates the coefficients $c_{\kappa,\lambda}$, and which is readily available. An alternative method of calculating zonal polynomials by computing sums of products of moments of independent normal random variables has been given by Kates (1980).

7.2.2 *A Fundamental Property*

Many results about zonal polynomials are proved with the help of a fundamental identity which has to do with averaging over the orthogonal group. This is given later in Theorem 7.2.5. Before getting to this we will look a little more closely at the differential form Δ_Y used in Definition 7.2.1 and at some related topics.

Let X be an $m\times m$ positive definite matrix and put

$$(ds)^2=\operatorname{tr}(X^{-1}dX\,X^{-1}dX) \tag{19}$$

where $dX=(dx_{ij})$. This is a (metric) differential form on the space $\mathcal{S}_m$ of $m\times m$ positive definite matrices which is invariant under the congruence transformation

$$X\rightarrow LXL' \tag{20}$$

for $L\in\mathcal{G}\ell(m,R)$, the group of $m\times m$ nonsingular real matrices. For then $dX\rightarrow L\,dXL'$, so that

$$\begin{aligned}\operatorname{tr}(X^{-1}dX\,X^{-1}dX)&\rightarrow\operatorname{tr}\big((LXL')^{-1}L\,dXL'(LXL')^{-1}L\,dXL'\big)\\&=\operatorname{tr}(X^{-1}dX\,X^{-1}dX).\end{aligned} \tag{21}$$

Now, put $n=m(m+1)/2$ and let $\mathbf{x}$ be the $n\times 1$ vector

$$\mathbf{x}=(x_{11},x_{12},\ldots,x_{1m},x_{22},\ldots,x_{2m},\ldots,x_{mm})'$$

consisting of the distinct elements in X. For notational convenience, relabel the components of $\mathbf{x}$ as $x_1,\ldots,x_n$. The differential form $(ds)^2$ is a quadratic form in the elements of the vector of differentials $d\mathbf{x}$ and can be written as

$$(ds)^2=\operatorname{tr}(X^{-1}dX\,X^{-1}dX)=d\mathbf{x}'G(\mathbf{x})\,d\mathbf{x}, \tag{22}$$

where $G(\mathbf{x})$ is an $n\times n$ nonsingular symmetric matrix. The reader is encouraged to write out $G(\mathbf{x})$ explicitly in the case $m=2$ (see Problem 7.2). Now define the differential operator Δ_X^* by

$$\Delta_X^* = \det G(\mathbf{x})^{-1/2} \sum_{j=1}^{n} \frac{\partial}{\partial x_j}\left[\det G(\mathbf{x})^{1/2} \sum_{i=1}^{n} g(\mathbf{x})^{ij}\frac{\partial}{\partial x_i}\right] \tag{23}$$

where $G(\mathbf{x})^{-1}=(g(\mathbf{x})^{ij})$. Denoting by $\partial/\partial\mathbf{x}$ the $n\times 1$ vector with components $\partial/\partial x_i$, we can write Δ_X^* as

$$\Delta_X^* = \det G(\mathbf{x})^{-1/2}\frac{\partial'}{\partial\mathbf{x}}\left[\det G(\mathbf{x})^{1/2}G(\mathbf{x})^{-1}\frac{\partial}{\partial\mathbf{x}}\right]. \tag{24}$$

This differential operator has the property that, like $(ds)^2$, it is invariant under the congruence transformation (20) for $L\in\mathcal{G}\ell(m,R)$; that is,

$$\Delta_{LXL'}^* = \Delta_X^*. \tag{25}$$

To show this, put $Z=LXL'$, let $\mathbf{z}$ be the $n\times 1$ vector of distinct elements of Z formed similarly to $\mathbf{x}$, and write

$$\mathbf{z} = T_L\mathbf{x} \tag{26}$$

where T_L (a function of L) is an $n\times n$ nonsingular matrix. It is easily verified that

$$\frac{\partial}{\partial\mathbf{x}} = T_L'\frac{\partial}{\partial\mathbf{z}},$$

so that

$$\frac{\partial}{\partial\mathbf{z}} = T_L'^{-1}\frac{\partial}{\partial\mathbf{x}}. \tag{27}$$

Since $(ds)^2$ is invariant under the transformation $X\to LXL'=Z$ it follows that

$$d\mathbf{x}'G(\mathbf{x})\,d\mathbf{x} = d\mathbf{z}'G(\mathbf{z})\,d\mathbf{z} = d\mathbf{x}'T_L'G(T_L\mathbf{x})\,T_L\,d\mathbf{x}, \tag{28}$$

where we have used (21), (26), and the fact that $d\mathbf{z}=T_L\,d\mathbf{x}$. This implies that

$$G(T_L\mathbf{x}) = T_L'^{-1}G(\mathbf{x})\,T_L^{-1}; \tag{29}$$

that is, under the transformation $X \to LXL'$ the matrix $G(\mathbf{x})$ defined by (22) is transformed as

$$(30) \qquad G(\mathbf{x}) \to T_L'^{-1} G(\mathbf{x}) T_L^{-1}.$$

By virtue of (24), (26), (27) and (29) it follows that

$$\begin{aligned}
\Delta_Z^* &= \Delta_{LXL'}^* \\
&= \det G(\mathbf{z})^{-1/2} \frac{\partial'}{\partial \mathbf{z}} \left[(\det G(\mathbf{z}))^{1/2} G(\mathbf{z})^{-1} \frac{\partial}{\partial \mathbf{z}} \right] \\
&= \det G(T_L\mathbf{x})^{-1/2} \frac{\partial'}{\partial \mathbf{x}} T_L^{-1} \left[(\det G(T_L\mathbf{x}))^{1/2} G(T_L\mathbf{x})^{-1} T_L'^{-1} \frac{\partial}{\partial \mathbf{x}} \right] \\
&= \det T_L \det G(\mathbf{x})^{-1/2} \frac{\partial'}{\partial \mathbf{x}} \left[T_L^{-1} (\det T_L^{-1}) (\det G(\mathbf{x}))^{1/2} T_L G(\mathbf{x})^{-1} T_L' T_L'^{-1} \frac{\partial}{\partial \mathbf{x}} \right] \\
&= \det G(\mathbf{x})^{-1/2} \frac{\partial'}{\partial \mathbf{x}} \left[(\det G(\mathbf{x}))^{1/2} G(\mathbf{x})^{-1} \frac{\partial}{\partial \mathbf{x}} \right] \\
&= \Delta_X^*,
\end{aligned}$$

proving the invariance of the differential operator Δ_X^*.

What does the operator Δ_Y of Definition 7.2.1 have to do with Δ_X^*? To answer this, let us see how Δ_X^* is transformed when we make a transformation from X to its latent roots and vectors. Put $X = HYH'$ where $H \in O(m)$ and $Y = \text{diag}(y_1, \ldots, y_m)$. In terms of H and Y the invariant differential form $(ds)^2$ given by (19) can be written

$$\begin{aligned}
(ds)^2 &= \text{tr}(X^{-1} dX X^{-1} dX) \\
&= \text{tr}[HY^{-1} H'(dH\, YH' + H\, dY\, H' + HY\, dH')HY^{-1} H' \\
&\quad \cdot (dH\, YH' + H\, dY\, H' + HY\, dH')]
\end{aligned}$$

On multiplying the terms on the right side and using the fact that the matrix $H'dH$ is skewsymmetric ($H'dH = -dH'H$), this becomes

$$(ds)^2 = \text{tr}(Y^{-1} dY\, Y^{-1} dY) - 2\text{tr}(d\Theta\, Y^{-1} d\Theta\, Y) + 2\text{tr}(d\Theta\, d\Theta),$$

where $d\Theta = (d\theta_{ij})$ denotes the matrix $H'dH$, or equivalently

$$(31) \qquad (ds)^2 = \sum_{i=1}^{m} \frac{(dy_i)^2}{y_i^2} + 2 \sum_{i<j}^{m} \frac{(y_i - y_j)^2}{y_i y_j} (d\theta_{ij})^2.$$

Putting $d\mathbf{y}=(dy_1,\dots,dy_m)'$ and $d\boldsymbol{\theta}=(d\theta_{12},d\theta_{13},\dots,d\theta_{m-1,m})'$ (so that $d\boldsymbol{\theta}$ contains the distinct elements of $d\Theta=H'dH$), we then have

$$(ds)^2=(d\mathbf{y}':d\boldsymbol{\theta}')G(\mathbf{y})\begin{pmatrix} d\mathbf{y}\\ d\boldsymbol{\theta}\end{pmatrix}$$

where

$$G(\mathbf{y})=\begin{bmatrix} y_1^{-2} & & & & & 0\\ & \ddots & & & & \\ & & y_m^{-2} & & & \\ & & & \dfrac{2(y_1-y_2)^2}{y_1y_2} & & \\ & & & & \ddots & \\ 0 & & & & & \dfrac{2(y_{m-1}-y_m)^2}{y_{m-1}y_m}\end{bmatrix}$$

In terms of the partial derivatives $\partial/\partial y_i$, $\partial/\partial\theta_{ij}$, the operator Δ_X^* is

$$\Delta_X^*=\Delta_{HYH'}^*=\det G(\mathbf{y})^{-1/2}\begin{pmatrix}\dfrac{\partial}{\partial\mathbf{y}}\\ \dfrac{\partial}{\partial\boldsymbol{\theta}}\end{pmatrix}'\left[\det G(\mathbf{y})^{1/2}G(\mathbf{y})^{-1}\begin{pmatrix}\dfrac{\partial}{\partial\mathbf{y}}\\ \dfrac{\partial}{\partial\boldsymbol{\theta}}\end{pmatrix}\right].$$

Substituting for $G(\mathbf{y})$ and simplifying, this is

$$\text{(32)}\qquad \Delta_X^*=\Delta_{HYH}^*=\sum_{i=1}^m y_i^2\frac{\partial^2}{\partial y_i^2}+\sum_{i=1}^m\sum_{\substack{j=1\\ j\neq i}}^m\frac{y_i^2}{y_i-y_j}\frac{\partial}{\partial y_i}$$
$$-\frac{1}{2}(m-3)\sum_{i=1}^m y_i\frac{\partial}{\partial y_i}+\frac{1}{4}\sum_{i<j}^m\frac{y_iy_j}{(y_i-y_j)^2}\frac{\partial^2}{\partial\theta_{ij}^2}$$
$$=\Delta_Y-\frac{1}{2}(m-3)E_Y+\frac{1}{4}\sum_{i<j}^m\frac{y_iy_j}{(y_i-y_j)^2}\frac{\partial^2}{\partial\theta_{ij}^2},$$

where Δ_Y is the differential operator (2) used in Definition 7.2.1 and E_Y is the Euler operator

$$\text{(33)}\qquad E_Y=\sum_{i=1}^m y_i\frac{\partial}{\partial y_i}.$$

Hence, apart from this latter operator, Δ_Y is the part of $\Delta^*_{HYH'}$ concerned with the roots $y_1,\dots,y_m$. Now,

$$C_\kappa(X)=C_\kappa(Y),$$

because the zonal polynomials are functions only of the latent roots and, since any homogeneous polynomial of degree k in $y_1,\dots,y_m$ is an eigenfunction of E_Y with eigenvalue k, it follows that

$$E_Y C_\kappa(Y)=kC_\kappa(Y). \tag{34}$$

Hence the effect of the operator Δ^*_X on $C_\kappa(X)$ is

$$\begin{aligned}\Delta^*_X C_\kappa(X)&=\Delta^*_{HYH'}C_\kappa(Y)\\ &=\left[\Delta_Y-\tfrac{1}{2}(m-3)E_Y+\frac{1}{4}\sum_{i<j}^{m}\frac{y_iy_j}{(y_i-y_j)^2}\frac{\partial^2}{\partial\theta_{ij}^2}\right]C_\kappa(Y)\\ &=\left[\rho_\kappa+k(m-1)-\tfrac{1}{2}k(m-3)\right]C_\kappa(Y)\\ &=\left[\rho_\kappa+\tfrac{1}{2}k(m+1)\right]C_\kappa(X),\end{aligned} \tag{35}$$

where we have used (32), (4), (34), and the fact that $C_\kappa(Y)$ is a function only of Y. In fact, we could have *defined* the zonal polynomial $C_\kappa(X)$ for $X>0$ in terms of the operator Δ^*_X rather than Δ_Y. Here the definition would be that $C_\kappa(X)(=C_\kappa(Y))$ is a symmetric homogeneous polynomial of degree k in the latent roots $y_1,\dots,y_m$ of X satisfying conditions (i) and (iii) of Definition 7.2.1 and such that $C_\kappa(X)$ is an eigenfunction of the differential operator Δ^*_X. The eigenvalue of Δ^*_X corresponding to $C_\kappa(X)$ is, from (35), equal to $\rho_\kappa+\frac{1}{2}k(m+1)$. This defines the zonal polynomials for positive definite matrices X, and since they are polynomials in the latent roots of X their definition can be extended to arbitrary (complex) symmetric matrices, and then to nonsymmetric matrices using (17).

We started out to prove a fundamental property of zonal polynomials. This is given in the following theorem.

THEOREM 7.2.5. If X_1 is a positive definite $m\times m$ matrix and X_2 is a symmetric $m\times m$ matrix, then

$$\int_{O(m)}C_\kappa(X_1HX_2H')(dH)=\frac{C_\kappa(X_1)C_\kappa(X_2)}{C_\kappa(I_m)} \tag{36}$$

where (dH) is the normalized invariant measure on $O(m)$.

Proof. Consider the integral on the left side of (36) as a function of X_2, say, $f_\kappa(X_2)$. Clearly $f_\kappa(X_2)=f_\kappa(QX_2Q')$ for all $Q\in O(m)$ so that $f_\kappa(X_2)$ is a symmetric function of X_2; in fact, a symmetric homogeneous polynomial of degree k. Suppose that X_2 is positive definite and apply the differential operator $\Delta^*_{X_2}$ to $f_\kappa(X_2)$. This gives

$$\Delta^*_{X_2} f_\kappa(X_2)=\int_{O(m)}\Delta^*_{X_2} C_\kappa(X_1 HX_2 H')(dH)$$

$$=\int_{O(m)}\Delta^*_{X_2} C_\kappa(X_1^{1/2} HX_2 H' X_1^{1/2})(dH)$$

$$=\int_{O(m)}\Delta^*_{X_2} C_\kappa(L X_2 L')(dH),$$

where $L=X_1^{1/2}H$. Using the invariance (25) of the operator Δ^* this is the same as

$$\Delta^*_{X_2} f_\kappa(X_2)=\int_{O(m)}\Delta^*_{LX_2L'} C_\kappa(L X_2 L')(dH)$$

$$=\left[\rho_\kappa+\tfrac{1}{2}k(m+1)\right]\int_{O(m)} C_\kappa(L X_2 L')(dH)$$

$$=\left[\rho_\kappa+\tfrac{1}{2}k(m+1)\right] f_\kappa(X_2),$$

where we have used (35) and the definition of $f_\kappa(X_2)$. By definition, $f_\kappa(X_2)$ must then be a multiple of the zonal polynomial $C_\kappa(X_2)$, $f_\kappa(X_2)=\lambda_\kappa C_\kappa(X_2)$. Putting $X_2=I_m$ and using the fact that $f_\kappa(I_m)=C_\kappa(X_1)$ shows that $\lambda_\kappa=C_\kappa(X_1)/C_\kappa(I_m)$. This proves (36) for $X_2>0$, and the desired result then follows for all (complex) symmetric X_2 by analytic continuation.

Theorem 7.2.5 plays a vital role in the evaluation of many integrals involving zonal polynomials. Some such integrals will be looked at in the next subsection.

We will now indicate the approach to zonal polynomials through group representation theory. Let V_k be the vector space of homogeneous polynomials $\phi(X)$ of degree k in the $n=m(m+1)/2$ different elements of the $m\times m$ positive definite matrix X. Corresponding to any congruence transformation

$$X\rightarrow LXL' \qquad [L\in \mathcal{GL}(m,R)],$$

we can define a linear transformation of the space V_k by

$$\phi \to T(L)\phi \quad : \quad (T(L)\phi)(X) = \phi(L^{-1}XL^{-1\prime}).$$

This transformation defines a *representation* of the real linear group $\mathcal{G}\ell(m, R)$ in the vector space V_k; that is, the mapping $L \to T(L)$ is a homomorphism from $\mathcal{G}\ell(m, R)$ to the group of linear transformations of V_k. To see this, note that

$$\begin{aligned}(T(L_1)(T(L_2)\phi))(X) &= (T(L_2)\phi)(L_1^{-1}XL_1^{-1\prime})\\ &= \phi(L_2^{-1}L_1^{-1}XL_1^{-1\prime}L_2^{-1\prime})\\ &= \phi((L_1L_2)^{-1}X(L_1L_2)^{-1\prime})\\ &= (T(L_1L_2)\phi)(X)\end{aligned}$$

for all X and ϕ so that

$$T(L_1L_2) = T(L_1)T(L_2).$$

Continuing, a subspace $V' \subset V_k$ is *invariant* if

$$T(L)V' \subset V'$$

for all $L \in \mathcal{G}\ell(m, R)$. If, in addition, V' contains no proper invariant subspaces, it is called an *irreducible* invariant subspace. The way in which the zonal polynomials arise is this. It can be shown that the space V_k (which is obviously invariant) decomposes into a direct sum of irreducible invariant subspaces V_κ

$$V_k = \bigoplus_{\kappa} V_\kappa,$$

where $\kappa = (k_1, k_2, \ldots, k_m)$, $k_1 \geq k_2 \geq \cdots \geq k_m \geq 0$, runs over all partitions of k into not more than m parts. The polynomial $(\operatorname{tr} X)^k \in V_k$ then has a unique decomposition

$$(\operatorname{tr} X)^k = \sum_{\kappa} C_\kappa(X)$$

into polynomials $C_\kappa(X) \in V_\kappa$, belonging to the respective invariant subspaces. The polynomial $C_\kappa(X)$ is the *zonal polynomial* corresponding to the partition κ; it is a symmetric homogeneous polynomial of degree k in the latent roots of X. The way in which we defined zonal polynomials in Definition 7.2.1 simply exploits a property that arises from the group representation theory. Because of its group-theoretic nature it is known that $C_\kappa(X)$ must be an eigenfunction of a certain differential operator called the Laplace–Beltrami operator. This is precisely the operator Δ_X^* given by (24) and, as we have seen, it leads directly to the operator Δ_Y used in Definition 7.2.1 when we write $X=HYH'$. For proofs, references, and much more detail, the reader is referred to James (1961b, 1964, 1968).

7.2.3. *Some Basic Integrals*

In this section we will evaluate some basic integrals involving zonal polynomials. The results here are due to Constantine (1963, 1966). Our starting point is the following lemma.

LEMMA 7.2.6. If $Y=\operatorname{diag}(y_1,\dots,y_m)$ and $X=(x_{ij})$ is an $m\times m$ positive definite matrix then

$$(37)\quad C_\kappa(XY)=d_\kappa y_1^{k_1}\cdots y_m^{k_m}x_{11}^{k_1-k_2}\det\begin{bmatrix}x_{11} & x_{12}\\ x_{21} & x_{22}\end{bmatrix}^{k_2-k_3}\cdots\det X^{k_m}$$

$$+\text{terms of lower weight in the } y\text{'s},$$

where $\kappa=(k_1,\dots,k_m)$ and d_κ is the coefficient of the term of highest weight in $C_\kappa(\cdot)$ [see (i) of Definition 7.2.1].

Proof. If A is an $m\times m$ symmetric matrix with latent roots $a_1,\dots,a_m$ we can write

$$(38)\quad C_\kappa(A)=d_\kappa a_1^{k_1}\cdots a_m^{k_m}+\text{terms of lower weight}$$

$$=d_\kappa a_1^{k_1-k_2}(a_1a_2)^{k_2-k_3}\cdots(a_1a_2\cdots a_m)^{k_m}+\cdots$$

$$=d_\kappa\left(\sum_{i=1}^m a_i\right)^{k_1-k_2}\left(\sum_{i<j}^m a_ia_j\right)^{k_2-k_3}\cdots(a_1a_2\cdots a_m)^{k_m}+\cdots$$

(since $C_\kappa(A)$ is symmetric in $a_1,\dots,a_m$)

$$=d_\kappa r_1^{k_1-k_2}r_2^{k_2-k_3}\cdots r_m^{k_m}+\cdots,$$

where r_j denotes the jth elementary symmetric function of $a_1,\ldots,a_m$; that is,

$$r_1 = a_1 + \cdots + a_m \tag{39}$$

$$r_2 = \sum_{i<j}^{m} a_i a_j$$

$$\vdots$$

$$r_m = a_1 \cdots a_m$$

Now, let $A_{i_1, i_2 \ldots i_k}$ denote the $k\times k$ matrix formed from A by deleting all but the $i_1,\ldots,i_k$th rows and columns and define the function $\text{tr}_k(\cdot)$ by

$$\text{tr}_k(A) = \sum_{1\le i_1 < i_2 \ldots < i_k \le m} \det A_{i_1, i_2 \ldots i_k}. \tag{40}$$

It is an easy matter to show that [see (xiii) of the Appendix, Section A7]

$$r_j = \text{tr}_j(A),$$

and using this in (38) gives

$$C_\kappa(A) = d_\kappa \text{tr}_1(A)^{k_1-k_2}\text{tr}_2(A)^{k_2-k_3}\cdots \text{tr}_m(A)^{k_m} + \cdots$$

$$= d_\kappa a_{11}^{k_1-k_2}\det\begin{bmatrix} a_{11} & a_{12} \\ a_{21} & a_{22}\end{bmatrix}^{k_2-k_3}\cdots(\det A)^{k_m} + \cdots.$$

Now, putting $A = XY$, so that $a_{ij} = x_{ij}y_j$, we have

$$C_\kappa(XY) = d_\kappa (y_1x_{11})^{k_1-k_2}\det\begin{bmatrix} y_1x_{11} & y_2x_{12} \\ y_1x_{21} & y_2x_{22}\end{bmatrix}^{k_2-k_3}\cdots\det(XY)^{k_m} + \cdots$$

$$= d_\kappa y_1^{k_1}\ldots y_m^{k_m} x_{11}^{k_1-k_2}\det\begin{bmatrix} x_{11} & x_{12} \\ x_{21} & x_{22}\end{bmatrix}^{k_2-k_3}\cdots(\det X)^{k_m} + \cdots,$$

which completes the proof.

A particular type of constant, called a *generalized hypergeometric coefficient*, will appear in the integrals that follow. If $\kappa = (k_1,\ldots,k_m)$ and α is a complex number we define $(\alpha)_\kappa$ by

$$(\alpha)_\kappa = \prod_{i=1}^{m} \left(\alpha - \tfrac{1}{2}(i-1)\right)_{k_i}, \tag{41}$$

where $(\alpha)_k=\alpha(\alpha+1)\cdots(\alpha+k-1),(\alpha)_0=1$. If $\operatorname{Re}(\alpha)>\frac{1}{2}(m-1)$ it is clear that

$$(\alpha)_\kappa=\pi^{m(m-1)/4}\frac{\prod_{i=1}^m\Gamma\left[\alpha+k_i-\frac{1}{2}(i-1)\right]}{\Gamma_m(\alpha)} \tag{42}$$

where $\Gamma_m(\alpha)$ is the multivariate gamma function (see Section 2.1.2).

THEOREM 7.2.7. Let Z be a complex symmetric $m\times m$ matrix with $\operatorname{Re}(Z)>0$ and let Y be a symmetric $m\times m$ matrix. Then

$$\begin{aligned}(43)\quad \int_{X>0}&\operatorname{etr}(-XZ)(\det X)^{a-(m+1)/2}C_\kappa(XY)(dX)\\&=(a)_\kappa\Gamma_m(a)(\det Z)^{-a}C_\kappa(YZ^{-1})\end{aligned}$$

for $\operatorname{Re}(a)>\frac{1}{2}(m-1)$. [Note that when $\kappa=(0)$, then $C_\kappa\equiv1$ and $(a)_\kappa\equiv1$, and (43) reduces to the result of Theorem 2.1.11.]

Proof. We will first prove the result for the special case $Z=I_m$. In this case it has to be shown that

$$\int_{X>0}\operatorname{etr}(-X)(\det X)^{a-(m+1)/2}C_\kappa(XY)(dX)=(a)_\kappa\Gamma_m(a)C_\kappa(Y). \tag{44}$$

Let $f(Y)$ denote the integral on the left side of (44); for any $H\in O(m)$ we have

$$f(HYH')=\int_{X>0}\operatorname{etr}(-X)(\det X)^{a-(m+1)/2}C_\kappa(XHYH')(dX). \tag{45}$$

Putting $U=H'XH$, so that $(dU)=(dX)$, this last integral becomes

$$\begin{aligned}(46)\qquad f(HYH')&=\int_{U>0}\operatorname{etr}(-U)(\det U)^{a-(m+1)/2}C_\kappa(UY)(dU)\\&=f(Y),\end{aligned}$$

so that f is a symmetric function of Y. Because of (45) and (46) we get, on integrating with respect to the normalized invariant measure (dH) on $O(m)$,

$$
\begin{aligned}
f(Y) &= \int_{O(m)} f(Y)(dH) \\
&= \int_{O(m)} f(HYH')(dH) \\
&= \int_{X>0} \operatorname{etr}(-X)(\det X)^{a-(m+1)/2} \int_{O(m)} C_\kappa(XHYH')(dH)(dX) \\
&= \int_{X>0} \operatorname{etr}(-X)(\det X)^{a-(m+1)/2} \frac{C_\kappa(X)C_\kappa(Y)}{C_\kappa(I_m)}(dX),
\end{aligned}
$$

where the last line follows from Theorem 7.2.5. From this we see that

(47) $$f(Y) = \frac{f(I_m)C_\kappa(Y)}{C_\kappa(I_m)}.$$

Since this is a symmetric homogeneous polynomial in the latent roots of Y it can be assumed without loss of generality that Y is diagonal, $Y = \operatorname{diag}(y_1,\dots,y_m)$. Using (i) of Definition 7.2.1 it then follows that

(48) $$f(Y) = \frac{f(I_m)}{C_\kappa(I_m)} d_\kappa y_1^{k_1}\dots y_m^{k_m} + \text{terms of lower weight},$$

where $\kappa = (k_1,\dots,k_m)$. On the other hand, using the result of Lemma 7.2.6 we get

$$
\begin{aligned}
f(Y) &= \int_{X>0} \operatorname{etr}(-X)(\det X)^{a-(m+1)/2} C_\kappa(XY)(dX) \\
&= d_\kappa y_1^{k_1}\dots y_m^{k_m} \int_{X>0} \operatorname{etr}(-X)(\det X)^{a-(m+1)/2} \\
&\quad \cdot x_{11}^{k_1-k_2} \det \begin{bmatrix} x_{11} & x_{12} \\ x_{21} & x_{22} \end{bmatrix}^{k_2-k_3} \dots (\det X)^{k_m}(dX) \\
&\quad + \text{terms of lower weight}.
\end{aligned}
$$

To evaluate this last integral, put $X=T'T$ where T is upper-triangular with positive diagonal elements. Then

$$\operatorname{tr} X=\sum_{i\le j}^{m} t_{ij}^2, \qquad x_{11}=t_{11}^2, \qquad \det\begin{bmatrix} x_{11} & x_{12} \\ x_{21} & x_{22}\end{bmatrix}=t_{11}^2 t_{22}^2,\ldots,$$

$$\det X=\prod_{i=1}^{m} t_{ii}^2,$$

and, from Theorem 2.1.9,

$$(dX)=2^m\prod_{i=1}^{m} t_{ii}^{m+1-i}\bigwedge_{i\le j}^{m} dt_{ij},$$

so that

$$(49)\quad f(Y)=d_\kappa y_1^{k_1}\cdots y_m^{k_m}\int\cdots\int_{t_{ij}}\exp\left(-\sum_{i\le j}^{m} t_{ij}^2\right)\prod_{i=1}^{m} t_{ii}^{2a+2k_i-i}$$

$$\times 2^m(dT)+\cdots$$

$$=d_\kappa y_1^{k_1}\cdots y_m^{k_m}\prod_{i<j}^{m}\left[\int_{-\infty}^{\infty}\exp(-t_{ij}^2)\,dt_{ij}\right]$$

$$\times\prod_{i=1}^{m}\left[\int_0^\infty e^{-t_{ii}^2}(t_{ii}^2)^{a+k_i-(i+1)/2}\,dt_{ii}^2\right]+\cdots$$

$$=d_\kappa y_1^{k_1}\cdots y_m^{k_m}\pi^{m(m-1)/4}\prod_{i=1}^{m}\Gamma\left[a+k_i-\tfrac{1}{2}(i-1)\right]+\cdots$$

$$=d_\kappa y_1^{k_1}\cdots y_m^{k_m}(a)_\kappa\Gamma_m(a)+\text{terms of lower weight},$$

where the last line follows from (42). Equating coefficients of $y_1^{k_1}\cdots y_m^{k_m}$ in (48) and (49) then shows that

$$\frac{f(I_m)}{C_\kappa(I_m)}=(a)_\kappa\Gamma_m(a),$$

and using this in (47) gives

$$f(Y)=(a)_\kappa\Gamma_m(a)C_\kappa(Y),$$

which establishes (44) and hence (43) for $Z=I_m$.

Now consider the integral (43) when $Z>0$ is real. Putting $V=Z^{1/2}XZ^{1/2}$ so that $(dV)=(\det Z)^{(m+1)/2}(dX)$, the left side of (43) becomes

$$(\det Z)^{-a}\int_{V>0}\operatorname{etr}(-V)(\det V)^{a-(m+1)/2}C_\kappa(Z^{-1/2}VZ^{-1/2}Y)(dV),$$

which is equal to

$$(a)_\kappa\Gamma_m(a)(\det Z)^{-a}C_\kappa(YZ^{-1})$$

by (44). Thus the theorem is true for real $Z>0$, and it follows for complex Z with $\operatorname{Re}(Z)>0$ by analytic continuation.

An interesting consequence of Theorem 7.2.7 is that a zonal polynomial has a reproductive property under expectation taken with respect to the Wishart distribution. This is made explicit in the following corollary.

COROLLARY 7.2.8. If A is $W_m(n,\Sigma)$ with $n>m-1$ and B is an arbitrary symmetric $m\times m$ (fixed) matrix then

$$E[C_\kappa(AB)]=2^k(\tfrac{1}{2}n)_\kappa C_\kappa(B\Sigma)$$

Proof. This follows immediately by multiplying $C_\kappa(AB)$ by the $W_m(n,\Sigma)$ density function for A given by Theorem 3.2.1 and integrating over $A>0$ using Theorem 7.2.7 with $Z=\frac{1}{2}\Sigma^{-1}$, $X=A$, $Y=B$, and $a=n/2$.

Taking $B=I_m$ in Corollary 7.2.8 shows that, if A is $W_m(n,\Sigma)$, then

$$E[C_\kappa(A)]=2^k(\tfrac{1}{2}n)_\kappa C_\kappa(\Sigma). \tag{50}$$

In particular, taking $\kappa=(1)$ we have $C_{(1)}(A)=\operatorname{tr}A$ so that

$$E(\operatorname{tr}A)=n\operatorname{tr}\Sigma,$$

a result we already know since $E(A)=n\Sigma$. In general, if $l_1,\dots,l_m$ denote the latent roots of A and $\kappa=(1,1,\dots,1)$ is a partition of k then

$$\begin{aligned}C_\kappa(A)&=d_\kappa l_1\dots l_k+\text{terms of lower weight}\\&=d_\kappa r_k(A),\end{aligned}$$

where $r_k(A)$ is the kth elementary symmetric function of $l_1,\dots,l_m$ [see (39)]. Similarly, for $\kappa=(1,\dots,1)$,

$$C_\kappa(\Sigma)=d_\kappa r_k(\Sigma)$$

where $r_k(\Sigma)$ is the kth elementary symmetric function of the latent roots $\lambda_1,\dots,\lambda_m$ of Σ. Corollary 7.2.8 then shows that

$$E[r_k(A)] = n(n-1)\dots(n-k+1)r_k(\Sigma) \tag{51}$$

for $k=1,\dots,m$.

A common method for evaluating integrals involves the use of multidimensional Laplace transforms.

DEFINITION 7.2.9. If $f(X)$ is a function of the positive definite $m\times m$ matrix X, the Laplace transform of $f(X)$ is defined to be

$$g(Z) = \int_{X>0} \operatorname{etr}(-XZ) f(X)(dX) \tag{52}$$

where $Z=U+iV$ is a complex symmetric matrix, U and V are real, and it is assumed that the integral is absolutely convergent in the right half-plane $\operatorname{Re}(Z)=U>U_0$ for some positive definite U_0.

The Laplace transform $g(Z)$ of $f(X)$ given in Definition 7.2.9 is an analytic function of Z in the half-plane $\operatorname{Re}(Z)>U_0$. If $g(Z)$ satisfies the conditions

$$\int |g(U+iV)|(dV) < \infty \tag{53}$$

and

$$\lim_{U\to\infty} \int |g(U+iV)|(dV) = 0, \tag{54}$$

where the integrals are over the space of all real symmetric matrices V, then the inverse formula

$$f(X) = \frac{2^{m(m-1)/2}}{(2\pi i)^{m(m+1)/2}} \int_{\operatorname{Re}(Z)>U_0>0} \operatorname{etr}(XZ) g(Z)(dZ) \tag{55}$$

holds. Here the integration is taken over $Z=U+iV$, with $U>U_0$ and fixed and V ranging over all real symmetric $m\times m$ matrices. Equivalently, given a function $g(Z)$ analytic in $\operatorname{Re}(Z)>U_0$ and satisfying (53) and (54), the inversion formula (55) defines a function $f(X)$ in $X>0$ which has $g(Z)$ as its Laplace transform.

The integrals (52) and (55) represent generalizations of the classical Laplace transform and inversion formulas to which they reduce when $m=1$. For more details and proofs in the general case the reader is referred to Herz (1955), page 479, and the references therein. For our purposes we will often prove that a certain equation is true by showing that both sides of the equation have the same Laplace transform and invoking the uniqueness of Laplace transforms.

Two examples of Laplace transforms have already been given. Theorem 2.1.11 (with $\Sigma^{-1}=2Z$) shows that the Laplace transform of

$$f_1(X)=(\det X)^{a-(m+1)/2}\quad [\mathrm{Re}(a)>\tfrac{1}{2}(m-1)]$$

is

$$g_1(Z)=\Gamma_m(a)(\det Z)^{-a},$$

while Theorem 7.2.7 (with $Y=I_m$) shows that the Laplace transform of

$$f_2(X)=(\det X)^{a-(m+1)/2}C_\kappa(X)\qquad \left[\mathrm{Re}(a)>\tfrac{1}{2}(m-1)\right]$$

is

$$g_2(Z)=(a)_\kappa\Gamma_m(a)(\det Z)^{-a}C_\kappa(Z^{-1}).$$

To apply the inversion formula (55), it would have to be shown that $g_1(Z)$ and $g_2(Z)$ satisfy conditions (53) and (54). This has been done for $g_1(Z)$ by Herz (1955) and for $g_2(Z)$ by Constantine (1963) and the reader is referred to these two papers for details. The inversion formula applied, for example, to $g_2(Z)$ shows that

$$(56)\qquad \frac{2^{m(m-1)/2}}{(2\pi i)^{m(m+1)/2}}\int_{\mathrm{Re}(Z)>0}\mathrm{etr}(XZ)(\det Z)^{-a}C_\kappa(Z^{-1})(dZ)$$

$$=\frac{1}{(a)_\kappa\Gamma_m(a)}(\det X)^{a-(m+1)/2}C_\kappa(X).$$

An important analog of the beta function integral is given in the following theorem.

THEOREM 7.2.10. If Y is a symmetric $m\times m$ matrix then

$$(57)\quad \int_{0<X<I_m}(\det X)^{a-(m+1)/2}\det(I_m-X)^{b-(m+1)/2}C_\kappa(XY)(dX)$$

$$=\frac{(a)_\kappa}{(a+b)_\kappa}\frac{\Gamma_m(a)\Gamma_m(b)}{\Gamma_m(a+b)}C_\kappa(Y)$$

for $\mathrm{Re}(a)>\frac{1}{2}(m-1)$, $\mathrm{Re}(b)>\frac{1}{2}(m-1)$.

Proof. Let $f(Y)$ denote the integral on the left side of (57); then, exactly as in the proof of Theorem 7.2.7,

$$f(Y)=f(HYH')\qquad \text{for all}\quad H\in O(m),$$

and hence

$$(58)\qquad f(Y)=\frac{f(I_m)}{C_\kappa(I_m)}C_\kappa(Y).$$

It remains to be shown that

$$f(I_m)=\frac{(a)_\kappa\Gamma_m(a)\Gamma_m(b)}{(a+b)_\kappa\Gamma_m(a+b)}C_\kappa(I_m),$$

or equivalently that

$$(59)\qquad (a+b)_\kappa\Gamma_m(a+b)f(I_m)=(a)_\kappa\Gamma_m(a)\Gamma_m(b)C_\kappa(I_m).$$

To establish this, note that the left side of (59) can be written as

$$(a+b)_\kappa\Gamma_m(a+b)f(I_m)$$

$$=\int_{W>0}\mathrm{etr}(-W)(\det W)^{a+b-(m+1)/2}\frac{f(I_m)}{C_\kappa(I_m)}C_\kappa(W)(dW)$$

(using Theorem 7.2.7)

$$=\int_{W>0}\mathrm{etr}(-W)(\det W)^{a+b-(m+1)/2}f(W)(dW)\qquad [\text{using (58)}]$$

$$=\int_{W>0}\mathrm{etr}(-W)(\det W)^{a+b-(m+1)/2}\int_{0<X<I}(\det X)^{a-(m+1)/2}$$

$$\cdot\det(I-X)^{b-(m+1)/2}C_\kappa(WX)(dX)(dW)\qquad [\text{by definition of } f(W)].$$

In the inner integral put $X=W^{-1/2}UW^{-1/2}$ with Jacobian $(dX)=(\det W)^{-(m+1)/2}(dU)$; then

$$
\begin{aligned}
&(a+b)_\kappa\Gamma_m(a+b)f(I_m)\\
&=\int_{W>0}\operatorname{etr}(-W)\int_{0<U<W}(\det U)^{a-(m+1)/2}\\
&\quad\cdot\det(W-U)^{b-(m+1)/2}C_\kappa(U)(dU)(dW)\\
&=\int_{U>0}\operatorname{etr}(-U)(\det U)^{a-(m+1)/2}C_\kappa(U)(dU)\\
&\quad\cdot\int_{V>0}\operatorname{etr}(-V)(\det V)^{b-(m+1)/2}(dV)\qquad(\text{on putting } V=W-U)\\
&=(a)_\kappa\Gamma_m(a)C_\kappa(I)\Gamma_m(b),
\end{aligned}
$$

where the last line follows Theorems 7.2.7 and Definition 2.1.10. This establishes (59) and completes the proof.

We have previously noted in Corollary 7.2.8 that a zonal polynomial has a reproductive property under expectation taken with respect to the Wishart distribution. A similar property also holds under expectation taken with respect to the multivariate beta distribution as the following corollary shows.

COROLLARY 7.2.11. If the matrix U has the $\text{Beta}_m(\frac{1}{2}n_1,\frac{1}{2}n_2)$ distribution of Definition 3.3.2 and B is a fixed $m\times m$ symmetric matrix then

$$
(60)\qquad E[C_\kappa(UB)]=\frac{(\frac{1}{2}n_1)_\kappa}{(\frac{1}{2}(n_1+n_2))_\kappa}C_\kappa(B)
$$

Proof. This follows immediately by multiplying $C_\kappa(UB)$ by the $\text{Beta}_m(\frac{1}{2}n_1,\frac{1}{2}n_2)$ density function for U given by Theorem 3.3.1 and integrating over $0<U<I_m$ using Theorem 7.2.10.

Taking $B=I_m$ in Corollary 7.2.11 shows that if U is $\text{Beta}_m(\frac{1}{2}n_1,\frac{1}{2}n_2)$ then

$$
(61)\qquad E[C_\kappa(U)]=\frac{(\frac{1}{2}n_1)_\kappa}{(\frac{1}{2}(n_1+n_2))_\kappa}C_\kappa(I_m).
$$

In particular, taking the partition $\kappa=(1,1,\dots,1)$ of k shows that

$$
(62)\quad E[r_\kappa(U)]=\frac{n_1(n_1-1)\cdots(n_1-k+1)}{(n_1+n_2)(n_1+n_2-1)\cdots(n_1+n_2-k+1)}\binom{m}{k},
$$

where $r_k(U)$ is the kth elementary symmetric function of the latent roots $u_1,\dots,u_m$ of U. The term

$$\binom{m}{k}$$

on the right side is the kth elementary symmetric function of the roots of I_m.

Our next result is proved with the help of the following lemma which is similar to Lemma 7.2.6 and whose proof is left as an exercise (see Problem 7.5).

LEMMA 7.2.12. If $Z=\mathrm{diag}(z_1,\dots,z_m)$ and $Y=(y_{ij})$ is an $m\times m$ positive definite matrix then

$$(63)\quad C_\kappa(Y^{-1}Z)=d_\kappa z_1^{k_1}\dots z_m^{k_m}y_{11}^{-(k_1-k_2)}\det\begin{bmatrix} y_{11} & y_{12}\\ y_{21} & y_{22}\end{bmatrix}^{-(k_2-k_3)}\dots\det Y^{-k_m}$$
$$+\text{terms of lower weight in the } z\text{'s},$$

where $\kappa=(k_1,\dots,k_m)$.

The following theorem should be compared with Theorem 7.2.7.

THEOREM 7.2.13. Let Z be a complex symmetric $m\times m$ matrix with $\mathrm{Re}(Z)>0$. Then

$$(64)\quad \int_{X>0}\mathrm{etr}(-XZ)(\det X)^{a-(m+1)/2}C_\kappa(X^{-1})(dX)$$
$$=\frac{(-1)^k\Gamma_m(a)}{(-a+\frac{1}{2}(m+1))_\kappa}(\det Z)^{-a}C_\kappa(Z)$$

for $\mathrm{Re}(a)>k_1+\frac{1}{2}(m-1)$, where $\kappa=(k_1,k_2,\dots,k_m)$.

Proof. First suppose that $Z>0$ is real. Let $f(Z)$ denote the integral on the left side of (64) and make the change of variables $X=Z^{-1/2}YZ^{-1/2}$, with Jacobian $(dX)=(\det Z)^{-(m+1)/2}(dY)$, to give

$$(65)\quad f(Z)=\int_{Y>0}\mathrm{etr}(-Y)(\det Y)^{a-(m+1)/2}C_\kappa(Y^{-1}Z)(dY)(\det Z)^{-a}.$$

Then, exactly as in the proof of Theorem 7.2.7,

$$(66)\qquad f(Z)=\frac{f(I_m)}{C_\kappa(I_m)}C_\kappa(Z)(\det Z)^{-a}.$$

Assuming without loss of generality that $Z=\operatorname{diag}(z_1,\ldots,z_m)$, it then follows, using (i) of Definition 7.2.1, that

$$(67)\quad f(Z)=\frac{f(I_m)}{C_\kappa(I_m)}(\det Z)^{-a}d_\kappa z_1^{k_1}\ldots z_m^{k_m}+\text{terms of lower weight.}$$

On the other hand, using the result of Lemma 7.2.12 in (65) gives

$$f(Z)=(\det Z)^{-a}d_\kappa z_1^{k_1}\ldots z_m^{k_m}\int_{Y>0}\operatorname{etr}(-Y)(\det Y)^{a-(m+1)/2}$$

$$\cdot y_{11}^{-(k_1-k_2)}\det\begin{bmatrix}y_{11}&y_{12}\\y_{21}&y_{22}\end{bmatrix}^{-(k_2-k_3)}\ldots\det Y^{-k_m}(dY)$$

$$+\text{terms of lower weight.}$$

To evaluate this last integral put $Y=T'T$ where T is upper-triangular with positive diagonal elements; then

$$\operatorname{tr}Y=\sum_{i\le j}^m t_{ij}^2,\qquad y_{11}=t_{11}^2,\qquad \det\begin{bmatrix}y_{11}&y_{12}\\y_{21}&y_{22}\end{bmatrix}=t_{11}^2t_{22}^2,\qquad\ldots,$$

$$\det Y=\prod_{i=1}^m t_{ii}^2$$

and, from Theorem 2.1.9,

$$(dY)=2^m\prod_{i=1}^m t_{ii}^{m+1-i}(dT).$$

Hence

(68)

$$f(Z)=(\det Z)^{-a}d_\kappa z_1^{k_1}\cdots z_m^{k_m}\left[\prod_{i<j}^m\int_{-\infty}^{\infty}\exp(-t_{ij}^2)\,dt_{ij}\right]$$

$$\cdot\left[\prod_{i=1}^m\int_0^\infty\exp(-t_{ii}^2)(t_{ii}^2)^{a-k_i-(i+1)/2}\,dt_{ii}^2\right]+\cdots$$

$$=(\det Z)^{-a}d_\kappa z_1^{k_1}\cdots z_m^{k_m}\pi^{m(m-1)/4}\prod_{i=1}^m\Gamma\left[a-k_i-\tfrac{1}{2}(i-1)\right]+\cdots$$

$$=(\det Z)^{-a}d_\kappa z_1^{k_1}\cdots z_m^{k_m}\frac{(-1)^k\Gamma_m(a)}{(-a+\frac{1}{2}(m+1))_\kappa}+\text{terms of lower weight.}$$

Equating coefficients of $z_1^{k_1}\dots z_m^{k_m}$ in (67) and (68) then gives

$$\frac{f(I_m)}{C_\kappa(I_m)}=\frac{(-1)^k\Gamma_m(a)}{(-a+\frac{1}{2}(m+1))_\kappa};$$

using this in (65) establishes the desired result for real $Z>0$, and it follows for complex Z with $\mathrm{Re}(Z)>0$ by analytic continuation.

7.3. HYPERGEOMETRIC FUNCTIONS OF MATRIX ARGUMENT

Many distributions of random matrices, and moments of test statistics, can be expressed in terms of functions known as *hypergeometric functions of matrix argument*, which involve series of zonal polynomials. These functions occur often in subsequent chapters.

Hypergeometric functions of a single variable have been introduced in Definition 1.3.1 as infinite power series. By analogy with this definition we will define hypergeometric functions of matrix argument.

DEFINITION 7.3.1. The hypergeometric functions of matrix argument are given by

$$(1)\qquad {}_pF_q(a_1,\dots,a_p;b_1,\dots,b_q;X)=\sum_{k=0}^{\infty}\sum_{\kappa}\frac{(a_1)_\kappa\cdots(a_p)_\kappa}{(b_1)_\kappa\cdots(b_q)_\kappa}\frac{C_\kappa(X)}{k!},$$

where $\sum_\kappa$ denotes summation over all partitions $\kappa=(k_1,\dots,k_m)$, $k_1\geq\cdots\geq k_m\geq 0$, of k, $C_\kappa(X)$ is the zonal polynomial of X corresponding to κ and the generalized hypergeometric coefficient $(a)_\kappa$ is given by

$$(2)\qquad (a)_\kappa=\prod_{i=1}^{m}\left(a-\tfrac{1}{2}(i-1)\right)_{k_i},$$

where $(a)_k=a(a+1)\dots(a+k-1)$, $(a)_0=1$. Here X, the argument of the function, is a complex symmetric $m\times m$ matrix and the parameters a_i, b_j are arbitrary complex numbers. No denominator parameter b_j is allowed to be zero or an integer or half-integer $\leq\frac{1}{2}(m-1)$ (otherwise some of the denominators in the series will vanish). If any numerator parameter a_i is a negative integer, say, $a_1=-n$, then the function is a polynomial of degree mn, because for $k\geq mn+1$, $(a_1)_\kappa=(-n)_\kappa=0$. The series converges for all X if $p\leq q$, it converges for $\|X\|<1$ if $p=q+1$, where $\|X\|$ denotes the maximum of the absolute values of the latent roots of X, and, unless it terminates, it diverges for all $X \neq 0$ if $p > q+1$. Finally, when $m=1$ the series (1) reduces to the classical hypergeometric function of Definition 1.3.1.

Two special cases of (1) are

$$(3) \qquad {}_0F_0(X) = \sum_{k=0}^{\infty} \sum_{\kappa} \frac{C_\kappa(X)}{k!} = \sum_{k=0}^{\infty} \frac{(\operatorname{tr} X)^k}{k!} = \operatorname{etr}(X),$$

where the second line follows from (iii) of Definition 7.2.1, and

$$(4) \qquad {}_1F_0(a; X) = \sum_{k=0}^{\infty} \sum_{\kappa} (a)_\kappa \frac{C_\kappa(X)}{k!} = \det(I_m - X)^{-a}, \qquad (\|X\| < 1)$$

a result which will be proved later in Corollary 7.3.5. Hence the ${}_1F_0$ series is a generalization of the usual binomial series.

We will see in later chapters that the hypergeometric functions given by Definition 7.3.1 appear in the density functions of matrix variates. The density functions of latent roots involve hypergeometric functions with two matrices as arguments. These are given by the following definition.

DEFINITION 7.3.2. The hypergeometric functions with the symmetric $m \times m$ matrices X and Y as arguments are given by

$$(5) \quad {}_pF_q^{(m)}(a_1,\dots,a_p; b_1,\dots,b_q; X, Y) = \sum_{k=0}^{\infty} \sum_{\kappa} \frac{(a_1)_\kappa \cdots (a_p)_\kappa}{(b_1)_\kappa \cdots (b_q)_\kappa} \frac{C_\kappa(X) C_\kappa(Y)}{k! C_\kappa(I_m)}$$

It is clear from Definition 7.3.2 that the order of X and Y is unimportant, that is

$${}_pF_q^{(m)}(a_1,\dots,a_p; b_1,\dots,b_q; X, Y) = {}_pF_q^{(m)}(a_1,\dots,a_p; b_1,\dots,b_q; Y, X).$$

Also, if one of the argument matrices is the identity this function reduces to the one-matrix function of Definition 7.3.1; that is

$$(6)\quad {}_pF_q^{(m)}(a_1,\dots,a_p;b_1,\dots,b_q;X,I_m)={}_pF_q(a_1,\dots,a_p;b_1,\dots,b_q;X)$$

The two-matrix functions ${}_pF_q^{(m)}$ can be obtained from the one-matrix function ${}_pF_q$ by averaging over the orthogonal group $O(m)$, as the following theorem shows.

THEOREM 7.3.3. If X is a positive definite $m\times m$ matrix and Y is a symmetric $m\times m$ matrix, then

$$(7)\quad \int_{O(m)} {}_pF_q(a_1,\dots,a_p;b_1,\dots,b_q;XHYH')(dH)$$

$$={}_pF_q^{(m)}(a_1,\dots,a_p;b_1,\dots,b_q;X,Y)$$

where (dH) denotes the normalized invariant measure on $O(m)$.

Proof. The result is immediate by expanding the integrand and integrating term by term using Theorem 7.2.5.

It was shown in Lemma 1.3.3 that the Laplace transform of a classical ${}_pF_q$ function is a ${}_{p+1}F_q$ function. A similar result is true in the matrix case.

THEOREM 7.3.4. If Z is a complex $m\times m$ symmetric matrix with $\mathrm{Re}(Z)>0$ and Y is a symmetric $m\times m$ matrix then

$$(8)\quad \int_{X>0}\mathrm{etr}(-XZ)(\det X)^{a-(m+1)/2}{}_pF_q(a_1,\dots,a_p;b_1,\dots,b_q;X)(dX)$$

$$=\Gamma_m(a)(\det Z)^{-a}{}_{p+1}F_q(a_1,\dots,a_p,a;b_1,\dots,b_q;Z^{-1})$$

and

(9)

$$\int_{X>0}\mathrm{etr}(-XZ)(\det X)^{a-(m+1)/2}{}_pF_q^{(m)}(a_1,\dots,a_p,b_1,\dots,b_q;X,Y)(dX)$$

$$=\Gamma_m(a)(\det Z)^{-a}{}_{p+1}F_q^{(m)}(a_1,\dots,a_p,a:b_1,\dots,b_q;Z^{-1},Y)$$

for $p<q$, $\mathrm{Re}(a)>\frac{1}{2}(m-1)$; or $p=q$, $\mathrm{Re}(a)>\frac{1}{2}(m-1)$, $\|Z^{-1}\|<1$ $(\|Y\|\leq 1)$.

Proof. Both (8) and (9) are immediately proved by expanding the ${}_pF_q$ functions in the integrands and integrating term by term using Theorem 7.2.7.

The generalization (4) of the binomial series is an immediate consequence.

COROLLARY 7.3.5.

$$ {}_1F_0(a;Z)=\det(I_m-Z)^{-a} \qquad (\|Z\|<1) $$

Proof. Replacing Z by Z^{-1} in (8), then making the change of variables $X=Z^{1/2}UZ^{1/2}$ with Jacobian $(dX)=(\det Z)^{(m+1)/2}(dU)$, and using ${}_0F_0(ZU)=\operatorname{etr}(ZU)$, gives

$$ {}_1F_0(a;Z)=\frac{1}{\Gamma_m(a)}\int_{U>0}\operatorname{etr}(-U)(\det U)^{a-(m+1)/2}\operatorname{etr}(ZU)(dU) $$

$$ =\frac{1}{\Gamma_m(a)}\int_{U>0}\operatorname{etr}(-U(I-Z))(\det U)^{a-(m+1)/2}(dU) $$

$$ =\det(I-Z)^{-a}, $$

where the last line follows from Theorem 2.1.11.

Theorem 7.3.4 shows that one can go from the ${}_pF_q$ function to the ${}_{p+1}F_q$ function by means of a Laplace transform (see Definition 7.2.9). There is also an inverse Laplace transformation which enables the ${}_pF_{q+1}$ functions to be found from the ${}_pF_q$ functions. Although we will not use the results explicitly in this book, we will state them for the sake of completeness. They are

$$ (10)\quad \frac{\Gamma_m(b)2^{m(m-1)/2}}{(2\pi i)^{m(m+1)/2}}\int_{\operatorname{Re}(Z)=U_0}\operatorname{etr}(XZ)(\det Z)^{-b}\,{}_pF_q\big(a_1,\dots,a_p;b_1,\dots,b_q;Z^{-1}\big)(dZ) $$

$$ =(\det X)^{b-(m+1)/2}\,{}_pF_{q+1}\big(a_1,\dots,a_p;b_1,\dots,b_q,b;X\big) $$

and

$$ (11)\quad \frac{\Gamma_m(b)2^{m(m-1)/2}}{(2\pi i)^{m(m+1)/2}}\int_{\operatorname{Re}(Z)=U_0}\operatorname{etr}(XZ)(\det Z)^{-b}\,{}_pF_q^{(m)}\big(a_1,\dots,a_p;b_1,\dots,b_q,Z^{-1},Y\big)(dZ) $$

$$ =(\det X)^{b-(m+1)/2}\,{}_pF_{q+1}^{(m)}\big(a_1,\dots,a_p;b_1,\dots,b_q,b;X,Y\big), $$

where the integrals are taken over all matrices $Z=U_0+iV$ for fixed positive definite U_0 and V arbitrary real symmetric. The reader can readily check that both (10) and (11) follow by expanding the ${}_pF_q$ functions in the integrands and integrating term by term using (56) of Section 7.2.

The hypergeometric functions of one-matrix argument were first introduced by Herz (1955), who started with the function ${}_0F_0(X)=\text{etr}(X)$ and then defined the general system of functions ${}_pF_q$ by means of the Laplace and inverse Laplace transforms (8) and (10). The zonal polynomial expansion for these functions given by Definition 7.3.1 was found by Constantine (1963).

7.4. SOME RESULTS ON SPECIAL HYPERGEOMETRIC FUNCTIONS

The hypergeometric functions of matrix argument which will occur in the distribution theory of subsequent chapters are ${}_0F_0$, ${}_1F_0$, ${}_0F_1$, ${}_1F_1$, and ${}_2F_1$. We have already seen that

$${}_0F_0(X)=\text{etr}(X)$$

and

$${}_1F_0(a;X)=\det(I-X)^{-a}.$$

The other three functions are, however, nontrivial. In this section we will derive some properties of these particular hypergeometric functions which will be useful later. The results here are due to Herz (1955).

Our first theorem gives a special integral representation for a ${}_0F_1$ function which will be useful in the derivation of the *noncentral* Wishart distribution in Chapter 10. The proof here is due to James.

THEOREM 7.4.1. If X is an $m\times n$ real matrix with $m\le n$ and $H=[H_1:H_2]\in O(n)$ where H_1 is $n\times m$ then

$$\text{(1)}\qquad \int_{O(n)}\text{etr}(XH_1)(dH)={}_0F_1\left(\tfrac{1}{2}n;\tfrac{1}{4}XX'\right)$$

where (dH) denotes the normalized invariant measure on $O(n)$.

Proof. It can be assumed without loss of generality in the proof that X has rank m (why?), so that $XX'>0$. Proving that (1) is true is equivalent to establishing that

(2)

$$\det(XX')^{(n-m-1)/2}\int_{O(n)} \operatorname{etr}(XH_1)(dH)=\det(XX')^{(n-m-1)/2}{}_0F_1(\tfrac{1}{2}n;\tfrac{1}{4}XX')$$

holds. The proof constructed here consists of showing that both sides of (2) have identical Laplace transforms. The Laplace transform of the left side of (2) is

$$g_1(Z)=\int_{XX'>0}\operatorname{etr}(-XX'Z)\det(XX')^{(n-m-1)/2}\int_{O(n)}\operatorname{etr}(XH_1)(dH)(d(XX'))$$

$$=\frac{\Gamma_m(\frac{1}{2}n)}{\pi^{mn/2}}\int_X\operatorname{etr}(-XX'Z)\int_{O(n)}\operatorname{etr}(XH_1)(dH)(dX),$$

on using Theorems 2.1.14 and 2.1.15. The first integral in the last line is over the space of $m\times n$ matrices X of rank m. Assuming $Z>0$ is real, put $X=Z^{-1/2}Y$ with Jacobian $(dX)=(\det Z)^{-n/2}(dY)$ (from Theorem 2.1.4) and interchange the order of integration of Y and H to give

(3)

$$g_1(Z)=\frac{\Gamma_m(\frac{1}{2}n)}{\pi^{mn/2}}\int_{O(n)}\int_Y\operatorname{etr}(-YY'+Z^{-1/2}YH_1)(dY)(dH)(\det Z)^{-n/2}$$

$$=(\det Z)^{-n/2}\operatorname{etr}(\tfrac{1}{4}Z^{-1})\frac{\Gamma_m(\frac{1}{2}n)}{\pi^{mn/2}}$$

$$\cdot\int_{O(n)}\int_Y\operatorname{etr}[-(Y-\tfrac{1}{2}Z^{-1/2}H_1')(Y-\tfrac{1}{2}Z^{-1/2}H_1')'](dY)(dH)$$

$$=(\det Z)^{-n/2}\operatorname{etr}(\tfrac{1}{4}Z^{-1})\Gamma_m(\tfrac{1}{2}n),$$

since

$$\frac{1}{\pi^{mn/2}}\operatorname{etr}[-(Y-M)(Y-M)']$$

is the density function of a matrix Y having the $N(M,\frac{1}{2}I_m\otimes I_n)$ distribution (see Theorem 3.1.1). Thus $g_1(Z)$ is equal to (3) for real $Z>0$, and by analytic continuation it equals (3) for complex symmetric Z with $\operatorname{Re}(Z)>0$.

Turning now to the right side of (2), the Laplace transform is

$$g_2(Z)=\int_{XX'>0}\operatorname{etr}(-XX'Z)\det(XX')^{(n-m-1)/2}{}_0F_1(\tfrac{1}{2}n;\tfrac{1}{4}XX')(d(XX'))$$

$$=\Gamma_m(\tfrac{1}{2}n)(\det Z)^{-n/2}{}_1F_1(\tfrac{1}{2}n;\tfrac{1}{2}n;\tfrac{1}{4}Z^{-1})$$

by Theorem 7.3.4. But the zonal polynomial expansion for ${}_1F_1$ makes it clear that

$$\begin{aligned}{}_1F_1(\tfrac{1}{2}n;\tfrac{1}{2}n;\tfrac{1}{4}Z^{-1})&={}_0F_0(\tfrac{1}{4}Z^{-1})\\&=\operatorname{etr}(\tfrac{1}{4}Z^{-1}),\end{aligned}$$

so that

$$g_2(Z)=\Gamma_m(\tfrac{1}{2}n)(\det Z)^{-n/2}\operatorname{etr}(\tfrac{1}{4}Z^{-1})$$

which is equal to $g_1(Z)$. The desired result now follows by uniqueness of Laplace transforms.

The next theorem generalizes two well-known integrals for the classical "confluent" hypergeometric function ${}_1F_1$ and the Gaussian hypergeometric function ${}_2F_1$.

THEOREM 7.4.2. The ${}_1F_1$ function has the integral representation

$$(4)\quad {}_1F_1(a;c;X)=\frac{\Gamma_m(c)}{\Gamma_m(a)\Gamma_m(c-a)}\int_{0<Y<I_m}\operatorname{etr}(XY)(\det Y)^{a-(m+1)/2}\cdot\det(I-Y)^{c-a-(m+1)/2}(dY),$$

valid for all symmetric X, $\operatorname{Re}(a)>\frac{1}{2}(m-1)$, $\operatorname{Re}(c)>\frac{1}{2}(m-1)$, and $\operatorname{Re}(c-a)>\frac{1}{2}(m-1)$, and the ${}_2F_1$ function has the integral representation

(5)

$${}_2F_1(a,b;c;X)=\frac{\Gamma_m(c)}{\Gamma_m(a)\Gamma_m(c-a)}\int_{0<Y<I_m}\det(I-XY)^{-b}(\det Y)^{a-(m+1)/2}\cdot\det(I-Y)^{c-a-(m+1)/2}(dY),$$

valid for $\operatorname{Re}(X)<I$, $\operatorname{Re}(a)>\frac{1}{2}(m-1)$, $\operatorname{Re}(c-a)>\frac{1}{2}(m-1)$.

Proof. To prove (4) expand

$$\operatorname{etr}(XY)=\sum_{k=0}^{\infty}\sum_{\kappa}\frac{C_\kappa(XY)}{k!}$$

and integrate term by term using Theorem 7.2.10. To prove (5) expand

$$\det(I-XY)^{-b}={}_1F_0(b;XY)$$

$$=\sum_{k=0}^{\infty}\sum_{\kappa}\frac{(b)_\kappa}{k!}C_\kappa(XY)$$

and integrate term by term using Theorem 7.2.10.

The Euler relation for the classical ${}_2F_1$ function has already been given by (17) of Section 5.1.3. This relation, and others, are generalized in the following theorem.

THEOREM 7.4.3.

(6) $${}_1F_1(a;c;X)=\operatorname{etr}(X)\,{}_1F_1(c-a;c;-X)$$

(7) $${}_2F_1(a,b;c;X)=\det(I-X)^{-b}\,{}_2F_1\left(c-a,b;c;-X(I-X)^{-1}\right)$$

$$=\det(I-X)^{c-a-b}\,{}_2F_1(c-a,c-b;c;X).$$

In the classical case $m=1$ the relation for ${}_1F_1$ is usually called the Kummer relation and those for ${}_2F_1$ the Euler relations. In the matrix case they can be established with the help of the integrals in Theorem 7.4.2; the proof is left as an exercise (see Problem 7.6).

Finally, let us note the confluence relations

(8) $$\lim_{b\to\infty}{}_2F_1\left(a,b;c;\frac{1}{b}X\right)={}_1F_1(a;c;X)$$

and

(9) $$\lim_{a\to\infty}{}_1F_1\left(a;c;\frac{1}{a}X\right)={}_0F_1(c;X)$$

which are an immediate consequence of the zonal polynomial expansions. Similar relations obviously also hold for the corresponding hypergeometric functions of two matrix arguments.

We have derived most of the integral results that we need concerning zonal polynomials and hypergeometric functions. Others will be derived in later chapters as the need arises.

7.5. PARTIAL DIFFERENTIAL EQUATIONS FOR HYPERGEOMETRIC FUNCTIONS

It will be seen that many density functions and moments can be expressed in terms of hypergeometric functions of matrix argument. Generally speaking, the zonal polynomial series for these functions converge extremely slowly and methods for approximating them have received a great deal of attention. One way of obtaining asymptotic results involves the use of differential equations for the hypergeometric functions; this method will be explained and used in subsequent chapters.

Differential equations satisfied by the classical hypergeometric functions are well-known; indeed, these functions are commonly defined as solutions of differential equations [see, for example, Erdélyi et al. (1953a)]. In this section we will give partial differential equations, from Muirhead (1970a) and Constantine and Muirhead (1972), satisfied by some hypergeometric functions of matrix argument. These differential equations will be expressed in terms of a number of differential operators in the latent roots $y_1,\ldots,y_m$ of the $m\times m$ symmetric matrix Y. The first of these is

$$(1)\qquad \Delta_Y=\sum_{i=1}^m y_i^2\frac{\partial^2}{\partial y_i^2}+\sum_{i=1}^m\sum_{\substack{j=1\\ j\neq i}}^m\frac{y_i^2}{y_i-y_j}\frac{\partial}{\partial y_i},$$

introduced in Definition 7.2.1. It was shown in Theorem 7.2.2 that

$$(2)\qquad \Delta_Y C_\kappa(Y)=[\rho_\kappa+k(m-1)]C_\kappa(Y),$$

where $\kappa=(k_1,\ldots,k_m)$ is a partition of k and

$$(3)\qquad \rho_\kappa=\sum_{i=1}^m k_i(k_i-i).$$

The other operators needed for the moment are

$$(4)\qquad E_Y=\sum_{i=1}^m y_i\frac{\partial}{\partial y_i},$$

$$(5)\qquad \varepsilon_Y=\sum_{i=1}^m\frac{\partial}{\partial y_i},$$

and

(6)
$$\delta_Y=\sum_{i=1}^m y_i\frac{\partial^2}{\partial y_i^2}+\sum_{i=1}^m\sum_{\substack{j=1\\ j\neq i}}^m\frac{y_i}{y_i-y_j}\frac{\partial}{\partial y_i}.$$

The operator E_Y has also appeared previously in (33) of Section 7.2.2 and its effect on $C_\kappa(Y)$ is given by

(7)
$$E_Y C_\kappa(Y)=kC_\kappa(Y).$$

To find the effect of the differential operators ε_Y and δ_Y on $C_\kappa(Y)$ we introduce the *generalized binomial expansion*

(8)
$$\frac{C_\kappa(I_m+Y)}{C_\kappa(I)}=\sum_{s=0}^{k}\sum_{\sigma}\binom{\kappa}{\sigma}\frac{C_\sigma(Y)}{C_\sigma(I)}$$

where the inner summation is over all partitions σ of the integer s. This defines the *generalized binomial coefficients*

$$\binom{\kappa}{\sigma}.$$

This generalization of the usual binomial expansion

$$(1+x)^k=\sum_{s=0}^{k}\binom{k}{s}x^s$$

was introduced by Constantine (1966), who tabulated the generalized binomial coefficients to $k=4$. These are given in Table 2. They have been tabulated to $k=8$ by Pillai and Jouris (1969). Now, corresponding to the partition $\kappa=(k_1,\dots,k_m)$ of k, let

(9)
$$\kappa_i=(k_1,\dots,k_{i-1},k_i+1,k_{i+1},\dots,k_m)$$

and

(10)
$$\kappa^{(i)}=(k_1,\dots,k_{i-1},k_i-1,k_{i+1},\dots,k_m),$$

whenever these partitions of $k+1$ and $k-1$ are *admissible*, that is whenever their parts are in non-increasing order. The following properties of the

Table 2. Generalized binomial coefficients $\binom{\kappa}{\sigma}$

$k=1$

κ \ σ	(0)	(1)
(1)	1	1

$k=2$

κ \ σ	(0)	(1)	(2)	(1,1)
(2)	1	2	1	0
(1,1)	1	2	0	1

$k=3$

κ \ σ	(0)	(1)	(2)	(1,1)	(3)	(2,1)	(1,1,1)
(3)	1	3	3	0	1	0	0
(2,1)	1	3	4/3	5/3	0	1	0
(1,1,1)	1	3	0	3	0	0	1

$k=4$

κ \ σ	(0)	(1)	(2)	(1,1)	(3)	(2,1)	(1,1,1)	(4)	(3,1)	(2,2)	(2,1,1)	(1,1,1,1)
(4)	1	4	6	0	4	0	0	1	0	0	0	0
(3,1)	1	4	11/3	7/3	6/5	14/5	0	0	1	0	0	0
(2,2)	1	4	8/3	10/3	0	4	0	0	0	1	0	0
(2,1,1)	1	4	5/3	13/3	0	5/2	3/2	0	0	0	1	0
(1,1,1,1)	1	4	0	6	0	0	4	0	0	0	0	1

Source: Reproduced from Constantine (1966) with the kind permission of the Institute of Mathematical Statistics.

generalized binomial coefficients are readily established:

(i) $\binom{\kappa}{(0)} \equiv 1$ for all κ.

(ii) $\binom{\kappa}{(1)} = k$ for any partition κ of k.

(iii) $\binom{\kappa}{\sigma} = 0$ if the partition σ has more non-zero parts than κ.

(iv) $\binom{\kappa}{\sigma} = 0$ if $\kappa > \sigma$.

(v) If κ and σ are both partitions of k then

$$\binom{\kappa}{\sigma} = \begin{cases} 1 & \text{if } \kappa = \sigma \\ 0 & \text{if } \kappa \neq \sigma. \end{cases}$$

(vi) **If κ is a partition of k and σ is a partition of $k-1$ then $\binom{\kappa}{\sigma}\neq 0$ only if $\sigma=\kappa^{(i)}$ for some i.**

The effects of the operators ε_Y and δ_Y on $C_\kappa(Y)$ are given in the following lemma.

LEMMA 7.5.1.

$$\varepsilon_Y \frac{C_\kappa(Y)}{C_\kappa(I)} = \sum_i \binom{\kappa}{\kappa^{(i)}} \frac{C_{\kappa^{(i)}}(Y)}{C_{\kappa^{(i)}}(I)} \tag{11}$$

and

$$\delta_Y \frac{C_\kappa(Y)}{C_\kappa(I)} = \sum_i \binom{\kappa}{\kappa^{(i)}} \left[k_i - 1 + \tfrac{1}{2}(m-i)\right] \frac{C_{\kappa^{(i)}}(Y)}{C_{\kappa^{(i)}}(I)} \tag{12}$$

where the summations are over all i such that $\kappa^{(i)}$ is admissible.

Proof. To prove (11) first note that, by (8),

$$\begin{aligned}\frac{C_\kappa(\lambda I+Y)}{C_\kappa(I)} &= \sum_{s=0}^{k} \lambda^{k-s} \sum_\sigma \binom{\kappa}{\sigma} \frac{C_\sigma(Y)}{C_\sigma(I)}\\ &= \frac{C_\kappa(Y)}{C_\kappa(I)} + \lambda \sum_i \binom{\kappa}{\kappa^{(i)}} \frac{C_{\kappa^{(i)}}(Y)}{C_{\kappa^{(i)}}(I)} + \text{terms involving higher powers of } \lambda.\end{aligned}$$

Hence

$$\begin{aligned}\varepsilon_Y \frac{C_\kappa(Y)}{C_\kappa(I)} &= \sum_{i=1}^{m} \frac{\partial}{\partial y_i} \frac{C_\kappa(Y)}{C_\kappa(I)}\\ &= \lim_{\lambda\to 0} \frac{C_\kappa(\lambda I+Y) - C_\kappa(Y)}{\lambda C_\kappa(I)}\\ &= \sum_i \binom{\kappa}{\kappa^{(i)}} \frac{C_{\kappa^{(i)}}(Y)}{C_{\kappa^{(i)}}(I)}.\end{aligned}$$

To prove (12), it is easily established that

$$\delta_Y = \tfrac{1}{2}(\varepsilon_Y \Delta_Y - \Delta_Y \varepsilon_Y) \tag{13}$$

(see Problem 7.9), and then (12) follows by applying ε_Y, Δ_Y to $C_\kappa(Y)$.

Two further sets of preliminary results are needed before we give differential equations for some one-matrix hypergeometric functions. These are contained in the following two lemmas.

LEMMA 7.5.2. Let $s_i = y_1^i + \cdots + y_m^i$, where $y_1, \dots, y_m$ are the latent roots of the symmetric $m \times m$ matrix Y. Then

$$s_1 \operatorname{etr}(Y) = \sum_{k=0}^{\infty} \sum_{\kappa} \frac{k C_\kappa(Y)}{k!}, \tag{14}$$

$$s_2 \operatorname{etr}(Y) = \sum_{k=0}^{\infty} \sum_{\kappa} \rho_\kappa \frac{C_\kappa(Y)}{k!}, \tag{15}$$

and

$$s_1 s_2 \operatorname{etr}(Y) = \sum_{k=0}^{\infty} \sum_{\kappa} (k-2)\rho_\kappa \frac{C_\kappa(Y)}{k!}, \tag{16}$$

where ρ_κ is given by (3).

Proof. To prove (14) we have

$$s_1 \operatorname{etr}(Y) = \sum_{k=0}^{\infty} \frac{s_1^{k+1}}{k!} = \sum_{k=0}^{\infty} \frac{k s_1^k}{k!} = \sum_{k=0}^{\infty} \sum_{\kappa} \frac{k C_\kappa(Y)}{k!},$$

where we have used

$$s_1^k = (\operatorname{tr} Y)^k = \sum_{\kappa} C_\kappa(Y).$$

Applying the operator $\Delta_Y - (m-1)E_Y$ to both sides of

$$\operatorname{etr}(Y) = \sum_{k=0}^{\infty} \sum_{\kappa} \frac{C_\kappa(Y)}{k!} \tag{17}$$

gives (15), and applying E_Y to both sides of (15) and collecting coefficients of $C_\kappa(Y)$ using (15) gives (16); the details are straightforward.

LEMMA 7.5.3.

$$\sum_{i} \binom{\kappa_i}{\kappa} C_{\kappa_i}(I) = m(k+1) C_\kappa(I). \tag{18}$$

$$\sum_{i} \binom{\kappa_i}{\kappa} \left[k_i - \tfrac{1}{2}(i-1)\right] C_{\kappa_i}(I) = k(k+1) C_\kappa(I). \tag{19}$$

$$\sum_{i} \binom{\kappa_i}{\kappa} \left[k_i - \tfrac{1}{2}(i-1)\right]^2 C_{\kappa_i}(I) = (k+1)\left[\rho_\kappa + \tfrac{1}{2}k(m+1)\right] C_\kappa(I). \tag{20}$$

Proof. We will sketch the proofs and the reader can fill in the details. Applying ε_Y to both sides of (17) and equating coefficients of $C_\kappa(Y)$ gives (18). Applying δ_Y to both sides of (14) and collecting coefficients of $C_\kappa(Y)$ using (14) gives (19). Applying δ_Y to both sides of (15) and collecting coefficients of $C_\kappa(Y)$ using (14), (15), (16), (18), and (19) gives (20).

We have now gathered enough ammunition to attack the problem of establishing differential equations for some one-matrix hypergeometric functions. We will start with the ${}_2F_1$ function. The classical ${}_2F_1(a,b;c;x)$ function satisfies the second order differential equation

$$x(1-x)\frac{d^2F}{dx^2}+\left[c-(a+b+1)x\right]\frac{dF}{dx}=abF;$$

see, for example, Erdélyi et al. (1953a), p. 56. In the matrix case a generalization of this is provided by the following theorem.

THEOREM 7.5.4. The function ${}_2F_1(a,b;c;Y)$ satisfies the partial differential equation

(21)

$$\delta_Y F+\left[c-\tfrac{1}{2}(m-1)\right]\varepsilon_Y F-\Delta_Y F-\left[a+b+1-\tfrac{1}{2}(m-1)\right]E_Y F=mabF.$$

Moreover, it is the unique solution subject to the condition that F has the form

$$F=\sum_{k=0}^{\infty}\sum_{\kappa}\alpha_\kappa C_\kappa(Y)\qquad(\alpha_{(0)}=1),$$

where the coefficients α_κ are independent of m.

Proof. It can be readily verified that substituting the series

$$F(Y)=\sum_{k=0}^{\infty}\sum_{\kappa}\alpha_\kappa C_\kappa(Y)\qquad(\alpha_{(0)}=1)\tag{22}$$

in the differential equation (21), applying each of the component differential operators to $C_\kappa(Y)$, and then equating coefficients of $C_\kappa(Y)$ on both sides gives

$$\sum_i\binom{\kappa_i}{\kappa}\left[c+k_i-\tfrac{1}{2}(i-1)\right]C_{\kappa_i}(I)\alpha_{\kappa_i}$$
$$=\left[mab+ka+kb+\rho_\kappa+\tfrac{1}{2}k(m+1)\right]C_\kappa(I)\alpha_\kappa.\tag{23}$$

This is a recurrence relation for the coefficients α_κ. We have to show that

$$\alpha_\kappa = \frac{(a)_\kappa (b)_\kappa}{(c)_\kappa k!}$$

is a solution of (23). Since

$$(a)_{\kappa_i} = (a)_\kappa\left[a + k_i - \tfrac{1}{2}(i-1)\right],$$

the problem reduces to showing that

$$\sum_i \binom{\kappa_i}{\kappa}\left[a+k_i-\tfrac{1}{2}(i-1)\right]\left[b+k_i-\tfrac{1}{2}(i-1)\right]C_{\kappa_i}(I)$$

$$= (k+1)\left[mab + \rho_\kappa + ka + kb + \tfrac{1}{2}k(m+1)\right]C_\kappa(I).$$

This, however, is a direct consequence of Lemma 7.5.3.

To establish the uniqueness claim, first put $\alpha_\kappa = \beta_\kappa/(c)_\kappa$, where $\beta_{(0)} = 1$. Then (23) becomes

$$(24)\quad \sum_i \binom{\kappa_i}{\kappa} C_{\kappa_i}(I)\beta_{\kappa_i} = \left[mab + ka + kb + \rho_\kappa + \tfrac{1}{2}k(m+1)\right]C_\kappa(I)\beta_\kappa,$$

and hence the β_κ do not depend on c. Now, from (18) of Section 7.2 we have

$$C_\kappa(I_m) = \frac{2^{2k}k!}{(2k)!}\left(\tfrac{1}{2}m\right)_\kappa \chi_\kappa$$

where

$$\chi_\kappa = \frac{(2k)!\prod_{i<j}^{p}(2k_i - 2k_j - i + j)}{\prod_{i=1}^{p}(2k_i + p - i)!}$$

with p being the number of nonzero parts of the partition κ. Note that χ_κ is defined for all partitions and is independent of m; the fact that $C_\kappa(I_m) \equiv 0$ if κ is a partition into more than m parts follows by noting that $(\frac{1}{2}m)_\kappa \equiv 0$. Since

$$\frac{C_{\kappa_i}(I)}{C_\kappa(I)} = \frac{(m+2k_i - i + 1)\chi_{\kappa_i}}{(2k+1)\chi_\kappa}$$

the recurrence relation (24) for the β_κ becomes

$$(25)\quad \sum_i \binom{\kappa_i}{\kappa}(m+2k_i-i+1)\chi_{\kappa_i}\beta_{\kappa_i} = (2k+1)\left[mab+ka+kb+\rho_\kappa+\tfrac{1}{2}k(m+1)\right]\chi_\kappa\beta_\kappa.$$

There are no restrictions here on the number of nonzero parts of κ, and the summation is over all i such that κ_i is admissible. Now assume that the β_κ (and hence the α_κ) are independent of m. Equating coefficients of m on both sides of (25) gives

$$(26)\qquad \sum_i \binom{\kappa_i}{\kappa}\chi_{\kappa_i}\beta_{\kappa_i} = (2k+1)\left(ab+\tfrac{1}{2}k\right)\chi_\kappa\beta_\kappa,$$

and equating constant terms gives

$$(27)\quad \sum_i \binom{\kappa_i}{\kappa}(2k_i-i+1)\chi_{\kappa_i}\beta_{\kappa_i} = (2k+1)\left[\rho_\kappa+ka+kb+\tfrac{1}{2}k\right]\chi_k\beta_\kappa.$$

As κ runs over all partitions of k (26) and (27) give equations in all the unknowns β corresponding to partitions of $k+1$, since any partition of $k+1$ can be expressed as κ_i for some i and some partition κ of k. The equations (26) and (27) determine the β_κ uniquely. With $\beta_{(0)}=1$, (26) gives $\beta_{(1)}=ab$. Next, with $\kappa=(1)$, (26) gives

$$2\beta_{(2)}+4\beta_{(1,1)} = 3\left(ab+\tfrac{1}{2}\right)ab$$

and (27) gives

$$4\beta_{(2)}-4\beta_{(1,1)} = 3\left(a+b+\tfrac{1}{2}\right)ab.$$

Solving these gives

$$\beta_{(2)} = \tfrac{1}{2}a(a+1)b(b+1) = \tfrac{1}{2}(a)_{(2)}(b)_{(2)}$$

and

$$\beta_{(1,1)} = \tfrac{1}{2}a\left(a-\tfrac{1}{2}\right)b\left(b-\tfrac{1}{2}\right) = \tfrac{1}{2}(a)_{(1,1)}(b)_{(1,1)}.$$

In general, letting $N(k)$ denote the number of partitions of k, (26) and (27) give $2N(k)$ equations in the $N(k+1)$ unknowns corresponding to the partitions of $k+1$. Since $2N(k) \geq N(k+1)$ there are more equations than unknowns. We know the equations are consistent since they are satisfied by $\beta_\kappa = (a)_\kappa (b)_\kappa / k!$. It is a straightforward matter to show that the $2N(k) \times N(k+1)$ matrix of coefficients formed from the left sides of (26) and (27) has rank $N(k+1)$ so that the equations have a unique solution.

Using Theorem 7.5.4 it is possible to prove a much stronger result than the one given there. The next theorem shows that the ${}_2F_1$ function is the unique solution of a *system* of partial differential equations.

THEOREM 7.5.5. The function ${}_2F_1(a, b; c; Y)$ is the unique solution of each of the m partial differential equations

$$(28)\quad y_i(1-y_i)\frac{\partial^2 F}{\partial y_i^2} + \left\{ c - \tfrac{1}{2}(m-1) - \left[a+b+1-\tfrac{1}{2}(m-1)\right] y_i + \frac{1}{2}\sum_{\substack{j=1\\ j\neq i}}^{m} \frac{y_i(1-y_i)}{y_i - y_j} \right\} \frac{\partial F}{\partial y_i} - \frac{1}{2}\sum_{\substack{j=1\\ j\neq i}}^{m} \frac{y_j(1-y_j)}{y_i - y_j}\frac{\partial F}{\partial y_j} = abF \qquad (i=1,\dots,m),$$

subject to the conditions that

(a) F is a symmetric function of $y_1,\dots,y_m$, and

(b) F is analytic at $Y=0$, and $F(0)=1$.

Proof. We will sketch the proof, which is lengthy. More complete details may be found in Muirhead (1970a). First note that any function which satisfies each of the m partial differential equations (28) also satisfies the equation obtained by summing them. It is readily verified that this sum is

$$\delta_Y F + \left[c - \tfrac{1}{2}(m-1)\right]\varepsilon_Y F - \Delta_Y F - \left[a+b+1-\tfrac{1}{2}(m-1)\right]E_Y F = mabF.$$

This is the differential equation of Theorem 7.5.4, and it is shown there that ${}_2F_1(a, b; c; Y)$ is the unique solution of this subject to F having the form

$$(29)\qquad F = \sum_{k=0}^{\infty}\sum_{\kappa} \alpha_\kappa C_\kappa(Y) \qquad (\alpha_{(0)}=1),$$

where the coefficients α_κ are independent of m. It suffices to show that the differential equations (28) have the same unique solution which can be expressed in the form (29).

We first demonstrate that the m equations (28) have the same unique solution subject to conditions (a) and (b) by transforming to a system of equations in terms of the elementary symmetric functions $r_1=\Sigma_{i=1}^m y_i$, $r_2=\Sigma_{i<j}^m y_i y_j, \ldots, r_m = y_1 y_2 \ldots y_m$, of $y_1, \ldots, y_m$. Let $r_j^{(1)}$ for $j = 1, 2, \ldots, m-1$ denote the jth elementeary symmetric function formed from $y_1, \ldots, y_m$, omitting y_i. Defining $r_0 = r_0^{(i)} = 1$, we have

$$(30) \qquad r_j = y_i r_{j-1}^{(i)} + r_j^{(i)} \qquad (j=1,\ldots,m-1).$$

Using this, it follows that

$$\frac{\partial}{\partial y_i} = \sum_{\nu=1}^{m} r_{\nu-1}^{(i)} \frac{\partial}{\partial r_\nu}$$

and

$$\frac{\partial^2}{\partial y_i^2} = \sum_{\mu,\nu=1}^{m} r_{\mu-1}^{(i)} r_{\nu-1}^{(i)} \frac{\partial^2}{\partial r_\mu \partial r_\nu}.$$

Substituting these in (28) and using (30), we find that the system (28) becomes

$$(31) \quad \sum_{\mu,\nu=1}^{m} \frac{\partial^2 F}{\partial r_\mu \partial r_\nu}\left\{\sum_{j=1}^{m} a_{\mu\nu}^{(j)}\left(r_{j-1}^{(i)} - r_j + r_j^{(i)}\right)\right\}$$
$$+\sum_{j=1}^{m}\left\{\left[c-\tfrac{1}{2}(j-1)\right]r_{j-1}^{(i)}+\left(a+b+1-\tfrac{1}{2}j\right)r_j^{(i)}-(a+b+1)r_j\right\}$$
$$\cdot\frac{\partial F}{\partial r_j} - abF = 0 \quad (i=1,\ldots,m),$$

where $a_{\mu\nu}^{(j)} = a_{\nu\mu}^{(j)}$ and, for $\mu \leq \nu$,

$$a_{\mu\nu}^{(j)} = \begin{cases} r_{\mu+\nu-j} & \text{for } 1 \leq j \leq \mu \\ 0 & \text{for } \mu < j \leq \nu \\ -r_{\mu+\nu-j} & \text{for } \nu < j \leq \mu+\nu \\ 0 & \text{for } \mu+\nu < j. \end{cases}$$

Any solution of (31) satisfies condition (a). In (31) we can equate coefficients of $r_{j-1}^{(i)}$ to zero for $j=1,\ldots,m$, giving the system of differential equations

$$(32)\quad \sum_{\mu,\nu=1}^{m} \frac{\partial^2 F}{\partial r_\mu \partial r_\nu}\left(a_{\mu\nu}^{(j)}+a_{\mu\nu}^{(j-1)}\right)+\left[c-\tfrac{1}{2}(j-1)\right]\frac{\partial F}{\partial r_j}$$

$$+\left[a+b+1-\tfrac{1}{2}(j-1)\right]\frac{\partial F}{\partial r_{j-1}}$$

$$-\delta_{1j}\left\{\sum_{\mu,\nu=1}^{m}\frac{\partial^2 F}{\partial r_\mu \partial r_\nu}\sum_{i=1}^{m} a_{\mu\nu}^{(i)} r_i+(a+b+1)\sum_{i=1}^{m} r_i\frac{\partial F}{\partial r_i}+abF\right\}=0$$

$$(j=1,\dots,m).$$

Now we put

$$(33)\qquad F(r_1,\dots,r_m)=\sum_{j_1,\dots,j_m=0}^{\infty}\gamma(j_1,\dots,j_m)r_1^{j_1}\cdots r_m^{j_m}$$

with $\gamma(0,\dots,0)=1$. Next, we introduce dictionary ordering for the coefficients $\gamma(j_1,\dots,j_m)$ on the basis of the indices arranged in the order $j_m, j_{m-1},\dots,j_2, j_1$. Substituting (33) in (32) with $j=m$ gives a recurrence relation which expresses $\gamma(j_1,\dots,j_m)$ in terms of coefficients whose last index is less than j_m, and by iteration $\gamma(j_1,\dots,j_m)$ can then be expressed in terms of coefficients whose last index is zero. Putting $r_m=0$ in the equation (32) with $j=m-1$, we can then express coefficients of the form $\gamma(j_1,\dots,j_{m-1},0)$ in terms of coefficients of the form $\gamma(t_1,\dots,t_{m-2},0,0)$. By repeating this procedure, all coefficients can be expressed in terms of $\gamma(0,\dots,0)$, which is 1. Hence all the coefficients $\gamma(j_1,\dots,j_m)$ in (33) are uniquely determined by the recurrence relations, and condition (b) is satisfied. Since each differential equation in (31) gives rise to the same system (32), it follows that each equation in the system (28) has the same unique solution F subject to conditions (a) and (b).

Next, note that the coefficients in the system (32) do not involve m explicitly so that the coefficients $\gamma(j_1,\dots,j_m)$ obtained from the recurrence relations will be functions of a, b, c, and j_i but will be independent of m. In fact, since $r_h\equiv 0$ for $h>m$ the system (32) can be formally extended to hold for all $i=1,2,\dots$ and the upper limit on the summations can be dropped. The coefficients γ in (33) are thus defined for any number of indices $j_1,\dots,j_h$ and are completely independent of m. Now, the series (33) could be rearranged as a series of zonal polynomials

$$F=\sum_{k=0}^{\infty}\sum_{\kappa}\alpha_{\kappa}C_{\kappa}(Y),\qquad \alpha_{(0)}=1. \tag{34}$$

Since the zonal polynomials when expressed in terms of the elementary symmetric functions $r_1,\ldots,r_m$ do not explicitly depend on m, the coefficients α_κ will be functions of a, b, c, and κ but not m. Since $C_\kappa(Y)\equiv 0$ for any partition into more than m nonzero parts, the α_κ can be defined for partitions into any number of parts and are completely independent of m. Hence the unique solution of (28) subject to (a) and (b) can be expressed as (34), where the coefficients α_κ are independent of m, and the proof is complete.

Theorem 7.5.5 also yields systems of partial differential equations satisfied by the ${}_1F_1$ and ${}_0F_1$ functions. These are given in the following theorem.

THEOREM 7.5.6. The function ${}_1F_1(a;c;Y)$ is the unique solution of each of the m partial differential equations in the system

(35)

$$y_i\frac{\partial^2 F}{\partial y_i^2}+\left\{c-\tfrac{1}{2}(m-1)-y_i+\frac{1}{2}\sum_{\substack{j=1\\ j\neq i}}^{m}\frac{y_i}{y_i-y_j}\right\}\frac{\partial F}{\partial y_i}-\frac{1}{2}\sum_{\substack{j=1\\ j\neq i}}^{m}\frac{y_j}{y_i-y_j}\frac{\partial F}{\partial y_j}=aF$$

$$(i=1,\ldots,m),$$

and the function ${}_0F_1(c;Y)$ is the unique solution of each of the m partial differential equations in the system

$$y_i\frac{\partial^2 F}{\partial y_i^2}+\left\{c-\tfrac{1}{2}(m-1)+\frac{1}{2}\sum_{\substack{j=1\\ j\neq i}}^{m}\frac{y_i}{y_i-y_j}\right\}\frac{\partial F}{\partial y_i}-\frac{1}{2}\sum_{\substack{j=1\\ j\neq i}}^{m}\frac{y_j}{y_i-y_j}\frac{\partial F}{\partial y_j}=F \tag{36}$$

$$(i=1,\ldots,m),$$

subject to the conditions that

(a) F is a symmetric function of $y_1,\ldots,y_m$, and

(b) F is analytic at $Y=0$, and $F(0)=1$.

Proof. Proofs similar to the proof of Theorem 7.5.5 can be constructed. However, the results follow directly from Theorem 7.5.5 using the confluence relations given by (8) and (9) of Section 7.4. Theorem 7.5.5 shows that subject to (a) and (b), the function ${}_2F_1(a,b;c;(1/b)Y)$ is the unique solution of each equation in the system

$$(37)\quad y_i\left(1-\frac{1}{b}y_i\right)\frac{\partial^2 F}{\partial y_i^2}+\left\{c-\tfrac{1}{2}(m-1)-\left[a+b+1-\tfrac{1}{2}(m-1)\right]\frac{1}{b}y_i\right.$$

$$\left.+\frac{1}{2}\sum_{\substack{j=1\\j\neq i}}^{m}\frac{y_i\left(1-\frac{1}{b}y_i\right)}{y_i-y_j}\right\}\frac{\partial F}{\partial y_i}-\frac{1}{2}\sum_{\substack{j=1\\j\neq i}}^{m}\frac{y_j\left(1-\frac{1}{b}y_j\right)}{y_i-y_j}\frac{\partial F}{\partial y_j}=aF.$$

Letting $b\to\infty$, ${}_2F_1(a,b;c;(1/b)Y)\to{}_1F_1(a;c;Y)$ and the system (37) tends to (35). Similarly, since ${}_1F_1(a;c;(1/a)Y)\to{}_0F_1(c;Y)$ as $a\to\infty$ the system (36) can be obtained using (35).

We now turn to the two-matrix hypergeometric functions given by Definition 7.3.2. To give differential equations satisfied by these functions we need to introduce two further differential operators, namely,

$$(38)\qquad \gamma_Y=\sum_{i=1}^{m}y_i^2\frac{\partial}{\partial y_i}$$

and

$$(39)\qquad \eta_Y\equiv\tfrac{1}{2}(\Delta_Y\gamma_Y-\gamma_Y\Delta_Y)$$

$$=\sum_{i=1}^{m}y_i^3\frac{\partial^2}{\partial y_i^2}+\sum_{i=1}^{m}\sum_{\substack{j=1\\j\neq i}}^{m}\frac{y_i^3}{y_i-y_j}\frac{\partial}{\partial y_i}+\tfrac{1}{2}(3-m)\gamma_Y$$

(see Problem 7.10). In order to obtain the effects of these operators on $C_\kappa(Y)$ we need the following lemma.

LEMMA 7.5.7.

$$(40)\qquad \sum_i\binom{\kappa_i}{\kappa}C_{\kappa_i}(Y)=(k+1)(\operatorname{tr}Y)C_\kappa(Y).$$

[Note that this reduces to (18) of Lemma 7.5.3 when $Y=I_m$.]

Proof. From Theorem 7.3.3 we have

$$(41)\qquad \int_{O(m)} \operatorname{etr}(XHYH')(dH)=\sum_{k=0}^{\infty}\sum_{\kappa}\frac{C_\kappa(X)C_\kappa(Y)}{k!C_\kappa(I_m)}.$$

Let $x_1,\dots,x_m$ denote the latent roots of X and apply the operator $\varepsilon_X=\sum_i\partial/\partial x_i$ to both sides to give, with the help of (11),

$$(42)\quad \int_{O(m)} \operatorname{tr}(HYH')\operatorname{etr}(XHYH')(dH)$$

$$=\sum_{k=0}^{\infty}\sum_{\kappa}\frac{C_\kappa(Y)}{k!}\sum_i\binom{\kappa}{\kappa^{(i)}}\frac{C_{\kappa^{(i)}}(X)}{C_{\kappa^{(i)}}(I)}.$$

Using (41) to evaluate the left side of (42), this becomes

$$(\operatorname{tr}Y)\sum_{k=0}^{\infty}\sum_{\kappa}\frac{C_\kappa(X)C_\kappa(Y)}{k!C_\kappa(I)}=\sum_{k=0}^{\infty}\sum_{\kappa}\frac{C_\kappa(Y)}{k!}\sum_i\binom{\kappa}{\kappa^{(i)}}\frac{C_{\kappa^{(i)}}(X)}{C_{\kappa^{(i)}}(I)}.$$

Equating coefficients of $C_\kappa(X)/C_\kappa(I)$ on both sides gives (40) and completes the proof.

The effects of γ_Y and η_Y on $C_\kappa(Y)$ are given in the following lemma.

LEMMA 7.5.8.

$$(43)\qquad \gamma_Y C_\kappa(Y)=\frac{1}{k+1}\sum_i\binom{\kappa_i}{\kappa}\left[k_i-\tfrac{1}{2}(i-1)\right]C_{\kappa_i}(Y)$$

and

$$(44)\quad \eta_Y C_\kappa(Y)=\frac{1}{k+1}\sum_i\binom{\kappa_i}{\kappa}\left[k_i-\tfrac{1}{2}(i-1)\right]\left[k_i-\tfrac{1}{2}(i-m)\right]C_{\kappa_i}(Y).$$

Proof. To prove (43), apply the operator Δ_Y to both sides of (40) and simplify using (40). To prove (44), apply Δ_Y and γ_Y to $C_\kappa(Y)$ and use (43). The details are straightforward and are left to the reader.

The results we have established enable us to give differential equations for some two-matrix hypergeometric functions. These results are from Constantine and Muirhead (1972). In what follows, X and Y are symmetric $m\times m$ matrices with latent roots $x_1,\dots,x_m$ and $y_1,\dots,y_m$, respectively. We start with the ${}_2F_1^{(m)}$ function.

THEOREM 7.5.9. The function ${}_2F_1^{(m)}(a,b;c;X,Y)$ is the unique solution of the partial differential equation

(45)

$$\delta_X F+\left[c-\tfrac{1}{2}(m-1)\right]\varepsilon_X F-\left[a+b-\tfrac{1}{2}(m-1)\right]\gamma_Y F-\eta_Y F=abF(\operatorname{tr}Y)$$

subject to the condition that F has the series expansion

$$F(X,Y)=\sum_{k=0}^{\infty}\sum_{\kappa}\alpha_\kappa\frac{C_\kappa(X)C_\kappa(Y)}{C_\kappa(I)}, \tag{46}$$

where $F(0,0)=1$; that is, $\alpha_{(0)}=1$.

Proof. Substitute the series (46) in the differential equation (45), apply each of the component differential operators to their respective zonal polynomials, and compare coefficients first of $C_\kappa(X)$ and then of $C_\kappa(Y)$ on both sides. It can readily be verified that this gives rise to the following recurrence relation for the α_κ:

(47)

$$(k+1)\left[c+k_i-\tfrac{1}{2}(i-1)\right]\alpha_{\kappa_i}=\left[a+k_i-\tfrac{1}{2}(i-1)\right]\left[b+k_i-\tfrac{1}{2}(i-1)\right]\alpha_\kappa.$$

The condition $\alpha_{(0)}=1$ used in (47) determines the α_κ uniquely as

$$\alpha_\kappa=\frac{(a)_\kappa(b)_\kappa}{(c)_\kappa k!},$$

and the proof is complete.

Theorem 7.5.9 yields partial differential equations for many other two-matrix hypergeometric functions. These are given in the following corollary.

COROLLARY 7.5.10.

(i) ${}_1F_1^{(m)}(a;c;X,Y)$ satisfies the differential equation

$$\delta_F F+\left[c-\tfrac{1}{2}(m-1)\right]\varepsilon_X F-\gamma_Y F=aF(\operatorname{tr}Y).$$

(ii) ${}_0F_1^{(m)}(c;X,Y)$ satisfies the differential equation

$$\Delta_X F+\left[c-\tfrac{1}{2}(m-1)\right]\varepsilon_X F=F(\operatorname{tr}Y).$$

(iii) ${}_1F_0^{(m)}(a; X, Y)$ satisfies the differential equation

$$\delta_X F - a\gamma_Y F - \eta_Y F = \tfrac{1}{2}a(m-1)F(\operatorname{tr} Y).$$

(iv) ${}_0F_0^{(m)}(X, Y)$ satisfies the differential equation

$$\delta_X F - \gamma_Y F = \tfrac{1}{2}a(m-1)F(\operatorname{tr} Y).$$

Proof. (i) follows from Theorem 7.5.9 via the confluence

$$\lim_{b\to\infty} {}_2F_1^{(m)}\left(a, b; c; X, \frac{1}{b}Y\right) = {}_1F_1^{(m)}(a; c; X, Y).$$

Similarly (ii) follows from (i) by confluence. Putting $b = c = (m-1)/2$ in Theorem 7.5.9 gives (iii), and putting $a = c = (m-1)/2$ in (i) gives (iv).

7.6. GENERALIZED LAGUERRE POLYNOMIALS

Having generalized the classical hypergeometric functions to functions of matrix argument, it is interesting to ask whether other classical special functions can be similarly generalized. The answer is that many of them can and have been, and the interested reader is referred to the references in Muirhead (1978). After the hypergeometric functions, the functions which appear to be most useful in multivariate analysis are generalizations of the classical Laguerre polynomials, which have been studied by Herz (1955) and Constantine (1966). The generalized Laguerre polynomials will be used in Sections 10.6.2, 10.6.4, and 11.3.4.

Let us first recall some facts about the classical Laguerre polynomials, one of the classical *orthogonal* polynomials. The Laguerre polynomial $L_k^\gamma(x)$ is given by

(1) $$L_k^\gamma(x) = (\gamma+1)_k \sum_{s=0}^{k} \binom{k}{s} \frac{(-x)^s}{(\gamma+1)_s}$$

for $\gamma > -1$. Various normalizations are used; here L_k^γ is normalized so that the coefficient of x^k is $(-1)^k$. Obviously $L_k^\gamma(x)$ is a polynomial of degree k in x, and the L_k^γ are orthogonal on $x > 0$ with respect to the weight function $e^{-x}x^\gamma$; in fact,

(2) $$\int_0^\infty e^{-x}x^\gamma L_k^\gamma(x) L_j^\gamma(x)\,dx = \delta_{jk}\, k!\,\Gamma(\gamma+1+k).$$

The basic generating function for the L_k^γ is

$$(3)\qquad (1-z)^{-\gamma-1}\exp\left(\frac{xz}{z-1}\right)=\sum_{k=0}^{\infty}\frac{L_k^\gamma(x)z^k}{k!}\qquad (|z|<1).$$

For proofs and other properties the reader is referred to Rainville (1967), Chapter 12, and Erdélyi et al. (1953b), Chapter 10. It should be noted that the polynomial in these references is $L_k^\gamma(x)/k!$ in our notation.

Each term in (1) has been generalized previously. In defining the hypergeometric functions of matrix argument, the powers of the variable were replaced by zonal polynomials and the coefficients $(a)_k$ by generalized hypergeometric coefficients $(a)_\kappa$ given by (2) of Section 7.3. The binomial coefficients have been generalized by the generalized binomial coefficients which appear in the generalized binomial expansion given by (8) of Section 7.5. Taking our cue from these we will proceed from the following definition, which should be compared with (1).

DEFINITION 7.6.1. The generalized Laguerre polynomial $L_\kappa^\gamma(X)$ of an $m\times m$ symmetric matrix X corresponding to the partition κ of k is

$$(4)\quad L_\kappa^\gamma(X)=(\gamma+p)_\kappa C_\kappa(I_m)\sum_{s=0}^{k}\sum_{\sigma}\binom{\kappa}{\sigma}\frac{C_\sigma(-X)}{(\gamma+p)_\sigma C_\sigma(I_m)}\qquad (\gamma>-1),$$

where the inner summation is over all partitions σ of the integer s and, throughout this section,

$$p=\tfrac{1}{2}(m+1).$$

Clearly $L_\kappa^\gamma(X)$ is a symmetric polynomial of degree k in the latent roots of X. Note that

$$(5)\qquad L_\kappa^\gamma(0)=(\gamma+p)_\kappa C_\kappa(I_m).$$

The following theorem gives the Laplace transform of $(\det X)^\gamma L_\kappa^\gamma(X)$ and is useful in the derivation of further results.

THEOREM 7.6.2. If Z is an $m\times m$ symmetric matrix with $\mathrm{Re}(Z)>0$ then

$$(6)\quad \int_{X>0}\operatorname{etr}(-XZ)(\det X)^\gamma L_\kappa^\gamma(X)(dX)$$

$$=(\gamma+p)_\kappa\Gamma_m(\gamma+p)(\det Z)^{-\gamma-p}C_\kappa(I-Z^{-1})$$

Proof. Substituting the right side of (4) for $L_\kappa^\gamma(X)$ in the integrand, and integrating using Theorem 7.2.7 shows that the left side of (6) is equal to

$$(\gamma+p)_\kappa C_\kappa(I_m)\sum_{s=0}^{k}\sum_{\sigma}\binom{\kappa}{\sigma}\frac{C_\sigma(-Z^{-1})}{C_\sigma(I)}\Gamma_m(\gamma+p)(\det Z)^{-\gamma-p}$$

$$=(\gamma+p)_\kappa\Gamma_m(\gamma+p)(\det Z)^{-\gamma-p}C_\kappa(I-Z^{-1})$$

using (8) of Section 7.5, and the proof is complete.

Our next result generalizes the generating function relation (3).

THEOREM 7.6.3. If $X>0$, then

$$(7)\quad \det(I-Z)^{-\gamma-p}{}_0F_0^{(m)}(-X,Z(I-Z)^{-1})$$
$$=\sum_{k=0}^{\infty}\sum_{\kappa}\frac{L_\kappa^\gamma(X)C_\kappa(Z)}{C_\kappa(I_m)k!}\qquad(\|Z\|<1).$$

Proof. The proof consists of considering both sides of (7) as functions of X and showing that they have the same Laplace transforms. First, multiply both sides of (7) by $(\det X)^\gamma$; we then must show that

$$(8)\quad \det(I-Z)^{-\gamma-p}(\det X)^\gamma{}_0F_0^{(m)}(-X,Z(I-Z)^{-1})$$
$$=\det X^\gamma\sum_{k=0}^{\infty}\sum_{\kappa}\frac{L_\kappa^\gamma(X)C_\kappa(Z)}{C_\kappa(I_m)k!}.$$

The Laplace transform of the left side of (8) is

$$g_1(W)=\det(I-Z)^{-\gamma-p}\int_{X>0}\operatorname{etr}(-XW)(\det X)^\gamma$$
$$\cdot{}_0F_0^{(m)}(-X,Z(I-Z)^{-1})(dX)$$
$$=\det(I-Z)^{-\gamma-p}\Gamma_m(\gamma+p)(\det W)^{-\gamma-p}{}_1F_0^{(m)}(\gamma+p;-W^{-1},Z(I-Z)^{-1})$$

by Theorem 7.3.4

$$=\Gamma_m(\gamma+p)(\det W)^{-\gamma-p}\det(I-Z)^{-\gamma-p}$$
$$\cdot\int_{O(m)}\det\left(I+H'W^{-1}HZ(I-Z)^{-1}\right)^{-\gamma-p}(dH),$$

using Theorem 7.3.3 and Corollary 7.3.5

$$= \Gamma_m(\gamma+p)(\det W)^{-\gamma-p} \int_{O(m)} \det(I - Z + H'W^{-1}HZ)^{-\gamma-p}(dH)$$

$$= \Gamma_m(\gamma+p)(\det W)^{-\gamma-p} \int_{O(m)} \det(I - H'(I-W^{-1})HZ)^{-\gamma-p}(dH)$$

$$= \Gamma_m(\gamma+p)(\det W)^{-\gamma-p} {}_1F_0^{(m)}(\gamma+p; I-W^{-1}, Z).$$

The Laplace transform of the right side of (8) is

$$g_2(W) = \sum_{k=0}^{\infty} \sum_{\kappa} \frac{C_\kappa(Z)}{C_\kappa(I_m)k!} \int_{X>0} \operatorname{etr}(-XW)(\det X)^{\gamma} L_\kappa^\gamma(X)(dX)$$

$$= \Gamma_m(\gamma+p)(\det W)^{-\gamma-p} \sum_{k=0}^{\infty} \sum_{\kappa} \frac{(\gamma+p)_\kappa}{k!} \frac{C_\kappa(Z)C_\kappa(I-W^{-1})}{C_\kappa(I_m)}$$

$$= \Gamma_m(\gamma+p)(\det W)^{-\gamma-p} {}_1F_0^{(m)}(\gamma+p; I-W^{-1}, Z),$$

which is equal to $g_1(W)$. The desired result now follows by uniqueness of Laplace transforms.

The integral expression for $L_\kappa^\gamma(X)$ in the next theorem is actually how Constantine (1966) defined the generalized Laguerre polynomials.

THEOREM 7.6.4. If X is a symmetric $m \times m$ matrix then

$$(9) \qquad \operatorname{etr}(-X)L_\kappa^\gamma(X) = \frac{1}{\Gamma_m(\gamma+p)} \int_{Y>0} \operatorname{etr}(-Y)(\det Y)^{\gamma} \cdot C_\kappa(Y) {}_0F_1(\gamma+p; -XY)(dY)$$

$$[\gamma > -1;\ p = (m+1)/2].$$

A proof of this result can be constructed by showing that both sides of (9) have the same Laplace transforms; the details are very similar to those in the proof of Theorem 7.6.3 and are left as an exercise (see Problem 7.18).

The final result we will present is the generalization of the orthogonality relation (2). Note that the following theorem says that the Laguerre polynomials are orthogonal with respect to a Wishart density function.

THEOREM 7.6.5. $L_\kappa^\gamma(X)$ and $L_\sigma^\gamma(X)$ are orthogonal on $X>0$ with respect to the weight function

$$W(X)=\operatorname{etr}(-X)(\det X)^\gamma,$$

unless $\kappa=\sigma$. Specifically,

$$(10)\quad \int_{X>0}\operatorname{etr}(-X)(\det X)^\gamma L_\kappa^\gamma(X)L_\sigma^\gamma(X)(dX)$$

$$=\delta_{\kappa\sigma}k!C_\kappa(I)\Gamma_m(\gamma+p)(\gamma+p)_\kappa$$

where

$$\delta_{\kappa\sigma}=\begin{cases}1 & \text{if } \kappa=\sigma\\ 0 & \text{if } \kappa\neq\sigma\end{cases}\quad\text{and}\quad p=\frac{m+1}{2}.$$

Proof. From the generating function (7) we have

$$(11)\quad \det(I-Z)^{-\gamma-p}\int_{O(m)}\operatorname{etr}\left(-XHZ(I-Z)^{-1}H'\right)(dH)$$

$$=\sum_{k=0}^{\infty}\sum_{\kappa}\frac{L_\kappa^\gamma(X)C_\kappa(Z)}{C_\kappa(I_m)k!}$$

Multiply both sides by $\operatorname{etr}(-X)(\det X)^\gamma C_\sigma(X)$, where σ is a partition of any integer s, and integrate over $X>0$. The left side of (11) becomes

$$(12)\quad \det(I-Z)^{-\gamma-p}\int_{O(m)}\int_{X>0}\operatorname{etr}\left(-X\left(I+HZ(I-Z)^{-1}H'\right)\right)$$

$$\cdot(\det X)^\gamma C_\sigma(X)(dX)(dH)$$

$$=\det(I-Z)^{-\gamma-p}\Gamma_m(\gamma+p)(\gamma+p)_\sigma\int_{O(m)}$$

$$\cdot\det\left(I+HZ(I-Z)^{-1}H'\right)^{-\gamma-p}(dH)\ C_\sigma(I-Z)$$

by Theorem 7.2.7

$$=\Gamma_m(\gamma+p)(\gamma+p)_\sigma C_\sigma(I-Z)$$

$$=\Gamma_m(\gamma+p)(\gamma+p)_\sigma(-1)^s C_\sigma(Z)+\text{terms of lower degree.}$$

The right side of (11) becomes

$$(13)\qquad \sum_{k=0}^{\infty}\sum_{\kappa}\frac{C_\kappa(Z)}{C_\kappa(I)k!}\int_{X>0}\operatorname{etr}(-X)(\det X)^{\gamma}C_\sigma(X)L_\kappa^{\gamma}(X)(dX).$$

Comparing coefficients of $C_\kappa(Z)$ on both sides shows that

$$\int_{X>0}\operatorname{etr}(-X)(\det X)^{\gamma}C_\sigma(X)L_\kappa^{\gamma}(X)(dX)=0$$

for $k\geq s$, unless $\kappa=\sigma$, so that $L_\kappa^{\gamma}(X)$ is orthogonal to all Laguerre polynomials of lower degree. Since, from Definition 7.6.1,

$$L_\sigma^{\gamma}(X)=(-1)^s C_\sigma(X)+\text{terms of lower degree}$$

it also follows that $L_\kappa^{\gamma}(X)$ is orthogonal to all Laguerre polynomials $L_\sigma^{\gamma}(X)$ of the same degree unless $\kappa=\sigma$. Putting $\kappa=\sigma$ and comparing coefficients of $C_\sigma(Z)$ in (12) and (13) gives

$$\int_{X>0}\operatorname{etr}(-X)(\det X)^{\gamma}C_\sigma(-X)L_\sigma^{\gamma}(X)(dX)$$
$$=s!C_\sigma(I)\Gamma_m(\gamma+p)(\gamma+p)_\sigma,$$

from which it follows that

$$\int_{X>0}\operatorname{etr}(-X)(\det X)^{\gamma}[L_\sigma(X)]^2(dX)=s!C_\sigma(I)\Gamma_m(\gamma+p)(\gamma+p)_\sigma$$

since

$$L_\sigma^{\gamma}(X)=C_\sigma(-X)+\text{terms of lower degree which integrate to zero.}$$

PROBLEMS

7.1. Using the recurrence relation given by (14) of Section 7.2 compute the coefficients of the monomial symmetric functions in all zonal polynomials of degree $k=4$ and $k=5$.

7.2. Let

$$X=\begin{bmatrix} x_1 & x_2 \\ x_2 & x_3 \end{bmatrix}$$

be a 2×2 positive definite matrix and put $\mathbf{x}=(x_1,x_2,x_3)'$; then

$$dX=\begin{bmatrix} dx_1 & dx_2 \\ dx_2 & dx_3 \end{bmatrix} \quad \text{and} \quad d\mathbf{x}=(dx_1,dx_2,dx_3)'.$$

(a) Show that the 3×3 matrix $G(\mathbf{x})$ satisfying

$$\operatorname{tr}(X^{-1}\,dXX^{-1}\,dX)=d\mathbf{x}'G(\mathbf{x})\,d\mathbf{x}$$

is

$$G(\mathbf{x})=\frac{1}{(x_1x_3-x_2^2)^2}\begin{bmatrix} x_3^2 & -2x_2x_3 & x_2^2 \\ -2x_2x_3 & 2(x_1x_3+x_2^2) & -2x_1x_2 \\ x_2^2 & -2x_1x_2 & x_1^2 \end{bmatrix}.$$

(b) Let Δ_X^* be the differential operator given by (23) or (24) of Section 7.2. Express Δ_X^* in terms of

$$\frac{\partial^2}{\partial x_i^2},\qquad \frac{\partial^2}{\partial x_i\,\partial x_j},\qquad \frac{\partial}{\partial x_i}.$$

(c) Put

$$Z=\begin{bmatrix} z_1 & z_2 \\ z_2 & z_3 \end{bmatrix}=LXL',$$

where $L=(l_{ij})$ is a 2×2 nonsingular matrix, and put $\mathbf{z}=(z_1,z_2,z_3)'$. Find the 3×3 matrix T_L such that $\mathbf{z}=T_L\mathbf{x}$, and verify directly that $G(T_L\mathbf{x})=T_L'^{-1}G(\mathbf{x})T_L^{-1}$.

7.3. If $g_1(Z)$ and $g_2(Z)$ are the Laplace transforms of $f_1(X)$ and $f_2(X)$ (see Definition 7.2.9) prove that $g_1(Z)g_2(Z)$ is the Laplace transform of the convolution

$$f(Y)=\int_{0<X<Y} f_1(X)f_2(Y-X)(dX).$$

7.4. Use the result of Problem 7.3 to prove Theorem 7.2.10 for $Y>0$. [*Hint*: Let $f(Y)$ denote the integral on the left side of (57) of Section 7.2 and show that $f(Y)=f(I)C_\kappa(Y)/C_\kappa(I)$. Putting $X=Y^{-1/2}VY^{-1/2}$ in the integral gives

$$f(Y)(\det Y)^{a+b-(m+1)/2}=\int_{0<V<Y}(\det V)^{a-(m+1)/2}$$

$$\cdot\det(Y-V)^{b-(m+1)/2}C_\kappa(V)(dV).$$

Now take Laplace transforms of both sides using the result of Problem 7.3 to evaluate the transform of the right side, and solve the resulting equation for $f(I)/C_\kappa(I)$.]

7.5. Prove Lemma 7.2.12.
[*Hint*: Note that if A^{-1} has latent roots $\alpha_1,\ldots,\alpha_m$ then

$$\begin{aligned}C_\kappa(A^{-1})&=d_\kappa\alpha_1^{k_1}\cdots\alpha_m^{k_m}+\text{terms of lower weight}\\&=d_\kappa(\alpha_1\alpha_2\cdots\alpha_m)^{k_1}(\alpha_2\cdots\alpha_m)^{k_2-k_1}\ldots\alpha_m^{k_m-k_{m-1}}+\cdots\\&=d_\kappa r_m^{k_1}r_{m-1}^{k_2-k_1}\cdots r_1^{k_m-k_{m-1}}+\cdots,\end{aligned}$$

where r_k is the kth elementary symmetric function of $\alpha_1,\ldots,\alpha_m$. Now use the fact that $r_j=\det A^{-1}\operatorname{tr}_{m-j}(A)$; see (xiv) of Section A7.]

7.6. Prove Theorem 7.4.3.

7.7. Show that

$${}_0F_0^{(m)}(I+X,Y)=\operatorname{etr}(Y)\,{}_0F_0^{(m)}(X,Y)$$

7.8. Suppose that $H_1\in V_{k,m}$, i.e., H_1 is $m\times k$ with $H_1'H_1=I_k$. Let (dH_1) be the normalized invariant measure on $V_{k,m}$, so that $\int_{V_{k,m}}(dH_1)=1$. If X is an $m\times m$ positive definite matrix prove that

$$\int_{V_{k,m}}\operatorname{etr}(XH_1H_1')(dH_1)={}_1F_1(\tfrac{1}{2}k;\tfrac{1}{2}m;X).$$

7.9. Prove (13) of Section 7.5.

7.10. Verify (39) of Section 7.5.

7.11. Calculate all the generalized binomial coefficients $\binom{\kappa}{\sigma}$ for partitions κ of $k=1,2,3$.

7.12. Prove that

$$\sum_{\sigma}\binom{\kappa}{\sigma}=\binom{k}{s},$$

where κ is a partition of k and the summation is over all partitions σ of s.

7.13. Prove that

$$\sum_{i}\binom{\sigma_i}{\sigma}\binom{\kappa}{\sigma_i}=(k-s)\binom{\kappa}{\sigma}\qquad(s<k).$$

[*Hint*: Start with $\sum_{i=1}^{m}x_i\,\partial/\partial x_i\,C_\kappa(X)=kC_\kappa(X)$, put $X=(I+Y)$, substitute the generalized binomial expansion of $C_\kappa(I+Y)$, and equate coefficients of $C_\kappa(Y)/C_\kappa(I)$.]

7.14. Prove that

$$C_\kappa(Y)\operatorname{etr}(Y)=k!\sum_{s=k}^{\infty}\sum_{\sigma}\binom{\sigma}{\kappa}\frac{C_\sigma(Y)}{s!}.$$

[*Hint*: Use the result of Problem 7.7.]

7.15. Prove that

$$\sum_{s=0}^{k}(-1)^s\sum_{\sigma}\binom{\kappa}{\sigma}\frac{(a)_\sigma}{(b)_\sigma}=\frac{(b-a)_\kappa}{(b)_\kappa}.$$

[*Hint*: Use the Kummer relation of Theorem 7.4.3 and the result of Problem 7.14.]

7.16. It is sometimes useful to express a product of two zonal polynomials in terms of other zonal polynomials (see, e.g., the proof of Lemma 10.6.1). Define constants $g_{\sigma\tau}^{\kappa}$ by

$$C_\sigma(Y)C_\tau(Y)=\sum_{\kappa}g_{\sigma\tau}^{\kappa}C_\kappa(Y)$$

where σ is a partition of s, τ is a partition of t, and κ runs over all partitions of $k=s+t$.

(a) Find all constants g for partitions of $s=2$, $t=1$.

(b) Prove that

$$\binom{k}{t}\sum_{\tau}g_{\sigma\tau}^{\kappa}=\binom{\kappa}{\sigma}.$$

[*Hint*: Use the result of Problem 7.7, expand $\operatorname{etr}(Y)$ as a zonal polynomial series, and equate coefficients of $C_\kappa(Y)$ on both sides.]

7.17. If $Y=\operatorname{diag}(y_1, y_2)$ prove that

$$ {}_2F_1(a,b;c;Y)=\sum_{k=0}^{\infty}\frac{(a)_k(c-a)_k(b)_k(c-b)_k}{(c)_{2k}(c-1/2)_k k!}(y_1y_2)^k $$

$$ \cdot {}_2F_1(a+k,b+k;c+2k;y_1+y_2-y_1y_2). $$

[*Hint*: Show that the right side satisfies the partial differential equations of Theorem 7.5.6.]

Using the confluence relations given by (8) and (9) of Section 7.4, obtain similar expressions for ${}_1F_1(a;c;Y)$ and ${}_0F_1(c;Y)$.

7.18. Prove Theorem 7.6.4.

7.19. Prove that

$$ \operatorname{etr}(Z)\,{}_0F_1^{(m)}(\gamma+p;X,-Z)=\sum_{k=0}^{\infty}\sum_{\kappa}\frac{L_\kappa^\gamma(X)C_\kappa(Z)}{(\gamma+p)_\kappa k!C_\kappa(I_m)} $$

for $X>0$, $Z>0$, $\gamma>-1$, $p=\frac{1}{2}(m+1)$.

7.20. Prove that

$$ \det(I-Z)^{-c}\,{}_1F_1^{(m)}\big(a;\gamma+p;X,-Z(I-Z)^{-1}\big) $$

$$ =\sum_{k=0}^{\infty}\sum_{\kappa}\frac{(a)_\kappa L_\kappa^\gamma(X)C_\kappa(Z)}{(\gamma+p)_\kappa k!C_\kappa(I_m)} $$

for $X>0$, $\|Z\|<1$, $\gamma>-1$, $p=\frac{1}{2}(m+1)$.

7.21. Prove that

$$ (1-t)^{-m(\gamma+p)}\operatorname{etr}\left[-\frac{t}{1-t}(X+Z)\right]{}_0F_1^{(m)}\left(\gamma+p;\frac{t}{(1-t)^2}X,Z\right) $$

$$ =\sum_{k=0}^{\infty}\sum_{\kappa}\frac{L_\kappa^\gamma(X)L_\kappa^\gamma(Z)t^k}{(\gamma+p)_\kappa k!C_\kappa(I_m)} $$

for $X>0$, $Z>0$, $\gamma>-1$, $p=\frac{1}{2}(m+1)$, $|t|<1$.

CHAPTER 8

Some Standard Tests on Covariance Matrices and Mean Vectors

8.1. INTRODUCTION

In this chapter we examine some standard likelihood ratio tests about the parameters of multivariate normal distributions. The null hypotheses considered in this chapter are

$$
\begin{aligned}
&H: \Sigma_1 = \Sigma_2 = \cdots = \Sigma_r \qquad &&\text{(Section 8.2)},\\
&H: \Sigma = \lambda I_m &&\text{(Section 8.3)},\\
&H: \Sigma = \Sigma_0 &&\text{(Section 8.4)}, \quad \text{and}\\
&H: \Sigma = \Sigma_0,\ \boldsymbol{\mu} = \boldsymbol{\mu}_0 &&\text{(Section 8.5)}.
\end{aligned}
$$

In each instance the likelihood ratio test is derived and invariance and unbiasedness properties are established. Moments of the test statistics are obtained and used to find asymptotic null and non-null distributions. The likelihood ratio test statistics are also compared briefly with other possible test statistics.

There are a number of other null hypotheses of interest about mean vectors and covariance matrices. Some of these will be treated later. These include testing equality of p mean vectors (Section 10.7), testing equality of p normal populations (Section 10.8), and testing independence of k sets of variables (Section 11.2).

8.2. TESTING EQUALITY OF r COVARIANCE MATRICES

8.2.1. The Likelihood Ratio Statistic and Invariance

In this section we consider testing the null hypothesis that the covariance matrices of r normal distributions are equal, given independent samples from these r populations. Let $\mathbf{X}_{i1},\ldots,\mathbf{X}_{iN_i}$ be independent $N_m(\boldsymbol{\mu}_i,\Sigma_i)$ random vectors ($i=1,\ldots,r$) and consider testing the null hypothesis

$$H:\Sigma_1=\cdots=\Sigma_r$$

against the alternative K which says that H is not true. In H the common covariance matrix is unspecified, as are the mean vectors. The assumption of equal covariance matrices is important in multivariate analysis of variance and discriminant analysis, as we shall see in Chapter 10. Let $\overline{\mathbf{X}}_i$ and A_i be, respectively, the mean vector and the matrix of sums of squares and products formed from the ith sample; that is,

$$\overline{\mathbf{X}}_i=\frac{1}{N_i}\sum_{j=1}^{N_i}\mathbf{X}_{ij},\qquad A_i=\sum_{j=1}^{N_i}(\mathbf{X}_{ij}-\overline{\mathbf{X}}_i)(\mathbf{X}_{ij}-\overline{\mathbf{X}}_i)'\qquad(i=1,\ldots,r),$$

and put

$$A=A_1+\cdots+A_r,\qquad N=N_1+\cdots+N_r.$$

The likelihood ratio test of H, first derived by Wilks (1932), is given in the following theorem.

THEOREM 8.2.1. The likelihood ratio test of size α of the null hypothesis $H:\Sigma_1=\cdots=\Sigma_r=\Sigma$, with Σ unspecified, rejects H if $\Lambda\le c_\alpha$, where

$$(1)\qquad \Lambda=\frac{\prod_{i=1}^r(\det A_i)^{N_i/2}}{(\det A)^{N/2}}\cdot\frac{N^{mN/2}}{\prod_{i=1}^r N_i^{mN_i/2}}$$

and c_α is chosen so that the size of the test is α.

Proof. Apart from a multiplicative constant the likelihood function based on the r independent samples is [see, for example, (8) of Section 3.1],

$$(2)\quad L(\boldsymbol{\mu}_1,\ldots,\boldsymbol{\mu}_r,\Sigma_1,\ldots,\Sigma_r)=\prod_{i=1}^r L_i(\boldsymbol{\mu}_i,\Sigma_i)$$

$$=\prod_{i=1}^r\Big\{(\det\Sigma_i)^{-N_i/2}\operatorname{etr}(-\tfrac{1}{2}\Sigma_i^{-1}A_i)$$

$$\cdot\exp\big[-\tfrac{1}{2}N_i(\overline{\mathbf{X}}_i-\boldsymbol{\mu}_i)'\Sigma_i^{-1}(\overline{\mathbf{X}}_i-\boldsymbol{\mu}_i)\big]\Big\}.$$

The likelihood ratio statistic is

$$(3)\qquad \Lambda=\frac{\sup_{\mu_1,\ldots,\mu_r,\Sigma} L(\mu_1,\ldots,\mu_r,\Sigma,\ldots,\Sigma)}{\sup_{\mu_1,\ldots,\mu_r,\Sigma_1,\ldots,\Sigma_r} L(\mu_1,\ldots,\mu_r,\Sigma_1,\ldots,\Sigma_r)}.$$

When the parameters are unrestricted, maximizing the likelihood is equivalent to maximizing each of the likelihoods in the product (2) and hence the denominator in (3) is

$$(4)\qquad L(\overline{\mathbf{X}}_1,\ldots,\overline{\mathbf{X}}_r,\hat{\Sigma}_1,\ldots,\hat{\Sigma}_r)=e^{-mN/2}\prod_{i=1}^{r}\left(N_i^{-mN_i/2}(\det A_i)^{-N_i/2}\right),$$

where $\hat{\Sigma}_i=N_i^{-1}A_i$. When the null hypothesis H: $\Sigma_1=\cdots=\Sigma_r=\Sigma$ is true, the likelihood function is, from (2)

$$L(\mu_1,\ldots,\mu_r,\Sigma,\ldots,\Sigma)=(\det\Sigma)^{-N/2}\operatorname{etr}(-\tfrac{1}{2}\Sigma^{-1}A)\cdot\prod_{i=1}^{r}\exp[-\tfrac{1}{2}N_i(\overline{\mathbf{X}}_i-\mu_i)'\Sigma^{-1}(\overline{\mathbf{X}}_i-\mu_i)],$$

which is maximized when $\mu_i=\overline{\mathbf{X}}_i$ and $\Sigma=N^{-1}A$. Hence the numerator in (3) is

$$(5)\qquad L(\overline{\mathbf{X}}_1,\ldots,\overline{\mathbf{X}}_r,N^{-1}A,\ldots,N^{-1}A)=N^{mN/2}e^{-mN/2}(\det A)^{-N/2}.$$

Using (4) and (5) in (3) then gives

$$\Lambda=\frac{\prod_{i=1}^{r}(\det A_i)^{N_i/2}}{(\det A)^{N/2}}\cdot\frac{N^{mN/2}}{\prod_{i=1}^{r}N_i^{mN_i/2}}$$

and the likelihood ratio test rejects H for small values of Λ, completing the proof.

We now look at the problem of testing equality of covariance matrices from an invariance point of view. Because it is somewhat simpler we will concentrate here on the case $r=2$, where we are testing H: $\Sigma_1=\Sigma_2$ against K: $\Sigma_1\neq\Sigma_2$. A sufficient statistic is $(\overline{\mathbf{X}}_1,\overline{\mathbf{X}}_2,A_1,A_2)$. Consider the group of nonsingular transformations

$$(6)\qquad G=\{(B,\mathbf{c},\mathbf{d});\ B\in\mathcal{Gl}(m,R),\ \mathbf{c}\in R^m,\ \mathbf{d}\in R^m\}$$

acting on the space $R^m \times R^m \times \mathbb{S}_m \times \mathbb{S}_m$ of points $(\overline{\mathbf{X}}_1, \overline{\mathbf{X}}_2, A_1, A_2)$ by

$$(B,\mathbf{c},\mathbf{d})(\overline{\mathbf{X}}_1, \overline{\mathbf{X}}_2, A_1, A_2) = (B\overline{\mathbf{X}}_1 + \mathbf{c},\, B\overline{\mathbf{X}}_2 + \mathbf{d},\, BA_1B',\, BA_2B'),$$

where the group operation is

$$(B_1,\mathbf{c}_1,\mathbf{d}_1)(B_2,\mathbf{c}_2,\mathbf{d}_2) = (B_1B_2,\, B_1\mathbf{c}_2 + \mathbf{c}_1,\, B_1\mathbf{d}_2 + \mathbf{d}_1).$$

The corresponding induced group of transformations (also G) on the parameter space of points $(\boldsymbol{\mu}_1, \boldsymbol{\mu}_2, \Sigma_1, \Sigma_2)$ is given by

$$(7) \qquad (B,\mathbf{c},\mathbf{d})(\boldsymbol{\mu}_1, \boldsymbol{\mu}_2, \Sigma_1, \Sigma_2) = (B\boldsymbol{\mu}_1 + \mathbf{c},\, B\boldsymbol{\mu}_2 + \mathbf{d},\, B\Sigma_1B',\, B\Sigma_2B')$$

and the testing problem is invariant under G, for the family of distributions of $(\overline{\mathbf{X}}_1, \overline{\mathbf{X}}_2, A_1, A_2)$ is invariant, as are the null and alternative hypotheses. Our next problem is to find a maximal invariant.

THEOREM 8.2.2. Under the group G of transformations (7) a maximal invariant is $(\delta_1,\dots,\delta_m)$, where $\delta_1 \geq \delta_2 \geq \cdots \geq \delta_m (>0)$ are the latent roots of $\Sigma_1\Sigma_2^{-1}$.

Proof. Let

$$\phi(\boldsymbol{\mu}_1, \boldsymbol{\mu}_2, \Sigma_1, \Sigma_2) = (\delta_1,\dots,\delta_m).$$

First note that ϕ is invariant, for the latent roots of

$$(B\Sigma_1B')(B\Sigma_2B')^{-1} = B\Sigma_1\Sigma_2^{-1}B^{-1}$$

are the same as those of $\Sigma_1\Sigma_2^{-1}$. To show it is maximal invariant suppose that

$$\phi(\boldsymbol{\mu}_1, \boldsymbol{\mu}_2, \Sigma_1, \Sigma_2) = \phi(\boldsymbol{\tau}_1, \boldsymbol{\tau}_2, \Gamma_1, \Gamma_2);$$

that is, $\Sigma_1\Sigma_2^{-1}$ and $\Gamma_1\Gamma_2^{-1}$ have the same latent roots $(\delta_1,\dots,\delta_m)$. By Theorem A9.9 there exist nonsingular matrices B_1 and B_2 such that

$$B_1\Sigma_1B_1' = \Delta, \qquad B_1\Sigma_2B_1' = I_m,$$
$$B_2\Gamma_1B_2' = \Delta, \qquad B_2\Gamma_2B_2' = I_m,$$

where

$$\Delta = \operatorname{diag}(\delta_1,\dots,\delta_m).$$

Then

$$\Gamma_1 = B_2^{-1}\Delta B_2'^{-1} = B_2^{-1}B_1\Sigma_1 B_1' B_2'^{-1} = B\Sigma_1 B'$$

and

$$\Gamma_2 = B_2^{-1}B_2'^{-1} = B_2^{-1}B_1\Sigma_2 B_1' B_2'^{-1} = B\Sigma_2 B',$$

where

$$B = B_2^{-1}B_1.$$

Putting $\mathbf{c} = -B\boldsymbol{\mu}_1 + \boldsymbol{\tau}_1$ and

$$\mathbf{d} = -B\boldsymbol{\mu}_2 + \boldsymbol{\tau}_2$$

we then have

$$(B,\mathbf{c},\mathbf{d})(\boldsymbol{\mu}_1,\boldsymbol{\mu}_2,\Sigma_1,\Sigma_2) = (\boldsymbol{\tau}_1,\boldsymbol{\tau}_2,\Gamma_1,\Gamma_2)$$

so that

$$(\boldsymbol{\mu}_1,\boldsymbol{\mu}_2,\Sigma_1,\Sigma_2) \sim (\boldsymbol{\tau}_1,\boldsymbol{\tau}_2,\Gamma_1,\Gamma_2) \qquad (\text{mod}\, G).$$

Hence $(\delta_1,\ldots,\delta_m)$ is a maximal invariant, and the proof is complete.

As a consequence of this theorem a maximal invariant under the group G acting on the sample space of the sufficient statistic $(\overline{\mathbf{X}}_1,\overline{\mathbf{X}}_2,A_1,A_2)$ is $(f_1,\ldots,f_m)$, where $f_1 \geq f_2 \geq \cdots \geq f_m (>0)$ are the latent roots of $A_1A_2^{-1}$. Any invariant test depends only on $f_1,\ldots,f_m$ and, from Theorem 6.1.12, the distribution of $f_1,\ldots,f_m$ depends only on $\delta_1,\ldots,\delta_m$, the latent roots of $\Sigma_1\Sigma_2^{-1}$. This distribution will be given explicitly in Theorem 8.2.8. Note that the likelihood ratio test is invariant, for

$$(8) \qquad \Lambda = \frac{N^{mN/2}}{N_1^{mN_1/2}N_2^{mN_2/2}} \frac{(\det A_1)^{N_1/2}(\det A_2)^{N_2/2}}{\det(A_1+A_2)^{(N_1+N_2)/2}}$$

$$= \frac{N^{mN/2}}{N_1^{mN_1/2}N_2^{mN_2/2}} \frac{\det(A_1A_2^{-1})^{N_1/2}}{\det(I+A_1A_2^{-1})^{(N_1+N_2)/2}}$$

$$= \frac{N^{mN/2}}{N_1^{mN_1/2}N_2^{mN_2/2}} \prod_{i=1}^{m} \frac{f_i^{N_1/2}}{(1+f_i)^{(N_1+N_2)/2}},$$

so that Λ is a function of $f_1,\ldots,f_m$. In terms of the latent roots of $\Sigma_1\Sigma_2^{-1}$ the null hypothesis is equivalent to

$$H: \delta_1 = \cdots = \delta_m = 1.$$

There is no uniformly most powerful invariant test and many other functions of $f_1,\ldots,f_m$ in addition to Λ have been proposed as test statistics. Some of these will be discussed in Section 8.2.8. For the most part, however, we will concentrate on the likelihood ratio approach.

8.2.2. *Unbiasedness and the Modified Likelihood Ratio Test*

The likelihood ratio test of Theorem 8.2.1 has the defect that, when the sample sizes $N_1,\ldots,N_r$ are not all equal, it is biased; that is, the probability of rejecting H when H is false can be smaller than the probability of rejecting H when H is true. This was first noted by Brown (1939) when $m=1$ (in which case the equality of r normal variances is being tested). We will establish the biasedness for general m using an argument due to Das Gupta (1969) for the case of $r=2$ populations. This involves the use of the following lemma.

LEMMA 8.2.3. Let Y be a random variable and $\delta(>0)$ a constant such that $\delta(N_2-1)Y/(N_1-1)$ has the F_{N_1-1,N_2-1} distribution, with $N_1<N_2$, and let

$$\beta(\delta) = P\left[\frac{Y^{N_1}}{(1+Y)^{N_1+N_2}} \geq k \,|\, \delta\right]$$

Then there exists a constant λ $(\lambda<1)$ independent of k such that

$$\beta(\delta) > \beta(1) \qquad \text{for all} \quad \delta \in (\lambda, 1).$$

Proof. Since the region

$$\frac{Y^{N_1}}{(1+Y)^{N_1+N_2}} \geq k$$

is equivalent to $y_2 \leq Y \leq y_1$, where

$$(9) \qquad \left(\frac{y_1}{y_2}\right)^{N_1} = \left(\frac{1+y_1}{1+y_2}\right)^{N_1+N_2},$$

it follows by integration of the F_{N_1-1,N_2-1} density function that

$$\beta(\delta)=C\int_{\delta y_2}^{\delta y_1}\frac{y^{(N_1-3)/2}}{(1+y)^{(N_1+N_2)/2-1}}\,dy,$$

where

$$C=\frac{\Gamma[\frac{1}{2}(N_1+N_2)-1]}{\Gamma[\frac{1}{2}(N_1-1)]\Gamma[\frac{1}{2}(N_2-1)]}.$$

Differentiating with respect to δ gives

$$\beta'(\delta)=C\delta^{(N_1-3)/2}\left[\frac{y_1^{(N_1-1)/2}}{(1+\delta y_1)^{(N_1+N_2)/2-1}}-\frac{y_2^{(N_1-1)/2}}{(1+\delta y_2)^{(N_1+N_2)/2-1}}\right].$$

It then follows that $\beta'(\delta)\gtreqless 0$ according as

$$\left(\frac{y_1}{y_2}\right)^{(N_1-1)/2}\gtreqless\left(\frac{1+\delta y_1}{1+\delta y_2}\right)^{(N_1+N_2)/2-1}.$$

Using (9) we then have that $\beta'(\delta)\gtreqless 0$ according as

(10) $$\left(\frac{1+y_1}{1+y_2}\right)^{\lambda}\gtreqless\frac{1+\delta y_1}{1+\delta y_2},$$

where

(11) $$\lambda=\frac{(N_1+N_2)(N_1-1)}{N_1(N_1+N_2-2)}$$

and $\lambda<1$. It now follows from (10) that there exists δ_0 such that $\beta'(\delta)\gtreqless 0$ according as $\delta\lesseqgtr\delta_0$, where $\delta_0<1$. Now, since the function $g(x)=(1+\lambda x)/(1+x)^{\lambda}$ is increasing in x we have $g(y_1)>g(y_2)$; from (10) this implies that $\beta'(\lambda)<0$. Hence $\delta_0<\lambda<1$. Consequently $\beta(\delta)>\beta(1)$ for all $\delta\in(\lambda,1)$, where λ does not depend on k.

We are now in a position to demonstrate that the likelihood ratio test for testing equality of two covariance matrices is biased. First note that rejecting H for small values of Λ given by (8) is equivalent to rejecting H for small values of

$$V=\frac{\Lambda^2 N_1^{mN_1}N_2^{mN_2}}{N^{mN}}=\frac{(\det A_1)^{N_1}(\det A_2)^{N_2}}{\det(A_1+A_2)^{N_1+N_2}}.$$

THEOREM 8.2.4. For testing $H:\Sigma_1=\Sigma_2$ against $K:\Sigma_1\neq\Sigma_2$ the likelihood ratio test having the critical region $V\leq k$ is biased.

Proof. Using an obvious invariance argument we can assume without loss of generality for power calculations that $\Sigma_2=I_m$ and $\Sigma_1=\Delta$, where Δ is diagonal. In particular, take

$$\Delta=\text{diag}(\delta,1,\dots,1).$$

Let $A_1=(a_{ij}^{(1)})$ and $A_2=(a_{ij}^{(2)})$, and define the random variable Z by

$$V=\frac{\left(a_{11}^{(1)}\right)^{N_1}\left(a_{11}^{(2)}\right)^{N_2}}{\left(a_{11}^{(1)}+a_{11}^{(2)}\right)^{N_1+N_2}}\cdot Z. \tag{12}$$

Then Z is independent of the first factor on the right side of (12), and its distribution does not depend on δ (use Theorem 3.2.10). Putting $Y=a_{11}^{(1)}/a_{11}^{(2)}$ so that $\delta^{-1}(N_2-1)Y/(N_1-1)$ is F_{N_1-1,N_2-1}, the first factor on the right side of (12) is $Y^{N_1}/(1+Y)^{N_1+N_2}$. Lemma 8.2.3 then shows that the power of the likelihood ratio test is less than its size if $\delta^{-1}\in(\lambda,1)$, where λ is given by (11), and the proof is complete.

Although unbiasedness is in no sense an optimal property, it is certainly a desirable one. It turns out that by modifying the likelihood ratio statistic slightly an unbiased test can be obtained. The modified likelihood ratio statistic, suggested by Bartlett (1937), is defined to be

$$\Lambda^*=\frac{\prod_{i=1}^r(\det A_i)^{n_i/2}}{(\det A)^{n/2}}\frac{n^{mn/2}}{\prod_{i=1}^r n_i^{mn_i/2}}, \tag{13}$$

where $n_i=N_i-1$ and $n=\sum_{i=1}^r n_i=N-r$. Note that Λ^* is obtained from Λ by replacing the sample sizes N_i by the corresponding degrees of freedom n_i. This is exactly the likelihood ratio statistic that is obtained by working with the likelihood function of $\Sigma_1,\dots,\Sigma_r$ specified by the joint marginal density function of $A_1,\dots,A_r$ (a product of Wishart densities), rather than the likelihood function specified by the original normally distributed variables. The modified likelihood ratio test then rejects $H:\Sigma_1=\cdots=\Sigma_r$ for small enough values of Λ^*. The unbiasedness of this test was established in the univariate case $m=1$ by Pitman (1939). If, in addition, $r=2$, this test is a uniformly most powerful unbiased test. The unbiasedness for general m and r was proved by Perlman (1980). Although his elegant proof is too lengthy to reproduce here, we will establish the unbiasedness for the case $r=2$ using an argument due to Sugiura and Nagao (1968).

THEOREM 8.2.5. For testing H: $\Sigma_1=\Sigma_2$ against K: $\Sigma_1\neq\Sigma_2$ the modified likelihood ratio test having the critical region

$$C=\left\{(A_1,A_2);\,A_1>0,A_2>0,\ \frac{(\det A_1)^{n_1/2}(\det A_2)^{n_2/2}}{\det(A_1+A_2)^{n/2}}\leq k_\alpha\right\}$$

is unbiased.

Proof. Using invariance we can assume without loss of generality that $\Sigma_2=I_m$ and $\Sigma_1=\Delta$, where $\Delta=\mathrm{diag}(\delta_1,\dots,\delta_m)$. The probability of the rejection region under K is

$$P_K(C)=c_{m,n_1}c_{m,n_2}\int_{(A_1,A_2)\in C}\mathrm{etr}\left[-\tfrac{1}{2}(\Delta^{-1}A_1+A_2)\right]$$

$$\cdot(\det A_1)^{(n_1-m-1)/2}(\det A_2)^{(n_2-m-1)/2}(dA_1)(dA_2)(\det\Delta)^{-n_1/2}$$

where

$$c_{m,n}=\left[2^{mn/2}\Gamma_m(\tfrac{1}{2}n)\right]^{-1}.$$

Now make the transformation $A_1=U_1$, $A_2=U_1^{1/2}U_2U_1^{1/2}$, where $U_1^{1/2}$ is the positive definite symmetric square root of U_1, so that $U_1^{1/2}U_1^{1/2}=U_1$. Then

$$(dA_1)(dA_2)=(\det U_1)^{(m+1)/2}(dU_1)(dU_2),$$

and

$$P_K(C)=c_{m,n_1}c_{m,n_2}\det\Delta^{-n_1/2}\int_{U_1>0}\int_{(I,U_2)\in C}\mathrm{etr}\left[-\tfrac{1}{2}(\Delta^{-1}+U_2)U_1\right]$$

$$\cdot(\det U_1)^{(n-m-1)/2}(\det U_2)^{(n_2-m-1)/2}(dU_2)(dU_1)$$

$$=\frac{c_{m,n_1}c_{m,n_2}}{c_{m,n_1+n_2}}(\det\Delta)^{-n_1/2}\int_{(I,U_2)\in C}(\det U_2)^{(n_2-m-1)/2}$$

$$\cdot\det(\Delta^{-1}+U_2)^{-n/2}(dU_2),$$

using Theorem 2.1.9. Now put $U_2=\Delta^{-1/2}V\Delta^{-1/2}$, with $(dU_2)=(\det\Delta)^{-(m+1)/2}(dV)$, so that

$$P_K(C)=d\int_{V\in C_1}(\det V)^{(n_2-m-1)/2}\cdot\det(I+V)^{-n/2}(dV),$$

where

$$C_1=\{V;\ V>0,\ (I,\Delta^{-1/2}V\Delta^{-1/2})\in C\}\quad\text{and}\quad d=c_{m,n_1}c_{m,n_2}c_{m,n_1+n_2}^{-1}.$$

Putting $C_2=\{V;\ V>0,\ (I,V)\in C\}$, so that $C_1=C_2$ when H is true (i.e., when $\Delta=I$), it then follows that

$$\begin{aligned}P_K(C)-P_H(C)&=d\left\{\int_{V\in C_1}-\int_{V\in C_2}\right\}(\det V)^{(n_2-m-1)/2}\det(I+V)^{-n/2}(dV)\\&=d\left\{\int_{V\in C_1-C_1\cap C_2}-\int_{V\in C_2-C_1\cap C_2}\right\}\\&\qquad\cdot(\det V)^{(n_2-m-1)/2}\det(I+V)^{-n/2}(dV).\end{aligned}$$

Now, for $V\in C_2-C_1\cap C_2$ we have

$$(\det V)^{(n_2-m-1)/2}\det(I+V)^{-n/2}\le k_\alpha(\det V)^{-(m+1)/2},$$

and for $V\in C_1-C_1\cap C_2$

$$(\det V)^{(n_2-m-1)/2}\det(I+V)^{-n/2}>k_\alpha(\det V)^{-(m+1)/2},$$

and hence

$$\begin{aligned}P_K(C)-P_H(C)&\ge dk_\alpha\left\{\int_{V\in C_1-C_1\cap C_2}-\int_{V\in C_2-C_1\cap C_2}\right\}\\&\qquad\cdot(\det V)^{-(m+1)/2}(dV)\\&=dk_\alpha\left\{\int_{V\in C_1}-\int_{V\in C_2}\right\}\\&\qquad\cdot(\det V)^{-(m+1)/2}(dV)\\&=0,\end{aligned}$$

since

$$\int_{V\in C_1}(\det V)^{-(m+1)/2}(dV)=\int_{V\in C_2}(\det V)^{-(m+1)/2}(dV);$$

this is easily proved by making the transformation $W=\Delta^{-1/2}V\Delta^{-1/2}$ in the integral on the left. We have used the fact that

$$\int_{V\in C_1}(\det V)^{(n_2-m-1)/2}\det(I+V)^{-n/2}(dV)<\infty$$

[because this integral is bounded above by $d^{-1}P_K(C)$], from which it follows that for any subset C^* of C_1

$$\infty>\int_{V\in C^*}(\det V)^{(n_2-m-1)/2}\det(I+V)^{-n/2}(dV)$$

$$\geq k_\alpha\int_{V\in C^*}(\det V)^{-(m+1)/2}(dV);$$

this implies that

$$\int_{V\in C_1}(\det V)^{-(m+1)/2}(dV)<\infty,$$

which has been used implicitly above. We have thus shown that $P_K(C)\geq P_H(C)$ and the proof is complete.

8.2.3. *Central Moments of the Modified Likelihood Ratio Statistic*

Information about the distribution of the modified likelihood ratio statistic Λ^* can be obtained from a study of its moments. In this section we find the moments for general r when the null hypothesis H: $\Sigma_1=\cdots=\Sigma_r$ is true. For notational convenience, define the statistic

$$W=\Lambda^*\frac{\prod_{i=1}^r n_i^{mn_i/2}}{n^{mn/2}} \tag{14}$$

$$=\frac{\prod_{i=1}^r(\det A_i)^{n_i/2}}{(\det A)^{n/2}},$$

where $n=\sum_{i=1}^r n_i$. The moments of W are given in the following theorem.

THEOREM 8.2.6. When $H: \Sigma_1 = \cdots = \Sigma_r$ is true, the hth moment of W is

$$E(W^h) = \frac{\Gamma_m(\frac{1}{2}n)}{\Gamma_m[\frac{1}{2}n(1+h)]} \prod_{i=1}^{r} \frac{\Gamma_m[\frac{1}{2}n_i(1+h)]}{\Gamma_m(\frac{1}{2}n_i)} \tag{15}$$

Proof. Let Σ denote the common covariance matrix, so that the A_i are independent $W_m(n_i, \Sigma)$ matrices ($i=1,\dots,r$). There is no loss of generality in assuming that $\Sigma = I_m$, since W is invariant under the group of transformations $A_i \to BA_iB'$, where $B \in \mathcal{G}\ell(m, R)$. Hence,

$$E(W^h) = \left(\prod_{i=1}^{r} c_{m,n_i}\right) \int_{A_1>0} \cdots \int_{A_r>0} (\det A)^{-nh/2}$$

$$\cdot \prod_{i=1}^{r} \left\{ \operatorname{etr}(-\tfrac{1}{2}A_i)(\det A_i)^{n_i(1+h)/2-(m+1)/2} \right\} (dA_r) \cdots (dA_1),$$

where

$$c_{m,n} = \left[2^{mn/2}\Gamma_m(n/2)\right]^{-1}. \tag{16}$$

Consequently,

$$E(W^h) = \prod_{i=1}^{r} \frac{c_{m,n_i}}{c_{m,n_i(1+h)}} \cdot E\left[(\det A)^{-nh/2}\right], \tag{17}$$

where $A = \sum_{i=1}^{r} A_i$ and the A_i have independent $W_m[n_i(1+h), I_m]$ distributions ($i=1,\dots,r$). Hence A is $W_m[n(1+h), I_m]$ so that, using (15) of Section 3.2,

$$E\left[(\det A)^{-nh/2}\right] = 2^{-mnh/2} \frac{\Gamma_m(\frac{1}{2}n)}{\Gamma_m[\frac{1}{2}n(1+h)]}.$$

Substituting back in (17) and using (16) then gives the desired result.

The moments of W may be used to obtain exact expressions for the distribution of W, and hence Λ^*. Briefly the approach used is as follows. The *Mellin transform* of a function $f(x)$, defined for $x>0$, is

$$M(s) = \int_0^{\infty} f(x)x^{s-1}\,dx,$$

a function of the complex variable s. The function $f(x)$ is called the *inverse Mellin transform* of $M(s)$; see, for example, Erdélyi et al. (1954), Chapter 6. If X is a positive random variable with density function $f(x)$, the Mellin transform $M(s)$ gives the $(s-1)$th moment of X. Hence Theorem 8.2.6 gives the Mellin transform of W evaluated at $s=h+1$; that is,

$$M(h+1)=E(W^h).$$

The inverse Mellin transform gives the density function of W. There is, of course, nothing special about the central moments here; given the noncentral moments (when H is not true) the noncentral distribution of W can be obtained using the inverse Mellin transform approach. It turns out that exact distributions of many of the likelihood ratio criteria that we will look at, including W, can be expressed via this method in terms of two types of special functions known as G and H functions. For work in this direction the interested reader is referred to Khatri and Srivastava (1971), Pillai and Nagarsenker (1972), Mathai and Saxena (1978), and to useful survey papers by Pillai (1976, 1977) and references therein. Although theoretically interesting, we will not go into this further because in general the exact density functions of likelihood ratio statistics in multivariate analysis are so complicated that they appear to be of limited usefulness. It should be mentioned that there are often some special cases for which the distributions are quite tractable. Rather than list these, however, we will concentrate on asymptotic distributions; these turn out to be simple and easy to use.

8.2.4. *The Asymptotic Null Distribution of the Modified Likelihood Ratio Statistic*

In this section we derive an asymptotic expansion for the distribution of Λ^* as all sample sizes increase. We put $n_i=k_i n$ $(i=1,\dots,r)$, where $\sum_{i=1}^r k_i=1$ and assume that $k_i>0$ and that $n\to\infty$. The general theory of likelihood ratio tests [see, e.g., Rao (1973), Chapter 6] shows that when the null hypothesis H: $\Sigma_1=\cdots=\Sigma_r$ is true, the asymptotic distribution as $n\to\infty$ of $-2\log\Lambda$ (and $-2\log\Lambda^*$) is χ_f^2, where

$f=$ number of independent parameters estimated in the full parameter space $-$ number of independent parameters estimated under the null hypothesis

$$=rm+\tfrac{1}{2}rm(m+1)-\left[rm+\tfrac{1}{2}m(m+1)\right]$$

$$=\tfrac{1}{2}m(m+1)(r-1).$$

It turns out, however, that in this situation and many others convergence to the asymptotic χ^2 distribution can be improved by considering a particular multiple of $-2\log\Lambda^*$. We will first outline the general theory and then specialize to the modified likelihood ratio statistic Λ^*. For a much more detailed treatment than the one given here, the reader should refer to Box (1949), to whom the theory is due, and to Anderson (1958), Section 8.6.

Consider a random variable Z $(0\le Z\le 1)$ with moments

$$E(Z^h)=K\left[\frac{\prod_{j=1}^p y_j^{y_j}}{\prod_{k=1}^q x_k^{x_k}}\right]^h\frac{\prod_{k=1}^q\Gamma[x_k(1+h)+\xi_k]}{\prod_{j=1}^p\Gamma[y_j(1+h)+\eta_j]}, \tag{18}$$

where

$$\sum_{j=1}^p y_j=\sum_{k=1}^q x_k \tag{19}$$

and K is a constant such that $E(Z^0)=1$. In our applications we will have $x_k=a_k n$, $y_j=b_j n$, where a_k and b_j are constant and n is the asymptotic variable, usually total sample size or a simple function of it. Hence we will write $O(n)$ for $O(x_k)$ and $O(y_j)$. Now, from (18), the characteristic function of $-2\rho\log Z$, where ρ $(0\le\rho\le1)$ for the moment is arbitrary, is

$$\begin{aligned}\phi(t)&=E[e^{-2it\rho\log Z}]\\&=E[Z^{-2it\rho}]\\&=K\left[\frac{\prod_{j=1}^p y_j^{y_j}}{\prod_{k=1}^q x_k^{x_k}}\right]^{-2it\rho}\frac{\prod_{k=1}^q\Gamma[x_k(1-2it\rho)+\xi_k]}{\prod_{j=1}^p\Gamma[y_j(1-2it\rho)+\eta_j]}.\end{aligned} \tag{20}$$

Putting

$$\beta_k=(1-\rho)x_k,\qquad \varepsilon_j=(1-\rho)y_j, \tag{21}$$

it then follows that the cumulant generating function $\Psi(t)$ of $-2\rho\log Z$ can be written as

$$\Psi(t)=\log\phi(t)=g(t)-g(0), \tag{22}$$

where

$$g(t)=2it\rho\left[\sum_{k=1}^{q} x_k\log x_k-\sum_{j=1}^{p} y_j\log y_j\right] \tag{23}$$

$$+\sum_{k=1}^{q}\log\Gamma[\rho x_k(1-2it)+\beta_k+\xi_k]$$

$$-\sum_{j=1}^{p}\log\Gamma[\rho y_j(1-2it)+\varepsilon_j+\eta_j]$$

and we have used the fact that $-g(0)=\log K$ because $\Psi(0)=0$. We now expand the log gamma functions in $\Psi(t)$ for large x_k and y_j. For this we need the following asymptotic expansion due to Barnes [see Erdélyi et al. (1953a), page 48]:

$$\log\Gamma(z+a)=(z+a-\tfrac{1}{2})\log z-z+\tfrac{1}{2}\log 2\pi \tag{24}$$

$$+\frac{B_2(a)}{1.2}z^{-1}+\cdots+(-1)^{l+1}\frac{B_{l+1}(a)}{l(l+1)}z^{-l}$$

$$+O(z^{-l-1})\qquad (l=1,2,3,\ldots,|\arg z|<\pi).$$

In (24), $B_j(a)$ is the Bernoulli polynomial of degree j, defined as the coefficient of $z^j/j!$ in the expansion of $ze^{az}(e^z-1)^{-1}$; that is,

$$ze^{az}(e^z-1)^{-1}=\sum_{j=0}^{\infty}B_j(a)\frac{z^j}{j!}\qquad (|z|<2\pi). \tag{25}$$

The first few Bernoulli polynomials are [see Erdélyi et al. (1953a), page 36]

$$B_0(a)=1,\quad B_1(a)=a-\tfrac{1}{2},\quad B_2(a)=a^2-a+\tfrac{1}{6}, \tag{26}$$

$$B_3(a)=a^3-\tfrac{3}{2}a^2+\tfrac{1}{2}a,\quad B_4(a)=a^4-2a^3+a^2-\tfrac{1}{30}.$$

Using (24) to expand the log gamma functions in $g(t)$ and $g(0)$ for large x_k and y_j we obtain, after some simplification

$$\Psi(t)=-\tfrac{1}{2}f\log(1-2it)+\sum_{\alpha=1}^{l}\omega_\alpha[(1-2it)^{-\alpha}-1]+O(n^{-l-1}), \tag{27}$$

where

(28) $$f=-2\left[\sum_{k=1}^{q}\xi_k-\sum_{j=1}^{p}\eta_j-\tfrac{1}{2}(q-p)\right]$$

and

(29) $$\omega_\alpha=\frac{(-1)^{\alpha+1}}{\alpha(\alpha+1)}\left[\sum_{k=1}^{q}\frac{B_{\alpha+1}(\beta_k+\xi_k)}{(\rho x_k)^\alpha}-\sum_{j=1}^{p}\frac{B_{\alpha+1}(\varepsilon_j+\eta_j)}{(\rho y_j)^\alpha}\right].$$

If $1-\rho$ is not small for large n it follows from (21) that β_k and ε_j are of order n; for (27) to be valid we take ρ to depend on n so that $1-\rho=O(n^{-1})$. If we take $\rho=1, l=1$ and use the fact that $\omega_1=O(n^{-1})$, (27) becomes

$$\Psi(t)=-\tfrac{1}{2}f\log(1-2it)+O(n^{-1}).$$

Exponentiating gives the characteristic function of $-2\log Z$ as

$$\phi(t)=(1-2it)^{-f/2}[1+O(n^{-1})].$$

Using the fact that $(1-2it)^{-f/2}$ is the characteristic function of the χ_f^2 distribution, it follows that $P(-2\log Z\le x)=P(\chi_f^2\le x)+O(n^{-1})$, so that the error or remainder term is of order n^{-1}. The point of allowing a value of ρ other than unity is that the term of order n^{-1} in the expansion for $P(-2\rho\log Z\le x)$ can be made to vanish. Taking $l=1$ in (27) and using the fact that $B_2(a)=a^2-a+\frac{1}{6}$ we have for the cumulant generating function of $-2\rho\log Z$,

$$\Psi(t)=-\tfrac{1}{2}f\log(1-2it)+\omega_1[(1-2it)^{-1}-1]+O(n^{-2}),$$

where f is given by (28) and

$$\omega_1=\frac{1}{2\rho}\left\{\sum_{k=1}^{q}\left[(\beta_k+\xi_k)^2-(\beta_k+\xi_k)+\frac{1}{6}\right]x_k^{-1}\right.$$

$$\left.-\sum_{j=1}^{p}\left[(\varepsilon_j+\eta_j)^2-(\varepsilon_j+\eta_j)+\tfrac{1}{6}\right]y_j^{-1}\right\}$$

$$=\frac{1}{2\rho}\left[-(1-\rho)f+\sum_{k=1}^{q}x_k^{-1}(\xi_k^2-\xi_k+\tfrac{1}{6})-\sum_{j=1}^{p}y_j^{-1}\left(\eta_j^2-\eta_j+\frac{1}{6}\right)\right],$$

on using (21). If we now choose the value of ρ to be

$$\text{(30)} \qquad \rho = 1 - \frac{1}{f}\left[\sum_{k=1}^{q} x_k^{-1}\left(\xi_k^2 - \xi_k + \frac{1}{6}\right) - \sum_{j=1}^{p} y_j^{-1}\left(\eta_j^2 - \eta_j + \frac{1}{6}\right)\right]$$

it then follows that $\omega_1 \equiv 0$. With this value of ρ we then have

$$\phi(t) = (1-2it)^{-f/2}\left[1 + O(n^{-2})\right],$$

and hence

$$\text{(31)} \qquad P(-2\rho\log Z \le x) = P\left(\chi_f^2 \le x\right) + O(n^{-2}).$$

This means that if the χ_f^2 distribution is used as an approximation to the distribution of $-2\rho\log Z$, the error involved is of order n^{-2}. If we also include the term of order n^{-2} we have, from (27) with $\omega_1 = 0$,

$$\Psi(t) = -\tfrac{1}{2}f\log(1-2it) + \omega_2\left[(1-2it)^{-2} - 1\right] + O(n^{-3}),$$

so that

(32)

$$P(-2\rho\log Z \le x) = P\left(\chi_f^2 \le x\right) + \omega_2\left[P\left(\chi_{f+4}^2 \le x\right) - P\left(\chi_f^2 \le x\right)\right] + O(n^{-3}).$$

We now return to the modified likelihood ratio statistic Λ^* for testing $H: \Sigma_1 = \cdots = \Sigma_r$. Putting $n_i = k_i n$, where $\sum_{i=1}^r k_i = 1$, it follows from Theorem 8.2.6 that

(33)

$$\begin{aligned} E[\Lambda^{*h}] &= \frac{n^{mnh/2}}{\prod_{i=1}^r n_i^{mn_ih/2}} E(W^h) \\ &= \frac{n^{mnh/2}}{\prod_{i=1}^r n_i^{mn_ih/2}} \frac{\Gamma_m(\frac{1}{2}n)}{\Gamma_m[\frac{1}{2}n(1+h)]} \prod_{i=1}^r \frac{\Gamma_m[\frac{1}{2}n_i(1+h)]}{\Gamma_m(\frac{1}{2}n_i)} \\ &= K\left[\frac{\prod_{j=1}^m(\frac{1}{2}n)^{n/2}}{\prod_{j=1}^m\prod_{i=1}^r(\frac{1}{2}nk_i)^{nk_i/2}}\right]^h \frac{\prod_{i=1}^r\prod_{j=1}^m\Gamma[\frac{1}{2}nk_i(1+h) - \frac{1}{2}(j-1)]}{\prod_{j=1}^m\Gamma[\frac{1}{2}n(1+h) - \frac{1}{2}(j-1)]}, \end{aligned}$$

where K is a constant not involving h. Comparing this with (18), we see that it has the same form with $p=m$; $q=rm$; $y_j=\frac{1}{2}n$; $\eta_j=-\frac{1}{2}(j-1)$; $x_k=\frac{1}{2}nk_i$, with $k=(i-1)m+1,\ldots,im$ $(i=1,\ldots,r)$; $\xi_k=-\frac{1}{2}(j-1)$, with $k=j$, $m+j,\ldots,(r-1)m+j$ $(j=1,\ldots,m)$.

The degrees of freedom in the limiting χ^2 distribution are, from (28),

$$(34)\qquad f=-2\left[\sum_{k=1}^{q}\xi_k-\sum_{j=1}^{p}\eta_j-\tfrac{1}{2}(q-p)\right]$$

$$=-2\left[\sum_{i=1}^{r}\sum_{j=1}^{m}\tfrac{1}{2}(1-j)-\sum_{j=1}^{m}\tfrac{1}{2}(1-j)-\tfrac{1}{2}m(r-1)\right]$$

$$=r\sum_{j=1}^{m}(j-1)-\sum_{j=1}^{m}(j-1)+m(r-1)$$

$$=(r-1)\left(\tfrac{1}{2}m(m+1)-m\right)+m(r-1)$$

$$=\tfrac{1}{2}m(m+1)(r-1),$$

as previously noted. The value of ρ which makes the term of order n^{-1} vanish in the asymptotic expansion of the distribution of $-2\rho\log\Lambda^*$ is, from (30),

$$(35)\qquad \rho=1-\frac{1}{f}\left[\sum_{k=1}^{q}x_k^{-1}\left(\xi_k^2-\xi_k+\frac{1}{6}\right)-\sum_{j=1}^{p}y_j^{-1}\left(\eta_j^2-\eta_j+\frac{1}{6}\right)\right]$$

$$=1-\frac{1}{f}\left\{\sum_{i=1}^{r}\sum_{j=1}^{m}\frac{2}{nk_i}\left[\frac{1}{4}(j-1)^2+\frac{1}{2}(j-1)+\frac{1}{6}\right]\right.$$

$$\left.-\sum_{j=1}^{m}\frac{2}{n}\left[\frac{1}{4}(j-1)^2+\frac{1}{2}(j-1)+\frac{1}{6}\right]\right\}$$

$$=1-\frac{2}{fn}\left\{\sum_{j=1}^{m}\left[\frac{1}{4}(j-1)^2+\frac{1}{2}(j-1)+\frac{1}{6}\right]\right\}\left(\sum_{i=1}^{r}\frac{1}{k_i}-1\right)$$

$$=1-\frac{2}{fn}\left[\sum_{j=1}^{m}\left(\frac{1}{4}j^2-\frac{1}{12}\right)\right]\left(\sum_{i=1}^{r}\frac{1}{k_i}-1\right)$$

$$=1-\frac{m}{12fn}(2m^2+3m-1)\left(\sum_{i=1}^{r}\frac{1}{k_i}-1\right)$$

$$=1-\frac{(2m^2+3m-1)}{6(m+1)(r-1)n}\left(\sum_{i=1}^{r}\frac{1}{k_i}-1\right).$$

With this value of ρ the term of order n^{-2} in the expansion is, from (32), (29) and (26)

$$\omega_2 = -\frac{1}{6\rho^2}\left\{\sum_{k=1}^{q} x_k^{-2}\left[(\beta_k+\xi_k)^3 - \frac{3}{2}(\beta_k+\xi_k)^2 + \frac{1}{2}(\beta_k+\xi_k)\right]\right.$$

$$\left. - \sum_{j=1}^{p} y_j^{-2}\left[(\varepsilon_j+\eta_j)^3 - \frac{3}{2}(\varepsilon_j+\eta_j)^2 + \frac{1}{2}(\varepsilon_j+\eta_j)\right]\right\},$$

which, after lengthy algebraic manipulation, reduces to

(36)

$$\omega_2 = \frac{m(m+1)}{48(n\rho)^2}\left\{(m-1)(m+2)\left(\sum_{i=1}^{r}\frac{1}{k_i^2}-1\right) - 6(r-1)[n(1-\rho)]^2\right\}.$$

Now define

$$M = \rho n = n - \frac{(2m^2+3m-1)}{6(m+1)(r-1)}\left(\sum_{i=1}^{r}\frac{1}{k_i}-1\right), \tag{37}$$

which we will use as the asymptotic variable, and put

(38)

$$\begin{aligned}\gamma &= M^2\omega_2 \\ &= \frac{m(m+1)}{48}\left\{(m-1)(m+2)\left(\sum_{i=1}^{r}\frac{1}{k_i^2}-1\right) - 6(r-1)[n(1-\rho)]^2\right\}.\end{aligned}$$

We now have obtained the following result.

THEOREM 8.2.7. When the null hypothesis $H: \Sigma_1 = \cdots = \Sigma_r$ is true the distribution of $-2\rho\log\Lambda^*$, where ρ is given by (35), can be expanded for large $M = \rho n$ as

$$P(-2\rho\log\Lambda^* \le x) = P(\chi_f^2 \le x) + \frac{\gamma}{M^2}\left[P(\chi_{f+4}^2 \le x) - P(\chi_f^2 \le x)\right] + O(M^{-3}) \tag{39}$$

where $f = m(m+1)(r-1)/2$ and γ is given by (38).

An approximate test of size α of H based on the modified likelihood ratio statistic is to reject H if $-2\rho\log\Lambda^* > c_f(\alpha)$, where $c_f(\alpha)$ denotes the upper $100\alpha\%$ point of the χ_f^2 distribution. The error in this approximation is of order n^{-2}. More accurate p-values can be obtained using the term of order n^{-2} from Theorem 8.2.7. For a detailed discussion of this and other approximations to the distribution of Λ^*, the interested reader is referred to

Table 3. Upper 5 percentage points of $-2\log\Lambda^*$, where Λ^* is the modified likelihood ratio statistic for testing equality of r covariance matrices (equal sample sizes)[a]

	$m=2$				$m=3$			
n_0 \ r	2	3	4	5	2	3	4	5
3	12.19	18.70	24.55	30.09				
4	10.70	16.65	22.00	27.07	22.41	35.00	46.58	57.68
5	9.97	15.63	20.73	25.56	19.19	30.52	40.95	50.95
6	9.53	15.02	19.97	24.66	17.57	28.24	38.06	47.49
7	9.24	14.62	19.46	24.05	16.59	26.84	36.29	45.37
8	9.04	14.33	19.10	23.62	15.93	25.90	35.10	43.93
9	8.88	14.11	18.83	23.30	15.46	25.22	34.24	42.90
10	8.76	13.94	18.61	23.05	15.11	24.71	33.59	42.11
11	8.67	13.81	18.44	22.84	14.83	24.31	33.08	41.50
12	8.59	13.70	18.31	22.68	14.61	23.99	32.67	41.00
13	8.52	13.60	18.19	22.54	14.43	23.73	32.33	40.60
14	8.47	13.53	18.09	22.43	14.28	23.50	32.05	40.26
15	8.42	13.46	18.01	22.33	14.15	23.32	31.81	39.97
16	8.38	13.40	17.94	22.24	14.04	23.16	31.60	39.72
17	8.35	13.35	17.87	22.16	13.94	23.02	31.42	39.50
18	8.32	13.31	17.82	22.10	13.86	22.89	31.27	39.31
19	8.28	13.27	17.77	22.04	13.79	22.79	31.13	39.15
20	8.26	13.23	17.72	21.98	13.72	22.69	31.00	39.00
25	8.17	13.10	17.56	21.78	13.48	22.33	30.55	38.44
30	8.11	13.01	17.45	21.65	13.32	22.10	30.25	38.09
35	8.07	12.95	17.37	21.56	13.21	21.94	30.04	37.83
40	8.03	12.90	17.31	21.49	13.13	21.82	29.89	37.65
45	8.01	12.87	17.27	21.44	13.07	21.73	29.77	37.51
50	7.99	12.84	17.23	21.40	13.02	21.66	29.68	37.39
55	7.97	12.82	17.20	21.36	12.98	21.60	29.60	37.30
60	7.96	12.80	17.18	21.33	12.94	21.55	29.54	37.23
120	7.89	12.69	17.05	21.18	12.77	21.28	29.20	36.82

[a]Here, r = number of covariance matrices; n = one less than common sample size; m = number of variables.

Source: Adapted from Davis and Field (1971) and Lee et al. (1977), with the kind permission of the Commonwealth Scientific and Industrial Research Organization (C.S.I.R.O.), Australia, North-Holland Publishing Company, and the authors.

Table 3 (*Continued*)

n_0 \ r	$m=4$: 2	3	4	5	$m=5$: 2	3	4	5
5	35.39	56.10	75.36	93.97				
6	30.07	48.63	65.90	82.60	51.14	81.99	110.92	138.98
7	27.31	44.69	60.90	76.56	43.40	71.06	97.03	122.22
8	25.61	42.24	57.77	72.77	39.29	65.15	89.45	113.03
9	24.45	40.56	55.62	70.17	36.70	61.39	84.62	107.17
10	23.62	39.34	54.05	68.27	34.92	58.78	81.25	103.06
11	22.98	38.41	52.85	66.81	33.62	56.85	78.75	100.03
12	22.48	37.67	51.90	65.66	32.62	55.37	76.83	97.68
13	22.08	37.08	51.13	64.73	31.83	54.19	75.30	95.81
14	21.75	36.59	50.50	63.96	31.19	53.23	74.05	94.29
15	21.47	36.17	49.96	63.31	30.67	52.44	73.02	93.02
16	21.23	35.82	49.51	62.75	30.21	51.76	72.14	91.95
17	21.03	35.52	49.12	62.28	29.83	51.19	71.39	91.03
18	20.86	35.26	48.77	61.86	29.51	50.69	70.74	90.24
19	20.70	35.02	48.47	61.49	29.22	50.26	70.17	89.54
20	20.56	34.82	48.21	61.17	28.97	49.88	69.67	88.93
25	20.06	34.06	47.23	59.98	28.05	48.49	67.85	86.70
30	19.74	33.58	46.60	59.22	27.48	47.61	66.71	85.29
35	19.52	33.25	46.17	58.69	27.09	47.01	65.92	84.33
40	19.36	33.01	45.85	58.30	26.81	46.57	65.35	83.62
45	19.23	32.82	45.61	58.00	26.59	46.24	64.91	83.08
50	19.14	32.67	45.42	57.77	26.42	45.98	64.56	82.66
55	19.06	32.55	45.26	57.58	26.28	45.77	64.28	82.31
60	18.99	32.45	45.13	57.42	26.17	45.59	64.05	82.03
120	18.64	31.92	44.44	56.57	25.57	44.66	62.83	80.52

Box (1949), Davis (1971), Davis and Field (1971), and Krishnaiah and Lee (1979). The upper 5 percentage points of the distribution of $-2\log\Lambda^*$ have been tabulated for equal sample sizes ($n_0=n_i$, with $i=1,\ldots,r$) and for various values of m and r by Davis and Field (1971) and Krishnaiah and Lee (1979); some of these are given in Table 3.

8.2.5. *Noncentral Moments of the Modified Likelihood Ratio Statistic when $r=2$*

In this section we will obtain the moments in general of Λ^* for the case $r=2$ where the equality of two covariance matrices is being tested. These will be used in the next section to obtain asymptotic expansions for the non-null distribution of Λ^* from which the power of the modified likelihood ratio test can be computed. Recall that when $r=2$

$$W=\Lambda^*\cdot\frac{n_1^{mn_1/2}n_2^{mn_2/2}}{n^{mn/2}} \tag{40}$$

$$=\frac{(\det A_1)^{n_1/2}(\det A_2)^{n_2/2}}{\det(A_1+A_2)^{n/2}}$$

$$=\prod_{i=1}^{m}\frac{f_i^{n_1/2}}{(1+f_i)^{n/2}},$$

where $f_1\geq f_2\geq\cdots\geq f_m(>0)$ are the latent roots of $A_1A_2^{-1}$. We start by giving the joint distribution of $f_1,\dots,f_m$ in the following theorem due to Constantine (see James, 1964).

THEOREM 8.2.8. If A_1 is $W_m(n_1,\Sigma_1)$, A_2 is $W_m(n_2,\Sigma_2)$, with $n_1>m-1$, $n_2>m-1$, and A_1 and A_2 are independent then the joint probability density function of $f_1,\dots,f_m$, the latent roots of $A_1A_2^{-1}$, is

$$\prod_{i=1}^{m}\delta_i^{-n_1/2}\,{}_1F_0^{(m)}\left(\tfrac{1}{2}n;-\Delta^{-1},F\right)\frac{\pi^{m^2/2}\Gamma_m\left(\tfrac{1}{2}n\right)}{\Gamma_m\left(\tfrac{1}{2}m\right)\Gamma_m\left(\tfrac{1}{2}n_1\right)\Gamma_m\left(\tfrac{1}{2}n_2\right)} \tag{41}$$

$$\cdot\prod_{i=1}^{m}f_i^{(n_1-m-1)/2}\prod_{i<j}^{m}(f_i-f_j)\qquad(f_1>f_2>\cdots>f_m>0),$$

where $n=n_1+n_2$, $F=\operatorname{diag}(f_1,\dots,f_m)$, $\Delta=\operatorname{diag}(\delta_1,\dots,\delta_m)$, with $\delta_1,\dots,\delta_m$ being the latent roots of $\Sigma_1\Sigma_2^{-1}$, and ${}_1F_0^{(m)}$ is a two-matrix hypergeometric function (see Section 7.3).

Proof. Using a (by now) familiar invariance argument, we can assume without loss of generality that $\Sigma_1=\Delta$ and $\Sigma_2=I_m$. The joint density function of A_1 and A_2 is then

$$\frac{(\det\Delta)^{-n_1/2}}{2^{mn/2}\Gamma_m\left(\tfrac{1}{2}n_1\right)\Gamma_m\left(\tfrac{1}{2}n_2\right)}\operatorname{etr}\left(-\tfrac{1}{2}\Delta^{-1}A_1\right)(\det A_1)^{(n_1-m-1)/2}$$

$$\cdot\operatorname{etr}\left(-\tfrac{1}{2}A_2\right)(\det A_2)^{(n_2-m-1)/2}.$$

Now make the transformation

$$\tilde{F}=U^{1/2}A_2^{-1}U^{1/2},\qquad U=A_1$$

and note that the latent roots of $A_1A_2^{-1}$ are the latent roots of $\tilde{F}$. The Jacobian is given by

$$(dA_1)(dA_2)=(\det U)^{(m+1)/2}(\det\tilde{F})^{-(m+1)}(dU)(d\tilde{F}),$$

and hence the joint density function of $\tilde{F}$ and U is

$$\frac{(\det\Delta)^{-n_1/2}}{2^{mn/2}\Gamma_m(\frac{1}{2}n_1)\Gamma_m(\frac{1}{2}n_2)}\operatorname{etr}[-\tfrac{1}{2}U(\Delta^{-1}+\tilde{F}^{-1})]$$
$$\cdot(\det U)^{(n-m-1)/2}(\det\tilde{F})^{-(n_2+m+1)/2}.$$

Integrating this over $U>0$ using Theorem 2.1.11 then shows that $\tilde{F}$ has the density function

$$\frac{\Gamma_m(\frac{1}{2}n)}{\Gamma_m(\frac{1}{2}n_1)\Gamma_m(\frac{1}{2}n_2)}(\det\Delta)^{-n_1/2}(\det\tilde{F})^{(n_1-m-1)/2}\det(I+\Delta^{-1}\tilde{F})^{-n/2}.$$

Using Theorem 3.2.17 it now follows that the joint density function of $f_1,\dots,f_m$ may be expressed in the form

$$(42)\quad \frac{\pi^{m^2/2}}{\Gamma_m(\frac{1}{2}m)}\frac{\Gamma_m(\frac{1}{2}n)}{\Gamma_m(\frac{1}{2}n_1)\Gamma_m(\frac{1}{2}n_2)}(\det\Delta)^{-n_1/2}\prod_{i=1}^{m}f_i^{(n_1-m-1)/2}\prod_{i<j}^{m}(f_i-f_j)$$
$$\cdot\int_{O(m)}\det(I+\Delta^{-1}HFH')^{-n/2}(dH).$$

The desired result (41) is now obtained using Corollary 7.3.5 and Theorem 7.3.3.

The null distribution of $f_1,\dots,f_m$ (that is, the distribution when $\Sigma_1=\Sigma_2$) follows easily from Theorem 8.2.8.

COROLLARY 8.2.9. When $\Sigma_1=\Sigma_2$ the joint density function of $f_1,\dots,f_m$, the latent roots of $A_1A_2^{-1}$, is

(43)

$$\frac{\pi^{m^2/2}\Gamma_m(\frac{1}{2}n)}{\Gamma_m(\frac{1}{2}m)\Gamma_m(\frac{1}{2}n_1)\Gamma_m(\frac{1}{2}n_2)}\prod_{i=1}^{m}\left[f_i^{(n_1-m-1)/2}(1+f_i)^{-n/2}\right]\prod_{i<j}^{m}(f_i-f_j).$$

Proof. When $\Sigma_1 = \Sigma_2$ we have $\Delta = I_m$ and the ${}_1F_0^{(m)}$ function in (41) is

$$\begin{aligned}{}_1F_0^{(m)}(\tfrac{1}{2}n; -\Delta^{-1}, F) &= {}_1F_0(\tfrac{1}{2}n; -F)\\ &= \det(I+F)^{-n/2}\\ &= \prod_{i=1}^{m}(1+f_i)^{-n/2},\end{aligned}$$

completing the proof.

The alert reader will recall that the null distribution (43) of $f_1,\dots,f_m$ has already been essentially derived in Theorem 3.3.4, where the distribution of the latent roots $u_1,\dots,u_m$ of $A_1(A_1+A_2)^{-1}$ is given. Corollary 8.2.9 follows immediately from Theorem 3.3.4 on putting $f_i = u_i/(1-u_i)$.

The zonal polynomial series for the hypergeometric function in (41) may not converge for all values of Δ and F, but the integral in (42) is well-defined. A convergent series can be obtained using the following result of Khatri (1967).

LEMMA 8.2.10.

$$\int_{O(m)} \det(I+\Delta^{-1}HFH')^{-n/2}(dH)$$

$$= \det(I+\lambda F)^{-n/2}\,{}_1F_0^{(m)}\left(\tfrac{1}{2}n; I-(\lambda\Delta)^{-1}, \lambda F(I+\lambda F)^{-1}\right),$$

where λ is any non-negative number such that the zonal polynomial series for the ${}_1F_0^{(m)}$ function on the right converges for all F.

Proof. Since

$$\det(I+\Delta^{-1}HFH')^{-n/2} = \det(I+\lambda F)^{-n/2}$$

$$\cdot\det\left[I-(I-\lambda^{-1}\Delta^{-1})H(\lambda F)(I+\lambda F)^{-1}H'\right]^{-n/2},$$

the desired result follows by integrating over $O(m)$.

We now turn our attention to the moments of the statistic W defined by (40), a multiple of the modified likelihood ratio statistic Λ^*. These can be expressed in terms of the ${}_2F_1$ one-matrix hypergeometric function (see Sections 7.3 and 7.4) as the following theorem due to Sugiura (1969a) shows.

THEOREM 8.2.11. The hth moment of W is

$$(44)\quad E(W^h)=\frac{\Gamma_m(\frac{1}{2}n)\Gamma_m[\frac{1}{2}n_1(1+h)]\Gamma_m[\frac{1}{2}n_2(1+h)]}{\Gamma_m(\frac{1}{2}n_1)\Gamma_m(\frac{1}{2}n_2)\Gamma_m[\frac{1}{2}n(1+h)]}\cdot(\det\Delta)^{n_1h/2}{}_2F_1\big(\tfrac{1}{2}nh,\tfrac{1}{2}n_1(1+h);\tfrac{1}{2}n(1+h);I-\Delta\big).$$

Proof. Let $\tilde{F}$ be a random matrix having the density function

$$\frac{\Gamma_m(\frac{1}{2}n)}{\Gamma_m(\frac{1}{2}n_1)\Gamma_m(\frac{1}{2}n_2)}(\det\Delta)^{-n_1/2}(\det\tilde{F})^{(n_1-m-1)/2}\det(I+\tilde{F})^{-n/2}\cdot{}_1F_0^{(m)}\big(\tfrac{1}{2}n;I-\Delta^{-1},\tilde{F}(I+\tilde{F})^{-1}\big)\qquad(\tilde{F}>0).$$

The latent roots $f_1,\dots,f_m$ of $A_1A_2^{-1}$ have the same distribution as the latent roots of $\tilde{F}$, as shown by Theorem 8.2.8 and Lemma 8.2.10 (with $\lambda=1$). Hence the hth moment of

$$W=\prod_{i=1}^{m}\frac{f_i^{n_1/2}}{(1+f_i)^{n/2}}=(\det\tilde{F})^{n_1/2}\det(I+\tilde{F})^{-n/2}$$

is given by

$$E(W^h)=\frac{\Gamma_m(\frac{1}{2}n)}{\Gamma_m(\frac{1}{2}n_1)\Gamma_m(\frac{1}{2}n_2)}(\det\Delta)^{-n_1/2}\int_{\tilde{F}>0}(\det\tilde{F})^{n_1(1+h)/2-(m+1)/2}\cdot\det(I+\tilde{F})^{-n(1+h)/2}{}_1F_0^{(m)}\big(\tfrac{1}{2}n;I-\Delta^{-1},\tilde{F}(I+\tilde{F})^{-1}\big)(d\tilde{F}).$$

Putting $U=(I+\tilde{F})^{-1/2}\tilde{F}(I+\tilde{F})^{-1/2}$ and using the zonal polynomial series for ${}_1F_0^{(m)}$, we get

$$E(W^h)=\frac{\Gamma_m(\frac{1}{2}n)}{\Gamma_m(\frac{1}{2}n_1)\Gamma_m(\frac{1}{2}n_2)}(\det\Delta)^{-n_1/2}\int_{\tilde{F}>0}(\det\tilde{F})^{n_1(1+h)/2-(m+1)/2}\cdot\det(I+\tilde{F})^{-n(1+h)/2}{}_1F_0^{(m)}\big(\tfrac{1}{2}n;I-\Delta^{-1},\tilde{F}(I+\tilde{F})^{-1}\big)(d\tilde{F}).$$

$$=\frac{\Gamma_m(\frac{1}{2}n)}{\Gamma_m(\frac{1}{2}n_1)\Gamma_m(\frac{1}{2}n_2)}(\det\Delta)^{-n_1/2}\sum_{k=0}^{\infty}\sum_{\kappa}\frac{(\frac{1}{2}n)_\kappa}{k!}\frac{C_\kappa(I-\Delta^{-1})}{C_\kappa(I_m)}\cdot\int_{0<U<I}(\det U)^{n_1(1+h)/2-(m+1)/2}\cdot\det(I-U)^{n_2(1+h)/2-(m+1)/2}C_\kappa(U)(dU).$$

(Here we are assuming that $\max_i|1-\delta_i^{-1}|<1$, i.e., $\max_i\delta_i=\delta_1>\frac{1}{2}$. This restriction can be dropped at the end.) Using Theorem 7.2.10 to evaluate this integral gives

$$E(W^h)=\frac{\Gamma_m(\frac{1}{2}n)}{\Gamma_m(\frac{1}{2}n_1)\Gamma_m(\frac{1}{2}n_2)}\frac{\Gamma_m[\frac{1}{2}n_1(1+h)]\Gamma_m[\frac{1}{2}n_2(1+h)]}{\Gamma_m[\frac{1}{2}n(1+h)]}$$

$$\cdot(\det\Delta)^{-n_1/2}\sum_{k=0}^{\infty}\sum_{\kappa}\frac{(\frac{1}{2}n)_\kappa(\frac{1}{2}n_1(1+h))_\kappa}{(\frac{1}{2}n(1+h))_\kappa}\frac{C_\kappa(I-\Delta^{-1})}{k!}$$

$$=\frac{\Gamma_m(\frac{1}{2}n)\Gamma_m[\frac{1}{2}n_1(1+h)]\Gamma_m[\frac{1}{2}n_2(1+h)]}{\Gamma_m(\frac{1}{2}n_1)\Gamma_m(\frac{1}{2}n_2)\Gamma_m[\frac{1}{2}n(1+h)]}$$

$$\cdot(\det\Delta)^{-n_1/2}{}_2F_1\left(\tfrac{1}{2}n,\tfrac{1}{2}n_1(1+h);\tfrac{1}{2}n(1+h);I-\Delta^{-1}\right).$$

The desired result now follows on using the Euler relation given in Theorem 7.4.3.

As mentioned earlier these moments could be used in principle to obtain exact expressions for the distribution of W; see, for instance, Khatri and Srivastava (1971). We concentrate here on asymptotic distributions.

8.2.6. Asymptotic Non-null Distributions of the Modified Likelihood Ratio Statistic when $r=2$

The power function of the modified likelihood ratio test of size α is $P(-2\rho\log\Lambda^*\geq k_\alpha^*|\delta_1,\dots,\delta_m)$, where ρ is given by (35) and k_α^* is the upper $100\alpha\%$ point of the distribution of $-2\rho\log\Lambda^*$ when $H:\Sigma_1=\Sigma_2$ is true. This is a function of the latent roots $\delta_1,\dots,\delta_m$ of $\Sigma_1\Sigma_2^{-1}$ (and, of course, n_1, n_2, and m). We have already seen that an approximation for k_α^* is $c_f(\alpha)$, the upper $100\alpha\%$ point of the χ_f^2 distribution, with $f=m(m+1)/2$. In this section we investigate ways of approximating the power function.

The asymptotic non-null distribution of a likelihood ratio statistic (or, in this case, a modified one) depends in general upon the type of alternative being considered. Here we will consider three different alternatives. To discuss these, we must recall the notation introduced in Section 8.2.4. Put $n_i=k_in$ with $k_i>0$ $(i=1,2)$, where $k_1+k_2=1$, and assume that $n\to\infty$. Instead of using n as the asymptotic variable, we will use $M=\rho n$, as in Theorem 8.2.7. We will consider asymptotic distributions of $-2\rho\log\Lambda^*$ as $M\to\infty$.

The three alternatives discussed here are

$$K:\quad \Sigma_1 \neq \Sigma_2,$$

$$K_M:\quad \Sigma_2^{-1/2}\Sigma_1\Sigma_2^{-1/2} = I_m + \frac{1}{M}\Theta,$$

and

$$K_M^*:\quad \Sigma_2^{-1/2}\Sigma_1\Sigma_2^{-1/2} = I_m + \frac{1}{M^{1/2}}\Theta,$$

where Θ is a fixed matrix. The alternative K is referred to as a fixed (or general) alternative, while K_M and K_M^* are sequences of *local* alternatives since they approach the null hypothesis $H: \Sigma_1 = \Sigma_2$ as $M \to \infty$. It is more convenient to express these alternative hypotheses in terms of the matrix $\Delta = \operatorname{diag}(\delta_1, \ldots, \delta_m)$ of latent roots of $\Sigma_1\Sigma_2^{-1}$; they are clearly equivalent to

$$K:\quad \Delta \neq I_m,$$

$$K_M:\quad \Delta = I_m + \frac{1}{M}\Omega,$$

and

$$K_M^*:\quad \Delta = I_m + \frac{1}{M^{1/2}}\Omega,$$

where Ω is a fixed diagonal matrix, $\Omega = \operatorname{diag}(\omega_1, \ldots, \omega_m)$.

The asymptotic distributions of Λ^* are different for each of these three cases. We will first state the results and prove and expand on them later. Under the fixed alternative K the asymptotic distribution as $M \to \infty$ of

$$\frac{-2\rho}{M^{1/2}}\log \Lambda^* + M^{1/2}\log\frac{(\det \Delta)^{k_1}}{\det(k_1\Delta + k_2 I)}$$

is $N(0, \tau^2)$, where

$$\tau^2 = \tfrac{1}{2}k_1k_2\sum_{i=1}^{m}\left(\frac{\delta_i - 1}{k_1\delta_i + k_2}\right)^2.$$

This normal approximation could be used for computing approximate powers of the modified likelihood ratio test for large deviations from the null hypothesis $H:\Delta=I$. Note, however, that the asymptotic variance $\tau^2\rightarrow 0$ as all $\delta_i\rightarrow 1$ so that the normal approximation can not be expected to be much good for alternatives close to I. This is where the sequences of local alternatives K_M and K_M^* give more accurate results. Under the sequence K_M of local alternatives the asymptotic distribution of $-2\rho\log\Lambda^*$ is χ_f^2, where $f=m(m+1)/2$, the same as the asymptotic distribution when H is true. For the sequence K_M^* of local alternatives, under which $\Delta\rightarrow I$ at a slower rate than under K_M, the asymptotic distribution of $-2\rho\log\Lambda^*$ is noncentral $\chi_f^2(\nu)$, where the noncentrality parameter is $\nu=k_1k_2\operatorname{tr}\Omega^2/2$.

Asymptotic expansions for the distributions in these three cases can be obtained as in Section 8.2.4 from expansions of the characteristic functions for large M. When H is not true, however, the characteristic functions involve a ${}_2F_1$ function of matrix argument (see Theorem 8.2.11) which must also be expanded asymptotically. There are a number of methods available for doing this; in this section we will concentrate on an approach which uses the partial differential equations of Theorem 7.5.5 satisfied by the ${}_2F_1$ function. The first correction term will be derived in asymptotic series for the distributions under each of the three alternatives. Under K and K_M^* this term is of order $M^{-1/2}$, while under K_M it is of order M^{-1}.

We consider first the general alternative $K:\Delta\neq I$. Define the random variable Y by

$$Y=\frac{-2\rho}{M^{1/2}}\log\Lambda^*+M^{1/2}\log\frac{(\det\Delta)^{k_1}}{\det(k_1\Delta+k_2I)},\tag{45}$$

and note that the asymptotic mean of $(-2\rho/M)\log\Lambda^*$ is $-\log[(\det\Delta)^{k_1}/\det(k_1\Delta+k_2I)]$ and its asymptotic variance is of order M^{-1}, so that Y has, asymptotically, zero mean and constant variance. The characteristic function of Y is

$$\begin{aligned}g(M,t,\Delta)&=E(e^{itY})\\&=\left[\frac{(\det\Delta)^{k_1}}{\det(k_1\Delta+k_2I)}\right]^{itM^{1/2}}E(\Lambda^{*-2it\rho/M^{1/2}})\\&=\left[\frac{(\det\Delta)^{k_1}}{\det(k_1\Delta+k_2I)}\right]^{itM^{1/2}}(k_1^{k_1}k_2^{k_2})^{itM^{1/2}m}\cdot E(W^{-2it\rho/M^{1/2}}),\end{aligned}$$

where W is given by (40). Using Theorem 8.2.11 to evaluate this last expectation then shows that $g(M,t,\Delta)$ can be expressed in the form

$$(46) \qquad g(M,t,\Delta)=G_1(M,t)G_2(M,t,\Delta),$$

where

$$(47) \quad G_1(M,t)=\left(k_1^{k_1}k_2^{k_2}\right)^{itM^{1/2}m}\frac{\Gamma_m\left(\frac{1}{2}M+\varepsilon_1\right)\Gamma_m\left(\frac{1}{2}k_1M-M^{1/2}k_1it+\gamma_1\right)}{\Gamma_m\left(\frac{1}{2}M-M^{1/2}it+\varepsilon_1\right)\Gamma_m\left(\frac{1}{2}k_1M+\gamma_1\right)}$$

$$\cdot\frac{\Gamma_m\left(\frac{1}{2}k_2M-M^{1/2}k_2it+\varepsilon_1-\gamma_1\right)}{\Gamma_m\left(\frac{1}{2}k_2M+\varepsilon_1-\gamma_1\right)}$$

and

(48)

$$G_2(M,t,\Delta)=\det(k_1\Delta+k_2I)^{-itM^{1/2}}$$

$$\cdot {}_2F_1\left(-itM^{1/2},\tfrac{1}{2}k_1M-M^{1/2}k_1it+\gamma_1;\tfrac{1}{2}M-M^{1/2}it+\varepsilon_1;I-\Delta\right)$$

with

$$\varepsilon_1=\tfrac{1}{2}n(1-\rho)=\frac{2m^2+3m-1}{12(m+1)}\left(\frac{1}{k_1}+\frac{1}{k_2}-1\right);$$

then

$$(49) \qquad \gamma_1=k_1\varepsilon_1.$$

Here $g(M,t,\Delta)$ has been written in terms of the asymptotic variable $M=\rho n$. Using (24) to expand the gamma functions for large M, it is a straightforward task to show that

$$(50) \qquad G_1(M,t)=1+\frac{m(m+1)it}{M^{1/2}}+O(M^{-1}).$$

An expansion for $G_2(M,t,\Delta)$, the other factor in $g(M,t,\Delta)$, follows from the following theorem, where for convenience we have put $R=I-\Delta=\text{diag}(r_1,\dots,r_m)$.

THEOREM 8.2.12. The function $G_2(M,t,I-R)$ defined by (48) can be expanded as

$$(51)\qquad G_2(M,t,I-R)=\exp(-k_1k_2\sigma_2t^2)\left[1+\frac{Q_1(R)}{M^{1/2}}+O(M^{-1})\right],$$

where

$$(52)\qquad Q_1(R)=-\tfrac{1}{2}itk_1k_2(1+4t^2)\sigma_2+\tfrac{4}{3}(it)^3k_1k_2(k_1-k_2)\sigma_3$$
$$-2(it)^3k_1^2k_2^2\sigma_4-\tfrac{1}{2}k_1k_2it\sigma_1^2$$

with

$$(53)\qquad \sigma_j=\operatorname{tr}\left[R(I-k_1R)^{-1}\right]^j=\sum_{i=1}^m\frac{r_i^j}{(1-k_1r_i)^j}.$$

Proof. We outline a proof using the partial differential equations satisfied by the ${}_2F_1$ function. From Theorem 7.5.5 the function

$${}_2F_1(-itM^{1/2},\tfrac{1}{2}k_1M-M^{1/2}k_1it+\gamma_1;\tfrac{1}{2}M-M^{1/2}it+\varepsilon_1;R),$$

which is part of $G_2(M,t,I-R)$, is the unique solution of the system of partial differential equations

$$r_j(1-r_j)\frac{\partial^2F}{\partial r_j^2}+\left\{\tfrac{1}{2}M-M^{1/2}it+\varepsilon_1-\tfrac{1}{2}(m-1)-\left[\tfrac{1}{2}k_1M-M^{1/2}it(1+k_1)\right.\right.$$
$$\left.\left.+\gamma_1+1-\frac{1}{2}(m-1)\right]r_j+\frac{1}{2}\sum_{\alpha\neq j}\frac{r_j(1-r_j)}{r_j-r_\alpha}\right\}\frac{\partial F}{\partial r_j}$$
$$-\frac{1}{2}\sum_{\alpha\neq j}\frac{r_\alpha(1-r_\alpha)}{r_j-r_\alpha}\frac{\partial F}{\partial r_\alpha}=-M^{1/2}it(\tfrac{1}{2}k_1M-M^{1/2}k_1it+\gamma_1)F$$
$$(j=1,\dots,m)$$

subject to the conditions that F be a symmetric function of $r_1,\dots,r_m$ which is analytic at $R=0$, with $F(0)=1$. From this system a system of differential equations satisfied by the function $H(R)\equiv\log G_2(M,t,I-R)$ can be obtained. The energetic reader can verify, after lengthy but straightforward algebraic manipulation, that this system is

$$(54)\quad r_j(1-r_j)\left[\frac{\partial^2 H}{\partial r_j^2}+\left(\frac{\partial H}{\partial r_j}\right)^2\right]+\Big\{\tfrac{1}{2}M-M^{1/2}it+\varepsilon_1-\tfrac{1}{2}(m-1)$$

$$-\left[\tfrac{1}{2}k_1M-M^{1/2}it(1+k_1)+\gamma_1+1-\tfrac{1}{2}(m-1)\right]r_j$$

$$\left.-2M^{1/2}k_1it\frac{r_j(1-r_j)}{1-k_1r_j}+\frac{1}{2}\sum_{\alpha\neq j}\frac{r_j(1-r_j)}{r_j-r_\alpha}\right\}\frac{\partial H}{\partial r_j}$$

$$-\frac{1}{2}\sum_{\alpha\neq j}\frac{r_\alpha(1-r_\alpha)}{r_j-r_\alpha}\frac{\partial H}{\partial r_\alpha}=M\frac{k_1k_2(it)^2r_j}{(1-k_1r_j)^2}$$

$$-M^{1/2}\left[\frac{k_1k_2itr_j}{(1-k_1r_j)^2}+\frac{k_1k_2it}{(1-k_1r_j)^2}\sum_{\alpha\neq j}\frac{r_\alpha}{1-k_1r_\alpha}\right]\qquad(j=1,\ldots,m).$$

The function $H(R)$ is the unique solution of each of these partial differential equations, subject to the conditions that $H(R)$ be a symmetric function of $r_1,\ldots,r_m$ which is analytic at $R=0$ with $H(0)=0$. In (54) we now substitute the series

$$H(R)=\sum_{k=0}^{\infty}\frac{Q_k(R)}{M^{k/2}}=Q_0+\frac{Q_1}{M^{1/2}}+\cdots,$$

where (a) $Q_k(R)$ is symmetric in $r_1,\ldots,r_m$ and (b) $Q_k(0)=0$ for $k=0,1,2,\ldots$, and equate coefficients of like powers of M on both sides. Equating the coefficients of M shows that $Q_0(R)$ satisfies the system

$$\frac{\partial Q_0}{\partial r_j}=-\frac{2k_1k_2t^2r_j}{(1-k_1r_j)^3}\qquad(j=1,\ldots,m).$$

Integrating with respect to r_j and using conditions (a) and (b) gives

$$(55)\qquad Q_0=-k_1k_2\sigma_2t^2,$$

where σ_2 is defined by (53). Equating the coefficients of $M^{1/2}$ on both sides of (54) and using (55) yields the system for $Q_1(R)$ as

$$\frac{\partial Q_1}{\partial r_j} = \frac{-2itk_1k_2r_j}{(1-k_1r_j)^3} + \frac{4(it)^3k_1k_2r_j}{(1-k_1r_j)^3} - \frac{4(it)^3k_1k_2^2r_j^2}{(1-k_1r_j)^4}$$

$$+ \frac{4(it)^3k_1^2k_2r_j^2}{(1-k_1r_j)^4} - \frac{8k_1^2k_2^2(it)^3r_j^3}{(1-k_1r_j)^5}$$

$$- \frac{k_1k_2it}{(1-k_1r_j)^2} \sum_{\alpha\neq j} \frac{r_\alpha}{1-k_1r_\alpha} \qquad (j=1,\dots,m),$$

the solution of which subject to conditions (a) and (b) is the function $Q_1(R)$ given by (52). We now have

$$\begin{aligned} G_2(M,t,I-R) &= \exp H(R) \\ &= \exp\left[Q_0(R) + \frac{Q_1(R)}{M^{1/2}} + O(M^{-1})\right] \\ &= \exp\left(Q_0(R)\right)\left[1 + \frac{Q_1(R)}{M^{1/2}} + O(M^{-1})\right], \end{aligned}$$

completing the proof.

An asymptotic expansion to order $M^{-1/2}$ of the distribution of Y is an immediate consequence of Theorem 8.2.12.

THEOREM 8.2.13. Under the fixed alternative K: $\Delta \neq I$, the distribution function of the random variable Y given by (45) can be expanded asymptotically up to terms of order $M^{-1/2}$ as

$$(56) \qquad P\left(\frac{Y}{\tau} \leq x\right) = \Phi(x) + \frac{1}{M^{1/2}}\left[a_1\phi(x) - a_2\phi^{(2)}(x)\right] + O(M^{-1})$$

where Φ and ϕ denote the standard normal distribution and density functions respectively, and

$$(57) \qquad \tau^2 = \tfrac{1}{2}k_1k_2\sigma_2,$$

$$(58) \qquad a_1 = \frac{1}{2\tau}\left[k_1k_2(\sigma_2+\sigma_1^2) - m(m+1)\right],$$

$$(59) \qquad a_2 = \frac{2k_1k_2}{\tau^3}\left[\sigma_2 + \frac{2}{3}(k_1-k_2)\sigma_3 - k_1k_2\sigma_4\right],$$

with

$$\sigma_j = \sum_{i=1}^{m} \left(\frac{1-\delta_i}{k_1\delta_i + k_2} \right)^j.$$

Proof. Putting $R = I - \Delta$ in Theorem 8.2.12 and using (50) and (46) shows that $g(M, t/\tau, \Delta)$, the characteristic function of Y/τ, can be expanded as

$$g(M, t/\tau, \Delta) = e^{-t^2/2}\left[1 + P_1(\Delta)/M^{1/2} + O(M^{-1})\right],$$

where

$$P_1(\Delta) = -ita_1 + (it)^3 a_2$$

with a_1 and a_2 given by (58) and (59). The desired result (56) now follows by straightforward inversion of this characteristic function, where we use

$$\int_{-\infty}^{\infty} e^{itx}\phi^{(k)}(x)\,dx = -(it)^k e^{-t^2/2} \qquad \text{for } k = 1, 3.$$

We now turn our attention to the sequence of local alternatives

$$K_M:\quad \Delta = I + \frac{1}{M}\Omega,$$

where $\Omega = \text{diag}(\omega_1, \ldots, \omega_m)$ is fixed. Under K_M the characteristic function of $-2\rho\log\Lambda^*$ is, using Theorem 8.2.11,

(60)
$$\begin{aligned}\phi(M, t, \Omega) &= E(\Lambda^{*-2it\rho})\\ &= \left(k_1^{k_1}k_2^{k_2}\right)^{itmM} E(W^{-2it\rho})\\ &= \phi(M, t, 0)G_3(M, t, \Omega),\end{aligned}$$

where

(61)
$$G_3(M, t, \Omega) = \det\left(I + \frac{1}{M}\Omega\right)^{-k_1Mit} \cdot {}_2F_1\left(-Mit, \tfrac{1}{2}k_1M(1-2it) + \gamma_1; \tfrac{1}{2}M(1-2it) + \varepsilon_1; -\frac{1}{M}\Omega\right)$$

with γ_1 and ε_1 given by (49), and $\phi(M,t,0)$ is the characteristic function of $-2\rho\log\Lambda^*$ when H is true ($\Omega=0$), obtained from (33) by putting $r=2$ and $h=-2\rho it$. From Theorem 8.2.7 we know that

$$\phi(M,t,0)=(1-2it)^{-f/2}+O(M^{-2}), \tag{62}$$

where $f=m(m+1)/2$. It remains to expand $G_3(M,t,\Omega)$ for large M. Because it will be useful in another context (see Section 11.2.6) we will generalize the function $G_3(M,t,\Omega)$ a little by introducing some more parameters. The term of order M^{-1} will be found in an asymptotic expansion of the function

$$(63)\qquad G(M,\Omega)=\det\left(I+\frac{1}{M}\Omega\right)^{\alpha_0 M+\alpha_1}$$
$$\cdot {}_2F_1\left(\beta_0 M+\beta_1,\gamma_0 M+\gamma_1;\varepsilon_0 M+\varepsilon_1;-\frac{1}{M}\Omega\right),$$

where $\alpha_i,\beta_i,\gamma_i,\varepsilon_i$ $(i=0,1)$ are arbitrary parameters independent of M and nonzero for $i=0$. Note that $G_3(M,t,\Omega)$ is the special case of $G(M,\Omega)$ obtained by putting $\alpha_0=-k_1 it,\alpha_1=0,\beta_0=-it,\beta_1=0,\gamma_0=\frac{1}{2}k_1(1-2it),\varepsilon_0=\frac{1}{2}(1-2it)$ and letting γ_1 and ε_1 be given by (49). Later we will see that with different values of the parameters $G(M,\Omega)$ is also part of another characteristic function of interest. The following theorem gives the term of order M^{-1} in an asymptotic expansion of $G(M,\Omega)$.

THEOREM 8.2.14. The function $G(M,\Omega)$ defined by (63) has the expansion

$$G(M,\Omega)=\exp\left[\left(\alpha_0-\frac{\beta_0\gamma_0}{\varepsilon_0}\right)\sigma_1\right]\left[1+\frac{P_1(\Omega)}{M}+O(M^{-2})\right] \tag{64}$$

where

$$(65)\qquad P_1(\Omega)=\frac{\sigma_1}{\varepsilon_0^2}\left(\alpha_1\varepsilon_0^2-\beta_0\gamma_1\varepsilon_0-\beta_1\gamma_0\varepsilon_0+\beta_0\gamma_0\varepsilon_1\right)$$
$$+\frac{\sigma_2}{2\varepsilon_0^3}\left(\beta_0^2\gamma_0\varepsilon_0+\beta_0\gamma_0^2\varepsilon_0-\beta_0^2\gamma_0^2-\alpha_0\varepsilon_0^3\right)$$

with

$$\sigma_j=\operatorname{tr}\Omega^j=\sum_{i=1}^m\omega_i^j.$$

Proof. A proof similar to that of Theorem 8.2.12 can be constructed here. Starting with the system of partial differential equations satisfied by the ${}_2F_1$ function in $G(M,\Omega)$ a system for the function $H(M,\Omega)=\log G(M,\Omega)$ can be readily derived. This is found to be

$$
\begin{aligned}
(66)\quad & \omega_j\left(1+\frac{1}{M}\omega_j\right)^2\left[\frac{\partial^2 H}{\partial\omega_j^2}+\left(\frac{\partial H}{\partial\omega_j}\right)^2\right]+\Big\{\varepsilon_0 M+\varepsilon_1-\tfrac{1}{2}(m-1)\\
&+\omega_j\left[\beta_0+\gamma_0+\varepsilon_0-2\alpha_0+\frac{1}{M}(\beta_1+\gamma_1+\varepsilon_1-2\alpha_1+2-m)\right]\\
&+\omega_j^2\left[\frac{1}{M}(\beta_0+\gamma_0-2\alpha_0)+\frac{1}{M^2}\left(\beta_1+\gamma_1+1-\tfrac{1}{2}(m-1)-2\alpha_1\right)\right]\\
&+\frac{1}{2}\sum_{l\neq j}\frac{\omega_j\left(1+\frac{1}{M}\omega_j\right)^2}{\omega_j-\omega_l}\Big\}\frac{\partial H}{\partial\omega_j}\\
&-\frac{1}{2}\sum_{l\neq j}\frac{\omega_l\left(1+\frac{1}{M}\omega_l\right)\left(1+\frac{1}{M}\omega_j\right)}{\omega_j-\omega_l}\frac{\partial H}{\partial\omega_l}\\
&=M(\alpha_0\varepsilon_0-\beta_0\gamma_0)+(\alpha_0\varepsilon_1+\alpha_1\varepsilon_0-\beta_0\gamma_1-\gamma_0\beta_1)+\frac{1}{M}(\alpha_1\varepsilon_1-\beta_1\gamma_1)\\
&+\omega_j\Big[\alpha_0\gamma_0+\alpha_0\beta_0-\beta_0\gamma_0-\alpha_0^2+\frac{1}{M}(\alpha_1\gamma_0+\alpha_1\beta_0+\alpha_0\gamma_1+\alpha_0\beta_1-2\alpha_0\alpha_1\\
&-\beta_1\gamma_0-\beta_0\gamma_1)+\frac{1}{M^2}(-\beta_1\gamma_1-\alpha_1^2+\alpha_1\beta_1+\alpha_1\gamma_1)\Big]\qquad (j=1,\dots,m).
\end{aligned}
$$

The function $H(M,\Omega)$ is the unique solution of each of these partial differential equations subject to the condition that $H(M,\Omega)$ be a symmetric function of $\omega_1,\dots,\omega_m$ which is analytic at $\Omega=0$ with $H(M,0)=0$. In (66) we substitute

$$H(M,\Omega)=\sum_{k=0}^{\infty}\frac{P_k(\Omega)}{M^k}=P_0(\Omega)+\frac{P_1(\Omega)}{M}+\cdots,$$

where (a) $P_k(\Omega)$ is symmetric in $\omega_1,\dots,\omega_m$ and (b) $P_k(0)=0$ for $k=0,1,2,\dots$. Equating coefficients of M on both sides gives

$$\frac{\partial P_0}{\partial \omega_j}=\alpha_0-\frac{\beta_0\gamma_0}{\varepsilon_0} \qquad (j=1,\dots,m)$$

the unique solution of which is, subject to conditions (a) and (b),

$$P_0(\Omega)=\left(\alpha_0-\frac{\beta_0\gamma_0}{\varepsilon_0}\right)\sigma_1, \tag{67}$$

where $\sigma_1=\operatorname{tr}(\Omega)$. Similarly, equating constant terms on both sides and using (67) gives a system of equations for P_1, the solution of which subject to (a) and (b) is the function $P_1(\Omega)$ given by (65). Hence

$$\begin{aligned} G(M,\Omega)&=\exp H(M,\Omega)\\ &=\exp\left[P_0(\Omega)+\frac{P_1(\Omega)}{M}+O(M^{-2})\right]\\ &=\exp[P_0(\Omega)]\left[1+\frac{P_1(\Omega)}{M}+O(M^{-2})\right], \end{aligned}$$

which is the desired result.

An asymptotic expansion to order M^{-1} of the distribution function of $-2\rho\log\Lambda^*$ under K_M now follows easily.

THEOREM 8.2.15. Under the sequence of local alternatives

$$K_M:\Delta=I+\frac{1}{M}\Omega,$$

the distribution function of $-2\rho\log\Lambda^*$ can be expanded as

(68)

$$\begin{aligned} P(-2\rho\log\Lambda^*\le x)&=P(\chi_f^2\le x)\\ &\quad+\frac{k_1k_2\sigma_2}{4M}\left[P(\chi_{f+2}^2\le x)-P(\chi_f^2\le x)\right]+O(M^{-2}), \end{aligned}$$

where $f=m(m+1)/2$ and $\sigma_2=\operatorname{tr}\Omega^2=\sum_{i=1}^m\omega_i^2$.

Proof. In Theorem 8.2.14 put $\alpha_0=-k_1it$, $\alpha_1=0$, $\beta_0=-it$, $\beta_1=0$, $\gamma_0=\frac{1}{2}k_1(1-2it)$, and $\varepsilon_0=\frac{1}{2}(1-2it)$, and let γ_1 and ε_1 be given by (49); it then follows, using the resulting expansion and (62) in (60), that the characteristic function $\phi(M,t,\Omega)$ of $-2\rho\log\Lambda^*$ under K_M has the expansion

$$\phi(M,t,\Omega)=(1-2it)^{-f/2}\left[1+\frac{k_1k_2\sigma_2}{4M}\left(\frac{1}{1-2it}-1\right)+O(M^{-1})\right].$$

Inverting this expansion term by term, using the fact that $(1-2it)^{-l/2}$ is the characteristic function of the χ_l^2 distribution, gives (68) and completes the proof.

Finally, we consider the other sequence of local alternatives K_M^*: $\Delta=I+(1/M^{1/2})\Omega$, under which $\Delta\to I$ at a slower rate than under K_M. In this case the characteristic function of $-2\rho\log\Lambda^*$ can be written from (60) as

$$\phi(M,t,M^{-1/2}\Omega)=\phi(M,t,0)G_3(M,t,M^{-1/2}\Omega). \tag{69}$$

The following theorem gives the term of order $M^{-1/2}$ in an expansion for $G_3(M,t,M^{-1/2}\Omega)$.

THEOREM 8.2.16. The function $G_3(M,t,M^{-1/2}\Omega)$ defined by (63) can be expanded as

$$G_3(M,t,M^{-1/2}\Omega)=\exp\left[\frac{itk_1k_2}{2(1-2it)}\sigma_2\right]\left[1+\frac{T_1(\Omega)}{M^{1/2}}+O(M^{-1})\right], \tag{70}$$

where

$$T_1(\Omega)=\frac{1}{6}k_1k_2\sigma_3\left[\frac{1-2k_1}{(1-2it)^2}-\frac{3k_2}{(1-2it)}+2-k_1\right] \tag{71}$$

with $\sigma_j=\sum_{i=1}^m\omega_i^j$.

Proof. The partial differential equation argument used in the proofs of Theorems 8.2.12 and 8.2.14 should be familiar to the reader by now. The function $H(M,\Omega)=\log G_3(M,t,M^{-1/2}\Omega)$ is the unique solution of each differential equation in the system

(72)

$$\omega_j\left(1+\frac{\omega_j}{M^{1/2}}\right)^2\left[\frac{\partial^2 H}{\partial\omega_j^2}+\left(\frac{\partial H}{\partial\omega_j}\right)^2\right]+\left\{\tfrac{1}{2}(1-2it)M+\varepsilon_1-\tfrac{1}{2}(m-1)\right.$$

$$+M^{1/2}\omega_j\left[\left(\tfrac{1}{2}-2it+\tfrac{1}{2}k_1+k_1it\right)+\frac{1}{M}(\gamma_1+\varepsilon_1+2-m)\right]$$

$$+\omega_j^2\left[\left(\tfrac{1}{2}k_1-it+k_1it\right)+\frac{1}{M}\left(\gamma_1+1-\tfrac{1}{2}(m-1)\right)\right]$$

$$\left.+\frac{1}{2}\sum_{\alpha\neq j}\frac{\omega_j\left(1+\frac{\omega_j}{M^{1/2}}\right)^2}{\omega_j-\omega_\alpha}\right\}\frac{\partial H}{\partial\omega_j}$$

$$-\frac{1}{2}\sum_{\alpha\neq j}\frac{\omega_\alpha\left(1+\frac{\omega_\alpha}{M^{1/2}}\right)\left(1+\frac{\omega_j}{M^{1/2}}\right)}{\omega_j-\omega_\alpha}\frac{\partial H}{\partial\omega_\alpha}$$

$$=\tfrac{1}{2}Mk_1k_2it\omega_j+\gamma_1k_2it\omega_j\qquad(j=1,\ldots,m)$$

subject to $H(\Omega)$ being a symmetric function of $\omega_1,\ldots,\omega_m$, analytic at $\Omega=0$ with $H(0)=0$. Substituting

$$H(M,\Omega)=\sum_{k=0}^{\infty}\frac{T_k(\Omega)}{M^{k/2}}$$

in (72) and equating coefficients, first of M and then $M^{1/2}$, yields differential equations for $T_0(\Omega)$ and $T_1(\Omega)$, the solutions of which are

$$T_0(\Omega)=\frac{itk_1k_2}{2(1-2it)}\sigma_2$$

and the function $T_1(\Omega)$ given by (71). Exponentiation of H to give G_3 then completes the proof.

An expansion of the distribution of $-2\rho\log\Lambda^*$ under K_M^* is an immediate consequence.

THEOREM 8.2.17. Under the sequence of local alternatives K_M^*: $\Delta = I + (1/M^{1/2})\Omega$ the distribution function of $-2\rho\log\Lambda^*$ can be expanded in terms of noncentral χ^2 probabilities as

(73)

$$P(-2\rho\log\Lambda^* \le x) = P\left(\chi_f^2(\nu)\le x\right) + \frac{k_1k_2\sigma_3}{6M^{1/2}}$$
$$\cdot\left[(1-2k_1)P\left(\chi_{f+4}^2(\nu)\le x\right) - 3k_2P\left(\chi_{f+2}^2(\nu)\le x\right)\right.$$
$$\left.+(2-k_1)P\left(\chi_f^2(\nu)\le x\right)\right] + O(M^{-1}),$$

where $f = m(m+1)/2$, $\nu = \frac{1}{2}k_1k_2\sigma_1$ and $\sigma_j = \operatorname{tr}\Omega^j = \sum_{i=1}^m \omega_i^j$.

Proof. Using Theorem 8.2.16 and (62) in (69) shows that the characteristic function $\phi(M,t,M^{-1/2}\Omega)$ of $-2\rho\log\Lambda^*$ under K_M^* has the expansion

$$\phi(M,t,M^{-1/2}\Omega) = \exp\left[\frac{itk_1k_2}{2(1-2it)}\sigma_2\right](1-2it)^{-f/2}$$
$$\cdot\left\{1+\frac{T_1(\Omega)}{M^{1/2}} + O(M^{-1})\right\},$$

where $T_1(\Omega)$ is given by (71). The desired expansion (73) is obtained by inverting this term by term using the fact that $\exp[\frac{1}{2}itk_1k_2\sigma_2/(1-2it)](1-2it)^{-f/2}$ is the characteristic function of the noncentral $\chi_f^2(\nu)$ distribution, where the noncentrality parameter is $\nu = \frac{1}{2}k_1k_2\sigma_2$.

For actual power calculations and for further terms in the asymptotic expansions presented here, the interested reader should see Sugiura (1969a, 1974), Pillai and Nagarsenker (1972), and Subrahmaniam (1975). Expansions of the distribution of the modified likelihood ratio statistic Λ^* in the more general setting where the equality of r covariance matrices is being tested have been derived by Nagao (1970, 1974).

8.2.7. The Asymptotic Null Distribution of the Modified Likelihood Ratio Statistic for Elliptical Samples

It is important to understand how inferences based on the assumption of multivariate normality are affected if this assumption is violated. In this section we sketch the derivation of the asymptotic null distribution of Λ^* for testing $H: \Sigma_1 = \Sigma_2$ when the two samples come from the same elliptical distribution with kurtosis parameter κ (see Section 1.6).

When testing H the modified likelihood ratio statistic (assuming normality) is

$$\Lambda^* = \frac{(\det S_1)^{n_1/2}(\det S_2)^{n_2/2}}{(\det S)^{n/2}},$$

where S_1 and S_2 are the two sample covariance matrices and $S = n^{-1}(n_1 S_1 + n_2 S_2)$, with $n = n_1 + n_2$. Let $n_i = k_i n$ $(i = 1,2)$, with $k_1 + k_2 = 1$, and write $S_i = \Sigma_0 + (nk_i)^{-1/2} Z_i$, where Σ_0 denotes the common value of Σ_1 and Σ_2. Then $-2\log\Lambda^*$ has the following expansion when H is true:

$$\begin{aligned}-2\log\Lambda^* &= \tfrac{1}{2}k_2 \operatorname{tr}(Z_1\Sigma_0^{-1})^2 + \tfrac{1}{2}k_1 \operatorname{tr}(Z_2\Sigma_0^{-1})^2 \\ &\quad - (k_1k_2)^{1/2}\operatorname{tr}(Z_1\Sigma_0^{-1}Z_2\Sigma_0^{-1}) + O_p(n^{-1/2}).\end{aligned}$$

Now assume that the two samples are drawn from the same elliptical distribution with kurtosis parameter κ. We can then write

$$\frac{-2\log\Lambda^*}{1+\kappa} = \mathbf{v}'\mathbf{v} + O_p(n^{-1/2}),$$

where

$$\mathbf{v} = (1+\kappa)^{-1/2}(\Sigma_0 \otimes \Sigma_0)^{-1/2}\left[(\tfrac{1}{2}k_2)^{1/2}\mathbf{z}_1 - (\tfrac{1}{2}k_1)^{1/2}\mathbf{z}_2\right]$$

with $\mathbf{z}_i = \operatorname{vec}(Z_i')$ $(i = 1,2)$; see Problem 8.2(a).

The asymptotic distribution of $\mathbf{v}$, as $n \to \infty$, is $N_{m^2}(\mathbf{0}, V)$, where

$$V = \frac{\kappa}{2(1+\kappa)} \sum_{i,j=1}^{m} (E_{ii} \otimes E_{jj}) + P, \tag{74}$$

with

$$P = \frac{1}{2}\left[I_{m^2} + \sum_{i,j=1}^{m} (E_{ij} \otimes E_{ij}')\right]$$

and E_{ij} being an $m \times m$ matrix with 1 in the (i, j) position and 0's elsewhere; see Problem 8.2(b). The rank of V is $f=\frac{1}{2}m(m+1)$. Let $\alpha_1,\ldots,\alpha_f$ denote the nonzero latent roots of V and let $H \in O(m^2)$ be such that

$$HVH'=\begin{bmatrix} \alpha_1 & & & & & 0 \\ & \ddots & & & & \\ & & \alpha_f & & & \\ & & & 0 & & \\ & & & & \ddots & \\ 0 & & & & & 0 \end{bmatrix}=D.$$

Putting $\mathbf{u}=H\mathbf{v}$ we then have

$$\frac{-2\log\Lambda^*}{1+\kappa}=\mathbf{u}'\mathbf{u}+O_p(n^{-1/2}),$$

where the asymptotic distribution of $\mathbf{u}$ is $N_{m^2}(\mathbf{0}, D)$. Summarizing, we have the following result.

THEOREM 8.2.18. Consider the modified likelihood ratio statistic Λ^* (assuming normality) for testing $H: \Sigma_1=\Sigma_2$. If the two samples come from the same elliptical distribution with kurtosis parameter κ then the asymptotic null distribution of $-2\log\Lambda^*/(1+\kappa)$ is the distribution of $\sum_{i=1}^{f}\alpha_i X_i$, where $f=\frac{1}{2}m(m+1)$, $X_1,\ldots,X_f$ are independent χ_1^2 variables, and $\alpha_1,\ldots,\alpha_f$ are the nonzero latent roots of the matrix V given by (74).

Three points are worth noting. First, if κ is unknown and is estimated by a consistent estimate $\hat{\kappa}$ then the limiting null distribution of $-2\log\Lambda^*/(1+\hat{\kappa})$ is the same as that of $-2\log\Lambda^*/(1+\kappa)$. Second, if the two samples are normally distributed we have $\kappa=0$ and the asymptotic covariance matrix V is equal to P which is idempotent. This shows that $-2\log\Lambda^*$ has an asymptotic $\chi^2_{m(m+1)/2}$ distribution, a result derived previously in Section 8.2.4. Third, the asymptotic distribution of $-2\log\Lambda^*/(1+\kappa)$ may differ substantially from $\chi^2_{m(m+1)/2}$, suggesting that the test based on Λ^* should be used with great caution, if at all, if the two samples are non-normal.

8.2.8. *Other Test Statistics*

We have been concentrating on the likelihood and modified likelihood ratio statistics, but a number of other invariant test statistics have also been proposed for testing the null hypothesis $H: \Sigma_1=\Sigma_2$. In terms of the latent roots $f_1 > \cdots > f_m$ of $A_1 A_2^{-1}$ these include

$$L_1 = \operatorname{tr}(A_1 A_2^{-1}) = \sum_{i=1}^m f_i,$$

$$L_2 = \operatorname{tr}\left[A_1(A_1+A_2)^{-1}\right] = \sum_{i=1}^m \frac{f_i}{1+f_i},$$

$$L_3 = \det A_2(A_1+A_2)^{-1} = \prod_{i=1}^m \frac{1}{1+f_i},$$

$$L_4 = \tfrac{1}{2} n_1 \operatorname{tr}\left(\frac{n}{n_1} A_1 A^{-1} - I\right)^2 + \tfrac{1}{2} n_2 \operatorname{tr}\left(\frac{n}{n_2} A_2 A^{-1} - I\right)^2$$

$$= \tfrac{1}{2} n \frac{n_1}{n_2} \sum_{i=1}^m \left(\frac{\frac{n_2}{n_1} f_i - 1}{1+f_i} \right)^2 \qquad (A = A_1 + A_2),$$

as well as the largest and smallest roots f_1 and f_m. Both the statistic

$$W = \prod_{i=1}^m \frac{f_i^{n_1/2}}{(1+f_i)^{n/2}},$$

(a multiple of the modified likelihood ratio statistic Λ^*) and L_3 are special cases of a more general class of statistics defined for arbitrary a and b by

$$L(a,b) = \frac{\det(A_1 A_2^{-1})^a}{\det(I + A_1 A_2^{-1})^b}$$

$$= \prod_{i=1}^m f_i^a (1+f_i)^{-b}.$$

Various properties of this class of statistics have been investigated by Pillai and Nagarsenker (1972) and Das Gupta and Giri (1973).

If H is true the roots f_i should be close to n_1/n_2 and any statistic which measures the deviation of the f_i from n_1/n_2 (regardless of sign) could be used for testing H against all alternatives K: $\Delta \neq I_m$. Both W and L_4 fall into this category, as does a test based on both f_1 and f_m. If we consider the one-sided alternative

$$K_1: \delta_i \ge 1 \qquad \left(i=1,\ldots,m;\quad \sum_{i=1}^{m} \delta_i > m\right),$$

then it is reasonable to reject H in favor of K_1 if the latent roots $f_1,\ldots,f_m$ of $A_1A_2^{-1}$ are "large" in some sense. Hence in this testing problem we reject H for large values of L_1, L_2, and f_1 and for small values of L_3. A comparison of the powers of these four one-sided tests was carried out by Pillai and Jayachandran (1968) for the bivariate case $m=2$. They concluded that for small deviations from H, or for large deviations but when δ_1 and δ_2 are close, the test based on L_2 appears to be generally better than that based on L_3, while L_3 is better than L_1. The reverse ordering appears to hold for large deviations from H with $\delta_1 - \delta_2$ large. The largest root f_1 has lower power than the other three statistics except when δ_1 is the only deviant root.

In most circumstances it is unlikely that one knows what the alternatives are, so it is probably more sensible to use a test which has reasonable power properties for all alternatives such as the modified likelihood test, or a test which rejects H for large values of L_4 (Nagao, 1973a), or one which rejects H if $f_1 > f_1^0$, $f_m < f_m^0$ (Roy, 1953). Asymptotic distributions of L_4 have been derived by Nagao (1973a, 1974). For reviews of other results concerning these tests, the reader is referred to Pillai (1976, 1977) and Giri (1977).

8.3. THE SPHERICITY TEST

8.3.1. *The Likelihood Ratio Statistic; Invariance and Unbiasedness*

Let $\mathbf{X}_1,\ldots,\mathbf{X}_N$ be independent $N_m(\boldsymbol{\mu},\Sigma)$ random vectors and consider testing the null hypothesis $H: \Sigma = \lambda I_m$ against the alternative $K: \Sigma \ne \lambda I_m$, where λ is unspecified. The null hypothesis H is called the *hypothesis of sphericity* since when it is true the contours of equal density in the normal distribution are spheres. We first look at this testing problem from an invariance point of view. A sufficient statistic is $(\overline{\mathbf{X}}, A)$, where

$$\overline{\mathbf{X}} = N^{-1}\sum_{i=1}^{N} \mathbf{X}_i \quad \text{and} \quad A = \sum_{i=1}^{N} (\mathbf{X}_i - \overline{\mathbf{X}})(\mathbf{X}_i - \overline{\mathbf{X}})'.$$

Consider the group of transformations given by

$$(1) \qquad \overline{\mathbf{X}} \to aH\overline{\mathbf{X}} + \mathbf{b} \quad \text{and} \quad A \to a^2HAH',$$

where $a \neq 0$, $H \in O(m)$, and $\mathbf{b} \in R^m$; this induces the transformations

$$\mu \to aH\mu + \mathbf{b} \quad \text{and} \quad \Sigma \to a^2 H \Sigma H' \tag{2}$$

on the parameter space and it is clear that the problem of testing H against K is invariant under this group, for the family of distributions of $(\overline{\mathbf{X}}, A)$ is invariant, as are the null and alternative hypotheses. The next problem is to find a maximal invariant under the action of this group. This is done in the next theorem whose proof is straightforward and left as an exercise (see Problem 8.6).

THEOREM 8.3.1. Under the group of transformations (2) a maximal invariant is

$$\left(\frac{\lambda_1}{\lambda_m}, \frac{\lambda_2}{\lambda_m}, \ldots, \frac{\lambda_{m-1}}{\lambda_m}\right),$$

where $\lambda_1 \geq \lambda_2 \geq \cdots \geq \lambda_m \ (>0)$ are the latent roots of Σ.

As a consequence of this theorem a maximal invariant under the group of transformations (1) of the sample space of the sufficient statistic $(\overline{\mathbf{X}}, A)$ is $(a_1/a_m, \ldots, a_{m-1}/a_m)$, where $a_1 > a_2 > \cdots > a_m > 0$ are the latent roots of the Wishart matrix A. Any invariant test statistic can be written in terms of these ratios and from Theorem 6.1.12 the distribution of the a_i/a_m $(i = 1, \ldots, m-1)$ depends only on λ_i/λ_m $(i = 1, \ldots, m-1)$. There is, however, no uniformly most powerful invariant test and the choice now of a particular test may be somewhat arbitrary. The most commonly used invariant test in this situation is the likelihood ratio test, first derived by Mauchly (1940); this is given in the following theorem.

THEOREM 8.3.2. Let $\mathbf{X}_1, \ldots, \mathbf{X}_N$ be independent $N_m(\mu, \Sigma)$ random vectors and put

$$A = nS = \sum_{i=1}^{N} (\mathbf{X}_i - \overline{\mathbf{X}})(\mathbf{X}_i - \overline{\mathbf{X}})' \qquad (n = N-1).$$

The likelihood ratio test of size α of $H: \Sigma = \lambda I_m$, where λ is unspecified, rejects H if

$$V \equiv \frac{\det A}{\left(\frac{1}{m}\operatorname{tr} A\right)^m} = \frac{\det S}{\left(\frac{1}{m}\operatorname{tr} S\right)^m} \leq k_\alpha, \tag{3}$$

where k_α is chosen so that the size of the test is α.

Proof. Apart from a multiplicative constant the likelihood function is

$$L(\mu,\Sigma)=(\det\Sigma)^{-N/2}\operatorname{etr}(-\tfrac{1}{2}\Sigma^{-1}A)\exp[-\tfrac{1}{2}N(\overline{X}-\mu)'\Sigma^{-1}(\overline{X}-\mu)];$$

see, for example, (8) of Section 3.1. The likelihood ratio statistic is

$$(4)\qquad \Lambda=\frac{\sup_{\mu\in R^m,\lambda>0}L(\mu,\lambda I_m)}{\sup_{\mu\in R^m,\Sigma>0}L(\mu,\Sigma)}.$$

The denominator in (4) is

$$(5)\qquad \sup_{\mu,\Sigma}L(\mu,\Sigma)=L(\overline{X},\hat{\Sigma})=N^{mN/2}e^{-mN/2}(\det A)^{-N/2},$$

where $\hat{\Sigma}=N^{-1}A$, while the numerator in (4) is

(6)

$$\begin{aligned}\sup_{\mu,\lambda}L(\mu,\lambda I_m)&=\sup_{\mu,\lambda}\lambda^{-mN/2}\operatorname{etr}\left(-\frac{1}{2\lambda}A\right)\exp\left[-\frac{N}{2\lambda}(\overline{X}-\mu)'(\overline{X}-\mu)\right]\\&=\sup_{\lambda}\lambda^{-mN/2}\operatorname{etr}\left(-\frac{1}{2\lambda}A\right)\\&=\left(\frac{1}{mN}\operatorname{tr}A\right)^{-mN/2}e^{-mN/2}.\end{aligned}$$

Using (5) and (6) in (4) we then get

$$\Lambda^{2/N}=\frac{\det A}{\left(\frac{1}{m}\operatorname{tr}A\right)^m}=V.$$

The likelihood ratio test is to reject H if the likelihood ratio statistic Λ is small; noting that this is equivalent to rejecting H if V is small completes the proof.

The statistic

$$(7)\qquad V=\frac{\det A}{\left(\frac{1}{m}\operatorname{tr}A\right)^m}$$

is commonly called the *ellipticity statistic*; note that

$$V=\frac{\prod_{i=1}^{m}a_i}{\left(\frac{1}{m}\sum_{i=1}^{m}a_i\right)^m}=\frac{\prod_{i=1}^{m}l_i}{\left(\frac{1}{m}\sum_{i=1}^{m}l_i\right)^m},$$

where $a_1,\ldots,a_m$ are the latent roots of A and $l_1,\ldots,l_m$ are the latent roots of the sample covariance matrix $S=n^{-1}A$, so that

$$V=\left[\frac{\prod_{i=1}^{m}l_i^{1/m}}{\frac{1}{m}\sum_{i=1}^{m}l_i}\right]^m,$$

that is, $V^{1/m}$ is the ratio of the geometric mean of $l_1,\ldots,l_m$ to the arithmetic mean. If the null hypothesis H is true it is clear that V should be close to 1. Note also that V is invariant, for

$$V=\frac{\prod_{i=1}^{m-1}\dfrac{a_i}{a_m}}{\left[\dfrac{1}{m}\left(1+\sum_{i=1}^{m-1}\dfrac{a_i}{a_m}\right)\right]^m}.$$

Obviously, in order to determine the constant k_α in Theorem 8.3.2 and to calculate powers of the likelihood ratio test we need to know the distribution of V. Some distributional results will be obtained in a moment. Before getting to these we will demonstrate that the likelihood ratio test has the satisfying property of unbiasedness. This is shown in the following theorem, first proved by Gleser (1966). The proof given here is due to Sugiura and Nagao (1968).

THEOREM 8.3.3. For testing $H: \Sigma=\lambda I_m$ against $K: \Sigma\neq\lambda I_m$, where λ is an unspecified positive number, the likelihood ratio test having the rejection or critical region

$$C=\left\{A>0;\quad V=\frac{\det A}{\left(\frac{1}{m}\operatorname{tr}A\right)^m}\leq k_\alpha\right\} \tag{8}$$

is unbiased.

Proof. By invariance we can assume without loss of generality that $\Sigma=\text{diag}(\lambda_1,\ldots,\lambda_m)$. The probability of the rejection region C under K can be written

$$P_K(C)=c_{m,n}\int_{A\in C}(\det\Sigma)^{-n/2}\operatorname{etr}(-\tfrac{1}{2}\Sigma^{-1}A)(\det A)^{(n-m-1)/2}(dA)$$

where $c_{m,n}=[2^{mn/2}\Gamma_m(\frac{1}{2}n)]^{-1}$ and $n=N-1$. Putting $U=\Sigma^{-1/2}A\Sigma^{-1/2}$ this becomes

$$P_K(C)=c_{m,n}\int_{\Sigma^{1/2}U\Sigma^{1/2}\in C}\operatorname{etr}(-\tfrac{1}{2}U)(\det U)^{(n-m-1)/2}(dU).$$

Now put $U=v_{11}V_0$ where V_0 is the symmetric matrix given by

$$V_0=\begin{bmatrix}1 & v_{12} & \cdots & v_{1m}\\ v_{12} & v_{22} & \cdots & v_{2m}\\ \vdots & \vdots & & \vdots\\ v_{1m} & v_{2m} & \cdots & v_{mm}\end{bmatrix}. \tag{9}$$

It is easily verified that

$$(dU)=v_{11}^{m(m+1)/2-1}dv_{11}(dV_0)$$

(i.e., the Jacobian is $v_{11}^{m(m+1)/2-1}$), so that

$$P_K(C)=c_{m,n}\int_{v_{11}\Sigma^{1/2}V_0\Sigma^{1/2}\in C}$$

$$\cdot\operatorname{etr}(-\tfrac{1}{2}v_{11}V_0)v_{11}^{mn/2-1}(\det V_0)^{(n-m-1)/2}dv_{11}(dV_0)$$

$$=2^{mn/2}\Gamma(\tfrac{1}{2}mn)c_{m,n}\int_{\Sigma^{1/2}V_0\Sigma^{1/2}\in C}(\det V_0)^{(n-m-1)/2}(\operatorname{tr}V_0)^{-mn/2}(dV_0),$$

where we have used the fact that the region C is invariant for the transformation $U\to cU$ for any $c>0$ and have integrated with respect to v_{11} from 0 to ∞. Now let

$$C^*=\{V_0>0;\ V_0\ \text{ has the form (9) and }\ \Sigma^{1/2}V_0\Sigma^{1/2}\in C\}.$$

Then, putting $b=2^{mn/2}\Gamma(\frac{1}{2}mn)c_{m,n}$, we have

$$P_K(C)-P_H(C)=b\left\{\int_{V_0\in C^*}-\int_{V_0\in C}\right\}(\det V_0)^{(n-m-1)/2}(\operatorname{tr} V_0)^{-mn/2}(dV_0)$$

$$=b\left\{\int_{V_0\in C^*-C\cap C^*}-\int_{V_0\in C-C\cap C^*}\right\}(\det V_0)^{(n-m-1)/2}(\operatorname{tr} V_0)^{-mn/2}(dV_0).$$

Now, for $V_0\in C^*-C\cap C^*$ we have $(\operatorname{tr} V_0)^{-mn/2}>k_\alpha^{n/2}m^{-mn/2}(\det V_0)^{-n/2}$ and for $V_0 \in C - C \cap C^*$, $(\det V_0)^{(n-m-1)/2} \le (\det V_0)^{-(m+1)/2}(\operatorname{tr} V_0)^{mn/2}k_\alpha^{n/2}m^{-mn/2}$, so that

$$P_K(C)-P_H(C)\ge bk_\alpha^{n/2}m^{-mn/2}\left\{\int_{V_0\in C^*-C\cap C^*}-\int_{V_0\in C-C\cap C^*}\right\}$$

$$\cdot(\det V_0)^{-(m+1)/2}(dV_0)$$

$$=bk_\alpha^{n/2}m^{-n/2}\left\{\int_{V_0\in C^*}-\int_{V_0\in C}\right\}(\det V_0)^{-(m+1)/2}(dV_0)$$

$$=0,$$

where we have used the fact that

$$\int_{V_0\in C^*}(\det V_0)^{-(m+1)/2}(dV_0)=\int_{V_0\in C}(\det V_0)^{-(m+1)/2}(dV_0);$$

this is easily proved by making the transformation $W_0=\lambda_1^{-1}\Sigma^{1/2}V_0\Sigma^{1/2}$ in the integral on the left side. We have thus shown that

$$P_K(C)\ge P_H(C);$$

(i.e., that the likelihood ratio test is unbiased), and the proof is complete.

Along the same lines it is worth mentioning a somewhat stronger result. With $\lambda_1\ge\cdots\ge\lambda_m$ being the latent roots of Σ, Carter and Srivastava (1977) have proved that the power function of the likelihood ratio test, $P_K(C)$, where C is given by (8), is a monotone nondecreasing function of $\delta_k=\lambda_k/\lambda_{k+1}$ for any k, $1\le k\le m-1$, while the remaining $m-2$ ratios $\delta_i=\lambda_i/\lambda_{i+1}$, with $i=1,\dots,m-1$, $i\ne k$, are held fixed.

Finally we note here that testing the null hypothesis of sphericity is equivalent to testing the null hypothesis $H\colon \Gamma=\lambda\Gamma_0$, where Γ_0 is a known positive definite matrix, given independent observations $y_1, \ldots, y_N$ distributed as $N_m(\tau, \Gamma)$. To see this, let B be an $m \times m$ nonsingular matrix $[B \in \mathcal{G}\ell(m, R)]$ such that $B\Gamma_0 B' = I_m$, and put $\mu = B\tau$, $\Sigma = B\Gamma B'$, $\mathbf{X}_i = B\mathbf{y}_i$ (with $i = 1, \ldots, N$). Then $\mathbf{X}_1, \ldots, \mathbf{X}_N$ are independent $N_m(\mu, \Sigma)$ vectors and the null hypothesis becomes H: $\Sigma = \lambda I_m$. It is easily verified that, in terms of the y-variables, the ellipticity statistic is

$$(10) \qquad V = \frac{\det\left(\Gamma_0^{-1} A_y\right)}{\left[\frac{1}{m} \operatorname{tr}\left(\Gamma_0^{-1} A_y\right)\right]^m},$$

where

$$A_y = \sum_{i=1}^N (\mathbf{y}_i - \bar{\mathbf{y}})(\mathbf{y}_i - \bar{\mathbf{y}})' = B^{-1} A B^{-1\prime}.$$

8.3.2. *Moments of the Likelihood Ratio Statistic*

Information about the distribution of the ellipticity statistic V can be obtained from a study of its moments. In order to find these we will need the distribution of $\operatorname{tr} A$, where A is $W_m(n, \Sigma)$. When $m = 2$ this is a mixture of gamma distributions, as we saw in Problem 3.12. In general it can be expressed as an infinite sum of the zonal polynomials introduced in Chapter 7. The result is given in the following theorem, from Khatri and Srivastava (1971).

THEOREM 8.3.4. If A is $W_m(n, \Sigma)$ with $n > m - 1$ the distribution function of $\operatorname{tr} A$ can be expressed in the form

$$(11) \qquad P(\operatorname{tr} A \le x) = \det(\lambda^{-1}\Sigma)^{-n/2} \sum_{k=0}^{\infty} \frac{1}{k!} P\left(\chi^2_{mn+2k} \le \frac{x}{\lambda}\right) \cdot \sum_{\kappa} \left(\tfrac{1}{2}n\right)_\kappa C_\kappa(I - \lambda\Sigma^{-1})$$

where $0 < \lambda < \infty$ is arbitrary. The second summation is over all partitions $\kappa = (k_1, k_2, \ldots, k_m)$, $k_1 \ge \cdots \ge k_m \ge 0$ of the integer k, $C_\kappa(\cdot)$ is the zonal polynomial corresponding to κ, and

$$(a)_\kappa = \prod_{i=1}^m \left(a - \tfrac{1}{2}(i-1)\right)_{k_i},$$

with

$$(x)_k = x(x+1)\cdots(x+k-1).$$

Proof. The moment generating function of $\operatorname{tr} A$ is

$$\begin{aligned}
\phi(t) &= E[\operatorname{etr}(tA)] \\
&= \frac{(\det\Sigma)^{-n/2}}{2^{mn/2}\Gamma_m(\frac{1}{2}n)} \int_{A>0} \operatorname{etr}[-\tfrac{1}{2}A(\Sigma^{-1}-2tI)](\det A)^{(n-m-1)/2}(dA) \\
&= (\det\Sigma)^{-n/2}\det(\Sigma^{-1}-2tI)^{-n/2} \\
&= \det(I-2t\Sigma)^{-n/2},
\end{aligned}$$

where the integral is evaluated using Theorem 2.1.11. In order to invert this moment-generating function it helps to expand it in terms of recognizable moment generating functions. For $0<\lambda<\infty$ write

$$\begin{aligned}
\phi(t) &= \det(I-2t\Sigma)^{-n/2} \\
&= (1-2t\lambda)^{-mn/2}\det(\lambda^{-1}\Sigma)^{-n/2}\det\left[I-\frac{1}{1-2t\lambda}(I-\lambda\Sigma^{-1})\right]^{-n/2} \\
&= (1-2t\lambda)^{-mn/2}\det(\lambda^{-1}\Sigma)^{-n/2}{}_1F_0\left(\tfrac{1}{2}n;\frac{1}{1-2t\lambda}(I-\lambda\Sigma^{-1})\right),
\end{aligned}$$

where the last equality follows by Corollary 7.3.5. The zonal polynomial expansion for the ${}_1F_0$ function [see (4) of Section 7.3] converges only if the absolute value of the maximum latent root of the argument matrix is less than 1. Hence if λ and t satisfy

$$\|I-\lambda\Sigma^{-1}\|<|1-2t\lambda|,$$

where $\|X\|$ denotes the maximum of the absolute values of the latent roots of X, we can write

$$(12)\quad \phi(t) = (1-2t\lambda)^{-mn/2}\det(\lambda^{-1}\Sigma)^{-n/2}\sum_{k=0}^{\infty}\sum_{\kappa}\frac{(\frac{1}{2}n)_\kappa C_\kappa(I-\lambda\Sigma^{-1})}{(1-2t\lambda)^k k!}.$$

Using the fact that $(1-2t\lambda)^{-r}$ is the moment-generating function of the gamma distribution with parameters r and 2λ and density function

$$(13)\qquad g_{r,2\lambda}(u)=\frac{e^{-u/2\lambda}u^{r-1}}{(2\lambda)^r\Gamma(r)}\qquad (u>0),$$

the moment-generating function (12) can be inverted term by term to give the density function of $\operatorname{tr}A$ as

(14)

$$f_{\operatorname{tr}A}(u)=\det(\lambda^{-1}\Sigma)^{-n/2}\sum_{k=0}^{\infty}\frac{1}{k!}g_{mn/2+k,2\lambda}(u)\sum_{\kappa}(\tfrac{1}{2}n)_{\kappa}C_{\kappa}(I-\lambda\Sigma^{-1})$$

$$(u>0),$$

valid for $0<\lambda<\infty$. Integrating with respect to u from 0 to x then gives the expression (11) for $P(\operatorname{tr}A\le x)$ and the proof is complete.

The parameter λ in (11) is open to choice; a sensible thing to do is to choose λ so that the series converges rapidly. We will not go into the specifics except to say that a value that appears to be close to optimal is $\lambda=2\lambda_1\lambda_m/(\lambda_1+\lambda_m)$, where λ_1 and λ_m are respectively the largest and smallest latent roots of Σ. For details see Khatri and Srivastava (1971) and the references therein. Note that when $\Sigma=\lambda I_m$, Theorem 8.1.13 reduces to

$$P(\operatorname{tr}A\le x)=P\left(\chi^2_{mn}\le\frac{x}{\lambda}\right);$$

that is, $\lambda^{-1}\operatorname{tr}A$ has the χ^2_{mn} distribution, a result previously derived in Theorem 3.2.20.

We now return to the problem of finding the moments of the ellipticity statistic V. These are given in the following theorem, also from Khatri and Srivastava (1971).

THEOREM 8.3.5. If $V=\det A/[(1/m)\operatorname{tr}A]^m$, where A is $W_m(n,\Sigma)$ (with $n>m-1$), then

(15)

$$E(V^h)=m^{mh}\frac{\Gamma(\tfrac{1}{2}mn)\Gamma_m(\tfrac{1}{2}n+h)}{\Gamma(\tfrac{1}{2}mn+mh)\Gamma_m(\tfrac{1}{2}n)}$$

$$\cdot\det(\lambda\Sigma^{-1})^{n/2}\sum_{k=0}^{\infty}\frac{(\tfrac{1}{2}mn)_k}{(\tfrac{1}{2}mn+mh)_k k!}\sum_{\kappa}(\tfrac{1}{2}n+h)_{\kappa}C_{\kappa}(I-\lambda\Sigma^{-1})$$

where λ $(0<\lambda<\infty)$ is arbitrary.

Proof. Using the $W_m(n,\Sigma)$ density function and the definition of V we have

$$(16)\quad E(V^h)=\frac{m^{mh}(\det\Sigma)^{-n/2}}{2^{mn/2}\Gamma_m(\frac{1}{2}n)}\int_{A>0}\operatorname{etr}(-\tfrac{1}{2}\Sigma^{-1}A)(\det A)^{(n+2h-m-1)/2}$$

$$\cdot(\operatorname{tr}A)^{-mh}(dA)$$

$$=(2m)^{mh}\frac{\Gamma_m(\frac{1}{2}n+h)}{\Gamma_m(\frac{1}{2}n)}(\det\Sigma)^h E\left[(\operatorname{tr}A)^{-mh}\right],$$

where the matrix A in this last expectation has the $W_m(n+2h,\Sigma)$ distribution. In this case it follows from Theorem 8.1.3 [see (14)] that the density function of $\operatorname{tr}A$ can be expressed as

$$f_{\operatorname{tr}A}(u)=\det(\lambda^{-1}\Sigma)^{-n/2-h}\sum_{k=0}^{\infty}\frac{1}{k!}g_{mn/2+mh+k,2\lambda}(u)$$

$$\cdot\sum_{\kappa}(\tfrac{1}{2}n+h)_\kappa C_\kappa(I-\lambda\Sigma^{-1}),$$

where $0<\lambda<\infty$ and $g_{a,b}(u)$ denotes the gamma (a,b) density function given by (13). Using the fact that the $(-mh)$th moment of the gamma distribution with parameters $\frac{1}{2}mn+mh+k$ and 2λ is $(2\lambda)^{-mh}\Gamma(\frac{1}{2}mn+k)/\Gamma(\frac{1}{2}mn+mh+k)$, we find that

$$E\left[(\operatorname{tr}A)^{-mh}\right]=\det(\lambda^{-1}\Sigma)^{-n/2-h}(2\lambda)^{-mh}\sum_{k=0}^{\infty}\frac{1}{k!}\frac{\Gamma(\frac{1}{2}mn+k)}{\Gamma(\frac{1}{2}mn+mh+k)}$$

$$\cdot\sum_{\kappa}(\tfrac{1}{2}n+h)_\kappa C_\kappa(I-\lambda\Sigma^{-1}).$$

Substituting this in (16) gives the desired result and completes the proof.

The expression (15) for the moments of V simplifies considerably when the null hypothesis $H:\Sigma=\lambda I_m$ is true.

COROLLARY 8.3.6. When $\Sigma=\lambda I_m$ the moments of the ellipticity statistic V are given by

$$(17)\qquad E(V^h)=m^{mh}\frac{\Gamma(\frac{1}{2}mn)}{\Gamma(\frac{1}{2}mn+mh)}\frac{\Gamma_m(\frac{1}{2}n+h)}{\Gamma_m(\frac{1}{2}n)}.$$

The moments of V may be used to obtain exact expressions for the density function using the Mellin transform approach described briefly in Section 8.2.3. For work in this direction, see Khatri and Srivastava (1971). Except in some special cases the exact distributions are extremely complicated, and we prefer here to concentrate on asymptotic distributions.

8.3.3. *The Asymptotic Null Distribution of the Likelihood Ratio Statistic*

Replacing h in Corollary 8.3.6 by $nh/2$, where $n=N-1$, shows that when $H:\Sigma=\lambda I_m$ is true, the hth moment of $W=V^{n/2}=\Lambda^{n/N}$ is

$$(18)\qquad E(W^h)=E(V^{nh/2})$$
$$=m^{mnh/2}\frac{\Gamma(\frac{1}{2}mn)}{\Gamma[\frac{1}{2}mn(1+h)]}\frac{\Gamma_m[\frac{1}{2}n(1+h)]}{\Gamma_m(\frac{1}{2}n)}$$
$$=Km^{mnh/2}\frac{\prod_{k=1}^{m}\Gamma[\frac{1}{2}n(1+h)-\frac{1}{2}(k-1)]}{\Gamma[\frac{1}{2}mn(1+h)]},$$

where K is a constant not involving h. This has the same form as (18) of Section 8.2.4, with $p=1$; $q=m$; $x_k=\frac{1}{2}n$; $\xi_k=-\frac{1}{2}(k-1)$, with $k=1,\dots,m$; $y_1=\frac{1}{2}mn$; $\eta_1=0$. The degrees of freedom in the limiting χ^2 distribution are, from (28) of Section 8.2.4,

$$(19)\qquad f=-2\left[\sum_{k=1}^{q}\xi_k-\sum_{j=1}^{p}\eta_j-\tfrac{1}{2}(q-p)\right]$$
$$=-2\left[\sum_{k=1}^{m}\tfrac{1}{2}(1-k)-\tfrac{1}{2}(m-1)\right]$$
$$=\tfrac{1}{2}(m+2)(m-1).$$

The value of ρ which makes the term of order n^{-1} vanish in the asymptotic expansion of the distribution of $-2\rho\log W$ is, from (30) of Section 8.2.4,

$$(20)\qquad \rho=1-\frac{1}{f}\left[\sum_{k=1}^{q}x_k^{-1}\left(\xi_k^2-\xi_k+\frac{1}{6}\right)-\sum_{j=1}^{p}y_j^{-1}\left(\eta_j^2-\eta_j+\frac{1}{6}\right)\right]$$
$$=1-\frac{1}{fn}\left\{\sum_{k=1}^{m}\left(\tfrac{1}{2}k^2-\frac{1}{6}\right)-\frac{1}{3m}\right\}$$
$$=1-\frac{2m^2+m+2}{6mn}.$$

With this value of ρ it is then found, using (29) of Section 8.2.4, that

$$\omega_2 = \frac{(m-1)(m-2)(m+2)(2m^3+6m^2+3m+2)}{288m^2n^2\rho^2}. \tag{21}$$

Hence we have obtained the following result, first given explicitly by Anderson (1958), Section 10.7.4.

THEOREM 8.3.7. When the null hypothesis $H: \Sigma = \lambda I_m$ is true, the distribution function of $-2\rho\log W$, where ρ is given by (20), can be expanded for large $M = \rho n$ as

(22)

$$P(-2\rho\log W \le x) = P(-n\rho\log V \le x) = P\left(-2\frac{n}{N}\rho\log\Lambda \le x\right)$$
$$= P\left(\chi_f^2 \le x\right) + \frac{\gamma}{M^2}\left[P\left(\chi_{f+4}^2 \le x\right) - P\left(\chi_f^2 \le x\right)\right] + O(M^{-3}),$$

where $f = (m+2)(m-1)/2$ and $\gamma = (n\rho)^2\omega_2 = M^2\omega_2$, with ω_2 given by (21).

Table 4, taken from Nagarsenker and Pillai (1973a), gives the lower 5 and 1 percentage points of the distribution of V.

8.3.4. Asymptotic Non-Null Distributions of the Likelihood Ratio Statistic

The power function of the likelihood ratio test of size α is $P(-n\rho\log V \ge k_\alpha^* | \lambda_1, \dots, \lambda_m)$, where ρ is given by (20) and k_α^* denotes the upper $100\alpha\%$ point of the distribution of $-n\rho\log V$ when $H: \Sigma = \lambda I_m$ is true. This is a function of the latent roots $\lambda_1, \dots, \lambda_m$ of Σ. We have seen that an approximation for k_α^* is $c_f(\alpha)$, the upper $100\alpha\%$ point of the χ_f^2 distribution, with $f = (m+2)(m-1)/2$. In this section we investigate ways of approximating the power function.

We noted earlier, in Section 8.2.6, that the asymptotic non-null distributions of likelihood ratio statistics depend upon the type of alternative being considered. Here again we consider three alternatives: a fixed alternative $K: \Sigma \ne \lambda I_m$, and two sequences of local alternatives expresssed in terms of the asymptotic variable $M = \rho n$ as

$$K_M: \quad \Sigma = \lambda\left(I_m + \frac{1}{M}\Omega\right)$$

Table 4. Lower 5 and 1 percentage points of the ellipticity statistic V for testing sphericity ($\Sigma = \lambda I$)[a]: $\alpha = .05$

$N \backslash m$	4	5	6	7	8	9	10
5	$0.0^{4}9528$						
6	$0.0^{2}3866$	$0.0^{4}2578$					
7	0.01687	$0.0^{2}1262$	$0.0^{5}7479$				
8	0.03866	$0.0^{2}6400$	$0.0^{3}4267$	$0.0^{5}2284$			
9	0.06640	0.01650	$0.0^{2}2553$	$0.0^{3}1473$	$0.0^{6}7219$		
10	0.09739	0.03110	$0.0^{2}7004$	$0.0^{3}9434$	$0.0^{4}5149$	$0.0^{6}2326$	
11	0.1297	0.04919	0.01435	$0.0^{2}2950$	$0.0^{3}3631$	$0.0^{4}1817$	$0.0^{7}7722$
12	0.1621	0.06970	0.02433	$0.0^{2}6524$	$0.0^{2}1233$	$0.0^{3}1397$	$0.0^{5}6455$
13	0.1938	0.09174	0.03653	0.01179	$0.0^{2}2924$	$0.0^{3}5114$	$0.0^{4}5370$
14	0.2244	0.1146	0.05051	0.01870	$0.0^{2}5613$	$0.0^{2}1295$	$0.0^{3}2107$
15	0.2535	0.1378	0.06583	0.02712	$0.0^{2}9379$	$0.0^{2}2629$	$0.0^{3}5667$
16	0.2812	0.1608	0.08210	0.03682	0.01423	$0.0^{2}4616$	$0.0^{2}1214$
17	0.3074	0.1835	0.09900	0.04761	0.02011	$0.0^{2}7314$	$0.0^{2}2235$
18	0.3321	0.2058	0.1163	0.05927	0.02693	0.01074	$0.0^{2}3692$
19	0.3533	0.2273	0.1337	0.07161	0.03460	0.01489	$0.0^{2}5630$
20	0.3772	0.2482	0.1511	0.08446	0.04299	0.01973	$0.0^{2}8071$
22	0.4173	0.2876	0.1854	0.1111	0.06154	0.03129	0.01448
24	0.4530	0.3240	0.2185	0.1383	0.08178	0.04494	0.02282
26	0.4848	0.3575	0.2501	0.1654	0.1030	0.06022	0.03287
28	0.5134	0.3882	0.2800	0.1920	0.1248	0.07667	0.04435
30	0.5390	0.4164	0.3081	0.2178	0.1467	0.09392	0.05698
34	0.5833	0.4663	0.3594	0.2665	0.1898	0.1296	0.08468
42	0.6508	0.5453	0.4442	0.3515	0.2697	0.2006	0.1444
50	0.6998	0.6046	0.5106	0.4211	0.3389	0.2660	0.2035
60	0.7447	0.6603	0.5749	0.4910	0.4112	0.3376	0.2715
80	0.8037	0.7354	0.6641	0.5916	0.5196	0.4499	0.3840
100	0.8406	0.7835	0.7228	0.6597	0.5955	0.5317	0.4694
140	0.8842	0.8413	0.7948	0.7453	0.6935	0.6405	0.5870
200	0.9179	0.8868	0.8525	0.8153	0.7757	0.7342	0.6913
300	0.9447	0.9234	0.8996	0.8734	0.8452	0.8151	0.7833

[a] Here, m = number of variables; N = sample size

Source: Reproduced from Nagarsenker and Pillai (1973a) with the kind permission of Academic Press, Inc., and the authors.

and

$$K_M^*: \quad \Sigma = \lambda\left(I_m + \frac{1}{M^{1/2}}\Omega\right),$$

where Ω is a fixed matrix. By invariance we can assume that both Σ and Ω are diagonal, with $\Omega = \text{diag}(\omega_1, \dots, \omega_m)$. The asymptotic distributions are

Table 4 (*Continued*): $\alpha = .01$

$N \backslash m$	4	5	6	7	8	9	10
5	$0.0^{5}3665$						
6	$0.0^{3}6904$	$0.0^{6}9837$					
7	$0.0^{2}5031$	$0.0^{3}2184$	$0.0^{6}2970$				
8	0.01503	$0.0^{2}1828$	$0.0^{4}7187$	$0.0^{7}8604$			
9	0.03046	$0.0^{2}6123$	$0.0^{3}6758$	$0.0^{4}2424$	$0.0^{7}2760$		
10	0.05010	0.01361	$0.0^{2}2498$	$0.0^{3}2520$	$0.0^{5}8306$	$0.0^{8}9216$	
11	0.07258	0.02416	$0.0^{2}6033$	$0.0^{2}1017$	$0.0^{4}9438$	$0.0^{5}2879$	$0.0^{8}3573$
12	0.09679	0.03730	0.01148	$0.0^{2}2646$	$0.0^{3}4120$	$0.0^{4}3544$	$0.0^{5}1004$
13	0.1218	0.05248	0.01880	$0.0^{2}5369$	$0.0^{2}1149$	$0.0^{3}1663$	$0.0^{4}1332$
14	0.1471	0.06915	0.02782	$0.0^{2}9296$	$0.0^{2}2476$	$0.0^{3}4943$	$0.0^{4}6681$
15	0.1721	0.08685	0.03830	0.01444	$0.0^{2}4516$	$0.0^{2}1126$	$0.0^{3}2108$
16	0.1966	0.10518	0.04998	0.02073	$0.0^{2}7343$	$0.0^{2}2160$	$0.0^{3}5065$
17	0.2204	0.1239	0.06261	0.02807	0.01098	$0.0^{2}3669$	$0.0^{2}1018$
18	0.2434	0.1426	0.07595	0.03635	0.01542	$0.0^{2}5707$	$0.0^{2}1804$
19	0.2655	0.1613	0.08982	0.04541	0.02062	$0.0^{2}8300$	$0.0^{2}2914$
20	0.2867	0.1797	0.1040	0.05514	0.02652	0.01146	$0.0^{2}4386$
22	0.3264	0.2156	0.1330	0.07612	0.04017	0.01940	$0.0^{2}8498$
24	0.3626	0.2497	0.1620	0.09845	0.05580	0.02933	0.01421
26	0.3956	0.2819	0.1904	0.1215	0.07287	0.04095	0.02145
28	0.4257	0.3120	0.2180	0.1447	0.09092	0.05392	0.03007
30	0.4531	0.3402	0.2445	0.1677	0.1096	0.06795	0.03989
34	0.5013	0.3910	0.2940	0.2125	0.1475	0.09805	0.06231
42	0.5769	0.4741	0.3789	0.2939	0.2211	0.1611	0.1136
50	0.6331	0.5383	0.4475	0.3632	0.2876	0.2221	0.1671
60	0.6856	0.6001	0.5157	0.4348	0.3594	0.2912	0.2311
80	0.7558	0.6852	0.6129	0.5408	0.4705	0.4034	0.3411
100	0.8006	0.7407	0.6782	0.6144	0.5505	0.4879	0.4275
140	0.8541	0.8085	0.7598	0.7088	0.6562	0.6028	0.5495
200	0.8961	0.8626	0.8262	0.7874	0.7465	0.7040	0.6605
300	0.9297	0.9066	0.8811	0.8535	0.8239	0.7927	0.7600

different for these three cases. We first look at the two sequences of local alternatives.

THEOREM 8.3.8. (a) Under the sequence of local alternatives K_M: $\Sigma = \lambda[I_m + (1/M)\Omega]$, the distribution function of $-M\log V$ can be expanded as

$$P(-M\log V \le x) = P(\chi_f^2 \le x) + \frac{1}{4M}\left(\sigma_2 - \frac{\sigma_1^2}{m}\right)\left[P(\chi_{f+2}^2 \le x)\right.$$
$$\left. - P(\chi_f^2 \le x)\right] + O(M^{-2}).$$

(b) Under the sequence of local alternatives $K_M^*: \Sigma = \lambda[I_m + (1/M^{1/2})\Omega]$, the distribution function of $-M\log V$ can be expanded as

$$P(-M\log V \le x) = P[\chi_f^2(\delta) \le x]$$
$$+\frac{1}{M^{1/2}}\left\{\left(\frac{\sigma_3}{3} - \frac{\sigma_1\sigma_2}{2m} + \frac{\sigma_1^3}{6m^2}\right)P[\chi_f^2(\delta) \le x]\right.$$
$$+\left(\frac{\sigma_1\sigma_2}{m} - \frac{\sigma_3}{2} - \frac{\sigma_1^3}{2m^2}\right)P[\chi_{f+2}^2(\delta) \le x]$$
$$\left.+\frac{1}{6}\left(\sigma_3 - \frac{\sigma_1\sigma_2}{3m} + \frac{2\sigma_1^3}{m^2}\right)P[\chi_{f+4}^2(\delta) \le x]\right\}$$
$$+O(M^{-1}).$$

Here $f = \frac{1}{2}(m+2)(m-1)$, $\sigma_j = \operatorname{tr}\Omega^j$ and in (b) the noncentrality parameter is $\delta = \frac{1}{2}(\sigma_2 - \sigma_1^2/m)$.

Proof. Both (a) and (b) can be proved by expanding the characteristic function of $-M\log V$. Here only a proof of (a) will be presented. Using Theorem 8.3.5, the characteristic function of $-M\log V$ under K_M is

$$(23)\qquad \phi(M,t,\Omega) = E(V^{-Mit})$$
$$= \phi(M,t,0)\det\left(I + \frac{1}{M}\Omega\right)^{-(M/2+\varepsilon)} G(M,t,\Omega),$$

where

(24)

$$G(M,t,\Omega) = \sum_{k=0}^{\infty} \frac{(\frac{1}{2}mN + m\varepsilon)_k}{[\frac{1}{2}mN(1-2it)+m\varepsilon]_k k!}\sum_{\kappa}\frac{[\frac{1}{2}N(1-2it)+\varepsilon]_\kappa}{\left(C_\kappa\left[\Omega\left(I - \frac{1}{N}\Omega\right)^{-1}\right]\right)^{-1}}$$

with

$$\varepsilon = \tfrac{1}{2}(n-M) = \tfrac{1}{2}n(1-\rho) = \frac{2m^2+m+2}{12m},$$

and $\phi(M,t,0)$ is the characteristic function of $-M\log V$ when H is true ($\Omega = 0$), obtained from Corollary 8.3.6 by putting $h = -Mit$. From Theorem

8.3.7 we know that

$$\phi(M,t,0)=(1-2it)^{-f/2}+O(M^{-2}), \tag{25}$$

where $f=\frac{1}{2}(m+2)(m-1)$. Consider next the determinant term in (23). Taking logs and expanding gives

$$\begin{aligned}\log\det\left(I+\frac{1}{M}\Omega\right)^{-(M/2+\varepsilon)} &= -(\tfrac{1}{2}M+\varepsilon)\sum_{i=1}^{m}\log\left(1+\frac{\omega_i}{M}\right)\\ &= -(\tfrac{1}{2}M+\varepsilon)\sum_{i=1}^{m}\left(\frac{\omega_i}{M}-\frac{\omega_i^2}{2M^2}+\cdots\right)\\ &= -\tfrac{1}{2}\sigma_1+\frac{1}{M}(\tfrac{1}{4}\sigma_2-\varepsilon\sigma_1)+O(M^{-2}),\end{aligned}$$

where $\sigma_j=\operatorname{tr}\Omega^j$, and hence

$$\det\left(I+\frac{1}{M}\Omega\right)^{-(M/2+\varepsilon)}=e^{-\sigma_1/2}\left[1+\frac{1}{M}(\tfrac{1}{4}\sigma_2-\varepsilon\sigma_1)+O(M^{-2})\right]. \tag{26}$$

It remains to expand the function $G(M,t,\Omega)$ for large M. The reader may recall that a partial differential equation approach was used in Section 8.2.6 for a similar problem. Here, however, although G slightly resembles a ${}_2F_1$ function, no differential equation is known for it. One way of expanding G, and the method which will be used here, is to expand each term in the series and then sum. This amounts to formally rearranging the series in terms of powers of M^{-1}. It is easily verified that

$$\begin{aligned}\frac{(\frac{1}{2}mM+m\varepsilon)_k}{(\frac{1}{2}mM(1-2it)+m\varepsilon)_k} &= (1-2it)^{-k}\left\{1+\frac{1}{mM}[k(k-1)+2km\varepsilon]\right.\\ &\quad\left.\cdot\left[1-\frac{1}{1-2it}\right]+O(M^{-2})\right\}\end{aligned} \tag{27}$$

and that, with $\kappa=(k_1,\ldots,k_m)$,

(28)

$$\frac{(\frac{1}{2}M(1-2it)+\varepsilon)_\kappa}{M^k}=2^{-k}(1-2it)^k\left[1+\frac{1}{(1-2it)M}(\rho_\kappa+2\varepsilon k)+O(M^{-2})\right],$$

where

$$\rho_\kappa = \sum_{i=1}^{m} k_i(k_i - i).$$

Multiplying (27) and (28) and substituting in (24) then gives

$$(29)\quad G(M,t,\Omega) = \sum_{k=0}^{\infty}\sum_{\kappa}\frac{C_\kappa(R)}{k!}\left\{1 + \frac{1}{mM(1-2it)}\left[m\rho_\kappa - k(k-1)\right] + \frac{1}{mN}\left[k(k-1) + 2km\varepsilon\right] + O(M^{-2})\right\},$$

where

$$R = \tfrac{1}{2}\Omega\left(I + \frac{1}{M}\Omega\right)^{-1}.$$

We now sum, using the fact that

$$(30)\quad \sum_{k=0}^{\infty}\sum_{\kappa}\frac{C_\kappa(R)}{k!} = \operatorname{etr}(R)$$

and applying the formulas

$$(31)\quad \sum_{k=0}^{\infty}\sum_{\kappa}k\frac{C_\kappa(R)}{k!} = \operatorname{etr}(R)\operatorname{tr}R$$

and

$$(32)\quad \sum_{k=0}^{\infty}\sum_{\kappa}\rho_\kappa\frac{C_\kappa(R)}{k!} = \operatorname{etr}(R)\operatorname{tr}R^2,$$

which were proved in Lemma 7.5.2. Also needed is the formula

$$(33)\quad \sum_{k=0}^{\infty}\sum_{\kappa}k^2\frac{C_\kappa(R)}{k!} = \operatorname{etr}(R)\left[\operatorname{tr}R + (\operatorname{tr}R)^2\right],$$

which is readily established by applying the differential operator $E =$

$\Sigma_{i=1}^m r_i \partial/\partial r_i$ to (31). We get

$$(34)\qquad G(M,t,\Omega)=\operatorname{etr}(R)\left\{1+\frac{1}{mM(1-2it)}\left[m\operatorname{tr}R^2-(\operatorname{tr}R)^2\right]\right.$$
$$\left.+\frac{1}{mM}\left[(\operatorname{tr}R)^2+2m\varepsilon\operatorname{tr}R\right]+O(M^{-2})\right\}.$$

We now use

$$\operatorname{etr}(R)=\operatorname{etr}\left[\tfrac{1}{2}\Omega\left(I+\frac{1}{M}\Omega\right)^{-1}\right]=\operatorname{etr}(\tfrac{1}{2}\Omega)\left[1-\frac{\sigma_2}{2M}+O(M^{-2})\right],$$
$$\operatorname{tr}R=\tfrac{1}{2}\sigma_1+O(M^{-1}),\qquad (\operatorname{tr}R)^2=\tfrac{1}{4}\sigma_1^2+O(M^{-1}),$$
$$\operatorname{tr}R^2=\tfrac{1}{4}\sigma_2+O(M^{-1})$$

in (34) and multiply the resulting expansion by the expansions (25) and (26) to give

$$\phi(M,t,\Omega)=(1-2it)^{-f/2}$$
$$\times\left\{1+\frac{1}{4M}\left(\sigma_2-\frac{\sigma_1^2}{m}\right)\left(\frac{1}{1-2it}-1\right)+O(M^{-2})\right\}.$$

Inverting this expansion for the characteristic function of $-M\log V$ then establishes (a). Part (b) can be proved in a similar way, although the details are nowhere near as straightforward.

Finally, we consider the asymptotic distribution of the likelihood ratio statistic under general alternatives. This is given in the following theorem, whose proof is omitted.

THEOREM 8.3.9. Under the fixed alternative $K:\Sigma\neq\lambda I$ the distribution of the random variable

$$Y=M^{1/2}\left[-\log V+\log\frac{\det\Sigma}{\left(\frac{1}{m}\operatorname{tr}\Sigma\right)^m}\right]$$

can be expanded as

$$P\left(\frac{Y}{\tau}\leq x\right)=\Phi(x)-\frac{1}{M^{1/2}}\left[a_1\phi(x)+a_2\phi^{(2)}(x)\right]+O(M^{-1}),$$

where $\Phi(x)$ and $\phi(x)$ denote the standard normal distribution and density functions, respectively, and

$$\tau^2=2m(t_2-1)$$

$$a_1=\tfrac{1}{2}\tau(m^2+m-2t_2)$$

$$a_2=\frac{2m}{3\tau^3}\left[2t_3-3t_2+1-3(t_2-1)^2\right],$$

with

$$t_j=m^{j-1}\frac{\operatorname{tr}\Sigma^j}{(\operatorname{tr}\Sigma)^j}.$$

For derivations and further terms in the asymptotic expansions presented in this section, the interested reader should see Sugiura (1969b), Nagao (1970, 1973b) and Khatri and Srivastava (1974).

8.3.5. *The Asymptotic Null Distribution of the Likelihood Ratio Statistic for an Elliptical Sample*

As an attempt to understand the effect of non-normality on the distribution of V we examine what happens when the sample comes from an elliptical distribution. In

$$V=\frac{\det S}{\left(\dfrac{1}{m}\operatorname{tr}S\right)^m},$$

where S is the sample covariance matrix, we substitute $S=\lambda(I_m+n^{-1/2}Z)$. Then, when $H:\Sigma=\lambda I_m$ is true, $-n\log V$ can be expanded as

$$\begin{aligned}-n\log V&=\tfrac{1}{2}\operatorname{tr}(Z^2)-\frac{1}{2m}(\operatorname{tr}Z)^2+O_p(n^{-1/2})\\&=\sum_{i<j}^{m}z_{ij}^2+\tfrac{1}{2}\sum_{i=1}^{m}(z_{ii}-\bar{z})^2+O_p(n^{-1/2})\\&=\mathbf{u}'B\mathbf{u}+O_p(n^{-1/2}),\end{aligned}$$

where

$$Z=(z_{ij}), \qquad \bar{z}=m^{-1}\sum_{i=1}^{m} z_{ii},$$

$$\mathbf{u}'=\left(\frac{z_{11}}{2^{1/2}},\ldots,\frac{z_{mm}}{2^{1/2}},z_{12},\ldots,z_{1m},z_{23},\ldots,z_{m-1,m}\right)$$

and

$$B=\begin{bmatrix} I_m-m^{-1}\mathbf{1}\mathbf{1}' & 0 \\ 0 & I_{m(m-1)/2}\end{bmatrix}$$

with $\mathbf{1}=(1,\ldots,1)'\in R^m$. Now suppose that the sample is drawn from an elliptical distribution with kurtosis parameter κ. The asymptotic distribution of $\mathbf{u}$, as $n\to\infty$, is $N_{m(m+1)/2}(\mathbf{0},\Gamma)$, where

$$\Gamma=\begin{bmatrix} (1+\kappa)I_m+\frac{1}{2}\kappa\mathbf{1}\mathbf{1}' & 0 \\ 0 & (1+\kappa)I_{m(m-1)/2}\end{bmatrix}.$$

Since $B\Gamma B=(1+\kappa)B$ and rank $(B\Gamma)=\frac{1}{2}(m-1)(m+2)$ it follows that the asymptotic distribution of $-n\log V/(1+\kappa)$ is $\chi^2_{(m-1)(m+2)/2}$. We summarize this result in the following theorem.

THEOREM 8.3.10. If the sample is drawn from an elliptical distribution with kurtosis parameter κ then the asymptotic distribution of $-n\log V/(1+\kappa)$, when $H:\Sigma=\lambda I_m$ is true, is $\chi^2_{(m-1)(m+2)/2}$.

When the sample is normally distributed $\kappa=0$, and this result agrees with that derived in Section 8.3.3. Note also that κ can be replaced by a consistent estimate without affecting the asymptotic distribution. A Monte Carlo study carried out by C. M. Waternaux (unpublished) indicates that the usual test statistic $-n\log V$ should be used with extreme caution, if at all, for testing $H:\Sigma=I_m$ when the underlying population is elliptical with longer tails than the normal distribution. For example 200 samples of size $N=200$ drawn from a six-variate contaminated normal distribution with $\sigma=2$, $\varepsilon=.3$, and $\Sigma=I_6$ gave an observed significance level for the test based on $-n\log V$, of 50%, for a nominal value of 5%. On the other hand, a test based on $-(n\log V)/(1+\hat{\kappa})$, where $\hat{\kappa}$ is a consistent estimate of κ, yielded an observed significance level of 7.5% for the same nominal level of 5%.

8.3.6. *Other Test Statistics*

A number of the other invariant test statistics have been proposed for testing the null hypothesis $H: \Sigma = \lambda I_m$. These include the two statistics

$$L_1 = \frac{a_1}{a_m},$$

the ratio of the largest to the smallest latent root of A, and

$$L_2 = \frac{\operatorname{tr} A^2}{(\operatorname{tr} A)^2} = \frac{a_1^2 + \cdots + a_m^2}{(a_1 + \cdots + a_m)^2}.$$

The null hypothesis is rejected if either of these statistics is large enough. Note that a test based on L_1 is equivalent to a test based on

$$L_3 = \frac{a_1 a_m}{[\frac{1}{2}(a_1 + a_m)]^2} = \frac{4L_1}{(1+L_1)^2}.$$

The exact null distribution of L_1 has been given by Sugiyama (1970) as a complicated series involving zonal polynomials, and it has been further studied by Waikar and Schuurmann (1973) and Krishnaiah and Schuurmann (1974); in these last two papers upper percentage points of the distribution may be found. The test based on L_2 has the optimal property of being a locally best (most powerful) invariant test. What this means is that for every $\lambda > 0$ and for every other invariant test L (say) there is a neighborhood $N(\lambda I_m)$ of λI_m such that the power of L_2 is no less than the power of L. For a proof of this and for distributional results associated with L_2, the reader is referred to John (1971, 1972) and Sugiura (1972b). No extensive power comparisons of the tests based on V, L_1, and L_2 (and others) have been carried out. In view of the fact that the asymptotic nonnull distributions of L_2 are similar to those of V (see Sugiura (1972b), it seems unlikely that there is much difference between these two.

8.4. TESTING THAT A COVARIANCE MATRIX EQUALS A SPECIFIED MATRIX

8.4.1. *The Likelihood Ratio Test and Invariance*

Let $\mathbf{X}_1, \dots, \mathbf{X}_N$ be independent $N_m(\boldsymbol{\mu}, \Sigma)$ random vectors and consider testing the null hypothesis $H: \Sigma = \Sigma_0$, where Σ_0 is a specified positive definite matrix, against $K: \Sigma \neq \Sigma_0$. An argument similar to that used at the end of Section 8.3.1 shows that this is equivalent to testing $H: \Sigma = I$ against $K: \Sigma \neq I$. All results in this section will be written in terms of the latter formulation.

We first look at this testing problem from an invariance point of view. Consider the affine group of transformations

$$(1) \qquad \mathcal{A}\ell^*(m,R) = \{(H,\mathbf{c});\, H \in O(m), \mathbf{c} \in R^m\}$$

(a subgroup of the full affine group $\mathcal{A}\ell(m,R)$ given by (14) of Section 6.3) acting on the sample space of the sufficient statistic $(\overline{\mathbf{X}}, A)$, where

$$\overline{\mathbf{X}} = N^{-1} \sum_{i=1}^{N} \mathbf{X}_i \quad \text{and} \quad A = \sum_{i=1}^{N} (\mathbf{X}_i - \overline{\mathbf{X}})(\mathbf{X}_i - \overline{\mathbf{X}})',$$

by

$$(2) \qquad \overline{\mathbf{X}} \rightarrow H\overline{\mathbf{X}} + \mathbf{c} \quad \text{and} \quad A \rightarrow HAH'.$$

The induced group of transformations on the parameter space is given by

$$(3) \qquad \boldsymbol{\mu} \rightarrow H\boldsymbol{\mu} + \mathbf{c} \quad \text{and} \quad \Sigma \rightarrow H\Sigma H',$$

and the testing problem is invariant, for the family of distributions of $(\overline{\mathbf{X}}, A)$ is invariant, as are the null and alternative hypotheses. A maximal invariant is given in the following theorem.

THEOREM 8.4.1. Under the group of transformations (3) a maximal invariant is $(\lambda_1, \dots, \lambda_m)$, where $\lambda_1 \geq \lambda_2 \geq \cdots \geq \lambda_m$ (>0) are the latent roots of Σ.

Proof. Let $\phi(\boldsymbol{\mu}, \Sigma) = (\lambda_1, \dots, \lambda_m)$, and first note that $\phi(\boldsymbol{\mu}, \Sigma)$ is invariant, because for $H \in O(m)$, $H\Sigma H'$ and Σ have the same latent roots. To show it is maximal invariant, suppose that

$$\phi(\boldsymbol{\mu}, \Sigma) = \phi(\boldsymbol{\tau}, \Gamma),$$

i.e., Σ and Γ have the same latent roots $\lambda_1, \dots, \lambda_m$. Choose $H_1 \in O(m)$, $H_2 \in O(m)$ such that $H_1 \Sigma H_1' = \Delta$, $H_2 \Gamma H_2' = \Delta$, where

$$\Delta = \operatorname{diag}(\lambda_1, \dots, \lambda_m).$$

Then

$$\Gamma = H_2' H_1 \Sigma H_1' H_2 = H\Sigma H',$$

where

$$H = H_2' H_1 \in O(m).$$

Putting $\mathbf{c} = -H\boldsymbol{\mu} + \boldsymbol{\tau}$, we then have

$$H\boldsymbol{\mu} + \mathbf{c} = \boldsymbol{\tau} \quad \text{and} \quad H\Sigma H' = \Gamma$$

so that

$$(\boldsymbol{\mu}, \Sigma) \sim (\boldsymbol{\tau}, \Gamma) \qquad [\operatorname{mod} \mathcal{AL}^*(m, R)],$$

and the proof is complete.

It follows from this that a maximal invariant under the group $\mathcal{AL}^*(m, R)$ acting on the sample space of the sufficient statistic $(\overline{\mathbf{X}}, A)$ is $(a_1, \dots, a_m)$, where $a_1 > \cdots > a_m > 0$ are the latent roots of the Wishart matrix A. Any invariant test depends only on $a_1, \dots, a_m$, and from Theorem 6.1.12 the distribution of $a_1, \dots, a_m$ depends only on $\lambda_1, \dots, \lambda_m$, the latent roots of Σ. This has previously been noted in the discussion following Theorem 3.2.18. The likelihood ratio test, given in the following theorem due to Anderson (1958), is an invariant test.

THEOREM 8.4.2. The likelihood ratio test of size α of $H: \Sigma = I_m$ rejects H if $\Lambda \le c_\alpha$, where

$$\Lambda = \left(\frac{e}{N}\right)^{mN/2} \operatorname{etr}\left(-\tfrac{1}{2}A\right)(\det A)^{N/2} \tag{4}$$

and c_α is chosen so that the size of the test is α.

Proof. The likelihood ratio statistic is

$$\Lambda = \frac{\sup_{\boldsymbol{\mu}} L(\boldsymbol{\mu}, I_m)}{\sup_{\boldsymbol{\mu}, \Sigma} L(\boldsymbol{\mu}, \Sigma)} \tag{5}$$

where

$$L(\boldsymbol{\mu}, \Sigma) = (\det \Sigma)^{-N/2} \operatorname{etr}\left(-\tfrac{1}{2}\Sigma^{-1}A\right) \exp\left[-\tfrac{1}{2}N(\overline{\mathbf{X}} - \boldsymbol{\mu})'\Sigma^{-1}(\overline{\mathbf{X}} - \boldsymbol{\mu})\right].$$

The numerator in (5) is found by putting $\boldsymbol{\mu} = \overline{\mathbf{X}}$, and the denominator by putting $\boldsymbol{\mu} = \overline{\mathbf{X}}$, $\Sigma = N^{-1}A$. Substitution of these values gives the desired result.

8.4.2. Unbiasedness and the Modified Likelihood Ratio Test

The likelihood ratio test given in Theorem 8.4.2 is biased. This is well known in the case $m=1$ and was established in general by Das Gupta (1969) with the help of the following lemma.

LEMMA 8.4.3. Let Y be a random variable and $\sigma^2>0$ a constant such that Y/σ^2 has the χ_p^2 distribution. For $r>0$, let

$$\beta(\sigma^2)=P\left[Y^r\exp(-\tfrac{1}{2}Y)\ge k\,|\,\sigma^2\right].$$

Then

$$\frac{d\beta(\sigma^2)}{d\sigma^2}\;\substack{>\\=\\<}\;0 \quad\text{according as}\quad \sigma^2\;\substack{<\\=\\>}\;\frac{2r}{p}.$$

Proof. Since the region

$$Y^r\exp(-\tfrac{1}{2}Y)\ge k$$

is equivalent to $y_1\le Y\le y_2$, where

(6) $$y_1^r\exp(-\tfrac{1}{2}y_1)=y_2^r\exp(-\tfrac{1}{2}y_2)=k,$$

it follows by integration of the χ_p^2 density function that

$$\beta(\sigma^2)=C\int_{y_1/\sigma^2}^{y_2/\sigma^2}e^{-y/2}y^{p/2-1}\,dy.$$

where $C=[2^{p/2}\Gamma(\frac{1}{2}p)]^{-1}$. Differentiating with respect to σ^2 gives

$$\beta'(\sigma^2)=C(\sigma^2)^{-p/2-1}\left[\exp(-y_1/2\sigma^2)y_1^{p/2}-\exp(-y_2/2\sigma^2)y_2^{p/2}\right].$$

It follows that $\beta'(\sigma^2)\substack{>\\=\\<}0$ according as

$$\exp(-y_1/2\sigma^2)y_1^{p/2}\;\substack{>\\=\\<}\;\exp(-y_2/2\sigma^2)y_2^{p/2},$$

i.e., according as

(7) $$\sigma^2\;\substack{<\\=\\>}\;\frac{y_2-y_1}{p\ln(y_2/y_1)}.$$

Using (6) the right side of (7) is easily seen to be $2r/p$, and the proof is complete.

The following corollary is an immediate consequence.

COROLLARY 8.4.4. With the assumptions and notation of Lemma 8.4.3, $\beta(\sigma^2)$ decreases monotonically as $|\sigma^2-1|$ increases, if $2r=p$.

We are now in a position to demonstrate that the likelihood ratio test is biased. Note that rejecting H for small values of Λ is equivalent to rejecting H for small values of

$$V=\operatorname{etr}(-\tfrac{1}{2}A)(\det A)^{N/2} \tag{8}$$

THEOREM 8.4.5. For testing $H:\Sigma=I$ against $K:\Sigma\neq I$, the likelihood ratio test having the critical region $V\le c$ is biased.

Proof. By invariance we can assume without loss of generality that $\Sigma=\operatorname{diag}(\lambda_1,\dots,\lambda_m)$. The matrix $A=(a_{ij})$ has the $W_m(n,\Sigma)$ distribution, with $n=N-1$. Write

$$V=\left[\frac{\det A}{\prod_{i=1}^m a_{ii}}\right]^{N/2}\prod_{i=1}^m\left[\exp(-\tfrac{1}{2}a_{ii})a_{ii}^{N/2}\right].$$

Now, the random variables $\det A/\prod_{i=1}^m a_{ii}$ and a_{ii} (with $i=1,\dots,m$) are all independent, the first one having a distribution which does not depend on $(\lambda_1,\dots,\lambda_m)$, while a_{ii}/λ_i is χ^2_n (see the proof of Theorem 5.1.3). From Lemma 8.4.3 it follows that there exists a constant $\lambda_m^*\in(1,N/n)$ such that

$$P\left[a_{mm}^{N/2}\exp(-\tfrac{1}{2}a_{mm})\ge k\,|\,\lambda_m=1\right]<P\left[a_{mm}^{N/2}\exp(-\tfrac{1}{2}a_{mm})\ge k\,|\,\lambda_m=\lambda_m^*\right]$$

for any k. The desired result now follows when we evaluate $P(V\ge c)$ by conditioning on $a_{11},\dots,a_{m-1,m-1}$ and $\det A/\prod_{i=1}^m a_{ii}$.

By modifying the likelihood ratio statistic slightly, an unbiased test is obtained. The modified likelihood ratio statistic is

$$\Lambda^*=\left(\frac{e}{n}\right)^{mn/2}\operatorname{etr}(-\tfrac{1}{2}A)(\det A)^{n/2} \tag{9}$$

and is obtained from Λ by replacing the sample size N by the degrees of freedom n. This is exactly the likelihood ratio statistic that is obtained by working with the likelihood function for Σ specified by the Wishart density for A instead of the likelihood function specified by the original normally distributed sample. The modified likelihood ratio test then rejects $H:\Sigma=I$ for small enough values of Λ^*, or equivalently, of

$$V^*=\operatorname{etr}(-\tfrac{1}{2}A)(\det A)^{n/2}. \tag{10}$$

That this test is unbiased is a consequence of the stronger result in the following theorem, from Nagao (1967) and Das Gupta (1969).

THEOREM 8.4.6. The power function of the modified likelihood ratio test with critical region $V^* \leq c$ increases monotonically as $|\lambda_i - 1|$ increases for each $i = 1, \ldots, m$.

Proof. Corollary 8.4.4 shows that $P[a_{mm}^{n/2}\exp(-\frac{1}{2}a_{mm}) \leq k|\lambda_m]$ increases monotonically as $|\lambda_m - 1|$ increases. The desired result now follows using a similar argument to that used in the proof of Theorem 8.4.5.

8.4.3. *Moments of the Modified Likelihood Ratio Statistic*

Distributional results associated with Λ^* can be obtained via a study of its moments given by Anderson (1958).

THEOREM 8.4.7. The hth moment of the modified likelihood ratio statistic Λ^* given by (9) is

$$(11)\qquad E(\Lambda^{*h}) = \left(\frac{2e}{n}\right)^{mnh/2} \frac{(\det \Sigma)^{nh/2}}{\det(I + h\Sigma)^{n(1+h)/2}} \frac{\Gamma_m(\frac{1}{2}n(1+h))}{\Gamma_m(\frac{1}{2}n)}.$$

Proof. Using the $W_m(n, \Sigma)$ density function, the definition of Λ^*, and Theorem 2.1.9 we have

$$E(\Lambda^{*h}) = \left(\frac{e}{n}\right)^{mnh/2} \frac{(\det \Sigma)^{-n/2}}{2^{mn/2}\Gamma_m(\frac{1}{2}n)} \int_{A>0} \operatorname{etr}[-\tfrac{1}{2}(hI + \Sigma^{-1})A]$$

$$\cdot(\det A)^{n(1+h)/2-(m+1)/2}(dA)$$

$$= \left(\frac{e}{n}\right)^{mnh/2} \frac{(\det \Sigma)^{-n/2}}{2^{mn/2}\Gamma_m(\frac{1}{2}n)} \frac{2^{mn(1+h)/2}\Gamma_m(\frac{1}{2}n(1+h))}{\det(hI + \Sigma^{-1})^{n(1+h)/2}}$$

$$= \left(\frac{2e}{n}\right)^{mnh/2} \frac{(\det \Sigma)^{nh/2}}{\det(I + h\Sigma)^{n(1+h)/2}} \frac{\Gamma_m(\frac{1}{2}n(1+h))}{\Gamma_m(\frac{1}{2}n)},$$

completing the proof.

COROLLARY 8.4.8. When the null hypothesis $H: \Sigma = I$ is true the hth moment of Λ^* is

$$(12)\qquad E(\Lambda^{*h}) = \left(\frac{2e}{n}\right)^{mnh/2} (1+h)^{-mn(1+h)/2} \frac{\Gamma_m(\frac{1}{2}n(1+h))}{\Gamma_m(\frac{1}{2}n)}.$$

These null moments have been used by Nagarsenker and Pillai (1973b) to derive expressions for the exact distribution of Λ^*. These have been used to compute the upper 5 and 1 percentage points of the distribution of $-2\log\Lambda^*$ for $m=4(1)10$ and various values of n. The table of percentage points in Table 5 is taken from Davis and Field (1971).

8.4.4. *The Asymptotic Null Distribution of the Modified Likelihood Ratio Statistic*

The null moments of Λ^* given by Corollary 8.4.8 are not of the form (18) of Section 8.2.4. Nevertheless, it is still possible to find a constant ρ so that the term of order n^{-1} vanishes in an asymptotic expansion of the distribution of $-2\rho\log\Lambda^*$. The result is given in the following theorem.

THEOREM 8.4.9. When the null hypothesis H: $\Sigma=I$ is true the distribution of $-2\rho\log\Lambda^*$ can be expanded as

$$P(-2\rho\log\Lambda^*\le x)=P(\chi_f^2\le x)+\frac{\gamma}{M^2}\left[P(\chi_{f+4}^2\le x)-P(\chi_f^2\le x)\right]+O(M^{-3})$$

where

$$(13)\qquad \rho=1-\frac{2m^2+3m-1}{6n(m+1)},\qquad M=\rho n,\qquad f=\tfrac{1}{2}m(m+1)$$

and

$$\gamma=\frac{m}{288(m+1)}(2m^4+6m^3+m^2-12m-13).$$

Proof. With $M=\rho n=n-(2m^2+3m-1)/6(m+1)=n-\alpha$, say, the characteristic function of $-2\rho\log\Lambda^*$ is, from Corollary 8.4.8,

(14)

$$g_1(t)=E[\Lambda^{*-2\rho it}]=\left(\frac{2e}{M+\alpha}\right)^{-mMit}\left[1-2it+\frac{2it\alpha}{M+\alpha}\right]^{-mM(1-2it)/2-\alpha m/2}\cdot\frac{\Gamma_m[\frac{1}{2}M(1-2it)+\frac{1}{2}\alpha]}{\Gamma_m[\frac{1}{2}(M+\alpha)]}$$

The desired result is an immediate consequence of expanding each of the terms in $\log g_1(t)$ for large M, where (24) of Section 8.2.4 is used to expand the gamma functions.

Table 5. Upper 5 and 1 percentage points of $-2\log \Lambda^*$, where Λ^* is the modified likelihood ratio statistic for testing that a covariance matrix equals a specified matrix[a]

m	2		3		4		5		6	
n	5%	1%	5%	1%	5%	1%	5%	1%	5%	1%
6	8.9415	13.0019	15.805	21.229						
7	8.7539	12.7231	15.1854	20.358	24.06	30.75				
8	8.6198	12.5246	14.7676	19.7750	23.002	29.32				
9	8.5193	12.3761	14.4663	19.3577	22.278	28.357	32.47	39.97		
10	8.4411	12.2608	14.2387	19.0439	21.749	27.657	31.36	38.55		
11	8.3786	12.1687	14.0605	18.7992	21.3456	27.1268	30.549	37.51	42.08	
12	8.3274	12.0935	13.9173	18.6029	21.0276	26.7102	29.922	36.710	40.92	48.96
13	8.2847	12.0308	12.7995	18.4421	20.7702	26.3743	29.424	36.079	40.02	47.84
14	8.2486	11.9778	13.7010	18.3078	20.5576	26.0975	29.0182	35.567	39.303	46.96
15	8.2177	11.9325	13.6174	18.1940	20.3789	25.8655	28.6812	35.1435	38.714	46.234
16	8.1909	11.8932	13.5456	18.0964	20.2266	25.6681	28.3967	34.7866	38.222	45.632
17	8.1674	11.8588	13.4832	18.0116	20.0953	25.4982	28.1532	34.4818	37.806	45.122
18	8.1467	11.8285	13.4284	17.9373	19.9808	25.3503	27.9425	34.2185	37.4475	44.686
19	8.1283	11.8015	13.3800	17.8718	19.8801	25.2204	27.7582	33.9886	37.1365	44.3069
20	8.1118	11.7774	13.3369	17.8134	19.7909	25.1054	27.5958	33.7862	36.8638	43.9754
21	8.0970	11.7557	13.2983	17.7611	19.7113	25.0029	27.4514	22.6066	36.6227	43.6827
22	8.0835	11.7361	13.2634	17.7141	19.6398	24.9109	27.3224	33.4461	36.4080	43.4224
23	8.0713	11.7183	13.2319	17.6715	19.5753	24.8279	27.2062	33.3018	36.2155	43.1892
24	8.0602	11.7020	13.2032	17.6327	19.5167	24.7527	27.1011	33.1713	36.0420	42.9792
25	8.0500	11.6871	13.1769	17.5973	19.4633	24.6841	27.0056	33.0529	35.8847	42.7890
26	8.0406	11.6734	13.1529	17.5648	19.4144	24.6214	26.9184	32.9448	35.7415	42.6160
27	8.0319	11.6608	13.1307	17.5349	19.3695	24.5638	26.8385	32.8458	35.6106	42.4579
28	8.0239	11.6490	13.1102	17.5073	19.3281	24.5107	26.7650	32.7547	35.4904	42.3128
29	8.0164	11.6382	13.0912	17.4817	19.2899	34.4617	26.6971	32.6707	35.3797	42.1793
30	8.0095	11.6280	13.0735	17.4579	19.2543	24.4162	26.6342	32.5930	35.2774	42.0559
35	7.9809	11.5864	13.0012	17.3606	19.1094	24.2307	26.3788	32.2774	34.8635	41.5575
40	7.9597	11.5554	12.9478	17.2887	19.0029	24.0946	26.1924	32.0474	34.5632	41.1965
45	7.9432	11.5314	12.9067	17.2335	18.9214	23.9906	26.0503	31.8723	34.3353	40.9229
50	7.9301	11.5124	12.8741	17.1898	18.8570	23.9085	25.9384	31.7345	34.1565	40.7083
60	7.9106	11.4840	12.8257	17.1249	18.7617	23.7870	25.7734	31.5315	33.8937	40.3935
120	7.8624	11.4139	12.7071	16.9660	18.5300	23.4922	25.3751	31.0424	33.2642	39.6405
∞	7.8147	11.3449	12.5916	16.8119	18.3070	23.2093	24.9958	30.5779	32.6705	38.9321

Table 5 (*Continued*)

m	7		8		9		10	
n	5%	1%	5%	1%	5%	1%	5%	1%
15	50 70	59.38	64.94					
16	49.90	58.41	63.66	73.39				
17	49.221	57.60	62.60	72.14				
18	48.645	56.911	61.71	71.08	76.86			
19	48.149	56.318	60.95	70.19	75.72	86.11		
20	47.716	55.801	60.290	69.41	74.75	84.97	91.28	
21	47.336	55.348	59.714	68.733	73.90	83.99	90.06	101.30
22	46.9982	54.947	59.206	68.138	73.16	83.13	89.01	100.08
23	46.6971	54.5889	58.754	67.609	72.502	82.37	88.08	99.02
24	46.4266	54.2678	58.350	67.137	71.918	81.697	87.25	98.08
25	46.1823	53.9780	57.986	66.712	71.394	81.092	86.52	97.23
26	45.9605	53.7151	57.6573	66.328	70.922	80.548	85.855	96.48
27	45.7582	53.4756	57.3580	65.9787	70.494	80.054	85.258	95.799
28	45.5730	53.2563	57.0847	65.6601	70 104	79.605	84.716	95.181
29	45.4027	53.0549	56.8340	65.3681	69.747	79.195	84.221	94.618
30	45.2456	52.8693	56.6032	65.0996	69.4190	78.818	83.768	94.103
35	44.6133	52.1228	55.6790	64.0253	68.1134	77.3197	81.9731	92.064
40	44.1575	51.5856	55.0172	63.2576	67.1852	76.2565	80.7067	90.6287
45	43.8132	51.1804	54.5197	62.6812	66.4910	75.4625	79.7646	89.5624
50	43.5440	50.8637	54.1321	62.2325	65.9521	74.8466	79.0361	88.7388
60	43.1499	50.4008	53.5669	61 5790	65.1694	73.9533	77.9823	87.5489
120	42.2125	49.3019	52.2325	60.0394	63.3356	71.8646	75.5328	84.7885
∞	41.3371	48.2782	50.9985	58.6192	61.6562	69.9568	72.3115	82.2921

[a]Here, m = number of variables; n = sample size minus one.

Source: Reproduced from Davis and Field (1971) with the kind permission of the Commonwealth Scientific and Industrial Research Organization (C.S.I.R.O.), Australia, and the authors.

This expansion was given by Davis (1971), who also derived further terms and used his expansion to compute upper percentage of the distribution of $-2\log\Lambda^*$; see also Korin (1968) for earlier calculations. Nagarsenker and Pillai (1973b) have compared exact percentage points with the approximate ones of Davis and Korin; it appears that the latter are quite accurate.

8.4.5. Asymptotic Non-Null Distributions of the Modified Likelihood Ratio Statistic

The power function of the modified likelihood ratio test of size α is $P(-2\rho\log\Lambda^*\geq k_\alpha^*|\lambda_1,\ldots,\lambda_m)$, where ρ is given by (13) and k_α^* is the upper $100\alpha\%$ point of the null distribution of $-2\rho\log\Lambda^*$. This is a function of $\lambda_1,\ldots,\lambda_m$, the latent roots of Σ. An approximation for k_α^* is $c_f(\alpha)$, the upper $100\alpha\%$ point of the χ_f^2 distribution with $f=\frac{1}{2}m(m+1)$. The error in this approximation is of order M^{-2}, where $M=\rho n$. Here we give approximations to the power function.

We consider the three different alternatives

$$K:\Sigma\neq I,$$

$$K_M:\Sigma=I+\frac{1}{M}\Omega,$$

and

$$K_M^*:\Sigma=I+\frac{1}{M^{1/2}}\Omega,$$

where Ω is a fixed matrix. By invariance it can be assumed without loss of generality that both Σ and Ω are diagonal, $\Sigma=\text{diag}(\lambda_1,\ldots,\lambda_m)$ and $\Omega=\text{diag}(\omega_1,\ldots,\omega_m)$. Here K is a fixed alternative and K_M, K_M^* are sequences of local alternatives. We consider first these latter two.

THEOREM 8.4.10. (a) Under the sequence of local alternatives K_M: $\Sigma=I+(1/M)\Omega$ the distribution function of $-2\rho\log\Lambda^*$ can be expanded as

$$P(-2\rho\log\Lambda^*\leq x)=P(\chi_f^2\leq x)+\frac{\sigma_2}{4M}\left[P(\chi_{f+2}^2\leq x)-P(\chi_f^2\leq x)\right]+O(M^{-2}),$$

where

$$f=\tfrac{1}{2}m(m+1),\qquad M=\rho n,\qquad \sigma_j=\operatorname{tr}\Omega^j.$$

(b) Under the sequence of local alternatives K_M^*: $\Sigma = I + (1/M^{1/2})\Omega$ the distribution function of $-2\rho\log\Lambda^*$ can be expanded as

$$P(-2\rho\log\Lambda^* \le x) = P\left(\chi_f^2(\delta) \le x\right) + \frac{\sigma_3}{6M^{1/2}} P$$
$$\cdot\left[\left(\chi_{f+4}^2(\delta) > x\right) - 3P\left(\chi_{f+2}^2(\delta) \le x\right) + 2P\left(\chi_f^2(\delta) \le x\right)\right] + O(M^{-1}),$$

where $\sigma_j = \operatorname{tr}\Omega^j$ and the noncentrality parameter is $\delta = \frac{1}{2}\sigma_2$.

Proof. As to part (a), under the sequence of alternatives K_M the characteristic function of $-2\rho\log\Lambda^*$ is, from Theorem 8.4.7,

$$g(t,\Omega) = g_1(t)\frac{\det\left(I + \dfrac{1}{M}\Omega\right)^{-Mit}}{\det\left(I - \dfrac{2it\rho}{1-2it\rho}\dfrac{1}{M}\Omega\right)^{M(1-2it)/2+\alpha/2}},$$

where $\alpha = n - M$ and $g_1(t)$ is the characteristic function of $-2\rho\log\Lambda^*$ given by (14). Using the formula

$$(15) \qquad -\log\det\left(I - \frac{1}{M}Z\right) = \frac{1}{M}\operatorname{tr}Z + \frac{1}{2M^2}\operatorname{tr}Z^2 + \frac{1}{3M^3}\operatorname{tr}Z^3 + \cdots$$

to expand the determinant terms gives

$$g(t,\Omega) = (1-2it)^{-f/2}\left[1 + \frac{\sigma_2}{4M}\left(\frac{1}{1-2it} - 1\right) + O(M^{-2})\right],$$

whence the desired result.

As to part (b) under the sequence of alternatives K_M^* the characteristic function of $-2\rho\log\Lambda^*$ is $g(t, M^{1/2}\Omega)$, using the above notation. Again, using (15) this is expanded as

$$g(t, M^{1/2}\Omega) = (1-2it)^{-f/2}\exp\left(\frac{it\sigma_2}{1-2it}\right)\cdot\left\{1 + \frac{\sigma_3}{6M^{1/2}}\left[(1-2it)^{-2}\right.\right.$$
$$\left.\left. - 3(1-2it)^{-1} + 2\right] + O(M^{-1})\right\},$$

and inversion completes the proof.

We turn now to the asymptotic behavior of the modified likelihood ratio statistic under fixed alternatives.

THEOREM 8.4.11. Under the fixed alternative $K:\Sigma\neq I$ the distribution of the random variable

$$Y=M^{-1/2}\{-2\rho\log\Lambda^*-M[\operatorname{tr}(\Sigma-I)-\log\det\Sigma]\}$$

may be expanded as

$$P\left(\frac{Y}{\tau}\le x\right)=\Phi(x)-\frac{1}{6M^{1/2}}\left[c_1\phi(x)+c_2\phi^{(2)}(x)\right]+O(M^{-1}),$$

where $\Phi(x)$ and $\phi(x)$ denote the standard normal distribution and density functions, respectively, and

$$\tau^2=2\operatorname{tr}(\Sigma-I)^2$$

$$c_1=\frac{3}{\tau}m(m+1)$$

$$c_2=\frac{4}{\tau^3}(m+2t_3-3t_2),$$

with

$$t_j=\operatorname{tr}\Sigma^j.$$

The proof follows on using Theorem 8.4.7 to obtain the characteristic function of Y/τ and expanding this for large M. The details are left as an exercise (see Problem 8.8).

The expansions in this section are similar to ones given by Sugiura (1969a, b) for the distribution of $-2\log\Lambda^*$ (not $-2\rho\log\Lambda^*$). Sugiura also gives further terms in his expansions, as well as some power calculations.

8.4.6. The Asymptotic Null Distribution of the Modified Likelihood Ratio Statistic for an Elliptical Sample

Here we examine how the asymptotic null distribution of Λ^* is affected when the sample is drawn from an elliptical distribution. In

$$\Lambda^*=e^{mn/2}\operatorname{etr}(-\tfrac{1}{2}nS)(\det S)^{n/2},$$

where S is the sample covariance matrix, we substitute $S=I_m+n^{-1/2}Z$.

Then when $H: \Sigma = I_m$ is true, $-2\log \Lambda^*$ can be expanded as

$$-2\log \Lambda^* = \tfrac{1}{2}\operatorname{tr}(Z^2) + O_p(n^{-1/2})$$

$$= \mathbf{u}'\mathbf{u} + O_p(n^{-1/2})$$

where $\mathbf{u}' = (z_{11}/2^{1/2}, \ldots, z_{mm}/2^{1/2}, z_{12}, \ldots, z_{1m}, z_{23}, \ldots, z_{m-1,m})$. Now suppose that the observations are drawn from an elliptical distribution with kurtosis parameter κ. Then the asymptotic distribution of $\mathbf{u}$, as $n \to \infty$, is $N_{m(m+1)/2}(0, \Gamma)$, where

$$\Gamma = \begin{bmatrix} (1+\kappa)I_m + \tfrac{1}{2}\kappa \mathbf{1}\mathbf{1}' & 0 \\ 0 & (1+\kappa)I_{m(m-1)/2} \end{bmatrix}$$

with $\mathbf{1} = (1, \ldots, 1)' \in R^m$. The latent roots of Γ are $1 + \kappa + \tfrac{1}{2}m\kappa$ and $1 + \kappa$ repeated $\tfrac{1}{2}(m-1)(m+2)$ times. Diagonalizing Γ by an orthogonal matrix and using an obvious argument establishes the result given in the following theorem.

THEOREM 8.4.12. If the sample is drawn from an elliptical distribution with kurtosis parameter κ then the asymptotic distribution of $-2\log \Lambda^*/(1+\kappa)$ is the distribution of the random variable

$$\left[1 + \frac{\kappa m}{2(1+\kappa)}\right] X_1 + X_2,$$

where X_1 is χ_1^2, X_2 is $\chi^2_{(m-1)(m+2)/2}$, and X_1 and X_2 are independent.

When $\kappa = 0$ this result reduces to the asymptotic distribution of $-2\log \Lambda^*$ under normality, namely, $\chi^2_{m(m+1)/2}$. The further that κ is away from zero the greater the divergence of the asymptotic distribution from $\chi^2_{m(m+1)/2}$, so great care should be taken in using the test based on Λ^* for non-normal data.

8.4.7. *Other Test Statistics*

In addition to the modified likelihood ratio test a test based on a_1 and a_m, the largest and smallest latent roots of A, was proposed by Roy (1957). The test is to reject $H: \Sigma = I_m$ if $a_1 > a_1^0$ or $a_m < a_m^0$. Various methods for choosing a_1^0, a_m^0 have been suggested; see Thompson (1962), Hanumara and Thompson (1968), and Clem et al. (1973). An alternative test, rejecting for large values of a_1 alone, was examined by Sugiyama (1972) and Muirhead

(1974). Further discussion of this test may be found in Section 9.7. Sugiura (1972b) and John (1971) have shown that for testing $H:\Sigma=I$ against $K_{\text{l}}: \Sigma>I$ the test which rejects H for large values of $\operatorname{tr}A$ is locally best invariant.

8.5. TESTING SPECIFIED VALUES FOR THE MEAN VECTOR AND COVARIANCE MATRIX

8.5.1. *The Likelihood Ratio Test*

Let $\mathbf{X}_1,\dots,\mathbf{X}_N$ be independent $N_m(\boldsymbol{\mu},\Sigma)$ random vectors and consider testing the null hypothesis $H:\boldsymbol{\mu}=\boldsymbol{\mu}_0, \Sigma=\Sigma_0$ against $K:\boldsymbol{\mu}\neq\boldsymbol{\mu}_0$ or $\Sigma\neq\Sigma_0$. An argument similar to that used at the end of Section 8.3.1 shows that this is equivalent to testing $H:\boldsymbol{\mu}=\mathbf{0}, \Sigma=I_m$. We will express all results in this section in terms of this latter formulation.

It is clear that this testing problem is invariant under the orthogonal group $O(m)$ acting on the space of the sufficient statistic $(\overline{\mathbf{X}}, A)$, where

$$\overline{\mathbf{X}}=N^{-1}\sum_{i=1}^{N}\mathbf{X}_i,\qquad A=\sum_{i=1}^{N}(\mathbf{X}_i-\overline{\mathbf{X}})(\mathbf{X}_i-\overline{\mathbf{X}})',$$

by

$$(1)\qquad \overline{\mathbf{X}}\to H\overline{\mathbf{X}},\qquad A\to HAH'\qquad [H\in O(m)]$$

The induced group of transformations on the parameter space is given by

$$(2)\qquad \boldsymbol{\mu}\to H\boldsymbol{\mu},\qquad \Sigma\to H\Sigma H'.$$

Various invariants under the transformation (2) include $\boldsymbol{\mu}'\boldsymbol{\mu}$, $\boldsymbol{\mu}'\Sigma\boldsymbol{\mu}$, $\boldsymbol{\mu}'\Sigma^{-1}\boldsymbol{\mu}$, and $(\lambda_1,\dots,\lambda_m)$, the latent roots of Σ. The problem of finding a maximal invariant is left as an exercise (see Problem 8.10).

The likelihood ratio test is given in the following theorem from T. W. Anderson (1958).

THEOREM 8.5.1. The likelihood ratio test of size α of $H:\boldsymbol{\mu}=\mathbf{0}, \Sigma=I_m$ rejects H if $\Lambda\le c_\alpha$, where

$$(3)\qquad \Lambda=\left(\frac{e}{N}\right)^{mN/2}(\det A)^{N/2}\operatorname{etr}(-\tfrac{1}{2}A)\exp(-\tfrac{1}{2}N\overline{\mathbf{X}}'\overline{\mathbf{X}}).$$

Proof. The likelihood ratio statistic is

$$(4)\qquad \Lambda=\frac{L(\mathbf{0},I)}{\sup_{\boldsymbol{\mu},\Sigma}L(\boldsymbol{\mu},\Sigma)},$$

where

$$L(\boldsymbol{\mu},\Sigma)=(\det\Sigma)^{-N/2}\operatorname{etr}(-\tfrac{1}{2}\Sigma^{-1}A)\exp\left[-\tfrac{1}{2}N(\overline{\mathbf{X}}-\boldsymbol{\mu})'\Sigma^{-1}(\overline{\mathbf{X}}-\boldsymbol{\mu})\right]$$

The denominator in (4) is found by putting $\boldsymbol{\mu}=\overline{\mathbf{X}}$ and $\Sigma=N^{-1}A$, and the desired result is immediate.

The unbiasedness of the likelihood ratio test has been established by Sugiura and Nagao (1968) and Das Gupta (1969). The proof given here is due to Sugiura and Nagao.

THEOREM 8.5.2. For testing H: $\boldsymbol{\mu}=\mathbf{0}$, $\Sigma=I$ against K: $\boldsymbol{\mu}\neq\mathbf{0}$ or $\Sigma\neq I$, the likelihood ratio test having the critical region

$$C=\left\{(\overline{\mathbf{X}},A);\overline{\mathbf{X}}\in R^m,A>0,(\det A)^{N/2}\operatorname{etr}(-\tfrac{1}{2}A)\exp(-\tfrac{1}{2}N\overline{\mathbf{X}}'\overline{\mathbf{X}})\le k_\alpha\right\}$$

is unbiased.

Proof. Without loss of generality it can be assumed that $\Sigma=\operatorname{diag}(\lambda_1,\dots,\lambda_m)$. The probability of the critical region under K can be written

$$P_K(C)=c_{m,n}\left(\frac{N}{2\pi}\right)^{m/2}\int_{(\overline{\mathbf{X}},A)\in C}(\det\Sigma)^{-N/2}(\det A)^{(N-m-2)/2}$$

$$\cdot\operatorname{etr}\left\{-\tfrac{1}{2}\Sigma^{-1}\left[A+N(\overline{\mathbf{X}}-\boldsymbol{\mu})(\overline{\mathbf{X}}-\boldsymbol{\mu})'\right]\right\}(d\overline{\mathbf{X}})(dA)$$

where

$$c_{m,n}=\left[2^{mn/2}\Gamma_m(\tfrac{1}{2}n)\right]^{-1}\quad\text{and}\quad n=N-1.$$

Now put $U=\Sigma^{-1/2}A\Sigma^{-1/2}$ and $\overline{\mathbf{Y}}=\Sigma^{-1/2}(\overline{\mathbf{X}}-\boldsymbol{\mu})$; then

$$(d\overline{\mathbf{X}})(dA)=(\det\Sigma)^{(m+2)/2}(d\overline{\mathbf{Y}})(dU)$$

and

$$P_K(C)=c_{m,n}\left(\frac{N}{2\pi}\right)^{m/2}\int_{(\overline{\mathbf{Y}},U)\in C^*}(\det U)^{(N-m-2)/2}$$

$$\cdot\operatorname{etr}\left[-\tfrac{1}{2}(U+N\overline{\mathbf{Y}}\overline{\mathbf{Y}}')\right](d\overline{\mathbf{Y}})(dU),$$

where

$$C^*=\left\{(\overline{\mathbf{Y}},U);\overline{\mathbf{Y}}\in R^m,U>0,(\Sigma^{1/2}\overline{\mathbf{Y}}+\boldsymbol{\mu},\Sigma^{1/2}U\Sigma^{1/2})\in C\right\}.$$

Note that when H is true the region C^* is equal to C. It follows that, with

$$b = c_{m,n}\left(\frac{N}{2\pi}\right)^{m/2},$$

$$P_K(C) - P_H(C) = b\left\{\int_{(\bar{\mathbf{Y}},U)\in C^*} - \int_{(\bar{\mathbf{Y}},U)\in C}\right\}(\det U)^{(N-m-2)/2}$$

$$\cdot \operatorname{etr}\left[-\tfrac{1}{2}(U + N\bar{\mathbf{Y}}\bar{\mathbf{Y}}')\right](d\bar{\mathbf{Y}})(dU)$$

$$= b\left\{\int_{(\bar{\mathbf{Y}},U)\in C^* - C\cap C^*} - \int_{(\bar{\mathbf{Y}},U)\in C - C\cap C^*}\right\}(\det U)^{(N-m-2)/2}$$

$$\cdot \operatorname{etr}\left[-\tfrac{1}{2}(U + N\bar{\mathbf{Y}}\bar{\mathbf{Y}}')\right](d\bar{\mathbf{Y}})(dU).$$

Now, for $(\bar{\mathbf{Y}},U)\in C - C\cap C^*$ we have

$$\operatorname{etr}\left[-\tfrac{1}{2}(U + N\bar{\mathbf{Y}}\bar{\mathbf{Y}}')\right] \le k_\alpha(\det U)^{-N/2},$$

while for $(\bar{\mathbf{Y}},U)\in C^* - C\cap C^*$

$$\operatorname{etr}\left[-\tfrac{1}{2}(U + N\bar{\mathbf{Y}}\bar{\mathbf{Y}}')\right] > k_\alpha(\det U)^{-N/2},$$

and hence

$$P_K(C) - P_H(C) \ge bk_\alpha\left\{\int_{(\bar{\mathbf{Y}},U)\in C^*} - \int_{(\bar{\mathbf{Y}},U)\in C}\right\}(\det U)^{-(m+2)/2}(d\bar{\mathbf{Y}})(dU)$$

$$=0,$$

since

$$\int_{(\bar{\mathbf{Y}},U)\in C^*}(\det U)^{-(m+2)/2}(d\bar{\mathbf{Y}})(dU) = \int_{(\bar{\mathbf{Y}},U)\in C}(\det U)^{-(m+2)/2}(d\bar{\mathbf{Y}})(dU);$$

This is easily seen by making the transformation

$$\bar{\mathbf{Z}} = \Sigma^{1/2}\bar{\mathbf{Y}} + \boldsymbol{\mu}, \qquad V = \Sigma^{1/2}U\Sigma^{1/2}$$

in the integral on the left. We have thus shown that

$$P_K(C) \ge P_H(C),$$

and the proof is complete.

8.5.2. *Moments of the Likelihood Ratio Statistic*

Distributional results associated with Λ can be obtained from the moments.

THEOREM 8.5.3. The hth moment of Λ is

$$(5)\qquad E(\Lambda^h)=\left(\frac{2e}{N}\right)^{mNh/2}\frac{\Gamma_m[\frac{1}{2}(n+Nh)]}{\Gamma_m(\frac{1}{2}n)}\frac{(\det\Sigma)^{Nh/2}}{\det(I+h\Sigma)^{N(1+h)/2}}$$

$$\cdot\exp\left\{-\tfrac{1}{2}Nh\boldsymbol{\mu}'\left[I-h(\Sigma^{-1}+hI)^{-1}\right]\boldsymbol{\mu}\right\},$$

where $n=N-1$.

Proof. Using the independence of A and $\overline{\mathbf{X}}$ we have

$$(6)\qquad E(\Lambda^h)=\left(\frac{e}{N}\right)^{mNh/2}\frac{(\det\Sigma)^{-n/2}}{2^{mn/2}\Gamma_m(\frac{1}{2}n)}\int_{A>0}\operatorname{etr}\left[-\tfrac{1}{2}(hI+\Sigma^{-1})A\right]$$

$$\cdot(\det A)^{(n+Nh-m-1)/2}(dA)\left(\frac{N}{2\pi}\right)^{m/2}(\det\Sigma)^{-1/2}$$

$$\cdot\int_{R^m}\exp\left[-\frac{N}{2}(\overline{\mathbf{X}}-\boldsymbol{\mu})'\Sigma^{-1}(\overline{\mathbf{X}}-\boldsymbol{\mu})-\tfrac{1}{2}Nh\overline{\mathbf{X}}'\overline{\mathbf{X}}\right](d\overline{\mathbf{X}})$$

The first integral on the right is equal to

$$(7)\qquad \det(hI+\Sigma^{-1})^{-(n+Nh)/2}\Gamma_m\left[\tfrac{1}{2}(n+Nh)\right]2^{m(n+Nh)/2},$$

while the second integral can be written as

$$(8)\quad \exp(\tfrac{1}{2}Nh\boldsymbol{\mu}'\boldsymbol{\mu})\int_{R^m}\exp\left[-\tfrac{1}{2}N(\overline{\mathbf{X}}-\boldsymbol{\mu})'(\Sigma^{-1}+hI)(\overline{\mathbf{X}}-\boldsymbol{\mu})\right]$$

$$\cdot\exp(-Nh\overline{\mathbf{X}}'\boldsymbol{\mu})(d\overline{\mathbf{X}})$$

$$=\exp\left(\frac{h}{2}N\boldsymbol{\mu}'\boldsymbol{\mu}\right)\left(\frac{2\pi}{N}\right)^{m/2}\det(\Sigma^{-1}+hI)^{-1/2}E\left[\exp(-Nh\overline{\mathbf{X}}'\boldsymbol{\mu})\right],$$

where this last expectation is taken with respect to $\overline{\mathbf{X}}$ distributed as

$$N_m\left(\boldsymbol{\mu},\frac{1}{N}(\Sigma^{-1}+hI)^{-1}\right).$$

Since

$$(9)\qquad E\left[\exp(-Nh\bar{\mathbf{X}}'\boldsymbol{\mu})\right]=\exp\left[-Nh\boldsymbol{\mu}'\boldsymbol{\mu}+\tfrac{1}{2}Nh^2\boldsymbol{\mu}'(\Sigma^{-1}+hI)^{-1}\boldsymbol{\mu}\right]$$

(see, for example, Theorem 1.2.5), the desired result now follows by substitution of (9) in (8), then of (7) and (8) in (6).

COROLLARY 8.5.4. When the null hypothesis $H: \boldsymbol{\mu}=\mathbf{0},\ \Sigma=I$ is true, the hth moment of Λ is

$$(10)\qquad E(\Lambda^h)=\left(\frac{2e}{N}\right)^{mNh/2}(1+h)^{-mN(1+h)/2}\frac{\Gamma_m\left[\frac{1}{2}(n+Nh)\right]}{\Gamma_m\left(\frac{1}{2}n\right)}.$$

These null moments have been used by Nagarsenker and Pillai (1974) to derive expressions for the exact distribution of Λ and hence to compute the upper 5 and 1 percentage points of the distribution of $-2\log\Lambda$ for $m=2(1)6$ and $N=4(1)20,\ 20(2)40,\ 40(5)100$. These are given in Table 6.

8.5.3. The Asymptotic Null Distribution of the Likelihood Ratio Statistic

The null moments of Λ given in Corollary 8.5.4 are not of the form (18) of Section 8.2.4. However it is still possible to find a constant ρ so that the term of order N^{-1} vanishes in an asymptotic expansion of the distribution of $-2\rho\log\Lambda$, as the following theorem from Davis (1971) shows. The proof is very similar to that of Theorem 8.4.9 and is left as an exercise (see Problem 8.11).

THEOREM 8.5.5. When the null hypothesis $H: \boldsymbol{\mu}=\mathbf{0},\ \Sigma=I_m$ is true, the distribution of $-2\rho\log\Lambda$ can be expanded as

$$(11)\qquad P(-2\rho\log\Lambda\le x)=P(\chi_f^2\le x)+\frac{\gamma}{M^2}\left[P(\chi_{f+4}^2\le x)-P(\chi_f^2\le x)\right]+O(M^{-3}),$$

where

$$(12)\qquad \rho=1-\frac{2m^2+9m+11}{6N(m+3)},\qquad M=\rho N,\qquad f=\tfrac{1}{2}m(m+3),$$

and

$$\gamma=\frac{m}{288(m+3)}(2m^4+18m^3+49m^2+36m-13).$$

Table 6. Upper 5 and 1 percentage points of $-2\log\Lambda$, where Λ is the likelihood ratio statistic for testing specified values for the mean vector and covariance matrix[a]: $\alpha = .05$

$N \backslash m$	2	3	4	5	6
4	17.381				
5	15.352	27.706			
6	14.318	24.431	39.990		
7	13.689	22.713	35.307	54.261	
8	13.265	21.646	32.787	48.039	70.475
9	12.960	20.915	31.190	44.610	62.660
10	12.729	20.382	30.080	42.400	58.222
11	12.549	19.975	29.261	40.843	55.321
12	12.404	19.655	28.631	39.683	53.254
13	12.285	19.396	28.131	38.782	51.698
14	12.186	19.181	27.723	38.061	50.480
15	12.101	19.002	27.384	37.470	49.499
16	12.029	18.848	27.098	36.977	48.691
17	11.966	18.716	26.854	36.559	48.013
18	11.911	18.601	26.642	36.200	47.436
19	11.862	18.499	26.457	35.888	46.938
20	11.819	18.410	26.294	35.614	46.504
22	11.745	18.258	26.019	35.157	45.785
24	11.684	18.134	25.797	34.790	45.212
26	11.633	18.031	25.614	34.489	44.745
28	11.591	17.944	25.460	34.237	44.357
30	11.554	17.870	25.329	34.023	44.029
32	11.522	17.806	25.215	33.840	43.748
34	11.494	17.750	25.117	33.681	43.505
36	11.469	17.701	25.030	33.541	43.292
38	11.447	17.657	24.954	33.417	43.105
40	11.427	17.618	24.885	33.307	42.938
45	11.386	17.536	24.742	33.079	42.594
50	11.353	17.471	24.630	32.900	42.324
55	11.327	17.419	24.539	32.755	42.107
60	11.305	17.375	24.465	32.636	41.929
65	11.286	17.339	24.402	32.537	41.780
70	11.271	17.308	24.348	32.452	41.654
75	11.257	17.281	24.302	32.379	41.546
80	11.245	17.258	24.262	32.316	41.451
85	11.235	17.237	24.227	32.261	
90	11.225	17.219	24.196	32.211	
95	11.217	17.203	24.168		
100	11.210	17.188	24.143		

Table 6 (*Continued*): $\alpha = .01$

$N \backslash m$	2	3	4	5	6
4	24.087				
5	21.114	36.308			
6	19.625	31.682	50.512		
7	18.729	29.318	44.073	66.728	
8	18.129	27.871	40.713	58.348	84.937
9	17.700	26.890	38.621	53.885	74.530
10	17.377	26.180	37.184	51.063	68.874
11	17.125	25.642	36.133	49.100	65.244
12	16.923	25.219	35.328	47.650	62.690
13	16.758	24.878	34.692	46.531	60.784
14	16.620	24.597	34.176	45.639	59.302
15	16.503	24.361	33.748	44.911	58.114
16	16.403	24.161	33.388	44.305	57.139
17	16.316	23.988	33.080	43.793	56.324
18	16.239	23.838	32.814	43.353	55.631
19	16.172	23.706	32.582	43.973	55.035
20	16.112	23.589	32.378	42.639	54.517
22	16.010	23.392	32.035	42.083	53.658
24	15.927	23.231	31.758	41.637	52.977
26	15.857	23.098	31.529	41.272	52.422
28	15.798	22.986	31.337	40.967	51.961
30	15.747	22.890	31.174	40.708	51.573
32	15.703	22.807	31.033	40.486	51.240
34	15.665	22.734	30.911	40.294	50.953
36	15.631	22.671	30.803	40.125	50.701
38	15.601	22.614	30.708	39.976	50.480
40	15.574	22.564	30.623	39.844	50.284
45	15.517	22.458	30.447	39.568	49.877
50	15.473	22.375	30.308	39.353	49.559
55	15.436	22.307	30.196	39.179	49.303
60	15.406	22.252	30.103	39.036	49.094
65	15.381	22.205	30.025	38.916	48.919
70	15.359	22.165	29.959	38.815	48.770
75	15.341	22.131	29.903	38.727	48.643
80	15.324	22.101	29.853	38.651	48.532
85	15.310	22.074	29.810	38.585	
90	15.297	22.051	29.771	38.526	
95	15.286	22.030	29.737		
100	15.276	22.011	29.706		

[a]Here, m = number of variables; N = sample size.

Source: Reproduced from Nagarsenker and Pillai (1974) with the kind permission of Academic Press, Inc., and the authors.

8.5.4. *Asymptotic Non-Null Distributions of the Likelihood Ratio Statistic*

Here we consider the asymptotic distributions of $-2\rho\log\Lambda$ under the three different alternatives

$$K: \boldsymbol{\mu}\neq 0 \quad \text{or} \quad \Sigma\neq I_m,$$

$$K_M: \boldsymbol{\mu}=\frac{1}{M}\boldsymbol{\tau}, \qquad \Sigma=I+\frac{1}{M}\Omega$$

and

$$K_M^*: \boldsymbol{\mu}=\frac{1}{M^{1/2}}\boldsymbol{\tau}, \qquad \Sigma=I+\frac{1}{M^{1/2}}\Omega,$$

where $M=\rho N$, with ρ given by (12), and Ω is a fixed matrix assumed diagonal without loss of generality. The two sequences of local alternatives K_M and K_M^* are considered first.

THEOREM 8.5.6. (a) Under the sequence of local alternatives

$$K_M: \boldsymbol{\mu}=\frac{1}{M}\boldsymbol{\tau}, \qquad \Sigma=I+\frac{1}{M}\Omega,$$

the distribution function of $-2\rho\log\Lambda$ can be expanded as

$$(13)\quad P(-2\rho\log\Lambda\le x)=P(\chi_f^2\le x)+\frac{1}{4M}(\sigma_2+2\boldsymbol{\tau}'\boldsymbol{\tau})\left[P(\chi_{f+2}^2\le x)-P(\chi_f^2\le x)\right]+O(M^{-2}),$$

where $f=\frac{1}{2}m(m+3)$, $M=\rho N$, and $\sigma_2=\operatorname{tr}\Omega^2$.

(b) Under the sequence of local alternatives

$$K_M^*: \boldsymbol{\mu}=\frac{1}{M^{1/2}}\boldsymbol{\tau}, \qquad \Sigma=I+\frac{1}{M^{1/2}}\Omega,$$

the distribution function of $-2\rho\log\Lambda$ can be expanded as

(14)

$$P(-2\rho\log\Lambda\le x)=P(\chi_f^2(\delta)\le x)+\frac{1}{6M^{1/2}}\left[(\sigma_3+3\boldsymbol{\tau}'\Omega\boldsymbol{\tau})P(\chi_{f+4}^2(\delta)\le x)\right.$$
$$\left.-(3\sigma_3+6\boldsymbol{\tau}'\Omega\boldsymbol{\tau})P(\chi_{f+2}^2(\delta)\le x)+(2\sigma_3+3\boldsymbol{\tau}'\Omega\boldsymbol{\tau})P(\chi_f^2(\delta)\le x)\right]+O(M^{-1}),$$

where $\sigma_j=\operatorname{tr}\Omega^j$ and the noncentrality parameter is $\delta=\frac{1}{2}\sigma_2+\boldsymbol{\tau}'\boldsymbol{\tau}$.

Proof. As to part (a), under the sequence of alternatives K_M the characteristic function of $-2\rho\log\Lambda$ is, from Theorem 8.5.3,

$$g(t,\tau,\Omega)=g_1(t)g_2(t,\tau,\Omega), \tag{15}$$

where $g_1(t)$ is the characteristic function of $-2\rho\log\Lambda$ when H is true obtained from Corollary 8.5.4 by putting $h=-2it\rho$, and

(16)

$$g_2(t,\tau,\Omega)=\frac{\det\left(I+\frac{1}{M}\Omega\right)^{-Mit}}{\det\left(I-\frac{2it\rho}{1-2it\rho}\frac{1}{M}\Omega\right)^{M(1-2it)/2+\alpha/2}}$$
$$\cdot\exp\left[\frac{it}{M}\tau'\tau-\frac{2t^2\rho}{(1+2it\rho)M}\tau'\left(I-\frac{2it\rho}{(1-2it\rho)M}\Omega\right)^{-1}\left(I+\frac{1}{M}\Omega\right)\tau\right]$$

with $\alpha=N-M$. From Theorem 8.5.5 it follows that

$$g_1(t)=(1-2it)^{-f/2}\left[1+O(M^{-2})\right]. \tag{17}$$

The ratio of determinants in g_2 can be expanded, as in the proof of Theorem 8.4.9, as

(18)

$$\frac{\det\left(I+\frac{1}{N}\Omega\right)^{-Mit}}{\det\left(I-\frac{2it\rho}{1-2it\rho}\frac{1}{M}\Omega\right)^{M(1-2it)/2+\alpha/2}}=1+\frac{\sigma_2}{4M}\left(\frac{1}{1-2it}-1\right)+O(M^{-2}),$$

where $\sigma_2=\operatorname{tr}\Omega^j$. The exponential term in g_2 can be expanded as

(19)

$$\exp\left[\frac{it}{M}\tau'\tau+\frac{2(it)^2\rho}{(1-2it\rho)M}\tau'\tau+O(M^{-2})\right]=\exp\left[\frac{it}{(1-2it)M}\tau'\tau+O(M^{-2})\right]$$
$$=1+\frac{it}{(1-2it)M}\tau'\tau+O(M^{-2}).$$

Multiplication of (17), (18), and (19) then shows that

$$g(t,\tau,\Omega)=(1-2it)^{-f/2}\left[1+\frac{1}{4M}(\sigma_2+2\tau'\tau)\left(\frac{1}{1-2it}-1\right)+O(M^{-2})\right],$$

and inverting this completes the proof.

As to part (b), under the sequence of alternatives K_M^* the characteristic function of $-2\rho\log\Lambda$ is $g(t, M^{1/2}\tau, M^{1/2}\Omega)$. The ratio of determinants here is expanded as

$$\exp\left[\frac{it\sigma_2}{2(1-2it)}\right]\left\{1+\frac{\sigma_3}{6M^{1/2}}\left[(1-2it)^{-2}-3(1-2it)^{-1}+2\right]+O(M^{-1})\right\}$$

(see the proof of Theorem 8.4.10), while the exponential term has the expansion

$$\exp\left(\frac{it}{1-2it}\tau'\tau\right)\left\{1+\frac{\tau'\Omega\tau}{2M^{1/2}}\left[(1-2it)^{-2}-2(1-2it)^{-1}+1\right]+O(M^{-1})\right\}.$$

Putting these together gives

$$g(t, M^{1/2}\tau, M^{1/2}\Omega)=(1-2it)^{-f/2}\exp\left[\frac{it}{1-2it}\left(\tfrac{1}{2}\sigma_2+\tau'\tau\right)\right]\left\{1+\frac{1}{6m^{1/2}}\right.$$

$$\left.\cdot\left[\frac{\sigma_3+3\tau'\Omega\tau}{(1-2it)^2}-\frac{3\sigma_3+6\tau'\Omega\tau}{1-2it}+2\sigma_3+3\tau'\Omega\tau\right]+O(M^{-1})\right\},$$

whence the desired result.

The next theorem describes the asymptotic behavior of the likelihood ratio statistic under fixed alternatives.

THEOREM 8.5.7. Under the alternative K: $\mu\neq 0$ or $\Sigma=I$, the distribution of the random variable

$$(20)\qquad Y=M^{-1/2}\{-2\rho\log\Lambda-M[\operatorname{tr}(\Sigma-I)-\log\det\Sigma+\mu'\mu]\}$$

may be expanded as

$$P\left(\frac{Y}{\beta}\le x\right)=\Phi(x)-\frac{1}{6M^{1/2}}\left[c_1\phi(x)+c_2\phi^{(2)}(x)\right]+O(M^{-1}),$$

where $\Phi(x)$ and $\phi(x)$ denote the standard normal distribution and density functions respectively and

$$\beta^2 = 2\,\mathrm{tr}(\Sigma - I)^2 + 4\mu'\Sigma\mu,$$

$$c_1 = \frac{3}{\beta} m(m+3),$$

$$c_2 = \frac{4}{\beta^3}\left(m + 6\mu'\Sigma^2\mu + 2t_3 - 3t_2\right),$$

with

$$t_j = \mathrm{tr}\,\Sigma^j.$$

The proof follows on using Theorem 8.5.3 to obtain the characteristic function of Y/β and expanding this for large M in the usual way. The details are left as an exercise (see Problem 8.12).

The expansions in these last two theorems are similar to ones given by Sugiura (1969a,b) for the distribution of $-2\log\Lambda$ (not $-2\rho\log\Lambda$). Sugiura also gives further terms in his expansions.

PROBLEMS

8.1. Let W be the statistic given by (14) of Section 8.2 when $r=2$, i.e., for testing $H: \Sigma_1 = \Sigma_2$. Using the null moments of W given in Theorem 8.2.6, show that, when $m=2$, W has the same distribution as $X^{n_1}(1-X)^{n_2}Y^{n_1+n_2}$, where X is beta(n_1-1, n_2-1), Y is beta$(n_1+n_2-2, 1)$, and X and Y are independent.

8.2. For testing $H: \Sigma_1 = \Sigma_2$ the modified likelihood ratio statistic is (assuming normality)

$$\Lambda^* = \frac{(\det S_1)^{n_1/2}(\det S_2)^{n_2/2}}{(\det S)^{n/2}},$$

where S_1 and S_2 are the two sample covariance matrices and $S = n^{-1}(n_1 S_1 + n_2 S_2)$, with $n = n_1 + n_2$. Suppose that H is true, and let Σ_0 denote the common value of Σ_1 and Σ_2. Let $n_i = k_i n$ $(i=1,2)$, with $k_1 + k_2 = 1$, and suppose $n \to \infty$. Put $Z_i = (nk_i)^{1/2}(S_i - \Sigma_0)$, with $i = 1,2$. (See Section 8.2.7.)

(a) Show that

$$-2\log\Lambda^* = \mu'\mu + O_p(n^{-1/2}),$$

where

$$\mu=(\Sigma_0\otimes\Sigma_0)^{-1/2}\left[\left(\tfrac{1}{2}k_2\right)^{1/2}\mathbf{z}_1-\left(\tfrac{1}{2}k_1\right)^{1/2}\mathbf{z}_2\right]$$

with $\mathbf{z}_i=\text{vec}(Z_i')$ $(i=1,2)$

(b) Suppose that the two samples are drawn from the same elliptical distribution with kurtosis parameter κ. Show that as $n\to\infty$ the asymptotic distribution of $\mathbf{v}=(1+\kappa)^{-1/2}\boldsymbol{\mu}$ is $N_{m^2}(\mathbf{0},V)$, where

$$V=\frac{\kappa}{2(1+\kappa)}\sum_{i,j=1}^{m}(E_{ij}\otimes E_{ij})+P,$$

with

$$P=\frac{1}{2}\left[I_{m^2}+\sum_{i,j=1}^{m}(E_{ij}\otimes E_{ij}')\right]$$

and E_{ij} being an $m\times m$ matrix with 1 in position (i,j) and 0's elsewhere.

[*Hint:* Use the result of Problem 3.3.] Show that the rank of V is $f=\frac{1}{2}m(m+1)$ and deduce the asymptotic distribution of $-2\log\Lambda^*/(1+\kappa)$ given in Theorem 8.2.18.

8.3. If A_1 is a non-negative definite $m\times m$ matrix and A_2 is a positive definite $m\times m$ matrix prove that for all $\boldsymbol{\alpha}\in R^m$

$$f_1\geq\frac{\boldsymbol{\alpha}'A_1\boldsymbol{\alpha}}{\boldsymbol{\alpha}'A_2\boldsymbol{\alpha}}\geq f_m,$$

where f_1 and f_m are the largest and smallest latent roots of $A_1A_2^{-1}$, respectively.

8.4. Suppose that A_1 is $W_m(n_1,\Sigma_1)$ and A_2 is $W_m(n_2,\Sigma_2)$ and A_1 and A_2 are independent. Let f_1 and f_m be the largest and smallest latent roots of $A_1A_2^{-1}$, respectively. Using the result of Problem 8.3 prove that

$$P(f_1\leq x)\leq\prod_{i=1}^{m}P\left(F_{n_1,n_2}\leq\frac{n_2x}{n_1\delta_i}\right)$$

and

$$P(f_m\leq x)\geq 1-\prod_{i=1}^{m}P\left(F_{n_1,n_2}\geq\frac{n_2x}{n_1\delta_i}\right),$$

where $\delta_1,\ldots,\delta_m$ are the latent roots of $\Sigma_1\Sigma_2^{-1}$.

[*Hint:* Use an initial invariance argument to express the problem in terms of the maximal invariant $(\delta_1,\ldots,\delta_m)$.]

8.5. Consider a sample of size N_i from a $N_m(\mu_i,\Sigma_i)$ distribution $(i=1,2,3)$.

(a) Show that the likelihood ratio statistic for testing H_0: $\Sigma_1=\Sigma_2=\Sigma_3$, given $\Sigma_1=\Sigma_2$, is

$$\Lambda_0=\frac{\det(A_1+A_2)^{(N_1+N_2)/2}(\det A_3)^{N_3/2}}{\det(A_1+A_2+A_2)^{(N_1+N_2+N_2)/2}}$$

$$\cdot\frac{(N_1+N_2+N_3)^{m(N_1+N_2+N_1)/2}}{(N_1+N_2)^{m(N_1+N_2)/2}N_3^{mN_3/2}}$$

(b) Let Λ_1 be the likelihood ratio statistic for testing H_1: $\Sigma_1=\Sigma_2$ (see Theorem 8.2.1). Show that Λ_0 and Λ_1 are independently distributed when $\Sigma_1=\Sigma_2=\Sigma_3$.

8.6. Prove Theorem 8.3.1.

8.7. Let V be the ellipticity statistic given by (7) of Section 8.3 for testing H: $\Sigma=\lambda I_m$.

(a) Show that when $m=2$ and H is not necessarily true the distribution function of V can be expressed in the form

$$P(V\le x)=\sum_{k=0}^{\infty} d_k I_x(\tfrac{1}{2}(n-1),k+1),$$

where $I_x(\alpha,\beta)$ denotes the incomplete beta function and d_k is the negative binomial probability

$$d_k=(-1)^k\binom{-\frac{1}{2}n}{k}p^{n/2}(1-p)^k \qquad (k=0,1,\ldots)$$

with $p=4\lambda_1\lambda_2/(\lambda_1+\lambda_2)^2$, where λ_1 and λ_2 are the latent roots of Σ.

[*Hint:* Show that

$$E(V^h)=\frac{2^{4h}\Gamma_2(\frac{1}{2}n+h)}{\Gamma_2(\frac{1}{2}n)}(\det\Sigma)^h E\left[(\operatorname{tr}A)^{-2h}\right]$$

and use the result of Problem 3.12 to evaluate the expectation on the right.]

(b) From (a) it follows that when $m=2$ and H is true, V is beta $(\frac{1}{2}(n-1),1))$. Show also that $V^{1/2}$ is beta$(n-1,1)$ and $-(n-1)\log V$ is χ_2^2.

8.8. Prove Theorem 8.4.11.

8.9. Suppose that $\mathbf{X}_1,\ldots,\mathbf{X}_N$ is a random sample from the $N_m(\boldsymbol{\mu},\Sigma)$ distribution. Derive the likelihood ratio statistic for testing $H: \boldsymbol{\mu}=\mathbf{0}, \Sigma=\sigma^2 I_m$, where $\sigma^2>0$ is unspecified, and find its moments. Show also that the likelihood ratio test is unbiased.

8.10. Show that a maximal invariant under the group of transformations (2) of Section 8.5 is $(\lambda_1,\ldots,\lambda_m, P\boldsymbol{\mu})$, where $\lambda_1,\ldots,\lambda_m$ are the latent roots of Σ and $P\in O(m)$ is such that $P\Sigma P'=\Lambda\equiv\text{diag}(\lambda_1,\ldots,\lambda_m)$.

8.11. Prove Theorem 8.5.5.

8.12. Prove Theorem 8.5.7.

CHAPTER 9

Principal Components and Related Topics

9.1. INTRODUCTION

In many practical situations observations are taken on a large number of correlated variables and in such cases it is natural to look at various ways in which the dimension of the problem (that is, the number of variables being studied) might be reduced, without sacrificing too much of the information about the variables contained in the covariance matrix. One such exploratory data-analytic technique, developed by Hotelling (1933), is principal components analysis. In this analysis the coordinate axes (representing the original variables) are rotated to give a new coordinate system representing variables having certain optimal variance properties. This is equivalent to making a special orthogonal transformation of the original variables. The first principal component (that is, the first variable in the transformed set) is the normalized linear combination of the original variables with maximum variance; the second principal component is the normalized linear combination having maximum variance out of all linear combinations uncorrelated with the first principal component, and so on. Hence principal components analysis is concerned with attempting to characterize or explain the variability in a vector variable by replacing it by a new variable with a smaller number of components with large variance.

It will be seen in Section 9.2 that principal components analysis is concerned fundamentally with the eigenstructure of covariance matrices, that is, with their latent roots and eigenvectors. The coefficients in the first principal component are, in fact, the components of the normalized eigenvector corresponding to the largest latent root, and the variance of the first principal component is this largest root. A common and often valid criticism of the technique is that it is not invariant under linear transformations of the variables since such transformations change the eigenstructure of the covariance matrix. Because of this, the choice of a particular coordinate system or units of measurement of the variables is very important; principal components analysis makes much more sense if all the variables are measured in the same units. If they are not, it is often recommended that principal components be extracted from the correlation matrix rather than the covariance matrix; in this case, however, questions of interpretation arise and problems of inference are exceedingly more complex [see, for example, T. W. Anderson (1963)], and will not be dealt with in this book.

This chapter is concerned primarily with results about the latent roots and eigenvectors of a covariance matrix formed from a normally distributed sample. Because of the complexity of exact distributions (see Sections 9.4 and 9.7) many of the results presented are asymptotic in nature. In Section 9.5 asymptotic joint distributions of the latent roots are derived, and these are used in Section 9.6 to investigate a number of inference procedures, primarily from T. W. Anderson (1963), of interest in principal components analysis. In Section 9.7 expressions are given for the exact distributions of the extreme latent roots of the covariance matrix.

9.2. POPULATION PRINCIPAL COMPONENTS

Let $\mathbf{X}$ be an $m\times 1$ random vector with mean $\boldsymbol{\mu}$ and positive definite covariance matrix Σ. Let $\lambda_1\ge\lambda_2\ge\cdots\ge\lambda_m$ (>0) be the latent roots of Σ and let $H=[\mathbf{h}_1\ldots\mathbf{h}_m]$ be an $m\times m$ orthogonal matrix such that

$$H'\Sigma H=\Lambda\equiv\operatorname{diag}(\lambda_1,\ldots,\lambda_m), \tag{1}$$

so that $\mathbf{h}_i$ is an eigenvector of Σ corresponding to the latent root λ_i. Now put $\mathbf{U}=H'\mathbf{X}=(U_1,\ldots,U_m)'$; then $\operatorname{Cov}(\mathbf{U})=\Lambda$, so that $U_1,\ldots,U_m$ are all uncorrelated, and $\operatorname{Var}(U_i)=\lambda_i$, $i=1,\ldots,m$. The components $U_1,\ldots,U_m$ of $\mathbf{U}$ are called the *principal components* of $\mathbf{X}$. The *first* principal component is $U_1=\mathbf{h}_1'\mathbf{X}$ and its variance is λ_1; the *second* principal component is $U_2=\mathbf{h}_2'\mathbf{X}$, with variance λ_2; and so on. Moreover, the principal components have the following optimality property. The first principal component U_1 is the normalized linear combination of the components of $\mathbf{X}$ with the largest possible variance, and this maximum variance is λ_1; then out of all normalized linear combinations of the components of $\mathbf{X}$ which are uncorrelated with U_1, the second principal component U_2 has maximum variance, namely, λ_2, and so on. In general, out of all normalized linear combinations which are uncorrelated with $U_1,\ldots,U_{k-1}$, the kth principal component U_k has maximum variance λ_k, with $k=1,\ldots,m$. We will prove this assertion in a moment. First note that the variance of an arbitrary linear function $\boldsymbol{\alpha}'\mathbf{X}$ of $\mathbf{X}$ is $\mathrm{Var}(\boldsymbol{\alpha}'\mathbf{X})=\boldsymbol{\alpha}'\Sigma\boldsymbol{\alpha}$ and that the condition that $\boldsymbol{\alpha}'\mathbf{X}$ be uncorrelated with the ith principal component $U_i=\mathbf{h}_i'\mathbf{X}$ is

$$\begin{aligned}0&=\mathrm{Cov}(\boldsymbol{\alpha}'\mathbf{X},\mathbf{h}_i'\mathbf{X})\\&=\boldsymbol{\alpha}'\Sigma\mathbf{h}_i\\&=\lambda_i\boldsymbol{\alpha}'\mathbf{h}_i,\end{aligned}$$

since $\Sigma\mathbf{h}_i=\lambda_i\mathbf{h}_i$, so that $\boldsymbol{\alpha}$ must be orthogonal to $\mathbf{h}_i$. The above optimality property of the principal components is a direct consequence of the following theorem.

THEOREM 9.2.1. Let $H=[\mathbf{h}_1,\dots,\mathbf{h}_m]\in O(m)$ be such that

$$H'\Sigma H=\Lambda=\mathrm{diag}(\lambda_1,\dots,\lambda_m),$$

where $\lambda_1\geq\cdots\geq\lambda_m$. Then

$$\lambda_k=\max_{\substack{\boldsymbol{\alpha}'\boldsymbol{\alpha}=1\\ \boldsymbol{\alpha}'\mathbf{h}_i=0\\ i=1,\dots,k-1}}\boldsymbol{\alpha}'\Sigma\boldsymbol{\alpha}=\mathbf{h}_k'\Sigma\mathbf{h}_k.$$

Proof. First note that with $\boldsymbol{\beta}=H'\boldsymbol{\alpha}=(\beta_1,\dots,\beta_m)'$ we have

$$\boldsymbol{\alpha}'\Sigma\boldsymbol{\alpha}=\boldsymbol{\alpha}'HH'\Sigma HH'\boldsymbol{\alpha}=\boldsymbol{\beta}'\Lambda\boldsymbol{\beta}=\sum_{i=1}^{m}\lambda_i\beta_i^2.$$

As a consequence, if $\boldsymbol{\alpha}'\boldsymbol{\alpha}=1$, so that $\boldsymbol{\beta}'\boldsymbol{\beta}=1$,

$$\boldsymbol{\alpha}'\Sigma\boldsymbol{\alpha}=\boldsymbol{\beta}'\Lambda\boldsymbol{\beta}\leq\lambda_1(\beta_1^2+\cdots+\beta_m^2)=\lambda_1,$$

with equality when $\boldsymbol{\beta}=(1,0,\dots,0)'$, i.e., when $\boldsymbol{\alpha}=\mathbf{h}_1$. Hence

$$\lambda_1=\max_{\boldsymbol{\alpha}'\boldsymbol{\alpha}=1}\boldsymbol{\alpha}'\Sigma\boldsymbol{\alpha}=\mathbf{h}_1'\Sigma\mathbf{h}_1.$$

Next, the condition that $\boldsymbol{\alpha}'\mathbf{h}_1=0$ is equivalent to $\boldsymbol{\beta}'H'\mathbf{h}_1=0$, that is, to $\beta_1=0$ so that, when this holds and when $\boldsymbol{\alpha}'\boldsymbol{\alpha}=\boldsymbol{\beta}'\boldsymbol{\beta}=1$ we have

$$\boldsymbol{\alpha}'\Sigma\boldsymbol{\alpha}=\boldsymbol{\beta}'\Lambda\boldsymbol{\beta}=\sum_{i=2}^{m}\lambda_i\beta_i^2\leq\lambda_2(\beta_2^2+\cdots+\beta_m^2)=\lambda_2$$

with equality when $\boldsymbol{\beta}=(0,1,0,\dots,0)'$, i.e., when $\boldsymbol{\alpha}=\mathbf{h}_2$. Hence

$$\lambda_2=\max_{\substack{\boldsymbol{\alpha}'\boldsymbol{\alpha}=1\\ \boldsymbol{\alpha}'\mathbf{h}_1=0}}\boldsymbol{\alpha}'\Sigma\boldsymbol{\alpha}=\mathbf{h}_2'\Sigma\mathbf{h}_2.$$

The rest of the proof follows in exactly the same way.

If the latent roots $\lambda_1,\dots,\lambda_m$ of Σ are *distinct*, the orthogonal matrix H which diagonalizes Σ is unique up to sign changes of the first element in each column so that the principal components $U_i=\mathbf{h}_i'\mathbf{X}$, $i=1,\dots,m$, are unique up to sign changes. If the latent roots are not all distinct, say

$$\underbrace{\lambda_1=\cdots=\lambda_{m_1}}_{\delta_1}>\underbrace{\lambda_{m_1+1}=\cdots=\lambda_{m_1+m_2}}_{\delta_2}>\cdots>\underbrace{\lambda_{m_1+\cdots+m_{r-1}+1}=\cdots=\lambda_m}_{\delta_r},$$

so that δ_j is a latent root of multiplicity m_j, $j=1,\dots,r$, with $\sum_{j=1}^r m_j=m$, then if $H\in O(m)$ diagonalizes Σ so does the orthogonal matrix

$$H\begin{bmatrix}P_1 & & & 0\\ & P_2 & & \\ & & \ddots & \\ 0 & & & P_r\end{bmatrix}\equiv HP,$$

say, where $P_i\in O(m_i)$, $i=1,\dots,r$, and hence the principal components are not unique. This, of course, does not affect the optimality property in terms of variance discussed previously.

If the random vector $\mathbf{X}$ has an *elliptical* distribution with covariance matrix Σ, the contours of equal probability density are ellipsoids and the principal components clearly represent a rotation of the coordinate axes to the principal axes of the ellipsoid. If Σ has multiple latent roots (it is easy to picture $\Sigma=\lambda I_2$), these principal axes are not unique.

Recall from Section 9.1 that what a principal components analysis attempts to do is "explain" the variability in $\mathbf{X}$. To do this, some overall measure of the "total variability" in $\mathbf{X}$ is required; two such measures are $\operatorname{tr}\Sigma$ and $\det\Sigma$, with the former being more commonly used since $\det\Sigma$ has the disadvantage of being very sensitive to any small latent roots even though the others may be large. Note that in transforming to principal components these measures of total variation are unchanged, for

$$\operatorname{tr}\Sigma=\operatorname{tr}H'\Sigma H=\operatorname{tr}\Lambda=\sum_{i=1}^{m}\lambda_i$$

and

$$\det \Sigma = \det H'\Sigma H = \det \Lambda = \prod_{i=1}^{m} \lambda_i .$$

Note also that $\lambda_1 + \cdots + \lambda_k$ is the variance of the first k principal components; in a principal components analysis the hope is that for some small k, $\lambda_1 + \cdots + \lambda_k$ is close to $\operatorname{tr} \Sigma$. If this is so, the first k principal components explain most of the variation in $\mathbf{X}$ and the remaining $m-k$ principal components contribute little, since these have small variances. Of course, in most practical situations, the covariance matrix Σ is unknown, and hence so are its roots and vectors. The next section deals with the estimation of principal components and their variances.

9.3. SAMPLE PRINCIPAL COMPONENTS

Suppose that the random vector $\mathbf{X}$ has the $N_m(\boldsymbol{\mu}, \Sigma)$ distribution and let $\mathbf{X}_1, \ldots, \mathbf{X}_N$ be a random sample of size $N = n+1$ on $\mathbf{X}$. Let S be the sample covariance matrix given by

$$A = nS = \sum_{i=1}^{N} (\mathbf{X}_i - \overline{\mathbf{X}})(\mathbf{X}_i - \overline{\mathbf{X}})'$$

and let $l_1 > \cdots > l_m$ be the latent roots of S. These are distinct with probability one and are estimates of the latent roots $\lambda_1 \geq \cdots \geq \lambda_m$ of Σ. Recall that $\lambda_1, \ldots, \lambda_m$ are the variances of the population principal components. Let $Q = [\mathbf{q}_1 \ldots \mathbf{q}_m]$ be an $m \times m$ orthogonal matrix such that

$$Q'SQ = L \equiv \operatorname{diag}(l_1, \ldots, l_m), \tag{1}$$

so that $\mathbf{q}_i$ is the normalized eigenvector of S corresponding to the latent root l_i; it represents an estimate of the eigenvector $\mathbf{h}_i$ of Σ given by the ith column of an orthogonal matrix H satisfying (1) of Section 9.2. The *sample principal components* are defined to be the components $\hat{U}_1, \ldots, \hat{U}_m$ of $\hat{\mathbf{U}} = Q'\mathbf{X}$. These are estimates of the population principal components given by $\mathbf{U} = H'\mathbf{X}$.

If we require that the first element in each column of H be non-negative the representation $\Sigma = H\Lambda H'$ is unique if the latent roots $\lambda_1, \ldots, \lambda_m$ of Σ are *distinct*. Similarly, with probability one, the sample covariance matrix S has the unique representation $S = QLQ'$, where the first element in each column of Q is nonnegative. The maximum likelihood estimates of λ_i and $\mathbf{h}_i$ are then, respectively, $\hat{\lambda}_i = nl_i/N$ and $\hat{\mathbf{h}}_i = \mathbf{q}_i$ for $i=1,\dots,m$; that is, the maximum likelihood estimates of Λ and H are $\hat{\Lambda}=(n/N)L$ and $\hat{H}=Q$. Note that $\hat{\mathbf{h}}_i$ is an eigenvector of the maximum likelihood estimate $\hat{\Sigma}=(n/N)S$ of Σ, corresponding to the latent root $\hat{\lambda}_i$. If, on the other hand, Σ has multiple roots then the maximum likelihood estimate of any multiple root is obtained by *averaging* the corresponding latent roots of $\hat{\Sigma}$ and the maximum likelihood estimate of the corresponding columns of H is not unique. These assertions are proved in the following theorem (from T. W. Anderson, 1963).

THEOREM 9.3.1. Suppose that the population covariance matrix Σ has latent roots $\delta_1 > \cdots > \delta_r$ with multiplicities $m_1,\dots,m_r$, respectively, and partition the orthogonal matrices H and Q as

$$H=\left[H_1 \vdots H_2 \vdots \dots \vdots H_r\right], \qquad Q=\left[Q_1 \vdots Q_2 \vdots \dots \vdots Q_r\right],$$

where H_i and Q_i are $m \times m_i$ matrices. Then the maximum likelihood estimate of δ_i is

$$\hat{\delta}_i = \frac{1}{m_i}\frac{n}{N}\sum_{j\in D_i} l_j \qquad (i=1,\dots,r), \tag{2}$$

where D_i is the set of integers $m_1+\cdots+m_{i-1}+1,\dots,m_1+\cdots+m_i$; and a maximum likelihood estimate of H_i is $\hat{H}_i = Q_i P_{ii}$, where P_{ii} is any $m_i \times m_i$ orthogonal matrix such that the first element in each column of $\hat{H}_i$ is nonnegative.

Proof. For notational simplicity we will give a proof in the case where there is one multiple root; the reader can readily generalize the argument that follows. Suppose, then, that the latent roots of Σ are

$$\lambda_1 > \cdots > \lambda_k > \lambda_{k+1} = \cdots = \lambda_m \quad (=\lambda);$$

that is, the largest k roots are distinct and the smallest root λ has multiplicity $m-k$. Ignoring the constant the likelihood function is [see (8) of Section 3.1]

$$L(\boldsymbol{\mu},\Sigma)=(\det\Sigma)^{-N/2}\operatorname{etr}\left(-\tfrac{1}{2}\Sigma^{-1}A\right)\exp\left[-\tfrac{1}{2}N(\overline{\mathbf{X}}-\boldsymbol{\mu})'\Sigma^{-1}(\overline{\mathbf{X}}-\boldsymbol{\mu})\right],$$

where $A=nS$. For each Σ, $L(\boldsymbol{\mu},\Sigma)$ is maximized when $\boldsymbol{\mu}=\overline{\mathbf{X}}$, so it remains to maximize the function

$$g(\Sigma)=\log L(\overline{\mathbf{X}},\Sigma)=-\tfrac{1}{2}N\log\det\Sigma-\tfrac{1}{2}\operatorname{tr}(\Sigma^{-1}A). \tag{3}$$

Putting $\Sigma = H\Lambda H'$ and $A = nS = nQLQ'$ where $\Lambda = \text{diag}(\lambda_1,\dots,\lambda_k,\lambda,\dots,\lambda)$, $L = \text{diag}(l_1,\dots,l_m)$ and H and Q are orthogonal, this becomes

$$
\begin{aligned}
g(\Sigma) &= -\tfrac{1}{2}N\sum_{i=1}^{k}\log\lambda_i - \tfrac{1}{2}N(m-k)\log\lambda - \tfrac{1}{2}n\,\text{tr}(H\Lambda^{-1}H'QLQ') \\
&= -\tfrac{1}{2}N\sum_{i=1}^{k}\log\lambda_i - \tfrac{1}{2}N(m-k)\log\lambda - \tfrac{1}{2}n\,\text{tr}(\Lambda^{-1}P'LP),
\end{aligned}
$$

where $P = Q'H \in O(m)$.

Now partition P as $P = [P_1 : P_2]$, where P_1 is $m\times k$ and P_2 is $m\times(m-k)$, and write Λ as

$$
\Lambda = \begin{bmatrix} \Lambda_1 & 0 \\ 0 & \lambda I_{m-k} \end{bmatrix},
$$

where $\Lambda_1 = \text{diag}(\lambda_1,\dots,\lambda_k)$, so that Λ_1 contains the distinct latent roots of Σ. Then

$$
\begin{aligned}
\text{tr}(\Lambda^{-1}P'LP) &= \text{tr}\left(\begin{bmatrix} \Lambda_1^{-1} & 0 \\ 0 & \lambda^{-1}I \end{bmatrix}\begin{bmatrix} P_1' \\ P_2' \end{bmatrix} L[P_1 P_2]\right) \\
&= \text{tr}(\Lambda_1^{-1}P_1'LP_1) + \frac{1}{\lambda}\text{tr}(LP_2P_2') \\
&= \frac{1}{\lambda}\sum_{i=1}^{m} l_i - \text{tr}\left(\left(\frac{1}{\lambda}I_k - \Lambda_1^{-1}\right)P_1'LP_1\right)
\end{aligned}
$$

where we have used the fact that $P_2P_2' = I - P_1P_1'$. Hence

$$
\begin{aligned}
(4)\qquad g(\Sigma) = &-\tfrac{1}{2}N\sum_{i=1}^{k}\log\lambda_i - \tfrac{1}{2}N(m-k)\log\lambda - \frac{n}{2\lambda}\sum_{i=1}^{m} l_i \\
&+ \tfrac{1}{2}n\,\text{tr}\left(\left(\frac{1}{\lambda}I_k - \Lambda_1^{-1}\right)P_1'LP_1\right).
\end{aligned}
$$

It is a straightforward matter to show that if $U = \text{diag}(u_1,\dots,u_k)$ with $u_1 > \cdots > u_k > 0$ and $V = \text{diag}(v_1,\dots,v_m)$, with $v_1 > \cdots > v_m > 0$, then for all $P_1 \in V_{k,m}$, the Stiefel manifold of $m\times k$ matrices P_1 with $P_1'P_1 = I_k$,

$$
\text{tr}(UP_1'VP_1) \le \sum_{i=1}^{k} u_i v_i,
$$

with equality only at the 2^k $m \times k$ matrices of the form

$$(5) \qquad P_1 = \begin{bmatrix} \pm 1 & & 0 \\ & \ddots & \\ 0 & & \pm 1 \\ \hdashline & 0 & \end{bmatrix}$$

(see Problem 9.4). Applying this result to the trace term in (4) with $U = \lambda^{-1} I_k - \Lambda_1^{-1}$ and $V = L$, it follows that this term is maximized with respect to P_1 when P_1 has the form (5), and the maximum value is

$$(6) \qquad \frac{n}{2\lambda} \sum_{i=1}^{k} l_i - \tfrac{1}{2} n \sum_{i=1}^{k} \frac{l_i}{\lambda_i}.$$

Since P is orthogonal it follows that the function $g(\Sigma)$ is maximized with respect to P when P has the form

$$\hat{P} = \left[\begin{array}{ccc:c} \pm 1 & & & 0 \\ & \ddots & & \\ & & \pm 1 & \\ \hdashline & 0 & & P_{22} \end{array}\right]$$

for any $P_{22} \in O(m-k)$, and then $\hat{H} = Q\hat{P}$ gives a maximum likelihood estimate of H. We now have, from (4) and (6),

$$(7) \qquad \max_{P \in O(m)} g(\Sigma) = -\frac{n}{2\lambda} \sum_{i=k+1}^{m} l_i - \tfrac{1}{2} n \sum_{i=1}^{k} \frac{l_i}{\lambda_i} - \tfrac{1}{2} N \sum_{i=1}^{k} \log \lambda_i$$
$$- \tfrac{1}{2} N(m-k) \log \lambda.$$

Straightforward differentiation now shows that the values of λ_i and λ which maximize this are

$$(8) \qquad \hat{\lambda}_i = \frac{n}{N} l_i \qquad (i = 1, \ldots, k)$$

and

$$(9) \qquad \hat{\lambda} = \frac{1}{m-k} \frac{n}{N} \sum_{j=k+1}^{m} l_j,$$

completing the proof.

Tractable expressions for the exact moments of the latent roots of S are unknown, but asymptotic expansions for some of these have been found by Lawley (1956). Lawley has shown that if λ_i is a distinct latent root of Σ the mean and variance of l_i can be expanded for large n as

$$(10) \qquad E(l_i)=\lambda_i+\frac{\lambda_i}{n}\sum_{\substack{j=1\\ j\neq i}}^{m}\frac{\lambda_j}{\lambda_i-\lambda_j}+O(n^{-2})$$

and

$$(11) \qquad \mathrm{Var}(l_i)=\frac{2\lambda_i^2}{n}\left[1-\frac{1}{n}\sum_{\substack{j=1\\ j\neq i}}^{m}\left(\frac{\lambda_j}{\lambda_i-\lambda_j}\right)^2\right]+O(n^{-3}).$$

9.4. THE JOINT DISTRIBUTION OF THE LATENT ROOTS OF A SAMPLE COVARIANCE MATRIX

In this and the following section we will derive expressions for the exact and asymptotic joint distributions of the latent roots of a covariance matrix formed from a normal sample. Let $l_1>\cdots>l_m$ be the latent roots of the sample covariance matrix S, where $A=nS$ has the $W_m(n,\Sigma)$ distribution. Recall that these roots are estimates of the variances of the population principal components. The exact joint density function of $l_1,\dots,l_m$ can be expressed in terms of the two-matrix ${}_0F_0$ hypergeometric function introduced in Section 7.3, having an expansion in terms of zonal polynomials. The result is given in the following theorem (from James, 1960).

THEOREM 9.4.1. Let nS have the $W_m(n,\Sigma)$ distribution, with $n>m-1$. Then the joint density function of $l_1,\dots,l_m$, the latent roots of S, can be expressed in the form

$$(1) \qquad \left(\frac{n}{2}\right)^{mn/2}\frac{\pi^{m^2/2}(\det\Sigma)^{-n/2}}{\Gamma_m(\frac{1}{2}n)\Gamma_m(\frac{1}{2}m)}\prod_{i=1}^{m}l_i^{(n-m-1)/2}\prod_{i<j}^{m}(l_i-l_j)$$

$$\cdot {}_0F_0^{(m)}(-\tfrac{1}{2}nL,\Sigma^{-1}) \qquad (l_1>\cdots>l_m>0),$$

where $L=\operatorname{diag}(l_1,\ldots,l_m)$ and

$$(2)\qquad {}_0F_0^{(m)}(-\tfrac{1}{2}nL,\Sigma^{-1})=\sum_{k=0}^{\infty}\sum_{\kappa}\frac{C_\kappa(-\frac{1}{2}nL)C_\kappa(\Sigma^{-1})}{k!\,C_\kappa(I_m)}$$

Proof. From Theorem 3.2.18 the joint density function of $l_1,\ldots,l_m$ is

$$(3)\qquad \left(\frac{n}{2}\right)^{mn/2}\frac{\pi^{m^2/2}(\det\Sigma)^{-n/2}}{\Gamma_m(\frac{1}{2}n)\Gamma_m(\frac{1}{2}m)}\prod_{i=1}^{m}l_i^{(n-m-1)/2}\prod_{i<j}^{m}(l_i-l_j)$$

$$\cdot\int_{O(m)}\operatorname{etr}(-\tfrac{1}{2}n\Sigma^{-1}HLH')(dH),$$

where (dH) is the normalized invariant measure on $O(m)$. (Note that in Theorem 3.2.18 the l_i are the latent roots of $A=nS$ so that l_i there must be replaced by nl_i.) The desired result now follows from (3) using Theorem 7.3.3 and the fact that

$$\operatorname{etr}(-\tfrac{1}{2}n\Sigma^{-1}HLH')={}_0F_0(-\tfrac{1}{2}n\Sigma^{-1}HLH')$$

[see (3) of Section 7.3].

It was noted in the discussion following Theorem 3.2.18 that the density function of $l_1,\ldots,l_m$ depends on the population covariance matrix Σ only through its latent roots. The zonal polynomial expansion (2) makes this obvious, since $C_\kappa(\Sigma^{-1})$ is a symmetric homogeneous polynomial in the latent roots of Σ^{-1}. We also noted in Corollary 3.2.19 that when $\Sigma=\lambda I_m$ the joint density function of the sample roots has a particularly simple form. For completeness we will restate the result here.

COROLLARY 9.4.2. When $\Sigma=\lambda I_m$ the joint density function of the latent roots $l_1,\ldots,l_m$ of the sample covariance matrix S is

(4)

$$\left(\frac{n}{2\lambda}\right)^{mn/2}\frac{\pi^{m^2/2}}{\Gamma_m(\frac{1}{2}n)\Gamma_m(\frac{1}{2}m)}\exp\left(-\frac{n}{2\lambda}\sum_{i=1}^{m}l_i\right)\prod_{i=1}^{m}l_i^{(n-m-1)/2}\prod_{i<j}^{m}(l_i-l_j)$$

$$(l_1>\cdots l_m>0)$$

The proof of this is a direct consequence of either (1) or (3), or it follows from Corollary 3.2.19 by replacing l_i there by nl_i.

The distribution of the sample latent roots when $\Sigma=\lambda I_m$ given by Corollary 9.4.2 is usually referred to as the *null* distribution; the distribution given in Theorem 9.4.1 for arbitrary positive definite Σ is called the *non-null* (or noncentral) distribution. If we write S as $S=QLQ'$, where the first element in each column of $Q\in O(m)$ is non-negative, in the null case when $\Sigma=\lambda I_m$ the matrix Q, whose columns are the eigenvectors of S, has the conditional Haar invariant distribution (as noted in the discussion following Corollary 3.2.19), that is, the distribution of an orthogonal $m\times m$ matrix having the invariant distribution on $O(m)$ conditional on the first element in each column being non-negative. Moreover the matrix Q is independently distributed of the latent roots $l_1,\dots,l_m$. Neither of these statements remains true in the non-null case.

9.5. ASYMPTOTIC DISTRIBUTIONS OF THE LATENT ROOTS OF A SAMPLE COVARIANCE MATRIX

The joint density function of the latent roots $l_1,\dots,l_m$ of the sample covariance matrix S given by Theorem 9.4.1 involves the hypergeometric function ${}_0F_0^{(m)}(-\frac{1}{2}nL,\Sigma^{-1})$ having an expansion in terms of zonal polynomials. If n is large, this zonal polynomial series converges very slowly in general. Moreover, it is difficult to obtain from this series any feeling for the behavior of the density function or an understanding of how the sample and population latent roots interact with each other. It often occurs in practical situations that one is dealing with a large sample size (so that n is large) and it makes sense to ask how the ${}_0F_0^{(m)}$ function behaves *asymptotically* for large n. It turns out that asymptotic representations for this function can be written in terms of elementary functions and sheds a great deal of light on the interaction between the sample and population roots.

The zonal polynomial expansion for ${}_0F_0^{(m)}$ given by (2) of Section 8.4 does not lend itself easily to the derivation of asymptotic results. Integral representations are generally the most useful tool for obtaining asymptotic results in analysis, so that here we will work with the integral

$$(1)\qquad {}_0F_0^{(m)}(-\tfrac{1}{2}nL,\Sigma^{-1})=\int_{O(m)}\operatorname{etr}(-\tfrac{1}{2}n\Sigma^{-1}HLH')(dH),$$

and examine its asymptotic behavior as $n\to\infty$. To do this we will make use of the following theorem, which gives a multivariate extension of Laplace's method for obtaining the asymptotic behavior of integrals. In this theorem, and subsequently, the notation "$a \sim b$ for large n" means that $a/b \to 1$ as $n \to \infty$.

THEOREM 9.5.1. Let D be a subset of R^p and let f and g be real-valued functions on D such that:

(i) f has an absolute maximum at an interior point ξ of D and $f(\xi) > 0$;

(ii) there exists a $k \geq 0$ such that $g(\mathbf{x})f(\mathbf{x})^k$ is absolutely integrable on D;

(iii) all partial derivatives

$$\frac{\partial f}{\partial x_i} \quad \text{and} \quad \frac{\partial^2 f}{\partial x_i \partial x_j} \qquad (i, j = 1, \ldots, p)$$

exist and are continuous in a neighborhood $N(\xi)$ of ξ;

(iv) there exists a constant $\gamma < 1$ such that

$$\left|\frac{f(\mathbf{x})}{f(\xi)}\right| < \gamma \quad \text{for all} \quad \mathbf{x} \in D - N(\xi);$$

(v) g is continuous in a neighborhood of ξ and $g(\xi) \neq 0$.

Then, for large n,

$$\int_D [f(\mathbf{x})]^n g(\mathbf{x})\, d\mathbf{x} \sim \left(\frac{2\pi}{n}\right)^{p/2} \frac{[f(\xi)]^n g(\xi)}{[\Delta(\xi)]^{1/2}}, \tag{2}$$

where $\Delta(\xi)$ denotes the Hessian of $-\log f$, namely,

$$\Delta(\xi) = \det \Omega(\xi), \qquad \Omega(\xi) = \left(\frac{-\partial^2 \log f(\xi)}{\partial \xi_i \partial \xi_j}\right). \tag{3}$$

For a rigorous proof of this very useful theorem, due originally to Hsu (1948), the reader is referred to Glynn (1977, 1980). The basic idea in the proof involves recognizing that for large n the major contribution to the integral will arise from a neighborhood of ξ and expanding f and g about ξ. We will sketch a heuristic proof. Write

$$\int_D [f(\mathbf{x})]^n g(\mathbf{x})\, d\mathbf{x} = [f(\xi)]^n \int_D g(\mathbf{x}) \exp\{n[\log f(\mathbf{x}) - \log f(\xi)]\}\, d\mathbf{x}.$$

In a neighborhood $N(\xi)$ of ξ, $\log f(\mathbf{x})-\log f(\xi)$ is approximately equal to $-\frac{1}{2}(\mathbf{x}-\xi)'\Omega(\xi)(\mathbf{x}-\xi)$, $g(\mathbf{x})$ is approximately equal to $g(\xi)$, and then, using (iv), n can be chosen sufficiently large so that the integral over $D-N(\xi)$ is negligible and hence the domain of integration can be extended to R^p. Thus for large n,

$$\int_D [f(\mathbf{x})]^n g(\mathbf{x})\,d\mathbf{x} \sim [f(\xi)]^n g(\xi)\cdot\int_{R^p}\exp\left[-\tfrac{1}{2}n(\mathbf{x}-\xi)'\Omega(\xi)(\mathbf{x}-\xi)\right]d\mathbf{x}$$

$$=\left(\frac{2\pi}{n}\right)^{p/2}\frac{[f(\xi)]^n g(\xi)}{[\Delta(\xi)]^{1/2}}.$$

Let us now return to the problem of finding the asymptotic behavior of the ${}_0F_0^{(m)}$ function in Theorem 9.4.1. It turns out that this depends fundamentally on the spread of the latent roots of the covariance matrix Σ. Different asymptotic results can be obtained by varying the multiplicities of these roots. Because it is somewhat simpler to deal with, we will first look at the case where the m latent roots of Σ are all distinct. The result is given in the following theorem, from G. A. Anderson (1965), where it is assumed without loss of generality that Σ is diagonal (since the ${}_0F_0^{(m)}$ function is a function only of the latent roots of the argument matrices).

THEOREM 9.5.2. If $\Sigma=\mathrm{diag}(\lambda_1,\dots,\lambda_m)$ and $L=\mathrm{diag}(l_1,\dots,l_m)$, where $\lambda_1>\cdots>\lambda_m>0$ and $l_1>\cdots>l_m>0$ then, for large n,

$$(4)\qquad {}_0F_0^{(m)}(-\tfrac{1}{2}nL,\Sigma^{-1})\sim\frac{\Gamma_m(\frac{1}{2}m)}{\pi^{m^2/2}}\exp\left(-\tfrac{1}{2}n\sum_{i=1}^{m}\frac{l_i}{\lambda_i}\right)\prod_{i<j}^{m}\left(\frac{2\pi}{nc_{ij}}\right)^{1/2},$$

where

$$(5)\qquad c_{ij}=\frac{(l_i-l_j)(\lambda_i-\lambda_j)}{\lambda_i\lambda_j}.$$

Proof. The proof is messy and disagreeable, in that it involves a lot of tedious algebraic manipulation; the ideas involved are, however, very simple. We will sketch the proof, leaving some of the details to the reader. The basic idea here is to write the ${}_0F_0^{(m)}$ function as a multiple integral to which the result of Theorem 9.5.1 can be applied. First, write

$${}_0F_0^{(m)}(-\tfrac{1}{2}nL,\Sigma^{-1})=\int_{O(m)}\mathrm{etr}(-\tfrac{1}{2}n\Sigma^{-1}HLH')(dH).$$

Here (dH) is the normalized invariant measure on $O(m)$; it is a little more convenient to work in terms of the unnormalized invariant measure

$$(H'dH)\equiv \bigwedge_{i<j}^{m} \mathbf{h}_j' d\mathbf{h}_i$$

(see Sections 2.1.4 and 3.2.5), equivalent to ordinary Lebesgue measure, regarding the orthogonal group $O(m)$ as a point set in Euclidean space of dimension $\frac{1}{2}m(m-1)$. These two measures are related by

$$(dH)=\frac{\Gamma_m(\frac{1}{2}m)}{2^m\pi^{m^2/2}}(H'dH)$$

[see (20) of Section 3.2.5], so that

(6) $$_0F_0^{(m)}(-\tfrac{1}{2}nL,\Sigma^{-1})=\frac{\Gamma_m(\frac{1}{2}m)}{2^m\pi^{m^2/2}}I(n),$$

where

(7) $$I(n)=\int_{O(m)}\operatorname{etr}(-\tfrac{1}{2}n\Sigma^{-1}HLH')(H'dH).$$

Note that this integral has the form

$$I(n)=\int_{O(m)}[f(H)]^n(H'dH),$$

where

(8) $$f(H)=\operatorname{etr}(-\tfrac{1}{2}\Sigma^{-1}HLH').$$

In order to apply Theorem 9.5.1 there are two things to be calculated, namely, the maximum value of $f(H)$ and the value of the Hessian of $-\log f$ at the maximum. Maximizing $f(H)$ is equivalent to minimizing

$$\phi(H)=\operatorname{tr}(\Sigma^{-1}HLH'),$$

and it is a straightforward matter to show that for all $H\in O(m)$,

(9) $$\operatorname{tr}(\Sigma^{-1}HLH')\geq\sum_{i=1}^{m}\frac{l_i}{\lambda_i},$$

with equality if and only if H is one of the 2^m matrices of the form

$$\begin{bmatrix} \pm 1 & & 0 \\ & \ddots & \\ 0 & & \pm 1 \end{bmatrix} \tag{10}$$

(see Problem 9.3). The function $f(H)$ thus has a maximum of $\exp[-\frac{1}{2}\Sigma_{i=1}^m(l_i/\lambda_i)]$ at each of the 2^m matrices (10). Theorem 9.5.1 assumes just one maximum point. The next step is to split $O(m)$ up into 2^m disjoint pieces, each containing exactly one of the matrices (10), and to recognize that the asymptotic behavior of each of the resulting integrals is the same. Hence for large n,

$$I(n) \sim 2^m \int_{N(I_m)} [f(H)]^n (H'dH), \tag{11}$$

where $N(I_m)$ is a neighborhood of the identity matrix I_m on the orthogonal manifold $O(m)$. Because the determinant of a matrix is a continuous function of the elements of the matrix we can assume that $N(I_m)$ contains only proper orthogonal matrices H (i.e., $\det H = 1$). This is important in the next step, which involves calculating the Hessian of $-\log f$, evaluated at $H = I_m$. This involves differentiating $\log f$ twice with respect to the elements of H. This is complicated by the fact that H has m^2 elements but only $\frac{1}{2}m(m-1)$ functionally independent ones. It helps at this stage to work in terms of a convenient parametrization of H. Any proper orthogonal $m \times m$ matrix H can be expressed as

$$H = \exp(U) \equiv I_m + U + \tfrac{1}{2}U^2 + \tfrac{1}{3!}U^3 + \cdots, \tag{12}$$

where U is an $m \times m$ skew-symmetric matrix (see Theorem A9.11). The $\frac{1}{2}m(m-1)$ elements of U provide a parametrization of H. The mapping $H \to U$ is a mapping from $O^+(m) \to R^{m(m-1)/2}$, where $O^+(m)$ is the subgroup of $O(m)$ consisting of proper orthogonal matrices. The image of $O^+(m)$ under this mapping is a bounded subset of $R^{m(m-1)/2}$. The Jacobian of this transformation is given by

$$\begin{aligned} (H'dH) &\equiv \bigwedge_{i<j}^{m} \mathbf{h}_j' d\mathbf{h}_i \\ &= \left[1 + O\left(u_{ij}^2\right)\right] \bigwedge_{i<j}^{m} du_{ij}, \end{aligned} \tag{13}$$

where $O(u_{ij}^r)$ denotes terms in the u_{ij} which are at least of order r (see Problem 9.11). Under the transformation $H=\exp(U)$, $N(I_m)$ is mapped into a neighborhood of $U=0$, say, $N^*(U=0)$, so that, using (13) in (11), we get

(14)

$$I(n)\sim 2^m\int_{N^*(U=0)}[f(\exp(U))]^n(1+\text{higher-order terms in } U)\prod_{i<j}^m du_{ij}.$$

Put

$$\psi(H)=\log f(H)=-\tfrac{1}{2}\operatorname{tr}(\Sigma^{-1}HLH')=-\frac{1}{2}\sum_{i,j=1}^m\frac{l_j}{\lambda_i}h_{ij}^2;$$

to calculate the Hessian, note that

$$-\frac{\partial^2\psi}{\partial u_{\alpha\beta}^2}=\sum_{i,j=1}^m\frac{l_j}{\lambda_i}\frac{\partial^2 h_{ij}}{\partial u_{\alpha\beta}^2}+\sum_{i,j=1}^m\frac{l_j}{\lambda_i}\left(\frac{\partial h_{ij}}{\partial u_{\alpha\beta}}\right)^2$$

and

$$-\frac{\partial^2\psi}{\partial u_{\alpha\beta}\partial u_{pq}}=\sum_{i,j=1}^m\frac{l_j}{\lambda_i}\frac{\partial^2 h_{ij}}{\partial u_{\alpha\beta}\partial u_{pq}}+\sum_{i,j=1}^m\frac{l_j}{\lambda_i}\frac{\partial h_{ij}}{\partial u_{\alpha\beta}}\frac{\partial h_{ij}}{\partial u_{pq}}$$

so that in order to find the Hessian Δ of $-\log f=-\psi$, we have to differentiate the elements of $H=\exp(U)$ at most twice and set $U=0$. Thus to calculate Δ we can use

$$H=U+\tfrac{1}{2}U^2.$$

It is then a simple matter to show that, at $U=0$,

$$-\frac{\partial^2\psi}{\partial u_{\alpha\beta}^2}=c_{\alpha\beta}\equiv\frac{(l_\alpha-l_\beta)(\lambda_\alpha-\lambda_\beta)}{\lambda_\alpha\lambda_\beta}$$

and

$$\frac{\partial^2\psi}{\partial u_{\alpha\beta}\partial u_{pq}}=\frac{\partial\psi}{\partial u_{\alpha\beta}}=0$$

so that, at $U=0$, the Hessian is

$$\Delta=\det\begin{bmatrix} c_{12} & & 0 \\ & \ddots & \\ 0 & & c_{m-1,m}\end{bmatrix}=\prod_{i<j}^{m} c_{ij}.$$

Hence applying Theorem 9.5.1 with $p=\frac{1}{2}m(m-1)$ to (14) shows that, for large n,

$$I(n)\sim 2^m \exp\left(-\tfrac{1}{2}n\sum_{i=1}^{m}\frac{l_i}{\lambda_i}\right)\prod_{i<j}^{m}\left(\frac{2\pi}{nc_{ij}}\right)^{1/2},$$

and substituting this asymptotic result for $I(n)$ in (6) gives the desired result and completes the proof.

Substituting the asymptotic formula (4) for ${}_0F_0^{(m)}$ back in (1) of Section 9.4 gives an asymptotic representation for the joint density function of the sample roots $l_1,\dots,l_m$ under the assumption that the population roots $\lambda_1,\dots,\lambda_m$ are all distinct. The result is, however, of somewhat limited use for statistical inference in principal components. One of the most commonly used procedures used in principal components analysis is to test whether the q smallest latent roots of Σ are equal. If they are then the variation in the last q dimensions is spherical and, if their common value is small compared with the other $m-q$ roots, then most of the variation in the sample is explained by the first $m-q$ principal components, and a reduction in dimensionality is achieved by considering only these components. We will investigate such a test later in Section 9.6; as a first step it makes sense to find an asymptotic representation for the ${}_0F_0^{(m)}$ function (and hence for the density function of $l_1,\dots,l_m$) under the *null hypothesis* that the smallest q latent roots of Σ are equal. Before doing this we need some preliminary results. First recall from Section 2.1.4 that if $H_1\in V_{k,m}$, the Stiefel manifold of $m\times k$ matrices with orthonormal columns, and we choose *any* $m\times(m-k)$ matrix H_2 (a function of H_1) so that $H=[H_1:H_2]\in O(m)$ then the unnormalized invariant measure on $V_{k,m}$ is

$$(H_1'dH_1)\equiv\bigwedge_{i=1}^{k}\bigwedge_{j=i+1}^{m}\mathbf{h}_j'd\mathbf{h}_i, \tag{15}$$

where $H=[\mathbf{h}_1,\dots,\mathbf{h}_k:\mathbf{h}_{k+1}\dots\mathbf{h}_m]$, and from Theorem 2.1.15,

$$\int_{V_{k,m}}(H_1'dH_1)=\frac{2^k\pi^{km/2}}{\Gamma_k(\frac{1}{2}m)}. \tag{16}$$

We will need the result contained in the following useful lemma given by Constantine and Muirhead (1976); given a function $f(H)$ of an $m\times m$ orthogonal matrix it enables us to first integrate over the last $m-k$ columns of H, the first k columns being *fixed*, and then to integrate over these k columns.

LEMMA 9.5.3.

(17)

$$\int_{O(m)} f(H_1,H_2)(H'dH)=\int_{H_1\in V_{k,m}}\int_{K\in O(m-k)} f(H_1,GK)(K'dK)(H_1'dH_1),$$

where $H=[H_1:H_2]$, H_1 is $m\times k$ and $G=G(H_1)$ is any $m\times(m-k)$ matrix with orthonormal columns orthogonal to those of H_1 (so that $GG'=I_m-H_1H_1'$).

Proof. For fixed H_1, the manifold $\mathcal{H}_2$, say, spanned by the columns of H_2 can be generated by orthogonal transformations of any fixed matrix G chosen so that $[H_1:G]$ is orthogonal; that is, any $H_2\in\mathcal{H}_2$ can be written as $H_2=GK$, and as H_2 runs over $\mathcal{H}_2$, K runs over $O(m-k)$, and the relationship is one-to-one. Writing

$$H=[H_1:H_2]=\left[\mathbf{h}_1\ldots\mathbf{h}_k\vdots\mathbf{h}_{k+1}\ldots\mathbf{h}_m\right]$$

and

$$K=[\mathbf{k}_1\ldots\mathbf{k}_{m-k}]$$

we have

$$d\mathbf{h}_{k+j}=G\,d\mathbf{k}_j\qquad (j=1,\ldots,m-k)$$

for fixed G. Now

$$\begin{aligned}(H'dH)&\equiv\bigwedge_{i<j}^{m}\mathbf{h}_j'd\mathbf{h}_i\\ &=\bigwedge_{i<j}^{k}\mathbf{h}_j'd\mathbf{h}_i\bigwedge_{j=1}^{m-k}\bigwedge_{i=1}^{k}\mathbf{h}_{k+j}'d\mathbf{h}_i\bigwedge_{i<j}^{m-k}\mathbf{h}_{k+j}'d\mathbf{h}_{k+i}\\ &=\bigwedge_{i<j}^{k}\mathbf{h}_j'd\mathbf{h}_i\bigwedge_{j=1}^{m-k}\bigwedge_{i=1}^{k}\mathbf{k}_j'G'd\mathbf{h}_i\bigwedge_{i<j}^{m-k}d\mathbf{k}_j'G'G\,d\mathbf{k}_i\\ &=\bigwedge_{i<j}^{k}\mathbf{h}_j'd\mathbf{h}_i\bigwedge_{j=1}^{m-k}\bigwedge_{i=1}^{k}k_j'G'd\mathbf{h}_i\bigwedge_{i<j}^{m-k}\mathbf{k}_j'd\mathbf{k}_i\\ &=(H_1'dH_1)(K'dK),\end{aligned}$$

using (15). This transformation of the measure $(H'dH)$ is to be interpreted as: first integrate over K for *fixed* H_1, and then integrate over H_1.

The following theorem, from James (1969), gives the asymptotic behavior of the ${}_0F_0^{(m)}$ function in Theorem 9.4.1 under the assumption or null hypothesis that the largest k latent roots of Σ are distinct and the smallest $q=m-k$ roots are equal. When $m=k$ it yields the result of Theorem 9.5.2 as a special case. Again, it is assumed without loss of generality that Σ is diagonal.

THEOREM 9.5.4. If $\Sigma=\text{diag}(\lambda_1,\dots,\lambda_k,\lambda,\dots,\lambda)$, where

$$\lambda_1>\cdots>\lambda_k>\lambda \tag{18}$$

and the smallest root λ is of multiplicity $m-k$, and $L=\text{diag}(l_1,\dots,l_m)$, where $l_1>\cdots>l_m>0$, then, for large n,

(19)

$$ {}_0F_0^{(m)}(-\tfrac{1}{2}nL,\Sigma^{-1})\sim\frac{\Gamma_k(\frac{1}{2}m)}{\pi^{km/2}}\exp\left(-\tfrac{1}{2}n\sum_{i=1}^{k}\frac{l_i}{\lambda_i}\right)$$

$$\cdot\exp\left(-\frac{n}{2\lambda}\sum_{i=k+1}^{m}l_i\right)\prod_{i<j}^{k}\left(\frac{2\pi}{nc_{ij}}\right)^{1/2}\prod_{i=1}^{k}\prod_{j=k+1}^{m}\left(\frac{2\pi}{nd_{ij}}\right)^{1/2},$$

where

$$c_{ij}=\frac{(l_i-l_j)(\lambda_i-\lambda_j)}{\lambda_i\lambda_j}\qquad (i,j=1,\dots,k) \tag{20}$$

and

$$d_{ij}=\frac{(l_i-l_j)(\lambda_i-\lambda)}{\lambda\lambda_i}\qquad (i=1,\dots,k;\ j=k+1,\dots,m). \tag{21}$$

Proof. The proof is similar to that of Theorem 9.5.2 but complicated by the fact that Σ has a multiple root. First, as in the proof of Theorem 9.5.2, write

$${}_0F_0^{(m)}(-\tfrac{1}{2}nL,\Sigma^{-1})=\frac{\Gamma_m(\frac{1}{2}m)}{2^m\pi^{m^2/2}}I(n), \tag{22}$$

where

(23) $$I(n)=\int_{O(m)}\text{etr}\left(-\tfrac{1}{2}n\Sigma^{-1}H'LH\right)(H'dH).$$

Now partition Σ and H as

$$\Sigma=\begin{bmatrix}\Sigma_1 & 0\\ 0 & \lambda I_{m-k}\end{bmatrix},\qquad \Sigma_1=\text{diag}(\lambda_1,\ldots,\lambda_k)$$

and $H=[H_1:H_2]$, where H_1 is $m\times k$. Then

$$\begin{aligned}\text{tr}(\Sigma^{-1}H'LH)&=\text{tr}(\Sigma_1^{-1}H_1'LH_1)+\text{tr}(\lambda^{-1}H_2'LH_2)\\ &=\text{tr}\left[(\Sigma_1^{-1}-\lambda^{-1}I_k)H_1'LH_1\right]+\text{tr}(\lambda^{-1}L),\end{aligned}$$

where we have used

$$\text{tr}(\lambda^{-1}H_2'LH_2)=\text{tr}(\lambda^{-1}LH_2H_2')$$

and the fact that $H_2H_2'=I-H_1H_1'$. Hence

$$I(n)=\exp\left(-\frac{n}{2\lambda}\sum_{i=1}^{m}l_i\right)\int_{O(m)}\text{etr}\left[-\tfrac{1}{2}n(\Sigma_1^{-1}-\lambda^{-1}I)H_1'LH_1\right](H'dH).$$

Applying Lemma 9.5.3 to this last integral gives

$$\begin{aligned}I(n)=\exp\left(-\frac{n}{2\lambda}\sum_{i=1}^{m}l_i\right)\int_{H_1\in V_{k,m}}\int_{K\in O(m-k)}&\text{etr}\left[-\tfrac{1}{2}n(\Sigma_1^{-1}-\lambda^{-1}I)H_1'LH_1\right]\\ &\cdot(K'dK)(H_1'dH_1).\end{aligned}$$

The integrand here is not a function of K, and using Corollary 2.1.16 we can integrate with respect to K to give

(25) $$I(n)=\frac{2^{m-k}\pi^{(m-k)^2/2}}{\Gamma_{m-k}\left[\frac{1}{2}(m-k)\right]}\exp\left(-\frac{n}{2\lambda}\sum_{i=1}^{m}l_i\right)J(n),$$

where

(26) $$J(n)=\int_{V_{k,m}}\text{etr}\left[-\tfrac{1}{2}n(\Sigma_1^{-1}-\lambda^{-1}I)H_1'LH_1\right](H_1'dH_1).$$

The proof from this point on becomes very similar to that of Theorem 9.5.2 (and even more disagreeable algebraically), and we will merely sketch it. The integral $J(n)$ is of the form

$$J(n)=\int_{V_{k,m}}[f(H_1)]^n(H_1'dH_1)$$

where

$$f(H_1)=\text{etr}\left[\tfrac{1}{2}(\lambda^{-1}I-\Sigma_1^{-1})H_1'LH_1\right],$$

so that in order to apply Theorem 9.5.1 to find the asymptotic behavior of $J(n)$ we have to find the maximum value of $f(H_1)$ and the Hessian of $-\log f$ at the maximum. Maximizing f is equivalent to maximizing

$$\phi(H_1)=\text{tr}\left[(\lambda^{-1}I-\Sigma_1^{-1})H_1'LH_1\right]$$

and, from Problem 9.4, it follows that for all $H_1\in V_{k,m}$,

(27) $$\phi(H_1)\le\frac{1}{\lambda}\sum_{i=1}^{k}l_i-\sum_{i=1}^{k}\frac{l_i}{\lambda_i}$$

with equality if and only if H_1 is one of the 2^k matrices of the form

(28) $$\begin{bmatrix}\pm1 & & 0\\ & \ddots & \\ 0 & & \pm1\\ \cdots & \cdots & \cdots\\ & 0 & \end{bmatrix}\qquad(m\times k).$$

Arguing as in the proof of Theorem 8.4.4 it follows that

$$J(n)\sim2^k\int_{N\left(\begin{bmatrix}I_k\\0\end{bmatrix}\right)}[f(H_1)]^n(H_1'dH_1),$$

where

$$N\left(\begin{bmatrix}I_k\\ \cdots\\0\end{bmatrix}\right)$$

denotes a neighborhood of the matrix

$$\begin{bmatrix}I_k\\ \cdots\\0\end{bmatrix}$$

on the Stiefel manifold $V_{k,n}$. Now let $[H_1:-]$ be an $m\times m$ orthogonal matrix whose first k columns are H_1. In the neighborhood above a parametrization of H_1 is given by [see James (1969)]

$$[H_1:-]=\exp\left(\begin{bmatrix} U_{11} & U_{12} \\ -U_{12}' & 0 \end{bmatrix}\right),$$

where U_{11} is a $k\times k$ skew-symmetric matrix and U_{12} is $k\times(m-k)$. The Jacobian of this transformation [cf. (13)] is given by

$$(H_1'dH_1)=\left(1+O\left(u_{ij}^2\right)\right)(dU_{11})(dU_{12}),$$

and the image of $N\left(\begin{bmatrix} I_k \\ \cdot\cdot \\ 0 \end{bmatrix}\right)$ under this transformation is a neighborhood, say, N^*, of $U_{11}=0$, $U_{12}=0$. Hence

$$J(n)\sim 2^k\int_{N^*} f^n\left[1+O\left(u_{ij}^2\right)\right](dU_{11})(dU_{12}).$$

To calculate the Hessian Δ of $-\log f$, put

$$\begin{aligned}\psi=\log f&=\tfrac{1}{2}\operatorname{tr}\left[\left(\lambda^{-1}I-\Sigma_1^{-1}\right)H_1'LH_1\right]\\ &=\frac{1}{2}\sum_{i=1}^{k}\sum_{j=1}^{m}\left(\lambda^{-1}-\lambda_i^{-1}\right)l_jh_{ji}^2,\end{aligned}$$

substitute for the h_{ji}'s in terms of the elements of U_{11} and U_{12}, and evaluate $\Delta=\det(-\partial^2\psi/\partial u_{ij}\partial u_{pq})$ at $U_{11}=0$ and $U_{12}=0$. We will omit the messy details. An application of Theorem 9.5.1 then gives the asymptotic behavior of $J(n)$ for large n as

$$(29)\qquad J(n)\sim 2^k\exp\left(\frac{n}{2\lambda}\sum_{i=1}^{k}l_i-\tfrac{1}{2}n\sum_{i=1}^{k}\frac{l_i}{\lambda_i}\right)$$

$$\cdot\prod_{i<j}^{k}\left(\frac{2\pi}{nc_{ij}}\right)^{1/2}\prod_{i=1}^{k}\prod_{j=k+1}^{m}\left(\frac{2\pi}{nd_{ij}}\right)^{1/2},$$

where c_{ij} and d_{ij} are given by (20) and (21). Substituting this for $J(n)$ in (25) and then the resulting expression for $I(n)$ in (22) gives the desired result on noting the easily proved fact that

$$\frac{\Gamma_m(\frac{1}{2}m)}{\Gamma_{m-k}[\frac{1}{2}(m-k)]}=\Gamma_k(\tfrac{1}{2}m)\pi^{-k(k-m)/2}.$$

The precise meaning of Theorem 9.5.4 is that, given $\varepsilon>0$, there exists $n_0\equiv n_0(\varepsilon,\Sigma,L)$ such that

$$\left|\frac{{}_0F_0^{(m)}(-\frac{1}{2}nL,\Sigma)}{h(n,L,\Sigma)}-1\right|<\varepsilon \qquad \text{for all} \quad n\geq n_0,$$

where $h(n,L,\Sigma)$ denotes the right side of (19). It is clear from the form of $h(n,L,\Sigma)$ that this does not hold uniformly in L or Σ; that is, n_0 cannot be chosen independently of L and Σ. However, it is possible to prove that it does hold uniformly on any set of $l_1,\dots,l_m$ $(l_1>\cdots>l_m>0)$ and $\lambda_1,\dots,\lambda_k,\lambda$ $(\lambda_1>\cdots>\lambda_k>\lambda>0)$ such that the l_i's are bounded away from one another and from zero, as are $\lambda_1,\dots,\lambda_k,\lambda$. The proof of this requires a more sophisticated version of Theorem 9.5.1 given by Glynn (1980).

Substitution of the asymptotic behavior (19) for ${}_0F_0^{(m)}$ in (1) of Section 9.4 yields an asymptotic representation for the joint density function of the sample roots $l_1,\dots,l_m$ when the population roots satisfy (18). The result is summarized in the following theorem.

THEOREM 9.5.5. Let $l_1,\dots,l_m$ be the latent roots of the sample covariance matrix S formed from a sample of size $N=n+1$ $(n\geq m)$ from the $N_m(\mu,\Sigma)$ distribution, and suppose the latent roots $\lambda_1,\dots,\lambda_m$ of Σ satisfy

(30) $$\lambda_1>\cdots>\lambda_k>\lambda_{k+1}=\cdots=\lambda_m \quad (=\lambda>0).$$

Then for large n an asymptotic representation for the joint density function of $l_1,\dots,l_m$ is

(31)

$$K_1\prod_{i=1}^{k}\left[l_i^{(n-m-1)/2}\exp\left(-\frac{nl_i}{2\lambda_i}\right)\right]\prod_{i<j}^{k}\left(\frac{l_i-l_j}{\lambda_i-\lambda_j}\right)^{1/2}\cdot\prod_{i=1}^{k}\prod_{j=k+1}^{m}\left(\frac{l_i-l_j}{\lambda_i-\lambda}\right)^{1/2}$$

$$\cdot\prod_{i=k+1}^{m}\left[l_i^{(n-m-1)/2}\exp\left(-\frac{nl_i}{2\lambda}\right)\right]\prod_{\substack{k+1\\ i<j}}^{m}(l_i-l_j) \qquad (l_1>\cdots>l_m>0),$$

where

$$(32) \qquad K_1 = \frac{\left(\frac{1}{2}n\right)^{mn/2-k(2m-k-1)/4}\pi^{m^2/2-k(k+1)/4}\Gamma_k\left(\frac{1}{2}m\right)}{\Gamma_m\left(\frac{1}{2}n\right)\Gamma_m\left(\frac{1}{2}m\right)}$$

$$\cdot \prod_{i=1}^{k} \lambda_i^{-(n-m+1)/2}\lambda^{-(m-k)(n-k)/2}.$$

This theorem has two interesting consequences.

COROLLARY 9.5.6. Suppose that the latent roots of Σ satisfy (30). For large n an asymptotic representation for the conditional density function of $l_{k+1},\ldots,l_m$, the $q=m-k$ smallest roots of S, given the k largest roots $l_1,\ldots,l_k$, is proportional to

(33)

$$\prod_{i=1}^{k}\prod_{j=k+1}^{m}(l_i-l_j)^{1/2}\cdot\prod_{i=k+1}^{m}\left[l_i^{(n-k-q-1)/2}\exp\left(-\frac{nl_i}{2\lambda}\right)\right]\prod_{\substack{k+1\\ i<j}}^{m}(l_i-l_j).$$

Note that this asymptotic conditional density function does not depend on $\lambda_1,\ldots,\lambda_k$, the k largest roots of Σ. Hence by conditioning on $l_1,\ldots,l_k$ the effects of these k largest population roots can be eliminated, at least asymptotically. In this sense $l_1,\ldots,l_k$ are *asymptotically sufficient* for $\lambda_1,\ldots,\lambda_k$. We can also see in (33) that the influence of the largest k sample roots l_i $(i=1,\ldots,k)$ in the asymptotic conditional distribution is felt through linkage factors of the form $(l_i-l_j)^{1/2}$.

COROLLARY 9.5.7. Suppose the latent roots of Σ satisfy

$$\lambda_1>\cdots>\lambda_k>\lambda_{k+1}=\cdots=\lambda_m \quad (=\lambda>0),$$

and put

$$(34) \qquad x_i=\left(\frac{n}{2}\right)^{1/2}\left(\frac{l_i-\lambda_i}{\lambda_i}\right) \qquad (i=1,\ldots,m).$$

Then the limiting joint density function of $x_1,\ldots,x_m$ as $n\to\infty$ is

$$(35) \qquad \prod_{i=1}^{k}\phi(x_i)\frac{\pi^{q(q-1)/4}}{2^{q/2}\Gamma_q\left(\frac{1}{2}q\right)}\exp\left(-\frac{1}{2}\sum_{j=k+1}^{m}x_j^2\right)\prod_{\substack{k+1\\ i<j}}^{m}(x_i-x_j),$$

where $q=m-k$ and $\phi(\cdot)$ denotes the standard normal density function.

This can be proved by making the change of variables (34) in (31) and letting $n\rightarrow\infty$. The details are left to the reader. Note that this shows that if λ_i is a *distinct* population root then x_i is asymptotically independent of x_j for $j\neq i$ and the limiting distribution of x_i is standard normal. This result was first observed by Girshick (1939) using the asymptotic theory of maximum likelihood estimates. In the more complicated case when Σ has multiple roots the definitive paper is that of T. W. Anderson (1963); Corollary 9.5.7 is a special case of a more general result of Anderson dealing with many multiple roots, although the derivation here is different.

It is interesting to look at the maximum likelihood estimates of the population latent roots obtained from the *marginal* distribution of the sample roots (rather than from the original normally distributed sample). The part of the joint density function of $l_1,\dots,l_m$ involving the population roots is

$$L^*=\prod_{i=1}^{m}\lambda_i^{-n/2}\,{}_0F_0^{(m)}\left(-\tfrac{1}{2}nL,\Sigma^{-1}\right),\tag{36}$$

which we will call the *marginal likelihood function*. When the population roots are all distinct (i.e., $l_1>\cdots>l_m>0$), Theorem 9.5.2 can be used to approximate this for large n, giving

$$L^*\approx K\cdot L_1L_2,\tag{37}$$

where

$$L_1=\prod_{i=1}^{m}\left[\lambda_i^{-n/2}\exp\left(-\frac{n}{2}\frac{l_i}{\lambda_i}\right)\right],$$

$$L_2=\prod_{i<j}^{m}\left(\frac{\lambda_i\lambda_j}{\lambda_i-\lambda_j}\right)^{1/2},$$

and K is a constant (depending on $n,l_1,\dots,l_m$, but not on $\lambda_1\dots\lambda_m$ and hence irrelevant for likelihood purposes). The values of the λ_i which maximize L_1 are

$$\hat\lambda_i=l_i\qquad(i=1,\dots,m),$$

that is, the usual sample roots. We have already noted in (10) of Section 9.3 that these are biased estimates of the λ_i, with bias terms of order n^{-1}. However, using the factor L_2 in the estimation procedure gives a bias correction. It is easy to show that the values of the λ_i which maximize L_1L_2

are

$$(38) \qquad \hat{\lambda}_i = l_i - \frac{1}{n} l_i \sum_{\substack{j=1 \\ j \neq i}}^{m} \frac{l_j}{l_i - l_j} + O(n^{-2}) \qquad (i=1,\ldots,m).$$

These estimates utilize information from other sample roots, adjacent ones of course having the most effect, and using (10) of Section 9.3 it follows easily that

$$(39) \qquad E(\hat{\lambda}_i) = \lambda_i + O(n^{-2}) \qquad (i=1,\ldots,m)$$

so that their bias terms are of order n^{-2}. This result was noted by G. A. Anderson (1965).

We have concentrated in this section on asymptotic distributions associated with the latent roots of a covariance matrix. The method used (Theorem 9.5.1) to derive these asymptotic distributions is useful in a variety of other situations as well. For further results and various extensions, particularly in the area of asymptotic *expansions*, the interested reader is referred to Muirhead (1978) and the references therein. We will conclude this section by stating without proof a theorem about the asymptotic distributions of the eigenvectors of S.

THEOREM 9.5.8. Suppose that the latent roots of Σ are $\lambda_1 \geq \cdots \geq \lambda_m > 0$, and let $\mathbf{h}_1 \ldots \mathbf{h}_m$ be the corresponding normalized eigenvectors. Let $\mathbf{q}_1, \ldots, \mathbf{q}_m$ be the normalized eigenvectors of the sample covariance matrix S corresponding to the latent roots $l_1 > \cdots > l_m > 0$. If λ_i is a *distinct* root then, as $n \to \infty$, $n^{1/2}(\mathbf{q}_i - \mathbf{h}_i)$ has a limiting m-variate normal distribution with mean $\mathbf{0}$ and covariance matrix

$$\Gamma = \lambda_i \sum_{\substack{j=1 \\ j \neq i}}^{m} \frac{\lambda_j}{(\lambda_i - \lambda_j)^2} \mathbf{h}_j \mathbf{h}_j'$$

and is asymptotically independent of l_i.

For a proof of this result the reader is referred to T. W. Anderson (1963).

9.6. SOME INFERENCE PROBLEMS IN PRINCIPAL COMPONENTS

In Section 8.3 we derived the likelihood ratio test of sphericity, that is, for testing the null hypothesis that all the latent roots of Σ are equal. If this hypothesis is accepted we conclude that the principal components all have the same variance and hence contribute equally to the total variation, so that no reduction in dimension is achieved by transforming to principal components. If the null hypothesis is rejected it is possible, for example, that the $m-1$ smallest roots are equal. If this is true and if their common value (or an estimate of it) is small compared with the largest root then most of the variation in the sample is explained by the first principal component, giving a substantial reduction in dimension. Hence it is reasonable to consider the null hypothesis that the $m-1$ smallest roots of Σ are equal. If this is rejected, we can test whether the $m-2$ smallest roots are equal, and so on. In practice then, we test sequentially the null hypotheses

$$(1) \qquad H_k: \lambda_{k+1} = \cdots = \lambda_m$$

for $k=0,1,\ldots,m-2$, where $\lambda_1 \geq \cdots \geq \lambda_m > 0$ are the latent roots of Σ. We saw in Section 8.3 that the likelihood ratio test of

$$H_0: \lambda_1 = \cdots = \lambda_m$$

is based on the statistic

$$(2) \qquad V_0 = \frac{\prod_{i=1}^m l_i}{\left(\frac{1}{m}\sum_{i=1}^m l_i\right)^m},$$

where $l_1 > \cdots > l_m$ are the latent roots of the sample covariance matrix S, and a test of asymptotic size α is to reject H_0 if

$$(3) \qquad -\left(n - \frac{2m^2+m+2}{6m}\right)\log V_0 > c\left(\alpha; \tfrac{1}{2}(m+2)(m-1)\right),$$

where $c(\alpha; r)$ denotes the upper $100\alpha\%$ point of the χ_r^2 distribution. When testing equality of a *subset* of latent roots the likelihood ratio statistic looks much the same as V_0, except that only those sample roots corresponding to the population roots being tested appear in the statistic. This is demonstrated in the following theorem from T. W. Anderson, (1963).

THEOREM 9.6.1. Given a sample of size N from the $N_m(\mu, \Sigma)$ distribution, the likelihood ratio statistic for testing the null hypothesis

$$H_k: \lambda_{k+1} = \cdots = \lambda_m \quad (=\lambda, \text{unknown})$$

is $\Lambda_k = V_k^{N/2}$, where

$$V_k \equiv \frac{\prod_{i=k+1}^m l_i}{\left(\frac{1}{m-k}\sum_{i=k+1}^m l_i\right)^{m-k}}. \tag{4}$$

Proof. This follows directly from the proof of Theorem 9.3.1. When H_k is true, the maximum value of the likelihood function is obtained from (7), (8), and (9) of Section 9.3 as

$$\exp\left(-\frac{n}{2\hat{\lambda}}\sum_{i=k+1}^{m} l_i - \tfrac{1}{2}n\sum_{i=1}^{k}\frac{l_i}{\hat{\lambda}_i}\right)\left(\prod_{i=1}^{k}\hat{\lambda}_i^{-N/2}\right)\hat{\lambda}^{-N(m-k)/2}, \tag{5}$$

where $n = N-1$, and

$$\hat{\lambda}_i = \frac{n}{N}l_i (i=1,\ldots,k), \quad \hat{\lambda} = \frac{1}{m-k}\frac{n}{N}\sum_{i=k+1}^{m} l_i$$

are the maximum likelihood estimates of the λ_i and λ under H_k. Substituting for these in (5) gives the maximum of the likelihood function under H_k as

$$\max_{H_k} L(\mu,\Sigma) = \left(\frac{N}{n}\right)^{mN/2}\left(\prod_{i=1}^{k} l_i^{-N/2}\right)\left(\frac{1}{m-k}\sum_{i=k+1}^{m} l_i\right)^{-N(m-k)/2} e^{-mN/2}.$$

When μ and Σ are unrestricted the maximum value of the likelihood function is given by

$$\max_{\mu,\Sigma} L(\mu,\Sigma) = \left(\frac{N}{n}\right)^{mN/2}\left(\prod_{i=1}^{m} l_i^{-N/2}\right)e^{-mN/2},$$

so that the likelihood ratio statistic for testing H_k is given by

$$\Lambda_k = \frac{\max_{H_k} L(\mu,\Sigma)}{\max_{\mu,\Sigma} L(\mu,\Sigma)} = \left[\frac{\prod_{i=k+1}^m l_i}{\left(\frac{1}{m-k}\sum_{i=k+1}^m l_i\right)^{m-k}}\right]^{N/2},$$

$$= V_k^{N/2},$$

where V_k is given by (4). Rejecting H_k for small values of Λ_k is equivalent to rejecting H_k for small values of V_k, and the proof is complete.

Let us now turn our attention to the asymptotic distribution of the statistic V_k when the null hypothesis H_k is true. It is convenient to put $q=m-k$ and

$$(6) \qquad \bar{l}_q=\frac{1}{q}\sum_{i=k+1}^{m} l_i,$$

the average of the smallest q latent roots of S, so that

$$V_k=\frac{\prod_{i=k+1}^{m} l_i}{\bar{l}_q^{\,q}}.$$

The general theory of likelihood ratio tests shows that, as $n\rightarrow\infty$, the asymptotic distribution of $-n\log V_k$ is $\chi^2_{(q+2)(q-1)/2}$ when H_k is true. An improvement over $-n\log V_k$ is the statistic

$$-\left(n-k-\frac{2q^2+q+2}{6q}\right)\log V_k \qquad (q=m-k)$$

suggested by Bartlett (1954). This should be compared with the test given by (3), to which it reduces when $k=0$, i.e., $q=m$. A further refinement in the multiplying factor was obtained by Lawley (1956) and James (1969). We will now indicate the approach used by James.

We noted in the discussion following Corollary 9.5.6 that when H_k is true the asymptotic conditional density function of $l_{k+1},\dots,l_m$, the q smallest latent roots of S, given the k largest roots $l_1,\dots,l_k$, does not depend on $\lambda_1,\dots,\lambda_k$, the k largest roots of Σ. In a test of H_k these k largest roots are *nuisance* parameters; the essential idea of James is that the effects of these nuisance parameters can be eliminated, at least asymptotically, by testing H_k using this conditional distribution.

If we put

$$(7) \qquad u_i=\frac{l_i}{\bar{l}_q} \qquad (i=k+1,\dots,m)$$

in the asymptotic conditional density function of $l_{k+1},\dots,l_m$, given $l_1,\dots,l_k$ in Corollary 9.5.6, then the asymptotic density function of $u_{k+1},\dots,u_{m-1}$, conditional on $l_1,\dots,l_k,\bar{l}_q$, follows easily as

$$(8) \qquad K_2\prod_{i=1}^{k}\prod_{j=k+1}^{m}(r_i-u_j)^{1/2}\prod_{i=k+1}^{m}u_i^{(n-k-q-1)/2}\prod_{\substack{k+1\\ i<j}}^{m}(u_i-u_j),$$

where $r_i = l_i/\bar{l}_q$ for $i=1,\dots,k$, and K_2 is a constant. Note that $\Sigma_{i=k+1}^{m} u_i = q$ and that

$$V_k = \prod_{i=k+1}^{m} \left(\frac{l_i}{\bar{l}_q} \right) = \prod_{i=k+1}^{m} u_i. \tag{9}$$

Put $T_k = -\log V_k$ so that the limiting distribution of nT_k is $\chi^2_{(q+2)(q-1)/2}$ when H_k is true. The appropriate multiplier of T_k can be obtained by finding its expected value. For notational convenience, let E_c denote expectation taken with respect to the conditional distribution (8) of $u_{k+1},\dots,u_{m-1}$ given $l_1,\dots,l_k,\bar{l}_q$ and let E_N denote expectation taken with respect to the "null" distribution

$$K_3 \prod_{i=k+1}^{m} u_i^{(n-k-q-1)/2} \prod_{\substack{k+1 \\ i<j}}^{m} (u_i - u_j), \tag{10}$$

where K_3 is constant, obtained from (8) by ignoring the linkage factor

$$\prod_{i=1}^{k} \prod_{j=k+1}^{m} (r_i - u_j)^{1/2}.$$

The following theorem gives the asymptotic result of Lawley (1956) together with the additional information about the accuracy of the χ^2 approximation provided by the means due to James (1969).

THEOREM 9.6.2. When the null hypothesis H_k is true the limiting distribution, as $n \to \infty$, of the statistic

$$P_k = -\left[n - k - \frac{2q^2+q+2}{6q} + \sum_{i=1}^{k} \frac{\bar{l}_q^2}{(l_i - \bar{l}_q)^2} \right] \log V_k \tag{11}$$

is $\chi^2_{(q+2)(q-1)/2}$, and

$$E_c(P_k) = \tfrac{1}{2}(q+2)(q-1) + O(n^{-2}). \tag{12}$$

Proof. We will merely sketch the details of the proof. First note that

$$(13)\qquad \begin{aligned} E_c(T_k) &= E_c\left(-\ln \prod_{i=k+1}^{m} u_i\right) \\ &= E_c\left[-\frac{\partial}{\partial h}\left(\prod_{i=k+1}^{m} u_i^h\right)_{h=0}\right] \\ &= -\frac{\partial}{\partial h}\left[E_c\left(\prod_{i=k+1}^{m} u_i^h\right)\right]_{h=0} \\ &= -\frac{\partial}{\partial h}\left[E_c(e^{-hT_k})\right]_{h=0}. \end{aligned}$$

We can interchange the order of differentiation and integration in (13) because in a neighborhood of $h=0$

$$h^{-1}\left(1-\prod_{i=k+1}^{m} u_i^h\right) \le 2\sum_{i=k+1}^{m} u_i = 2q.$$

Hence, in order to find $E_c(T_k)$ we will first obtain

$$(14)\qquad E_c(e^{-hT_k}) = E_c\left(\prod_{i=k+1}^{m} u_i^h\right).$$

This can obviously be done by finding

$$(15)\qquad E_N\left[\prod_{i=1}^{k}\prod_{j=k+1}^{m}(r_i-u_j)^{1/2}\cdot\exp(-hT_k)\right].$$

Now, when H_k is true,

$$1-u_j = O_p(n^{-1/2})$$

so that

$$\begin{aligned} (r_i-u_j)^{1/2} &= (r_i-1)^{1/2}\left(1+\frac{1-u_j}{r_i-1}\right)^{1/2} \\ &= (r_i-1)^{1/2}\left[1+\frac{1}{2}\frac{(1-u_j)}{(r_i-1)}-\frac{1}{8}\frac{(1-u_j)^2}{(r_i-1)^2}+O_p(n^{-3/2})\right]. \end{aligned}$$

Since $\Sigma_{j=k+1}^{m}(1-u_j)=0$, we get

$$(16)\quad \prod_{i=1}^{k}\prod_{j=k+1}^{m}(r_i-u_j)^{1/2}$$

$$=\prod_{i=1}^{k}(r_i-1)^{q/2}\left[1+\frac{1}{2(r_i-1)^2}\sum_{\substack{k+1\\ j<p}}^{m}(1-u_j)(1-u_p)+O_p(n^{-3/2})\right]$$

$$=\prod_{i=1}^{k}(r_i-1)^{q/2}\cdot\left\{1+\tfrac{1}{2}\alpha\left[\sum_{\substack{k+1\\ i<j}}^{m}u_iu_j-\binom{q}{2}\right]+O_p(n^{-3/2})\right\}$$

where $q=m-k$ and

$$(17)\qquad \alpha=\sum_{i=1}^{k}\frac{1}{(r_i-1)^2}=\sum_{i=1}^{k}\frac{\bar{l}_q^2}{(l_i-\bar{l}_q)^2}.$$

Substituting (16) in (15) it is seen that we need to evaluate

$$E_N\left[\left(\sum_{\substack{k+1\\ i<j}}^{m}u_iu_j\right)\exp(-hT_k)\right]=E_N\left[\left(\sum_{\substack{k+1\\ i<j}}^{m}u_iu_j\right)\prod_{i=k+1}^{m}u_i^h\right].$$

This problem is addressed in the following lemma.

LEMMA 9.6.3.

$$(18)\quad E_N\left[\left(\sum_{\substack{k+1\\ i<j}}^{m}u_iu_j\right)\exp(-hT_k)\right]=\binom{q}{2}\left(\frac{n-k-1+2h}{n-k+2/q+2h}\right)E_0(h),$$

where

$$(19)\qquad E_0(h)\equiv E_N[\exp(-hT_k)]=E_N\left(\prod_{i=k+1}^{m}u_i^h\right)$$

Proof. Since $u_i=l_i/\bar{l}_q$ for $i=k+1,\dots,m$, it follows that

$$\left(\sum_{\substack{k+1\\ i<j}}^{m}l_il_j\right)\prod_{i=k+1}^{m}l_i^h=\bar{l}_q^{qh+2}\left(\sum_{\substack{k+1\\ i<j}}^{m}u_iu_j\right)\prod_{i=k+1}^{m}u_i^h.$$

The "null" distribution of $l_{k+1},\ldots,l_m$ is the same as the distribution of the latent roots of a $q \times q$ covariance matrix S such that $(n-k)S$ has the $W_q(n-k, \lambda I_q)$ distribution, so that we will regard $l_{k+1},\ldots,l_m$ as the latent roots of S. All subsequent expectations involving l_i for $i = k+1,\ldots,m$ are taken with respect to this distribution. Put $n' = n - k$; then $(n'/\lambda)S$ is $W_q(n', I_q)$ and hence by Theorem 3.2.7, $(n'/\lambda)\operatorname{tr} S = (n'/\lambda)q\bar{l}_q$ is $\chi^2_{n'q}$, from which it follows easily that

$$(20) \qquad E(\bar{l}^r_q) = \left(\frac{1}{2}\frac{n'}{\lambda}q\right)^{-r}(\tfrac{1}{2}n'q)_r,$$

where $(x)_r = x(x+1)\cdots(x+r-1)$. Furthermore (see the proof of Theorem 3.2.20), $\bar{l}_q$ is independent of u_i, $i = k+1,\ldots,m$, and hence

$$(21) \qquad \frac{E_N\left[\left(\Sigma^m_{\substack{k+1 \\ i<j}}u_i u_j\right)\Pi^m_{i=k+1}u^h_i\right]}{E_N\left(\Pi^m_{i=k+1}u^h_i\right)} = \frac{E\left[\left(\Sigma^m_{\substack{k+1 \\ i<j}}l_i l_j\right)\Pi^m_{i=k+1}l^h_i\right]E(\bar{l}^{qh}_q)}{E(\bar{l}^{qh+2}_q)E\left(\Pi^m_{i=k+1}l^h_i\right)},$$

where we have used the fact that

$$\prod^m_{i=k+1}l_i = \bar{l}^q_q\prod^m_{i=k+1}u_i.$$

Two of the expectations on the right side of (21) can be evaluated using (20); it remains to calculate the other two. Now

$$\prod^m_{i=k+1}l_i = \det S$$

and

$$\sum^m_{\substack{k+1 \\ i<j}}l_i l_j = r_2(S),$$

the sum of second-order (2×2) principal minors of S. Since the principal minors all give the same expectation, we need only find the expectation involving the first one,

$$\Delta=\det\begin{bmatrix} s_{11} & s_{12} \\ s_{12} & s_{22} \end{bmatrix},$$

and multiply by

$$\binom{q}{2},$$

the number of them. Put $(n'/\lambda)S=T'T$, where $T=(t_{ij})$ is a $q\times q$ upper-triangular matrix; by Theorem 3.2.14, the t_{ii}^2 are independent $\chi^2_{n'-i+1}$ random variables $(i=1,\dots,q)$, from which it is easy to verify that

$$\text{(22)}\qquad E\left(\prod_{i=k+1}^{m} l_i^h\right)=E\left[(\det S)^h\right]$$
$$=\left(\frac{n'}{\lambda}\right)^{-qh}E\left(\prod_{i=1}^{q} t_{ii}^{2h}\right)$$
$$=\left(\frac{n'}{2\lambda}\right)^{-qh}\prod_{i=1}^{q}\left(\tfrac{1}{2}(n'-i+1)\right)_h$$

and

(23)

$$E\left[\left(\sum_{\substack{k+1\\ i<j}}^{m} l_i l_j\right)\prod_{i=k+1}^{m} l_i^h\right]=\binom{q}{2}E\left[\Delta(\det S)^h\right]$$
$$=\binom{q}{2}E\left[t_{11}^2 t_{22}^2\prod_{i=1}^{q} t_{ii}^{2h}\right]\left(\frac{n'}{\lambda}\right)^{-(qh+2)}$$
$$=\binom{q}{2}\left(\frac{1}{2}\frac{n'}{\lambda}\right)^{-(qh+2)}\left(\tfrac{1}{2}n'\right)_{h+1}\left(\tfrac{1}{2}(n'-1)\right)_{h+1}$$
$$\cdot\prod_{i=3}^{q}\left(\tfrac{1}{2}(n'-i+1)\right)_h.$$

Substituting (20), (22), and (23) for the expectations on the right side of (21) then gives

$$\frac{E_N\left[\left(\Sigma_{i<j}^{k+1} u_i u_j\right)\Pi_{i=k+1}^m u_i^h\right]}{E_N\left(\Pi_{i=k+1}^m u_i^h\right)} = \frac{\binom{q}{2}\left(\frac{1}{2}n'\right)_{h+1}\left(\frac{1}{2}(n'-1)\right)_{h+1} q^2 \left(\frac{1}{2}n'q\right)_{qh}}{\left(\frac{1}{2}n'\right)_h\left(\frac{1}{2}(n'-1)\right)_h\left(\frac{1}{2}n'q\right)_{qh+2}}$$

$$= \binom{q}{2}\frac{\left(\frac{1}{2}n'+h\right)\left(\frac{1}{2}n'-\frac{1}{2}+h\right)q^2}{\left(\frac{1}{2}n'q+qh\right)\left(\frac{1}{2}n'q+qh+1\right)}$$

$$= \binom{q}{2}\frac{\frac{1}{2}n'-\frac{1}{2}+h}{\frac{1}{2}n'+1/q+h}$$

$$= \binom{q}{2}\frac{n'-1+2h}{n'+2/q+2h}$$

$$= \binom{q}{2}\frac{n-k-1+2h}{n-k+2/q+2h},$$

which completes the proof of the lemma.

Returning now to our outline of the proof of Theorem 9.6.2 it follows from (15), (16), and Lemma 9.6.3 that

$$E_c\left(e^{-hT_k}\right) = \frac{\theta(h)}{\theta(0)} \tag{24}$$

where

$$\theta(h) = E_0(h) f(h),$$

with

$$f(h) = 1 + \tfrac{1}{2}\alpha\binom{q}{2}\left[\frac{n-k-1+2h}{n-k+\dfrac{2}{q}+2h} - 1\right]$$

Using (13) we have

$$E_c(T_k) = -\frac{\partial}{\partial h}\left[\frac{\theta(h)}{\theta(0)}\right]_{h=0} \tag{25}$$

$$= -E_0'(0) - \frac{f'(0)}{f(0)}$$

$$= -E_0'(0) - \frac{\alpha d}{n^2} + O(n^{-3}),$$

where $d=(q-1)(q+2)/2$ and α is given by (17). But $-E_0'(0)=E_N(T_k)$, and in the case of the null distribution (where $l_{k+1},\dots,l_m$ are regarded as the latent roots of a $q\times q$ sample covariance matrix S such that $(n-k)S$ is $W_q(n-k,\lambda I_q)$) we know from Section 8.3 that $[n-k-(2q^2+q+2)/6q]T_k$ has an asymptotic χ_d^2 distribution as $n\to\infty$, and the means agree to $O(n^{-2})$ so that

$$-E_0'(0)=\frac{d}{n-k-(2q^2+q+2)/6q}+O(n^{-3}).$$

Substituting this in (25) then gives

$$E_0(T_k)=\frac{d}{n-k-(2q^2+q+2/6q)}-\frac{\alpha d}{n^2}+O(n^{-3}),$$

from which it follows that if P_k is the statistic defined by (11) then

$$E_c(P_k)=d+O(n^{-2})$$

and the proof is complete.

It follows from Theorem 9.5.2 that if n is large an approximate test of size α of the null hypothesis

$$H_k:\lambda_{k+1}=\cdots=\lambda_m$$

is to reject H_k if $P_k>c(\alpha;(q+2)(q-1)/2)$, where P_k is given by (11), $q=m-k$ and $c(\alpha;r)$ is the upper $100\alpha\%$ point of the χ_r^2 distribution. Suppose that the hypotheses H_k, $k=0,1,\dots,m-1$ are tested sequentially and that for some k the hypothesis H_k is accepted and we are prepared to conclude that the $q=m-k$ smallest latent roots of Σ are equal. If their common value is λ and λ is negligible (compared with the other roots) we might decide to ignore the last q principal components and study only the first k components. One way of deciding whether λ is negligible, suggested by T. W. Anderson (1963), is to construct a one-sided confidence interval. An estimate of λ is provided by

$$\bar{l}_q=q^{-1}\sum_{i=k+1}^{m}l_i,$$

and it is easy to show, from Corollary 9.5.7 for example (see Problem 9.6), that as $n\to\infty$ the asymptotic distribution of $(\frac{1}{2}nq)^{1/2}(\bar{l}_q-\lambda)/\lambda$ is standard normal $N(0,1)$. Let z_α be the upper $100\alpha\%$ point of the $N(0,1)$ distribution, that is, such that $\Phi(z_\alpha)=1-\alpha$, where $\Phi(\cdot)$ denotes the standard normal distribution function. Then asymptotically,

$$P\left(\left(\frac{nq}{2}\right)^{1/2}\left(\frac{\bar{l}_q-\lambda}{\lambda}\right)\geq -z_\alpha\right)=1-\alpha,$$

which leads to a one-sided confidence interval for λ, namely,

$$\lambda\leq\frac{\bar{l}_q}{1-z_\alpha(2/nq)^{1/2}}$$

with asymptotic confidence coefficient $1-\alpha$. If the upper limit of this confidence interval is sufficiently small we might decide that λ is negligible and study only the first k principal components. It is also worth noting in passing that if we assume that λ_i is a distinct latent root the asymptotic normality of $(n/2)^{1/2}(l_i-\lambda_i)/\lambda_i$ guaranteed by Corollary 9.5.7 can be used to test the null hypothesis that λ_i is equal to some specified value and to construct confidence intervals for λ_i.

Even if we cannot conclude that some of the smallest latent roots of Σ are equal, it still may be possible that the variation explained by the last $q=m-k$ principal components, namely $\Sigma_{i=k+1}^m\lambda_i$, is small compared with the total variation $\Sigma_{i=1}^m\lambda_i$, in which case we might decide to study only the first k principal components. Thus it is of interest to consider the null hypothesis

$$H_k^*:\frac{\Sigma_{i=k+1}^m\lambda_i}{\Sigma_{i=1}^m\lambda_i}=h,$$

where h $(0<h<1)$ is a number to be specified by the experimenter. This can be tested using the statistic

$$M_k\equiv\sum_{i=k+1}^m l_i-h\sum_{i=1}^m l_i=-h\sum_{i=1}^k l_i+(1-h)\sum_{i=k+1}^m l_i.$$

Assuming the latent roots of Σ are distinct, Corollary 9.5.7 shows that the limiting distribution as $n\rightarrow\infty$ of

$$n^{1/2}\left[M_k+h\sum_{i=1}^k\lambda_i-(1-h)\sum_{i=k+1}^m\lambda_i\right]$$

is normal with mean 0 and variance

$$\tau^2 = 2h^2 \sum_{i=1}^{k} \lambda_i^2 + 2(1-h)^2 \sum_{i=k+1}^{m} \lambda_i^2.$$

Replacing λ_i by l_i $(i=1,\dots,m)$ in τ^2, this result can be used to construct an approximate test of H_k^* and to give confidence intervals for

$$\sum_{i=k+1}^{m} \lambda_i - h \sum_{i=1}^{m} \lambda_i.$$

Finally, let us derive an asymptotic test for a given principal component (also from T. W. Anderson, 1963). To be specific we will concentrate on the first component. Let H^{**} be the null hypothesis that the vector of coefficients $\mathbf{h}_1$ of the first principal component is equal to a specified $m \times 1$ vector $\mathbf{h}_1^0$, i.e.,

$$H^{**}: \mathbf{h}_1 = \mathbf{h}_1^0, \qquad \mathbf{h}_1^{0\prime}\mathbf{h}_1^0 = 1.$$

Recall that $\mathbf{h}_1$ is the eigenvector of Σ corresponding to the largest latent root λ_1; we will assume that λ_1 is a *distinct* root. A test of H^{**} can be constructed using the result of Theorem 9.5.8, namely, that if $\mathbf{q}_1$ is the normalized eigenvector of the sample covariance matrix S corresponding to the largest latent root l_1 of S then the asymptotic distribution of $\mathbf{y} = n^{1/2}(\mathbf{q}_1 - \mathbf{h}_1)$ is $N_m(\mathbf{0}, \Gamma)$, where

$$\Gamma = \sum_{i=2}^{m} \frac{\lambda_1 \lambda_i}{(\lambda_1 - \lambda_i)^2} \mathbf{h}_i \mathbf{h}_i'$$

$$= H_2 B^2 H_2',$$

with $H_2 = [\mathbf{h}_2 \dots \mathbf{h}_m]$ and

$$B^2 = \begin{bmatrix} \dfrac{\lambda_1\lambda_2}{(\lambda_1-\lambda_2)^2} & & & 0 \\ & \dfrac{\lambda_1\lambda_3}{(\lambda_1-\lambda_3)^2} & & \\ & & \ddots & \\ 0 & & & \dfrac{\lambda_1\lambda_m}{(\lambda_1-\lambda_m)^2} \end{bmatrix}$$

Note that the covariance matrix Γ in this asymptotic distribution is singular, as is to be expected. Put $\mathbf{z}=B^{-1}H_2'\mathbf{y}$; then the limiting distribution of $\mathbf{z}$ is $N_{m-1}(\mathbf{0}, I_{m-1})$, and hence the limiting distribution of $\mathbf{z}'\mathbf{z}$ is χ^2_{m-1}. Now note that

$$\mathbf{z}'\mathbf{z}=\mathbf{y}'H_2B^{-2}H_2'\mathbf{y}$$

and the matrix of this quadratic form in $\mathbf{y}$ is

(26)

$$H_2B^{-2}H_2' = H_2\begin{bmatrix} \frac{\lambda_1}{\lambda_2}-2+\frac{\lambda_2}{\lambda_1} & & & 0 \\ & \frac{\lambda_1}{\lambda_3}-2+\frac{\lambda_3}{\lambda_1} & & \\ & & \ddots & \\ 0 & & & \frac{\lambda_1}{\lambda_m}-2+\frac{\lambda_m}{\lambda_1}\end{bmatrix}H_2'$$

$$=\lambda_1H_2\begin{bmatrix}\frac{1}{\lambda_2} & & 0\\ & \ddots & \\ 0 & & \frac{1}{\lambda_m}\end{bmatrix}H_2'-2H_2H_2'$$

$$+\frac{1}{\lambda_1}H_2\begin{bmatrix}\lambda_2 & & 0\\ & \ddots & \\ 0 & & \lambda_m\end{bmatrix}H_2'$$

$$=\lambda_1\sum_{i=2}^{m}\frac{1}{\lambda_i}\mathbf{h}_i\mathbf{h}_i'-2H_2H_2'+\frac{1}{\lambda_1}\sum_{i=2}^{m}\lambda_i\mathbf{h}_i\mathbf{h}_i'.$$

Putting $\Lambda=\text{diag}(\lambda_1,\ldots,\lambda_m)$ and using

$$\Sigma=H\Lambda H'=\sum_{i=1}^{m}\lambda_i\mathbf{h}_i\mathbf{h}_i',$$

$$\Sigma^{-1}=H\Lambda^{-1}H'=\sum_{i=1}^{m}\frac{1}{\lambda_i}\mathbf{h}_i\mathbf{h}_i'$$

and

$$H_2 H_2' = I - \mathbf{h}_1 \mathbf{h}_1'$$

(26) becomes

$$\begin{aligned} H_2 B^{-2} H_2' &= \lambda_1 \left(\Sigma^{-1} - \frac{1}{\lambda_1} \mathbf{h}_1 \mathbf{h}_1' \right) - 2(I - \mathbf{h}_1 \mathbf{h}_1') + \frac{1}{\lambda_1} (\Sigma - \lambda_1 \mathbf{h}_1 \mathbf{h}_1') \\ &= \lambda_1 \Sigma^{-1} - 2I + \frac{1}{\lambda_1} \Sigma. \end{aligned}$$

Hence the limiting distribution of

$$\begin{aligned} (27) \qquad n(\mathbf{q}_1 - \mathbf{h}_1)' &\left(\lambda_1 \Sigma^{-1} - 2I + \frac{1}{\lambda_1} \Sigma \right) (\mathbf{q}_1 - \mathbf{h}_1) \\ &= n\mathbf{q}_1' \left(\lambda_1 \Sigma^{-1} - 2I + \frac{1}{\lambda_1} \Sigma \right) \mathbf{q}_1 \\ &= n \left(\lambda_1 \mathbf{q}_1' \Sigma^{-1} \mathbf{q}_1 + \frac{1}{\lambda_1} \mathbf{q}_1' \Sigma \mathbf{q}_1 - 2 \right) \end{aligned}$$

is χ^2_{m-1}. Since S, S^{-1}, and l_1 are consistent estimates of Σ, Σ^{-1}, and λ_1, they can be substituted for Σ, Σ^{-1}, and λ_1 in (27) without affecting the limiting distribution. Hence, when $H^{**}: \mathbf{h}_1 = \mathbf{h}_1^0$ is true, the limiting distribution of

$$\begin{aligned} W &\equiv n(\mathbf{q}_1 - \mathbf{h}_1^0)' \left(l_1 S^{-1} - 2I + \frac{1}{l_1} S \right) (\mathbf{q}_1 - \mathbf{h}_1^0) \\ &= n \left(l_1 \mathbf{h}_1^{0\prime} S^{-1} \mathbf{h}_1^0 + \frac{1}{l_1} \mathbf{h}_1^{0\prime} S \mathbf{h}_1^0 - 2 \right) \end{aligned}$$

is χ^2_{m-1}. It follows that a test of H^{**} of asymptotic size α is to reject H^{**} if $W > c(\alpha; m-1)$, where $c(\alpha; m-1)$ is the upper $100\alpha\%$ point of the χ^2_{m-1} distribution.

It should be pointed out that most inference procedures in principal components analysis are quite sensitive to departures from normality of the underlying distribution. For work in this direction the interested reader should see Waternaux (1976) and Davis (1977).

9.7. DISTRIBUTIONS OF THE EXTREME LATENT ROOTS OF A SAMPLE COVARIANCE MATRIX

In theory the marginal distribution of any latent root of the sample covariance matrix S, or of any subset of latent roots, can be obtained from the joint density function given in Theorem 9.4.1 by integrating with respect to the roots not under consideration. In general the integrals involved are not particularly tractable, even in the null case ($\Sigma=\lambda I_m$) of Corollary 9.4.2. A number of techniques have been developed in order to study the marginal distributions, and for a discussion of these the interested reader is referred to two useful surveys by Pillai (1976, 1977). We will concentrate here on the largest and smallest roots since expressions for their marginal distributions can be found using some of the theory presented in Chapter 7.

An expression for the distribution function of the largest root of S follows from the following theorem due to Constantine (1963).

THEOREM 9.7.1. If A is $W_m(n,\Sigma)$ $(n>m-1)$ and Ω is an $m\times m$ positive definite matrix ($\Omega>0$) then the probability that $\Omega-A$ is positive definite ($A<\Omega$) is

$$(1)\quad P(A<\Omega)=\frac{\Gamma_m[\frac{1}{2}(m+1)]}{\Gamma_m[\frac{1}{2}(n+m+1)]}\cdot\det(\tfrac{1}{2}\Sigma^{-1}\Omega)^{n/2}\,{}_1F_1(\tfrac{1}{2}n;\tfrac{1}{2}(n+m+1);-\tfrac{1}{2}\Sigma^{-1}\Omega),$$

where

$$(2)\qquad {}_1F_1(a;c;X)=\sum_{k=0}^{\infty}\sum_{\kappa}\frac{(a)_\kappa}{(c)_\kappa}\frac{C_\kappa(X)}{k!}.$$

Proof. Using the $W_m(n,\Sigma)$ density function for A, it follows that

$$P(A<\Omega)=\frac{1}{2^{mn/2}\Gamma_m(\frac{1}{2}n)(\det\Sigma)^{n/2}}\int_{0<A<\Omega}\operatorname{etr}(-\tfrac{1}{2}\Sigma^{-1}A)\cdot(\det A)^{(n-m-1)/2}(dA)$$

Putting $A=\Omega^{1/2}X\Omega^{1/2}$ so that $(dA)=(\det\Omega)^{(m+1)/2}(dX)$, this becomes

$$P(A<\Omega)=\frac{\det\left(\frac{1}{2}\Sigma^{-1}\Omega\right)^{n/2}}{\Gamma_m\left(\frac{1}{2}n\right)}\int_{0<X<I}\operatorname{etr}\left(-\tfrac{1}{2}\Omega^{1/2}\Sigma^{-1}\Omega^{1/2}X\right)\cdot(\det X)^{(n-m-1)/2}(dX)$$

$$=\frac{\det\left(\frac{1}{2}\Sigma^{-1}\Omega\right)^{n/2}}{\Gamma_m\left(\frac{1}{2}n\right)}\sum_{k=0}^{\infty}\sum_{\kappa}\frac{1}{k!}\int_{0<X<I}(\det X)^{(n-m-1)/2}\cdot C_\kappa\left(-\tfrac{1}{2}\Omega^{1/2}\Sigma^{-1}\Omega^{1/2}X\right)(dX),$$

where we have used the fact that

$$\operatorname{etr}(R)={}_0F_0(R)=\sum_{k=0}^{\infty}\sum_{\kappa}\frac{C_\kappa(R)}{k!}.$$

Using Theorem 7.2.10 to evaluate this last integral we get

$$P(A<\Omega)=\frac{\Gamma_m\left[\frac{1}{2}(m+1)\right]}{\Gamma_m\left[\frac{1}{2}(n+m+1)\right]}\det\left(\tfrac{1}{2}\Sigma^{-1}\Omega\right)^{n/2}\cdot\sum_{k=0}^{\infty}\sum_{\kappa}\frac{\left(\frac{1}{2}n\right)_\kappa}{\left(\frac{1}{2}(n+m+1)\right)_\kappa}\frac{C_\kappa\left(-\frac{1}{2}\Sigma^{-1}\Omega\right)}{k!},$$

and the proof is complete.

It is worth noting here that for $m\geq 2$, $P(A<\Omega)\neq 1-P(A>\Omega)$ because the set of A where neither of the relations $A<\Omega$ nor $A>\Omega$ holds is not of measure zero.

If S is the sample covariance matrix formed from a sample of size $N=n+1$ from the $N_m(\mu,\Sigma)$ distribution then $A=nS$ is $W_m(n,\Sigma)$ and an expression for the distribution function of the largest latent root of S follows immediately from Theorem 9.7.1. The result is given in the following corollary.

COROLLARY 9.7.2. If l_1 is the largest latent root of S, where $A=nS$ is $W_m(n,\Sigma)$, then the distribution function of l_1 can be expressed in the form

$$P_\Sigma(l_1<x)=\frac{\Gamma_m\left[\frac{1}{2}(m+1)\right]}{\Gamma_m\left[\frac{1}{2}(n+m+1)\right]}\det\left(\tfrac{1}{2}nx\Sigma^{-1}\right)^{n/2}\cdot{}_1F_1\left(\tfrac{1}{2}n;\tfrac{1}{2}(n+m+1);-\tfrac{1}{2}nx\Sigma^{-1}\right)\tag{3}$$

Proof. Note that the inequality $l_1 < x$ is equivalent to $S < xI$, i.e., to $A < nxI$. The result then follows by putting $\Omega = nxI$ in Theorem 9.7.1.

The problem of finding the distribution of the smallest latent root of S is more difficult. In the case when $r = \frac{1}{2}(n-m-1)$ is a positive integer, an expression for the distribution function in terms of a *finite* series of zonal polynomials follows from the following result of Khatri (1972).

THEOREM 9.7.3. Let A be $W_m(n, \Sigma)$ (with $n > m-1$), and let Ω be an $m \times m$ positive definite matrix ($\Omega > 0$). If $r = \frac{1}{2}(n-m-1)$ is a positive integer then

$$(4) \qquad P(A > \Omega) = \operatorname{etr}(-\tfrac{1}{2}\Sigma^{-1}\Omega) \sum_{k=0}^{mr} \sum_{\kappa}{}^{*} \frac{C_\kappa(\frac{1}{2}\Sigma^{-1}\Omega)}{k!},$$

where Σ_κ^* denotes summation over those partitions $\kappa = (k_1, \ldots, k_m)$ of k with $k_1 \le r$.

Proof. In

$$P(A > \Omega) = \frac{1}{2^{mn/2}\Gamma_m(\frac{1}{2}n)(\det \Sigma)^{n/2}} \int_{A > \Omega} \operatorname{etr}(-\tfrac{1}{2}\Sigma^{-1}A) \cdot (\det A)^{(n-m-1)/2}(dA)$$

put $A = \Omega^{1/2}(I+X)\Omega^{1/2}$ with $(dA) = (\det \Omega)^{(m+1)/2}(dX)$ to get

(5)

$$P(A > \Omega) = \frac{\operatorname{etr}(-\frac{1}{2}\Sigma^{-1}\Omega)\det(\frac{1}{2}\Sigma^{-1}\Omega)^{n/2}}{\Gamma_m(\frac{1}{2}n)} \cdot \int_{X > 0} \operatorname{etr}(-\tfrac{1}{2}\Omega^{1/2}\Sigma^{-1}\Omega^{1/2}X) \cdot (\det X)^{(n-m-1)/2}\det(I + X^{-1})^{(n-m-1)/2}(dX).$$

Now $\det(I + X^{-1})^{(n-m-1)/2}$ can be expanded in terms of zonal polynomials, and the series terminates because $r = (n-m-1)/2$ is a positive integer. By Corollary 7.3.5

$$\det(I + X^{-1})^r = {}_1F_0(-r; -X^{-1}) = \sum_{k=0}^{mr} \sum_{\kappa}{}^{*} \frac{(-r)_\kappa C_\kappa(X^{-1})}{k!},$$

because $(-r)_\kappa \equiv 0$ is any part of κ greater than r. Using this in (5) gives

$$P(A>\Omega)=\frac{\operatorname{etr}(-\frac{1}{2}\Sigma^{-1}\Omega)\det(\frac{1}{2}\Sigma^{-1}\Omega)^{n/2}}{\Gamma_m(\frac{1}{2}n)}\sum_{k=0}^{mr}\sum_{\kappa}{}^{*}\frac{(-r)_\kappa(-1)^k}{k!}$$

$$\cdot\int_{X>0}\operatorname{etr}(-\tfrac{1}{2}\Omega^{1/2}\Sigma^{-1}\Omega^{1/2}X)(\det X)^{(n-m-1)/2}C_\kappa(X^{-1})(dX).$$

For each partition $\kappa=(k_1,\ldots,k_m)$ in this sum we have $k_1\le r$; the desired result follows easily on using Theorem 7.2.13 to evaluate the last integral.

An immediate consequence of Theorem 9.7.3 is an expression for the distribution function of the smallest latent root of the sample covariance matrix S.

COROLLARY 9.7.4. If l_m is the smallest latent root of S, where $A=nS$ is $W_m(n,\Sigma)$ and if $r=\frac{1}{2}(n-m-1)$ is a positive integer then

$$P_\Sigma(l_m>x)=\operatorname{etr}(-\tfrac{1}{2}nx\Sigma^{-1})\sum_{k=0}^{mr}\sum_{\kappa}{}^{*}\frac{C_\kappa(\frac{1}{2}nx\Sigma^{-1})}{k!},\tag{6}$$

where Σ_κ^* denotes summation over those partitions $\kappa=(k_1,\ldots,k_m)$ of k with $k_1\le r$.

Proof. Note that the inequality $l_m>x$ is equivalent to $S>xI$, i.e., to $A>nxI$, and put $\Omega=nxI$ in Theorem 9.7.3.

In principle the distributional results in Corollaries 9.7.2 and 9.7.4 could be used to test hypotheses about Σ using statistics which are functions of the largest latent root l_1 or the smallest latent root l_m. Consider, for example, the null hypothesis $H\colon \Sigma=I_m$. The likelihood ratio test was considered in Section 8.4; an alternative test of size α based on the largest root l_1 is to reject H if $l_1>l(\alpha;m,n)$, where $l(\alpha;n,m)$ is the upper $100\alpha\%$ point of the distribution of l_1 when $\Sigma=I_m$, that is, such that $P_{I_m}(l_1>l(\alpha;m,n))=\alpha$. The power function of this test is then,

$$\beta(\Sigma)=P_\Sigma(l_1>l(\alpha;m,n)),$$

which depends on Σ only through its latent roots. These percentage points and powers could theoretically be computed using the distribution function for l_1 given in Corollary 9.7.2, and this has actually been done by Sugiyama (1972) for $m=2$ and 3. In general, however, this approach poses severe computational problems because the zonal polynomial series (2) for the $_1F_1$

hypergeometric function converges very slowly, even for small n and m. If n is large and λ_1, the largest root of Σ, is a distinct root an approximate test based on l_1 can be constructed using the asymptotic normality of $n^{1/2}(l_1-\lambda_1)$ guaranteed by Corollary 9.5.7. If n is small or moderate further terms in an asymptotic *series* can be used to get more accurate approximations; see Sugiura (1973b), Muirhead and Chikuse (1975b), and Muirhead (1974) for work in this direction. If λ_1 is not a distinct latent root of Σ the asymptotic distribution of l_1 is considerably more complicated (see Corollary 9.5.7). We will give it explicitly when $\Sigma=\lambda I_m$ and $m=2$ and 3, leaving the details as an exercise (see Problem 9.7). The distribution function of $t_1=(n/2)^{1/2}(l_1-\lambda)/\lambda$ can be expanded when $\Sigma=\lambda I_2$ as

(7) $$P(t_1\le x)=\Phi\left(x\sqrt{2}\right)-\sqrt{\pi}\,\phi(x)\Phi(x)+O(n^{-1/2})$$

and when $\Sigma=\lambda I_3$ as

(8)
$$P(t_1\le x)=\Phi\left(x\sqrt{2}\right)\Phi(x)-2x\phi(x)\Phi\left(x\sqrt{2}\right)-\pi^{-1/2}\phi\left(x\sqrt{3}\right)+O(n^{-1/2}),$$

where $\phi(\cdot)$ and $\Phi(\cdot)$ denote, respectively, the density and distribution function of the standard normal distribution. Further terms in asymptotic series for these two distribution functions may be found in Muirhead (1974).

Since the exact distributions of the extreme roots l_1 and l_m are computationally difficult and their asymptotic distributions depend fundamentally on the eigenstructure of Σ, it is occasionally useful to have quick, albeit rough, approximations for their distribution functions. The bounds in the following theorem could be used for this purpose.

THEOREM 9.7.5. If l_1 and l_m are the largest and smallest latent roots of S, where nS is $W_m(n,\Sigma)$, then

(9) $$P(l_1\le x)\le\prod_{i=1}^{m}P\left(\chi_n^2\le\frac{nx}{\lambda_i}\right)$$

and

(10) $$P(l_m\ge x)\le\prod_{i=1}^{m}P\left(\chi_n^2\ge\frac{nx}{\lambda_i}\right),$$

where $\lambda_1,\dots,\lambda_m$ are the latent roots of Σ.

Proof. Let $H\in O(m)$ be such that

$$H'\Sigma H=\Lambda=\mathrm{diag}(\lambda_1,\dots,\lambda_m),$$

and put $S^*=H'SH$ so that nS^* is $W_m(n,\Lambda)$. Since S and S^* have the same latent roots, l_1 and l_m are the extreme latent roots of S^*. We have already seen in Theorem 9.2.1 that for all $\boldsymbol{\alpha}\in R^m$ with $\boldsymbol{\alpha}'\boldsymbol{\alpha}=1$ we have

$$l_1\geq\boldsymbol{\alpha}'S^*\boldsymbol{\alpha},$$

and a similar proof can be used to show that

$$l_m\leq\boldsymbol{\alpha}'S^*\boldsymbol{\alpha}.$$

Taking $\boldsymbol{\alpha}$ to be the vectors $(1,0,\dots,0)'$, $(0,1,0,\dots,0)'$, and so on, then shows that

$$l_1\geq\max(s_{11}^*,\dots,s_{mm}^*)\tag{11}$$

and

$$l_m\leq\min(s_{11}^*,\dots,s_{mm}^*),\tag{12}$$

where $S^*=(s_{ij}^*)$. By Theorem 3.2.7 the random variables ns_{ii}^*/λ_i have independent χ_n^2 distributions for $i=1,\dots,m$ so that, using (11) and (12),

$$\begin{aligned}P(l_1\leq x)&\leq P(\max(s_{11}^*,\dots,s_{mm}^*)\leq x)\\&=\prod_{i=1}^m P(s_{ii}^*\leq x)\\&=\prod_{i=1}^m P\left(\chi_n^2\leq\frac{nx}{\lambda_i}\right)\end{aligned}$$

and

$$\begin{aligned}P(l_m\geq x)&\leq P(\min(s_{11}^*,\dots,s_{mm}^*)\geq x)\\&=\prod_{i=1}^m P(s_{ii}^*\geq x)\\&=\prod_{i=1}^m P\left(\chi_n^2\geq\frac{nx}{\lambda_i}\right).\end{aligned}$$

PROBLEMS

9.1. Suppose that the $m\times 1$ random vector $\mathbf{X}$ has covariance matrix

$$\Sigma=\sigma^2\begin{bmatrix} 1 & \rho & \rho & \cdots & \rho \\ \rho & 1 & \rho & \cdots & \rho \\ \vdots & & & & \\ \rho & \rho & \rho & \cdots & 1 \end{bmatrix}.$$

(a) Find the population principal components and their variances.

(b) Suppose a sample of size $N=n+1$ is taken on $\mathbf{X}$, and let $\hat{U}_1=\mathbf{q}_1'\mathbf{X}$ be the first sample principal component. Write $\mathbf{q}_1=(q_{11},\dots,q_{1m})'$, so that

$$\hat{U}_1=\sum_{i=1}^{m} q_{1i}X_i.$$

Using Theorem 9.5.8, show that the covariance matrix in the asymptotic distribution of $n^{1/2}\mathbf{q}_1$ is $\Gamma=(\gamma_{ij})$, where

$$\gamma_{ii}=\frac{[1+(m-1)\rho](m-1)(1-\rho)}{m^2\rho^2} \qquad (i=1,\dots,m)$$

and

$$\gamma_{ij}=\frac{-[1+(m-1)\rho](1-\rho)}{m^3\rho^2} \qquad (i\neq j).$$

Why would you expect the covariances to be negative?

9.2. Let Σ be a $m\times m$ positive definite covariance matrix, and consider the problem of approximating Σ by an $m\times m$ matrix Γ of rank r obtained by minimizing

$$\|\Sigma-\Gamma\|=\left[\sum_{i=1}^{m}\sum_{j=1}^{m}(\sigma_{ij}-\gamma_{ij})^2\right]^{1/2}.$$

(a) Show that

$$\|\Sigma-\Gamma\|^2=\operatorname{tr}(\Lambda-P)(\Lambda-P)',$$

where $H'\Sigma H=\Lambda=\operatorname{diag}(\lambda_1,\dots,\lambda_m)$, $\lambda_1\geq\cdots\geq\lambda_m\geq 0$, $H\in O(m)$, $P=H'\Gamma H$.

(b) Using (a), show that the matrix Γ of rank r which minimizes $\|\Sigma-\Gamma\|$ is

$$\Gamma=\sum_{i=1}^{r}\lambda_i \mathbf{h}_i\mathbf{h}_i',$$

where $H=[\mathbf{h}_1,\dots,\mathbf{h}_m]$.

9.3. Prove that if $A=\text{diag}(a_1,\dots,a_m)$, $a_1>a_2>\cdots>a_m>0$, and $B=\text{diag}(b_1,\dots,b_m)$, $0<b_1<b_2<\cdots<b_m$, then for all $H\in O(m)$,

$$\text{tr}(BHAH')\geq\sum_{i=1}^{m}a_ib_i$$

with equality if and only if H has the form $H=\text{diag}(\pm1,\pm1,\dots,\pm1)$.

9.4. Let $A=\text{diag}(a_1,\dots,a_m)$, with $a_1>a_2>\cdots>a_m>0$, and $B=\text{diag}(b_1,\dots,b_k)$, with $b_1>b_2>\cdots>b_k>0$. Show that for all $H_1\in V_{k,m}$

$$\text{tr}(BH_1'AH_1)\leq\sum_{i=1}^{k}a_ib_i$$

with equality if and only if H_1 has the form

$$H_1=\begin{bmatrix}\pm1 & & 0\\ & \ddots & \\ 0 & & \pm1\\ \hdashline & 0 & \end{bmatrix}.$$

9.5. Obtain Corollary 9.5.7 from Theorem 9.5.5

9.6. Let $\bar{l}_q=q^{-1}\Sigma_{i=k+1}^{m}l_i$, where $q=m-k$ and $l_{k+1}>\cdots>l_m$ are the q smallest latent roots of a sample covariance matrix S. Suppose that λ, the smallest latent root of Σ, has multiplicity q. Prove that as $n\to\infty$ the asymptotic distribution of $(nq/2\lambda^2)^{1/2}(\bar{l}_q-\lambda)$ is $N(0,1)$.

9.7. Establish equations (7) and (8) of Section 9.7.

9.8. Suppose that the latent roots $\lambda_1>\cdots>\lambda_m$ of Σ are all distinct; let l_1 be the largest latent root of S, and put $x_1=(n/2)^{1/2}(l_1/\lambda_1-1)$.

(a) Using Corollary 9.7.2 show that

$$P(x_1<x)=\frac{\Gamma_m(p)(\det R)^{n/2}}{\Gamma_m(\frac{1}{2}n+p)}\,{}_1F_1(\tfrac{1}{2}n;\tfrac{1}{2}n+p;-R)$$

where $p=\frac{1}{2}(m+1)$, $R=\text{diag}(r_1,\ldots,r_m)$, with $r_i=[\frac{1}{2}n+(\frac{1}{2}n)^{1/2}x]z_i$, $z_i=\lambda_1/\lambda_i$ $(i=1,\ldots,m)$. (Note that $z_1=1$ is a dummy variable.)

(b) Starting with the partial differential equations of Theorem 7.5.8 satisfied by the ${}_1F_1$ function, find a system satisfied by $P(x_1<x)$ in terms of derivatives with respect to $x, z_2,\ldots,z_m$.

(c) Assuming that $P(x_1<x)$ has an expansion of the form

$$P(x_1<x)=\Phi(x)+\frac{Q_1}{n^{1/2}}+\frac{Q_2}{n}+\cdots,$$

use the differential equations obtained in (b) and the boundary conditions $P(x_1<\infty)=1$, $P(x_1<-\infty)=0$ to show that

$$Q_1=-\left(\sum_{j=2}^{m}\frac{\lambda_j}{\lambda_1-\lambda_j}\right)2^{-1/2}\phi(x)+\frac{2^{1/2}(1-x^2)\phi(x)}{3},$$

where $\phi(x)$ denotes the standard normal density function.

9.9. Let $\mathbf{X}_1,\ldots,\mathbf{X}_N$ be independent $N_m(\boldsymbol{\mu},\Sigma)$ random variables. Find the likelihood ratio statistic Λ for testing that the smallest $q=m-k$ latent roots of Σ are equal to a specified value λ_0, and the asymptotic distribution of $-2\log\Lambda$.

9.10. Suppose that $\mathbf{X}$ is $N_m(\boldsymbol{\mu},\Sigma)$, where $\Sigma=\Gamma+\sigma^2 I_m$, with Γ being a non-negative definite matrix of rank r. Let $\lambda_1\geq\cdots\geq\lambda_m>0$ be the latent roots of Σ and $\gamma_1\geq\cdots\geq\gamma_r>0$ be the nonzero latent roots of Γ.

(a) Show that $\lambda_i=\gamma_i+\sigma^2$ (with $i=i,\ldots,r$) and $\lambda_j=\sigma^2$ (with $j=r+1,\ldots,m$). How are the latent vectors of Σ and Γ related?

(b) Given a sample of size N on $\mathbf{X}$, find the maximum likelihood estimate of σ^2.

9.11. Let $H=[\mathbf{h}_1\ldots\mathbf{h}_m]$ be a proper orthogonal $m\times m$ matrix and write $H=\exp(U)$, where U is an $m\times m$ skew-symmetric matrix. Establish equation (13) of Section 9.5.

CHAPTER 10

The Multivariate Linear Model

10.1. INTRODUCTION

In this chapter we consider the multivariate linear model. Before introducing this we review a few results about the familiar (univariate) linear model given by

$$\mathbf{y} = X\boldsymbol{\beta} + \boldsymbol{\varepsilon},$$

where $\mathbf{y}$ and $\boldsymbol{\varepsilon}$ are $n \times 1$ random vectors, X is a known $n \times p$ matrix of rank p (the full-rank case), and $\boldsymbol{\beta}$ is a $p \times 1$ vector of unknown parameters (regression coefficients). The vector $\mathbf{y}$ is a vector of n observations, and $\boldsymbol{\varepsilon}$ is an error vector. Under the assumption that $\boldsymbol{\varepsilon}$ is $N_n(\mathbf{0}, \sigma^2 I_n)$, where σ^2 is unknown [i.e., the errors are independent $N(0, \sigma^2)$ random variables]:

(i) the maximum likelihood estimates of $\boldsymbol{\beta}$ and σ^2 are

$$\hat{\boldsymbol{\beta}} = (X'X)^{-1}X'\mathbf{y}$$

and

$$\hat{\sigma}^2 = \frac{1}{n}(\mathbf{y} - X\hat{\boldsymbol{\beta}})'(\mathbf{y} - X\hat{\boldsymbol{\beta}});$$

(ii) $(\hat{\boldsymbol{\beta}}, \hat{\sigma}^2)$ is sufficient for $(\boldsymbol{\beta}, \sigma^2)$;
(iii) the maximum likelihood estimates $\hat{\boldsymbol{\beta}}$ and $\hat{\sigma}^2$ are independent; $\hat{\boldsymbol{\beta}}$ is $N_p(\boldsymbol{\beta}, \sigma^2(X'X)^{-1})$ and $n\hat{\sigma}^2/\sigma^2$ is χ^2_{n-p}; and
(iv) the likelihood ratio test of the null hypothesis H: $C\boldsymbol{\beta} = \mathbf{0}$, where C is

a known $r\times p$ matrix of rank r, rejects H for large values of

(1) $$F=\left(\frac{n-p}{rn\hat{\sigma}^2}\right)\hat{\beta}'C'\left[C(X'X)^{-1}C'\right]^{-1}C\hat{\beta}.$$

When H is true F has the $F_{r,n-p}$ distribution. Proofs of these assertions, which should be familiar to the reader, may be found, for example, in Graybill (1961), Searle (1971), and Seber (1977).

The multivariate linear model generalizes this model in the sense that it allows a vector of observations, given by the rows of a matrix Y, to correspond to the rows of the known matrix X. The multivariate model takes the form

(2) $$Y=X\mathbf{B}+E$$

where Y and E are $n\times m$ random matrices, X is a known $n\times p$ matrix, and $\mathbf{B}$ is an unknown $p\times m$ matrix of parameters called regression coefficients. We will assume throughout this chapter that X has rank p, that $n\geq m+p$, and that the rows of the error matrix E are independent $N_m(\mathbf{0},\Sigma)$ random vectors. Using the notation introduced in Chapter 3, this means that E is $N(0, I_n\otimes\Sigma)$ so that Y is $N(X\mathbf{B}, I_n\otimes\Sigma)$. We now find the maximum likelihood estimates of $\mathbf{B}$ and Σ and show that they are sufficient.

THEOREM 10.1.1. If Y is $N(X\mathbf{B}, I_n\otimes\Sigma)$ and $n\geq m+p$ the maximum likelihood estimates of $\mathbf{B}$ and Σ are

(3) $$\hat{\mathbf{B}}=(X'X)^{-1}X'Y$$

and

(4) $$\hat{\Sigma}=\frac{1}{n}(Y-X\hat{\mathbf{B}})'(Y-X\hat{\mathbf{B}}).$$

Moreover $(\hat{\mathbf{B}},\hat{\Sigma})$ is sufficient for $(\mathbf{B},\Sigma)$.

Proof. Since Y is $N(X\mathbf{B}, I_n\otimes\Sigma)$ the density function of Y is

$$(2\pi)^{-mn/2}(\det\Sigma)^{-n/2}\operatorname{etr}\left[-\tfrac{1}{2}(Y-X\mathbf{B})\Sigma^{-1}(Y-X\mathbf{B})'\right].$$

Noting that $X'(Y-X\hat{\mathbf{B}})=0$, it follows that the likelihood function can be written (ignoring the constant) as

$$L(\mathbf{B},\Sigma)=(\det\Sigma)^{-n/2}\operatorname{etr}\left[-\frac{n}{2}\Sigma^{-1}\hat{\Sigma}-\tfrac{1}{2}\Sigma^{-1}(\mathbf{B}-\hat{\mathbf{B}})'X'X(\mathbf{B}-\hat{\mathbf{B}})\right].$$

This shows immediately that $(\hat{\mathbf{B}},\hat{\Sigma})$ is sufficient for $(\mathbf{B},\Sigma)$. That $\hat{\mathbf{B}}$ and $\hat{\Sigma}$ are the maximum likelihood estimates follows using a proof similar to that of Theorem 3.1.5.

The next theorem shows that the maximum likelihood estimates are independently distributed and gives their distributions.

THEOREM 10.1.2. If Y is $N(X\mathbf{B}, I_n\otimes\Sigma)$ the maximum likelihood estimates $\hat{\mathbf{B}}$ and $\hat{\Sigma}$, given by (3) and (4), are independently distributed; $\hat{\mathbf{B}}$ is $N[\mathbf{B},(X'X)^{-1}\otimes\Sigma]$ and $n\hat{\Sigma}$ is $W_m(n-p,\Sigma)$.

Proof. Let H be an $n\times(n-p)$ matrix such that

$$X'H=0,\ H'H=I_{n-p},$$

so that the columns of H form an orthogonal basis for $R(X)^{\perp}$, the orthogonal complement of the range of X. Hence $HH'=I_n-X(X'X)^{-1}X'$. Now, put $Z=H'Y,(n-p\times m)$; then

$$Z'Z=Y'HH'Y=Y'\left(I_n-X(X'X)^{-1}X'\right)Y=n\hat{\Sigma}.$$

The distribution of the matrix

$$\begin{bmatrix}\hat{\mathbf{B}}\\ \cdots\\ Z\end{bmatrix}=\begin{bmatrix}(X'X)^{-1}X'\\ \cdots\cdots\\ H'\end{bmatrix}Y$$

is normal with mean

$$E\begin{bmatrix}\hat{\mathbf{B}}\\ \cdots\\ Z\end{bmatrix}=\begin{bmatrix}(X'X)^{-1}X'\\ \cdots\cdots\\ H'\end{bmatrix}X\mathbf{B}=\begin{bmatrix}\mathbf{B}\\ \cdots\\ 0\end{bmatrix}.$$

The covariance matrix is (see the example following Lemma 2.2.2)

$$\left(\begin{bmatrix}(X'X)^{-1}X'\\ \cdots\cdots\\ H'\end{bmatrix}\otimes I_m\right)(I_n\otimes\Sigma)\left(\left[X(X'X)^{-1}:H\right]\otimes I_m\right)=\begin{bmatrix}(XX)^{-1} & 0\\ 0 & I_{n-p}\end{bmatrix}\otimes\Sigma.$$

Hence $\hat{\mathbf{B}}$ is $N(\mathbf{B},(X'X)^{-1}\otimes\Sigma)$, Z is $N(0, I_{n-p}\otimes\Sigma)$ and $\hat{\mathbf{B}}$ and Z are independent. Since $n\hat{\Sigma}=Z'Z$ it follows from Definition 3.1.1 that $n\hat{\Sigma}$ is $W_m(n-p,\Sigma)$, and the proof is complete.

Note that the Wishart density for $n\hat{\Sigma}$ exists only if $n\geq m+p$. When $n<m+p$ Z has rank $n-p$ and $n\hat{\Sigma}=Z'Z$ is singular.

10.2. A GENERAL TESTING PROBLEM: CANONICAL FORM, INVARIANCE, AND THE LIKELIHOOD RATIO TEST

In this section we consider testing the null hypothesis H: $C\mathbf{B}=0$, where C is a known $r\times p$ matrix of rank r, against the alternative K: $C\mathbf{B}\neq 0$. This null hypothesis H is often referred to as the *general linear hypothesis*. With various choices for the matrix C this incorporates many hypotheses of interest. For example, partition $\mathbf{B}$ as

$$\mathbf{B}=\begin{bmatrix}\mathbf{B}_1\\ \mathbf{B}_2\end{bmatrix},$$

where $\mathbf{B}_1$ is $r\times m$ and $\mathbf{B}_2$ is $(p-r)\times m$. The null hypothesis that $\mathbf{B}_1=0$ is the same as $C\mathbf{B}=0$, with $C=[I_r:0]$. As a second example suppose that $\mathbf{y}_{i1},\dots,\mathbf{y}_{iq_i}$ are independent $N_m(\boldsymbol{\mu}_i,\Sigma)$ random vectors $(i=1,\dots,p)$ and consider testing that $\boldsymbol{\mu}_1=\cdots=\boldsymbol{\mu}_p$. This model, the single classification model, can be written in the form $Y=X\mathbf{B}+E$ with

$$Y=\begin{bmatrix}\mathbf{y}'_{11}\\ \vdots\\ \mathbf{y}'_{1q_1}\\ \mathbf{y}'_{21}\\ \vdots\\ \mathbf{y}'_{2q_2}\\ \vdots\\ \mathbf{y}'_{p1}\\ \vdots\\ \mathbf{y}'_{pq_p}\end{bmatrix},\qquad X=\begin{bmatrix}1&0&0&\dots&0\\ \vdots&\vdots&\vdots&\dots&\vdots\\ 1&0&0&\dots&0\\ 0&1&0&\dots&0\\ \vdots&\vdots&\vdots&\dots&\vdots\\ 0&1&0&\dots&0\\ \vdots&\vdots&\vdots&\dots&\vdots\\ 0&0&0&\dots&1\\ \vdots&\vdots&\vdots&\dots&\vdots\\ 0&0&0&\dots&1\end{bmatrix},\qquad \mathbf{B}=\begin{bmatrix}\boldsymbol{\mu}'_1\\ \boldsymbol{\mu}'_2\\ \vdots\\ \boldsymbol{\mu}'_p\end{bmatrix}.$$

Here Y is $n \times m$, with $n = \Sigma_{i=1}^{p} q_i$, X is $n \times p$, $\mathbf{B}$ is $p \times m$, and E is $N(0, I_n \otimes \Sigma)$. The hypothesis that the means are equal can be expressed as $C\mathbf{B} = 0$ with

$$C = \begin{bmatrix} 1 & 0 & \dots & 0 & -1 \\ 0 & 1 & \dots & 0 & -1 \\ \vdots & & & & \\ 0 & 0 & \dots & 1 & -1 \end{bmatrix}.$$

Here C is $r \times p$ with $r = p - 1$.

Returning to our general discussion we will see that by transforming the variables and parameters in the model $Y = X\mathbf{B} + E$ the problem can be assumed to be in the following form:

Let

$$Y^* = \begin{bmatrix} Y_1^* \\ Y_2^* \\ Y_3^* \end{bmatrix}$$

[where Y_1^* is $r \times m$, Y_2^* is $(p-r) \times m$, and Y_3^* is $(n-p) \times m$] be a random matrix whose rows are independent m-variate normal with common covariance matrix Σ and expectations given by

$$E(Y_1^*) = M_1, \qquad E(Y_2^*) = M_2, \qquad E(Y_3^*) = 0.$$

The null hypothesis $H: C\mathbf{B} = 0$ is equivalent to $H: M_1 = 0$. This form of the testing problem is generally referred to as the *canonical form*.

We now verify that we can express the problem in this particular form. First, write

$$X = Q\begin{bmatrix} I_p \\ 0 \end{bmatrix} D,$$

where $Q \in O(n)$ and $D \in \mathcal{G}\ell(p, R)$. Note that if Q is partitioned as $Q = [Q_1 : Q_2]$, where Q_1 is $n \times p$ and Q_2 is $n \times (n-p)$, then $X = Q_1 D$ and the columns of Q_1 form an orthogonal basis for the range of X. Now write $CD^{-1} = E[I_r : 0]P$, where $E \in \mathcal{G}\ell(r, R)$ and $P \in O(p)$, and transform Y to Y^* defined by

$$Y^* = \begin{bmatrix} P & 0 \\ 0 & I_{n-p} \end{bmatrix} Q'Y.$$

The rows of Y^* are independently normal with covariance matrix Σ and the expectation of Y^* is

$$E(Y^*)=\begin{bmatrix} P & 0 \\ 0 & I_{n-p} \end{bmatrix} Q'X\mathbf{B}$$

$$=\begin{bmatrix} P & 0 \\ 0 & I_{n-p} \end{bmatrix} Q'Q\begin{bmatrix} I_p \\ 0 \end{bmatrix} D\mathbf{B}$$

$$=\begin{bmatrix} P \\ 0 \end{bmatrix} D\mathbf{B}$$

$$=\begin{bmatrix} PD\mathbf{B} \\ 0 \end{bmatrix}$$

$$=\begin{bmatrix} M \\ 0 \end{bmatrix},$$

where $M=PD\mathbf{B}$. Partitioning Y^* and M as

$$Y^*=\begin{bmatrix} Y_1^* \\ Y_2^* \\ Y_3^* \end{bmatrix}, \qquad M=\begin{bmatrix} M_1 \\ M_2 \end{bmatrix}$$

where Y_1^* and M_1 are $r\times m$, Y_2^* and M_2 are $(p-r)\times m$, and Y_3^* is $(n-p)\times m$, we then have

$$E(Y_1^*)=M_1, \qquad E(Y_2^*)=M_2, \qquad E(Y_3^*)=0.$$

The null hypothesis H: $C\mathbf{B}=0$ is equivalent to

$$0=CD^{-1}D\mathbf{B}=E[I_r:0]PD\mathbf{B},$$

i.e.,

$$[I_r:0]M=0; \quad \text{hence} \quad M_1=0.$$

This form of the testing problem is due to Roy (1957) and T. W. Anderson (1958).

We now express the maximum likelihood estimates of $\mathbf{B}$ and Σ in terms of the transformed variables. First note that, if $P\in O(p)$ is partitioned as

$$P=\begin{bmatrix} P_1 \\ P_2 \end{bmatrix},$$

where P_1 is $r\times p$ and P_2 is $(p-r)\times p$, then

$$Y=Q\begin{bmatrix} P' & 0 \\ 0 & I_{n-p} \end{bmatrix} Y^* \tag{1}$$

$$=[Q_1:Q_2]\begin{bmatrix} P_1' & P_2' & 0 \\ 0 & 0 & I \end{bmatrix}\begin{bmatrix} Y_1^* \\ Y_2^* \\ Y_3^* \end{bmatrix}$$

$$=Q_1P_1'Y_1^*+Q_1P_2'Y_2^*+Q_2Y_3^*.$$

Then

$$\hat{\mathbb{B}}=(X'X)^{-1}X'Y \tag{2}$$

$$=(D'D)^{-1}D'Q_1'(Q_1P_1'Y_1^*+Q_1P_2'Y_2^*+Q_2Y_3^*)$$

$$=D^{-1}P_1'Y_1^*+D^{-1}P_2'Y_2^*$$

and

$$n\hat{\Sigma}=Y'Y-\hat{\mathbb{B}}'X'Y \tag{3}$$

$$=Y^{*\prime}Y^*-\left(Y_1^{*\prime}P_1D^{-1\prime}+Y_2^{*\prime}P_2D^{-1\prime}\right)D'Q_1'(Q_1P_1'Y_1^*+Q_1P_2'Y_2^*$$
$$+Q_2Y_3^*)$$

$$=Y^{*\prime}Y^*-(Y_1^{*\prime}Y_1^*+Y_2^{*\prime}Y_2^*)$$

$$=Y_3^{*\prime}Y_3^*,$$

where we have used

$$Q_1'Q_1=I,\qquad Q_1'Q_2=0,\qquad P_1P_1'=I,\qquad P_1P_2'=0.$$

The matrix version of the numerator of the F ratio in (1) of Section 10.1 is

$$\hat{\mathbb{B}}'C'\left[C(X'X)^{-1}C'\right]^{-1}C\hat{\mathbb{B}}$$

$$=Y_1^{*\prime}Y_1^*.$$

These quantities are often summarized in a generalization of the usual Analysis of Variance table, called a Multivariate Analysis of Variance table (or MANOVA table) in which the sums of squares are replaced by matrices of sums of squares (S.S.) and sums of products (S.P.).

MANOVA Table

Source of Variation due to	Degrees of freedom	S.S. and S.P.	Expectation
Regression or model	r	$A=Y_1^{*\prime}Y_1^*$	$r\Sigma+M_1'M_1$
Deviations or error	$n-p$	$B=Y_3^{*\prime}Y_3^*=n\hat{\Sigma}$	$(n-p)\Sigma$
Total (corrected)	$n-(p-r)$	$Y_1^{*\prime}Y_1^*+Y_3^{*\prime}Y_3^* = Y'Y-Y_2^{*\prime}Y_2^*$	

The matrices A and B defined as

$$(4) \qquad A=Y_1^{*\prime}Y_1^*, \qquad B=Y_3^{*\prime}Y_3^*=n\hat{\Sigma}$$

are called the matrices due to the hypothesis and due to error, respectively. We have already seen in Theorem 10.1.2 that B is $W_m(n-p,\Sigma)$ and hence $E(B)=(n-p)\Sigma$. We will see later that in general the matrix A has what is called a noncentral Wishart distribution (central when H: $M_1=0$ is true). Here we find its expectation. We have $A=Y_1^{*\prime}Y_1^*$, where Y_1^* is $N(M_1, I_r\otimes\Sigma)$. Since $Y_1^*-M_1$ is $N(0,I_r\otimes\Sigma)$ it follows that $(Y_1^*-M_1)'(Y_1^*-M_1)$ is $W_m(r,\Sigma)$, so that

$$\begin{aligned} r\Sigma &= E[(Y_1^*-M_1)'(Y_1^*-M_1)] \\ &= E(Y_1^{*\prime}Y_1^*-Y_1^{*\prime}M_1-M_1'Y_1^*+M_1'M_1) \\ &= E(Y_1^{*\prime}Y_1^*)-M_1'M_1-M_1'M_1+M_1'M_1 \\ &= E(Y_1^{*\prime}Y_1^*)-M_1'M_1. \end{aligned}$$

Hence

$$(5) \qquad E(A)=r\Sigma+M_1'M_1.$$

Note that when H is true, $M_1=0$, so that both A and B, divided by their respective degrees of freedom, provide unbiased estimates of Σ. In the univariate case $m=1$ the ratio of these mean squares has the $F_{r,n-p}$ distribution when H is true.

We next look at our testing problem from an invariance point of view. We have independent matrices Y_1^*, Y_2^*, Y_3^*, where

$$Y_1^* \text{ is } N(M_1,I_r\otimes\Sigma), \qquad Y_2^* \text{ is } N(M_2,I_{p-r}\otimes\Sigma),$$

$$Y_3^* \text{ is } N(0,I_{n-p}\otimes\Sigma),$$

and the null hypothesis is H: $M_1=0$. The joint density function of Y_1^*, Y_2^*, and Y_3^* is

$$(6)\quad (2\pi)^{-mn/2}(\det\Sigma)^{-n/2}$$
$$\cdot\operatorname{etr}\{-\tfrac{1}{2}\Sigma^{-1}[(Y_1^*-M_1)'(Y_1^*-M_1)+(Y_2^*-M_2)'(Y_2^*-M_2)+Y_3^{*\prime}Y_3^*]\},$$

from which it is apparent that a sufficient statistic is (Y_1^*, Y_2^*, B), with $B=Y_3^{*\prime}Y_3^*=n\hat{\Sigma}$. With $M_{m,n}$ denoting the space of $m\times n$ real matrices, consider the group of transformations

$$(7)\qquad G=\{(\Gamma,E,N);\ \ \Gamma\in O(r),\ \ E\in \mathcal{G}\ell(m,R),\ \ N\in M_{p-r,m}\}$$

acting on the space of the sufficient statistic by

$$(8)\qquad (\Gamma,E,N)(Y_1^*,Y_2^*,B)=(\Gamma Y_1^*E',\ \ Y_2^*E'+N,\ \ EBE')$$

where the group operation is

$$(9)\qquad (\Gamma_1,E_1,N_1)(\Gamma_2,E_2,N_2)=(\Gamma_1\Gamma_2,E_1E_2,N_2E_1'+N_1).$$

The corresponding induced group of transformations on the parameter space of points (M_1, M_2, Σ) is given by

$$(10)\qquad (\Gamma,E,N)(M_1,M_2,\Sigma)=(\Gamma M_1E',M_2E'+N,E\Sigma E')$$

and the problem of testing H: $M_1=0$ against K: $M_1\neq 0$ is invariant under G. A maximal invariant is given in the following theorem.

THEOREM 10.2.1. Under the group G of transformations (8) a maximal invariant is $(f_1,\dots,f_s)$, where $s=\operatorname{rank}(Y_1^*B^{-1}Y_1^{*\prime})$ and $f_1\geq\cdots\geq f_s>0$ are the nonzero latent roots of $Y_1^*B^{-1}Y_1^{*\prime}$.

Proof. Let

$$\phi(Y_1^*,Y_2^*,B)=(f_1,\dots,f_s).$$

First note that ϕ is invariant, for the latent roots of

$$(\Gamma Y_1^*E')(EBE')^{-1}(\Gamma Y_1^*E')'=\Gamma Y_1^*B^{-1}Y_1^{*\prime}\Gamma'$$

are the same as those of $Y_1^*B^{-1}Y_1^{*\prime}$. To show that it is maximal invariant, suppose that

$$\phi(Y_1^*,Y_2^*,B)=\phi(Z_1^*,Z_2^*,F),$$

i.e., $Y_1^*B^{-1}Y_1^{*\prime}$ and $Z_1^*F^{-1}Z_1^{*\prime}$ have the same nonzero latent roots $(f_1,\dots,f_s)$. By Theorem A9.2 there exist $r\times r$ orthogonal matrices H_1, H_2 such that

$$H_1Y_1^*B^{-1}Y_1^{*\prime}H_1' = H_2Z_1^*F^{-1}Z_1^{*\prime}H_2' = \begin{bmatrix} f_1 & & & & & 0 \\ & \ddots & & & & \\ & & f_s & & & \\ & & & 0 & & \\ & & & & \ddots & \\ 0 & & & & & 0 \end{bmatrix}.$$

Then

$$Z_1^*F^{-1}Z_1^{*\prime} = H_2'H_1Y_1^*B^{-1}Y_1^{*\prime}H_1'H_2$$

i.e.,

$$(Z_1^*F^{-1/2})(Z_1^*F^{-1/2})' = (H_2'H_1Y_1^*B^{-1/2})(H_2'H_1Y_1^*B^{-1/2})'.$$

By Theorem A9.5 there exists $H_3\in O(m)$ such that

$$H_3F^{-1/2}Z_1^{*\prime} = B^{-1/2}Y_1^{*\prime}H_1'H_2.$$

Hence

$$Z_1^* = \Gamma Y_1^*E'$$

with

$$\Gamma = H_2'H_1 \quad \text{and} \quad E = F^{1/2}H_3'B^{-1/2}.$$

Note that $EBE' = F$. Putting $N = Z_2^* - Y_2^*E'$ we then have

$$(\Gamma, E, N)(Y_1^*, Y_2^*, B) = (Z_1^*, Z_2^*, F)$$

so that

$$(Y_1^*, Y_2^*, B) \sim (Z_1^*, Z_2^*, F) (\operatorname{mod} G).$$

Hence $(f_1,\dots,f_s)$ is a maximal invariant and the proof is complete.

As a consequence of this theorem any invariant test depends only on $f_1,\dots,f_s$ and, from Theorem 6.1.12, the distribution of $f_1,\dots,f_s$ depends only on the nonzero latent roots of $M_1\Sigma^{-1}M_1'$. Note that

$$(11) \qquad s=\operatorname{rank}\left(Y_1^*B^{-1}Y_1^{*\prime}\right)$$
$$=\min(r,m).$$

Note also that when $r=1$ and $Y_1^*\equiv \mathbf{y}_1^{*\prime}$, $M_1\equiv \mathbf{m}_1'$, both $1\times m$, then a maximal invariant is $f_1=\mathbf{y}_1^{*\prime}B^{-1}\mathbf{y}_1^*$, a multiple of Hotelling's T^2 statistic (see Theorem 6.3.1). We have already seen in Theorem 6.3.4 that the test which rejects $H:\mathbf{m}_1=\mathbf{0}$ for large values of $\mathbf{y}_1^{*\prime}B^{-1}\mathbf{y}_1^*$ is a uniformly most powerful invariant test of $H:\mathbf{m}_1=\mathbf{0}$ against $K:\mathbf{m}_1\neq\mathbf{0}$. Note also that when $m=1$ (the univariate case) and $B=n\hat{\Sigma}=n\hat{\sigma}^2$ the maximal invariant is the nonzero latent root of

$$\frac{1}{n\hat{\sigma}^2}Y_1^*Y_1^{*\prime},$$

namely,

$$f_1=\frac{Y_1^{*\prime}Y_1^*}{n\hat{\sigma}^2},$$

which is a multiple of the usual F ratio used for testing H: $M_1=0$ [see (1) of Section 10.1]. The test based on this is also uniformly most powerful invariant (Problem 10.1).

In general, however, there is no uniformly most powerful invariant test and many functions of the latent roots of $Y_1^*B^{-1}Y_1^{*\prime}$ have been proposed as test statistics. The likelihood ratio test statistic (from Wilks, 1932), given in the following theorem, is one such statistic.

THEOREM 10.2.2. The likelihood ratio test of size α of H: $M_1=0$ against K: $M_1\neq 0$ rejects H if $\Lambda\leq c_\alpha$, where

$$\Lambda=\frac{(\det B)^{n/2}}{\det(A+B)^{n/2}}$$

with $A=Y_1^{*\prime}Y_1^*$, $B=Y_3^{*\prime}Y_3^*$, and c_α is chosen so that the size of the test is α.

Proof. Apart from a multiplicative factor the likelihood function based on the independent matrices Y_1^*, Y_2^*, and Y_3^*, where Y_1^* is $N(M_1, I_r\otimes\Sigma)$, Y_2^* is $N(M_2, I_{p-r}\otimes\Sigma)$ and Y_3^* is $N(0, I_{n-p}\otimes\Sigma)$ is [see (6)]

$$L(M_1,M_2,\Sigma)=(\det\Sigma)^{-n/2}\operatorname{etr}\left\{-\tfrac{1}{2}\Sigma^{-1}\left[(Y_1^*-M_1)'(Y_1^*-M_1)\right.\right.$$
$$\left.\left.+(Y_2^*-M_2)'(Y_2^*-M_2)+Y_3^{*\prime}Y_3^*\right]\right\}.$$

The likelihood ratio statistic is

$$(12) \qquad \Lambda = \frac{\sup_{M_2,\Sigma} L(0, M_2, \Sigma)}{\sup_{M_1, M_2, \Sigma} L(M_1, M_2, \Sigma)}$$

When $M_1 = 0$ the likelihood function is maximized when $M_2 = Y_2^*$ and

$$\Sigma = \frac{1}{n}(Y_1^{*\prime}Y_1 + Y_3^{*\prime}Y_3^*) = \frac{1}{n}(A+B)$$

so that the numerator in (12) is

$$(13) \qquad L\left(0, Y_2^*, \frac{1}{n}(A+B)\right) = \det\left[\frac{1}{n}(A+B)\right]^{-n/2} e^{-mn/2}.$$

When the parameters are unrestricted the likelihood function is maximized when

$$M_1 = Y_1^*, \quad M_2 = Y_2^*, \quad \text{and} \quad \Sigma = \frac{1}{n}Y_3^{*\prime}Y_3^* = \frac{1}{n}B$$

so that the denominator in (12) is

$$(14) \qquad L\left(Y_1^*, Y_2^*, \frac{1}{n}B\right) = \det\left(\frac{1}{n}B\right)^{-n/2} e^{-mn/2}.$$

Using (13) and (14) in (12) then gives

$$\Lambda = \frac{(\det B)^{n/2}}{\det(A+B)^{n/2}},$$

and the likelihood ratio test rejects H for small values of Λ, completing the proof.

For notational convenience, define the statistic

$$(15) \qquad W = \frac{\det B}{\det(A+B)} = \Lambda^{2/n}.$$

The likelihood ratio test is equivalent to rejecting H: $M_1 = 0$ for small values of W. Note that this is an invariant test for

$$W = \frac{\det B}{\det(Y_1^{*\prime}Y_1^* + B)}$$

$$= \det\left(I + Y_1^* B^{-1} Y_1^{*\prime}\right)^{-1}$$

$$= \prod_{i=1}^{s} (1+f_i)^{-1},$$

where $s = \min(r, m) = \operatorname{rank}(Y_1^* B^{-1} Y_1^{*\prime})$ and $f_1 \geq \cdots \geq f_s > 0$ are the nonzero latent roots of $Y_1^* B^{-1} Y_1^{*\prime}$. The distribution of W will be discussed in detail in Section 10.5. Other invariant test statistics include

$$T_0^2 = \operatorname{tr} AB^{-1} = \sum_{i=1}^{s} f_i$$

called the generalized T_0^2 statistic and suggested by Hotelling (1947) and Lawley (1938),

$$V = \operatorname{tr} A(A+B)^{-1} = \sum_{i=1}^{s} \frac{f_i}{1+f_i},$$

proposed by Pillai (1955), and f_1, the largest latent root of $Y_1^* B^{-1} Y_1^{*\prime}$, suggested by Roy (1953). Distributional results associated with these statistics will be given in Section 10.6. The joint distribution of $f_1, \ldots, f_s$ is given in Section 10.4 and can be used as a starting point for deriving distributional results about these statistics. Before getting to this we need to introduce the noncentral Wishart distribution, which is the distribution of the matrix $A = Y_1^{*\prime} Y_1^*$. This is done in the next section.

10.3. THE NONCENTRAL WISHART DISTRIBUTION

The noncentral Wishart distribution generalizes the noncentral χ^2 distribution in the same way that the usual or central Wishart distribution generalizes the χ^2 distribution. It forms a major building block for noncentral distributions.

DEFINITION 10.3.1. If $A = Z'Z$, where the $n \times m$ matrix Z is $N(M, I_n \otimes \Sigma)$ then A is said to have the noncentral Wishart distribution with n degrees of freedom, covariance matrix Σ, and matrix of noncentrality parameters $\Omega=\Sigma^{-1}M'M$. We will write that A is $W_m(n,\Sigma,\Omega)$.

Note that when $M=0$, so that $\Omega=0$, A is $W_m(n,\Sigma)$ (i.e., central Wishart), and when $m=1$ with $\Sigma=\sigma^2$, A/σ^2 is $\chi_n^2(\delta)$, with $\delta=M'M/\sigma^2$. We have already seen in (5) of Section 10.2 that

$$E(A)=n\Sigma+M'M=n\Sigma+\Sigma\Omega.$$

When $n<m$, A is singular and the $W_m(n,\Sigma,\Omega)$ distribution does not have a density function. The following theorem, which gives the density function of A when $n\geq m$, should be compared with Theorem 1.3.4, giving the noncentral χ^2 density function.

THEOREM 10.3.2. If the $n\times m$ matrix Z is $N(M,I_n\otimes\Sigma)$ with $n\geq m$ then the density function of $A=Z'Z$ is

$$(1)\quad \frac{1}{2^{mn/2}\Gamma_m(\frac{1}{2}n)(\det\Sigma)^{n/2}}$$

$$\cdot\operatorname{etr}(-\tfrac{1}{2}\Sigma^{-1}A)(\det A)^{(n-m-1)/2}\operatorname{etr}(-\tfrac{1}{2}\Omega)\,{}_0F_1(\tfrac{1}{2}n;\tfrac{1}{4}\Omega\Sigma^{-1}A)\qquad (A>0),$$

where $\Omega=\Sigma^{-1}M'M$.

Proof. The density of Z is

$$(2\pi)^{-mn/2}(\det\Sigma)^{-n/2}\operatorname{etr}(-\tfrac{1}{2}\Sigma^{-1}Z'Z)\operatorname{etr}(-\tfrac{1}{2}\Sigma^{-1}M'M)\operatorname{etr}(\Sigma^{-1}M'Z)(dZ).$$

Put $Z=H_1T$, where H_1 is $n\times m$, with $H_1'H_1=I_m$ and T being upper-triangular. Then $A=Z'Z=T'T$ and, from Theorem 2.1.14,

$$(dZ)=2^{-m}\det(Z'Z)^{(n-m-1)/2}(d(Z'Z))(H_1'dH_1)$$

so that the density becomes

$$2^{-m}(2\pi)^{-mn/2}(\det\Sigma)^{-n/2}\operatorname{etr}(-\tfrac{1}{2}\Sigma^{-1}Z'Z)\operatorname{etr}(-\tfrac{1}{2}\Sigma^{-1}M'M)$$

$$\cdot\det(Z'Z)^{(n-m-1)/2}\operatorname{etr}(\Sigma^{-1}M'H_1T)(d(Z'Z))(H_1'dH_1).$$

Now integrate with respect to $H_1\in V_{m,n}$, the Stiefel manifold of $n\times m$

matrices with orthonormal columns. From Lemma 9.5.3 we have

$$\int_{H_1\in V_{m,n}}\operatorname{etr}(\Sigma^{-1}M'H_1T)(H_1'dH_1)=\frac{\Gamma_{n-m}[\frac{1}{2}(n-m)]}{2^{n-m}\pi^{(n-m)^2/2}}$$

$$\cdot\int_{H_1\in V_{m,n}}\int_{K\in O(n-m)}\operatorname{etr}(\Sigma^{-1}M'H_1T)(K'dK)(H_1'dH_1)$$

$$=\frac{\Gamma_{n-m}[\frac{1}{2}(n-m)]}{2^{n-m}\pi^{(n-m)^2/2}}\int_{H\in O(n)}\operatorname{etr}(\Sigma^{-1}M'H_1T)(H'dH)$$

$$=\frac{2^m\pi^{mn/2}}{\Gamma_m(\frac{1}{2}n)}\int_{O(n)}\operatorname{etr}(\Sigma^{-1}M'H_1T)(dH).$$

Using Theorem 7.4.1 to evaluate this last integral then gives

$$\int_{H_1\in V_{m,n}}\operatorname{etr}(\Sigma^{-1}M'H_1T)(H_1'dH_1)=\frac{2^m\pi^{mn/2}}{\Gamma_m(\frac{1}{2}n)}\,{}_0F_1(\tfrac{1}{2}n;\tfrac{1}{4}T\Sigma^{-1}M'M\Sigma^{-1}T')$$

$$=\frac{2^m\pi^{mn/2}}{\Gamma_m(\frac{1}{2}n)}\,{}_0F_1(\tfrac{1}{2}n;\tfrac{1}{4}\Omega\Sigma^{-1}A)$$

where $\Omega=\Sigma^{-1}M'M$, $A=T'T$, and the proof is complete.

It should be noted that, although n is an integer ($\geq m$) in the derivation of the noncentral Wishart density function of Theorem 10.3.2, the function (1) is still a density function when n is any real number greater than $m-1$, so our definition of the noncentral Wishart distribution can be extended to cover noninteger degrees of freedom n, for $n>m-1$.

The noncentral Wishart distribution was first studied in detail by T. W. Anderson (1946), who obtained explicit forms of the density function when rank $(\Omega)=1,2$. Weibull (1953) found the distribution when rank $(\Omega)=3$. Herz (1955) expressed the distribution in terms of a ${}_0F_1$ function of matrix argument, and James (1961a) and Constantine (1963) gave the zonal polynomial expansion for it.

Recall that if A is $W_1(n,\sigma^2,\delta)$ then A/σ^2 is $\chi_n^2(\delta)$, so that the characteristic function of A is

$$(1-2it\sigma^2)^{-n/2}\exp(-\tfrac{1}{2}\delta)\exp\left[\tfrac{1}{2}\delta(1-2it\sigma^2)^{-1}\right].$$

The following theorem generalizes this result.

THEOREM 10.3.3. If A is $W_m(n,\Sigma,\Omega)$ then the characteristic function of A is

$$\text{(2)}\qquad \phi(\Theta)=E\left[\exp\left(i\sum_{j\le k}^{m}\theta_{jk}a_{jk}\right)\right]$$

$$=\det(I_m-i\Gamma\Sigma)^{-n/2}\operatorname{etr}(-\tfrac{1}{2}\Omega)\operatorname{etr}\left[\tfrac{1}{2}\Omega(I_m-i\Gamma\Sigma)^{-1}\right]$$

where

$$\Gamma=(\gamma_{ij})\; i,j=1,\dots,m,\qquad \text{with}\quad \gamma_{ij}=(1+\delta_{ij})\theta_{ij},\theta_{ji}=\theta_{ij},$$

and δ_{ij} is the Kronecker delta,

$$\delta_{ij}=\begin{cases}1 & \text{if}\quad i=j\\ 0 & \text{if}\quad i\neq j.\end{cases}$$

Proof. The characteristic function of A can be written as

$$\phi(\Theta)=E\left[\exp\left(\frac{i}{2}\sum_{j,k=1}^{m}(1+\delta_{jk})\theta_{jk}a_{jk}\right)\right]$$

$$=E\left[\operatorname{etr}\left(\frac{i}{2}A\Gamma\right)\right].$$

There are two cases to consider.

(i) First, suppose that n is a postive integer and write $A=Z'Z$ when Z is $N(M,I_n\otimes\Sigma)$, and $\Omega=\Sigma^{-1}M'M$. Let $\mathbf{z}_1,\dots,\mathbf{z}_n$ be the columns of Z' and $\mathbf{m}_1,\dots,\mathbf{m}_n$ be the columns of M'; then $\mathbf{z}_1,\dots,\mathbf{z}_n$ are independent, $\mathbf{z}_i$ is $N_m(\mathbf{m}_i,\Sigma)$, and $A=Z'Z=\sum_{j=1}^{n}\mathbf{z}_j\mathbf{z}_j'$. Hence

$$\phi(\Theta)=E\left[\operatorname{etr}\left(\frac{i}{2}Z'Z\Gamma\right)\right]=E\left[\operatorname{etr}\left(\frac{i}{2}\sum_{j=1}^{n}\mathbf{z}_j\mathbf{z}_j'\Gamma\right)\right]$$

$$=E\left[\exp\frac{i}{2}\sum_{j=1}^{n}\operatorname{tr}(\mathbf{z}_j\mathbf{z}_j'\Gamma)\right]$$

$$=E\left[\exp\frac{i}{2}\sum_{j=1}^{n}\mathbf{z}_j'\Gamma\mathbf{z}_j\right].$$

$$=\prod_{j=1}^{n}E\left[\exp\frac{i}{2}\mathbf{z}_j'\Gamma\mathbf{z}_j\right].$$

Put $\mathbf{y}_j=\Sigma^{-1/2}\mathbf{z}_j$; then $\mathbf{y}_j$ is $N_m(\boldsymbol{\tau}_j, I_m)$ with $\boldsymbol{\tau}_j=\Sigma^{-1/2}\mathbf{m}_j$, and

$$\phi(\Theta)=\prod_{j=1}^{n}E\left[\exp\frac{i}{2}\mathbf{y}_j'\Sigma^{1/2}\Gamma\Sigma^{1/2}\mathbf{y}_j\right].$$

Let $H\in O(m)$ be such that

$$H\Sigma^{1/2}\Gamma\Sigma^{1/2}H'=\operatorname{diag}(\lambda_1,\ldots,\lambda_m)=D_\Lambda,$$

where $\lambda_1,\ldots,\lambda_m$ are the latent roots of $\Sigma^{1/2}\Gamma\Sigma^{1/2}$. Put $\mathbf{u}_j=H\mathbf{y}_j$ then $\mathbf{u}_j$ is $N_m(\boldsymbol{\nu}_j, I_m)$ with $\boldsymbol{\nu}_j=H\boldsymbol{\tau}_j$ and

$$\phi(\Theta)=\prod_{j=1}^{n}E\left[\exp\frac{i}{2}\mathbf{u}_j'D_\Lambda\mathbf{u}_j\right]=\prod_{j=1}^{n}E\left[\exp\frac{i}{2}\sum_{k=1}^{m}\lambda_k u_{jk}^2\right]$$

where $\mathbf{u}_j=(u_{j1},\ldots,u_{jm})'$. Then

$$\phi(\Theta)=\prod_{j=1}^{n}\prod_{k=1}^{m}E\left[\exp\frac{i}{2}\lambda_k u_{jk}^2\right]=\prod_{j=1}^{n}\prod_{k=1}^{m}\left\{(1-i\lambda_k)^{-1/2}\exp\left(-\tfrac{1}{2}\nu_{jk}^2\right)\right.$$
$$\left.\cdot\exp\left[\tfrac{1}{2}\nu_{jk}^2(1-i\lambda_k)^{-1}\right]\right\},$$

where $\boldsymbol{\nu}_j=(\nu_{j1},\ldots,\nu_{jm})'$ and we have used the fact that the u_{jk}^2 are independent $\chi_1^2(\nu_{jk}^2)$ random variables. The desired result now follows by noting that

$$\begin{aligned}\prod_{j=1}^{n}\prod_{k=1}^{m}(1-i\lambda_k)&=\det(I_m-iD_\Lambda)^n\\&=\det\left(I_m-i\Sigma^{1/2}\Gamma\Sigma^{1/2}\right)^n\\&=\det(I-i\Gamma\Sigma)^n,\end{aligned}$$

$$\begin{aligned}\prod_{j=1}^{n}\prod_{k=1}^{m}\exp\left(-\tfrac{1}{2}\nu_{jk}^2\right)&=\prod_{j=1}^{n}\exp\left(-\tfrac{1}{2}\boldsymbol{\nu}_j'\boldsymbol{\nu}_j\right)\\&=\prod_{j=1}^{n}\exp\left(-\tfrac{1}{2}\boldsymbol{\tau}_j'\boldsymbol{\tau}_j\right)\\&=\prod_{j=1}^{n}\exp\left(-\tfrac{1}{2}\mathbf{m}_j'\Sigma^{-1}\mathbf{m}_j\right)\\&=\operatorname{etr}\left(-\tfrac{1}{2}\Sigma^{-1}M'M\right)\\&=\operatorname{etr}\left(-\tfrac{1}{2}\Omega\right),\end{aligned}$$

and

$$\prod_{j=1}^{n}\prod_{k=1}^{m}\exp\left[\tfrac{1}{2}\nu_{jk}^2(1-i\lambda_k)^{-1}\right]=\prod_{j=1}^{n}\exp\left[\tfrac{1}{2}\nu_j'(I-iD_\Lambda)^{-1}\nu_j\right]$$

$$=\prod_{j=1}^{n}\exp\left[\tfrac{1}{2}\mathbf{m}_j'\Sigma^{-1/2}(I-i\Sigma^{1/2}\Gamma\Sigma^{1/2})^{-1}\Sigma^{-1/2}\mathbf{m}_j\right]$$

$$=\operatorname{etr}\left[\tfrac{1}{2}M'M\Sigma^{-1}(\Sigma^{-1}-i\Gamma)^{-1}\Sigma^{-1}\right]$$

$$=\operatorname{etr}\left[\tfrac{1}{2}\Omega(I-i\Gamma\Sigma)^{-1}\right].$$

(ii) Now suppose that n is any real number with $n>m-1$. Then A has the density function (1) so that

$$\phi(\Theta)=E\left[\operatorname{etr}\left(\frac{i}{2}A\Gamma\right)\right]$$

$$=\frac{1}{2^{mn/2}\Gamma_m(\frac{1}{2}n)(\det\Sigma)^{n/2}}\int_{A>0}\operatorname{etr}\left[-\tfrac{1}{2}(\Sigma^{-1}-i\Gamma)A\right]$$

$$\cdot(\det A)^{(n-m-1)/2}{}_0F_1\left(\tfrac{1}{2}n;\tfrac{1}{4}\Omega\Sigma^{-1}A\right)(dA)$$

and the desired result then follows using Theorem 7.3.4.

The characteristic function can be used to derive various properties of the noncentral Wishart distribution. The following theorems should be compared with Theorems 3.2.4, 3.2.5, and 3.2.8, the central Wishart analogs.

THEOREM 10.3.4. If the $m\times m$ matrices $A_1,\ldots,A_r$ are all independent and A_i is $W_m(n_i,\Sigma,\Omega_i)$, $i=1,\ldots,r$, then $\sum_{i=1}^r A_i$ is $W_m(n,\Sigma,\Omega)$, with $n=\sum_{i=1}^r n_i$ and $\Omega=\sum_{i=1}^r\Omega_i$.

Proof. The characteristic function of $\sum_{i=1}^r A_i$ is the product of the characteristic functions of $A_1,\ldots,A_r$ and hence, with the notation of Theorem 10.3.3, is

$$\prod_{j=1}^{r}\left\{\det(I-i\Gamma\Sigma)^{-n_j/2}\operatorname{etr}\left(-\tfrac{1}{2}\Omega_j\right)\operatorname{etr}\left[\tfrac{1}{2}\Omega_j(I-i\Gamma\Sigma)^{-1}\right]\right\}$$

$$=\det(I-i\Gamma\Sigma)^{-n/2}\operatorname{etr}\left(-\tfrac{1}{2}\Omega\right)\operatorname{etr}\left[\tfrac{1}{2}\Omega(I-i\Gamma\Sigma)^{-1}\right],$$

which is the characteristic function of the $W_m(n,\Sigma,\Omega)$ distribution.

THEOREM 10.3.5. If the $n\times m$ matrix Z is $N(M, I_n\otimes\Sigma)$ and P is $k\times m$ of rank k then

$$PZ'ZP' \quad \text{is} \quad W_k\left(n, P\Sigma P', (P\Sigma P')^{-1}PM'MP'\right)$$

Proof. Note that $PZ'ZP'=(ZP')'(ZP')$ and ZP' is $N(MP', I_n\otimes P\Sigma P')$; the desired result is now an immediate consequence of Definition 10.3.1.

THEOREM 10.3.6. If A is $W_m(n,\Sigma,\Omega)$, where n is a positive integer and $\boldsymbol{\alpha}$ $(\neq \mathbf{0})$ is an $m\times 1$ fixed vector, then $\boldsymbol{\alpha}'A\boldsymbol{\alpha}/\boldsymbol{\alpha}'\Sigma\boldsymbol{\alpha}$ is $\chi_n^2(\delta)$, with $\delta=\boldsymbol{\alpha}'\Sigma\Omega\boldsymbol{\alpha}/\boldsymbol{\alpha}'\Sigma\boldsymbol{\alpha}$.

Proof. From Theorem 10.3.5 the distribution of $\boldsymbol{\alpha}'A\boldsymbol{\alpha}$ is $W_1(n,\boldsymbol{\alpha}'\Sigma\boldsymbol{\alpha},\boldsymbol{\alpha}'\Sigma\Omega\boldsymbol{\alpha}/\boldsymbol{\alpha}'\Sigma\boldsymbol{\alpha})$, which is the desired result.

Many other properties of the central Wishart distribution can be readily generalized to the noncentral Wishart. Here we will look at just two more. Recall that if A is $W_m(n,\Sigma)$ then the moments of the generalized variance $\det A$ are given by [see (14) of Section 3.2]

$$E\left[(\det A)^r\right]=(\det\Sigma)^r 2^{mr}\frac{\Gamma_m\left(\frac{1}{2}n+r\right)}{\Gamma_m\left(\frac{1}{2}n\right)}.$$

In the noncentral case the moments are given in the following result due to Herz (1955) and Constantine (1963).

THEOREM 10.3.7. If A is $W_m(n,\Sigma,\Omega)$ with $n\geq m$

$$E\left[(\det A)^r\right]=(\det\Sigma)^r 2^{mr}\frac{\Gamma_m\left(\frac{1}{2}n+r\right)}{\Gamma_m\left(\frac{1}{2}n\right)}\,{}_1F_1\left(-r;\tfrac{1}{2}n;-\tfrac{1}{2}\Omega\right). \tag{3}$$

(Note that this is a polynomial of degree mr if r is a positive integer.)

Proof. Using the noncentral Wishart density function, and Theorem 7.3.4 gives

$$\begin{aligned}E\left[(\det A)^r\right]&=\frac{\operatorname{etr}\left(-\frac{1}{2}\Omega\right)}{2^{mn/2}\Gamma_m\left(\frac{1}{2}n\right)(\det\Sigma)^{n/2}}\int_{A>0}\operatorname{etr}\left(-\tfrac{1}{2}\Sigma^{-1}A\right)\\&\quad\cdot\det A^{(n+2r-m-1)/2}\,{}_0F_1\left(\tfrac{1}{2}n;\tfrac{1}{4}\Omega\Sigma^{-1}A\right)(dA)\\&=(\det\Sigma)^r 2^{mr}\frac{\Gamma_m\left(\frac{1}{2}n+r\right)}{\Gamma_m\left(\frac{1}{2}n\right)}\operatorname{etr}\left(-\tfrac{1}{2}\Omega\right){}_1F_1\left(\tfrac{1}{2}n+r;\tfrac{1}{2}n;\tfrac{1}{2}\Omega\right).\end{aligned}$$

The desired result now follows using the Kummer relation in Theorem 7.4.3.

The moments of $\det A$ can be used to obtain asymptotic distributions. For work in this direction see Problems 10.2, 10.3, and 10.4.

The next theorem generalizes Bartlett's decomposition (Theorem 3.2.14) when the noncentrality matrix has rank 1.

THEOREM 10.3.8. Let the $n\times m$ $(n\geq m)$ matrix Z be $N(M, I_n\otimes I_m)$, where $M=[\mathbf{m}_1,\mathbf{0},\ldots,\mathbf{0}]$, so that $A=Z'Z$ is $W_m(n, I,\Omega)$, with $\Omega=\operatorname{diag}(\mathbf{m}_1'\mathbf{m}_1,0,\ldots,0)$. Put $A=T'T$, where T is an upper-triangular matrix with positive diagonal elements. Then the elements $t_{ij}(1\leq i\leq j\leq m)$ of T are all independent, t_{11}^2 is $\chi_n^2(\delta)$ with $\delta=\mathbf{m}_1'\mathbf{m}_1$, t_{ii}^2 is χ^2_{n-i+1} $(i=2,\ldots,m)$, and t_{ij} is $N(0,1)$ $(1\leq i<j\leq m)$.

Proof. With Ω having the above form, the density of A is

$$\frac{1}{2^{mn/2}\Gamma_m\left(\frac{1}{2}n\right)}\operatorname{etr}\left(-\tfrac{1}{2}A\right)(\det A)^{(n-m-1)/2}$$

$$\cdot\exp\left(-\tfrac{1}{2}\mathbf{m}_1'\mathbf{m}_1\right){}_0F_1\left(\tfrac{1}{2}n;\tfrac{1}{4}\mathbf{m}_1'\mathbf{m}_1a_{11}\right)(dA).$$

Since $A=T'T$ we have

$$\operatorname{tr}A=\operatorname{tr}T'T=\sum_{i\leq j}^{m}t_{ij}^2$$

$$\det A=\prod_{i=1}^{m}t_{ii}^2,\quad a_{11}=t_{11}^2$$

and, from Theorem 2.1.9

$$(dA)=2^m\prod_{i=1}^{m}t_{ii}^{m+1-i}\bigwedge_{i\leq j}^{m}dt_{ij}$$

so that the joint density function of the t_{ij} $(1\leq i\leq j\leq m)$ can be written in the form

$$\prod_{i<j}^{m}\left[\frac{1}{(2\pi)^{1/2}}\exp\left(-\tfrac{1}{2}t_{ij}^2\right)dt_{ij}\right]$$

$$\cdot\frac{1}{2^{n/2}\Gamma\left(\frac{1}{2}n\right)}\exp\left(-\tfrac{1}{2}t_{11}^2\right)\left(t_{11}^2\right)^{n/2-1}e^{-\delta/2}{}_0F_1\left(\tfrac{1}{2}n;\tfrac{1}{4}\delta t_{11}^2\right)dt_{11}^2$$

$$\cdot\prod_{i=2}^{m}\left[\frac{1}{2^{(n-i+1)/2}\Gamma\left[\frac{1}{2}(n-i+1)\right]}\exp\left(-\tfrac{1}{2}t_{ii}^2\right)\left(t_{ii}^2\right)^{(n-i-1)/2}dt_{ii}^2\right],$$

where $\delta=\mathbf{m}_1'\mathbf{m}_1$, which is the product of the marginal density functions for the elements of T stated in the theorem.

10.4. JOINT DISTRIBUTION OF THE LATENT ROOTS IN MANOVA

In this section we return to distribution problems associated with multivariate linear models. We saw in Section 10.2 that invariant test statistics for testing the general linear hypothesis $H: M_1=0$ against $K: M_1\neq 0$ are functions of $f_1,\dots,f_s$, where $s=\text{rank }(Y_1^*B^{-1}Y_1^{*\prime})=\min(r,m)$ and $f_1\geq\cdots\geq f_s$ are the nonzero latent roots of $Y_1^*B^{-1}Y_1^{*\prime}$. Here Y_1^* is an $r\times m$ matrix having the $N(M_1, I_r\otimes\Sigma)$ distribution, B is $W_m(n-p,\Sigma)$, and Y_1^* and B are independent. We are assuming throughout that $n\geq m+p$ so that the distribution of B is nonsingular. There are two cases to be considered, namely $s=m$ and $s=r$.

Case 1: $r\geq m$

In this case rank $(Y_1^*B^{-1}Y_1^{*\prime})=m$ and $f_1\geq\cdots\geq f_m$ (>0) are the nonzero latent roots of $Y_1^*B^{-1}Y_1^{*\prime}$ or, equivalently, the latent roots of AB^{-1}, where $A=Y_1^{*\prime}Y_1^*$. The distribution of A is $W_m(r,\Sigma,\Omega)$, with $\Omega=\Sigma^{-1}M_1'M_1$. The distribution of $f_1,\dots,f_m$ will follow from the next theorem, which is a generalization of the noncentral F distribution and should be compared with Theorem 1.3.6.

THEOREM 10.4.1. Let $\tilde{A}$ and $\tilde{B}$ be independent, where $\tilde{A}$ is $W_m(r,I,\tilde{\Omega})$ and $\tilde{B}$ is $W_m(n-p,I)$ with $r\geq m$, $n-p\geq m$. Then the density function of the matrix $\tilde{F}=\tilde{A}^{1/2}\tilde{B}^{-1}\tilde{A}^{1/2}$ is

$$\text{(1)}\qquad \operatorname{etr}\left(-\tfrac{1}{2}\tilde{\Omega}\right){}_1F_1\left(\tfrac{1}{2}(n+r-p);\tfrac{1}{2}r;\tfrac{1}{2}\tilde{\Omega}\tilde{F}(I+\tilde{F})^{-1}\right)$$
$$\cdot\frac{\Gamma_m\left[\frac{1}{2}(n+r-p)\right]}{\Gamma_m\left(\frac{1}{2}r\right)\Gamma_m\left[\frac{1}{2}(n-p)\right]}\frac{(\det\tilde{F})^{(r-m-1)/2}}{\det(I+\tilde{F})^{(n+r-p)/2}}\qquad(\tilde{F}>0).$$

Proof. The joint density of $\tilde{A}$ and $\tilde{B}$ is

$$\frac{2^{-m(n+r-p)/2}}{\Gamma_m\left(\frac{1}{2}r\right)\Gamma_m\left[\frac{1}{2}(n-p)\right]}\operatorname{etr}\left[-\tfrac{1}{2}(\tilde{A}+\tilde{B})\right](\det\tilde{A})^{(r-m-1)/2}$$
$$\cdot(\det\tilde{B})^{(n-p-m-1)/2}$$
$$\cdot\operatorname{etr}\left(-\tfrac{1}{2}\tilde{\Omega}\right){}_0F_1\left(\tfrac{1}{2}r;\tfrac{1}{4}\tilde{\Omega}\tilde{A}\right)(d\tilde{A})(d\tilde{B})$$

Now make the change of variables $\tilde{F}=\tilde{A}^{1/2}\tilde{B}^{-1}\tilde{A}^{1/2}$, $U=\tilde{A}$ so that

$$(d\tilde{A})(d\tilde{B})=(\det U)^{(m+1)/2}(\det\tilde{F})^{-(m+1)}(dU)(d\tilde{F}).$$

The joint density of U, $\tilde{F}$ is then

$$\operatorname{etr}(-\tfrac{1}{2}\tilde{\Omega})\,{}_0F_1(\tfrac{1}{2}r;\tfrac{1}{4}\tilde{\Omega}U)\frac{2^{-m(n+r-p)/2}}{\Gamma_m(\tfrac{1}{2}r)\Gamma_m[\tfrac{1}{2}(n-p)]}$$

$$\cdot\operatorname{etr}[-\tfrac{1}{2}(I+\hat{F}^{-1})U](\det U)^{(n+r-p-m-1)/2}(\det\hat{F})^{-(n-p+m+1)/2}(dU)(d\hat{F}).$$

Integrating with respect to U using

$$\int_{U>0}\operatorname{etr}[-\tfrac{1}{2}(I+\tilde{F}^{-1})U](\det U)^{(n+r-p-m-1)/2}\,{}_0F_1(\tfrac{1}{2}r;\tfrac{1}{4}\tilde{\Omega}U)(dU)$$

$$=2^{m(n+r-p)/2}\det(I+\tilde{F}^{-1})^{-(n+r-p)/2}\Gamma_m[\tfrac{1}{2}(n+r-p)]$$

$$\cdot{}_1F_1(\tfrac{1}{2}(n+r-p);\tfrac{1}{2}r;\tfrac{1}{2}\tilde{\Omega}(I+\tilde{F}^{-1})^{-1})$$

(from Theorem 7.3.4) gives the stated marginal density function for $\tilde{F}$.

This theorem is used to find the distribution of the latent roots $f_1,\dots,f_m$, due to Constantine (1963).

THEOREM 10.4.2. If $A=Y_1^{*\prime}Y_1^*$ and $B=Y_3^{*\prime}Y_3^*$, where Y_1^* and Y_3^* are $r\times m$ and $(n-p)\times m$ matrices independently distributed as $N(M_1, I_r\otimes\Sigma)$ and $N(0, I_{n-p}\otimes\Sigma)$, respectively, with $r\geq m$, $n-p\geq m$, then the joint density function of $f_1,\dots,f_m$, the latent roots of AB^{-1} is

$$(2)\quad \operatorname{etr}(-\tfrac{1}{2}\Omega)\,{}_1F_1^{(m)}(\tfrac{1}{2}(n+r-p);\tfrac{1}{2}r;\tfrac{1}{2}\Omega, F(I+F)^{-1})$$

$$\cdot\frac{\Gamma_m[\tfrac{1}{2}(n+r-p)]}{\Gamma_m(\tfrac{1}{2}r)\Gamma_m[\tfrac{1}{2}(n-p)]}\frac{\pi^{m^2/2}}{\Gamma_m(\tfrac{1}{2}m)}\prod_{i=1}^{m}\frac{f_i^{(r-m-1)/2}}{(1+f_i)^{(n+r-p)/2}}\prod_{i<j}^{m}(f_i-f_j)$$

$$(f_1>\cdots>f_m>0),$$

where $F=\operatorname{diag}(f_1,\dots,f_m)$ and $\Omega=\Sigma^{-1}M_1'M_1$.

Proof. Putting $U=Y_1^*\Sigma^{-1/2}$, $V=Y_3^*\Sigma^{-1/2}$, and $M^*=M_1\Sigma^{-1/2}$ it follows that U and V are independent, U is $N(M^*, I_r\otimes I_m)$, V is $N(0, I_{n-p}\otimes I_m)$, and $f_1,\dots,f_m$ are the latent roots of $\tilde{A}\tilde{B}^{-1}$, where $\tilde{A}=U'U$ and $\tilde{B}=V'V$, or equivalently of $\tilde{F}=\tilde{A}^{1/2}\tilde{B}^{-1}\tilde{A}^{1/2}$. The distributions of $\tilde{A}$ and $\tilde{B}$ are, respectively, $W_m(r,I,\tilde{\Omega})$ and $W_m(n-p,I)$ where $\tilde{\Omega}=M^{*\prime}M^{*}=\Sigma^{-1/2}M_1'M_1\Sigma^{-1/2}$. The proof is completed by making the transformation $\tilde{F}=HFH'$ in Theorem 10.4.1 and integrating over $H\in O(m)$ using Theorems 3.2.17 and 7.3.3., noting that the latent roots $\omega_1,\dots,\omega_m$ of $\Omega=\Sigma^{-1}M_1'M_1$ are the same as those of $\tilde{\Omega}$.

The reader should note that the distribution of $f_1,\dots,f_m$ depends only on $\omega_1,\dots,\omega_m$, the latent roots of $\Omega=\Sigma^{-1}M_1'M_1$. [Some of these, of course, may be zero. The number of nonzero roots is rank $(M_1\Sigma^{-1}M_1')$.] This is because the nonzero roots form a maximal invariant under the group of transformations discussed in Section 10.2.

The null distribution of $f_1,\dots,f_m$, i.e., the distribution when $M_1=0$, follows easily from Theorem 10.4.2 by putting $\Omega=0$.

COROLLARY 10.4.3. If A is $W_m(r,\Sigma)$, B is $W_m(n-p,\Sigma)$, with $r\geq m$, $n-p\geq m$, and A and B are independent then the joint density function of $f_1,\dots,f_m$, the latent roots of AB^{-1}, is

$$\frac{\pi^{m^2/2}\Gamma_m[\frac{1}{2}(n+r-p)]}{\Gamma_m(\frac{1}{2}m)\Gamma_m(\frac{1}{2}r)\Gamma_m[\frac{1}{2}(n-p)]}\prod_{i=1}^{m}\frac{f_i^{(r-m-1)/2}}{(1+f_i)^{(n+r-p)/2}}\prod_{i<j}^{m}(f_i-f_j) \tag{3}$$

$$(f_1>\cdots>f_m>0).$$

It is worth noting that the null distribution (3) of $f_1,\dots,f_m$ has already been essentially derived in Theorem 3.3.4, where the distribution of the latent roots $u_1,\ldots,u_m$ of $A(A+B)^{-1}$ is given. Corollary 10.4.3 follows immediately from Theorem 3.3.4 on putting $n_1=r$, $n_2=n-p$, and $f_i=u_i/(1-u_i)$.

Case 2: $r<m$

In this case rank $(Y_1^*B^{-1}Y_1^{*\prime})=r$ and $f_1\geq\cdots\geq f_r$ (>0) are the latent roots of $Y_1^*B^{-1}Y_1^{*\prime}$ or, equivalently, the nonzero latent roots of AB^{-1} where $A=Y_1^{*\prime}Y_1^*$. The distribution of A is noncentral Wishart, but it does not have a density function. The distribution of $f_1,\dots,f_r$ in this case will follow from the following theorem (from James, 1964), which gives what is sometimes called the "studentized Wishart" distribution.

THEOREM 10.4.4. If $A=Y_1^{*\prime}Y_1^*$ and $B=Y_3^{*\prime}Y_3^*$, where Y_1^* and Y_3^* are independent $r\times m$ and $(n-p)\times m$ matrices independently distributed as $N(M_1,I_r\otimes\Sigma)$ and $N(0,I_{n-p}\otimes\Sigma)$, respectively, with $n-p\geq m\geq r$, then the density function of $\tilde{F}=Y_1^*B^{-1}Y_1^{*\prime}$ is

$$(4)\qquad \operatorname{etr}\left(-\tfrac{1}{2}\tilde{\Omega}\right){}_1F_1\left(\tfrac{1}{2}(n+r-p);\tfrac{1}{2}m;\tfrac{1}{2}\tilde{\Omega}\tilde{F}(I+\tilde{F})^{-1}\right)$$

$$\cdot\frac{\Gamma_r[\frac{1}{2}(n+r-p)]}{\Gamma_r(\frac{1}{2}m)\Gamma_r[\frac{1}{2}(n+r-p-m)]}\frac{(\det\tilde{F})^{(m-r-1)/2}}{\det(I+\tilde{F})^{(n+r-p)/2}}\qquad(\tilde{F}>0),$$

where $\tilde{\Omega}=M_1\Sigma^{-1}M_1'$.

Proof. Putting $U=Y_1^*\Sigma^{-1/2}$, $V=Y_3^*\Sigma^{-1/2}$ and $M^*=M_1\Sigma^{-1/2}$, it follows that U and V are independent, U is $N(M^*,I_r\otimes I_m)$, V is $N(0,I_{n-p}\otimes I_m)$, and $\tilde{F}=U(V'V)^{-1}U'$. Now put

$$U'=H_1T=[H_1:H_2]\begin{bmatrix}T\\ \cdots\\ 0\end{bmatrix}=H\begin{bmatrix}T\\ \cdots\\ 0\end{bmatrix},$$

where H_1 is $m\times r$ with orthonormal columns ($H_1'H_1=I_r$), T is $r\times r$ upper-triangular and H_2 is any $m\times(m-r)$ matrix such that $H=[H_1:H_2]$ is orthogonal. Then

$$\begin{aligned}\tilde{F}&=[T':0]H'(V'V)^{-1}H\begin{bmatrix}T\\ \cdot\cdot\\ 0\end{bmatrix}\\&=[T':0](Z'Z)^{-1}\begin{bmatrix}T\\ \cdot\cdot\\ 0\end{bmatrix}\\&=T'\tilde{B}^{-1}T,\end{aligned}$$

where $Z=VH$ and $\tilde{B}^{-1}$ denotes the $r\times r$ matrix formed by the first r rows and columns of $(Z'Z)^{-1}$. The distribution of $Z'Z$ is $W_m(n-p,I_m)$ and, from Theorem 3.2.10, the distribution of $\tilde{B}$ is $W_r(n+r-p-m,I_r)$. The joint density of U and $\tilde{B}$ is

$$(2\pi)^{-mr/2}\operatorname{etr}\left[-\tfrac{1}{2}(U-M^*)'(U-M^*)\right]\frac{2^{-r(n+r-p-m)/2}}{\Gamma_r[\frac{1}{2}(n+r-p-m)]}\operatorname{etr}\left(-\tfrac{1}{2}\tilde{B}\right)$$

$$\cdot(\det\tilde{B})^{(n-p-m-1)/2}(dU)(d\tilde{B}).$$

Since $U'=H_1T$, where $H_1\in V_{r,m}$, the Stiefel manifold of $m\times r$ matrices with orthonormal columns, we have from Theorem 2.1.14 that

$$(dU)=2^{-r}\det(T'T)^{(m-r-1)/2}(d(T'T))(H_1'dH_1)$$

so that the joint density becomes

$$\frac{(2\pi)^{-mr/2}2^{-r(n+r-p-m)/2-r}}{\Gamma_r[\frac{1}{2}(n+r-p-m)]}\operatorname{etr}(-\tfrac{1}{2}T'T)\operatorname{etr}(-\tfrac{1}{2}M^{*\prime}M^*)\operatorname{etr}(H_1TM^*)$$

$$\cdot\operatorname{etr}(-\tfrac{1}{2}\tilde{B})(\det\tilde{B})^{(n-p-m-1)/2}\det(T'T)^{(m-r-1)/2}(d(T'T))(H_1'dH_1)(d\tilde{B})$$

Now integrate with respect to $H_1\in V_{r,m}$. The same argument as that used in the proof of Theorem 10.3.2 shows that

$$\int_{H_1\in V_{r,m}}\operatorname{etr}(H_1TM^*)(H_1'dH_1)=\frac{2^r\pi^{mr/2}}{\Gamma_r(\frac{1}{2}m)}\,{}_0F_1(\tfrac{1}{2}m;\tfrac{1}{4}M^*M^{*\prime}T'T)$$

and hence the joint density of $T'T$ and $\tilde{B}$ is

$$\frac{2^{-r(n+r-p)/2}}{\Gamma_r[\frac{1}{2}(n+r-p-m)]\Gamma_r(\frac{1}{2}m)}\operatorname{etr}(-\tfrac{1}{2}T'T)\det(T'T)^{(m-r-1)/2}$$

$$\cdot{}_0F_1(\tfrac{1}{2}m;\tfrac{1}{4}M^*M^{*\prime}T'T)$$

$$\cdot\operatorname{etr}(-\tfrac{1}{2}M^{*\prime}M^*)\operatorname{etr}(-\tfrac{1}{2}\tilde{B})\det\tilde{B}^{(n-p-m-1)/2}(d(T'T)(d\tilde{B}).$$

Now put $\tilde{F}=T'\tilde{B}^{-1}T, G=T'T$ with Jacobian given by

$$(d(T'T))(d\tilde{B})=(\det G)^{(r+1)/2}(\det\tilde{F})^{-(r+1)}(d\tilde{F})(dG).$$

The joint density of $\tilde{F}$ and G is then

$$\frac{2^{-r(n+r-p)/2}}{\Gamma_r[\frac{1}{2}(n+r-p-m)]\Gamma_r(\frac{1}{2}m)}\operatorname{etr}\left[-\tfrac{1}{2}(I+\tilde{F}^{-1})G\right](\det G)^{(n+r-p-r-1)/2}$$

$$\cdot{}_0F_1(\tfrac{1}{2}m;\tfrac{1}{4}M^*M^{*\prime}G)\operatorname{etr}(-\tfrac{1}{2}M^{*\prime}M^*)(\det\tilde{F})^{-(n-p+m+2r+1)/2}(d\tilde{F})(dG).$$

Integrating with respect to G using

$$\int_{G>0}\operatorname{etr}\left[-\tfrac{1}{2}(I+\tilde{F}^{-1})G\right](\det G)^{(n+r-p-r-1)/2}{}_0F_1\left(\tfrac{1}{2}m;\tfrac{1}{4}M^*M^{*\prime}G\right)(dG)$$

$$=2^{r(n+r-p)/2}\Gamma_r\left[\tfrac{1}{2}(n+r-p)\right]\det(I+\tilde{F}^{-1})^{-(n+r-p)/2}$$

$$\cdot{}_1F_1\left(\tfrac{1}{2}(n+r-p);\tfrac{1}{2}m;\tfrac{1}{2}M^*M^{*\prime}(I+\tilde{F}^{-1})^{-1}\right)$$

then gives the stated marginal density function for $\tilde{F}$.

The distribution of $f_1,\ldots,f_r$, the latent roots of $Y_1^*B^{-1}Y_1^{*\prime}$ can be easily obtained from Theorem 10.4.4 and is now given.

THEOREM 10.4.5. If B is $W_m(n-p,\Sigma)$ and $A=Y_1^{*\prime}Y_1^*$, where Y_1^* is $N(M_1,I_r\otimes\Sigma)$ and is independent of B, and $n-p\geq m\geq r$ then the joint density function of $f_1,\ldots,f_r$, the latent roots of $\tilde{F}=Y_1^*B^{-1}Y_1^{*\prime}$ or, equivalently, the nonzero latent roots of AB^{-1}, is

$$(5)\quad \operatorname{etr}\left(-\tfrac{1}{2}\tilde{\Omega}\right){}_1F_1^{(r)}\left(\tfrac{1}{2}(n+r-p);\tfrac{1}{2}m;\tfrac{1}{2}\tilde{\Omega},F(I+F)^{-1}\right)$$

$$\cdot\frac{\pi^{r^2/2}\Gamma_r\left[\tfrac{1}{2}(n+r-p)\right]}{\Gamma_r(\tfrac{1}{2}r)\Gamma_r(\tfrac{1}{2}m)\Gamma_r\left[\tfrac{1}{2}(n+r-p-m)\right]}\prod_{i=1}^{r}\frac{f_i^{(m-r-1)/2}}{(1+f_i)^{(n+r-p)/2}}\prod_{i<j}^{r}(f_i-f_j)$$

$$\cdot(f_1>\cdots>f_r>0),$$

where $F=\operatorname{diag}(f_1,\ldots,f_r)$ and $\tilde{\Omega}=M_1\Sigma^{-1}M_1'$.

Proof. Putting $\tilde{F}=HFH'$ in Theorem 10.4.4 and integrating over $H\in O(r)$ using Theorems 3.2.17 and 7.3.3 gives the desired result.

Putting $\Omega=0$ in Theorem 10.4.5 gives the null distribution of $f_1,\ldots,f_r$.

COROLLARY 10.4.6. If A is $W_m(r,\Sigma)$, B is $W_m(n-p,\Sigma)$, with $n-p\geq m\geq r$, and A and B are independent then the joint density function of $f_1,\ldots,f_r$, the nonzero latent roots of AB^{-1}, is

$$(6)\quad \frac{\pi^{r^2/2}\Gamma_r\left[\tfrac{1}{2}(n+r-p)\right]}{\Gamma_r(\tfrac{1}{2}r)\Gamma_r(\tfrac{1}{2}m)\Gamma_r\left[\tfrac{1}{2}(n+r-p-m)\right]}\prod_{i=1}^{r}\frac{f_i^{(m-r-1)/2}}{(1+f_i)^{(n+r-p)/2}}\prod_{i<j}^{r}(f_i-f_j)$$

$$(f_1>\cdots>f_r>0).$$

Let us now compare the distributions of the latent roots obtained in the two cases $n-p\geq m$, $r\geq m$ and $n-p\geq m\geq r$, i.e., compare Theorem 10.4.2

with Theorem 10.4.5 and Corollary 10.4.3 with Corollary 10.4.6. When $r=m$ they agree, and it is easy to check that the distributions in Case 2 $(n-p\geq m\geq r)$ can be obtained from the distributions in Case 1 $(n-p\geq m, r\geq m)$ by replacing m by r, r by m and $n-p$ by $n+r-p-m$, i.e., by making the substitutions

$$(7) \qquad m\rightarrow r, \qquad r\rightarrow m, \qquad n-p\rightarrow n+r-p-m.$$

One consequence of this is that the distribution of any invariant test statistic (i.e., function of the f_i) in Case 2 can be obtained from its distribution in Case 1 by using the substitution rule (7). In what follows we will concentrate primarily on Case 1.

10.5. DISTRIBUTIONAL RESULTS FOR THE LIKELIHOOD RATIO STATISTIC

10.5.1. Moments

In this section we return to the likelihood statistic for testing H: $M_1=0$ against K: $M_1\neq 0$. In Section 10.2 it was shown that the likelihood ratio test rejects H for small values of

$$W=\frac{\det B}{\det(A+B)}=\Lambda^{2/n},$$

where A is $W_m(r,\Sigma,\Omega)$, B is $W_m(n-p,\Sigma)$ and $\Omega=\Sigma^{-1}M_1'M_1$. We will assume here that $n-p\geq m, r\geq m$. In terms of the latent roots $f_1,\ldots,f_m$ of AB^{-1} the test statistic is $W=\prod_{i=1}^m(1+f_i)^{-1}$. The moments of W are given in the following theorem due to Constantine (1963).

THEOREM 10.5.1. The hth moment of W, when $n-p\geq m, r\geq m$, is

(1)

$$E(W^h)=\frac{\Gamma_m[\frac{1}{2}(n-p)+h]\Gamma_m[\frac{1}{2}(n+r-p)]}{\Gamma_m[\frac{1}{2}(n-p)]\Gamma_m[\frac{1}{2}(n+r-p)+h]}\,{}_1F_1(h;\tfrac{1}{2}(n+r-p)+h;-\tfrac{1}{2}\Omega)$$

where $\Omega=\Sigma^{-1}M_1'M_1$.

Proof. From Theorem 10.4.2 we have

$$E(W^h)=\frac{\Gamma_m[\frac{1}{2}(n+r-p)]\pi^{m^2/2}}{\Gamma_m(\frac{1}{2}r)\Gamma_m[\frac{1}{2}(n-p)]\Gamma_m(\frac{1}{2}m)}\operatorname{etr}(-\tfrac{1}{2}\Omega)$$

$$\cdot\int\cdots\int_{f_1>\cdots>f_m>0}\prod_{i=1}^m\left[f_i^{(r-m-1)/2}(1+f_i)^{-(n+2h-r-p)/2}\right]\prod_{i<j}^m(f_i-f_j)$$

$$\cdot{}_1F_1^{(m)}(\tfrac{1}{2}(n+r-p);\tfrac{1}{2}r;\tfrac{1}{2}\Omega,F(I+F)^{-1})\,df_m\ldots df_1$$

$$=\frac{\Gamma_m[\frac{1}{2}(n+r-p)]}{\Gamma_m(\frac{1}{2}r)\Gamma_m[\frac{1}{2}(n-p)]}\operatorname{etr}(-\tfrac{1}{2}\Omega)\int_{\tilde{F}>0}(\det\tilde{F})^{(r-m-1)/2}$$

$$\cdot\det(I+\tilde{F})^{-(n+2h+r-p)/2}{}_1F_1(\tfrac{1}{2}(n+r-p);\tfrac{1}{2}r;\tfrac{1}{2}\Omega(I+\tilde{F})^{-1/2}$$

$$F(I+\tilde{F})^{-1/2})(d\tilde{F}).$$

Putting $U=(I+\tilde{F})^{-1/2}\tilde{F}(I+\tilde{F})^{-1/2}$ so that

$$(d\tilde{F})=\det(I-U)^{-(m+1)}(dU)$$

and using the zonal polynomial series for the ${}_1F_1$ function gives

$$E(W^h)=\frac{\Gamma_m[\frac{1}{2}(n+r-p)]}{\Gamma_m(\frac{1}{2}r)\Gamma_m[\frac{1}{2}(n-p)]}\operatorname{etr}(-\tfrac{1}{2}\Omega)\int_{0<U<I}(\det U)^{(r-m-1)/2}$$

$$\cdot\det(I-U)^{(n+2h-p-m-1)/2}{}_1F_1(\tfrac{1}{2}(n+r-p);\tfrac{1}{2}r;\tfrac{1}{2}\Omega U)(dU)$$

$$=\frac{\Gamma_m[\frac{1}{2}(n+r-p)]}{\Gamma_m(\frac{1}{2}r)\Gamma_m[\frac{1}{2}(n-p)]}\operatorname{etr}(-\tfrac{1}{2}\Omega)\sum_{k=0}^{\infty}\sum_{\kappa}\frac{(\frac{1}{2}(n+r-p))_\kappa}{(\frac{1}{2}r)_\kappa}$$

$$\cdot\int_{0<U<I}(\det U)^{(r-m-1)/2}\det(I-U)^{(n+2h-p-m-1)/2}C_\kappa(\tfrac{1}{2}\Omega U)(dU).$$

Using Theorem 7.2.10 to evaluate this integral shows that

$$E(W^h)=\frac{\Gamma_m[\frac{1}{2}(n+r-p)]\Gamma_m[\frac{1}{2}(n-p)+h]}{\Gamma_m[\frac{1}{2}(n+r-p)+h]\Gamma_m[\frac{1}{2}(n-p)]}\operatorname{etr}(-\tfrac{1}{2}\Omega)$$

$$\cdot{}_1F_1(\tfrac{1}{2}(n+r-p);\tfrac{1}{2}(n+r-p)+h;\tfrac{1}{2}\Omega).$$

The desired result (1) now follows on using the Euler relation given in Theorem 7.4.3.

The moments of W when H: $M_1=0$ is true are obtained by putting $\Omega=0$.

COROLLARY 10.5.2. When $M_1=0$ (i.e., $\Omega=0$) the moments of W are given by

$$(2) \qquad E(W^h)=\frac{\Gamma_m[\frac{1}{2}(n-p)+h]\Gamma_m[\frac{1}{2}(n+r-p)]}{\Gamma_m[\frac{1}{2}(n-p)]\Gamma_m[\frac{1}{2}(n+r-p)+h]}.$$

It is worth emphasizing again that the expression for the moments given here assume that $r\geq m$. When $r<m$ the moments can be obtained using the substitution rule given by (7) of Section 10.4.

10.5.2. Null Distribution

When the null hypothesis H: $M_1=0$ is true, the likelihood ratio statistic W has the same distribution as a product of independent beta random variables. The result is given in the following theorem.

THEOREM 10.5.3. When H: $M_1=0$ is true and $r\geq m, n-p\geq m$, the statistic W has the same distribution as $\prod_{i=1}^m V_i$, where $V_1,\ldots,V_m$ are independent random variables and V_i is beta$(\frac{1}{2}(n-p-i+1),\frac{1}{2}r)$.

Proof. We have $W=\prod_{i=1}^m(1-u_i)$, where $u_1,\ldots,u_m$ are the latent roots of $A(A+B)^{-1}$. The distribution of $u_1,\ldots,u_m$, given in Theorem 3.3.3 with $n_1=r$ and $n_2=n-p$, is the distribution of the latent roots of a matrix U having the Beta$_m(\frac{1}{2}r,\frac{1}{2}(n-p))$ distribution, so that $W=\det(I-U)$. The distribution of $I-U$ is Beta$_m(\frac{1}{2}(n-p),\frac{1}{2}r)$. Putting $I-U=T'T$, where $T=(t_{ij})$ is upper-triangular gives $W=\prod_{i=1}^m t_{ii}^2$; from Theorem 3.3.2 the t_{ii}^2 are independent beta $(\frac{1}{2}(n-p-i+1),\frac{1}{2}r)$ random variables, and the proof is complete.

This result can, of course, also be obtained from the moments of W given in Corollary 10.5.2 by writing these as a product of moments of beta random variables. When $r\leq m$, the distribution of W is obtained from Theorem 10.5.3 using the substitution rule given by (7) of Section 10.4 and is given in the following corollary.

COROLLARY 10.5.4. When $n-p\geq m\geq r$, W has the same distribution as $\prod_{i=1}^r V_i$, where $V_1,\ldots,V_r$ are independent and V_i is beta$(\frac{1}{2}(n+r-p-m-i+1),\frac{1}{2}m)$.

Let us look briefly at two special cases. When $m=1$, Theorem 10.5.3 shows that W has the beta$(\frac{1}{2}(n-p),\frac{1}{2}r)$ distribution or, equivalently,

$$\frac{1-W}{W}\frac{n-p}{r}$$

is $F_{r,n-p}$. This is the usual F statistic for testing H: $M_1=0$ in the univariate setting. When $r=1$, Corollary 10.5.4 shows that W has the beta$(\frac{1}{2}(n+1-p-m),\frac{1}{2}m)$ distribution or, equivalently,

$$\frac{1-W}{W}\frac{n+1-p-m}{m}$$

is $F_{m,n+1-p-m}$. This is the null distribution of Hotelling's T^2 statistic given in Theorem 6.3.1.

In general it is not a simple matter to actually find the density function of a product of independent beta random variables. For some other special cases the interested reader should see T. W. Anderson (1958), Section 8.5.3, Srivastava and Khatri (1979), Section 6.3.6, and Problem 10.12.

10.5.3. *The Asymptotic Null Distribution*

Replacing h in Corollary 10.5.2 by $nh/2$ shows that the hth null moment of the likelihood ratio statistic $\Lambda=W^{n/2}$ is

$$(3)\qquad E(\Lambda^h)=E(W^{nh/2})$$
$$=\frac{\Gamma_m[\frac{1}{2}n(1+h)-\frac{1}{2}p]\Gamma_m[\frac{1}{2}(n+r-p)]}{\Gamma_m[\frac{1}{2}(n-p)]\Gamma_m[\frac{1}{2}n(1+h)+\frac{1}{2}(r-p)]}$$
$$=\frac{K\prod_{k=1}^{m}\Gamma[\frac{1}{2}n(1+h)+\frac{1}{2}(1-k-p)]}{\prod_{j=1}^{m}\Gamma[\frac{1}{2}n(1+h)+\frac{1}{2}(1-j+r-p)]},$$

where K is a constant not involving h. This has the same form as (18) of Section 8.2.4, where there we put

$$p=m,\quad q=m,\quad x_k=\tfrac{1}{2}n,\quad y_j=\tfrac{1}{2}n,\quad \xi_k=\tfrac{1}{2}(1-k-p),\quad \eta_j=\tfrac{1}{2}(1-j+r-p)$$
$$(k,j=1,\ldots,m).$$

The degrees of freedom in the limiting χ^2 distribution are, from (28) of Section 8.2.4,

$$f = -2\left[\sum_{k=1}^{m} \xi_k - \sum_{j=1}^{m} \eta_j\right]$$
$$= rm.$$

The value of ρ which makes the term of order n^{-1} vanish in the asymptotic expansion of the distribution of $-2\rho\log\Lambda$ is, from (30) of Section 8.2.4

$$(4) \qquad \rho = 1 - \frac{1}{f}\left[\frac{2}{n}\sum_{k=1}^{m}\left(\xi_k^2 - \xi_k - \eta_j^2 + \eta_j\right)\right]$$
$$= 1 - \frac{r}{2fn}\sum_{k=1}^{m}(-r+2k+2p)$$
$$= 1 - \frac{1}{n}\left[p - r + \tfrac{1}{2}(m+r+1)\right].$$

With this value of ρ it is then found, using (29) of Section 8.2.4, that

$$(5) \qquad \omega_2 = \frac{mr(m^2+r^2-5)}{48(\rho n)^2}.$$

Hence we have the following result, first given explicitly by T. W. Anderson (1958), Section 8.6.2.

THEOREM 10.5.5. When the null hypothesis H: $M_1 = 0$ is true the distribution function of $-2\rho\log\Lambda$, where ρ is given by (4), can be expanded for large $N = \rho n$ as

(6)

$$P(-2\rho\log\Lambda \le x) = P(-N\log W \le x)$$
$$= P(\chi_f^2 \le x) + \frac{\gamma}{N^2}\left[P(\chi_{f+4}^2 \le x) - P(\chi_f^2 \le x)\right] + O(N^{-3}),$$

where $f = mr$ and $\gamma = (n\rho)^2\omega_2 = N^2\omega_2$, with ω_2 given by (5).

Lee (1972) has shown that in an asymptotic series in terms of powers of N^{-1} for the distribution function of $-2\rho\log\Lambda$ only terms of even powers in N^{-1} are involved. For a detailed discussion of this and other work on the distribution of Λ the interested reader is referred to T. W. Anderson (1958), Rao (1951), Schatzoff (1966a), Pillai and Gupta (1969), Lee (1972), and Krishnaiah and Lee (1979).

Tables of upper $100\alpha\%$ points of $-2\rho\log\Lambda=-N\log W=-[n-p-\frac{1}{2}(m-r+1)]\log W$ for $\alpha=.100$, .050, .025, and .005 have been prepared for various values of m, r, and $n-p$ by Schatzoff (1966a), Pillai and Gupta (1969), Lee (1972), and Davis (1979). These are reproduced in Table 9 which is given at the end of the book. The function tabulated is a multiplying factor C which when multiplied by the χ^2_{mr} upper $100\alpha\%$ point gives the upper $100\alpha\%$ point of $-N\log W$. Each table represents a particular (m,r) combination, with arguments $M=n-p-m+1$ and significance level α. Note that since the distribution of W when $n-p\geq m\geq r$ is obtained from the distribution when $n-p\geq m, r\geq m$ by making the substitutions

$$m\to r,\quad r\to m,\quad n-p\to n+r-p-m,$$

it follows that m and r are interchangeable in the tables.

10.5.4. Asymptotic Non-null Distributions

The power function of the likelihood ratio test of size α is $P(-2\rho\log\Lambda\geq k_\alpha^*|\omega_1,\dots,\omega_m)$, where ρ is given by (4) and k_α^* denotes the upper $100\alpha\%$ point of the distribution of $-2\rho\log\Lambda$ when $H:\Omega=0$ is true. This depends on Ω only through its latent roots $\omega_1\geq\cdots\geq\omega_m\ (\geq 0)$ so that in this section, without loss of generality, it will be assumed that Ω is diagonal, $\Omega=\operatorname{diag}(\omega_1,\dots,\omega_m)$. Here we investigate ways of approximating the power function.

We consider first the general alternative $K:\Omega\neq 0$. From Theorem 10.5.1 the characteristic function of $-2\rho\log\Lambda$ under K is

$$(7)\qquad \begin{aligned} g(N,t,\Omega)&=E(\Lambda^{-2it\rho})\\ &=E(W^{-itN})\\ &=g(N,t,0)G(N,t,\Omega),\end{aligned}$$

where $g(N,t,0)$ is the characteristic function of $-2\rho\log\Lambda$ when H is true, obtained from (2) by putting $h=-2it\rho$ and $n=N+p-r+\frac{1}{2}(m+r+1)$, and

$$(8)\qquad G(N,t,\Omega)={}_1F_1(-itN;\tfrac{1}{2}N(1-2it)+\tfrac{1}{4}(r+m+1);-\tfrac{1}{2}\Omega).$$

From Theorem 10.5.5 we know that

$$(9)\qquad g(N,t,0)=(1-2it)^{-mr/2}+O(N^{-2}).$$

The following theorem gives the term of order N^{-1} in an expansion for $G(N,t,\Omega)$.

THEOREM 10.5.6. The function $G(N,t,\Omega)$ defined by (8) can be expanded as

$$(10)\qquad G(N,t,\Omega)=\exp\left(\frac{it\sigma_1}{1-2it}\right)\left[1+\frac{P_1(\Omega)}{N}+O(N^{-2})\right]$$

where $\sigma_j=\operatorname{tr}\Omega^j=\sum_{k=1}^m\omega_k^j$ and

$$(11)\qquad P_1(\Omega)=-\frac{1}{2}(r+m+1)\frac{it}{(1-2it)^2}\sigma_1-\frac{it}{(1-2it)^3}\frac{\sigma_2}{2}.$$

Proof. The proof given here is based on the partial differential equations satisfied by the ${}_1F_1$ function. Using Theorem 7.5.6 the function $H(N,\Omega)=\log G(N,t,\Omega)$ is found to satisfy the system of partial differential equations

$$(12)\qquad \omega_j\left[\frac{\partial^2 H}{\partial\omega_j^2}+\left(\frac{\partial H}{\partial\omega_j}\right)^2\right]$$
$$+\left[\tfrac{1}{2}N(1-2it)+\tfrac{1}{4}(r-m+3)+\tfrac{1}{2}\omega_j+\frac{1}{2}\sum_{l\neq j}\frac{\omega_j}{\omega_j-\omega_l}\right]\frac{\partial H}{\partial\omega_j}$$
$$-\frac{1}{2}\sum_{l\neq j}\frac{\omega_l}{\omega_j-\omega_l}\frac{\partial H}{\partial\omega_l}=\tfrac{1}{2}itN\qquad(j=1,\dots,m).$$

The function $H(N,\Omega)$ is the unique solution of each of these partial differential equations subject to the condition that $H(N,\Omega)$ be a symmetric function of $\omega_1,\dots,\omega_m$ which is analytic at $\Omega=0$ with $H(N,0)=0$. In (12) we substitute

$$H(N,\Omega)=\sum_{k=0}^{\infty}\frac{P_k(\Omega)}{N^k},$$

where (a) $P_k(\Omega)$ is symmetric in $\omega_1,\dots,\omega_m$ and (b) $P_k(0)=0$ for $k=$

$0,1,2,\ldots$. Equating coefficients of N on both sides gives

$$(1-2it)\frac{\partial P_0}{\partial \omega_j}=it \qquad (j=1,\ldots,m),$$

the unique solution of which is, subject to conditions (a) and (b),

$$P_0(\Omega)=\frac{it\sigma_1}{1-2it}, \tag{13}$$

where $\sigma_1=\operatorname{tr}\Omega$. Equating constant terms and using (13) yields the system for $P_1(\Omega)$ as

$$\frac{\partial P_1}{\partial \omega_j}=-\frac{1}{2}(r+m+1)\frac{it}{(1-2it)^2}-\frac{it}{(1-2it)^3}\omega_j \qquad (j=1,\ldots,m),$$

the solution of which is the function $P_1(\Omega)$ given by (11). Hence

$$\begin{aligned} G(N,t,\Omega)&=\exp H(N,\Omega)\\ &=\exp\left[P_0(\Omega)+\frac{P_1(\Omega)}{N}+O(N^{-2})\right]\\ &=\exp[P_0(\Omega)]\left[1+\frac{P_1(\Omega)}{N}+O(N^{-2})\right], \end{aligned}$$

and the proof is complete.

An asymptotic expansion to order N^{-1} of the distribution function of $-2\rho\log\Lambda$ under K now follows easily.

THEOREM 10.5.7. Under the fixed alternative $K:\Omega\neq 0$ the distribution function of $-2\rho\log\Lambda$ can be expanded for large $N=\rho n$ as

$$\begin{aligned} P(-2\rho\log\Lambda\le x)&=P\left(\chi_f^2(\sigma_1)\le x\right)\\ &\quad+\frac{1}{4N}\Big\{(m+r+1)\sigma_1P\left(\chi_{f+2}^2(\sigma_1)\le x\right)\\ &\quad-[(m+r+1)\sigma_1-\sigma_2]P\left(\chi_{f+4}^2(\sigma_1)\le x\right)\\ &\quad-\sigma_2P\left(\chi_{f+6}^2(\sigma_1)\le x\right)\Big\}+O(N^{-2}), \end{aligned} \tag{14}$$

where $f=mr$ and $\sigma_j=\operatorname{tr}\Omega^j$.

Proof. Using (9) and (10) in (7) shows that the characteristic function of $-2\rho\log\Lambda$ has the expansion

$$g(N,t,\Omega)=(1-2it)^{-f/2}\exp\left(\frac{it\sigma_1}{1-2it}\right)\left[1+\frac{P_1(\Omega)}{N}+O(N^{-2})\right],$$

where $P_1(\Omega)$ is given by (11). The desired expansion (14) is obtained by inverting this term by term.

A different limiting distribution for the likelihood ratio statistic can be obtained by assuming that $\Omega=O(n)$. Under this assumption the noncentrality parameters are increasing with n. A situation where this assumption is reasonable is the multivariate single classification model introduced briefly at the beginning of Section 10.2 and which will be considered in more detail in Section 10.7. In the notation of Section 10.2 the noncentrality matrix Ω turns out to have the form

$$\Omega=\Sigma^{-1}\sum_{i=1}^{p}q_i(\mu_i-\bar{\mu})(\mu_i-\bar{\mu})',$$

where $\bar{\mu}=n^{-1}\Sigma_{i=1}^{p}q_i\mu_i$, $n=\Sigma_{i=1}^{p}q_i$. Hence if $\mu_i=O(1)$ and $q_i\to\infty$ for fixed q_i/n $(i=1,\ldots,p)$ it follows that $\Omega=O(n)$. We will consider therefore the sequence of alternatives K_N: $\Omega=N\Delta$, where $N=\rho n$ and Δ is a fixed matrix. Without loss of generality Δ can be assumed diagonal, $\Delta=\text{diag}(\delta_1,\ldots,\delta_m)$. Define the random variable Y by

$$Y=\frac{-2\rho}{N^{1/2}}\log\Lambda-N^{1/2}\log\det(I+\Delta), \tag{15}$$

and note that asymptotically Y has zero mean and constant variance. The characteristic function of Y is, using Theorem 10.5.1,

$$\begin{aligned} h(N,t,\Delta)&=\det(I+\Delta)^{-N^{1/2}it}E(\Lambda^{-2\rho it/N^{1/2}}) \\ &=\det(I+\Delta)^{-N^{1/2}it}E(W^{-itN^{1/2}}) \\ &=C_1(N,t)C_2(N,t,\Delta) \end{aligned} \tag{16}$$

where

$$C_1(N,t)=\frac{\Gamma_m\left[\frac{1}{2}N-itN^{1/2}+\frac{1}{4}(m-r+1)\right]\Gamma_m\left[\frac{1}{2}N+\frac{1}{4}(m+r+1)\right]}{\Gamma_m\left[\frac{1}{2}N+\frac{1}{4}(m-r+1)\right]\Gamma_m\left[\frac{1}{2}N-itN^{1/2}+\frac{1}{4}(m+r+1)\right]} \tag{17}$$

and

$$(18)\quad C_2(N,t,\Delta)=\det(I+\Delta)^{-N^{1/2}it}\cdot {}_1F_1(-itN^{1/2};\tfrac{1}{2}N-itN^{1/2}+\tfrac{1}{4}(m+r+1);-\tfrac{1}{2}N\Delta).$$

Using (24) of Section 8.2 to expand the gamma functions for large N it is straightforward to show that

$$(19)\qquad C_1(N,t)=1+\frac{rmit}{N^{1/2}}+O(N^{-1}).$$

An expansion for $C_2(M,t,\Delta)$ is given in the following theorem.

THEOREM 10.5.8. The function $C_2(M,t,\Delta)$ defined by (18) can be expanded as

$$(20)\qquad C_2(M,t,\Delta)=\exp(-\tfrac{1}{2}\tau^2t^2)\left\{1+\frac{Q_1(\Delta)}{N^{1/2}}+O(N^{-1})\right\},$$

where

$$\tau^2=-2s_2-4s_1,$$

and

$$Q_1(\Delta)=\left[\tfrac{1}{2}s_2+\tfrac{1}{2}s_1^2+\tfrac{1}{2}(r+m+1)s_1\right]it-\left[2s_4+\tfrac{20}{3}s_3+8s_2+4s_1\right](it)^3$$

with

$$s_j=\operatorname{tr}\left[(I+\Delta)^{-1}-I\right]^j=\sum_{k=1}^{m}\left(\frac{-\delta_k}{1+\delta_k}\right)^j.$$

A proof of this theorem similar to that of Theorem 8.2.12 can be constructed using the partial differential equations satisfied by the ${}_1F_1$ hypergeometric function. The details are left as an exercise (see Problem 10.13).

An asymptotic expansion to order $N^{-1/2}$ of the distribution of Y is an immediate consequence of Theorem 10.5.8.

THEOREM 10.5.9. Under the sequence of alternatives K_N: $\Omega=N\Delta$ the distribution function of the random variable Y given by (15) can be expanded for large $N = \rho n$ as

$$(21) \quad P\left(\frac{Y}{\tau} \le x\right) = \Phi(x) - \frac{1}{N^{1/2}}\left[a_1\phi(x) - a_2\phi^{(2)}(x)\right] + O(N^{-1}),$$

where Φ and ϕ denote the standard normal distribution and density functions respectively, and

$$\tau^2 = -2s_2 - 4s_1,$$

$$(22) \qquad a_1 = \frac{1}{2\tau}\left[s_2 + s_1^2 + (r+m+1)s_1 + 2rm\right],$$

$$(23) \qquad a_2 = \frac{2}{3\tau^3}\left[3s_4 + 10s_3 + 12s_2 + 6s_1\right],$$

with

$$s_j = \sum_{i=1}^{m}\left(\frac{-\delta_i}{1+\delta_i}\right)^j.$$

Proof. From (19) and (20) it follows that $h(N, t/\tau, \Delta)$, the characteristic function of Y/τ, can be expanded as

$$h\left(N, \frac{t}{\tau}, \Delta\right) = \exp(-\tfrac{1}{2}t^2)\left[1 + \frac{1}{N^{1/2}}\left(a_1 it - a_2(it)^3\right) + O(N^{-1})\right]$$

with a_1 and a_2 given by (22) and (23). The desired result now follows by term by term inversion.

Further terms in the asymptotic expansions presented here have been obtained by Sugiura and Fujikoshi (1969), Sugiura (1973a), and Fujikoshi (1973).

10.6. OTHER TEST STATISTICS

10.6.1. Introduction

In this section we will derive some distributional results associated with three other invariant statistics used for testing the general linear hypothesis. In terms of the latent roots $f_1 > \cdots > f_m$ of AB^{-1}, where A is $W_m(r, \Sigma, \Omega)$ and B is independently $W_m(n-p, \Sigma)$ $(r \ge m, n-p \ge m)$, the three statistics are $T_0^2=\Sigma_{i=1}^m f_i$, $V=\Sigma_{i=1}^m f_i(1+f_i)^{-1}$, and f_1, the largest root. The null hypothesis is rejected for large values of these statistics.

10.6.2. The T_0^2 Statistic

The T_0^2 statistic given by $T_0^2=\Sigma_{i=1}^m f_i$ was proposed by Lawley (1938) and by Hotelling (1947) in connection with a problem dealing with the air-testing of bombsights. Here we derive expressions for the exact and asymptotic distributions of T_0^2.

An expression for the density function of T_0^2 in the general (noncentral) situation ($\Omega\neq 0$) has been obtained by Constantine (1966) as a series involving the generalized Laguerre polynomials introduced in Section 7.6. We will outline the derivation here. In this we will need the following lemma which provides an estimate for a Laguerre polynomial and which is useful for examining convergence of series involving Laguerre polynomials.

LEMMA 10.6.1. If $L_\kappa^\beta(X)$ denotes the generalized Laguerre polynomial of the $m\times m$ symmetric matrix X corresponding to the partition κ of k (see Definition 7.6.1) then

$$(1)\qquad |L_\kappa^\beta(X)|\le[\beta+\tfrac{1}{2}(m+1)]_\kappa C_\kappa(I)\,\mathrm{etr}(X)\qquad(\beta>-1).$$

Proof. We first assume that $\beta=-\frac{1}{2}$. From Theorem 7.6.4

$$L_\kappa^{-1/2}(X)=\frac{\mathrm{etr}(X)}{\Gamma_m(\frac{1}{2}m)}\int_{Y>0}\mathrm{etr}(-Y)(\det Y)^{-1/2}C_\kappa(Y)\,{}_0F_1(\tfrac{1}{2}m;-XY)(dY),$$

while from Theorem 7.4.1

$${}_0F_1(\tfrac{1}{2}m;-U)=\int_{O(m)}\mathrm{etr}(2iU^{1/2}H)(dH)$$

so that

$$|{}_0F_1(\tfrac{1}{2}m;-U)|\le 1.$$

Hence

$$(2)\quad |L_\kappa^{-1/2}(X)|\le\frac{\mathrm{etr}(X)}{\Gamma_m(\frac{1}{2}m)}\int_{Y>0}\mathrm{etr}(-Y)(\det Y)^{(m-m-1)/2}C_\kappa(Y)(dY)$$

$$=(\tfrac{1}{2}m)_\kappa C_\kappa(I)\,\mathrm{etr}(X),$$

where the integral on the right is evaluated using Theorem 7.2.7. This establishes (1) for $\beta=-\frac{1}{2}$. To prove it for general $\beta>-1$, we use the identity

$$\text{(3)} \qquad \frac{L_\kappa^\beta(X)}{k!C_\kappa(I)}=\sum_{\substack{t,n=0\\t+n=k}}^{k}\sum_\tau\sum_\nu\frac{(\beta-\gamma)_\tau}{t!n!}g_{\tau,\nu}^\kappa\frac{L_\nu^\gamma(X)}{C_\nu(I)},$$

where the summation is over all partitions τ of t and ν of n such that $t+n=k$ and $g_{\tau,\nu}^\kappa$ is the coefficient of $C_\kappa(Y)$ in $C_\tau(Y)C_\nu(Y)$; that is,

$$\text{(4)} \qquad C_\tau(Y)C_\nu(Y)=\sum_\kappa g_{\tau,\nu}^\kappa C_\kappa(Y).$$

To establish the identity (3) we start with the generating function given in Theorem 7.6.3 for the Laguerre polynomials, namely,

$$\det(I-Y)^{-\gamma-p}{}_0F_0^{(m)}(-X,Y(I-Y)^{-1})=\sum_{n=0}^\infty\sum_\nu\frac{L_\nu^\gamma(X)C_\nu(Y)}{C_\nu(I)n!},$$

where $p=\frac{1}{2}(m+1)$. Multiplying both sides by $\det(I-Y)^{-\beta+\gamma}$ the left side becomes

$$\det(I-Y)^{-\beta-p}{}_0F_0^{(m)}(-X,Y(I-Y)^{-1}),$$

which by Theorem 7.6.3 is equal to

$$\text{(5)} \qquad \sum_{k=0}^\infty\sum_\kappa\frac{L_\kappa^\beta(X)C_\kappa(Y)}{C_\kappa(I)k!}.$$

The coefficient of $C_\kappa(Y)$, the term of degree k in Y, is $L_\kappa^\beta(X)/C_\kappa(I)k!$. The right side becomes

$$\det(I-Y)^{-\beta+\gamma}\sum_{n=0}^\infty\sum_\nu\frac{L_\nu^\gamma(X)C_\nu(Y)}{C_\nu(I)n!},$$

which, using the zonal polynomial expansion for $\det(I-Y)^{-\beta+\gamma}$ (Corollary 7.3.5) is equal to

$$\text{(6)} \qquad \sum_{t=0}^\infty\sum_\tau\sum_{n=0}^\infty\sum_\nu(\beta-\gamma)_\tau\frac{C_\tau(Y)L_\nu^\gamma(X)C_\nu(Y)}{C_\nu(I)t!n!}.$$

The term of degree k in Y here is

$$\sum_{\tau}\sum_{\nu}(\beta-\gamma)_{\tau}\frac{C_{\tau}(Y)L_{\nu}^{\gamma}(X)C_{\nu}(Y)}{C_{\nu}(I)t!n!},$$

where $t+n=k$, and using (4) this is

$$\sum_{\tau}\sum_{\nu}\sum_{\kappa}(\beta-\gamma)_{\tau}\frac{L_{\nu}^{\gamma}(X)}{C_{\nu}(I)t!n!}g_{\tau,\nu}^{\kappa}C_{\kappa}(Y).$$

Equating coefficients of $C_{\kappa}(Y)$ in (5) and (6) hence establishes the identity (3). Now put $\gamma=-\frac{1}{2}$ in (3) and use the estimate (2) for $L_{\nu}^{-1/2}(X)$ to get

$$(7)\qquad |L_{\kappa}^{\beta}(X)|\le k!C_{\kappa}(I)\operatorname{etr}(X)\sum_{t+n=k}\sum_{\tau}\sum_{\nu}(\beta+\tfrac{1}{2})_{\tau}\frac{(\frac{1}{2}m)_{\nu}}{t!n!}g_{\tau,\nu}^{\kappa}.$$

It is easily seen that the sum on the right is the coefficient of $C_{\kappa}(Y)$ in the expansion of

$$\det(I-Y)^{-(\beta+1/2)}\det(I-Y)^{-m/2}=\det(I-Y)^{-(\beta+p)}$$

$$=\sum_{k=0}^{\infty}\sum_{\kappa}(\beta+p)_{\kappa}\frac{C_{\kappa}(Y)}{k!}\qquad\left[\text{where }\ p=\tfrac{1}{2}(m+1)\right].$$

Hence the sum on the right side of (7) is equal to $(\beta+p)_{\kappa}/k!$, and the proof is complete.

We are now in a position to derive an expression for the probability density function of T_0^2, valid over the range $0\le x<1$. The result is given in the next theorem.

THEOREM 10.6.2. If A is $W_m(r,\Sigma,\Omega)$, B is $W_m(n-p,\Sigma)$ and A and B are independent $(n-p\ge m, r\ge m)$ then the density function of $T_0^2=\operatorname{tr}AB^{-1}$ can be expressed as

$$(8)\qquad f_{T_0^2}(x)=\frac{\operatorname{etr}(-\frac{1}{2}\Omega)\Gamma_m[\frac{1}{2}(r+n-p)]}{\Gamma_m[\frac{1}{2}(n-p)]\Gamma(\frac{1}{2}mr)}x^{mr/2-1}$$

$$\cdot\sum_{k=0}^{\infty}\frac{(-x)^k}{(\frac{1}{2}mr)_k k!}\sum_{\kappa}[\tfrac{1}{2}(r+n-p)]_{\kappa}L_{\kappa}^{\gamma}(\tfrac{1}{2}\Omega)$$

$$(0\le x<1),$$

where $\gamma=\frac{1}{2}(r-m-1)$.

Proof. By invariance it can be assumed without loss of generality that $\Sigma = I$ and $\Omega = \mathrm{diag}(\omega_1,\ldots,\omega_m)$. The joint density function of A and B is

$$\frac{2^{-m(r+n-p)/2}}{\Gamma_m(\frac{1}{2}r)\Gamma_m[\frac{1}{2}(n-p)]}\operatorname{etr}(-\tfrac{1}{2}\Omega)\,{}_0F_1(\tfrac{1}{2}r;\tfrac{1}{4}\Omega A)$$
$$\cdot\operatorname{etr}(-\tfrac{1}{2}A)\operatorname{etr}(-\tfrac{1}{2}B)(\det A)^{(r-m-1)/2}(\det B)^{(n-p-m-1)/2}.$$

Hence the Laplace transform of $f_{T_0^2}(x)$, the density function of T_0^2, is

$$\begin{aligned} g(t) &= \int_0^\infty f_{T_0^2}(x)e^{-tx}\,dx \\ &= E\left[\operatorname{etr}(-tAB^{-1})\right] \\ &= \frac{2^{-m(r+n-p)/2}}{\Gamma_m(\frac{1}{2}r)\Gamma_m[\frac{1}{2}(n-p)]}\operatorname{etr}(-\tfrac{1}{2}\Omega) \\ &\quad\cdot\int_{B>0}\operatorname{etr}(-\tfrac{1}{2}B)(\det B)^{(n-p-m-1)/2}\int_{A>0}\operatorname{etr}\left[-\tfrac{1}{2}A(I+2tB^{-1})\right] \\ &\quad\cdot(\det A)^{(r-m-1)/2}\,{}_0F_1(\tfrac{1}{2}r;\tfrac{1}{4}\Omega A)(dA)(dB). \end{aligned}$$

Integrating over $A>0$ using Theorem 7.3.4 then gives

$$\text{(9)}\quad \begin{aligned} g(t) &= \frac{\operatorname{etr}(-\frac{1}{2}\Omega)}{2^{m(n-p)/2}\Gamma_m[\frac{1}{2}(n-p)]}\int_{B>0}\operatorname{etr}(-\tfrac{1}{2}B)(\det B)^{(n-p-m-1)/2} \\ &\quad\cdot\det(I+2tB^{-1})^{-r/2}\operatorname{etr}\left[\tfrac{1}{2}\Omega(I+2tB^{-1})^{-1}\right](dB) \\ &= \frac{\operatorname{etr}(-\frac{1}{2}\Omega)(2t)^{-mr/2}}{2^{m(n-p)/2}\Gamma_m[\frac{1}{2}(n-p)]}\int_{B>0}\operatorname{etr}(-B)(\det B)^{(r+n-p-m-1)/2} \\ &\quad\cdot\det\left(I+\frac{1}{2t}B\right)^{-r/2}\operatorname{etr}\left[\frac{1}{2}\Omega\frac{1}{2t}B\left(I+\frac{1}{2t}B\right)^{-1}\right](dB). \end{aligned}$$

No tractable expression for this last integral is known. However, we will see that we can perform the Laplace inversion first and then integrate over B. First note that if $h(\Omega)$ denotes the integral in (9) then $h(H\Omega H')=h(\Omega)$ for any $H\in O(m)$ [i.e., $h(\Omega)$ is a symmetric function of Ω]; hence replacing Ω by $H\Omega H'$ and integrating over $O(m)$ with respect to (dH), the normalized invariant measure on $O(m)$, we get

$$(10)\quad g(t)=\frac{\operatorname{etr}(-\frac{1}{2}\Omega)(2t)^{-mr/2}}{2^{m(n-p)/2}\Gamma_m[\frac{1}{2}(n-p)]}\int_{B>0}\operatorname{etr}(-B)(\det B)^{(r+n-p-m-1)/2}$$

$$\cdot\det\left(I+\frac{1}{2t}B\right)^{-r/2}{}_0F_0^{(m)}\left(\frac{1}{2}\Omega,\frac{1}{2t}B\left(I+\frac{1}{2t}B\right)^{-1}\right)(dB)$$

$$=\frac{\operatorname{etr}(-\frac{1}{2}\Omega)(2t)^{-mr/2}}{2^{m(n-p)/2}\Gamma_m[\frac{1}{2}(n-p)]}\int_{B>0}\operatorname{etr}(-B)(\det B)^{(r+n-p-m-1)/2}$$

$$\cdot\sum_{k=0}^{\infty}\sum_{\kappa}L_\kappa^{\gamma}(\tfrac{1}{2}\Omega)\frac{C_\kappa\left(-\frac{1}{2t}B\right)(dB)}{C_\kappa(I)k!},$$

where $\gamma=\frac{1}{2}(r-m-1)$ and we have used Theorem 7.6.3. By Lemma 10.6.1, the series in (10) is dominated termwise by the series

$$\operatorname{etr}(\tfrac{1}{2}\Omega)\sum_{k=0}^{\infty}\sum_{\kappa}(\tfrac{1}{2}r)_\kappa\frac{C_\kappa\left(-\frac{1}{2t}B\right)}{k!}=\operatorname{etr}(\tfrac{1}{2}\Omega)\det\left(I+\frac{1}{2t}B\right)^{-r/2}.$$

Hence for B fixed, $\operatorname{Re}(t)=c$ sufficiently large the series in (10) can be integrated term by term with respect to t, since this is true for

$$\det\left(I+\frac{1}{2t}B\right)^{-r/2}.$$

Using the easily proved fact that the Laplace transform of $x^{mr/2+k-1}(\frac{1}{2}mr)_k\Gamma(\frac{1}{2}mr)$ is $t^{-mr/2-k}$, it then follows that inversion of $g(t)$ yields an expression for $f_{T_0^2}(x)$, the density function of T_0^2, as

$$(11)\quad f_{T_0^2}(x)=\frac{\operatorname{etr}(-\frac{1}{2}\Omega)x^{mr/2-1}}{2^{m(r+n-p)/2}\Gamma_m[\frac{1}{2}(n-p)]\Gamma(\frac{1}{2}mr)}\int_{B>0}\operatorname{etr}(-\tfrac{1}{2}B)$$

$$\cdot(\det B)^{(r+n-p-m-1)/2}\sum_{k=0}^{\infty}\frac{(-x)^k}{(\frac{1}{2}mr)_k k!}$$

$$\cdot\sum_{\kappa}L_\kappa^{\gamma}(\tfrac{1}{2}\Omega)\frac{C_\kappa(\frac{1}{2}B)}{C_\kappa(I)}(dB).$$

Again using Lemma 10.6.1 the series in (11) is dominated termwise by the series

$$\operatorname{etr}(\tfrac{1}{2}\Omega)\sum_{k=0}^{\infty}\sum_{\kappa}\frac{(\frac{1}{2}r)_\kappa}{(\frac{1}{2}mr)_k}\frac{C_\kappa(-\frac{1}{2}xB)}{k!}$$

and, since $(\frac{1}{2}r)_\kappa/(\frac{1}{2}mr)_k \le 1$ for all m, this series is dominated termwise by the series

$$\operatorname{etr}(\tfrac{1}{2}\Omega)\sum_{k=0}^{\infty}\sum_{\kappa}\frac{C_\kappa(-\frac{1}{2}xB)}{k!}=\operatorname{etr}(\tfrac{1}{2}\Omega)\operatorname{etr}(-\tfrac{1}{2}xB).$$

Hence the series in (11) may be integrated term by term for $|x|<1$ using Theorem 7.2.7 to give (8) and complete the proof.

An expression for the null density function of T_0^2 is obtained by putting $\Omega=0$.

COROLLARY 10.6.3. When $\Omega=0$ the density function of T_0^2 is

(12)

$$f_{T_0^2}(x)=\frac{\Gamma_m[\frac{1}{2}(r+n-p)]}{\Gamma_m[\frac{1}{2}(n-p)]\Gamma(\frac{1}{2}mr)}x^{mr/2-1}\sum_{k=0}^{\infty}\frac{(-x)^k}{(\frac{1}{2}mr)_k k!}\sum_{\kappa}[\tfrac{1}{2}(r+n-p)]_\kappa \cdot(\tfrac{1}{2}r)_\kappa C_\kappa(I) \qquad (0\le x<1).$$

Proof. Putting $\Omega=0$ in Theorem 10.6.2 and using

$$L_\kappa^\gamma(0)=(\tfrac{1}{2}r)_\kappa C_\kappa(I) \qquad [\gamma=\tfrac{1}{2}(r-m-1)]$$

[see (5) of Section 7.6] completes the proof.

It is worth remarking that the distribution of T_0^2 for $r<m$ is obtained from Theorem 10.6.2 and Corollary 10.6.3 by making the substitutions

$$m\to r, \qquad r\to m, \qquad n-p\to n+r-p-m.$$

In view of the complexity of the exact distribution of T_0^2, approximations appear more useful. In a moment we will look at the asymptotic null distribution (as $n\to\infty$) of T_0^2. Before doing so we need to introduce another function of matrix argument. The reader will recall the ${}_1F_1$ confluent hypergeometric function of matrix argument introduced in Sections 7.3 and 7.4. As in the univariate case there is another type of confluent function Ψ, with an $m\times m$ symmetric matrix X as argument, defined by the integral representation

$$(13)\quad \Psi(a,c;X)=\frac{1}{\Gamma_m(a)}\int_{Y>0}\operatorname{etr}(-XY)(\det Y)^{a-(m+1)/2}\cdot\det(I+Y)^{c-a-(m+1)/2}(dY),$$

valid for $\operatorname{Re}(X)>0$, $\operatorname{Re}(a)>\frac{1}{2}(m-1)$. It will be shown later that this function satisfies the same system of partial differential equations as the function ${}_1F_1(a;c;X)$, namely, the system given in Theorem 7.5.6. First we obtain Ψ as a limiting function from the ${}_2F_1$ Gaussian hypergeometric function.

LEMMA 10.6.4.

$$(14)\quad \lim_{c\to\infty} {}_2F_1(a,b;c;I-cX^{-1})=(\det X)^b\Psi(b,b-a+\tfrac{1}{2}(m+1);X)$$

Proof. From Theorem 7.4.2 we have

$${}_2F_1(a,b;c;I-cX^{-1})=\frac{\Gamma_m(c)}{\Gamma_m(b)\Gamma_m(c-b)}\int_{0<Y<I}\det(I-(I-cX^{-1})Y)^{-a}\cdot(\det Y)^{b-(m+1)/2}\det(I-Y)^{c-b-(m+1)/2}(dY)$$

Putting $S=Y(I-Y)^{-1}$ this becomes

$${}_2F_1(a,b;c;I-cX^{-1})=\frac{\Gamma_m(c)}{\Gamma_m(b)\Gamma_m(c-b)}\int_{S>0}(\det S)^{b-(m+1)/2}\cdot\det(I+S)^{a-c}\det(I+cSX^{-1})^{-a}(dS)$$

Putting $Z=cS$ and then letting $c\to\infty$ gives

$$\begin{aligned}\lim_{c\to\infty} {}_2F_1(a,b;c;I-cX^{-1})&=\frac{1}{\Gamma_m(b)}\int_{Z>0}\operatorname{etr}(-Z)(\det Z)^{b-(m+1)/2}\cdot\det(I+ZX^{-1})^{-a}(dZ)\\&=(\det X)^b\Psi(b,b-a+\tfrac{1}{2}(m+1);X),\end{aligned}$$

where the last line follows by putting $U=ZX^{-1}$ and using (13). This completes the proof.

We now give a system of partial differential equations satisfied by Ψ; note that the system in the following theorem is the same as the system for ${}_1F_1$ given in Theorem 7.5.6.

THEOREM 10.6.5. The function $\Psi(a,c;X)$ is a solution of each of the partial differential equations

$$(15)\quad x_i\frac{\partial^2 F}{\partial x_i^2}+\left\{c-\tfrac{1}{2}(m-1)-x_i+\frac{1}{2}\sum_{j\neq i}\frac{x_i}{x_i-x_j}\right\}\frac{\partial F}{\partial x_i}-\frac{1}{2}\sum_{j\neq i}\frac{x_j}{x_i-x_j}\frac{\partial F}{\partial x_j}=aF\qquad (i=1,\dots,m),$$

where $x_1,\dots,x_m$ are the latent roots of X.

Proof. Using Theorem 7.5.5 it is readily found that $(\det X)^{-b}\,{}_2F_1(a,b;c;I-cX^{-1})$ satisfies the system

$$(16)\quad \frac{x_i^2}{c}\left(\frac{c}{x_i}-1\right)\frac{\partial^2 F}{\partial x_i^2}+\left\{b-a+1-x_i+\frac{x_i}{c}(a-b-1)-\frac{1}{2c}\cdot\sum_{j\neq i}\frac{x_ix_j\left(1-\dfrac{c}{x_i}\right)}{x_i-x_j}\right\}\frac{\partial F}{\partial x_i}+\frac{1}{2c}\sum_{j\neq i}\frac{x_j^2\left(1-\dfrac{c}{x_j}\right)}{x_i-x_j}\frac{\partial F}{\partial x_j}=\left(b-\frac{ab}{c}\right)F$$

$$(i=1,\dots,m).$$

Letting $c\to\infty$ the system (16) tends to the system

(17)

$$x_i\frac{\partial^2 F}{\partial x_i^2}+\left\{b-a+1-x_i+\frac{1}{2}\sum_{j\neq i}\frac{x_i}{x_i-x_j}\right\}\frac{\partial F}{\partial x_i}-\frac{1}{2}\sum_{j\neq i}\frac{x_j}{x_i-x_j}\frac{\partial F}{\partial x_j}=bF,$$

which by Lemma 10.6.4 must be satisfied by $\Psi(b,b-a+\frac{1}{2}(m+1);X)$. Noting that this is exactly the system satisfied by ${}_1F_1(b;b-a+\frac{1}{2}(m+1);X)$ (see Theorem 7.5.6) completes the proof.

We now return to the T_0^2 statistic and show that the Laplace transform $G_1(t)$ of the null density function of $(n-p)T_0^2$ can be expressed in terms of the function Ψ. For convenience put $n_0=n-p$. Using Corollary 10.4.3 we have

$$(18)\quad G_1(t)=E\left[\operatorname{etr}(-n_0tAB^{-1})\right]$$

$$=\frac{\pi^{m^2/2}\Gamma_m[\frac{1}{2}(r+n_0)]}{\Gamma_m(\frac{1}{2}m)\Gamma_m(\frac{1}{2}r)\Gamma_m(\frac{1}{2}n_0)}\int_{f_1>\cdots>f_m>0}\cdots\int\exp\left(-n_0t\sum_{i=1}^m f_i\right)$$

$$\cdot\prod_{i=1}^m\left[f_i^{(r-m-1)/2}(1+f_i)^{-(r+n_0)/2}\right]\prod_{i<j}^m(f_i-f_j)\,df_1\ldots df_m$$

$$=\frac{\Gamma_m[\frac{1}{2}(r+n_0)]}{\Gamma_m(\frac{1}{2}r)\Gamma_m(\frac{1}{2}n_0)}\int_{F>0}\operatorname{etr}(-n_0tF)(\det F)^{(r-m-1)/2}$$

$$\cdot\det(I+F)^{-(r+n_0)/2}(dF)$$

$$=\frac{\Gamma_m[\frac{1}{2}(r+n_0)]}{\Gamma_m(\frac{1}{2}n_0)}\Psi\left(\tfrac{1}{2}r,\tfrac{1}{2}(m+1-n_0);n_0tI\right),$$

where the last line follows from (13). Note that $G_1(0)=1$ because $G_1(t)=E[\operatorname{etr}(-n_0tAB^{-1})]$. Now let us find the limit of $G_1(t)$ as $n_0\to\infty$. Putting $T=\frac{1}{2}n_0F$ in the last integral of (18) gives

$$G_1(t)=\frac{\Gamma_m[\frac{1}{2}(r+n_0)]}{\Gamma_m(\frac{1}{2}r)\Gamma_m(\frac{1}{2}n_0)(\frac{1}{2}n_0)^{mr/2}}\int_{T>0}\operatorname{etr}(-2tT)(\det T)^{(r-m-1)/2}$$

$$\cdot\det\left(I+\frac{2}{n_0}T\right)^{-(r+n_0)/2}(dT).$$

Letting $n_0=n-p\to\infty$ then gives

$$(19)\quad \lim_{n_0\to\infty}G_1(t)=\frac{1}{\Gamma_m(\frac{1}{2}r)}\int_{T>0}\operatorname{etr}[-(1+2t)T](\det T)^{(r-m-1)/2}(dT)$$

$$=(1+2t)^{-rm/2}.$$

Since $(1+2t)^{-rm/2}$ is the Laplace transform of the χ^2_{rm} density function it follows that the asymptotic null distribution of $n_0T_0^2$ as $n_0\to\infty$ is χ^2_{rm}, the same as the asymptotic distribution of the likelihood ratio statistic (see Theorem 10.5.5). An asymptotic expansion for the distribution of $n_0 T_0^2$ can be obtained from an asymptotic expansion for $G_1(t)$. Because it will be useful in another context (see Section 10.6.3), we will generalize the function $G_1(t)$ a little. The term of order n_0^{-2} will be found in an asymptotic expansion of the function

$$G(R)=\frac{\Gamma_m\left(\alpha+\frac{\varepsilon}{2}n_0\right)}{\Gamma_m\left(\frac{\varepsilon}{2}n_0\right)}\Psi\left(\alpha,\beta-\frac{\varepsilon}{2}n_0;\frac{1}{2}n_0R\right), \tag{20}$$

where α, β are arbitrary parameters independent of n_0 and ε is either $+1$ or -1. Note that $G_1(t)$ is the special case of $G(R)$ obtained by putting $\alpha=\frac{1}{2}r$, $\beta=\frac{1}{2}(m+1)$, $\varepsilon=1$, and $R=2tI$. The following theorem gives an expansion, to terms of order n_0^{-2}, of $\log G(R)$.

THEOREM 10.6.6. The function $\log G(R)$, where $G(R)$ is given by (20), has the expansion

$$\log G(R)=\alpha\log\det(I-Y)+\frac{Q_1}{n_0}+\frac{Q_2}{n_0^2}+O(n_0^{-3}) \tag{21}$$

where $Y=I-(I+\varepsilon R)^{-1}$,

$$Q_1=\tfrac{1}{2}\varepsilon\alpha\left[\sigma_1^2+(2\alpha+1)\sigma_2-4\beta\sigma_1\right], \tag{22}$$

and

(23)

$$\begin{aligned}Q_2=\frac{1}{12}\alpha\big[&6\sigma_1^2\sigma_2+3(2\alpha+1)\sigma_2^2+12(2\alpha+1)\sigma_1\sigma_3+3(8\alpha^2+10\alpha+5)\sigma_4\\&-8\sigma_1^3-24(2\alpha+2\beta+1)\sigma_1\sigma_2-8(4\alpha^2+6\alpha\beta+6\alpha+3\beta+4)\sigma_3\\&+6(2\alpha+6\beta+1)\sigma_1^2+6(12\alpha\beta+4\beta^2+2\alpha+6\beta+3)\sigma_2\\&-48\beta^2\sigma_1\big],\end{aligned}$$

with $\sigma_j=\operatorname{tr}Y^j$.

Proof. The proof outlined here uses the partial differential equations satisfied by the Ψ function. From Theorem 10.6.5 it follows that $G(R)$

satisfies each partial differential equation in the system

$$(24)\quad r_i\frac{\partial^2 G}{\partial r_i^2}+\left[\beta-\frac{1}{2}\varepsilon n_0-\frac{1}{2}(m-1)-\frac{1}{2}n_0 r_i+\frac{1}{2}\sum_{j\neq i}\frac{r_i}{r_i-r_j}\right]\frac{\partial G}{\partial r_i}$$

$$-\frac{1}{2}\sum_{j\neq i}\frac{r_j}{r_i-r_j}\frac{\partial G}{\partial r_j}=\frac{1}{2}\alpha n_0 G\qquad (i=1,\dots,m)$$

where $r_1,\dots,r_m$ are the latent roots of R. An argument similar to that used in deriving (20) shows that

$$\lim_{n_0\to\infty} G(R)=\det(I+\varepsilon R)^{-\alpha}$$

$$=\det(I-Y)^{\alpha},$$

where $Y=I-(I+\varepsilon R)^{-1}$. Changing variables from R to Y and writing

$$G(R)=\det(I-Y)^{\alpha}\exp H(Y)$$

it is found that $H(Y)$ satisfies the system

$$(25)\qquad y_i(1-y_i)^2\left[\frac{\partial^2 H}{\partial y_i^2}+\left(\frac{\partial H}{\partial y_i}\right)^2\right]$$

$$+\Big\{\beta-\tfrac{1}{2}(m-1)-\tfrac{1}{2}\varepsilon n_0-y_i[\beta-\tfrac{1}{2}(m-5)+2\alpha]$$

$$+2y_i^2(\alpha+1)+\tfrac{1}{2}\sum_{j\neq i}\frac{y_i(1-y_i)(1-y_j)}{y_i-y_j}\Big\}\frac{\partial H}{\partial y_i}$$

$$-\tfrac{1}{2}\sum_{j\neq i}\frac{y_j(1-y_j)^2}{y_i-y_j}\frac{\partial H}{\partial y_j}=\alpha\beta-\alpha y_i(\alpha+1)-\frac{1}{2}\alpha\sum_{j\neq i}y_j$$

$$(i=1,\dots,m),$$

where $y_1,\dots,y_m$ are the latent roots of Y. In (25) we now substitute

$$H(Y)=\sum_{k=1}^{\infty}\frac{Q_k(Y)}{n_0^k}=\frac{Q_1}{n_0}+\frac{Q_2}{n_0^2}+\cdots,$$

where (a) $Q_k(Y)$ is symmetric in $y_1,\dots,y_m$ and (b) $Q_k(0)=0$ for $k=1,2,\dots$. Equating constant terms on both sides gives

$$\frac{\partial Q_1}{\partial y_i}=\varepsilon\alpha\left[\sum_{j\neq i} y_j+2y_i(\alpha+1)-2\beta\right]$$

whose solution, subject to (a) and (b), is the function Q_1 given by (22). Equating the coefficients of n_0^{-1} on both sides gives

$$(26)\quad \frac{\partial Q_2}{\partial y_i}=2\varepsilon\left\{y_i(1-y_i)^2\frac{\partial^2 Q_1}{\partial y_i^2}+\left[\beta-\tfrac{1}{2}(m-1)-y_i\left(\beta-\tfrac{1}{2}(m-5)+2\alpha\right)\right.\right.$$

$$\left.+2y_i^2(\alpha+1)+\frac{1}{2}\sum_{j\neq i}\frac{y_i(1-y_i)(1-y_j)}{y_i-y_j}\right]\frac{\partial Q_1}{\partial y_i}$$

$$\left.-\frac{1}{2}\sum_{j\neq i}\frac{y_j(1-y_j)^2}{y_i-y_j}\frac{\partial Q_1}{\partial y_j}\right\}.$$

Using Q_1 and its derivatives in (25) and then integrating gives the function Q_2 given by (23) and the proof is complete.

An asymptotic expansion of the null distribution of $n_0T_0^2$ now follows readily from Theorem 10.6.6.

THEOREM 10.6.7. The null distribution function of $n_0T_0^2$ can be expanded as

$$(27)\quad P\left(n_0T_0^2\le x\right)=P\left(\chi^2_{rm}\le x\right)+\frac{rm}{4n_0}\sum_{j=0}^{2}a_jP\left(\chi^2_{rm+2j}\le x\right)$$

$$+\frac{rm}{96n_0^2}\sum_{j=0}^{4}b_jP\left(\chi^2_{rm+2j}\le x\right)+O\left(n_0^{-3}\right),$$

where $n_0=n-p$; then

$a_0=r-m-1,$

$a_1=-2r,$

$a_2=m+r+1,$

$b_0=3m^3r-2m^2(3r^2-3r+4)+3m(r^3-2r^2+5r-4)-4(2r^2-3r-1),$

(28)

$$b_1 = 12mr^2(m-r+1),$$

$$b_2 = -6\left[m^3r + 2m^2r - 3mr(r^2+1) - 4r(2r+1)\right]$$

$$b_3 = -4\left[m^2(3r^2+4) + 3m(r^3+r^2+8r+4) + 8(2r^2+3r+2)\right],$$

$$b_4 = 3\left[m^3r + 2m^2(r^2+r+4) + m(r^3+2r^2+21r+20)\right.$$
$$\left. + 4(2r^2+5r+5)\right].$$

Proof. Putting $\alpha = \frac{1}{2}r$, $\beta = \frac{1}{2}(m+1)$, $\varepsilon = 1$, and $R = 2tI$ in Theorem 10.6.6 yields an expansion for $\log G_1(t)$, where $G_1(t)$ is the Laplace transform of the density function of $n_0 T_0^2$ given by (18). Note that with $R = 2tI$ we have $Y = (2t/1+2t)I$ so that

$$\sigma_j = \operatorname{tr} Y^j = \left(\frac{2t}{1+2t}\right)^j m = \left(1 - \frac{1}{1+2t}\right)^j m.$$

Exponentiation of the resulting expansion gives

$$G_1(t) = (1+2t)^{-mr/2}\left[1 + \frac{rm}{4n_0}\sum_{j=0}^{2} a_j(1+2t)^{-j}\right.$$
$$\left. + \frac{rm}{96n_0^2}\sum_{j=0}^{4} b_j(1+2t)^{-j} + O(n_0^{-3})\right],$$

where the a_j and b_j are given by (28). The desired result now follows by inverting this expansion.

The term of order n_0^{-3} in the expansion (27) is also implicit in Muirhead (1970b); see also Ito (1956), Davis (1968), and Fujikoshi (1970). Asymptotic expansions for the distribution function of $n_0 T_0^2$ in the non-null case in terms of noncentral χ^2 distributions have been obtained by Siotani (1957), (1971), Ito (1960), Fujikoshi (1970), and Muirhead (1972b). Percentage points of the null distribution of T_0^2 may be found in Pillai and Sampson (1959), Davis (1970, 1980), and Hughes and Saw (1972). For further references concerning distributional results for T_0^2 see the survey papers by Pillai (1976, 1977).

10.6.3. The V Statistic

The statistic

$$V=\operatorname{tr}A(A+B)^{-1}=\sum_{i=1}^{m}\frac{f_i}{1+f_i}$$

was suggested by Pillai (1955, 1956). Its exact non-null distribution over the range $0<V<1$ has been found by Khatri and Pillai (1968) as a complicated zonal polynomial series. Here we will concentrate on the asymptotic null distribution as $n\to\infty$.

We start by finding the Laplace transform $G_2(t)$ of the null density function of n_0V, where $n_0=n-p$. Putting $u_i=f_i/(1+f_i)$ (with $i=1,\ldots,m$), we have $V=\Sigma_{i=1}^m u_i$ and $u_1,\ldots,u_m$ as the latent roots of a matrix U having the $\text{Beta}_m(\frac{1}{2}r,\frac{1}{2}n_0)$ distribution (see the discussion following Corollary 10.4.3). Hence

$$\begin{aligned}(29)\qquad G_2(t)&=E[\operatorname{etr}(-n_0tU)]\\&=\frac{\Gamma_m[\frac{1}{2}(r+n_0)]}{\Gamma_m(\frac{1}{2}r)\Gamma_m(\frac{1}{2}n_0)}\int_{0<U<I}\operatorname{etr}(-n_0tU)\det U^{(r-m-1)/2}\\&\quad\cdot\det(I-U)^{(n_0-m-1)/2}(dU)\\&={}_1F_1\left(\tfrac{1}{2}r;\tfrac{1}{2}(r+n_0);-n_0tI\right),\end{aligned}$$

where Theorem 7.4.2 has been used to evaluate this integral. This result has been given by James (1964). Note that as $n_0\to\infty$

$$G_2(t)\to{}_1F_0\left(\tfrac{1}{2}r;-2tI\right)=(1+2t)^{-mr/2},$$

and hence the limiting null distribution of n_0V is χ^2_{rm}. The following theorem gives an asymptotic expansion for the distribution of n_0V.

THEOREM 10.6.7. The null distribution function of n_0V can be expanded as

$$\begin{aligned}(30)\qquad P(n_0V\le x)&=P(\chi^2_{rm}\le x)+\frac{rm}{4n_0}\sum_{j=0}^{2}c_jP(\chi^2_{rm+2j}\le x)\\&\quad+\frac{rm}{96n_0^2}\sum_{j=0}^{4}d_jP(\chi^2_{rm+2j}\le x)+O(n_0^{-3}),\end{aligned}$$

where $n_0 = n - p$; then

$$
\begin{aligned}
c_0 &= r - m - 1,\\
c_1 &= 2(m+1),\\
c_2 &= -(r+m+1),\\
d_0 &= 3m^3r - 2m^2(3r^2 - 3r + 4) + 3m(r^3 - 2r^2 + 5r - 4) - 4(2r^2 - 3r - 1),\\
d_1 &= -12mr\left[m^2 - m(r-2) - (r-1)\right],\\
d_2 &= 6\left[3m^3r + 2m^2(3r+4) - m(r^3 - 7r - 16) + 4(r+2)\right],\\
d_3 &= -4\left[3m^3r + m^2(3r^2 + 6r + 16) + 3m(r^2 + 9r + 12) + 4(r^2 + 6r + 7)\right],\\
d_4 &= 3\left[m^3r + 2m^2(r^2 + r + 4) + m(r^3 + 2r^2 + 21r + 20) + 4(2r^2 + 5r + 5)\right].
\end{aligned}
\tag{31}
$$

Proof. From Theorem 7.5.6 it follows that the function ${}_1F_1(\alpha; \beta - \frac{1}{2}\varepsilon n_0; \frac{1}{2}n_0 R)$ satisfies the same system of partial differential equations (24) as the function $G(R)$ given by (20). Hence an expansion for $\log G_2(t)$ follows from Theorem 10.6.6 by putting $\alpha = \frac{1}{2}r$, $\beta = \frac{1}{2}r$, $\varepsilon = -1$ and $R = -2tI$. Exponentiation of this expansion yields

$$
G_2(t) = (1+2t)^{-mr/2}\left[1 + \frac{rm}{4n_0}\sum_{j=0}^{2} c_j(1+2t)^{-j} + \frac{rm}{96n_0^2}\sum_{j=0}^{4} d_j(1+2t)^{-j} + O(n_0^{-3})\right], \tag{32}
$$

where the a_j and b_j are given by (31). The expansion (30) now follows by inversion of (32).

The term of order n_0^{-3} in the expansion is also implicit in Muirhead (1970b); see also Fujikoshi (1970) and Lee (1971b). Asymptotic expansions for the distribution function of n_0V in the non-null case have been obtained by Fujikoshi (1970) and Lee (1971b). For further references concerning distributional results for V the interested reader is referred to the survey papers of Pillai (1976, 1977).

10.6.4. The Largest Root

Roy (1953) proposed a test of the general linear hypothesis based on f_1, the largest latent root of AB^{-1}. The following theorem due to Khatri (1972) gives the exact distribution of f_1 in a special case as a finite series of Laguerre polynomials.

THEOREM 10.6.8. If A is $W_m(r,\Sigma,\Omega)$, B is $W_m(n-p,\Sigma)$ $(r\geq m, n-p\geq m)$ and A and B are independent and if $t=\frac{1}{2}(n-p-m-1)$ is an integer then the distribution function of f_1, the largest latent root of AB^{-1}, may be expressed as

(33)

$$P(f_1\leq x)=\left(\frac{x}{1+x}\right)^{mr/2}\operatorname{etr}\left[-\frac{1}{2(1+x)}\Omega\right]\sum_{k=0}^{mt}\sum_{\kappa}^{*}\frac{L_\kappa^\gamma\left(-\frac{1}{2}\frac{x}{1+x}\Omega\right)}{(1+x)^k k!},$$

where $\gamma=\frac{1}{2}(r-m-1)$ and Σ_κ^* denotes summation over those partitions $\kappa=(k_1,\dots,k_m)$ of k with largest part $k_1\leq t$.

Proof. Without loss of generality it can be assumed that $\Sigma=I$ and Ω is diagonal. Noting that the region $f_1\leq x$ is equivalent to the region $B>(1/x)A$, with $A>0$, we have, by integration of the joint density function of A and B,

(34)

$$P(f_1\leq x)=\frac{2^{-m(r+n-p)/2}\operatorname{etr}(-\frac{1}{2}\Omega)}{\Gamma_m(\frac{1}{2}r)\Gamma_m[\frac{1}{2}(n-p)]}\int_{A>0}\operatorname{etr}(-\tfrac{1}{2}A)(\det A)^{(r-m-1)/2}$$

$$\cdot{}_0F_1(\tfrac{1}{2}r;\tfrac{1}{4}\Omega A)\int_{B>\frac{1}{x}A}\operatorname{etr}(-\tfrac{1}{2}B)(\det B)^{(n-p-m-1)/2}(dB)(dA).$$

Let $\mathcal{J}$ denote the inner integral in (34); putting $T=B-(1/x)A$ in this yields

(35) $$\mathcal{J}=\int_{T>0}\operatorname{etr}(-\tfrac{1}{2}T)\det\left(T+\frac{1}{x}A\right)^t(dT)\operatorname{etr}\left(-\frac{1}{2x}A\right)$$

where $t=\frac{1}{2}(n-p-m-1)$. Now

$$\det\left(T+\frac{1}{x}A\right)^t$$

can be expanded in terms of zonal polynomials and the series terminates because t is a positive integer. We have

$$\begin{aligned}\det\left(T+\frac{1}{x}A\right)^t&=(\det T)^t\det\left(I+\frac{1}{x}AT^{-1}\right)^t\\&=(\det T)^t{}_1F_0\left(-t;-\frac{1}{x}AT^{-1}\right)\\&=(\det T)^t\sum_{k=0}^{mt}\sum_{\kappa}{}^*\frac{(-t)_\kappa C_\kappa\left(-\frac{1}{x}AT^{-1}\right)}{k!}\end{aligned}$$

because $(-t)_\kappa\equiv 0$ if any part of κ is greater than t. Using this in (35) gives

$$\begin{aligned}\mathcal{I}&=\operatorname{etr}\left(-\frac{1}{2x}A\right)\sum_{k=0}^{mt}\sum_{k=0}{}^*\frac{(-t)_\kappa\left(-\frac{1}{x}\right)^k}{k!}\\&\quad\cdot\int_{T>0}\operatorname{etr}(-\tfrac{1}{2}T)(\det T)^tC_\kappa(AT^{-1})(dT).\end{aligned}$$

Putting $X^{-1}=A^{1/2}T^{-1}A^{1/2}$ with $(dT)=(\det A)^{(m+1)/2}(dX)$ gives

$$\begin{aligned}\mathcal{I}&=\operatorname{etr}\left(-\frac{1}{2x}A\right)(\det A)^{t+(m+1)/2}\sum_{k=0}^{mt}\sum_{\kappa}{}^*\frac{(-t)_\kappa\left(-\frac{1}{x}\right)^k}{k!}\\&\quad\cdot\int_{X>0}\operatorname{etr}(-\tfrac{1}{2}AX)(\det X)^tC_\kappa(X^{-1})(dX)\end{aligned}$$

For each partition $\kappa=(k_1,\ldots,k_m)$ in this sum we have $k_1\le t$; using Theorem 7.2.13 to evaluate this last integral gives

$$\mathcal{I}=\operatorname{etr}\left(-\frac{1}{2x}A\right)2^{(n-p)m/2}\Gamma_m[\tfrac{1}{2}(n-p)]\sum_{k=0}^{mt}\sum_{\kappa}{}^*\frac{C_\kappa(A)}{k!(2x)^k}.\tag{36}$$

Using this in (34) we get

$$\begin{aligned}P(f_1\le x)&=\frac{2^{-rm/2}}{\Gamma_m(\frac{1}{2}r)}\operatorname{etr}(-\tfrac{1}{2}\Omega)\sum_{k=0}^{mt}\sum_{\kappa}{}^*\frac{1}{k!(2x)^k}\int_{A>0}\operatorname{etr}\left(-\frac{1}{2}\frac{x+1}{x}A\right)\\&\quad\cdot(\det A)^{(r-m-1)/2}C_\kappa(A){}_0F_1(\tfrac{1}{2}r;\tfrac{1}{4}\Omega A)(dA).\end{aligned}$$

Putting

$$U=\frac{1}{2}\frac{x+1}{x}A,$$

this then becomes

$$P(f_1\le x)=\left(\frac{x}{x+1}\right)^{mr/2}\frac{\operatorname{etr}(-\frac{1}{2}\Omega)}{\Gamma_m(\frac{1}{2}r)}\sum_{k=0}^{mt}\sum_{\kappa}{}^{*}\frac{1}{k!(x+1)^k}$$

$$\cdot\int_{U>0}\operatorname{etr}(-U)(\det U)^{(r-m-1)/2}C_\kappa(U)\,{}_0F_1\left(\frac{1}{2}r;\frac{1}{2}\frac{x}{x+1}\Omega U\right)(dU),$$

and the desired result now follows on using Theorem 7.6.4 to evaluate this last integral.

An expression for the null distribution function of f_1 follows by putting $\Omega=0$ and using

$$L_\kappa^\gamma(0)=(\tfrac{1}{2}r)_\kappa C_\kappa(I).$$

This gives the following corollary.

COROLLARY 10.6.9. When $\Omega=0$ and $t=\frac{1}{2}(n-p-m-1)$ is a positive integer, the distribution function of f_1 is

(37) $$P(f_1\le x)=\left(\frac{x}{1+x}\right)^{mr/2}\sum_{k=0}^{mt}\sum_{\kappa}{}^{*}\frac{(\frac{1}{2}r)_\kappa C_\kappa(I)}{(1+x)^k k!}.$$

A quick approximation to the distribution function of f_1 is the upper bound in the following theorem.

THEOREM 10.6.10.

(38) $$P(f_1\le x)\le\prod_{i=1}^{m}P\left(F_{r,n-p}(\omega_i)\le\frac{n-p}{r}x\right)$$

where $\omega_1,\dots,\omega_m$ are the latent roots of Ω.

Proof. By invariance it can be assumed that $\Sigma=I$ and $\Omega=\operatorname{diag}(\omega_1,\dots,\omega_m)$. Putting $A=(a_{ij})$ and $B=(b_{ij})$, it then follows that the a_{ii} and the b_{ii} are all independent, with b_{ii} having the χ^2_{n-p} distribution and a_{ii} having the $\chi_r^2(\omega_i)$ distribution (from Corollary 10.3.5). Hence the $(n-p)a_{ii}/rb_{ii}$ have independent $F_{r,n-p}(\omega_i)$ distributions $(i=1,\dots,m)$. We now use the fact that for all $\alpha \in R^m$,

$$f_1 \geq \frac{\alpha' A \alpha}{\alpha' B \alpha} \geq f_m$$

(see Problem 10.15). Taking α to be the vectors $(1,0,\ldots,0)'$, $(0,1,0,\ldots,0)'$, and so on, shows that

$$f_1 \geq \max\left(\frac{a_{11}}{b_{11}},\ldots,\frac{a_{mm}}{b_{mm}}\right).$$

Hence

$$\begin{aligned} P(f_1 \leq x) &\leq P\left(\max\left(\frac{a_{11}}{b_{11}},\ldots,\frac{a_{mm}}{b_{mm}}\right) \leq x\right) \\ &= \prod_{i=1}^{m} P\left(\frac{a_{ii}}{b_{ii}} \leq x\right) \\ &= \prod_{i=1}^{m} P\left(F_{r,n-p}(\omega_i) \leq \frac{n-p}{r}x\right). \end{aligned}$$

This upper bound is exact when $m=1$; some calculations by Muirhead and Chikuse (1975b) when $m=2$ in the linear case when $\omega_2=0$ indicate that as a quick approximation to the exact probability the bound (38) appears quite reasonable.

Upper percentage points of f_1 (in the null case) have been given by Heck (1960), Pillai and Bantegui (1959), and Pillai (1964, 1965, 1967). For further papers concerned with f_1 the interested reader is referred to the surveys of Pillai (1976, 1977).

10.6.5. Power Comparisons

Power comparisons of the four tests we have considered, namely, tests based on W, T_0^2, V, and f_1, have been carried out by a number of authors (see Mikhail, 1965; Schatzoff, 1966b; Pillai and Jayachandran, 1967; Fujikoshi, 1970; and Lee, 1971b). The consensus is that the differences between W, T_0^2, and V are very slight; if the ω_i's are very unequal then T_0^2 appears to be more powerful than W, and W more powerful than V. The reverse is true if the ω_i's are close. This conclusion was reached by Pillai and Jayachandran (1967) when $m=2$ and by Fujikoshi (1970) for $m=3$ and Lee (1971b) for $m=3,4$. Lee (1971b) further notes that in the region $\operatorname{tr}\Omega=$constant, the power of V varies the most while that of T_0^2 is most nearly constant; the power of W is intermediate between the two. Pillai and Jayachandran (1967) have noted that in general the largest root f_1 has lower power than the other tests when there is more than one nonzero noncentrality parameter ω_i.

The tests based on W, T_0^2, V, and f_1 are all unbiased. For details the interested reader is referred to Das Gupta et al. (1964) and to Perlman and Olkin (1980). Perlman and Olkin have shown that if $u_1,\ldots,u_m$ denote the latent roots of $A(A+B)^{-1}$ then any test whose acceptance region has the form $\{g(u_1,\ldots,u_m)\le c\}$, where g is nondecreasing in each argument, is unbiased.

10.7. THE SINGLE CLASSIFICATION MODEL

10.7.1. Introduction

The multivariate single classification or one-way analysis of variance model is concerned with testing the equality of the mean vectors of p m-variate normal distributions with common covariance matrix Σ, given independent samples from these distributions. Here we examine this model in order to illustrate some of the foregoing theory and because it leads naturally into the area of multiple discriminant analysis.

Let $\mathbf{y}_{i1},\ldots,\mathbf{y}_{iq_i}$ be independent $N_m(\boldsymbol{\mu}_i,\Sigma)$ random vectors ($i=1,\ldots,p$). It was noted at the beginning of Section 10.2 that this model can be written in the form $Y=X\mathbb{B}+E$ with

$$Y'=\left[\mathbf{y}_{11}\cdots\mathbf{y}_{1q_1}\mathbf{y}_{21}\cdots\mathbf{y}_{2q_2}\cdots\mathbf{y}_{p1}\cdots\mathbf{y}_{pq_p}\right],$$

$$\mathbb{B}'=\left[\boldsymbol{\mu}_1\cdots\boldsymbol{\mu}_p\right],$$

$$X=\begin{bmatrix} 1 & 0 & \cdots & 0\\ \vdots & \vdots & & \vdots\\ 1 & 0 & & 0\\ 0 & 1 & & 0\\ \vdots & \vdots & & \vdots\\ 0 & 1 & & 0\\ \vdots & \vdots & & \vdots\\ 0 & 0 & & 1\\ \vdots & \vdots & & \vdots\\ 0 & 0 & & 1 \end{bmatrix}.$$

Here Y is $n\times m$, with $n=\Sigma_{i=1}^{p}q_i$, X is $n\times p$, $\mathbf{B}$ is $p\times m$, and E is $N(0, I_n\otimes\Sigma)$. The null hypothesis

$$H:\boldsymbol{\mu}_1=\cdots=\boldsymbol{\mu}_p$$

is equivalent to $H: C\mathbf{B}=0$, with the $(p-1)\times p$ matrix C being

$$C=\begin{bmatrix}1 & 0 & \dots & 0 & -1\\ 0 & 1 & & 0 & -1\\ \vdots & \vdots & & \vdots & \vdots\\ 0 & 0 & & 1 & -1\end{bmatrix}.$$

It is useful for the reader to follow the steps given in Section 10.2 involved in reducing this model to canonical form. Here we will give the final result, leaving the details as an exercise.

Let

$$\bar{\mathbf{y}}_i=\frac{1}{q_i}\sum_{j=1}^{q_i}\mathbf{y}_{ij}\quad\text{and}\quad\bar{\mathbf{y}}=\frac{1}{n}\sum_{i=1}^{p}q_i\bar{\mathbf{y}}_i,$$

so that $\bar{\mathbf{y}}_i$ is the sample mean of the q_i observations in the ith sample $(i=1,\dots,p)$ and $\bar{\mathbf{y}}$ is the sample mean of all observations. The matrices due to the hypothesis and error (usually called the between-classes and within-classes matrices) are, respectively,

(1) $$A=\sum_{i=1}^{p}q_i(\bar{\mathbf{y}}_i-\bar{\mathbf{y}})(\bar{\mathbf{y}}_i-\bar{\mathbf{y}})'$$

and

(2) $$B=\sum_{i=1}^{p}\sum_{j=1}^{q_i}(\mathbf{y}_{ij}-\bar{\mathbf{y}}_i)(\mathbf{y}_{ij}-\bar{\mathbf{y}}_i)'.$$

These matrices are, of course, just matrix generalizations of the usual between-classes and within-classes sums of squares that occur in the analysis of variance table for the univariate single classification model. The matrices A and B are independently distributed: B is $W_m(n-p,\Sigma)$ and A is $W_m(p-1,\Sigma,\Omega)$, where the noncentrality matrix is

(3) $$\Omega=\Sigma^{-1}\sum_{i=1}^{p}q_i(\boldsymbol{\mu}_i-\bar{\boldsymbol{\mu}})(\boldsymbol{\mu}_i-\bar{\boldsymbol{\mu}})'\quad\text{with}\quad\bar{\boldsymbol{\mu}}=\frac{1}{n}\sum_{i=1}^{p}q_i\boldsymbol{\mu}_i.$$

The null hypothesis $H: \mu_1 = \cdots = \mu_p$ is equivalent to $H: \Omega = 0$. See the accompanying MANOVA table.

Variation	d.f.	S.S. & S.P.	Distribution	Expectation
Between classes	$p-1$	A	$W_m(p-1, \Sigma, \Omega)$	$(p-1)\Sigma + \Sigma\Omega$
Within classes	$n-p$	B	$W_m(n-p, \Sigma)$	$(n-p)\Sigma$
Total (corrected)	$n-1$	$A+B$		

We have noted in Section 10.2 that invariant test statistics for testing H are functions of the nonzero latent roots of AB^{-1}. The likelihood ratio test rejects H for small values of

$$W = \frac{\det B}{\det(A+B)} = \prod_{i=1}^{s} (1+f_i)^{-1}, \tag{4}$$

where $s = \min(p-1, m)$ and $f_1 \geq \cdots \geq f_s > 0$ are the nonzero latent roots of AB^{-1}. The distributions of these roots and of W have been derived in Sections 10.4 and 10.5. It is also worth noting that the diagonal elements of the matrices in the MANOVA table are the appropriate entries in the univariate analysis of variance, i.e., if $\mu_i = (\mu_{i1}, \ldots, \mu_{im})'$ (with $i = 1, \ldots, p$) then the analysis of variance table for testing.

$$H_j^*: \mu_{1j} = \mu_{2j} = \cdots = \mu_{pj}$$

is as shown in the tabulation, with $A = (a_{ij})$, $B = (b_{ij})$, $\Sigma = (\sigma_{ij})$. Here a_{jj} and b_{jj} are independent, b_{jj}/σ_{jj} is χ^2_{n-p}, and, if H_j^* is true, a_{jj}/σ_{jj} is χ^2_{p-1} so that

$$F = \frac{n-p}{p-1} \frac{a_{jj}}{b_{jj}},$$

the usual ratio of mean squares, has the $F_{p-1, n-p}$ distribution when H_j^* is true. If the null hypothesis H that the mean vectors are all equal is rejected looking at the univariate tables for $j = 1, \ldots, m$ can often provide useful information as to why H has been rejected. It should, however, be remembered that these m F-tests are not independent.

Variation	d.f.	S.S.
Between classes	$p-1$	a_{jj}
Within classes	$n-p$	b_{jj}
Total (corrected)	$n-1$	$a_{jj} + b_{jj}$

10.7.2. Multiple Discriminant Analysis

Suppose now that the null hypothesis

$$H: \boldsymbol{\mu}_1 = \cdots = \boldsymbol{\mu}_p$$

is rejected and we conclude that there are differences between the mean vectors. An interesting question to ask is: Is there a linear combination $\mathbf{l}'\mathbf{y}$ of the variables which "best" discriminates between the p groups? To answer this, suppose that a univariate single classification analysis is performed on an arbitrary linear combination $\mathbf{l}'\mathbf{y}$ of the original variables. The data are given in the accompanying table. All the random variables in this table are

Group 1	Group 2 ···	Group p
$\mathbf{l}'\mathbf{y}_{11}$	$\mathbf{l}'\mathbf{y}_{21}$	$\mathbf{l}'\mathbf{y}_{p1}$
$\vdots$	$\vdots$	$\vdots$
$\mathbf{l}'\mathbf{y}_{1q_1}$	$\mathbf{l}'\mathbf{y}_{2q_2}$	$\mathbf{l}'\mathbf{y}_{pq_p}$

independent and the observations in group i are a random sample of size q_i from the $N(\mathbf{l}'\boldsymbol{\mu}_i, \mathbf{l}'\Sigma\mathbf{l})$ distribution. The analysis of variance table for testing equality of the means of these p normal distributions, i.e., for testing

$$\mathbf{l}'\boldsymbol{\mu}_1 = \cdots = \mathbf{l}'\boldsymbol{\mu}_p$$

is shown next.

Variation	d.f.	S.S.	Distribution	Expectation
Between-classes	$p-1$	$\mathbf{l}'A\mathbf{l}$	$(\mathbf{l}'\Sigma\mathbf{l})\chi^2_{p-1}(\delta)$	$(p-1)\mathbf{l}'\Sigma\mathbf{l}+\mathbf{l}'\Sigma\Omega\mathbf{l}$
Within-classes	$n-p$	$\mathbf{l}'B\mathbf{l}$	$(\mathbf{l}'\Sigma\mathbf{l})\chi^2_{n-p}$	$(n-p)\mathbf{l}'\Sigma\mathbf{l}$
Total (corrected)	$n-1$	$\mathbf{l}'(A+B)\mathbf{l}$		

In this table A and B are the between-classes and within-classes matrices given by (1) and (2) which appear in the multivariate analysis of variance. The noncentrality parameter in the noncentral χ^2 distribution for $\mathbf{l}'A\mathbf{l}/\mathbf{l}'\Sigma\mathbf{l}$ is

$$\delta = \frac{\mathbf{l}'\Sigma\Omega\mathbf{l}}{\mathbf{l}'\Sigma\mathbf{l}} \tag{5}$$

(see Theorem 10.3.5). Let us now ask: What vector $\mathbf{l}$ best discriminates between the p groups in the sense that it maximizes

(6) $$f(\mathbf{l})=\frac{\mathbf{l}'A\mathbf{l}}{\mathbf{l}'B\mathbf{l}},$$

i.e., maximizes the ratio of the between-classes S.S. to the within-classes S.S.? We attack this problem by differentiating $f(\mathbf{l})$. We have

$$df=\frac{2d\mathbf{l}'A\mathbf{l}}{\mathbf{l}'B\mathbf{l}}-\frac{2(\mathbf{l}'A\mathbf{l})(d\mathbf{l}'B\mathbf{l})}{(\mathbf{l}'B\mathbf{l})^2}$$

and equating coefficients of $d\mathbf{l}'$ to zero gives

$$\frac{1}{\mathbf{l}'B\mathbf{l}}\left(A-\frac{\mathbf{l}'A\mathbf{l}}{\mathbf{l}'B\mathbf{l}}B\right)\mathbf{l}=\mathbf{0}$$

or, equivalently,

$$(A-f(\mathbf{l})B)\mathbf{l}=\mathbf{0}.$$

This equation has a nonzero solution for $\mathbf{l}$ if and only if

(7) $$\det(A-f(\mathbf{l})B)=0.$$

The nonzero solutions of this equation are $f_1>\cdots>f_s$, the nonzero latent roots of AB^{-1}, where $s=\min(p-1,m)$. The distributions of these roots and functions of them have been derived in Sections 10.4, 10.5, and 10.6. Corresponding to the root f_i let $\mathbf{l}_i$ be of a solution of

(8) $$(A-f_iB)\mathbf{l}_i=\mathbf{0}.$$

The vector $\mathbf{l}_1$ corresponding to the largest root f_1 gives what is often called the principal discriminant function $\mathbf{l}_1'\mathbf{y}$. The vectors $\mathbf{l}_i$ corresponding to the other roots f_i give "subsidiary" discriminant functions $\mathbf{l}_i'\mathbf{y}$, $i=2,\ldots,s$. The vectors $\mathbf{l}_1,\ldots,\mathbf{l}_s$ are, of course, all orthogonal to one another. The roots $f_1,\ldots,f_s$ provide a measure of the discriminating ability of the discriminant functions $\mathbf{l}_1'\mathbf{y},\ldots,\mathbf{l}_s'\mathbf{y}$. We have shown that $\mathbf{l}_1$ maximizes the ratio

$$f(\mathbf{l})=\frac{\mathbf{l}'A\mathbf{l}}{\mathbf{l}'B\mathbf{l}}$$

and the maximum value is f_1. Then out of all vectors which are orthogonal to $\mathbf{l}_1$, $\mathbf{l}_2$ maximizes $f(\mathbf{l})$ and the maximum value is f_2, and so on.

The next question to be answered is: How many of the discriminant functions are actually useful? It is of course possible that some of the roots f_i are quite small compared with the larger ones, in which case it is natural to claim that most of the discrimination, at least for practical purposes, can be achieved by using the first few discriminant functions. The problem now is to decide how many of the roots f_i are significantly large.

Let us write the equation

$$\det(A - fB) = 0$$

in the form

$$(9) \qquad \det\left[(n-p-m-1)B^{-1}A - (p-1)I_m - \hat{\omega}I_m\right] = 0,$$

where

$$(10) \qquad \hat{\omega} = (n-p-m-1)f - p + 1.$$

Now note that $(n-p-m-1)B^{-1}A-(p-1)I_m$ is an unbiased estimate of the noncentrality matrix Ω given by (3). This is easily seen from the independence of A and B, using

$$E(B^{-1}) = (n-p-m-1)^{-1}\Sigma^{-1},$$

$$E(A) = (p-1)\Sigma + \Sigma\Omega.$$

Consequently the solutions $\hat{\omega}$ of (9) are estimates of the solutions ω of

$$(11) \qquad \det(\Omega - \omega I_m) = 0,$$

i.e., they are estimates of the latent roots of the noncentrality matrix Ω. Let $\omega_1 \geq \cdots \geq \omega_m \geq 0$ be the latent roots of Ω, and $\hat{\omega}_1 \geq \cdots \geq \hat{\omega}_m$ be the latent roots of $(n-p-m-1)B^{-1}A-(p-1)I_m$. If the rank of Ω is k then $\omega_{k+1} = \cdots = \omega_m = 0$ and their estimates $\hat{\omega}_{k+1}, \cdots, \hat{\omega}_m$ should also be close to zero, at least if n is large. Since

$$(12) \qquad \hat{\omega}_i = (n-p-m-1)f_i - p + 1,$$

the discriminating ability of the discriminant functions $\mathbf{l}_1'\mathbf{y},\ldots,\mathbf{l}_s'\mathbf{y}$ can be measured by the $\hat{\omega}_i$. We can then say that a discriminant function $\mathbf{l}_i'\mathbf{y}$ is not useful for discrimination if $\hat{\omega}_i$ is not significantly different from zero. Hence, in practice, determining the rank of the noncentrality matrix Ω is important. This is, in fact, the dimension of the space in which the p group means lie.

To see this note that, using (3),

$$\operatorname{rank}\Omega=\operatorname{rank}\left[q_1^{1/2}(\mu_1-\bar{\mu})\dots q_p^{1/2}(\mu_p-\bar{\mu})\right]\begin{bmatrix} q_1^{1/2}(\mu_1-\bar{\mu})' \\ \vdots \\ q_p^{1/2}(\mu_p-\bar{\mu})' \end{bmatrix}$$

$$=\operatorname{rank}\left[\mu_1-\bar{\mu}\dots\mu_p-\bar{\mu}\right]\begin{bmatrix} q_1^{1/2} & & 0 \\ & \ddots & \\ 0 & & q_p^{1/2} \end{bmatrix}$$

$$=\operatorname{rank}\left[\mu_1-\bar{\mu}\dots\mu_p-\bar{\mu}\right].$$

Testing the null hypothesis that the p mean vectors are equal is equivalent to testing

$$H_0: \Omega=0.$$

If this is rejected it is possible that the $m-1$ smallest roots of Ω are zero [i.e., $\operatorname{rank}(\Omega)=1$], in which case only the principal discriminant function $\mathbf{l}_1'\mathbf{y}$ is useful for discrimination. Hence it is reasonable to consider the null hypothesis that the $m-1$ smallest roots of Ω are zero. If this is rejected we can test whether the $m-2$ smallest roots are zero, and so on. In practice then, we test the sequence of null hypotheses

$$H_k: \omega_{k+1}=\cdots=\omega_m=0 \tag{13}$$

for $k=0,1,\cdots,m-1$, where $\omega_1\geq\cdots\geq\omega_m\geq 0$ are the latent roots of Ω. We have seen in Section 10.2 that the likelihood ratio test of H_0: $\Omega=0$ is based on the statistic

$$W_0=\prod_{i=1}^{s}(1+f_i)^{-1} \tag{14}$$

where $s=\min(p-1,m)$ and $f_1>\cdots>f_s>0$ are the nonzero latent roots of AB^{-1}. The likelihood ratio test of H_k is based on the statistic

$$W_k=\prod_{i=k+1}^{s}(1+f_i)^{-1}; \tag{15}$$

see, for example. T. W. Anderson (1951) and Fujikoshi (1974a). In order to derive the asymptotic null distribution of W_k we will first give an asymptotic representation for the joint density function of $f_1,\ldots,f_s$; this is done in the next subsection.

10.7.3. *Asymptotic Distributions of Latent Roots in MANOVA*

Here we consider the asymptotic distribution of the latent roots of AB^{-1} for large sample size n. We will assume now that $p \geq m+1$ so that, with probability 1, AB^{-1} has m nonzero latent roots $f_1 > \cdots > f_m > 0$. It is a little easier, from the point of view of notation, to consider the latent roots $1 > u_1 > \cdots > u_m > 0$ of $A(A+B)^{-1}$; these two sets of roots are related by $f_i = u_i/(1-u_i)$. For convenience we will put

$$n_1 = p-1, \qquad n_2 = n-p, \tag{16}$$

so that n_1 and n_2 are, respectively, the degrees of freedom for the between-classes and within-classes entries in the MANOVA table and $n_1 \geq m$, $n_2 \geq m$. From Theorem 10.4.2 (with $r = n_1$, $U = F(I+F)^{-1}$) the joint density function of $u_1,\ldots,u_m$ is

$$\begin{aligned} (17)\quad & \operatorname{etr}(-\tfrac{1}{2}\Omega)\,{}_1F_1^{(m)}\big(\tfrac{1}{2}(n_1+n_2);\tfrac{1}{2}n_1;\tfrac{1}{2}\Omega,U\big) \\ & \cdot \frac{\pi^{m^2/2}\Gamma_m\big[\tfrac{1}{2}(n_1+n_2)\big]}{\Gamma_m(\tfrac{1}{2}n_1)\Gamma_m(\tfrac{1}{2}n_2)\Gamma_m(\tfrac{1}{2}m)} \\ & \cdot \prod_{i=1}^{m}\Big[u_i^{(n_1-m-1)/2}(1-u_i)^{(n_2-m-1)/2}\Big]\prod_{i<j}^{m}(u_i-u_j) \\ & \qquad\qquad (1 > u_1 > \cdots > u_m > 0), \end{aligned}$$

where $U = \operatorname{diag}(u_1,\ldots,u_m)$. The noncentrality matrix Ω is given by (3). If $\mu_i = O(1)$ and $q_i \to \infty$ for fixed q_i/n $(i=1,\ldots,p)$, it follows that $\Omega = O(n)$. Hence a reasonable approach from an asymptotic viewpoint is to put $\Omega = n_2\Theta$, where Θ is a fixed matrix, and let $n_2 \to \infty$. Because the density function (17) depends only on the latent roots of Ω, both Ω and Θ can be assumed diagonal without loss of generality:

$$\begin{aligned} (18)\qquad \Omega &= \operatorname{diag}(\omega_1,\ldots,\omega_m), \qquad \omega_1 \geq \cdots \geq \omega_m \geq 0 \\ \Theta &= \operatorname{diag}(\theta_1,\ldots,\theta_m), \qquad \theta_1 \geq \cdots \geq \theta_m \geq 0 \end{aligned}$$

The null hypothesis $H_k: \omega_{k+1} = \cdots = \omega_m = 0$ is equivalent to $H_k: \theta_{k+1} =$

$\cdots=\theta_m=0$. The following theorem due to Glynn (1977, 1980) gives the asymptotic behavior of the ${}_1F_1^{(m)}$ function in (17) when this null hypothesis is true.

THEOREM 10.7.1. If $U=\text{diag}(u_1,\dots,u_m)$ with $1>u_1>\cdots>u_m>0$ and $\Theta=\text{diag}(\theta_1,\dots,\theta_m)$ with $\theta_1>\cdots>\theta_k>\theta_{k+1}=\cdots=\theta_m=0$ then, as $n_2\to\infty$,

$$(19)\quad {}_1F_1^{(m)}\left(\tfrac{1}{2}(n_1+n_2);\tfrac{1}{2}n_1;\tfrac{1}{2}n_2\Theta,U\right)$$

$$\sim K_{n_2}\prod_{i=1}^{k}\prod_{\substack{j=1\\ i<j}}^{m}\left[(\theta_i-\theta_j)(u_i-u_j)\right]^{-1/2}$$

$$\cdot\prod_{i=1}^{k}\left[\xi_i^{n_2+(n_1-m-1)/2}(\theta_iu_i+4)^{-1/4}(\theta_iu_i)^{(m-n_1)/4}\right]$$

$$\cdot\exp\left[\tfrac{1}{4}n_2\sum_{i=1}^{k}(\theta_iu_i)^{1/2}\xi_i\right],$$

where

$$(20)\qquad \xi_i=(\theta_iu_i)^{1/2}+(\theta_iu_i+4)^{1/2}$$

and

(21)

$$K_{n_2}=\frac{e^{-kn_2/2}n_2^{(k/2)(n_2-m+1/2+k/2)}2^{(k/2)(2m-k/2-3n_2-n_1-3/2)}\Gamma_k\left(\tfrac{1}{2}n_1\right)\Gamma_k\left(\tfrac{1}{2}m\right)}{\Gamma_k\left[\tfrac{1}{2}(n_1+n_2)\right]\pi^{k(k+1)/4}}.$$

The proof is somewhat similar to those of Theorems 9.5.2 and 9.5.3 but even more messy and disagreeable. The basic idea is to express the ${}_1F_1^{(m)}$ function as a multiple integral to which the result of Theorem 9.5.1 can be applied. We will sketch the development of this multiple integral; for the rest of the analysis, involving the maximization of the integrand and the calculation of the Hessian term, the interested reader should see Glynn (1980). First, write

$${}_1F_1^{(m)}\left(\tfrac{1}{2}(n_1+n_2);\tfrac{1}{2}n_1;\tfrac{1}{2}n_2\Theta,U\right)$$

$$=\int_{O(m)}{}_1F_1\left(\tfrac{1}{2}(n_1+n_2);\tfrac{1}{2}n_1;\tfrac{1}{2}n_2\Theta H'UH\right)(dH).$$

Here (dH) is the normalized invariant measure on $O(m)$; it is convenient to work in terms of the unnormalized invariant measure

$$(H'dH)=\bigwedge_{i<j}^{m}\mathbf{h}_j'd\mathbf{h}_i$$

(see Sections 2.1.4 and 3.2.5). These two measures are related by

$$(dH)=\frac{\Gamma_m(\frac{1}{2}m)}{2^m\pi^{m^2/2}}(H'dH),$$

so that

$${}_1F_1^{(m)}(\tfrac{1}{2}(n_1+n_2);\tfrac{1}{2}n_1;\tfrac{1}{2}n_2\Theta,U)=\frac{\Gamma_m(\frac{1}{2}m)}{2^m\pi^{m^2/2}}I_1(n_2),$$

where

$$I_1(n_2)=\int_{O(m)}{}_1F_1(\tfrac{1}{2}(n_1+n_2);\tfrac{1}{2}n_1;\tfrac{1}{2}n_2\Theta H'UH)(H'dH).$$

Now partition Θ and H as

$$\Theta=\begin{bmatrix}\Theta_1 & 0\\ 0 & 0\end{bmatrix},\qquad \Theta_1=\text{diag}(\theta_1,\ldots,\theta_k)$$

and $H=[H_1:H_2]$, where H_1 is $m\times k$, H_2 is $m\times(m-k)$. Then

$$I_1(n_2)=\int_{O(m)}{}_1F_1(\tfrac{1}{2}(n_1+n_2);\tfrac{1}{2}n_1;\tfrac{1}{2}n_2\Theta_1H_1'UH_1)(H'dH).$$

Applying Lemma 9.5.3 to this last integral gives

$$I_1(n_2)=\int_{H_1\in V_{k,m}}\int_{K\in O(m-k)}{}_1F_1(\tfrac{1}{2}(n_1+n_2);$$
$$\tfrac{1}{2}n_1;\tfrac{1}{2}n_2\Theta_1H_1'UH_1)(K'dK)(H_1'dH_1).$$

The integrand here is not a function of K, and using Corollary 2.1.14 we can integrate with respect to K to give

$$I_1(n_2)=\frac{2^{m-k}\pi^{(m-k)^2/2}}{\Gamma_{m-k}[\frac{1}{2}(m-k)]}I_2(n_2),$$

where

$$I_2(n_2)=\int_{H_1\in V_{k,m}} {}_1F_1\left(\tfrac{1}{2}(n_1+n_2);\tfrac{1}{2}n_1;\tfrac{1}{2}n_2\Theta_1 H_1'UH_1\right)(H_1'dH_1).$$

Now, for $n_1+n_2>k-1$ the ${}_1F_1$ function in this integrand may be expressed as the Laplace transform of a ${}_0F_1$ function; using Theorem 7.3.4 we obtain

$$I_2(n_2)=\frac{1}{\Gamma_k\left[\frac{1}{2}(n_1+n_2)\right]}\int_{H_1\in V_{k,m}}\int_{X>0}\operatorname{etr}(-X)(\det X)^{(n_1+n_2-k-1)/2}$$

$$\cdot {}_0F_1\left(\tfrac{1}{2}n_1;\tfrac{1}{2}n_2X^{1/2}\Theta_1^{1/2}H_1'UH_1\Theta_1^{1/2}X^{1/2}\right)(dX)(H_1'dH_1),$$

where X is a $k\times k$ positive definite matrix. Using the integral representation for the ${}_0F_1$ function given in Theorem 7.4.1 and transforming to the unnormalized measure $(Q'dQ)$ on $O(n_1)$ now shows that

$$I_2(n_2)=\frac{\Gamma_{n_1}\left(\frac{1}{2}n_1\right)}{2^{n_1}\pi^{n_1^2/2}\Gamma_k\left[\frac{1}{2}(n_1+n_2)\right]}I_3(n_2),$$

where

$$I_3(n_2)=\int_{H_1\in V_{k,m}}\int_{X>0}\int_{Q\in O(n_1)}\operatorname{etr}(-X)(\det X)^{(n_1+n_2-k-1)/2}$$

$$\cdot\operatorname{etr}\left(2^{1/2}n_2^{1/2}\left[X^{1/2}\Theta_1^{1/2}H_1'U^{1/2}:0\right]Q_1\right)(Q'dQ)(dX)(H_1'dH_1)$$

where $Q\in O(n_1)$ is partitioned as $Q=[Q_1:Q_2]$ with Q_1 being $n_1\times k$ and the 0 matrix in $[X^{1/2}\Theta_1^{1/2}H_1'U^{1/2}:0]$ is the $k\times(n_1-k)$ zero matrix. Applying Lemma 9.5.3 and Corollary 2.1.14 to the integral involving Q gives

$$I_3(n_2)=\frac{2^{n_1-k}\pi^{(n_1-k)^2/2}}{\Gamma_{n-k}\left[\frac{1}{2}(n_1-k)\right]}I_4(n_2)$$

where

$$I_4(n_2)=\int_{H_1\in V_{k,m}}\int_{X>0}\int_{Q_1\in V_{k,n_1}}\operatorname{etr}(-X)(\det X)^{(n_1+n_2-k-1)/2}$$

$$\cdot\operatorname{etr}\left(2^{1/2}n_2^{1/2}\left[X^{1/2}\Theta_1^{1/2}H_1'U^{1/2}:0\right]Q_1\right)(Q_1'dQ_1)(dX)(H_1'dH_1).$$

Now put $X=\frac{1}{2}n_2G'V^2G$, where $V=\text{diag}(v_1,\dots,v_k)$ with $v_1>\cdots>v_k>0$ and $G\in O(k)$. Using the Jacobian in the proof of Theorem 3.2.17, this integral then becomes

$$I_4(n_2)=\left(\tfrac{1}{2}n_2\right)^{k(n_1+n_2)/2}\int_{H_1\in V_{k,m}}\int_{G\in O(k)}\int_{V\in D_v}\int_{Q_1\in V_{k,n_1}}$$

$$\cdot\text{etr}\left(-\tfrac{1}{2}n_2V^2\right.$$

$$\left.+n_2\left[G'VG\Theta_1^{1/2}H_1'U^{1/2}:0\right]Q_1\right)$$

$$\cdot(\det V)^{n_1+n_2-k}\prod_{i<j}^{k}\left(v_i^2-v_j^2\right)(Q_1'dQ_1)(dV)(G'dG)(H_1'dH_1),$$

where

$$D_v=\{(v_1,\dots,v_k);\,v_1>v_2>\cdots>v_k>0\}.$$

Finally, making the transformations $E_1=H_1G'$, $P_1\in Q_1G'$, $G=G$ gives

$$I_4(n_2)=\left(\tfrac{1}{2}n_2\right)^{k(n_1+n_2)/2}\int_{\Lambda}[f(\mathbf{x})]^{n_2}g(\mathbf{x})\,d\mathbf{x}, \tag{22}$$

where

$$f(\mathbf{x})=\text{etr}\left(-\tfrac{1}{2}V^2+\left[VG\Theta_1^{1/2}G'E_1'U^{1/2}:0\right]P_1\right)\det V,$$

$$g(\mathbf{x})=(\det V)^{n_1-k}\prod_{i<j}^{k}\left(v_i^2-v_j^2\right),$$

$$\Lambda=V_{k,m}\times O(k)\times D_v\times V_{k,n_1}$$

$$d\mathbf{x}=(E_1'dE_1)(G'dG)(dV)(P_1'dP_1),$$

with

$$E_1\in V_{k,m},\qquad G\in O(k),\qquad V\in D_v,\qquad P_1\in V_{k,n_1}.$$

The easy part of the proof is over. The integral (22) is in the right form for an application of Theorem 9.5.1. In order to use this to find the asymptotic behavior of $I_4(n_2)$ as $n_2\to\infty$ we have to find the maximum value of $f(\mathbf{x})$ and the Hessian of $-\log f$ at the maximum. This has been done by Glynn (1980), to whom the reader is referred for details.

Glynn (1980) has proved a stronger result than that stated in Theorem 10.7.1, namely, that the asymptotic approximation stated there holds uniformly on any set of $u_1,\dots,u_m$ and $\theta_1,\dots,\theta_k$ such that the u_i's are strictly bounded away from one another and from 0 and 1, and the θ_i's are bounded and are bounded away from one another and from 0.

Substitution of the asymptotic behavior (19) for ${}_1F_1^{(m)}$ in (17) yields an asymptotic representation for the joint density function of the sample roots $u_1,\dots,u_m$. This result is summarized in the following theorem.

THEOREM 10.7.2. An asymptotic representation for large n_2 of the joint density function of the latent roots $1>u_1>\cdots>u_m>0$ of $A(A+B)^{-1}$ when $\Omega=n_2\Theta$ with

$$\Theta=\operatorname{diag}(\theta_1,\dots,\theta_m)\qquad(\theta_1>\cdots>\theta_k>\theta_{k+1}=\cdots=\theta_m=0)$$

is

$$(23)\quad C_{n_2}\prod_{i=1}^{k}\left[\xi_i^{n_2+(n_1-m+1)/2}(\theta_iu_i+4)^{-1/4}(\theta_iu_i)^{(m-n_1)/4}\right]$$

$$\cdot\prod_{i=1}^{k}\left[u_i^{(n_1-m-1)/2}(1-u_i)^{(n_2-m-1)/2}\right]\prod_{i<j}^{k}\left(\frac{u_i-u_j}{\theta_i-\theta_j}\right)^{1/2}$$

$$\cdot\exp\left[\tfrac{1}{4}n_2\sum_{i=1}^{k}(\theta_iu_i)^{1/2}\xi_i\right]\prod_{i=1}^{k}\prod_{j=k+1}^{m}(u_i-u_j)^{1/2}$$

$$\cdot\prod_{i=k+1}^{m}\left[u_i^{(n_1-m-1)/2}(1-u_i)^{(n_2-m-1)/2}\right]\prod_{\substack{k+1\\ i<j}}^{m}(u_i-u_j),$$

where

$$(24)\qquad \xi_i=(\theta_iu_i)^{1/2}+(\theta_iu_i+4)^{1/2}$$

and

(25)

$$C_{n_2}=K_{n_2}\frac{\pi^{m^2/2}\Gamma_m\left[\tfrac{1}{2}(n_1+n_2)\right]}{\Gamma_m(\tfrac{1}{2}n_1)\Gamma_m(\tfrac{1}{2}n_2)\Gamma_m(\tfrac{1}{2}m)}\exp\left[-\tfrac{1}{2}n_2\left(k+\sum_{i=1}^{k}\theta_i\right)\right]\prod_{i=1}^{k}\theta_i^{-(m-k)/2}$$

with K_{n_2} given by (21).

This theorem has two interesting consequences.

COROLLARY 10.7.3. Under the conditions of Theorem 10.7.2 the asymptotic conditional density function for large n_2 of $u_{k+1},\ldots,u_m$, the $q=m-k$ smallest latent roots of $A(A+B)^{-1}$, given the k largest roots $u_1,\ldots,u_k$, is

$$(26)\qquad K\prod_{i=1}^{k}\prod_{j=k+1}^{m}(u_i-u_j)^{1/2}$$

$$\cdot\prod_{i=k+1}^{m}\left[u_i^{(n_1-k-q-1)/2}(1-u_i)^{(n_2-k-q-1)/2}\right]\prod_{\substack{k+1\\ i<j}}^{m}(u_i-u_j),$$

where K is a constant.

Note that this asymptotic conditional density function does not depend on $\theta_1,\ldots,\theta_k$, the k nonzero population roots. Hence by conditioning on $u_1,\ldots,u_k$ the effects of these k population roots can be eliminated, at least asymptotically. In this sense $u_1,\ldots,u_k$ are asymptotically sufficient for $\theta_1,\ldots,\theta_k$. We can also see in (26) that the influence of the k largest sample roots $u_1,\ldots,u_k$ in the asymptotic conditional distribution is felt through linkage factors of the form $(u_i-u_j)^{1/2}$.

COROLLARY 10.7.4. Assume that the conditions of Theorem 10.7.2 hold and put

$$(27)\qquad \begin{aligned} x_i&=\frac{n_2^{1/2}}{\sigma_i}\left(\frac{u_i}{1-u_i}-\theta_i\right) &&(\text{for}\quad i=1,\ldots,k)\\ x_j&=\frac{n_2u_j}{1-u_j} &&(\text{for}\quad j=k+1,\ldots,m),\end{aligned}$$

where

$$\sigma_i=[2\theta_i(2+\theta_i)]^{1/2}\qquad (i=1,\ldots,k).$$

Then the limiting joint density function of $x_1,\ldots,x_m$ as $n_2\to\infty$ is

$$(28)\qquad \prod_{i=1}^{k}\phi(x_i)\cdot\frac{\pi^{q^2/2}}{2^{q(n_1-k)/2}\Gamma_q(\frac{1}{2}q)\Gamma_q[\frac{1}{2}(n_1-k)]}$$

$$\cdot\prod_{j=k+1}^{m}\left[x_j^{(n_1-k-q-1)/2}e^{-x_j/2}\right]\prod_{\substack{k+1\\ i<j}}^{m}(x_i-x_j),$$

where $q=m-k$ and $\phi(\cdot)$ denotes the standard normal density function.

This result, due originally to Hsu (1941a), can be proved by making the change of variables (27) in (23) and letting $n_2 \to \infty$. Note that this shows that if θ_i is a distinct nonzero population root then x_i is asymptotically independent of x_j for $j \neq i$ and the limiting distribution of x_i is standard normal. Note also that $x_{k+1},\ldots,x_m$ (corresponding to the θ_j's equal to zero) are dependent and their asymptotic distribution is the same as the distribution of the roots of a $q \times q$ matrix having the $W_q(n_1-k, I_q)$ distribution.

For other asymptotic approaches to the distribution of the latent roots $u_1,\ldots,u_m$ the interested reader is referred to Constantine and Muirhead (1976), Muirhead (1978), and Chou and Muirhead (1979).

10.7.4. Determining the Number of Useful Discriminant Functions

It was noted in Section 10.7.2 that it is of interest to test the sequence of null hypotheses

$$H_k: \omega_{k+1} = \cdots = \omega_m = 0$$

for $k=0,1,\ldots,m-1$, where $\omega_1 \geq \cdots \geq \omega_m \geq 0$ are the latent roots of the noncentrality matrix Ω. If H_k is true then the rank of Ω is k and this is the number of useful discriminant functions. The likelihood ratio test rejects H_k for small values of the statistic

$$\text{(29)} \qquad W_k = \prod_{i=k+1}^{m} (1+f_i)^{-1} = \prod_{i=k+1}^{m} (1-u_i)$$

where $f_1 > \cdots > f_m > 0$ are the latent roots of AB^{-1} and $1 > u_1 > \cdots > u_m > 0$ are the latent roots of $A(A+B)^{-1}$. We are assuming here, as in the last subsection, that $n_1 \geq m$ and $n_2 \geq m$, where $n_1 = p-1$ and $n_2 = n-p$ are the between-classes and within-classes degrees of freedom, respectively. The asymptotic distribution as $n_2 \to \infty$ of $-n_2 \log W_k$ is $\chi^2_{(m-k)(n_1-k)}$ when H_k is true. An improvement over $-n_2 \log W_k$ is the statistic $-[n_2+\frac{1}{2}(n_1-m-1)]\log W_k$ suggested by Bartlett (1947). The multiplying factor here is exactly that given by Theorem 10.5.5, where it was shown that $-[n_2+\frac{1}{2}(n_1-m-1)]\log W_0$ has an asymptotic $\chi^2_{mn_1}$ distribution when $H_0: \Omega = 0$ is true. A further refinement in the multiplying factor was obtained by Chou and Muirhead (1979) and Glynn (1980). We will now indicate their approach.

We noted in Corollary 10.7.3 that the asymptotic conditional density function of $u_{k+1},\ldots,u_m$ given $u_1,\ldots,u_k$ is

$$(30)\qquad K\prod_{i=1}^{k}\prod_{j=k+1}^{m}(u_i-u_j)^{1/2}$$

$$\cdot\prod_{i=k+1}^{m}\left[u_i^{(n_1-k-q-1)/2}(1-u_i)^{(n_2-k-q-1)/2}\right]\prod_{\substack{k+1\\ i<j}}^{m}(u_i-u_j),$$

where $q=m-k$ and K is a constant. Put

$$(31)\qquad T_k=-\log W_k$$

so that the limiting distribution of n_2T_k is $\chi^2_{(m-k)(n_1-k)}$ when H_k is true. The appropriate multiplier of T_k can be obtained by finding its expected value. For notational convenience let E_c denote expectation taken with respect to the conditional distribution (30) of $u_{k+1},\ldots,u_m$ given $u_1,\ldots,u_k$ and let E_N denote expectation taken with respect to the "null" distribution

$$(32)\quad K_1\prod_{i=k+1}^{m}\left[u_i^{(n_1-k-q-1)/2}(1-u_i)^{(n_2-k-q-1)/2}\right]\prod_{\substack{k+1\\ i<j}}^{m}(u_i-u_j)$$

obtained from (30) by ignoring the linkage factor

$$\prod_{i=1}^{k}\prod_{j=k+1}^{m}(u_i-u_j)^{1/2}.$$

This distribution is just the distribution of the latent roots of a $q\times q$ matrix U having the $\text{Beta}_q(\frac{1}{2}(n_1-k),\frac{1}{2}(n_2-k))$ distribution (see Theorem 3.3.4). The following theorem gives the asymptotic distribution of the likelihood ratio statistic with additional information about the accuracy of the χ^2 approximation.

THEOREM 10.7.5. When the null hypothesis H_k is true the asymptotic distribution as $n_2\to\infty$ of the statistic

$$(33)\qquad L_k=-\left[n_2-k+\tfrac{1}{2}(n_1-m-1)+\sum_{i=1}^{k}u_i^{-1}\right]\log W_k$$

is $\chi^2_{(m-k)(n_1-k)}$, and

$$E_c(L_k)=(m-k)(n_1-k)+O(n_2^{-2}). \tag{34}$$

Proof. We will sketch the proof, which is rather similar to that of Theorem 9.6.2. First note that, with T_k defined by (31),

$$E_c(T_k)=-\frac{\partial}{\partial h}\{E_c[\exp(-hT_k)]\}_{h=0} \tag{35}$$

so that, in order to find $E_c(T_k)$ we will first obtain

$$E_c[\exp(-hT_k)]=E_c\left[\prod_{i=k+1}^{m}(1-u_i)^h\right]. \tag{36}$$

This can obviously be done by finding

$$E_N\left[\prod_{i=1}^{k}\prod_{j=k+1}^{m}\left(1-\frac{u_j}{u_i}\right)^{1/2}\cdot\exp(-hT_k)\right]. \tag{37}$$

Now, when H_k is true we can write

$$\prod_{i=1}^{k}\prod_{j=k+1}^{m}\left(1-\frac{u_j}{u_i}\right)^{1/2}=1-\tfrac{1}{2}\alpha\sum_{j=k+1}^{m}u_j+O_p(n_2^{-2}) \tag{38}$$

where

$$\alpha=\sum_{i=1}^{k}u_i^{-1}. \tag{39}$$

Substituting (38) in (37) it is seen that we need to evaluate

$$E_N\left[\exp(-hT_k)\sum_{j=k+1}^{m}u_j\right]. \tag{40}$$

This problem is addressed in the following lemma.

LEMMA 10.7.6.

$$E_N\left[\exp(-hT_k)\sum_{j=k+1}^{m}u_j\right]=\frac{(m-k)(n_1-k)}{n_1+n_2-2k+2h}E_0(h), \tag{41}$$

where

$$E_0(h)=E_N[\exp(-hT_k)]. \tag{42}$$

Proof. Let $v_i=1-u_{k+i}$ $(i=1,\dots,m-k)$. The null distribution of $v_1,\dots,v_{m-k}$ is the same as the distribution of the latent roots of a $q\times q$ $(q=m-k)$ matrix $V=(v_{ij})$ having the $\text{Beta}_q(\frac{1}{2}(n_2-k),\frac{1}{2}(n_1-k))$ distribution. Note that

$$\sum_{j=k+1}^{m} u_j=\text{tr}(I_q-V).$$

Since the diagonal elements of I_q-V all have the same expectation, we need only find the expectation of the first element $\Delta=1-v_{11}$ and multiply the result by $m-k$. Put $V=T'T$, where $T=(t_{ij})$ is a $q\times q$ upper-triangular matrix. By Theorem 3.3.3, $t_{11},\dots,t_{qq}$ are all independent and t_{ii}^2 has a $\text{beta}(\frac{1}{2}(n_2-k-i+1),\frac{1}{2}(n_1-k))$ distribution, and $\Delta=1-t_{11}^2$. Hence

$$\begin{aligned}
E_N\left[\exp(-hT_k)\sum_{j=k+1}^{m}u_j\right]&=(m-k)E_N\left[(1-t_{11}^2)\prod_{j=k+1}^{m}(1-u_j)^h\right]\\
&=(m-k)E_N\left[(1-t_{11}^2)\prod_{i=1}^{m-k}t_{ii}^{2h}\right]\\
&=(m-k)E_N\left[\prod_{i=1}^{m-k}t_{ii}^{2h}\right]-(m-k)E_N\left[t_{11}^2\prod_{i=1}^{m-k}t_{ii}^{2h}\right].
\end{aligned}$$

It is easily shown that

$$E_N\left[t_{11}^2\prod_{i=1}^{m-k}t_{ii}^2\right]=\frac{n_1+n_2-k-n_1+2h}{n_1+n_2-2k+2h}E_0(h)$$

and hence

$$E_N\left[\exp(-hT_k)\sum_{j=k+1}^{m}u_j\right]=\frac{(m-k)(n_1-k)}{n_1+n_2-2k+2h}E_0(h),$$

completing the proof of the lemma.

Returning to our outline of the proof of Theorem 10.7.5, it follows from (36), (37), (38), and Lemma 10.7.6 that

$$E_c[\exp(-hT_k)] = \frac{\theta(h)}{\theta(0)}, \tag{43}$$

where

$$\theta(h) = E_0(h) f(h) \tag{44}$$

with

$$f(h) = 1 - \frac{\alpha(m-k)(n_1-k)}{2(n_1+n_2-2k+2h)}. \tag{45}$$

Using (35), we have

$$\begin{aligned} E_c(T_k) &= -\frac{\partial}{\partial h}\left[\frac{\theta(h)}{\theta(0)}\right]_{h=0} \\ &= -E_0'(0) - \frac{\alpha(m-k)(n_1-k)}{(n_1+n_2-2k)^2} + O(n_2^{-3}). \end{aligned} \tag{46}$$

But $-E_0'(0) = E_N(T_k)$, and in the case of the null distribution we know that $[n_2 - k + \frac{1}{2}(n_1 - m - 1)]T_k$ has an asymptotic $\chi^2_{(m-k)(n_1-k)}$ distribution and the means agree to $O(n_2^{-2})$ so that

$$-E_0'(0) = \frac{(m-k)(n_1-k)}{n_2 - k + \frac{1}{2}(n_1-m-1)} + O(n_2^{-3}). \tag{47}$$

Hence it follows that

$$E_c(T_k) = \frac{(m-k)(n_1-k)}{n_2 - k + \frac{1}{2}(n_1-m-1) + \alpha + O(n_2^{-1})}, \tag{48}$$

from which it is seen that if L_k is the statistic defined by (33) then

$$E_c(L_k) = (m-k)(n_1-k) + O(n_2^{-2}),$$

and the proof is complete.

The multiplying factor in Theorem 10.7.5 is approximately that suggested by Bartlett (1947), namely, $n_2+\frac{1}{2}(n_1-m-1)$, if the observed values of $u_1,\ldots,u_k$ are all close to 1; in this case α is approximately equal to k.

It follows from Theorem 10.7.5 that if n_2 is large an approximate test of size α of the null hypothesis

$$H_k:\omega_{k+1}=\cdots=\omega_m=0$$

is to reject H_k if $L_k>c(\alpha;(m-k)(n_1-k))$, where L_k is given by (33) and $c(\alpha;r)$ is the upper $100\alpha\%$ point of the χ_r^2 distribution.

Let us now suppose that the hypotheses H_k, $k=0,1,\ldots,m-1$ are tested sequentially and that for some k the hypothesis H_k is accepted and we are prepared to conclude that there are k useful discriminant functions $\mathbf{l}_1'\mathbf{y},\ldots,\mathbf{l}_k'\mathbf{y}$, where $\mathbf{l}_1,\ldots,\mathbf{l}_k$ are solutions of (8) associated with the largest k latent roots $f_1,\ldots,f_k$ of AB^{-1}. How should a new observation $\mathbf{y}_0$ be assigned to one of the p groups? Let $L=[\mathbf{l}_1\ldots\mathbf{l}_k]'$ and put $\mathbf{x}_0=L\mathbf{y}_0,\mathbf{x}_1=L\bar{\mathbf{y}}_1,\ldots,\mathbf{x}_p=L\bar{\mathbf{y}}_p$. The distance between $\mathbf{y}_0$ and $\bar{\mathbf{y}}_i$ based on the new system of coordinates $\mathbf{l}_1,\ldots,\mathbf{l}_k$ is

$$d_i=\|\mathbf{x}_0-\mathbf{x}_i\| \quad i=1,\ldots,p.$$

A simple classification rule is then to assign $\mathbf{y}_0$ to the ith group if $\mathbf{x}_0$ is closer to $\mathbf{x}_i$ than to any other $\mathbf{x}_j$, i.e., if

$$d_i=\min(d_1,\ldots,d_p).$$

10.7.5. *Discrimination Between Two Groups*

In Section 10.7.2 we noted that the number of discriminant functions is equal to $s=\min(p-1,m)$, where p denotes the number of groups and m the number of variables in each observation. Hence when $p=2$ there is only one discriminant function. The reader can readily check that the nonzero latent root of AB^{-1} is

$$f_1=(\bar{\mathbf{y}}_1-\bar{\mathbf{y}}_2)'B^{-1}(\bar{\mathbf{y}}_1-\bar{\mathbf{y}}_2), \tag{49}$$

where $\bar{\mathbf{y}}_1$ and $\bar{\mathbf{y}}_2$ are the sample means of the two groups. A solution of the equation $(A-f_1B)\mathbf{l}_1=0$ is

$$\mathbf{l}_1=S^{-1}(\bar{\mathbf{y}}_1-\bar{\mathbf{y}}_2), \tag{50}$$

where $S=(n-2)^{-1}B$ $(n=q_1+q_2)$, which is unbiased for Σ. The discriminant function is then

$$(\bar{\mathbf{y}}_1 - \bar{\mathbf{y}}_2)'S^{-1}\mathbf{y}. \tag{51}$$

This is an estimate of a population discriminant function due to Fisher (1936) which is appropriate when all the parameters are known and which can be obtained in the following way. Suppose we have a new observation $\mathbf{y}_0$ which belongs to one of the two populations. The problem is to decide which one. If $\mathbf{y}_0$ belongs with the ith group, i.e., is drawn from the $N_m(\boldsymbol{\mu}_i, \Sigma)$ population, then its density function is

$$f_i(\mathbf{y}_0) = (2\pi)^{-m/2}(\det \Sigma)^{-1/2}\exp\left[-\tfrac{1}{2}(\mathbf{y}_0 - \boldsymbol{\mu}_i)'\Sigma^{-1}(\mathbf{y}_0 - \boldsymbol{\mu}_i)\right] \qquad (i = 1,2).$$

An intuitively appealing procedure is to assign $\mathbf{y}_0$ to the $N_m(\boldsymbol{\mu}_1, \Sigma)$ population if the likelihood function corresponding to this population is large enough compared with the likelihood function corresponding to the $N_m(\boldsymbol{\mu}_2, \Sigma)$ population, i.e., if

$$\frac{f_1(\mathbf{y}_0)}{f_2(\mathbf{y}_0)} \geq c,$$

where c is a constant. This inequality is readily seen to be equivalent to

$$(\boldsymbol{\mu}_1 - \boldsymbol{\mu}_2)'\Sigma^{-1}\mathbf{y}_0 + \tfrac{1}{2}(\boldsymbol{\mu}_1 + \boldsymbol{\mu}_2)'\Sigma^{-1}(\boldsymbol{\mu}_1 - \boldsymbol{\mu}_2) \geq k = \log c \tag{52}$$

The function

$$g(\mathbf{y}_0) = (\boldsymbol{\mu}_1 - \boldsymbol{\mu}_2)'\Sigma^{-1}\mathbf{y}_0 \tag{53}$$

is called Fisher's discriminant function. If

$$g(\mathbf{y}_0) \geq k_1 = k - \tfrac{1}{2}(\boldsymbol{\mu}_1 + \boldsymbol{\mu}_2)'\Sigma^{-1}(\boldsymbol{\mu}_1 - \boldsymbol{\mu}_2) \tag{54}$$

$\mathbf{y}_0$ is assigned to the $N_m(\boldsymbol{\mu}_1, \Sigma)$ population; otherwise, to the $N_m(\boldsymbol{\mu}_2, \Sigma)$ population. There are, of course, two errors associated with this classification rule. The probability that $\mathbf{y}_0$ is misclassified as belonging to the $N_m(\boldsymbol{\mu}_2, \Sigma)$ population when it belongs to the $N_m(\boldsymbol{\mu}_1, \Sigma)$ population is

$$\begin{aligned} \alpha_1 &= P_{(\boldsymbol{\mu}_1, \Sigma)}\left((\boldsymbol{\mu}_1 - \boldsymbol{\mu}_2)'\Sigma^{-1}\mathbf{y}_0 < k_1\right) \\ &= \Phi\left(\frac{k_1 - (\boldsymbol{\mu}_1 - \boldsymbol{\mu}_2)'\Sigma^{-1}\boldsymbol{\mu}_1}{d}\right), \end{aligned} \tag{55}$$

where

$$d^2=(\boldsymbol{\mu}_1-\boldsymbol{\mu}_2)'\Sigma^{-1}(\boldsymbol{\mu}_1-\boldsymbol{\mu}_2). \tag{56}$$

This follows because if $\mathbf{y}_0$ is $N_m(\boldsymbol{\mu}_1,\Sigma)$ then $(\boldsymbol{\mu}_1-\boldsymbol{\mu}_2)'\Sigma^{-1}\mathbf{y}_0$ is $N((\boldsymbol{\mu}_1-\boldsymbol{\mu}_2)'\Sigma^{-1}\boldsymbol{\mu}_1, d^2)$. The number d is a measure of the distance between $\boldsymbol{\mu}_1$ and $\boldsymbol{\mu}_2$ suggested by Mahalanobis (1930). Similarly the probability that $\mathbf{y}_0$ is misclassified as belonging to the $N_m(\boldsymbol{\mu}_1,\Sigma)$ population when it belongs to the $N_m(\boldsymbol{\mu}_2,\Sigma)$ population is

$$\begin{aligned}\alpha_2&=P_{(\boldsymbol{\mu}_2,\Sigma)}((\boldsymbol{\mu}_1-\boldsymbol{\mu}_2)'\Sigma^{-1}\mathbf{y}_0\geq k_1)\\&=\Phi\left(\frac{(\boldsymbol{\mu}_1-\boldsymbol{\mu}_2)'\Sigma^{-1}\boldsymbol{\mu}_2-k_1}{d}\right).\end{aligned} \tag{57}$$

Suppose now we assume that $\alpha_1=\alpha_2$, i.e., that the probabilities of the two misclassification errors are equal. It then follows that

$$k_1=\tfrac{1}{2}(\boldsymbol{\mu}_1-\boldsymbol{\mu}_2)'\Sigma^{-1}(\boldsymbol{\mu}_1+\boldsymbol{\mu}_2), \tag{58}$$

and substituting this value for k_1 in α_1 and α_2 we have

$$\alpha_1=\alpha_2=\Phi(-\tfrac{1}{2}d), \tag{59}$$

which is a decreasing function of d. Hence the procedure is to compute Fisher's discriminant function $g(\mathbf{y}_0)$ given by (53) and to assign $\mathbf{y}_0$ to the $N_m(\boldsymbol{\mu}_1,\Sigma)$ population if $g(\mathbf{y}_0)\geq k_1$ and to the $N_m(\boldsymbol{\mu}_2,\Sigma)$ population if $g(\mathbf{y}_0)<k_1$.

In most practical situations the parameters $\boldsymbol{\mu}_1$, $\boldsymbol{\mu}_2$, and Σ are unknown and have to be estimated. A reasonable estimate of $g(\mathbf{y}_0)$ is

$$\hat{g}(\mathbf{y}_0)=(\bar{\mathbf{y}}_1-\bar{\mathbf{y}}_2)'S^{-1}\mathbf{y}_0 \tag{60}$$

in which $\boldsymbol{\mu}_1$, $\boldsymbol{\mu}_2$, and Σ have been replaced by unbiased estimates. This is the (sample) discriminant function introduced at the beginning of this subsection [see (50)] and which arises from the theory of Section 10.7.2. The constant k_1 given by (58) can be similarly estimated by

$$\hat{k}_1=\tfrac{1}{2}(\bar{\mathbf{y}}_1-\bar{\mathbf{y}}_2)'S^{-1}(\bar{\mathbf{y}}_1+\bar{\mathbf{y}}_2), \tag{61}$$

so that the procedure is now to assign $\mathbf{y}_0$ to the $N_m(\boldsymbol{\mu}_1,\Sigma)$ population if the statistic

$$W(\mathbf{y}_0)=\hat{g}(\mathbf{y}_0)-\hat{k}_1 \tag{62}$$

is ≥ 0 and to the $N_m(\boldsymbol{\mu}_2, \Sigma)$ population if $W(\mathbf{y}_0)<0$. The statistic $W(\mathbf{y}_0)$ was suggested by T. W. Anderson (1958) and is known as Anderson's classification statistic. The probability of misclassification is no longer the same because $\hat{g}(\mathbf{y}_0)$ is no longer normal and $\hat{k}_1$ is random. A great deal of work has been done on the problem of estimating probabilities of misclassification associated with the classification statistic $W(\mathbf{y}_0)$ and other procedures. We will not delve into this problem; the interested reader is referred to a review of such work in Kshirsagar (1972), Chapter 6. It is, however, worth pointing out that the asymptotic distribution of $W(\mathbf{y}_0)$, as q_1 and $q_2 \to \infty$, is normal with variance equal to d^2. The mean is $(\boldsymbol{\mu}_1 - \boldsymbol{\mu}_2)'\Sigma^{-1}\boldsymbol{\mu}_1 - k_1$ if $\mathbf{y}_0$ belongs to the $N_m(\boldsymbol{\mu}_1, \Sigma)$ population and $(\boldsymbol{\mu}_1 - \boldsymbol{\mu}_2)'\Sigma^{-1}\boldsymbol{\mu}_2 - k_1$ if $\mathbf{y}_0$ belongs to the $N_m(\boldsymbol{\mu}_2, \Sigma)$ population. Hence asymptotically the probability of misclassification is the same as before.

The subjects of discriminant analysis and classification have been very widely studied. For various approaches and generalizations as to these subjects useful references are T. W. Anderson (1958), Chapter 6; Kshirsagar (1972), Chapters 6 and 9; and Srivastava and Khatri (1979), Chapter 8.

10.8. TESTING EQUALITY OF p NORMAL POPULATIONS

10.8.1. The Likelihood Ratio Statistic and Moments

In Section 10.7 we considered the problem of testing whether the mean vectors of p normal distributions are equal under the assumption that the distributions have a common covariance matrix. A closely related problem arises when we drop this assumption and test the null hypothesis

$$H: \quad \boldsymbol{\mu}_1 = \cdots = \boldsymbol{\mu}_p, \qquad \Sigma_1 = \cdots = \Sigma_p,$$

i.e. that the p m-variate normal distributions are *identical*. We are given $\mathbf{y}_{i1},\ldots,\mathbf{y}_{iN_i}$, a random sample from the $N_m(\boldsymbol{\mu}_i, \Sigma_i)$ distribution $(i=1,\ldots,p)$. The null hypothesis treated in Section 10.7 is

$$H_0: \quad \boldsymbol{\mu}_1 = \cdots = \boldsymbol{\mu}_p \qquad (\text{given} \quad \Sigma_1 = \cdots = \Sigma_p),$$

and the likelihood ratio statistic for testing H_0 is (see Theorem 10.2.2)

$$\Lambda_0 = \frac{(\det B)^{N/2}}{\det(A+B)^{N/2}} \tag{1}$$

where

$$(2) \qquad A=\sum_{i=1}^{p} N_i(\bar{\mathbf{y}}_i-\bar{\mathbf{y}})(\bar{\mathbf{y}}_i-\bar{\mathbf{y}})'$$

and

$$(3) \qquad B=\sum_{i=1}^{p}\sum_{j=1}^{N_i}(\mathbf{y}_{ij}-\bar{\mathbf{y}}_i)(\mathbf{y}_{ij}-\bar{\mathbf{y}}_i)',$$

with

$$\bar{\mathbf{y}}_i=\frac{1}{N_i}\sum_{j=1}^{N_i}\mathbf{y}_{ij}, \qquad \bar{\mathbf{y}}=\frac{1}{N}\sum_{i=1}^{p}N_i\bar{\mathbf{y}}_i, \qquad N=\sum_{i=1}^{p}N_i.$$

Now recall from Section 8.2 that the likelihood ratio statistic for testing the null hypothesis

$$H_1:\ \Sigma_1=\cdots=\Sigma_p$$

is

$$(4) \qquad \Lambda_1=\frac{N^{mN/2}}{\prod_{i=1}^{p}N_i^{mN_i/2}}\,\frac{\prod_{i=1}^{p}(\det B_i)^{N_i/2}}{(\det B)^{N/2}}$$

where

$$(5) \qquad B_i=\sum_{j=1}^{N_i}(\mathbf{y}_{ij}-\bar{\mathbf{y}}_i)(\mathbf{y}_{ij}-\bar{\mathbf{y}}_i)'$$

and $B=\Sigma_{j=1}^{p}B_j$, as in (3). The likelihood ratio statistic for testing H is the product of Λ_1 and Λ_2 and is given in the following theorem from Wilks (1932).

THEOREM 10.8.1. The likelihood ratio statistic for testing

$$H:\ \boldsymbol{\mu}_1=\cdots=\boldsymbol{\mu}_p, \qquad \Sigma_1=\cdots=\Sigma_p$$

is

$$\Lambda = \frac{\prod_{i=1}^{p}(\det B_i)^{N_i/2}}{\det(A+B)^{N/2}} \cdot \frac{N^{mN/2}}{\prod_{i=1}^{p} N_i^{mN_i/2}} \tag{6}$$

where A, B, and B_i are given by (2), (3), and (5), respectively.

Proof. Let $L(\mu_1,\dots,\mu_p,\ \Sigma_1,\dots,\Sigma_p)$ denote the likelihood function formed from the p independent normal samples. Then

$$\Lambda = \frac{\sup_{\mu,\Sigma} L(\mu,\dots,\mu,\ \Sigma,\dots,\Sigma)}{\sup_{\mu_1,\dots,\mu_p,\ \Sigma_1,\dots,\Sigma_p} L(\mu_1,\dots,\mu_p,\ \Sigma_1,\dots,\Sigma_p)}$$

$$= \frac{\sup_{\mu,\Sigma} L(\mu,\dots,\mu,\ \Sigma,\dots,\Sigma)}{\sup_{\mu_1,\dots,\mu_p,\Sigma} L(\mu_1,\dots,\mu_p,\ \Sigma,\dots,\Sigma)}$$

$$\cdot \frac{\sup_{\mu_1,\dots,\mu_p,\Sigma} L(\mu_1,\dots,\mu_p,\ \Sigma,\dots,\Sigma)}{\sup_{\mu_1,\dots,\mu_p,\ \Sigma_1,\dots,\Sigma_p} L(\mu_1,\dots,\mu_p,\ \Sigma_1,\dots,\Sigma_p)}$$

$$= \Lambda_0 \Lambda_1,$$

where Λ_0 and Λ_1 are the likelihood ratio statistics for testing the null hypothesis H_0 and H_1. The desired result now follows by substituting the values of Λ_0 and Λ_1 given by (1) and (4).

It follows that the likelihood ratio test of H rejects H for small values of the statistic

$$\lambda = \frac{\prod_{i=1}^{p}(\det B_i)^{N_i/2}}{\det(A+B)^{N/2}} = \Lambda \cdot \frac{\prod_{i=1}^{p} N_i^{mN_i/2}}{N^{mN/2}}. \tag{7}$$

T. W. Anderson (1958), Section 10.3, suggested the use of a modified test statistic Λ^* in which the sample sizes N_i are replaced by $n_i = N_i - 1$ and N is replaced by $n = \sum_{i=1}^{p} n_i = N - p$, i.e.,

$$\Lambda^* = \frac{\prod_{i=1}^{p}(\det B_i)^{n_i/2}}{\det(A+B)^{n/2}} \frac{n^{mn/2}}{\prod_{i=1}^{p} n_i^{mn_i/2}}. \tag{8}$$

However, Perlman (1980) has shown that it is the likelihood ratio test itself, not the modified test, which is unbiased for testing H.

The moments of λ are very difficult to obtain explicitly, except in some special cases. The following theorem gives the moments when the covariance matrices are all assumed equal.

THEOREM 10.8.2. When $\Sigma_1=\cdots=\Sigma_p$ $(=\Sigma)$ the hth moment of λ is

$$E(\lambda^h)=\frac{\Gamma_m[\frac{1}{2}(N-1)]}{\Gamma_m[\frac{1}{2}N(1+h)-\frac{1}{2}]}\prod_{i=1}^{p}\frac{\Gamma_m[\frac{1}{2}N_i(1+h)-\frac{1}{2}]}{\Gamma_m[\frac{1}{2}(N_i-1)]}\cdot{}_1F_1(\tfrac{1}{2}Nh;\tfrac{1}{2}N(1+h)-\tfrac{1}{2};-\tfrac{1}{2}\Omega), \tag{9}$$

where

$$\Omega=\Sigma^{-1}\sum_{i=1}^{p}N_i(\mu_i-\bar{\mu})(\mu_i-\bar{\mu})' \tag{10}$$

with

$$\bar{\mu}=\frac{1}{N}\sum_{i=1}^{p}N_i\mu_i.$$

Proof. The matrices B_i have independent $W_m(n_i,\Sigma)$ distributions ($i=1,\ldots,p$), where $n_i=N_i-1$, and the matrix A is independently distributed as $W_m(p-1,\Sigma,\Omega)$, where Ω is given by (10) (see Section 10.7). Hence

$$\begin{aligned}E(\lambda^h)=&\left(\prod_{i=1}^{p}c_{m,n_i}\right)c_{m,p-1}\operatorname{etr}(-\tfrac{1}{2}\Omega)\int_{B_1>0}\cdots\int_{B_p>0}\int_{A>0}\\&\cdot\det\left(A+\sum_{i=1}^{p}B_i\right)^{-Nh/2}\\&\cdot\prod_{i=1}^{p}\left[\operatorname{etr}(-\tfrac{1}{2}\Sigma^{-1}B_i)(\det B_i)^{n_i(1+h)/2+h/2-(m+1)/2}\right]\\&\cdot\operatorname{etr}(-\tfrac{1}{2}\Sigma^{-1}A)(\det A)^{(p-1-m-1)/2}{}_0F_1(\tfrac{1}{2}(p-1);\tfrac{1}{4}\Omega\Sigma^{-1}A)\\&\cdot(dA)(dB_p)\ldots(dB_1),\end{aligned}$$

where

$$c_{m,n}=\left[2^{mn/2}\Gamma_m(\tfrac{1}{2}n)(\det\Sigma)^{n/2}\right]^{-1}$$

Consequently,

$$E(\lambda^h)=\left(\prod_{i=1}^{p}\frac{c_{m,n_i}}{c_{m,n_i(1+h)+h}}\right)E\left[\det\left(A+\sum_{i=1}^{p}B_i\right)^{-Nh/2}\right],$$

where now the B_i have independent $W_m(n_i(1+h)+h,\Sigma)$ distributions $(i=1,\ldots,p)$ and are independent of A which is $W_m(\text{p}-1,\Sigma,\Omega)$. Hence, from Theorem 10.3.3, the matrix $\tilde{A}=A+\Sigma_{i=1}^{p}B_i$ is $W_m(N(1+h)-1,\Sigma,\Omega)$ so that, using Theorem 10.3.6 we have

$$E(\lambda^h)=\left(\prod_{i=1}^{p}\frac{c_{m,n_i}}{c_{m,n_i(1+h)+h}}\right)E\left[(\det\tilde{A})^{-Nh/2}\right]$$

$$=\left(\prod_{i=1}^{p}\frac{c_{m,n_i}}{c_{m,n_i(1+h)+h}}\right)\cdot(\det\Sigma)^{-Nh/2}2^{-Nhm/2}\frac{\Gamma_m[\frac{1}{2}(N-1)]}{\Gamma_m[\frac{1}{2}N(1+h)-\frac{1}{2}]}$$

$$\cdot{}_1F_1(\tfrac{1}{2}Nh;\tfrac{1}{2}N(1+h)-\tfrac{1}{2};-\tfrac{1}{2}\Omega)$$

$$=\frac{\Gamma_m[\frac{1}{2}(N-1)]}{\Gamma_m[\frac{1}{2}N(1+h)-\frac{1}{2}]}\prod_{i=1}^{p}\frac{\Gamma_m[\frac{1}{2}N_i(1+h)-\frac{1}{2}]}{\Gamma_m[\frac{1}{2}(N_i-1)]}$$

$$\cdot{}_1F_1(\tfrac{1}{2}Nh;\tfrac{1}{2}N(1+h)-\tfrac{1}{2};-\tfrac{1}{2}\Omega),$$

and the proof is complete.

COROLLARY 10.8.3. When the null hypothesis $H:\boldsymbol{\mu}_1=\cdots=\boldsymbol{\mu}_p$, $\Sigma_1=\cdots=\Sigma_p$ is true the hth moment of λ is

$$(11)\qquad E(\lambda^h)=\frac{\Gamma_m[\frac{1}{2}(N-1)]}{\Gamma_m[\frac{1}{2}N(1+h)-\frac{1}{2}]}\prod_{i=1}^{p}\frac{\Gamma_m[\frac{1}{2}N_i(1+h)-\frac{1}{2}]}{\Gamma_m[\frac{1}{2}(N_i-1)]}$$

10.8.2. The Asymptotic Null Distribution of the Likelihood Ratio Statistic

Putting $N_i = k_i N$ where $\sum_{i=1}^{p} k_i = 1$ it follows from Corollary 10.8.3 that when H is true

(12)
$$E(\Lambda^h) = \frac{N^{mNh/2}}{\prod_{i=1}^{p} N_i^{mN_i h/2}} E(\lambda^h)$$

$$= \frac{N^{mNh/2}}{\prod_{i=1}^{p} N_i^{mN_i h/2}} \frac{\Gamma_m[\frac{1}{2}(N-1)]}{\Gamma_m[\frac{1}{2}N(1+h)-\frac{1}{2}]} \prod_{i=1}^{p} \frac{\Gamma_m[\frac{1}{2}N_i(1+h)-\frac{1}{2}]}{\Gamma_m[\frac{1}{2}(N_i-1)]}$$

$$= K\left[\frac{\prod_{j=1}^{m}(\frac{1}{2}N)^{N/2}}{\prod_{j=1}^{m}\prod_{i=1}^{p}(\frac{1}{2}k_i N)^{k_i N/2}}\right]^h \frac{\prod_{i=1}^{p}\prod_{j=1}^{m}\Gamma[\frac{1}{2}Nk_i(1+h)-\frac{1}{2}j]}{\prod_{j=1}^{m}\Gamma[\frac{1}{2}N(1+h)-\frac{1}{2}j]},$$

where K is a constant not involving h. This has the same form as (18) of Section 8.2, where we put $p = m$, $q = mp$, $y_j = \frac{1}{2}N$, $\eta_j = -\frac{1}{2}j (j = 1, \ldots, m)$:

$$x_k = \tfrac{1}{2}Nk_i, \qquad k = (i-1)m+1, \ldots, im \qquad (i = 1, \ldots, p)$$

$$\xi_k = -\tfrac{1}{2}j, \qquad k = j, m+j, \ldots, (p-1)m+j \qquad (j = 1, \ldots, m).$$

The degrees of freedom in the limiting χ^2 distribution are, using (28) of Section 8.2,

(13)
$$f = -2\left[\sum_{k=1}^{q}\xi_k - \sum_{j=1}^{p}\eta_j - \tfrac{1}{2}(q-p)\right]$$

$$= \sum_{i=1}^{p}\sum_{j=1}^{m} j - \sum_{j=1}^{m} j + m(p-1)$$

$$= \tfrac{1}{2}m(p-1)(m+3).$$

The value of ρ which makes the term of order N^{-1} vanish in the asymptotic expansion of the distribution of $-2\rho\log\Lambda$ is, using (30) of Section 8.2

(14)
$$\rho = \frac{1}{f}\left[\sum_{i=1}^{p}\sum_{j=1}^{m}\frac{2}{Nk_i}\left(\tfrac{1}{4}j^2+\tfrac{1}{2}j+\tfrac{1}{6}\right) - \sum_{j=1}^{m}\frac{2}{N}\left(\tfrac{1}{4}j^2+\tfrac{1}{2}j+\tfrac{1}{6}\right)\right]$$

$$= 1 - \frac{2m^2+9m+11}{6(p-1)(m+3)N}\left(\sum_{i=1}^{p}\frac{1}{k_i} - 1\right).$$

The term of order N^{-2} can also be obtained by the theory of Section 8.2 giving the following result.

THEOREM 10.8.4. When the null hypothesis $H: \boldsymbol{\mu}_1 = \cdots = \boldsymbol{\mu}_p$, $\Sigma_1 = \cdots = \Sigma_p$ is true the distribution of $-2\rho\log\Lambda$, where ρ is given by (14), can be expanded for large $M = \rho N$ as

$$(15) \qquad P(-2\rho\log\Lambda \le x) = P(\chi_f^2 \le x) + \frac{\gamma}{M^2}\left[P(\chi_{f+4}^2 \le x) - P(\chi_f^2 \le x)\right] + O(M^{-3}),$$

where $f = \frac{1}{2}m(p-1)(m+3)$ and

$$\gamma = \frac{1}{288}\left[6m(m+1)(m+2)(m+3)\left(\sum_{i=1}^{p}\frac{1}{k_i^2} - 1\right) - \frac{(2m^2+9m+11)^2(2m-1)}{(p-1)m(m+3)}\left(\sum_{i=1}^{p}\frac{1}{k_i} - 1\right)^2\right].$$

A similar type of expansion, including the term of order M^{-2}, was given by T. W. Anderson (1958), Section 10.5, but for his modified likelihood ratio statistic.

An approximate test of size α of H is to reject H if $-2\rho\log\Lambda > c_f(\alpha)$, where $c_f(\alpha)$ denotes the upper $100\alpha\%$ point of the χ_f^2 distribution. The error in this approximation is of order N^{-2}.

The modified likelihood ratio statistic Λ^* given by (8) has been studied more extensively than Λ. Chang et al. (1977) have calculated the upper 5 percentage points of the distribution of $-2\log\Lambda^*$ for $n_i = n_0$ $(i = 1,\ldots,p)$, $p = 2(1)8$, $m = 1(1)4$. These are given in Table 7, in which $M_0 = n_0 - m$.

10.8.3. An Asymptotic Non-null Distribution of the Likelihood Ratio Statistic

The power function of the likelihood ratio test of size α is $P(-2\rho\log\Lambda \ge k_\alpha^* \mid \boldsymbol{\mu}_1,\ldots,\boldsymbol{\mu}_p, \Sigma_1,\ldots,\Sigma_p)$ where ρ is given by (14) and k_α^* denotes the upper $100\alpha\%$ point of the distribution of $-2\rho\log\Lambda$ when H is true. We will now derive the asymptotic distribution of $-2\rho\log\Lambda$ in a special case. We consider the sequence of local alternatives

$$(16) \qquad K_N: \quad \boldsymbol{\mu}_i = \bar{\boldsymbol{\mu}} + \frac{1}{N_i^{1/2}}\boldsymbol{\delta}_i \qquad (i = 1,\ldots,p)$$

$$\Sigma_1 = \cdots = \Sigma_p$$

Table 7. Upper 5 percentage points of $-2\log\Lambda^*$, where Λ^* is the modified likelihood ratio statistic for testing equality of p normal populations (equal sample sizes)[a]

$m=1$

M_0	$p=2$	$p=3$	$p=4$	$p=5$	$p=6$	$p=7$	$p=8$
1	6.96	10.39	13.42	16.26	19.00	21.66	24.26
2	6.68	10.13	13.18	16.03	18.78	21.45	24.06
3	6.52	9.99	13.04	15.91	18.66	21.34	23.95
4	6.42	9.89	12.96	15.83	18.59	21.27	23.89
5	6.36	9.83	12.90	15.78	18.54	21.23	23.85
6	6.31	9.78	12.85	15.74	18.51	21.19	23 83
7	6.27	9.75	12.82	15.71	18.48	21.17	23.80
8	6.24	9.72	12.80	15.68	18.46	21.15	23.79
9	6.21	9.69	12.78	15.67	18.45	21.14	23.78
10	6.19	9.68	12.76	15.65	18.43	21.13	23.77
11	6.18	9.66	12.75	15.64	18.42	21.12	23.76
12	6.16	9.65	12.73	15.63	18.41	21.11	23.75
13	6.15	9.64	12.72	15.62	18.40	21.10	23.75
14	6.14	9.63	12.71	15.62	18.40	21.10	23.75
15	6.13	9.62	12.71	15.61	18.39	21.09	23.74
16	6.12	9.61	12.70	15.60	18.39	21.09	23.74
17	6.12	9.60	12.70	15.59	18.38	21 09	23.73
18	6.11	9.60	12.69	15.59	18.38	21.08	23.73
19	6.10	9.59	12.68	15.59	18.37	21.08	23.73
20	6.10	9.59	12.68	15.58	18.37	21.08	23.72
25	6.08	9.57	12.66	15.57	18.36	21.07	23.72
30	6.06	9.56	12.65	15.56	18.35	21.06	23.71

$m=2$

M_0	$p=2$	$p=3$	$p=4$	$p=5$	$p=6$	$p=7$	$p=8$
1	15.74	24.25	32.03	39.46	46.70	53.78	60 77
2	14.19	22.26	29.63	36.70	43.56	50.29	56.91
3	13.41	21.27	28.46	35.33	42.02	48.58	55.04
4	12.94	20.67	27.75	34.53	41.12	47.58	53.95
5	12.63	20.28	27.29	34.00	40.52	46.92	53.22
6	12.41	20.00	26.96	33.62	40.10	46.46	52.71
7	12.24	19.78	26.71	33.34	39.79	46.11	52.34
8	12.12	19.62	26.51	33.12	39.54	45.84	52.04
9	12.01	19.49	26.36	32.94	39.35	45.63	51.81
10	11.93	19.38	26.24	32.81	39.19	45.46	51.62
11	11.86	19.29	26.13	32.69	39.06	45.31	51.47
12	11.79	19.22	26.04	32.58	38.95	45.19	51.34
13	11.74	19.15	25.97	32.50	38.86	45.09	51.23
14	11.69	19.09	25.90	32.43	38.77	44.99	51.13
15	11.66	19.04	25.84	32.36	38.70	44.92	51.04
16	11.63	19.00	25.79	32.31	38.64	44.85	50.97
17	11.59	18.96	25.75	32.26	38.59	44.79	50.90
18	11.56	18.93	25.72	32.21	38.54	44.73	50.85
19	11.54	18.90	25.67	32.17	38.49	44.69	50.79
20	11.52	18.87	25.65	32.14	38.45	44.64	50.74
25	11.43	18.76	25.52	32.00	38.29	44.47	50.55
30	11.37	18.69	25.44	31.90	38.19	44.36	50.43

Table 7 (*Continued*)

			$m=3$				
M_0	$p=2$	$p=3$	$p=4$	$p=5$	$p=6$	$p=7$	$p=8$
1	27.27	42.89	57.37	71.35	85.01	98.45	111.75
2	23.95	38.43	51.87	64.83	77.51	89.99	102.31
3	22.26	35.15	49.05	61.51	73.68	85.67	97.51
4	21.22	34.75	47.33	59.47	71.35	83.03	94.58
5	20.53	33.80	46.17	58.10	69.78	81.27	92.62
6	20.03	33.12	45.33	57.11	68.64	79.98	91.20
7	19.66	32.60	44.69	56.37	67.79	79.03	90.13
8	19.35	32.20	44.20	55.79	67.12	78.27	89.31
9	19.12	31.88	43.80	55.32	66.58	77.68	88.64
10	18.92	31.61	43.47	54.93	66.15	77.18	88.09
11	18.76	31.39	43.20	54.61	65.77	76.77	87.64
12	18.62	31.20	42.97	54.33	65.46	76.41	87.25
13	18.50	31.03	42.76	54.10	65.20	76.11	86.91
14	18.40	30.90	42.59	53.90	64.97	75.85	86.62
15	18.31	30.77	42.44	53.72	64.75	75.62	86.36
16	18.22	30.66	42.31	53.56	64.58	75.42	86.15
17	18.15	30.56	42.18	53.42	64.42	75.24	85.94
18	18.09	30.48	42.07	53.29	64.27	75.08	85.76
19	18.03	30.39	41.98	53.18	64.14	74.94	85.60
20	17.98	30.32	41.89	53.08	64.03	74.81	85.45
25	17.78	30.05	41.55	52.69	63.58	74.30	84.90
30	17.64	29.86	41.33	52.41	63.26	73.96	84.52

			$m=4$				
M_0	$p=2$	$p=3$	$p=4$	$p=5$	$p=6$	$p=7$	$p=8$
1	41.57	66.34	89.52	111.98	134.01	155.75	177.24
2	36.13	58.85	80.15	100.78	121.03	141.02	160.76
3	33.27	54.89	75.18	94.85	114.15	133.18	152.03
4	31.50	52.42	72.07	91.13	109.84	128.28	146.54
5	30.28	50.72	69.94	88.58	106.86	124.92	142.79
6	29.40	49.48	68.37	86.71	104.70	122.47	140.05
7	26.72	48.53	67.18	85.28	103.04	120.57	137.94
8	28.19	47.78	66.24	84.16	101.74	119.10	136.28
9	27.76	47.17	65.48	83.24	100.68	117.91	134.95
10	27.41	46.67	64.84	82.49	99.80	116.91	133.84
11	27.11	46.25	64.31	81.85	99.08	116.07	132.90
12	26.85	45.89	63.86	81.31	98.44	115.36	132.11
13	26.64	45.58	63.46	80.84	97.90	114.75	131.43
14	26.45	45.31	63.13	80.43	97.44	114.22	130.83
15	26.28	45.06	62.83	80.08	97.02	113.74	130.30
16	26.13	44.86	62.56	79.75	96.65	113.32	129.83
17	26.01	44.66	62.32	79.48	96.32	112.95	129.42
18	25.88	44.50	62.10	79.23	96.03	112.62	129.04
19	25.77	44.34	61.92	78.99	95.76	112.32	128.71
20	25.68	44.20	61.74	78.78	95.52	112.04	128.40
25	25.30	43.67	61.06	77.98	94.58	110.99	127.23
30	25.05	43.30	60.59	77.43	93.94	110.26	126.42

[a]Here, p = number of populations; m = number of variables; n_0 = one less than common sample size ($n_0 = n_i$; $i = 1, \ldots, p$); $M_0 = n_0 - m$

Source. Reproduced from Chang et al. (1977) with the kind permission of North-Holland Publishing Company and the authors.

under which the covariance matrices are all equal and the mean vectors approach a common value.

As before we assume that $N_i = k_i N$ with $k_i > 0$ $(i = 1,\dots,p)$ and $\sum_{i=1}^{p} k_i = 1$ and let $M = \rho N \to \infty$. Using Theorem 10.8.2 the characteristic function of $-2\rho\log\Lambda$ under K_N may be expressed as

$$(17) \qquad \phi_N(t;\Omega) = \phi_N(t;0)\,{}_1F_1\left(-Mit;\tfrac{1}{2}M(1-2it)+\alpha;-\tfrac{1}{2}\Omega\right)$$

where

$$(18) \qquad \Omega = \Sigma^{-1}\sum_{i=1}^{p}\delta_i\delta_i',$$

$$(19) \qquad \begin{aligned}\alpha &= \tfrac{1}{2}(N-m) - \tfrac{1}{2}\\ &= \tfrac{1}{2}N(1-\rho) - \tfrac{1}{2}\\ &= \frac{2m^2+9m+11}{12(p-1)(m+3)}\left(\sum_{i=1}^{p}\frac{1}{k_i}-1\right) - \frac{1}{2}\end{aligned}$$

and $\phi_N(t;0)$ is the characteristic function of $-2\rho\log\Lambda$ when H is true, obtained from (12) by putting $h = -2it\rho$. From Theorem 10.8.4 we know that

$$\phi_N(t;0) = (1-2it)^{-f/2} + O(M^{-2})$$

where $f = \frac{1}{2}m(p-1)(m+3)$. An asymptotic expansion for the ${}_1F_1$ function in (17) was obtained in Theorem 10.5.6, where we there replace N by M and r by $4\alpha - m - 1$. Theorem 10.5.6 then shows that

$$(20) \qquad \begin{aligned}&{}_1F_1\left(-Mit;\tfrac{1}{2}M(1-2it)+\alpha;-\tfrac{1}{2}\Omega\right)\\ &= \exp\left(\frac{it\sigma_1}{1-2it}\right)\left[1+\frac{P_1(\Omega)}{M}+O(M^{-2})\right],\end{aligned}$$

where $\sigma_j = \operatorname{tr}\Omega^j$ and

$$(21) \qquad P_1(\Omega) = -2\alpha\frac{it}{(1-2it)^2}\sigma_1 - \frac{it}{(1-2it)^3}\frac{\sigma_2}{2}.$$

Hence $\phi_N(t;\Omega)$ can be expanded as

$$(22)\quad \phi_N(t;\Omega)=(1-2it)^{-f/2}\exp\left(\frac{it\sigma_1}{1-2it}\right)\left[1+\frac{P_1(\Omega)}{M}+O(M^{-2})\right]$$

Inverting this gives the following theorem.

THEOREM 10.8.5. Under the sequence of local alternatives K_N given by (16), the distribution function of $-2\rho\log\Lambda$ can be expanded for large $M=\rho N$ as

$$(23)\quad \begin{aligned} P(-2\rho\log\Lambda\le x)=P\left(\chi_f^2(\sigma_1)\le x\right)+\frac{1}{4M}\Big\{&4\alpha\sigma_1 P\left(\chi_{f+2}^2(\sigma_1)\le x\right)\\ &-[4\alpha\sigma_1-\sigma_2]P\left(\chi_{f+4}^2(\sigma_1)\le x\right)\\ &-\sigma_2 P\left(\chi_{f+6}^2(\sigma_1)\le x\right)\Big\}+O(M^{-2}), \end{aligned}$$

where $f=\frac{1}{2}(p-1)m(m+3)$, $\sigma_j=\operatorname{tr}\Omega^j$, with Ω and α given by (18) and (19).

For Anderson's modified statistic a similar expansion has been obtained by Fujikoshi (1970) and an expansion in terms of normal distributions has been obtained by Nagao (1972).

PROBLEMS

10.1. In the univariate linear model consider testing $H: C\boldsymbol{\beta}=\mathbf{0}$ against $K: C\boldsymbol{\beta}\neq\mathbf{0}$, where C is a specified $p\times r$ matrix of rank r. Show that the test which rejects H for large values of the statistic F given by (1) of Section 10.1 is a uniformly most powerful invariant test.

10.2. Suppose that $A=nS$ has the $W_m(n,\Sigma,\Omega)$ distribution. Show that as $n\to\infty$ the asymptotic distribution of $(n/2m)^{1/2}\log(\det S/\det\Sigma)$ is $N(0,1)$ (see Fujikoshi, 1968).

10.3. Suppose that $A=nS$ has the $W_m(n,\Sigma,\Omega)$ distribution, where $\Omega=n\Delta$ with Δ a fixed $m\times m$ matrix. Show that as $n\to\infty$ the asymptotic distribution of

$$n^{1/2}\left\{\frac{\det S}{\det\Sigma}-\det(I+\Delta)\right\}$$

is $N\{0,2\det(I+\Delta)^2\operatorname{tr}[(I+2\Delta)(I+\Delta)^{-2}]\}$.

10.4. Suppose that $A=nS$ has the $W_m(n,\Sigma,\Omega)$ distribution, where $\Omega=n^{1/2}\Delta$, with Δ a fixed $m\times m$ matrix. Show that as $n\rightarrow\infty$ the asymptotic distribution of

$$\left(\frac{n}{2m}\right)^{1/2}\log\frac{\det S}{\det\Sigma}-\left(\frac{2}{m}\right)^{1/2}\operatorname{tr}\Delta$$

is $N(0,1)$ (see Fujikoshi, 1970).

10.5. If A is $W_m(n,\Sigma,\Omega)$ show that the characteristic function of $\operatorname{tr}A$ is

$$E[\exp(it\operatorname{tr}A)]=\det(I-2it\Sigma)^{-n/2}\operatorname{etr}\left[-\tfrac{1}{2}\Omega+\tfrac{1}{2}\Omega(I-2it\Sigma)^{-1}\right]$$

10.6. Using the result of Problem 10.5, show that if $A=nS$ is $W_m(n,\Sigma,\Omega)$ then as $n\rightarrow\infty$ the asymptotic distribution of

$$\left[\frac{n}{2\operatorname{tr}(\Sigma^2)}\right]^{1/2}(\operatorname{tr}S-\operatorname{tr}\Sigma)$$

is $N(0,1)$ (see Fujikoshi, 1970).

10.7. If $A=nS$ is $W_m(n,\Sigma,\Omega)$ and $\Omega=n\Delta$, where Δ is a fixed $m\times m$ matrix, show that as $n\rightarrow\infty$ the asymptotic distribution of

$$\left[\frac{n}{2\operatorname{tr}\Sigma^2(I+2\Delta)}\right]^{1/2}[\operatorname{tr}S-\operatorname{tr}\Sigma(I+\Delta)]$$

is $N(0,1)$.

10.8. If the $n\times m$ matrix Z is $N(M,I_n\otimes\Sigma)$, so that $A=Z'Z$ is $W_m(n,\Sigma,\Omega)$ with $\Omega=\Sigma^{-1}M'M$, prove that

$$\operatorname{Cov}(\operatorname{vec}(A))=(I_{m^2}+K)[n(\Sigma\otimes\Sigma)+\Sigma\otimes(M'M)+(M'M)\otimes\Sigma],$$

where K is the commutation matrix defined in Problem 3.2.

10.9. If A is $W_m(n,\Sigma,\Omega)$ with $n>m-1$ show that the density function of $w_1,\ldots,w_m$, the latent roots of $\Sigma^{-1}A$, is

$$\operatorname{etr}(-\tfrac{1}{2}\Omega)\,{}_0F_1^{(m)}(\tfrac{1}{2}n;\tfrac{1}{4}\Omega,W)\cdot\frac{\pi^{m^2/2}}{2^{mn/2}\Gamma_m(\tfrac{1}{2}n)\Gamma_m(\tfrac{1}{2}m)}$$

$$\cdot\exp\left(-\tfrac{1}{2}\sum_{i=1}^{m}w_i\right)\prod_{i=1}^{m}w_i^{(n-m-1)/2}\prod_{i<j}^{m}(w_i-w_j)\qquad(w_1>\cdots>w_m>0)$$

where $W=\operatorname{diag}(w_1,\dots,w_m)$. Why does this distribution depend only on the latent roots of Ω?

10.10. If A is $W_m(n,\Sigma,\Omega)$ where A,Σ,Ω are partitioned as

$$A=\begin{bmatrix} A_{11} & A_{12} \\ A_{21} & A_{22} \end{bmatrix},\qquad \Sigma=\begin{bmatrix} \Sigma_{11} & \Sigma_{12} \\ \Sigma_{21} & \Sigma_{22} \end{bmatrix},\qquad \Omega=\begin{bmatrix} \Omega_{11} & \Omega_{12} \\ \Omega_{21} & \Omega_{22} \end{bmatrix}$$

where A_{11}, Σ_{11}, and Ω_{11} are $k\times k$, show that the marginal distribution of A_{11} is $W_k(n,\Sigma_{11},\Sigma_{11}^{-1}(\Sigma_{11}\Omega_{11}+\Sigma_{12}\Omega_{21}))$.

10.11. Suppose that the $n\times m$ $(n\geq m)$ matrix Z is $N(M,I_n\otimes I_m)$, where $M=[\mathbf{m}_1\mathbf{0}\dots\mathbf{0}]$, so that $A=Z'Z$ is $W_m(n,I,\Omega)$, with $\Omega=\operatorname{diag}(\mathbf{m}_1'\mathbf{m}_1,0,\dots,0)$. Partition A as

$$A=\begin{bmatrix} a_{11} & \mathbf{a}_{12}' \\ \mathbf{a}_{12} & A_{22} \end{bmatrix}\begin{matrix} 1 \\ m-1 \end{matrix}$$
$$\qquad\quad 1 \qquad m-1$$

and put $A_{22\cdot 1}=A_{22}-a_{11}^{-1}\mathbf{a}_{12}\mathbf{a}_{12}'$. Prove that:

(a) $A_{22\cdot 1}$ is $W_{m-1}(n-1,I_{m-1})$ and is independent of $\mathbf{a}_{12}$ and a_{11}.

(b) The conditional distribution of $\mathbf{a}_{12}$ given a_{11} is $N_{m-1}(\mathbf{0},a_{11}I_{m-1})$.

(c) a_{11} is $\chi_n^2(\delta)$, with $\delta=\mathbf{m}_1'\mathbf{m}_1$.

(See Theorem 3.2.10.)

10.12. Consider the moments of the likelihood ratio statistic W given by (2) of Section 10.5. Using the duplication formula

$$\Gamma(a+1)\Gamma(a+\tfrac{1}{2})=\pi^{1/2}2^{-2a}\Gamma(2a+1),$$

show that, when m is even, $m=2k$ say, these moments can be written as

$$E(W^h)=\prod_{i=1}^{k}\frac{\Gamma(r+n-p+1-2i)\Gamma(n-p+1-2i+2h)}{\Gamma(r+n-p+1-2i+2h)\Gamma(n-p+1-2i)}.$$

Hence, show that W has the same distribution as $\prod_{i=1}^{k}U_i^2$, where $U_1,\dots,U_k$ are independent and U_i is beta$(n-p+1-2i,r)$. Show also that if m is odd, $m=2k+1$, W has the same distribution as $\prod_{i=1}^{k}Y_i^2Y_{k+1}$, where $Y_1,\dots,Y_{k+1}$ are independent, with Y_i having the beta$(n-p+1-2i,r)$ distribution $(i=1,\dots,k)$ and Y_{k+1} having the beta$(\frac{1}{2}(n-p+1-m),\frac{1}{2}r)$ distribution.

10.13. Prove Theorem 10.5.8.

10.14. Let $T_0^2 = \operatorname{tr}(AB^{-1})$, where A and B are independent with A having the $W_m(r, \Sigma, \Omega)$ distribution and B having the $W_m(n-p, \Sigma)$ distribution (see Section 10.6.2), with $r \geq m$, $n-p \geq m$. Using the joint density function of A and B show that if $n-p \geq 2k+m-1$ the kth moment of T_0^2 is

$$E\left[(T_0^2)^k\right] = (-1)^k \sum_{\kappa} \frac{L_\kappa^\gamma(-\frac{1}{2}\Omega)}{\left(\frac{1}{2}(m+1-n+p)\right)_\kappa} \qquad \left[\gamma = \tfrac{1}{2}(r-m-1)\right],$$

where the summation is over all partitions κ of k.

10.15. If f_m is the smallest latent root of AB^{-1}, where A is $W_m(r, \Sigma, \Omega)$ and B is independently $W_m(n-p, \Sigma)$, show that

$$P(f_m \geq x) \leq \prod_{i=1}^{m} P\left(F_{r,n-p}(\omega_i) \leq \frac{n-p}{r} x\right),$$

where $\omega_1, \ldots, \omega_m$ are the latent roots of Ω. [*Hint*: Use the result of Problem 8.3.]

10.16. For the single classification model of Section 10.7 show that the steps involved in reducing it to canonical form (see Section 10.2) lead to the between-classes and within-classes matrices A and B given by (1) and (2) of Section 10.7. Show also that the noncentrality matrix Ω is given by (3) of Section 10.7.

10.17. Obtain Corollary 10.7.4 from Theorem 10.7.2.

10.18. The generalized MANOVA model (GMANOVA) (Potthoff and Roy, 1964; Gleser and Olkin, 1970; Fujikoshi, 1974b; Kariya, 1978). Let Y be a $n \times m$ matrix whose rows have independent m-variate normal distributions with unknown covariance matrix Ψ and where

$$E(Y) = X_1 \mathbf{B} X_2;$$

Here X_1 is a known $n \times p$ matrix of rank $p \leq n$; X_2 is a known $q \times m$ matrix of rank $q \leq m$; and $\mathbf{B}$ is a $p \times q$ matrix of unknown parameters. This is known as the GMANOVA model. When $X_2 = I_m$, $q = m$, it reduces to the classical MANOVA model introduced in Section 10.1. When $p = 1$, $X_1 = \mathbf{1} = (1, 1, \ldots, 1)'$, the model is usually called the "growth curves" model.

Consider the problem of testing the null hypothesis H: $X_3 \mathbf{B} X_4 = 0$ against K: $X_3 \mathbf{B} X_4 \neq 0$, where X_3 is a known $u \times p$ matrix of rank $u \leq p$ and X_4 is a known $q \times v$ matrix of rank $v \leq q$.

(a) Show that by transforming Y the problem can be expressed in the following canonical form: Let Z be a random $n \times m$ matrix whose rows have independent m-variate normal distributions with covariance matrix Σ, partitioned as

$$Z=\begin{bmatrix} Z_{11} & Z_{12} & Z_{13} \\ Z_{21} & Z_{22} & Z_{23} \\ Z_{31} & Z_{32} & Z_{33} \end{bmatrix}\begin{matrix} u \\ p-u \\ n-p \end{matrix},$$
$$\qquad\quad \begin{matrix} q-v & v & m-q \end{matrix}$$

with

$$E(Z)=\begin{bmatrix} \Theta_{11} & \Theta_{12} & 0 \\ \Theta_{21} & \Theta_{22} & 0 \\ 0 & 0 & 0 \end{bmatrix}\begin{matrix} u \\ p-u \\ n-p \end{matrix}$$
$$\qquad\quad \begin{matrix} q-v & v & m-q \end{matrix}$$

The null hypothesis $H: X_3 \mathbb{B} X_4 = 0$ is equivalent to $H: \Theta_{12} = 0$. [*Hint*: Write

$$X_1 = H_1\begin{bmatrix} I_p \\ \cdot\cdot \\ 0 \end{bmatrix}L_1, \qquad X_2 = L_2[I_q : 0]H_2,$$

where $H_1 \in O(n)$, $H_2 \in O(m)$, $L_1 \in \mathcal{G}\ell(p, R)$, $L_2 \in \mathcal{G}\ell(q, R)$, and put

$$Y^* = H_1' Y H_2', \qquad \mathbb{B}^* = L_1 \mathbb{B} L_2, \qquad \Psi^* = H_1' \Psi H_2'.$$

Express $E(Y^*), \operatorname{Cov}(Y^*)$ in terms of $\mathbb{B}^*$ and Ψ^*. Next, write

$$X_3 L_1^{-1} = L_3[I_u : 0]H_3, \qquad L_2^{-1} X_4 = H_4\begin{bmatrix} 0 \\ \cdots \\ I_v \end{bmatrix}L_4,$$

where $H_3 \in O(p)$, $H_4 \in O(q)$, $L_3 \in \mathcal{G}\ell(u, R)$, $L_4 \in \mathcal{G}\ell(v, R)$, and put

$$Z = \begin{bmatrix} H_3 & 0 \\ 0 & I_{n-p} \end{bmatrix} Y^* \begin{bmatrix} H_4 & 0 \\ 0 & I_{m-q} \end{bmatrix}.$$

Show that

$$E(Z) = \begin{bmatrix} \Theta & 0 \\ 0 & 0 \end{bmatrix}\begin{matrix} p \\ n-p \end{matrix},$$
$$\qquad\quad \begin{matrix} q & m-q \end{matrix}$$

where $\Theta = H_3 \mathbf{B}^* H_4$. Partitioning Θ as

$$\Theta = \begin{bmatrix} \Theta_{11} & \Theta_{12} \\ \Theta_{21} & \Theta_{22} \end{bmatrix} \begin{matrix} u \\ p-u \end{matrix}, \qquad \text{columns: } q-v, \; v$$

show that $H: X_3 \mathbf{B} X_4 = 0$ is equivalent to $H: \Theta_{12} = 0$. Letting Σ be the covariance matrix of each of the rows of Z, express Σ in terms of Ψ^*.]

(b) Put

$$Z_1 = \begin{bmatrix} Z_{11} & Z_{12} & Z_{13} \\ Z_{21} & Z_{22} & Z_{23} \end{bmatrix}, \qquad Z_2 = \begin{bmatrix} Z_{31} & Z_{32} & Z_{33} \end{bmatrix}.$$

Show that a sufficient statistic is (Z_1, B), where $B = Z_2' Z_2$. State the distributions of Z_1 and B.

(c) Partition B and Σ as

$$B = \begin{bmatrix} B_{11} & B_{12} & B_{13} \\ B_{21} & B_{22} & B_{23} \\ B_{31} & B_{32} & B_{33} \end{bmatrix} \begin{matrix} q-v \\ v \\ m-q \end{matrix}, \qquad \Sigma = \begin{bmatrix} \Sigma_{11} & \Sigma_{12} & \Sigma_{13} \\ \Sigma_{21} & \Sigma_{22} & \Sigma_{23} \\ \Sigma_{31} & \Sigma_{32} & \Sigma_{33} \end{bmatrix} \begin{matrix} q-v \\ v \\ m-q \end{matrix}.$$
$$\text{columns: } q-v, \; v, \; m-q$$

For ease of notation, put $m_1 = q - v$, $m_2 = v$, $m_3 = m - q$ so that B_{ij} and Σ_{ij} are $m_i \times m_j$ matrices, and put $n_1 = u$, $n_2 = p - u$. Consider the group of transformations

$$G = \left\{ (A, F);\; A = \begin{bmatrix} A_{11} & 0 & 0 \\ A_{21} & A_{22} & 0 \\ A_{31} & A_{32} & A_{33} \end{bmatrix} \in \mathcal{G}\ell(m, R), \right.$$

$$A_{ii} \in \mathcal{G}\ell(m_i, R),\; i = 1, 2, 3,$$

$$\left. F = \begin{bmatrix} F_{11} & 0 & 0 \\ F_{21} & F_{22} & 0 \end{bmatrix} \begin{matrix} n_1 \\ n_2 \end{matrix} \right\} \qquad \text{columns: } m_1, \; m_2, \; m_3$$

acting on the sample space of the sufficient statistic by

$$(Z_1, B) \overset{(A,F)}{\rightarrow} (Z_1 A + F, A'BA).$$

This transformation induces the transformation

$$\left(\begin{bmatrix}\Theta_{11} & \Theta_{12} & 0\\ \Theta_{21} & \Theta_{22} & 0\end{bmatrix}, \Sigma\right) \rightarrow \left(\begin{bmatrix}\Theta_{11} & \Theta_{12} & 0\\ \Theta_{21} & \Theta_{22} & 0\end{bmatrix} A + F, A'\Sigma A\right)$$

in the parameter space. Show that the problem of testing $H: \Theta_{12}=0$ against $K: \Theta_{12} \neq 0$ is invariant under G.

(d) Prove that, if $n_1 \le m_2, p \le m_3$, a maximal invariant in the sample space under the group G is $g(Z_1, B) = (g_1(Z_1, B), g_2(Z_1, B))$ where

$$g_1(Z_1, B) = [Z_{12} \quad Z_{13}] \begin{bmatrix} B_{22} & B_{23}\\ B_{32} & B_{33}\end{bmatrix}^{-1} [Z_{12} \quad Z_{13}]'$$

$$g_2(Z_1, B) = \begin{bmatrix} Z_{13}\\ Z_{23}\end{bmatrix} B_{33}^{-1} \begin{bmatrix} Z_{13}\\ Z_{23}\end{bmatrix}'.$$

[*Hint*: First show that $g_1(Z_1, B)$ and $g_2(Z_1, B)$ are invariant. Next, consider any invariant function $h(Z_1, B)$, i.e., any h satisfying

$$h(Z_1, B) = h(Z_1 A + F, A'BA)$$

for all $(A, F) \in G$. It suffices to show that $h(Z_1, B)$ depends on (Z_1, B) only through $g_1(Z_1, B)$ and $g_2(Z_1, B)$. First, note that there exists a matrix

$$T = \begin{bmatrix} T_{11} & 0 & 0\\ T_{21} & T_{22} & 0\\ T_{31} & T_{32} & T_{33}\end{bmatrix} \in \mathcal{G}\ell(m, R)$$

with $T_{ii} \in \mathcal{G}\ell(m_i, R)$ $(i=1,2,3)$, such that $B^{-1} = TT'$. (Why?) Let

$$H = \begin{bmatrix} H_{11} & 0 & 0\\ 0 & H_{22} & 0\\ 0 & 0 & H_{33}\end{bmatrix},$$

where $H_{ii} \in O(m_i)$ $(i=1,2,3)$, be an arbitrary orthogonal $m \times m$ matrix, and put $A_H = TH$. Then A_H has the same form as the matrix A in (c), and for all F of the form given in (c) and all H of the form above

$$h(Z_1A_H+F,I_m)=h(Z_1A_H+F,A'_HBA_H)=h(Z_1,B).$$

This shows that $h(Z_1,B)$ depends on (Z_1,B) only through Z_1A_H+F. By writing this matrix out, show that F can be chosen so that $h(Z_1,B)$ is a function only of the matrices

$$[Z_{12}\quad Z_{13}]\begin{bmatrix} T_{22} \\ T_{32} \end{bmatrix}H_{22} \quad \text{and} \quad \begin{bmatrix} Z_{13} \\ Z_{23} \end{bmatrix}T_{33}\,H_{33}.$$

Now show that H_{22} and H_{33} can be chosen so that

$$[Z_{12}\quad Z_{13}]\begin{bmatrix} T_{22} \\ T_{32} \end{bmatrix}H_{22}=[g_1(Z_1,B):0]$$

and

$$\begin{bmatrix} Z_{13} \\ Z_{23} \end{bmatrix}T_{33}H_{33}=[g_2(Z_1,B):0]].$$

(e) Prove that a maximal invariant in the parameter space under the group induced by G is

$$\Delta=[\Theta_{12}\quad 0]\begin{bmatrix} \Sigma_{22} & \Sigma_{23} \\ \Sigma_{32} & \Sigma_{33} \end{bmatrix}^{-1}\begin{bmatrix} \Theta'_{12} \\ 0 \end{bmatrix}$$

$$=\Theta_{12}(\Sigma_{22}-\Sigma_{23}\Sigma_{33}^{-1}\Sigma_{32})^{-1}\Theta'_{12}$$

(f) Show that the problem of testing $H:\Theta_{12}=0$ against $K:\Theta_{12}\neq 0$ is also invariant under the larger group of transformations

$$G^*=\left\{(Q,A,F);\quad Q=\begin{bmatrix} Q_1 & 0 \\ 0 & Q_2 \end{bmatrix}\in O(p),\qquad Q_i\in O(n_i),\right.$$

$$\left. i=1,2,\quad (A,F)\in G\right\},$$

acting on the sample space of the sufficient statistic by

$$(Z_1,B)\overset{(Q,A,F)}{\rightarrow}(QZ_1A+F,A'BA).$$

(The group G is isomorphic to the subgroup $\{(I_p,A,F);(A,F)\in G\}$ of the group G^*.)

(g) A tractable maximal invariant under G^* in the sample space is difficult to characterize. Show, however, that a maximal invariant in the parameter space under the group induced by G^* is $(\delta_1,\dots,\delta_{n_1})$, where $\delta_1\geq\cdots\geq\delta_{n_1}$ are the latent roots of the matrix Δ given in (e).

(h) Show that the likelihood ratio statistic for testing $H:\Theta_{12}=0$ against $K:\Theta_{12}\neq 0$ is $\Lambda=W^{n/2}$, where

$$W=\frac{\det\left(I+Z_{13}B_{33}^{-1}Z_{13}'\right)}{\det\left(I+[Z_{12}\quad Z_{13}]\begin{bmatrix}B_{22}&B_{23}\\B_{32}&B_{33}\end{bmatrix}^{-1}[Z_{12}\quad Z_{13}]'\right)}$$

(i) Show that the statistic W in (h) can be expressed in the form

$$W=\frac{\det B_{22\cdot 3}}{\det(X'X+B_{22\cdot 3})},$$

where $B_{22\cdot 3}=B_{22}-B_{23}B_{33}^{-1}B_{32}$ and

$$X=\left(I+Z_{13}B_{33}^{-1}Z_{13}'\right)^{-1/2}\left(Z_{12}-Z_{13}B_{33}^{-1}B_{32}\right).$$

Show also that W is invariant under the group G^*.

(j) Show that, given (Z_{13},B_{33}), the conditional distribution of X is $N((I+Z_{13}B_{33}^{-1}Z_{13}')^{-1/2}\Theta_{12},I_{n_1}\otimes\Sigma_{22\cdot 3})$, where $\Sigma_{22\cdot 3}=\Sigma_{22}-\Sigma_{23}\Sigma_{33}^{-1}\Sigma_{32}$. Show also that $B_{22\cdot 3}$ is $W_{m_2}(n-p-m_3,\Sigma_{22\cdot 3})$ and that $B_{22\cdot 3}$ is independent of X and $Z_{13}B_{33}^{-1}Z_{13}'$.

(k) When the null hypothesis $H:\Theta_{12}=0$ is true, $X'X$ is $W_{m_2}(n_1,\Sigma_{22\cdot 3})$. Using Corollary 10.5.2, write down the null moments of W and use Theorem 10.5.5 to approximate the null distribution of Λ. [The moments of W under $K:\Theta_{12}\neq 0$ can be expressed in terms of a ${}_2F_2$ hypergeometric function having the matrix $-\frac{1}{2}\Delta$ as argument, where Δ is given in (e). For a derivation, as well as for asymptotic non-null distributions, see Fujikoshi (1974b).]

CHAPTER 11

Testing Independence Between k Sets of Variables and Canonical Correlation Analysis

11.1. INTRODUCTION

In this chapter we begin in Section 11.2 by considering how to test the null hypothesis that k vectors, jointly normally distributed, are independent. The likelihood ratio test is derived and central moments of the test statistic are obtained, from which the null distribution and the asymptotic null distribution are found. For $k=2$ noncentral moments of the test statistic are given and used to find asymptotic non-null distributions.

Testing independence between two sets of variables is very closely related to an exploratory data-analytic technique known as canonical correlation analysis, which is considered in Section 11.3. This technique is concerned with replacing the variables in the two sets by new variables, some of which are highly correlated; in essence it is concerned with reducing the correlation structure between the two sets of variables to the simplest possible form by means of linear transformations on each.

11.2. TESTING INDEPENDENCE OF k SETS OF VARIABLES

11.2.1. The Likelihood Ratio Statistic and Invariance

In this section we consider testing the null hypothesis that k vectors, jointly normally distributed, are independent. Suppose that $\mathbf{X}$ is $N_m(\boldsymbol{\mu}, \Sigma)$ and that

$\mathbf{X}$, $\boldsymbol{\mu}$ and Σ are partitioned as

$$\mathbf{X}'=(\mathbf{X}_1'\mathbf{X}_2'\ldots\mathbf{X}_k'),\qquad \boldsymbol{\mu}'=(\boldsymbol{\mu}_1'\boldsymbol{\mu}_2'\ldots\boldsymbol{\mu}_k')$$

and

$$\Sigma=\begin{bmatrix}\Sigma_{11} & \Sigma_{12} & \cdots & \Sigma_{1k}\\ \Sigma_{21} & \Sigma_{22} & \cdots & \Sigma_{2k}\\ \vdots & & & \\ \Sigma_{k1} & \Sigma_{k2} & \cdots & \Sigma_{kk}\end{bmatrix},$$

where $\mathbf{X}_i$ and $\boldsymbol{\mu}_i$ are $m_i\times 1$ and Σ_{ii} is $m_i\times m_i$ $(i=1,\ldots,k)$, with $\sum_{i=1}^k m_i=m$. We wish to test the null hypothesis H that the subvectors $\mathbf{X}_1,\ldots,\mathbf{X}_k$ are independent, i.e.,

$$H:\Sigma_{ij}=0\qquad (i,j=1,\ldots,k;\ i\neq j),$$

against the alternative K that H is not true. Let $\overline{\mathbf{X}}$ and S be, respectively, the sample mean vector and covariance matrix formed from a sample of $N=n+1$ observations on $\mathbf{X}$, and let $A=nS$ and partition $\overline{\mathbf{X}}$, and A as

$$(1)\qquad \overline{\mathbf{X}}'=(\overline{\mathbf{X}}_1'\overline{\mathbf{X}}_2'\ldots\overline{\mathbf{X}}_k'),\qquad A=\begin{bmatrix}A_{11} & A_{12} & \cdots & A_{1k}\\ A_{21} & A_{22} & \cdots & A_{2k}\\ \vdots & & & \\ A_{k1} & A_{k2} & \cdots & A_{kk}\end{bmatrix},$$

where $\overline{\mathbf{X}}_i$ is $m_i\times 1$ and A_{ii} is $m_i\times m_i$. The likelihood ratio test of H (from Wilks, 1935) is given in the following theorem.

THEOREM 11.2.1. The likelihood ratio test of level α for testing the null hypothesis H of independence rejects H if $\Lambda\le c_\alpha$, where

$$(2)\qquad \Lambda=\frac{(\det A)^{N/2}}{\prod_{i=1}^{k}(\det A_{ii})^{N/2}}$$

and c_α is chosen so that the significance level of the test is α.

Proof. Apart from a multiplicative constant the likelihood function is

$$L(\boldsymbol{\mu},\Sigma)=(\det\Sigma)^{-N/2}\operatorname{etr}(-\tfrac{1}{2}\Sigma^{-1}A)\exp[-\tfrac{1}{2}N(\overline{\mathbf{X}}-\boldsymbol{\mu})'\Sigma^{-1}(\overline{\mathbf{X}}-\boldsymbol{\mu})]$$

and

$$\sup_{\mu,\Sigma} L(\mu,\Sigma)=L(\overline{\mathbf{X}},\hat{\Sigma})=N^{mN/2}e^{-mN/2}(\det A)^{-N/2},$$

where $\hat{\Sigma}=N^{-1}A$. When the null hypothesis H is true Σ has the form

$$\Sigma=\Sigma^*=\begin{bmatrix}\Sigma_{11} & 0 & \cdots & 0\\ 0 & \Sigma_{22} & \cdots & 0\\ \vdots & & \ddots & \\ 0 & 0 & & \Sigma_{kk}\end{bmatrix}, \tag{3}$$

so that the likelihood function becomes

$$L(\mu,\Sigma^*)=\prod_{i=1}^{k} L_i(\mu_i,\Sigma_{ii}),$$

where

$$L_i(\mu_i,\Sigma_{ii})=(\det\Sigma_{ii})^{-N/2}\operatorname{etr}\left(-\tfrac{1}{2}\Sigma_{ii}^{-1}A_{ii}\right)$$

$$\cdot\exp\left[-\tfrac{1}{2}N(\overline{\mathbf{X}}_i-\mu_i)'\Sigma_{ii}^{-1}(\overline{\mathbf{X}}_i-\mu_i)\right].$$

Hence it follows that

$$\begin{aligned}\sup_{\mu,\Sigma^*} L(\mu,\Sigma^*)&=\prod_{i=1}^{k}\sup_{\mu_i,\Sigma_{ii}} L_i(\mu_i,\Sigma_{ii})\\ &=\prod_{i=1}^{k} L_i(\overline{\mathbf{X}}_i,\hat{\Sigma}_{ii})\\ &=N^{mN/2}e^{-mN/2}\prod_{i=1}^{k}(\det A_{ii})^{-N/2},\end{aligned}$$

where $\hat{\Sigma}_{ii}=N^{-1}A_{ii}$. Consequently, the likelihood ratio statistic is

$$\Lambda=\frac{\sup_{\mu,\Sigma^*}L(\mu,\Sigma^*)}{\sup_{\mu,\Sigma}L(\mu,\Sigma)}=\frac{(\det A)^{N/2}}{\prod_{i=1}^{k}(\det A_{ii})^{N/2}}$$

and the likelihood ratio test rejects H for small values of Λ, completing the proof.

We now look at the problem of testing independence from an invariance point of view. Because of its importance and because it is more tractable we will concentrate here on the case $k=2$ where we are testing the independence of $\mathbf{X}_1$ $(m_1\times 1)$ and $\mathbf{X}_2$ $(m_2\times 1)$, i.e., we are testing $H: \Sigma_{12}=0$ against $K: \Sigma_{12}\neq 0$. We will assume, without loss of generality, that $m_1 \le m_2$. A sufficient statistic is $(\overline{\mathbf{X}}, A)$ where $\overline{\mathbf{X}}'=(\overline{\mathbf{X}}_1', \overline{\mathbf{X}}_2')$ and

$$A=\begin{bmatrix} A_{11} & A_{12} \\ A_{21} & A_{22} \end{bmatrix}.$$

Consider the group of transformations

$$(4) \quad G=\{(B,\mathbf{c});\, B=\text{diag}(B_{11},B_{22}),\, B_{ii}\in \mathcal{GL}(m_i,R)\ (i=1,2), \mathbf{c}\in R^m\}$$

acting on the space $R^m\times \mathbb{S}_m$ of points $(\overline{\mathbf{X}}, A)$ by

$$(5) \qquad (B,\mathbf{c})(\overline{\mathbf{X}},A)=(B\overline{\mathbf{X}}+\mathbf{c},\, BAB')$$

i.e.,

$$\overline{\mathbf{X}}\rightarrow B\overline{\mathbf{X}}+\mathbf{c}, \qquad A_{ij}\rightarrow B_{ii}A_{ij}B_{jj}' \qquad (i,j=1,2).$$

The corresponding induced group of transformations (also G) on the parameter space of points $(\boldsymbol{\mu},\Sigma)$ is given by

$$(6) \qquad (B,\mathbf{c})(\boldsymbol{\mu},\Sigma)=(B\boldsymbol{\mu}+\mathbf{c},\, B\Sigma B'),$$

and the testing problem is invariant under G, for the family of distributions of $(\overline{\mathbf{X}}, A)$ is invariant as are the null and alternative hypotheses.

Our next problem is to find a maximal invariant.

THEOREM 11.2.2. Under the group of transformations (6) a maximal invariant is $(\rho_1^2,\rho_2^2,\ldots,\rho_{m_1}^2)$, where $(1\ge)\rho_1^2\ge\rho_2^2\ge\cdots\ge\rho_{m_1}^2$ (≥ 0) are the latent roots of $\Sigma_{11}^{-1}\Sigma_{12}\Sigma_{22}^{-1}\Sigma_{21}$. [Some of these may be zero; the maximum number of nonzero roots is rank (Σ_{12}).]

Proof. Let $\phi(\boldsymbol{\mu},\Sigma)=(\rho_1^2,\ldots,\rho_{m_1}^2)$. First note that ϕ is invariant, for the latent roots of

$$(B_{11}\Sigma_{11}B_{11}')^{-1}(B_{11}\Sigma_{12}B_{22}')(B_{22}\Sigma_{22}B_{22}')^{-1}(B_{22}\Sigma_{21}B_{11}')$$
$$=B_{11}'^{-1}\Sigma_{11}^{-1}\Sigma_{12}\Sigma_{22}^{-1}\Sigma_{21}B_{11}'$$

are the same as those of $\Sigma_{11}^{-1}\Sigma_{12}\Sigma_{22}^{-1}\Sigma_{21}$. To show it is maximal invariant, suppose

$$\phi(\mu,\Sigma)=\phi(\tau,\Gamma)$$

i.e.,

$$\Sigma_{11}^{-1}\Sigma_{12}\Sigma_{22}^{-1}\Sigma_{21} \quad \text{and} \quad \Gamma_{11}^{-1}\Gamma_{12}\Gamma_{22}^{-1}\Gamma_{21}$$

have the same latent roots $\rho_1^2,\dots,\rho_{m_1}^2$ or, equivalently, $(\Sigma_{11}^{-1/2}\Sigma_{12}\Sigma_{22}^{-1/2})(\Sigma_{11}^{-1/2}\Sigma_{12}\Sigma_{22}^{-1/2})'$ and $(\Gamma_{11}^{-1/2}\Gamma_{12}\Gamma_{22}^{-1/2})(\Gamma_{11}^{-1/2}\Gamma_{12}\Gamma_{22}^{-1/2})'$ have the same latent roots $\rho_1^2,\dots,\rho_{m_1}^2$. By Theorem A9.10 there exist $H\in O(m_1)$ and $Q\in O(m_2)$ such that

$$H\Sigma_{11}^{-1/2}\Sigma_{12}\Sigma_{22}^{-1/2}Q'=\tilde{P}$$

where

$$\tilde{P}=\begin{bmatrix} \rho_1 & & 0 & \vdots & \\ & \ddots & & \vdots & 0 \\ 0 & & \rho_{m_1} & \vdots & \end{bmatrix} \qquad (m_1\times m_2).$$

Putting $D_{11}=H\Sigma_{11}^{-1/2}$ and $D_{22}=Q\Sigma_{22}^{-1/2}$, it then follows that $D_{11}\Sigma_{11}D_{11}'=I_{m_1}$, $D_{22}\Sigma_{22}D_{22}'=I_{m_2}$, and $D_{11}\Sigma_{12}D_{22}'=\tilde{P}$. Hence with $D=\operatorname{diag}(D_{11},D_{22})$ we have

$$D\Sigma D'=\begin{bmatrix} I_{m_1} & \tilde{P} \\ \tilde{P}' & I_{m_2}\end{bmatrix}. \tag{7}$$

A similar argument shows that there exist nonsingular $m_1\times m_1$ and $m_2\times m_2$ matrices E_{11} and E_{22}, respectively, such that

$$E\Gamma E'=\begin{bmatrix} I_{m_1} & \tilde{P} \\ \tilde{P}' & I_{m_2}\end{bmatrix}$$

where $E=\operatorname{diag}(E_{11},E_{22})$. Hence $\Gamma=E^{-1}D\Sigma D'E'^{-1}=B\Sigma B'$, where

$$B=E^{-1}D=\begin{bmatrix} E_{11}^{-1}D_{11} & 0 \\ 0 & E_{22}^{-1}D_{22}\end{bmatrix}.$$

Putting $\mathbf{c} = -B\boldsymbol{\mu} + \boldsymbol{\tau}$, we then have

$$(B,\mathbf{c})(\boldsymbol{\mu},\Sigma) = (\boldsymbol{\tau},\Gamma).$$

Hence $(\rho_1^2,\dots,\rho_{m_1}^2)$ is a maximal invariant, and the proof is complete.

As a consequence of this theorem a maximal invariant under the group G acting on the sample space of the sufficient statistic $(\overline{\mathbf{X}}, A)$ is $(r_1^2,\dots,r_{m_1}^2)$, where $r_1^2 > \cdots > r_{m_1}^2 > 0$ are the latent roots of $A_{11}^{-1}A_{12}A_{22}^{-1}A_{21}$. Any invariant test depends only on $r_1^2,\dots,r_{m_1}^2$ and, from Theorem 6.1.12, the distribution of $r_1^2,\dots,r_{m_1}^2$ depends only on $\rho_1^2,\dots,\rho_{m_1}^2$. Their positive square roots $r_1,\dots,r_{m_1}$ and $\rho_1,\dots,\rho_{m_1}$ are called, respectively, the *sample* and *population canonical correlation coefficients.* Canonical correlation analysis is a technique aimed at investigating the correlation structure between two sets of variables; we will examine this in detail in Section 11.3. Note that the likelihood ratio test of Theorem 11.2.1 is invariant, for

$$\begin{aligned}\Lambda &= \left(\frac{\det A}{\det A_{11}\det A_{22}}\right)^{N/2}\\ &= \det\left(I - A_{11}^{-1}A_{12}A_{22}^{-1}A_{21}\right)^{N/2}\\ &= \prod_{i=1}^{m_1}\left(1-r_i^2\right)^{N/2},\end{aligned}$$

so that Λ is a function of $r_1^2,\dots,r_{m_1}^2$. In terms of the population canonical correlation coefficients the null hypothesis is equivalent to

$$H:\ \rho_1 = \cdots = \rho_{m_1} = 0.$$

We have already studied the case $k=2$, $m_1=1$; here $r_1 = R$ and $\rho_1 = \overline{R}$ are, respectively, the sample and multiple correlation coefficients between X_1 and the $m_2 = m-1$ variables in $\mathbf{X}_2$. In Theorem 6.2.2 it was shown that the likelihood ratio test of H: $\overline{R}=0$ against K: $\overline{R}\neq 0$ is a uniformly most powerful invariant test under the group G. In general, however, there is no uniformly most powerful invariant test and other functions of $r_1^2,\dots,r_{m_1}^2$ in addition to Λ have been proposed as test statistics. Some of these will be discussed in Section 11.2.8.

The likelihood ratio test was shown by Narain (1950) to be unbiased. For the case $k=2$, Anderson and Das Gupta (1964) established a somewhat stronger result, namely, that the power function of the likelihood ratio test increases monotonically as each population canonical correlation coefficient ρ_i increases; see also Perlman and Olkin (1980).

11.2.2. Central Moments of the Likelihood Ratio Statistic

Information about the distribution of the likelihood ratio statistic Λ can be obtained from a study of its moments. In this section we find the moments for general k when the null hypothesis $H: \Sigma_{ij}=0$ $(i, j=1,\ldots,k, i\neq j)$ is true. For notational convenience we define the statistic

$$W=\Lambda^{2/N}=\frac{\det A}{\prod_{i=1}^{k}\det A_{ii}}. \tag{8}$$

The moments of W are given in the following theorem.

THEOREM 11.2.3. When H is true, the hth moment of W is

$$E(W^h)=\frac{\Gamma_m(\frac{1}{2}n+h)}{\Gamma_m(\frac{1}{2}n)}\prod_{i=1}^{k}\frac{\Gamma_{m_i}(\frac{1}{2}n)}{\Gamma_{m_i}(\frac{1}{2}n+h)}, \tag{9}$$

where $n=N-1$.

Proof. When H is true, Σ has the form Σ^* given by (3). There is no loss of generality in assuming that $\Sigma^*=I_m$ since W is invariant under the group of transformations $\Sigma\rightarrow B\Sigma B'$, where $B=\text{diag}(B_{11},\ldots,B_{kk})$, with $B_{ii}\in \mathfrak{gl}(m_i, R)$, $i=1,\ldots,k$. Hence, with $c_{m,n}=[2^{mn/2}\Gamma_m(\frac{1}{2}n)]^{-1}$ we have

$$\begin{aligned}E(W^h)&=c_{m,n}\int_{A>0}\prod_{i=1}^{k}(\det A_{ii})^{-h}\operatorname{etr}(-\tfrac{1}{2}A)(\det A)^{(n+2h-m-1)/2}(dA)\\&=\frac{c_{m,n}}{c_{m,n+2h}}E\left[\prod_{i=1}^{k}(\det A_{ii})^{-h}\right],\end{aligned}$$

where the matrix A in this last expectation has the $W_m(n+2h, I_m)$ distribution. Consequently, $A_{11},\ldots,A_{kk}$ are independent, and A_{ii} is $W_{m_i}(n+2h, I_{m_i})$, $i=1,\ldots,k$, so that, using (15) of Section 3.2,

$$\begin{aligned}E(W^h)&=\frac{c_{m,n}}{c_{m,n+2h}}\prod_{i=1}^{k}E\left[(\det A_{ii})^{-h}\right]\\&=\frac{c_{m,n}}{c_{m,n+2h}}\prod_{i=1}^{k}2^{-m_ih/2}\frac{\Gamma_{m_i}(\frac{1}{2}n)}{\Gamma_{m_i}(\frac{1}{2}n+h)}\\&=\frac{\Gamma_m(\frac{1}{2}n+h)}{\Gamma_m(\frac{1}{2}n)}\prod_{i=1}^{k}\frac{\Gamma_{m_i}(\frac{1}{2}n)}{\Gamma_{m_i}(\frac{1}{2}n+h)},\end{aligned}$$

and the proof is complete.

11.2.3. *The Null Distribution of the Likelihood Ratio Statistic*

When the null hypothesis H is true, the statistic W has the same distribution as a product of independent beta random variables. The result is given in the following theorem.

THEOREM 11.2.4. When H is true, W has the same distribution as

$$\prod_{i=2}^{k}\prod_{j=1}^{m_i} V_{ij},$$

where the V_{ij} are independent random variables and V_{ij} is beta$(\frac{1}{2}(n+1-m_i^*-j),\frac{1}{2}m_i^*)$ with $m_i^*=\sum_{l=1}^{i-1}m_l$.

Proof. Starting with the result of Theorem 11.2.3 we have

$$\begin{aligned}
E(W^h)&=\frac{\Gamma_m(\frac{1}{2}n+h)}{\Gamma_m(\frac{1}{2}n)}\prod_{i=1}^{k}\frac{\Gamma_{m_i}(\frac{1}{2}n)}{\Gamma_{m_i}(\frac{1}{2}n+h)}\\
&=\prod_{i=1}^{m}\frac{\Gamma[\frac{1}{2}(n+1-i)+h]}{\Gamma[\frac{1}{2}(n+1-i)]}\prod_{i=1}^{k}\prod_{j=1}^{m_i}\frac{\Gamma[\frac{1}{2}(n+1-j)]}{\Gamma[\frac{1}{2}(n+1-j)+h]}\\
&=\prod_{i=m_1+1}^{m}\frac{\Gamma[\frac{1}{2}(n+1-i)+h]}{\Gamma[\frac{1}{2}(n+1-i)]}\prod_{i=2}^{k}\prod_{j=1}^{m_i}\frac{\Gamma[\frac{1}{2}(n+1-j)]}{\Gamma[\frac{1}{2}(n+1-j)+h]}\\
&=\prod_{i=2}^{k}\prod_{j=1}^{m_i}\frac{\Gamma[\frac{1}{2}(n+1-m_i^*-j)+h]\Gamma[\frac{1}{2}(n+1-j)]}{\Gamma[\frac{1}{2}(n+1-m_i^*-j)]\Gamma[\frac{1}{2}(n+1-j)+h]}.
\end{aligned}$$

Since the hth moment of a random variable having the beta(α,β) distribution is $\Gamma(\alpha+h)\Gamma(\alpha+\beta)/\Gamma(\alpha)\Gamma(\alpha+\beta+h)$, it follows that

$$E(W^h)=\prod_{i=2}^{k}\prod_{j=1}^{m_i}E(V_{ij}^h)$$

where V_{ij} has the beta$(\frac{1}{2}(n+1-m_i^*-j),\frac{1}{2}m_i^*)$ distribution. Because W is bounded its moments uniquely determine its distribution, and the proof is complete.

The important case $k=2$, where the independence of two sets of variables is being tested, merits special attention. In this case the hth moment of

W is

$$(10)\quad E(W^h)=\prod_{i=1}^{m}\frac{\Gamma[\frac{1}{2}(n+1-i)+h]}{\Gamma[\frac{1}{2}(n+1-i)]}\prod_{i=1}^{2}\prod_{j=1}^{m_i}\frac{\Gamma[\frac{1}{2}(n+1-j)]}{\Gamma[\frac{1}{2}(n+1-j)+h]}$$

$$=\prod_{i=m_2+1}^{m}\frac{\Gamma[\frac{1}{2}(n+1-i)+h]}{\Gamma[\frac{1}{2}(n+1-i)]}\prod_{j=1}^{m_1}\frac{\Gamma[\frac{1}{2}(n+1-j)]}{\Gamma[\frac{1}{2}(n+1-j)+h]}$$

$$=\prod_{i=1}^{m_1}\frac{\Gamma[\frac{1}{2}(n+1-m_2-i)+h]\Gamma[\frac{1}{2}(n+1-i)]}{\Gamma[\frac{1}{2}(n+1-m_2-i)]\Gamma[\frac{1}{2}(n+1-i)+h]}$$

$$=\frac{\Gamma_{m_1}[\frac{1}{2}(N-m_2-1)+h]\Gamma_{m_1}[\frac{1}{2}(N-1)]}{\Gamma_{m_1}[\frac{1}{2}(N-m_2-1)]\Gamma_{m_1}[\frac{1}{2}(N-1)+h]}.$$

These moments have exactly the same form as the moments of the statistic W used for testing the general linear hypothesis given in Corollary 10.5.2, where there we make the substitutions

$$m\to m_1,\qquad r\to m_2,\qquad n-p\to N-m_2-1.$$

It hence follows from Theorem 10.5.3, or from the moments (10), that W has the same distribution as $\prod_{i=1}^{m_1}V_i$, where $V_1,\dots,V_{m_1}$ are independent, with V_i having the beta$(\frac{1}{2}(N-m_2-i),\frac{1}{2}m_2)$ distribution. If, in addition, $m_1=1$, $m_2=m-1$, this shows that W has the beta$(\frac{1}{2}(N-m),\frac{1}{2}(m-1))$ distribution. In this case $W=1-R^2$, where R is the sample multiple correlation coefficient between X_1 and the variables in $\mathbf{X}_2$, so that the result agrees with Theorem 5.2.2.

In general it is not an easy matter to find expressions for the probability density function of W. For some special cases the interested reader is referred to T. W. Anderson (1958), Section 9.4.2, and Srivastava and Khatri (1979), Section 7.5.3.

11.2.4. The Asymptotic Null Distribution of the Likelihood Ratio Statistic

Replacing h in Theorem 11.2.3 by $\frac{1}{2}Nh$ shows that when H is true the hth moment of $\Lambda=W^{N/2}$ is

$$(11)\qquad E(\Lambda^h)=\frac{\Gamma_m[\frac{1}{2}N(1+h)-\frac{1}{2}]}{\Gamma_m[\frac{1}{2}(N-1)]}\prod_{i=1}^{k}\frac{\Gamma_{m_i}[\frac{1}{2}(N-1)]}{\Gamma_{m_i}[\frac{1}{2}N(1+h)-\frac{1}{2}]}$$

$$=K\frac{\prod_{i=1}^{m}\Gamma[\frac{1}{2}N(1+h)-\frac{1}{2}i]}{\prod_{i=1}^{k}\prod_{j=1}^{m_i}\Gamma[\frac{1}{2}N(1+h)-\frac{1}{2}j]},$$

where K is a constant not involving h. This has the same form as (18) of Section 8.2.4 with $p=m$, $q=m$, $x_l=\frac{1}{2}N$, $\xi_l=-\frac{1}{2}l$ $(l=1,\ldots,m)$, $y_l=\frac{1}{2}N$, $\eta_j=-\frac{1}{2}j$ $(j=1,\ldots,m_i;\ i=1,\ldots,k)$. The degrees of freedom in the limiting χ^2 distribution are, from (28) of Section 8.2.4,

$$(12)\qquad f=-2\left[\sum_{l=1}^{q}\xi_l-\sum_{j=1}^{p}\eta_j-\tfrac{1}{2}(q-p)\right]$$

$$=\sum_{l=1}^{m}l-\sum_{i=1}^{k}\sum_{j=1}^{m_i}j$$

$$=\frac{1}{2}\left(m^2-\sum_{i=1}^{k}m_i^2\right)=\sum_{i<j}^{k}m_im_j.$$

The value of ρ (not to be confused with the population canonical correlation coefficient ρ_i) which makes the term of order n^{-1} vanish in the asymptotic expansion of the distribution of $-2\rho\log\Lambda$ is, from (30) of Section 8.2.4,

$$(13)\qquad \rho=1-\frac{1}{f}\left[\sum_{l=1}^{q}x_l^{-1}\left(\xi_l^2-\xi_l+\frac{1}{6}\right)-\sum_{j=1}^{p}y_j^{-1}\left(\eta_j^2-\eta_j+\frac{1}{6}\right)\right]$$

$$=1-\frac{1}{f}\left[\sum_{l=1}^{m}\frac{2}{N}\left(\frac{1}{4}l^2+\frac{1}{2}l+\frac{1}{6}\right)-\sum_{i=1}^{k}\sum_{j=1}^{m_i}\frac{2}{N}\left(\frac{1}{4}j^2+\frac{1}{2}j+\frac{1}{6}\right)\right]$$

$$=1-\frac{2\left(m^3-\sum_{i=1}^{k}m_i^3\right)+9\left(m^2-\sum_{i=1}^{k}m_i^2\right)}{6N\left(m^2-\sum_{i=1}^{k}m_i^2\right)}.$$

With this value of ρ it is then found, using (29) of Section 8.2.4, that the term of order N^{-2} in the expansion is

(14)

$$\omega_2=\frac{1}{\rho^2N^2}\left[\frac{1}{48}\left(m^4-\sum_{i=1}^{k}m_i^4\right)-\frac{5}{96}\left(m^2-\sum_{i=1}^{k}m_i^2\right)-\frac{m^3-\sum_{i=1}^{k}m_i^3}{72\left(m^2-\sum_{i=1}^{k}m_i^2\right)}\right].$$

Hence we have the following result (from Box, 1949).

THEOREM 11.2.5. When the null hypothesis $H: \Sigma_{ij}=0$ $(i, j=1,\dots,k, i \neq j)$, is true the distribution function of $-2\rho\log\Lambda$, where ρ is given by (13), can be expanded for large $M=\rho N$ as

$$(15)\quad P(-2\rho\log\Lambda\le x)=P(-N\rho\log W\le x)$$
$$=P(\chi_f^2\le x)+\frac{\gamma}{M^2}\left[P(\chi_{f+4}^2\le x)-P(\chi_f^2\le x)\right]$$
$$+O(M^{-3}),$$

where f is given by (12) and $\gamma=(N\rho)^2\omega_2=M^2\omega_2$, with ω_2 given by (14).

In the important case $k=2$ we have $f=m_1m_2$; then

$$(16)\qquad \rho=1-\frac{m_1+m_2+3}{2N}$$
$$\omega_2=\frac{m_1m_2}{48(\rho N)^2}(m_1^2+m_2^2-5),$$

and the resulting expansion for the distribution function of $-2\rho\log\Lambda$ agrees with that in Theorem 10.5.5, where we make the substitutions

$$(17)\quad m\to m_1,\qquad r\to m_2,\qquad n\to N,\qquad p\to m_2+1,\qquad N\to M.$$

An approximate test of significance level α is to reject H if $-2\rho\log\Lambda > c_f(\alpha)$, where $c_f(\alpha)$ denotes the upper $100\alpha\%$ point of the χ_f^2 distribution. The error in the approximation is of order N^{-2}.

For the case $k=2$, Table 9 gives upper $100\alpha\%$ points of the distribution of $-2\rho\log\Lambda$ for $\alpha=.1$, .05, .025, and .005, after the substitutions (16) and $M\to N-m_1-m_2$ have been made. The function tabulated is a multiplying factor C which when multiplied by $c_{m_1m_2}(\alpha)$, the upper $100\alpha\%$ point of the $\chi_{m_1m_2}^2$ distribution, gives the upper $100\alpha\%$ point of $-2\rho\log\Lambda$.

For testing independence between $k>2$ sets of variables Davis and Field (1971) have prepared tables of upper $100\alpha\%$ points of $-2\rho\log\Lambda$ for $\alpha=.05$, .01 and for various values of the m_i, $i=1,\dots,k$. These are reproduced in Table 8. The function tabulated is a multiplying factor C which when multiplied by the upper $100\alpha\%$ point of the χ_f^2 distribution, where f is given by (12), yields the upper $100\alpha\%$ point of $-2\rho\log\Lambda$. (A "partition" in the table gives the values of $m_1, m_2, m_3,\dots$.)

11.2.5. Noncentral Moments of the Likelihood Ratio Statistic when k = 2

In this section we will obtain the moments in general of Λ for the case $k=2$ where the independence of two subvectors $\mathbf{X}_1,\mathbf{X}_2$ of sizes $m_1\times 1, m_2\times 1$

Table 8. χ^2 adjustments to the likelihood ratio statistic for testing independence: factor C for upper percentiles of $-2\rho \log \Lambda$ (see Section 11.2.4)[a]

Partitions	2,1,1		3,1,1		2,2,1		4,1,1		3,2,1		Partitions
N	5%	1%	5%	1%	5%	1%	5%	1%	5%	1%	N
5	1.07	1.08	1.32	1.40	1.29						5
6	1.034	1.042	1.12	1.152	1.109	1.13					6
7	1.020	1.025	1.067	1.0813	1.058	1.071	1.18	1.21	1.15	1.18	7
8	1.0135	1.0168	1.042	1.0509	1.036	1.044	1.100	1.12	1.083	1.10	8
9	1.0097	1.0119	1.0291	1.0350	1.0250	1.0300	1.0646	1.077	1.0536	1.063	9
10	1.0072	1.0089	1.0213	1.0256	1.0182	1.0218	1.0454	1.054	1.0376	1.044	10
11	1.0056	1.0069	1.0162	1.0195	1.0139	1.0166	1.0338	1.0399	1.0279	1.033	11
12	1.0045	1.0055	1.0128	1.0154	1.0110	1.0131	1.0261	1.0308	1.0216	1.0252	12
13	1.0037	1.0045	1.0104	1.0124	1.0088	1.0105	1.0209	1.0245	1.0172	1.0201	13
14	1.0031	1.0038	1.0086	1.0103	1.0073	1.0087	1.0170	1.0200	1.0140	1.0163	14
15	1.0026	1.0032	1.0072	1.0086	1.0061	1.0073	1.0142	1.0166	1.0117	1.0136	15
16	1.0022	1.0027	1.0061	1.0073	1.0052	1.0062	1.0120	1.0141	1.0099	1.0115	16
17	1.0019	1.0024	1.0053	1.0063	1.0045	1.0053	1.0103	1.0120	1.0084	1.0098	17
18	1.0177	1.0021	1.0046	1.0055	1.0039	1.0046	1.0089	1.0104	1.0073	1.0085	18
19	1.0015	1.0018	1.0040	1.0048	1.0034	1.0041	1.0078	1.0091	1.0064	1.0074	19
20	1.0013	1.0016	1.0036	1.0043	1.0030	1.0036	1.0069	1.0080	1.0056	1.0065	20
21	1.0012	1.0014	1.0032	1.0038	1.0027	1.0032	1.0061	1.0072	1.0050	1.0058	21
22	1.0011	1.0013	1.0029	1.0034	1.0024	1.0029	1.0055	1.0064	1.0045	1.0052	22
23	1.0010	1.0012	1.0026	1.0031	1.0022	1.0026	1.0049	1.0058	1.0040	1.0047	23
24	1.0009	1.0011	1.0023	1.0028	1.0020	1.0024	1.0044	1.0052	1.0036	1.0042	24
25	1.0008	1.0010	1.0021	1.0026	1.0018	1.0021	1.0040	1.0047	1.0033	1.0038	25
30	1.0005	1.0007	1.0014	1.0017	1.0012	1.0014	1.0027	1.0031	1.0022	1.0025	30
35	1.0004	1.0005	1.0010	1.0012	1.0009	1.0010	1.0019	1.0022	1.0015	1.0018	35
40	1.0003	1.0004	1.0008	1.0009	1.0006	1.0008	1.0014	1.0016	1.0011	1.0013	40
45	1.0002	1.0003	1.0006	1.0007	1.0005	1.0006	1.0011	1.0013	1.0009	1.0010	45
50	1.0002	1.0002	1.0005	1.0006	1.0004	1.0005	1.0009	1.0010	1.0007	1.0008	50
55	1.0001	1.0002	1.0004	1.0005	1.0003	1.0004	1.0007	1.0008	1.0006	1.0007	55
60	1.0001	1.0001	1.0003	1.0004	1.0003	1.0003	1.0006	1.0007	1.0005	1.0006	60
120	1.0000	1.0000	1.0001	1.0001	1.0001	1.0001	1.0001	1.0002	1.0001	1.0001	120
∞	1.0000	1.0000	1.0000	1.0000	1.0000	1.0000	1.0000	1.0000	1.0000	1.0000	∞
χ^2_f	11.0705	15.0863	14.0671	18.4753	15.5073	20.0902	16.9190	21.6660	19.6751	24.7250	χ^2_f

Table 8 (*Continued*)

Partitions	1^4		1^5		1^6		1^7		1^8		Partitions
N	5%	1%	5%	1%	5%	1%	5%	1%	5%	1%	N
6	1.034	1 043	1.115	1.14							6
7	1.021	1.0255	1.062	1.08							7
8	1 0136	1.0169	1.039	1.047							8
9	1.0097	1.0120	1.0267	1.032	1.055	1.06	1.10	1.12			9
10	1.0073	1.0089	1.0195	1.0233	1.0386	1.045	1 07	1.08	1.12	1.13	10
11	1.0056	1.0069	1.0149	1.0177	1.0288	1.033	1.049	1.057	1.08	1.09	11
12	1.0045	1.0055	1.0118	1.0139	1.0223	1.026	1.037	1.043	1.059	1.07	12
13	1.0037	1.0045	1.0095	1.0112	1.0178	1.0205	1.0293	1.033	1.045	1.051	13
14	1.0031	1.0037	1.0079	1.0093	1.0145	1.0167	1.0236	1.027	1.036	1 040	14
15	1.0026	1.0032	1.0066	1.0078	1.0121	1.0139	1.0195	1.0220	1.0293	1.033	15
16	1.0022	1.0027	1.0056	1.0066	1.0102	1.0117	1.0163	1.0184	1.0243	1.027	16
17	1.0019	1.0023	1.0048	1.0057	1.0088	1.0101	1.0139	1.0157	1.0205	1.0229	17
18	1.0017	1.0020	1.0042	1.0049	1.0076	1.0087	1.0120	1.0135	1.0176	1.0195	18
19	1.0015	1.0018	1.0037	1.0043	1.0066	1.0076	1.0104	1.0117	1 0152	1.0169	19
20	1.0013	1.0016	1 0033	1.0038	1.0059	1.0067	1.0092	1.0103	1.0133	1.0148	20
21	1 0012	1.0014	1.0029	1.0034	1.0052	1.0060	1.0081	1.0091	1.0117	1.0130	21
22	1.0011	1.0013	1.0026	1.0031	1.0047	1.0053	1.0072	1.0081	1.0104	1.0116	22
23	1.0010	1.0012	1.0024	1.0028	1.0042	1.0048	1.0065	1.0073	1.0093	1.0103	23
24	1.0009	1.0010	1.0021	1.0025	1.0038	1.0043	1.0059	1.0066	1.0084	1.0093	24
25	1.0008	1.0010	1.0020	1.0023	1.0035	1.0039	1.0053	1.0060	1.0076	1.0084	25
30	1.0005	1.0006	1.0013	1.0015	1 0023	1.0026	1.0035	1.0039	1.0049	1.0054	30
35	1.0004	1.0005	1.0009	1.0011	1.0016	1.0018	1.0025	1.0027	1.0035	1.0038	35
40	1 0003	1.0003	1.0007	1.0008	1.0012	1.0014	1.0018	1.0020	1.0026	1 0028	40
50	1.0002	1.0002	1.0004	1.0005	1.0007	1.0008	1 0011	1.0012	1.0016	1.0017	50
60	1 0001	1.0001	1.0003	1.0003	1.0005	1.0006	1.0008	1.0008	1.0010	1.0012	60
90	1.0001	1.0001	1.0001	1.0001	1.0002	1.0002	1.0003	1.0004	1.0004	1.0005	90
120	1.0000	1.0000	1.0001	1.0001	1.0001	1.0001	1.0002	1.0002	1.0002	1.0003	120
∞	1.0000	1.0000	1.0000	1.0000	1.0000	1 0000	1.0000	1.0000	1.0000	1.0000	∞
χ_f^2	12.5916	16.8119	18.3070	23.2093	24.9958	30.5779	32.6705	38.9321	41.3372	48.2782	χ_f^2

[a]Here, m = number of variables; N = sample size; m_i = number of variables in ith set, $i = 1, \ldots, k$ partition $= m_1, m_2, m_3, \ldots$;

$$C = \frac{\text{level for } -2\rho \log \Lambda}{\text{level for } \chi^2 \text{ of } f \text{ degrees of freedom}};$$

$$f = \frac{1}{2}\left(m^2 - \sum_{i=1}^{k} m_i^2\right).$$

Source: Reproduced from Davis and Field (1971) with the kind permission of the Commonwealth Scientific and Industrial Research Organization (C.S.I.R.O), Australia, and the authors.

$(m_1+m_2=m)$ is being tested. These will be used in the next section to derive asymptotic non-null distributions of Λ. We assume without loss of generality that $m_1 \le m_2$. Recall that in this case

$$W=\Lambda^{2/N}=\frac{\det A}{(\det A_{11})(\det A_{22})}=\prod_{i=1}^{m_1}(1-r_i^2),$$

where $r_1^2,\dots,r_{m_1}^2$, the squares of the sample canonical correlation coefficients, are the latent roots of $A_{11}^{-1}A_{12}A_{22}^{-1}A_{21}$. The hth moment of W can be expressed in terms of the ${}_2F_1$ one-matrix hypergeometric function (see Sections 7.3 and 7.4), as the following theorem from Sugiura and Fujikoshi (1969) shows.

THEOREM 11.2.6. The hth moment of W is

$$(18)\qquad E(W^h)=\frac{\Gamma_{m_1}(\frac{1}{2}n)\Gamma_{m_1}[\frac{1}{2}(n-m_2)+h]}{\Gamma_{m_1}(\frac{1}{2}n+h)\Gamma_{m_1}[\frac{1}{2}(n-m_2)]}$$

$$\cdot\det(I-P^2)^{n/2}{}_2F_1(\tfrac{1}{2}n,\tfrac{1}{2}n;\tfrac{1}{2}n+h;P^2),$$

where $n=N-1$ and $P^2=\operatorname{diag}(\rho_1^2,\dots,\rho_{m_1}^2)$, with $\rho_1^2,\dots,\rho_{m_1}^2$, the squares of the population canonical correlation coefficients, being the latent roots of $\Sigma_{11}^{-1}\Sigma_{12}\Sigma_{22}^{-1}\Sigma_{21}$.

Proof. We start with the $W_m(n,\Sigma)$ distribution for A. By invariance (see the proof of Theorem 11.2.2) we can assume without loss of generality that

$$\Sigma=\begin{bmatrix} I_{m_1} & \tilde{P} \\ \tilde{P}' & I_{m_2}\end{bmatrix},\qquad \tilde{P}=\begin{bmatrix} \rho_1 & & 0 & \vdots & 0 \\ & \ddots & & \vdots & \\ 0 & & \rho_{m_1} & \vdots & \end{bmatrix}=[P\,\vdots\,0],$$

where $\tilde{P}$ is $m_1\times m_2$. Write A as $A=Z'Z$ where Z is $N(0,I_n\otimes\Sigma)$ and partition Z as $Z=[Y\,\vdots\,X]$, where Y is $n\times m_1$ and X is $n\times m_2$. Then

$$A=Z'Z=\begin{bmatrix} Y'Y & Y'X \\ X'Y & X'X\end{bmatrix}=\begin{bmatrix} A_{11} & A_{12} \\ A_{21} & A_{22}\end{bmatrix},$$

and $W=\prod_{i=1}^{m_1}(1-r_i^2)$, where $r_1^2,\dots,r_{m_1}^2$ are the latent roots of $(Y'Y)^{-1}Y'X(X'X)^{-1}X'Y$, i.e., the solutions of the equation

$$(19)\qquad \det\left(Y'X(X'X)^{-1}X'Y-r^2Y'Y\right)=0.$$

We first condition on X. The conditional distribution of Y given X is $N(X\Sigma_{22}^{-1}\Sigma_{21}, I_n \otimes \Phi)$, where

$$\Phi = \Sigma_{11} - \Sigma_{12}\Sigma_{22}^{-1}\Sigma_{21} = I - \tilde{P}\tilde{P}' = \operatorname{diag}\left(1-\rho_1^2,\ldots,1-\rho_{m_1}^2\right)$$
$$=(I-P^2)$$

and

$$\Sigma_{22}^{-1}\Sigma_{21} = \tilde{P}'.$$

Hence the conditional density function of Y given X is

$$(2\pi)^{-nm_1/2}\det(I-P^2)^{-n/2}\operatorname{etr}\left[-\tfrac{1}{2}\Phi^{-1}(Y-X\tilde{P}')'(Y-X\tilde{P}')\right].$$

Now X is $n \times m_2$ of rank m_2 (with probability 1) and so there exists $H \in O(n)$ such that

$$HX = \begin{bmatrix} X_1 \\ \cdots \\ 0 \end{bmatrix},$$

where X_1 is a nonsingular $m_2 \times m_2$ matrix. Putting $T = HY$ equation (19) becomes

$$\det\left(T'\begin{bmatrix} X_1 \\ \cdots \\ 0 \end{bmatrix}(X_1'X_1)^{-1}[X_1' : 0']T - r^2 T'T\right) = 0$$

i.e.,

$$\det\left(T'\begin{bmatrix} I_{m_2} & 0 \\ 0 & 0 \end{bmatrix}T - r^2 T'T\right) = 0. \tag{20}$$

Partitioning T as

$$T = \begin{bmatrix} U \\ \cdots \\ V \end{bmatrix},$$

where U is $m_2 \times m_1$ and V is $(n-m_2) \times n_1$, (20) becomes

$$\det\left(U'U - r^2(U'U + V'V)\right) = 0. \tag{21}$$

Under the above transformations we have

$$\begin{aligned}
\operatorname{tr}\Phi^{-1}(Y-X\tilde{P}')'(Y-X\tilde{P}')&=\operatorname{tr}\Phi^{-1}(HY-HX\tilde{P}')'(HY-HX\tilde{P}')\\
&=\operatorname{tr}\Phi^{-1}\left(T-\begin{bmatrix}X_1\tilde{P}'\\ \cdots\\ 0\end{bmatrix}\right)'\left(T-\begin{bmatrix}X_1\tilde{P}'\\ \cdots\\ 0\end{bmatrix}\right)\\
&=\operatorname{tr}\Phi^{-1}\begin{bmatrix}U-X_1\tilde{P}'\\ \cdots\cdots\\ V\end{bmatrix}'\begin{bmatrix}U-X_1\tilde{P}'\\ \cdots\cdots\\ V\end{bmatrix}\\
&=\operatorname{tr}\Phi^{-1}(U-X_1\tilde{P}')'(U-X_1\tilde{P}')+\operatorname{tr}\Phi^{-1}V'V,
\end{aligned}$$

and hence the conditional density function of U and V given X is

$$\begin{aligned}
(2\pi)^{-nm_1/2}&\det(I-P^2)^{-n/2}\operatorname{etr}\left[-\tfrac{1}{2}\Phi^{-1}(U-X_1\tilde{P}')'(U-X_1\tilde{P}')\right]\\
&\cdot\operatorname{etr}\left(-\tfrac{1}{2}\Phi^{-1}V'V\right).
\end{aligned}$$

This shows that, conditional on X, U and V are independent, U is $N(X_1\tilde{P}', I_{m_2}\otimes\Phi)$ and V is $N(0, I_{n-m_2}\otimes\Phi)$, and hence, given X, $V'V$ is $W_{m_1}(n-m_2,\Phi)$, $U'U$ is noncentral $W_{m_1}(m_2,\Phi,\Omega)$, where the noncentrality matrix Ω is

$$\Omega=\Phi^{-1}\tilde{P}X_1'X_1\tilde{P}'=\Phi^{-1}\tilde{P}X'X\tilde{P}',$$

and $U'U$, $V'V$ are independent. In terms of U and V the statistic W is

$$W=\prod_{i=1}^{m_1}(1-r_i^2)=\frac{\det(V'V)}{\det(U'U+V'V)}.$$

This is the same as the likelihood ratio criterion for testing the general linear hypothesis (see Section 10.2) and hence the conditional moments of W given $X'X$ can be obtained from Theorem 10.5.1, where we put $r=m_2$, $m=m_1$, $n-p=n-m_2$. This gives

$$\begin{aligned}
E(W^h|X'X)=&\frac{\Gamma_{m_1}\left[\frac{1}{2}(n-m_2)+h\right]\Gamma_{m_1}\left(\frac{1}{2}n\right)}{\Gamma_{m_1}\left[\frac{1}{2}(n-m_2)\right]\Gamma_{m_1}\left(\frac{1}{2}n+h\right)}\\
&\cdot{}_1F_1\left(h;\tfrac{1}{2}n+h;-\tfrac{1}{2}\Phi^{-1}\tilde{P}X'X\tilde{P}'\right).
\end{aligned}$$

Since the matrix $X'X$ is $W_{m_2}(n, I)$ it now follows that

$$E(W^h) = E\left[E\left[W^h | X'X\right]\right]$$

$$= \frac{\Gamma_{m_1}\left[\frac{1}{2}(n-m_2)+h\right]\Gamma_{m_1}\left(\frac{1}{2}n\right)}{\Gamma_{m_1}\left[\frac{1}{2}(n-m_2)\right]\Gamma_m\left(\frac{1}{2}n+h\right)} \frac{2^{-mn_2/2}}{\Gamma_{m_2}\left(\frac{1}{2}n\right)} \int_{X'X>0}$$

$$\cdot \operatorname{etr}\left(-\tfrac{1}{2}X'X\right)(\det X'X)^{(n-m_2-1)/2}$$

$$\cdot {}_1F_1\left(h; \tfrac{1}{2}n+h; -\tfrac{1}{2}\Phi^{-1}\tilde{P}X'X\tilde{P}'\right)(d(X'X))$$

$$= \frac{\Gamma_{m_1}\left[\frac{1}{2}(n-m_2)+h\right]\Gamma_{m_1}\left(\frac{1}{2}n\right)}{\Gamma_{m_1}\left[\frac{1}{2}(n-m_2)\right]\Gamma_{m_1}\left(\frac{1}{2}n+h\right)} {}_2F_1\left(h, \tfrac{1}{2}n; \tfrac{1}{2}n+h; -\Phi^{-1}\tilde{P}\tilde{P}'\right),$$

where the integral has been evaluated using Theorem 7.3.4. The desired result now follows if we use the Euler relation of Theorem 7.4.3, namely,

$${}_2F_1\left(h, \tfrac{1}{2}n; \tfrac{1}{2}n+h; -\Phi^{-1}\tilde{P}\tilde{P}'\right)$$

$$= \det\left(I+\Phi^{-1}\tilde{P}\tilde{P}'\right)^{-n/2} {}_2F_1\left(\tfrac{1}{2}n, \tfrac{1}{2}n; \tfrac{1}{2}n+h; \Phi^{-1}\tilde{P}\tilde{P}'\left(I+\Phi^{-1}\tilde{P}\tilde{P}'\right)^{-1}\right),$$

and note that

$$\Phi^{-1}\tilde{P}\tilde{P}' = (I-P^2)^{-1}P^2 = \operatorname{diag}\left(\frac{\rho_1^2}{1-\rho_1^2}, \dots, \frac{\rho_{m_1}^2}{1-\rho_{m_1}^2}\right),$$

so that

$$\det\left(I+\Phi^{-1}\tilde{P}\tilde{P}'\right) = \det(I-P^2)^{-1}$$

and

$$\Phi^{-1}\tilde{P}\tilde{P}'\left(I+\Phi^{-1}\tilde{P}\tilde{P}'\right)^{-1} = P^2.$$

11.2.6. Asymptotic Non-null Distributions of the Likelihood Ratio Statistic when $k=2$

The power function of the likelihood ratio test of level α is $P(-2\rho\log\Lambda > k_\alpha^*)$, where ρ is given by (16) and k_α^* is the upper $100\alpha\%$ point of the distribution of $-2\rho\log\Lambda$ when $H: \Sigma_{12}=0$ is true. This is a function of

$\rho_1^2,\ldots,\rho_{m_1}^2$, the latent roots of $\Sigma_{11}^{-1}\Sigma_{12}\Sigma_{22}^{-1}\Sigma_{21}$. (We are assuming, as in the previous section, that $m_1 \le m_2$.) It has been shown that an approximation for k_α^* is $c_f(\alpha)$, the upper $100\alpha\%$ point of the χ_f^2 distribution, with $f = m_1 m_2$. The error in this approximation is of order M^{-2}, where $M = \rho N$. In this section we investigate ways of approximating the power function.

Letting $P^2 = \text{diag}(\rho_1^2,\ldots,\rho_{m_1}^2)$, we consider the three different alternatives

$$K:\quad P^2 \ne 0,$$

$$K_M:\quad P^2 = \frac{1}{M}\Omega,$$

and

$$K_M^*:\quad P^2 = \frac{1}{M^2}\Omega,$$

where $\Omega = \text{diag}(\omega_1,\ldots,\omega_{m_1})$ is fixed. Here K is a fixed alternative and K_M, K_M^* are sequences of local alternatives. We begin by looking at the asymptotic distribution of $-2\rho\log\Lambda$ under the sequence K_M.

THEOREM 11.2.7. Under the sequence of local alternatives K_M: $P^2 = (1/M)\Omega$ the distribution function of $-2\rho\log\Lambda$ can be expanded as

$$\begin{aligned}(22)\quad P(-2\rho\log\Lambda \le x) &= P\big(\chi_f^2(\sigma_1) \le x\big)\\ &\quad - \frac{1}{4M}\Big\{\big[\sigma_2 + (m+1)\sigma_1\big]P\big(\chi_f^2(\sigma_1) \le x\big)\\ &\quad - 2(m+1)\sigma_1 P\big(\chi_{f+2}^2(\sigma_1) \le x\big)\\ &\quad - \big[2\sigma_2 - (m+1)\sigma_1\big]P\big(\chi_{f+4}^2(\sigma_1) \le x\big)\\ &\quad + \sigma_2 P\big(\chi_{f+6}^2(\sigma_1) \le x\big)\Big\} + O(M^{-2}),\end{aligned}$$

where $f = m_1 m_2$, $\sigma_j = \text{tr}\,\Omega^j = \omega_1^j + \cdots + \omega_{m_1}^j$, and $m = m_1 + m_2$.

Proof. Under K_M the characteristic function of $-2\rho\log\Lambda$ is, using Theorem 11.2.6,

$$\begin{aligned}(23)\quad \phi(M,t,\Omega) &= E\big[\Lambda^{-2it\rho}\big]\\ &= E\big[W^{-Mit}\big]\\ &= \phi(M,t,0)G(M,t,\Omega),\end{aligned}$$

where

(24)

$$G(M,t,\Omega)=\det\left(I-\frac{1}{M}\Omega\right)^{(M+\delta)/2}$$

$$\cdot {}_2F_1\left(\frac{1}{2}(M+\delta),\frac{1}{2}(M+\delta);\frac{1}{2}M(1-2it)+\frac{1}{2}\delta;\frac{1}{M}\Omega\right)$$

with $\delta=\frac{1}{2}(m+1)$, and $\phi(M,t,0)$ is the characteristic function of $-2\rho\log\Lambda$ when H is true ($\Omega=0$) obtained from (10) by putting $h=-Mit$. From Theorem 11.2.5 we know that

$$\phi(M,t,0)=(1-2it)^{-f/2}+O(M^{-2}), \tag{25}$$

where $f=m_1m_2$. It remains to expand $G(M,t,\Omega)$ for large M. This has already been done in Theorem 8.2.14. If we there put $\alpha_0=\beta_0=\gamma_0=\frac{1}{2}$, $\alpha_1=\beta_1=\gamma_1=\varepsilon_1=\frac{1}{2}\delta$, $\varepsilon_0=\frac{1}{2}(1-2it)$, and replace Ω by $-\Omega$, Theorem 8.2.14 shows that $G(M,t,\Omega)$ may be expanded as

$$G(M,t,\Omega)=\exp\left(\frac{it\sigma_1}{1-2it}\right)\left\{1-\frac{1}{4M}\left[\sigma_2+(m+1)\sigma_1-\frac{2(m+1)\sigma_1}{1-2it}\right.\right. \tag{26}$$

$$\left.\left.-\frac{2\sigma_2-(m+1)\sigma_1}{(1-2it)^2}+\frac{\sigma_2}{(1-2it)^3}\right]\right\}+O(M^{-2})$$

where $\sigma_j=\operatorname{tr}\Omega^j$. Multiplying (25) and (26) then gives an expansion for the characteristic function $\phi(M,t,\Omega)$ which when inverted term by term gives (22) and completes the proof.

We consider next the sequence of local alternatives K_M^*: $P^2=(1/M^2)\Omega$ under which $P^2\to 0$ at a faster rate than under K_M. In this case the characteristic function of $-2\rho\log\Lambda$ can be written from (23) as

$$\phi\left(M,t,\frac{1}{M}\Omega\right)=\phi(M,t,0)G\left(M,t,\frac{1}{M}\Omega\right). \tag{27}$$

The partial differential equations of Theorem 7.5.5 can be used to expand $G(M,t,M^{-1}\Omega)$ for large M (as, for example, in the proofs of Theorems 8.2.12 and 8.2.14). It is readily found that

$$G\left(M,t,\frac{1}{M}\Omega\right)=1+\frac{it\sigma_1}{M(1-2it)}+O(M^{-2}), \tag{28}$$

where $\sigma_1 = \operatorname{tr}\Omega$. Multiplying (25) and (28) and inverting the resulting expansion then gives the following result.

THEOREM 11.2.8. Under the sequence of local alternatives K_M^*: $P^2 = (1/M^2)\Omega$ the distribution function of $-2\rho\log\Lambda$ can be expanded as

$$(29)\quad P(-2\rho\log\Lambda \le x) = P(\chi_f^2 \le x) + \frac{\sigma_1}{2M}\cdot\left[P(\chi_{f+2}^2 \le x) - P(\chi_f^2 \le x)\right] + O(M^{-2}),$$

where $f = m_1 m_2$ and $\sigma_1 = \operatorname{tr}\Omega$.

Finally, we consider the general alternative K: $P^2 \ne 0$. Define the random variable Y by

$$(30)\qquad Y = \frac{-2\rho\log\Lambda}{M^{1/2}} + M^{1/2}\log\det(I - P^2).$$

The characteristic function of Y is, using Theorems 11.2.6 and 7.4.3,

$$(31)\qquad \begin{aligned} g(M,t,P^2) &= E(e^{itY}) \\ &= \det(I-P^2)^{itM^{1/2}} E\left[\Lambda^{-2it/M^{1/2}}\right] \\ &= \det(I-P^2)^{itM^{1/2}} E\left[W^{-itM^{1/2}}\right] \\ &= G_1(M,t)G_2(M,t,P^2), \end{aligned}$$

where

$$(32)\quad G_1(M,t) = \frac{\Gamma_{m_1}\left[\frac{1}{2}(M+\delta)\right]\Gamma_{m_1}\left[\frac{1}{2}M - itM^{1/2} + \frac{1}{4}(m_1 - m_2 + 1)\right]}{\Gamma_{m_1}\left[\frac{1}{2}M + \frac{1}{4}(m_1 - m_2 + 1)\right]\Gamma_{m_1}\left[\frac{1}{2}(M+\delta) - M^{1/2}it\right]}$$

and

$$(33)\quad G_2(M,t,P^2) = {}_2F_1\left(-M^{1/2}it, -M^{1/2}it; \tfrac{1}{2}(M+\delta) - M^{1/2}it; P^2\right),$$

where $\delta = \frac{1}{2}(m_1 + m_2 + 1)$. Using (24) of Section 8.2 to expand the gamma functions for large M it is straightforward to show that

$$(34)\qquad G_1(M,t) = 1 + \frac{m_1 m_2 it}{M^{1/2}} + O(M^{-1}).$$

A function very similar to $G_2(M,t,P^2)$ has been expanded for large M in

Theorem 8.2.12. The same technique used there shows that

(35)
$$G_2(M,t,P^2)=\exp(-2t^2\sigma_1)\cdot\left\{1+\frac{1}{M^{1/2}}\left[m_1m_2it+4(it)^3(\sigma_1-\sigma_2)\right]+O(M^{-1})\right\},$$

where $\sigma_j=\operatorname{tr}P^{2j}$. Putting $\tau^2=4\sigma_1$ it then follows from (34) and (35) that $g(M,t/\tau,P^2)$, the characteristic function of Y/τ, can be expanded as

(36)
$$g\left(M,\frac{t}{\tau},P^2\right)=\exp(-\tfrac{1}{2}t^2)\cdot\left\{1+\frac{1}{M^{1/2}}\left[\frac{m_1m_2it}{\tau}+\frac{4(it)^3}{\tau^3}(\sigma_1-\sigma_2)\right]\right\}+O(M^{-1}).$$

Inverting this expansion then gives the following result.

THEOREM 11.2.9. Under the fixed alternative K: $P^2\neq 0$, the distribution function of the random variable Y given by (30) can be expanded as

(37)
$$P\left(\frac{Y}{\tau}\le x\right)=\Phi(x)-\frac{1}{M^{1/2}}\cdot\left[\frac{m_1m_2}{\tau}\phi(x)+\frac{4}{\tau^3}(\sigma_1-\sigma_2)\phi^{(2)}(x)\right]+O(M^{-1}),$$

where Φ and ϕ denote the standard normal distribution and density functions, respectively, and $\sigma_j=\operatorname{tr}P^{2j}$, $\tau^2=4\sigma_1$.

For further terms in the asymptotic expansions presented here the interested reader should see Sugiura (1969a), Sugiura and Fujikoshi (1969), Lee (1971a), and Muirhead (1972a). For work in the more general setting where the independence of $k>2$ sets of variables is being tested see Nagao (1972).

11.2.7. The Asymptotic Null Distribution of the Likelihood Ratio Statistic for Elliptical Samples

In order to understand the effect of non-normality on the distribution of Λ we examine the asymptotic null distribution of Λ for testing

$$H:\Sigma=\Sigma^*=\operatorname{diag}(\Sigma_{11},\Sigma_{22},\dots,\Sigma_{kk}),$$

where Σ_{ii} is $m_i \times m_i$, when the sample comes from an elliptical distribution. This null hypothesis, of course, does not specify independence between the k sets of variables unless the sample is normally distributed. We have

$$\Lambda = \frac{(\det A)^{N/2}}{\prod_{i=1}^{k} (\det A_{ii})^{N/2}} = \frac{(\det S)^{N/2}}{\prod_{i=1}^{k} (\det S_{ii})^{N/2}},$$

where $S = n^{-1}A$ is the sample covariance matrix partitioned similarly to A as in (1). Writing $S = \Sigma^* + N^{-1/2}Z$ and partitioning Z similarly to A and S, the statistic $-2\log\Lambda$ can be expanded when H is true as

$$-2\log\Lambda = \sum_{i<j}^{k} \operatorname{tr}\left(Z_{ij}\Sigma_{jj}^{-1}Z_{ji}\Sigma_{ii}^{-1}\right) + O_p(N^{-1/2})$$

$$= \sum_{i<j}^{k} \mathbf{z}_{ij}'\left(\Sigma_{ii}\otimes\Sigma_{jj}\right)^{-1}\mathbf{z}_{ij} + O_p(N^{-1/2}),$$

where $\mathbf{z}_{ij} = \operatorname{vec}(Z_{ij}')$. Now assume that the observations are drawn from an elliptical distribution with kurtosis parameter κ, in which case the $\mathbf{z}_{ij}$'s $(i<j)$ are all asymptotically independent and the asymptotic distribution, as $N \to \infty$, of $\mathbf{z}_{ij}$ is $N_{m_i m_j}(\mathbf{0}, (1+\kappa)(\Sigma_{ii}\otimes\Sigma_{jj}))$. This leads immediately to the following result.

THEOREM 11.2.10. Let Λ be the likelihood ratio statistic for testing $H: \Sigma = \Sigma^* = \operatorname{diag}(\Sigma_{11}, \dots, \Sigma_{kk})$, assuming normality. If the sample is drawn from an elliptical distribution with kurtosis parameter κ then the asymptotic null distribution of $-2(\log\Lambda)/(1+\kappa)$ is χ_f^2, where

$$f = \sum_{i<j}^{k} m_i m_j = \tfrac{1}{2}\left(m^2 - \sum_{i=1}^{k} m_i^2\right).$$

Two points are worth noting. First, if the sample is normally distributed $\kappa = 0$, and this result agrees with that derived in Section 11.2.4. Secondly, if κ is unknown and is estimated by a consistent estimate $\hat{\kappa}$ then the limiting null distribution of $-2(\log\Lambda)/(1+\hat{\kappa})$ is also χ_f^2. Monte Carlo studies carried out by Muirhead and Waternaux (1980) for the case $k=2$ indicate that the usual test statistic $-2\log\Lambda$ should be used with extreme care, if at all, for testing $H: \Sigma = \Sigma^*$ when the underlying population is elliptical with longer tails than the normal distribution. For example, 200 samples of size $N = 100$ drawn from the elliptical 7-variate t distribution on 5 degrees of freedom $(m_1 = 3, m_2 = 4)$ when $\Sigma = I_7$ gave an observed significance level for the test based on $-2\log\Lambda$, of 56% for a nominal level of 10%. On the other hand, a test based on $-2(\log\Lambda)/(1+\hat{\kappa})$, where $\hat{\kappa}$ is a consistent estimate of κ, yielded an observed significance level of 10.5% for the same nominal level of 10%. For further details about the Monte Carlo study, and for a method of estimating κ the interested reader is referred to Muirhead and Waternaux (1980).

11.2.8. Other Test Statistics

A number of other invariant test statistics have been proposed when $k=2$ for testing the null hypothesis $H: \Sigma_{12}=0$. In terms of the latent roots $r_1^2 > \cdots > r_{m_1}^2$ of $S_{11}^{-1}S_{12}S_{22}^{-1}S_{21}$ these include

$$L_1 = \sum_{i=1}^{m_1} \frac{r_i^2}{1-r_i^2}$$

and

$$L_2 = \sum_{i=1}^{m_1} r_i^2$$

and the largest root r_1^2. We reject H for large values of these three statistics. A comparison of the powers of the tests based on Λ, L_1, L_2, and r_1^2 was carried out by Pillai and Jayachandran (1968) for the case $m_1=2$. They concluded that for small deviations from H, or for large deviations when ρ_1^2 and ρ_2^2 are close, the test based on L_2 appears to have higher power than that based on Λ, while Λ has higher power than L_1. The reverse ordering appears to hold for large deviations from H with $\rho_1^2-\rho_2^2$ large. The largest root r_1^2 has lower power than the other three except when ρ_1^2 is the only deviant root.

An expression for the distribution function of r_1^2 will be obtained in Section 11.3.4. Asymptotic expansions for the distributions of L_1 and L_2 have been obtained by Lee (1971a). For a survey of other results concerning these tests the reader is referred to Pillai (1976, 1977).

11.3. CANONICAL CORRELATION ANALYSIS

11.3.1. Introduction

When observations are taken on a large number of correlated variables it is natural to look at various ways in which the number of variables might be reduced without sacrificing too much information. When the variables are regarded as belonging to a single set of variables a principal components analysis (Chapter 9) is often insightful. When the variables fall naturally into two sets an important exploratory technique is *canonical correlation analysis*, developed by Hotelling (1936). This analysis is concerned with reducing the correlation structure between two sets of variables $\mathbf{X}$ and $\mathbf{Y}$ to the simplest possible form by means of linear transformations on $\mathbf{X}$ and $\mathbf{Y}$. The first canonical variables U_1, V_1 are the two linear functions $U_1 = \boldsymbol{\alpha}_1'\mathbf{X}, V_1 = \boldsymbol{\beta}_1'\mathbf{Y}$ having the maximum correlation subject to the condition that $\mathrm{Var}(U_1) = \mathrm{Var}(V_1) = 1$; the second canonical variables U_2, V_2 are the two linear functions $U_2 = \boldsymbol{\alpha}_2'\mathbf{X}, V_2 = \boldsymbol{\beta}_2'\mathbf{Y}$ having maximum correlation subject to the conditions that U_2 and V_2 are uncorrelated with both U_1 and V_1 and have unit variance, and so on. When the two sets of variables are large it is often the case that the first few canonical variables exhibit high correlations compared with the remaining canonical variables. When this occurs it is natural, at least as an exploratory device, to restrict attention to the first few canonical variables. In essence, then, canonical correlation analysis is concerned with attempting to characterize the correlation structure between two sets of variables by replacing them with two new sets with a smaller number of variables which are pairwise highly correlated.

11.3.2. Population Canonical Correlation Coefficients and Canonical Variables

Suppose that $\mathbf{X}$ and $\mathbf{Y}$ are, respectively, $p \times 1$ and $q \times 1$ random vectors having covariance matrix

$$\text{(1)} \qquad \mathrm{Cov}\begin{pmatrix} \mathbf{X} \\ \mathbf{Y} \end{pmatrix} = \Sigma = \begin{bmatrix} \Sigma_{11} & \Sigma_{12} \\ \Sigma_{21} & \Sigma_{22} \end{bmatrix},$$

where Σ_{11} is $p \times p$ and Σ_{22} is $q \times q$. We will assume without loss of generality that $p \le q$. Let $k = \mathrm{rank}(\Sigma_{12})$. From Theorem A9.10 there exist $H \in O(p), Q \in O(q)$ such that

$$\text{(2)} \qquad \Sigma_{11}^{-1/2}\Sigma_{12}\Sigma_{22}^{-1/2} = H'\tilde{P}Q$$

where

$$\text{(3)} \qquad \tilde{P} = \left[\begin{array}{ccc|c} \rho_1 & & 0 & \\ & \ddots & & 0 \\ 0 & & \rho_k & \\ \hline & 0 & & 0 \end{array}\right] \qquad (p \times q)$$

with $\rho_1, \ldots, \rho_k$ $(1 \geq \rho_1 \geq \cdots \geq \rho_k > 0)$ being the positive square roots of $\rho_1^2, \ldots, \rho_k^2$, the nonzero latent roots of $\Sigma_{11}^{-1}\Sigma_{12}\Sigma_{22}^{-1}\Sigma_{21}$. Putting

$$(4) \qquad L_1 = H\Sigma_{11}^{-1/2}, \qquad L_2 = Q\Sigma_{22}^{-1/2},$$

it then follows that

$$(5) \qquad L_1\Sigma_{11}L_1' = I_p, \; L_2\Sigma_{22}L_2' = I_q \quad \text{and} \quad L_1\Sigma_{12}L_2' = \tilde{P}.$$

Putting $\mathbf{U} = L_1\mathbf{X}$, $\mathbf{V} = L_2\mathbf{Y}$ so that

$$(6) \qquad \begin{pmatrix} \mathbf{U} \\ \mathbf{V} \end{pmatrix} = \begin{bmatrix} L_1 & 0 \\ 0 & L_2 \end{bmatrix} \begin{pmatrix} \mathbf{X} \\ \mathbf{Y} \end{pmatrix} = L \begin{pmatrix} \mathbf{X} \\ \mathbf{Y} \end{pmatrix},$$

where $L = \text{diag}(L_1, L_2)$, we then have

$$(7) \qquad \text{Cov}\begin{pmatrix} \mathbf{U} \\ \mathbf{V} \end{pmatrix} = L\Sigma L' = \begin{bmatrix} I_p & \tilde{P} \\ \tilde{P}' & I_q \end{bmatrix}.$$

Hence, by means of linear transformations on $\mathbf{X}$ and $\mathbf{Y}$ the correlation structure implicit in the covariance matrix Σ has been reduced to a form involving only the parameters $\rho_1, \ldots, \rho_k$. This reduction has already been carried out in Theorem 11.2.2. The parameters $\rho_1, \ldots, \rho_k$ whose squares are the nonzero latent roots of $\Sigma_{11}^{-1}\Sigma_{12}\Sigma_{22}^{-1}\Sigma_{21}$ or, equivalently, the nonzero latent roots of $\Sigma_{22}^{-1}\Sigma_{21}\Sigma_{11}^{-1}\Sigma_{12}$, are called the population canonical correlation coefficients. The covariance matrix (7) of $\mathbf{U}, \mathbf{V}$ is called a canonical form for Σ under the group of transformations

$$\Sigma \to B\Sigma B',$$

where $B = \text{diag}(B_{11}, B_{22})$ with B_{11} and B_{22} being nonsingular $p \times p$ and $q \times q$ matrices, respectively, since it involves only a maximal invariant under the group (see Theorem 11.2.2). Letting $\mathbf{U}' = (U_1, U_2, \ldots, U_p)$ and $\mathbf{V}' = (V_1, \ldots, V_q)$ the variables U_i, V_i are called the ith canonical variables, $i = 1, \ldots, p$. That these variables have the optimal correlation properties mentioned in Section 11.3.1 will be established later. Note that, from (5)

$$\begin{aligned} (8) \qquad L_1\Sigma_{12}\Sigma_{22}^{-1}\Sigma_{21}L_1' &= (L_1\Sigma_{12}L_2')(L_2\Sigma_{22}L_2')^{-1}(L_2\Sigma_{21}L_1') \\ &= \tilde{P}I_q\tilde{P}' \\ &= \tilde{P}\tilde{P}' = \text{diag}(\rho_1^2, \ldots, \rho_k^2, 0, \ldots, 0) \qquad (p \times p). \end{aligned}$$

Hence, if $L_1' = [\mathbf{l}_1 \dots \mathbf{l}_p]$ it follows that $\mathbf{l}_i$ $(i = 1,\dots,k)$ is a solution of the equation

(9) $$(\Sigma_{12}\Sigma_{22}^{-1}\Sigma_{21} - \rho_i^2\Sigma_{11})\mathbf{l}_i = \mathbf{0}$$

normalized so that $\mathbf{l}_i'\Sigma_{11}\mathbf{l}_i = 1$. If the ρ_i's are distinct $\mathbf{l}_1,\dots,\mathbf{l}_k$ are unique apart from sign. Similarly,

(10) $$\begin{aligned} L_2\Sigma_{21}\Sigma_{11}^{-1}\Sigma_{12}L_2' &= (L_2\Sigma_{21}L_1')(L_1\Sigma_{11}L_1')^{-1}(L_1\Sigma_{12}L_2') \\ &= \tilde{P}'I_p\tilde{P} \\ &= \tilde{P}'\tilde{P} = \operatorname{diag}(\rho_1^2,\dots,\rho_k^2,0,\dots,0) \qquad (q \times q), \end{aligned}$$

so that if $L_2' = [\mathbf{l}_1^* \dots \mathbf{l}_q^*]$ it follows that $\mathbf{l}_i^*$ $(i = 1,\dots,k)$ is a solution of the equation

(11) $$(\Sigma_{21}\Sigma_{11}^{-1}\Sigma_{12} - \rho_i^2\Sigma_{22})\mathbf{l}_i^* = \mathbf{0},$$

normalized so that $\mathbf{l}_i^{*\prime}\Sigma_{22}\mathbf{l}_i^* = 1$. Again, if the ρ_i's are distinct $\mathbf{l}_1^*,\dots,\mathbf{l}_k^*$ are unique apart from sign. Once the signs have been set for $\mathbf{l}_1,\dots,\mathbf{l}_k$ they are determined for $\mathbf{l}_1^*,\dots,\mathbf{l}_k^*$ by the requirement that

(12) $$\mathbf{l}_i'\Sigma_{12}\mathbf{l}_j^* = \rho_i\delta_{i,j} \quad (i,j = 1,\dots,k),$$

which follows from (5). Note also that

(13) $$L_1\Sigma_{12}\Sigma_{22}^{-1} = (L_1\Sigma_{12}L_2')(L_2\Sigma_{22}L_2')^{-1}L_2 = \tilde{P}L_2$$

so that

(14) $$\Sigma_{22}^{-1}\Sigma_{21}\mathbf{l}_i = \rho_i\mathbf{l}_i^* \qquad (i = 1,\dots,k)$$

and

(15) $$\Sigma_{21}\mathbf{l}_i = \mathbf{0} \qquad (i = k+1,\dots,p),$$

and similarly, since

(16) $$\Sigma_{11}^{-1}\Sigma_{12}L_2' = L_1'(L_1\Sigma_{11}L_1')^{-1}(L_1\Sigma_{12}L_2') = L_1'\tilde{P},$$

it follows that

$$\Sigma_{11}^{-1}\Sigma_{12}\mathbf{l}_i^* = \rho_i \mathbf{l}_i \qquad (i=1,\dots,k) \tag{17}$$

and

$$\Sigma_{12}\mathbf{l}_i^* = \mathbf{0} \qquad (i=k+1,\dots,q). \tag{18}$$

In one matrix equation (14) and (17) become

$$\begin{bmatrix} -\rho_i\Sigma_{11} & \Sigma_{12} \\ \Sigma_{21} & -\rho_i\Sigma_{22} \end{bmatrix} \begin{pmatrix} \mathbf{l}_i \\ \mathbf{l}_i^* \end{pmatrix} = \mathbf{0} \qquad (i=1,\dots,k). \tag{19}$$

The canonical variables have the following optimality property. The first canonical variables $U_1 = \mathbf{l}_1'\mathbf{X}, V_1 = \mathbf{l}_1^{*\prime}\mathbf{Y}$ are linear combinations of the components of $\mathbf{X}$ and $\mathbf{Y}$, respectively, with unit variance having the largest possible correlation, and this correlation is ρ_1; then out of all linear combinations of the components of $\mathbf{X}$ and $\mathbf{Y}$ which are uncorrelated with both U_1 and V_1 and have unit variance the second canonical variables are most highly correlated, and the correlation is ρ_2, and so on. In general, out of all linear combinations of $\mathbf{X}$ and $\mathbf{Y}$ with unit variance which are uncorrelated with every one of $U_1,\dots,U_{j-1},V_1,\dots,V_{j-1}$, the jth canonical variables U_j, V_j have maximum correlation ρ_j, $j=1,\dots,k$. We will prove this assertion in a moment. First note that the correlation between two arbitrary linear functions $\boldsymbol{\alpha}'\mathbf{X}$ and $\boldsymbol{\beta}'\mathbf{Y}$ with unit variance is

$$\operatorname{Corr}(\boldsymbol{\alpha}'\mathbf{X}, \boldsymbol{\beta}'\mathbf{Y}) = \frac{\boldsymbol{\alpha}'\Sigma_{12}\boldsymbol{\beta}}{(\boldsymbol{\alpha}'\Sigma_{11}\boldsymbol{\alpha}\boldsymbol{\beta}'\Sigma_{22}\boldsymbol{\beta})^{1/2}} = \boldsymbol{\alpha}'\Sigma_{12}\boldsymbol{\beta}. \tag{20}$$

The condition that $\boldsymbol{\alpha}'\mathbf{X}$ be uncorrelated with $U_i = \mathbf{l}_i'\mathbf{X}$ is

$$\begin{aligned} 0 &= \boldsymbol{\alpha}'\Sigma_{11}\mathbf{l}_i \\ &= \frac{1}{\rho_i}\boldsymbol{\alpha}'\Sigma_{12}\mathbf{l}_i^*, \end{aligned} \tag{21}$$

using (17), and hence

$$0 = \boldsymbol{\alpha}'\Sigma_{12}\mathbf{l}_i^* \tag{22}$$

so that $\boldsymbol{\alpha}'\mathbf{X}$ and $V_i = \mathbf{l}_i^{*\prime}\mathbf{Y}$ are uncorrelated. Similarly the condition that $\boldsymbol{\beta}'\mathbf{Y}$

be uncorrelated with $V_i = \mathbf{l}_i^{*\prime}\mathbf{Y}$ is

$$(23)\qquad 0 = \boldsymbol{\beta}'\Sigma_{22}\mathbf{l}_i^* = \frac{1}{\rho_i}\boldsymbol{\beta}'\Sigma_{12}\mathbf{l}_i^*$$

using (14), and hence

$$(24)\qquad 0 = \mathbf{l}_i'\Sigma_{12}\boldsymbol{\beta},$$

so that $\boldsymbol{\beta}'\mathbf{Y}$ and $U_i = \mathbf{l}_i'\mathbf{X}$ are uncorrelated. The above optimality property of the canonical variables is a consequence of the following theorem.

THEOREM 11.3.1. Let Σ be partitioned as in (1) where Σ_{12} has rank k and let $\rho_1^2,\ldots,\rho_k^2$ $(\rho_1 \geq \cdots \geq \rho_k > 0)$ be the nonzero latent roots of $\Sigma_{11}^{-1}\Sigma_{12}\Sigma_{22}^{-1}\Sigma_{21}$. Then

$$(25)\qquad \rho_j = \sup \boldsymbol{\alpha}'\Sigma_{12}\boldsymbol{\beta} = \mathbf{l}_j'\Sigma_{12}\mathbf{l}_j^*,$$

where the supremum is taken over all $\boldsymbol{\alpha}\in R^p$, $\boldsymbol{\beta}\in R^q$ satisfying $\boldsymbol{\alpha}'\Sigma_{11}\boldsymbol{\alpha} = 1$, $\boldsymbol{\beta}'\Sigma_{22}\boldsymbol{\beta} = 1$, $\boldsymbol{\alpha}'\Sigma_{11}\mathbf{l}_i = 0$, $\boldsymbol{\beta}'\Sigma_{22}\mathbf{l}_i^* = 0$ $(i = 1,\ldots,j-1)$.

Proof. Putting $\boldsymbol{\gamma} = \Sigma_{11}^{1/2}\boldsymbol{\alpha}$ and $\boldsymbol{\delta} = \Sigma_{22}^{1/2}\boldsymbol{\beta}$ we have

$$(26)\qquad \begin{aligned}\boldsymbol{\alpha}'\Sigma_{12}\boldsymbol{\beta} &= \boldsymbol{\gamma}'\Sigma_{11}^{-1/2}\Sigma_{12}\Sigma_{22}^{-1/2}\boldsymbol{\delta}\\ &\leq (\boldsymbol{\gamma}'\boldsymbol{\gamma})^{1/2}(\boldsymbol{\delta}'\Sigma_{22}^{-1/2}\Sigma_{21}\Sigma_{11}^{-1}\Sigma_{12}\Sigma_{22}^{-1/2}\boldsymbol{\delta})^{1/2}\end{aligned}$$

by the Cauchy–Schwarz inequality. Now from (4) and (10) we have

$$Q\Sigma_{22}^{-1/2}\Sigma_{21}\Sigma_{11}^{-1}\Sigma_{12}\Sigma_{22}^{-1/2}Q' = \operatorname{diag}(\rho_1^2,\ldots,\rho_k^2,0,\ldots,0) \qquad (q\times q)$$

so that

$$\begin{aligned}\Sigma_{22}^{-1/2}\Sigma_{21}\Sigma_{11}^{-1}\Sigma_{12}\Sigma_{22}^{-1/2} &= Q'\operatorname{diag}(\rho_1^2,\ldots,\rho_k^2,0,\ldots,0)Q\\ &= \sum_{i=1}^{k}\rho_i^2\mathbf{b}_i\mathbf{b}_i',\end{aligned}$$

where $\mathbf{b}_1,\ldots,\mathbf{b}_q$ are the columns of Q'. Using this in (26), together with

$\gamma'\gamma=1$, we have

$$\begin{aligned}\alpha'\Sigma_{12}\beta &\le \left[\sum_{i=1}^{k}\rho_i^2(\delta'\mathbf{b}_i)^2\right]^{1/2}\\ &\le \rho_1\left[\sum_{i=1}^{k}(\delta'\mathbf{b}_i)^2\right]^{1/2}\\ &= \rho_1\left[\delta'\sum_{i=1}^{k}\mathbf{b}_i\mathbf{b}_i'\delta\right]^{1/2}\\ &= \rho_1\left[\delta'\left(I-\sum_{i=k+1}^{q}\mathbf{b}_i\mathbf{b}_i'\right)\delta\right]^{1/2}\\ &= \rho_1\left[\delta'\delta-\sum_{i=k+1}^{q}(\delta'\mathbf{b}_i)^2\right]^{1/2}\\ &\le \rho_1.\end{aligned}$$

From (12) we have $\mathbf{l}_1'\Sigma_{12}\mathbf{l}_1^*=\rho_1$, and hence

$$\rho_1=\sup\alpha'\Sigma_{12}\beta=\mathbf{l}_1'\Sigma_{12}\mathbf{l}_1^*,$$

where the supremum is taken over all $\alpha\in R^p$, $\beta\in R^q$ with $\alpha'\Sigma_{11}\alpha=1$, $\beta'\Sigma_{22}\beta=1$. Note that from (12) this is attained when $\beta=\mathbf{l}_1^*=\Sigma_{22}^{-1/2}\mathbf{b}_1$ and $\alpha=\mathbf{l}_1=\Sigma_{11}^{-1/2}\mathbf{h}_1$, where $\mathbf{h}_1,\ldots,\mathbf{h}_p$ are the columns of H. Next, again putting $\gamma=\Sigma_{11}^{1/2}\alpha$, $\delta=\Sigma_{22}^{1/2}\beta$, we have by the same argument

$$\alpha'\Sigma_{12}\beta\le\left[\sum_{i=1}^{k}\rho_i^2(\delta'\mathbf{b}_i)^2\right]^{1/2}.$$

Now, when $\beta'\Sigma_{22}\mathbf{l}_1^*=0$ we have $\delta'\mathbf{b}_1=0$ and hence

$$\begin{aligned}\alpha'\Sigma_{12}\beta &\le \left[\sum_{i=2}^{k}\rho_i^2(\delta'\mathbf{b}_i)^2\right]^{1/2}\\ &\le \rho_2\left[(\delta'\mathbf{b}_2)^2+\cdots+(\delta'\mathbf{b}_k)^2\right]^{1/2}\\ &\le \rho_2.\end{aligned}$$

From (12) we have $\mathbf{l}_2'\Sigma_{12}\mathbf{l}_2^* = \rho_2$, where $\mathbf{l}_2'\Sigma_{11}\mathbf{l}_1 = \mathbf{h}_2'\mathbf{h}_1 = 0$. Hence

$$\rho_2 = \sup \boldsymbol{\alpha}'\Sigma_{12}\boldsymbol{\beta} = \mathbf{l}_2'\Sigma_{12}\mathbf{l}_2^*,$$

where the supremum is taken over all $\boldsymbol{\alpha} \in R^p$, $\boldsymbol{\beta} \in R^q$ with $\boldsymbol{\alpha}'\Sigma_{11}\boldsymbol{\alpha} = 1$, $\boldsymbol{\beta}'\Sigma_{22}\boldsymbol{\beta} = 1$, $\boldsymbol{\alpha}'\Sigma_{11}\mathbf{l}_1 = 0$, $\boldsymbol{\beta}'\Sigma_{22}\mathbf{l}_1^* = 0$. The rest of the proof follows using a similar and obvious argument.

It is worth noting that the canonical correlation coefficients can be interpreted as multiple correlation coefficients. From (8) we have

$$\rho_i^2 = \mathbf{l}_i'\Sigma_{12}\Sigma_{22}^{-1}\Sigma_{21}\mathbf{l}_i$$

$$= \frac{(\mathbf{l}_i'\Sigma_{12})\Sigma_{22}^{-1}(\Sigma_{21}\mathbf{l}_i)}{\mathbf{l}_i'\Sigma_{11}\mathbf{l}_i}.$$

Noting that $\Sigma_{21}\mathbf{l}_i$ is the vector of covariances between $U_i = \mathbf{l}_i'\mathbf{X}$ and $\mathbf{Y}$ and that $\text{Var}(U_i) = \mathbf{l}_i'\Sigma_{11}\mathbf{l}_i$ this shows that ρ_i is the multiple correlation cofficience between U_i and $\mathbf{Y}$. A similar argument also shows that ρ_i is the multiple correlation coefficient between $V_i = \mathbf{l}_i^{*\prime}\mathbf{Y}$ and $\mathbf{X}$.

11.3.3. Sample Canonical Correlation Coefficients and Canonical Variables

In most practical applications the covariance matrix Σ is unknown, and hence so are the canonical correlations and canonical variables. These then have to be estimated. Suppose that $\hat{\Sigma}$ is the maximum likelihood estimate of Σ formed from a sample of size N observations on $(\mathbf{X}', \mathbf{Y}')'$ drawn from a $N_{p+q}(\boldsymbol{\mu}, \Sigma)$ distribution, and put $A = N\hat{\Sigma}$ and $S = n^{-1}A$ where $n = N-1$. Partition A and S similarly to Σ as

$$(27) \qquad A = \begin{bmatrix} A_{11} & A_{12} \\ A_{21} & A_{22} \end{bmatrix}, \qquad S = \begin{bmatrix} S_{11} & S_{12} \\ S_{21} & S_{22} \end{bmatrix},$$

where A_{11} and S_{11} are $p \times p$ and A_{22} and S_{22} are $q \times q$. Assuming, as before, that $p \le q$, let $r_1^2, \dots, r_p^2$ be the latent roots of $S_{11}^{-1}S_{12}S_{22}^{-1}S_{21}$ $(1 > r_1^2 > \cdots > r_p^2 > 0)$. (These are the same as the latent roots of $A_{11}^{-1}A_{12}A_{22}^{-1}A_{21}$.) These are distinct and nonzero with probability 1 and are estimates of the latent roots of $\Sigma_{11}^{-1}\Sigma_{12}\Sigma_{22}^{-1}\Sigma_{21}$ (some of which may be zero). Their positive square roots $r_1, \dots, r_p (1 > r_1 > \cdots > r_p > 0)$ are called the sample canonical correlation coefficients. The ith population canonical variables $U_i = \mathbf{l}_i'\mathbf{X}$, $V_i = \mathbf{l}_i^{*\prime}\mathbf{Y}$ are estimated by $\hat{U}_i = \hat{\mathbf{l}}_i'\mathbf{X}$, $\hat{V}_i = \hat{\mathbf{l}}_i^{*\prime}\mathbf{Y}$, called the ith sample canonical variables, where $\hat{\mathbf{l}}_i$ and $\hat{\mathbf{l}}_i^*$ satisfy equations similar to those satisfied by $\mathbf{l}_i$ and $\mathbf{l}_i^*$ with Σ replaced by S and ρ_i by r_i. Hence from (9), $\hat{\mathbf{l}}_i$, for $i=1,\ldots,p$, is a solution of the equation

$$(28)\qquad (S_{12}S_{22}^{-1}S_{21}-r_i^2S_{11})\hat{\mathbf{l}}_i=\mathbf{0},$$

normalized so that $\hat{\mathbf{l}}_i'S_{11}\hat{\mathbf{l}}_i=1$. Similarly, from (11), $\hat{\mathbf{l}}_i^*$ is a solution of

$$(29)\qquad (S_{21}S_{11}^{-1}S_{12}-r_i^2S_{22})\hat{\mathbf{l}}_i^*=\mathbf{0},$$

normalized so that $\hat{\mathbf{l}}_i^{*\prime}S_{22}\hat{\mathbf{l}}_i^*=1$. Equations (14), (17), (18), and (19) become, respectively,

$$(30)\qquad S_{22}^{-1}S_{21}\hat{\mathbf{l}}_i=r_i\hat{\mathbf{l}}_i^* \qquad (i=1,\ldots,p),$$

$$(31)\qquad S_{11}^{-1}S_{12}\hat{\mathbf{l}}_i^*=r_i\hat{\mathbf{l}}_i \qquad (i=1,\ldots,p),$$

$$(32)\qquad S_{12}\hat{\mathbf{l}}_i^*=\mathbf{0} \qquad (i=p+1,\ldots,q),$$

and

$$(33)\qquad \begin{bmatrix} -r_iS_{11} & S_{12} \\ S_{21} & -r_iS_{22}\end{bmatrix}\begin{pmatrix}\hat{\mathbf{l}}_i \\ \hat{\mathbf{l}}_i^*\end{pmatrix}=\mathbf{0} \qquad (i=1,\ldots,p).$$

Note that r_i is the sample multiple correlation coefficient between $\hat{U}_i$ and $\mathbf{Y}$, and also between $\hat{V}_i$ and $\mathbf{X}$.

Tractable expressions for the exact moments of $r_1,\ldots,r_p$ are unknown but asymptotic expansions for some of these have been found by Lawley (1959). If $k=\text{rank}(\Sigma_{12})$ and ρ_i^2 is a simple nonzero latent root of $\Sigma_{11}^{-1}\Sigma_{12}\Sigma_{22}^{-1}\Sigma_{21}$ then

(34)

$$E(r_i)=\rho_i+\frac{1-\rho_i^2}{2n\rho_i}\left[p+q-2-\rho_i^2+2(1-\rho_i^2)\sum_{\substack{j=1\\ j\neq i}}^{k}\frac{\rho_j^2}{\rho_i^2-\rho_j^2}\right]+O(n^{-2}),$$

and

$$(35)\qquad \text{Var}(r_i)=\frac{(1-\rho_i^2)^2}{n}+O(n^{-2}).$$

11.3.4. *Distributions of the Sample Canonical Correlation Coefficients*

We have noted in Section 11.2 that invariant test statistics used for testing the hypothesis of independence between two vectors of dimensions p and q, respectively ($p \le q$), are functions of the squares of the sample canonical correlation coefficients $r_1^2,\ldots,r_p^2$. The exact joint distribution of $r_1^2,\ldots,r_p^2$ can be expressed in terms of the two-matrix ${}_2F_1$ hypergeometric function introduced in Section 7.3, having an expansion in terms of zonal polynomials. The result is given in the following theorem due to Constantine (1963).

THEOREM 11.3.2. Let A have the $W_{p+q}(n,\Sigma)$ distribution where $p \le q$, $n \ge p+q$ and Σ and A are partitioned as in (1) and (27). Then the joint probability density function of $r_1^2,\ldots,r_p^2$, the latent roots of $A_{11}^{-1}A_{12}A_{22}^{-1}A_{21}$, is

$$(36)\qquad \prod_{i=1}^{p}\left(1-\rho_i^2\right)^{n/2} {}_2F_1^{(p)}\left(\tfrac{1}{2}n,\tfrac{1}{2}n;\tfrac{1}{2}q;P^2,R^2\right)$$

$$\cdot\frac{\pi^{p^2/2}}{\Gamma_p\left(\frac{1}{2}p\right)}\frac{\Gamma_p\left(\frac{1}{2}n\right)}{\Gamma_p\left[\frac{1}{2}(n-q)\right]\Gamma_p\left(\frac{1}{2}q\right)}$$

$$\cdot\prod_{i=1}^{p}\left[\left(r_i^2\right)^{(q-p-1)/2}\left(1-r_i^2\right)^{(n-p-q-1)/2}\right]$$

$$\cdot\prod_{i<j}^{p}\left(r_i^2-r_j^2\right)\qquad \left(1>r_1^2>\cdots>r_p^2>0\right),$$

where $\rho_1^2,\ldots,\rho_p^2$ are the latent roots of $\Sigma_{11}^{-1}\Sigma_{12}\Sigma_{22}^{-1}\Sigma_{21}$ (some of which may be zero), $P^2=\operatorname{diag}(\rho_1^2,\ldots,\rho_p^2)$, and $R^2=\operatorname{diag}(r_1^2,\ldots,r_p^2)$.

Proof. Most of the work involved in the proof has already been carried out in the proof of Theorem 11.2.6. In that proof with $m_1=p$, $m_2=q$, we saw that $r_1^2,\ldots,r_p^2$ are the solutions of the equation

$$(37)\qquad \det\left(U'U-r^2(U'U+V'V)\right)=0.$$

We also saw that, conditional on $X'X$, which is $W_q(n,I_q)$, the random matrices $U'U$ and $V'V$ are independent, with $U'U$ having the $W_p(q,\Phi,\Omega)$ distribution and $V'V$ having the $W_p(n-q,\Phi)$ distribution, where $\Phi=I-P^2$ and

$$\Omega=\Phi^{-1}\tilde{P}X'X\tilde{P}'$$

with

$$\tilde{P}=\begin{bmatrix} \rho_1 & & 0 & \vdots \\ & \ddots & & \vdots\ 0 \\ 0 & & \rho_p & \vdots \end{bmatrix} \qquad (p\times q).$$

Hence conditional on $X'X$ the density function of the latent roots $r_1^2,\ldots,r_p^2$ of $(U'U)(U'U+V'V)^{-1}$ follows from the density function of the latent roots $f_1,\ldots,f_p$ of $(U'U)(V'V)^{-1}$ given in Theorem 10.4.2 by putting $f_i=r_i^2/(1-r_i^2)$, $r=q$, $n-p=n-q$, $m=p$, and is

$$\operatorname{etr}\left(-\tfrac{1}{2}\Phi^{-1}\tilde{P}X'X\tilde{P}'\right){}_1F_1^{(p)}\left(\tfrac{1}{2}n;\tfrac{1}{2}q;\tfrac{1}{2}\Phi^{-1}\tilde{P}X'X\tilde{P}',R^2\right)$$

$$\cdot\frac{\pi^{p^2/2}\Gamma_p\left(\frac{1}{2}n\right)}{\Gamma_p\left(\frac{1}{2}p\right)\Gamma_p\left[\frac{1}{2}(n-q)\right]\Gamma_p\left(\frac{1}{2}q\right)}\prod_{i=1}^{p}\left[\left(r_i^2\right)^{(q-p-1)/2}\left(1-r_i^2\right)^{(n-p-q-1)/2}\right]$$

$$\cdot\prod_{i<j}^{p}\left(r_i^2-r_j^2\right).$$

Multiplying this by the $W_q(n,I_q)$ density function for $X'X$ gives the joint density function of $r_1^2,\ldots,r_p^2$ and $X'X$ as

$$\operatorname{etr}\left(-\tfrac{1}{2}\Phi^{-1}\tilde{P}X'X\tilde{P}'\right){}_1F_1^{(p)}\left(\tfrac{1}{2}n;\tfrac{1}{2}q;\tfrac{1}{2}\Phi^{-1}\tilde{P}X'X\tilde{P}',R^2\right)\frac{2^{-nq/2}}{\Gamma_q\left(\frac{1}{2}n\right)}\operatorname{etr}\left(-\tfrac{1}{2}X'X\right)$$

$$\cdot\det(X'X)^{(n-q-1)/2}\frac{\pi^{p^2/2}\Gamma_p\left(\frac{1}{2}n\right)}{\Gamma_p\left(\frac{1}{2}p\right)\Gamma_p\left[\frac{1}{2}(n-q)\right]\Gamma_p\left(\frac{1}{2}q\right)}\prod_{i=1}^{p}\left[\left(r_i^2\right)^{(q-p-1)/2}\right.$$

$$\left.\cdot\left(1-r_i^2\right)^{(n-p-q-1)/2}\right]\prod_{i<j}^{p}\left(r_i^2-r_j^2\right).$$

We now integrate with respect to $X'X$ using Theorem 7.3.4 to show

$$\frac{2^{-nq/2}}{\Gamma_q\left(\frac{1}{2}n\right)}\int_{X'X>0}\operatorname{etr}\left[-\tfrac{1}{2}\left(I+\tilde{P}'\Phi^{-1}\tilde{P}\right)X'X\right]\det(X'X)^{(n-q-1)/2}$$

$${}_1F_1^{(p)}\left(\tfrac{1}{2}n;\tfrac{1}{2}q;\tfrac{1}{2}\tilde{P}'\Phi^{-1}\tilde{P}X'X,R^2\right)(d(X'X))$$

$$=\det\left(I+\Phi^{-1}\tilde{P}\tilde{P}'\right)^{-n/2}{}_2F_1^{(p)}\left(\tfrac{1}{2}n,\tfrac{1}{2}n;\tfrac{1}{2}q;\Phi^{-1}\tilde{P}\left(I+\tilde{P}'\Phi^{-1}\tilde{P}\right)^{-1}\tilde{P}',R^2\right).$$

The desired result now follows if we note that

$$\Phi^{-1}\tilde{P}\tilde{P}'=\operatorname{diag}\left(\frac{\rho_1^2}{1-\rho_1^2},\dots,\frac{\rho_p^2}{1-\rho_p^2}\right),$$

so that

$$\det(I+\Phi^{-1}\tilde{P}\tilde{P}')^{-n/2}=\prod_{i=1}^{p}(1-\rho_i^2)^{n/2},$$

and that

$$\Phi^{-1}\tilde{P}(I+\tilde{P}'\Phi^{-1}\tilde{P})^{-1}\tilde{P}'=\operatorname{diag}(\rho_1^2,\dots,\rho_p^2)=P^2.$$

The reader should note that the distribution of $r_1^2,\dots,r_p^2$ depends only on $\rho_1^2,\dots,\rho_p^2$. [Some of these may be zero. The number of nonzero ρ_i is rank (Σ_{12}).] This is because the nonzero ρ_i^2 form a maximal invariant under the group of transformations discussed in Section 11.2.1.

The null distribution of $r_1^2,\dots,r_p^2$, i.e., the distribution when $\rho_1^2=\cdots=\rho_p^2=0$ $(\Sigma_{12}=0)$ follows easily from Theorem 11.3.2 by putting $P^2=0$.

COROLLARY 11.3.3. When $P^2=0$ the joint density function of $r_1^2,\dots,r_p^2$, the latent roots of $A_{11}^{-1}A_{12}A_{22}^{-1}A_{21}$, is

$$(38)\qquad \frac{\pi^{p^2/2}\Gamma_p(\frac{1}{2}n)}{\Gamma_p(\frac{1}{2}p)\Gamma_p[\frac{1}{2}(n-q)]\Gamma_p(\frac{1}{2}q)}\prod_{i=1}^{p}\left[(r_i^2)^{(q-p-1)/2}(1-r_i^2)^{(n-p-q-1)/2}\right]$$

$$\cdot\prod_{i<j}^{p}(r_i^2-r_j^2)\qquad (1>r_1^2>\cdots>r_p^2>0).$$

It is worth noting that the null distribution (38) could also have been derived using Theorem 3.3.4. In the proof of Theorem 11.3.2 we noted that $r_1^2,\dots,r_p^2$ are the latent roots of $U'U(U'U+V'V)^{-1}$, where, if $P^2=0$, $U'U$ and $V'V$ are independent, $U'U$ is $W_p(q,I_p)$, and $V'V$ is $W_p(n-q,I)$. Corollary 11.3.3 then follows immediately from Theoren 3.3.4 on putting $n_1=q$, $n_2=n-q$, and $m=p$.

In theory the marginal distribution of any single canonical correlation coefficient, or of any subset of $r_1^2,\dots,r_p^2$ can be obtained from Theorem 11.3.2. In general, however, the integrals involved are not particularly tractable, even in the null case of Corollary 11.3.3. The square of the largest sample canonical correlation coefficient r_1^2 is of some interest as this can be used for testing independence (see Section 11.2.8). In the case when $t=\frac{1}{2}(n-p-q-1)$ is a positive integer an expression can be obtained for the distribution function of r_1^2 as a finite series of zonal polynomials. The result is given in the following theorem due to Constantine.

THEOREM 11.3.4. Suppose that the assumptions of Theorem 11.3.2 hold and that $t=\frac{1}{2}(n-p-q-1)$ is a positive integer. Then the distribution function of r_1^2 may be expressed as

(39)

$$P(r_1^2\le x)=x^{pq/2}\prod_{i=1}^{p}\left(\frac{1-\rho_i^2}{1-x\rho_i^2}\right)^{n/2}$$

$$\cdot\sum_{k=0}^{pt}\sum_{\kappa}{}^{*}\frac{(1-x)^k}{k!}\left(\frac{1}{2}q\right)_{\kappa}C_{\kappa}(I_p)\sum_{s=0}^{k}\sum_{\sigma}\binom{\kappa}{\sigma}\frac{(\frac{1}{2}n)_{\sigma}C_{\sigma}(B)}{(\frac{1}{2}q)_{\sigma}C_{\sigma}(I_p)},$$

where

$$B=\operatorname{diag}\left(\frac{x\rho_1^2}{1-x\rho_1^2},\ldots,\frac{x\rho_p^2}{1-x\rho_p^2}\right).$$

Here Σ^* denotes summation over those partitions $\kappa=(k_1,\ldots,k_p)$ of k with largest part $k_1\le t$;

$$\binom{\kappa}{\sigma}$$

is the generalized binomial coefficient defined by (8) of Section 7.5; and $(\alpha)_\kappa$ is the generalized hypergeometric coefficient given by (2) of Section 7.3.

Proof. As in the proof of Theorem 11.3.2 we start with $r_1^2,\ldots,r_p^2$ being the latent roots of $U'U(U'U+V'V)^{-1}$ where, conditional on $X'X$, which is $W_q(n,I_q)$, $U'U$ and $V'V$ are independent with $U'U$ being $W_p(q,\Phi,\Omega)$ and $V'V$ being $W_p(n-q,\Phi)$, where $\Phi=I-P^2$, $\Omega=\Phi^{-1}\tilde{P}X'X\tilde{P}'$, and

$$\tilde{P}=\left[\begin{array}{ccc:c}\rho_1 & & 0 & \\ & \ddots & & 0\\ 0 & & \rho_p & \end{array}\right]\qquad(p\times q).$$

Hence, conditional on $X'X$, the distribution function of the largest latent root r_1^2 of $U'U(U'U+V'V)^{-1}$ follows from the distribution function of the largest root f_1 of $(U'U)(V'V)^{-1}$ given in Theorem 10.6.8 by replacing x there by $x/(1-x)$ and putting $r=q$, $n-p=n-q$, $m=p$. This shows that

(40) $$P(r_1^2\le x|X'X)=x^{pq/2}\operatorname{etr}\left[-\tfrac{1}{2}(1-x)\Phi^{-1}\tilde{P}X'X\tilde{P}'\right]$$
$$\cdot\sum_{k=0}^{pt}\sum_{\kappa}{}^{*}L_\kappa^\gamma\left(-\tfrac{1}{2}x\Phi^{-1}\tilde{P}X'X\tilde{P}'\right)\frac{(1-x)^k}{k!},$$

where $\gamma=\frac{1}{2}(q-p-1)$ and L_κ^γ denotes the Laguerre polynomial corresponding to the partition κ of k (see Section 7.6). To find the unconditional distribution function of r_1^2 we multiply (40) by the $W_q(n,I_q)$ density function for $X'X$ and integrate with respect to $X'X$. This gives

$$P(r_1^2\le x)=\frac{x^{pq/2}2^{-nq/2}}{\Gamma_q(\frac{1}{2}n)}\sum_{k=0}^{pt}\sum_{\kappa}{}^{*}\frac{(1-x)^k}{k!}$$
$$\cdot\int_{X'X>0}\operatorname{etr}\left[-\tfrac{1}{2}X'X\left(I+(1-x)\tilde{P}'\Phi^{-1}\tilde{P}\right)\right]\det(X'X)^{(n-q-1)/2}$$
$$\cdot L_\kappa^\gamma\left(-\tfrac{1}{2}x\Phi^{-1}\tilde{P}X'X\tilde{P}'\right)(d(X'X)).$$

Using the zonal polynomial series for L_κ^γ (see (4) of Section 7.6) this becomes

$$P(r_1^2\le x)=\frac{x^{pq/2}2^{-nq/2}}{\Gamma_q(\frac{1}{2}n)}\sum_{k=0}^{pt}\sum_{\kappa}{}^{*}\frac{(1-x)^k}{k!}\left(\tfrac{1}{2}q\right)_\kappa C_\kappa(I_p)$$
$$\cdot\sum_{s=0}^{k}\sum_{\sigma}\binom{\kappa}{\sigma}\frac{\left(\frac{1}{2}x\right)^s}{\left(\frac{1}{2}q\right)_\sigma C_\sigma(I_p)}\int_{X'X>0}\operatorname{etr}\left[-\tfrac{1}{2}X'X\left(I+(1-x)\tilde{P}'\Phi^{-1}\tilde{P}\right)\right]$$
$$\cdot\det(X'X)^{(n-q-1)/2}C_\sigma\left(\Phi^{-1}\tilde{P}X'X\tilde{P}'\right)(d(X'X))$$
$$=x^{pq/2}\sum_{k=0}^{pt}\sum_{\kappa}{}^{*}\frac{(1-x)^k}{k!}\left(\tfrac{1}{2}q\right)_\kappa C_\kappa(I_p)\sum_{s=0}^{k}\sum_{\sigma}\binom{\kappa}{\sigma}\frac{\left(\frac{1}{2}n\right)_\sigma}{\left(\frac{1}{2}q\right)_\sigma}$$
$$\cdot\det\left(I+(1-x)\tilde{P}'\Phi^{-1}\tilde{P}\right)^{-n/2}\frac{C_\sigma\left(x\tilde{P}'\Phi^{-1}\tilde{P}\left[(1-x)\tilde{P}'\Phi^{-1}\tilde{P}+I\right]^{-1}\right)}{C_\sigma(I_p)},$$

where the integral has been evaluated using Theorem 7.2.7. The desired result now follows on noting that

$$x\tilde{P}'\Phi^{-1}\tilde{P}\left[(1-x)\tilde{P}'\Phi^{-1}\tilde{P}+I\right]^{-1}=\operatorname{diag}\left(\frac{x\rho_1^2}{1-x\rho_1^2},\dots,\frac{x\rho_p^2}{1-x\rho_p^2},0,\dots,0\right)$$

$$=\begin{bmatrix} B & 0 \\ 0 & 0 \end{bmatrix} \qquad (q\times q).$$

and that

$$\det\left(I+(1-x)\tilde{P}'\Phi^{-1}\tilde{P}\right)^{-n/2}=\prod_{i=1}^{p}\left(\frac{1-\rho_i^2}{1-x\rho_i^2}\right)^{n/2}.$$

When $\rho_1=\cdots=\rho_p=0$ the distribution function of r_1^2 in Theorem 11.3.4 simplifies considerably.

COROLLARY 11.3.4. When $\rho_1=\cdots=\rho_p=0$ and $t=\frac{1}{2}(n-p-q-1)$ is a positive integer, the distribution function of r_1^2 may be expressed as

$$P(r_1^2\le x)=x^{pq/2}\sum_{k=0}^{pt}\sum_{\kappa}{}^{*}\frac{(1-x)^k}{k!}\left(\tfrac{1}{2}q\right)_\kappa C_\kappa(I_p). \tag{41}$$

11.3.5. Asymptotic Distributions of the Sample Canonical Correlation Coefficients

The ${}_2F_1^{(p)}$ function in the density function of $r_1^2,\dots,r_p^2$ converges very slowly for large n and it is difficult to obtain from the zonal polynomial series any feeling for the behavior of the density function or an understanding of how the sample and population canonical correlation coefficients interact with each other. It makes sense to ask how the ${}_2F_1^{(p)}$ function behaves asymptotically for large n. It turns out that an asymptotic representation for the function involves only elementary functions and tells a great deal about the interaction between the sample and population coefficients.

One of the most commonly used procedures in canonical correlation analysis is to test whether the smallest $p-k$ population canonical correlation coefficients are zero. If they are, then the correlation structure between the two sets of variables is explained by the first k canonical variables and a reduction in dimensionality is achieved by considering these canonical variables as new variables. We will investigate such a test later in Section 11.3.6. The following theorem gives the asymptotic behavior of the ${}_2F_1^{(p)}$ function under the null hypothesis that the smallest $p-k$ population canonical correlation coefficients are zero.

THEOREM 11.3.5. If $R^2=\text{diag}(r_1^2,\ldots,r_p^2)$, where $r_1^2>\cdots>r_p^2>0$ and $P^2=\text{diag}\,(\rho_1^2,\ldots,\rho_k^2,0,\ldots,0)$ $(p\times p)$, where $1>\rho_1^2>\cdots\rho_k^2>0$ then, as $n\to\infty$,

(42)

$$ {}_2F_1^{(p)}(\tfrac{1}{2}n,\tfrac{1}{2}n;\tfrac{1}{2}q;P^2,R^2)\sim K_1\prod_{i=1}^{k}\left[(1-r_i\rho_i)^{-n+(p+q+1)/2}(r_i\rho_i)^{(p-q)/2}\right] $$

$$ \cdot\prod_{i=1}^{k}\prod_{\substack{j=1\\ i<j}}^{p}c_{ij}^{-1/2}, $$

where

(43) $$ c_{ij}=(r_i^2-r_j^2)(\rho_i^2-\rho_j^2)\qquad(i=1,\ldots,k;\ j=1,\ldots,p) $$

and

(44) $$ K_1=(\tfrac{1}{2}n)^{-k(p+q-k-1)/2}\pi^{-k(k+1)/2}\Gamma_k(\tfrac{1}{2}q)\Gamma_k(\tfrac{1}{2}p)2^{-k}. $$

For a proof of this theorem the interested reader is referred to Glynn and Muirhead (1978) and Glynn (1980). The proof involves writing the ${}_2F_1^{(p)}$ function as a multiple integral and applying the result of Theorem 9.5.1. The multiple integral is similar to (22) of Section 10.7.3 for the ${}_1F_1^{(m)}$ function but involves even more steps.

Substitution of the asymptotic behavior (42) for ${}_2F_1^{(p)}$ in (36) yields an asymptotic representation for the joint density function of $r_1^2,\ldots,r_p^2$. The result is summarized in the following theorem.

THEOREM 11.3.6. An asymptotic representation for large n of the joint density function of $r_1^2,\ldots,r_p^2$ when the population canonical correlation coefficients satisfy

(45) $$ 1>\rho_1>\cdots>\rho_k>\rho_{k+1}=\cdots=\rho_p=0 $$

is

$$(46)\quad K_2 \prod_{i=1}^{k}\left[(1-r_i\rho_i)^{-n+(p+q+1)/2}(r_i^2)^{(q-p)/4-1/2}(1-r_i^2)^{(n-p-q-1)/2}\right]$$

$$\cdot \prod_{i<j}^{k}\left(\frac{r_i^2-r_j^2}{\rho_i^2-\rho_j^2}\right)^{1/2}\prod_{i=1}^{k}\prod_{j=k+1}^{p}(r_i^2-r_j^2)^{1/2}$$

$$\cdot \prod_{i=k+1}^{p}\left[(r_i^2)^{(q-p-1)/2}(1-r_i^2)^{(n-q-p-1)/2}\right]\prod_{\substack{k+1\\i<j}}^{p}(r_i^2-r_j^2),$$

where

$$(47)\quad K_2=K_1\frac{\Gamma_p(\frac{1}{2}n)\pi^{p^2/2}}{\Gamma_p(\frac{1}{2}p)\Gamma_p[\frac{1}{2}(n-q)]\Gamma_p(\frac{1}{2}q)}\prod_{i=1}^{k}\left[(1-\rho_i^2)^{n/2}\rho_i^{k-(p+q)/2}\right]$$

with K_1 given by (44).

This theorem has two interesting consequences.

COROLLARY 11.3.7. Under the conditions of Theorem 11.3.6 the asymptotic conditional density function for large n of $r_{k+1}^2,\ldots,r_p^2$, the squares of the smallest $p-k$ sample canonical correlation coefficients, given the k largest coefficients $r_1^2,\ldots,r_k^2$, is

(48)

$$K\prod_{i=1}^{k}\prod_{j=k+1}^{p}(r_i^2-r_j^2)^{1/2}\prod_{i=k+1}^{p}\left[(r_i^2)^{(q-p-1)/2}(1-r_i^2)^{(n-q-p-1)/2}\right]$$

$$\cdot \prod_{\substack{k+1\\i<j}}^{p}(r_i^2-r_j^2),$$

where K is a constant.

Note that this asymptotic conditional density function does not depend on $\rho_1^2,\ldots,\rho_k^2$, the nonzero population coefficients, so that $r_1^2,\ldots,r_k^2$ are asymptotically sufficient for $\rho_1^2,\ldots,\rho_k^2$.

COROLLARY 11.3.8. Assume that the conditions of Theorem 11.3.6 hold and put

$$x_i=\frac{n^{1/2}(r_i^2-\rho_i^2)}{2\rho_i(1-\rho_i^2)}\qquad (\text{for}\quad i=1,\dots,k)$$

$$x_j=nr_j^2\quad (\text{for}\quad j=k+1,\dots,p). \tag{49}$$

Then the limiting joint density function of $x_1,\dots,x_p$ as $n\to\infty$ is

$$\prod_{i=1}^{k}\phi(x_i)\cdot\frac{\pi^{(p-k)^2/2}\exp\left(-\frac{1}{2}\sum_{j=k+1}^{p}x_j\right)}{2^{(p-k)(q-k)/2}\Gamma_{p-k}[\frac{1}{2}(q-k)]\Gamma_{p-k}[\frac{1}{2}(p-k)]} \tag{50}$$

$$\cdot\prod_{j=k+1}^{p}x_j^{(q-p-1)/2}\prod_{\substack{k+1\\ i<j}}^{p}(x_i-x_j),$$

where $\phi(\cdot)$ denotes the standard normal density function.

This result, due originally to P. L. Hsu (1941b), can be proved by making the change of variables (49) in (46) and letting $n\to\infty$. Note that this shows that asymptotically the x_i's corresponding to distinct nonzero ρ_i's are marginally standard normal, independent of all x_j, $j\neq i$, while the x_j's corresponding to zero population canonical correlation coefficients are non-normal and dependent, and their asymptotic distribution is the same as the distribution of the latent roots of a $(p-k)\times(p-k)$ matrix having the $W_{p-k}(q-k, I_{p-k})$ distribution.

It is interesting to look at the maximum likelihood estimates of the population coefficients obtained from the marginal distribution of the sample coefficients. The part of the joint density function of $r_1^2,\dots,r_p^2$ involving the population coefficients is, from Theorem 11.3.2,

$$L^*=\prod_{i=1}^{p}(1-\rho_i^2)^{n/2}{}_2F_1^{(p)}(\tfrac{1}{2}n,\tfrac{1}{2}n;\tfrac{1}{2}q;P^2,R^2), \tag{51}$$

called the marginal likelihood function. When the population coefficients are all disinct and nonzero $(1>\rho_1>\cdots>\rho_p>0)$, Theorem 11.3.5 (with $k=p$) can be used to approximate L^* for large n, giving

$$L^*\sim K\cdot L_1L_2 \tag{52}$$

where

$$L_1=\prod_{i=1}^{p}\left[(1-\rho_i^2)^{n/2}(1-r_i\rho_i)^{-n}\right],$$

$$L_2=\prod_{i=1}^{p}\left[(1-r_i\rho_i)^{(p+q+1)/2}\rho_i^{(p-q)/2}\right]\prod_{i<j}^{p}\left(\rho_i^2-\rho_j^2\right)^{-1/2},$$

and K is a constant (depending on $n, r_1^2,\ldots,r_p^2$, but not on $\rho_1,\ldots,\rho_p$ and hence irrelevant for likelihood purposes). The values of the ρ_i which maximize L_1 are

$$\hat{\rho}_i=r_i \qquad (i=1,\ldots,p),$$

i.e., the usual maximum likelihood estimates. The values of the ρ_i which maximize L_1L_2 are

$$(53)\quad \hat{\hat{\rho}}_i=r_i-\frac{1-r_i^2}{2nr_i}\left[p+q-2+r_i^2+2(1-r_i^2)\sum_{\substack{j=1\\j\neq i}}^{p}\frac{r_j^2}{r_i^2-r_j^2}\right]+O(n^{-2})$$

$$(i=1,\ldots,p).$$

These estimates utilize information from other sample coefficients, adjacent ones having the most effect. It is natural to apply Fisher's z transformation in the canonical correlation case. Lawley (1959) noted that, as estimates of the parameters $\xi_i=\tanh^{-1}\rho_i$, the statistics $z_i=\tanh^{-1}r_i$ fail to stabilize the mean and variance to any marked extent. In fact z_i has a bias term of order n^{-1}. The estimate

$$(54)\qquad \hat{z}_i=\tanh^{-1}\hat{\hat{\rho}}_i=\tfrac{1}{2}\log\frac{1+\hat{\hat{\rho}}_i}{1-\hat{\hat{\rho}}_i}$$

fares much better. Substituting (53) for $\hat{\hat{\rho}}_i$ in (54) it is easily shown that

$$(55)\quad \hat{z}_i=z_i-\frac{1}{2nr_i}\left[p+q-2+r_i^2+2(1-r_i^2)\sum_{\substack{j=1\\j\neq i}}^{p}\frac{r_j^2}{r_i^2-r_j^2}\right]+O(n^{-2})$$

and using (34) and (35) the mean and variance of $\hat{z}_i$ are

$$(56)\qquad E(\hat{z}_i)=\xi_i+O(n^{-2})$$

and

$$\text{Var}(\hat{z}_i) = \frac{1}{n} + O(n^{-2}). \tag{57}$$

Hence Fisher's z transformation applied to the maximum marginal likelihood estimates $\hat{\rho}_i$ not only stabilizes the variance to order n^{-1} but also provides a correction for bias.

11.3.6. Determining the Number of Useful Canonical Variables

In Section 11.2 we derived the likelihood ratio test of independence of two sets of variables **X** and **Y** where **X** is $p \times 1$ and **Y** is $q \times 1$, $p \le q$, i.e., for testing the null hypothesis that $\rho_1 = \cdots = \rho_p = 0$ $(\Sigma_{12} = 0)$. If this is accepted there are clearly no useful canonical variables. If it is rejected it is possible that $\rho_1 > \rho_2 = \cdots = \rho_p = 0$ [rank $(\Sigma_{12}) = 1$], in which case only the first canonical variables are useful. If this is tested and rejected, we can test whether the smallest $p-2$ population canonical correlation coefficients are zero, and so on. In practice, then, we test the sequence of null hypotheses

$$H_k : \rho_{k+1} = \cdots = \rho_p = 0$$

for $k = 0, 1, \ldots, p-1$. We saw in Section 11.2 that the likelihood ratio test of $H_0 : \rho_1 = \cdots = \rho_p = 0$ is based on the statistic

$$W_0 = \prod_{i=1}^{p} (1 - r_i^2)$$

where $r_1^2, \ldots, r_p^2$ $(1 > r_1^2 > \cdots > r_p^2 > 0)$ are the squares of the sample canonical correlation coefficients and a test of asymptotic level α is to reject H_0 if

$$-\left[n - \tfrac{1}{2}(p+q+1)\right] \log W_0 > c_f(\alpha) \tag{58}$$

where $c_f(\alpha)$ is the upper $100\alpha\%$ point of the χ_f^2 distribution, with $f = pq$. Fujikoshi (1974a) has shown that the likelihood ratio test of H_k rejects H_k for small values of the statistic

$$W_k = \prod_{i=k+1}^{p} (1 - r_i^2). \tag{59}$$

The asymptotic distribution as $n \to \infty$ of $-n \log W_k$ is $\chi^2_{(p-h)(q-k)}$ when H_k is true. An improvement over $-n \log W_k$ is the statistic $-[n - \frac{1}{2}(p + q +$

$1)]\log W_k$ suggested by Bartlett (1938, 1947). The multiplying factor here is the same as that in (58) used for testing H_0. A further refinement to the multiplying factor was obtained by Lawley (1959) and Glynn and Muirhead (1978). We will now indicate the approach taken by Glynn and Muirhead.

We noted in Corollary 11.3.7 that the asymptotic conditional density function of $r_{k+1}^2,\dots,r_p^2$ given $r_1^2,\dots,r_k^2$ is

(60)

$$K\prod_{i=1}^{k}\prod_{j=k+1}^{p}\left(r_i^2-r_j^2\right)^{1/2}\prod_{i=k+1}^{p}\left[\left(r_i^2\right)^{(q-p-1)/2}\left(1-r_i^2\right)^{(n-q-p-1)/2}\right]$$

$$\cdot\prod_{k+1}^{p}\left(r_i^2-r_j^2\right),$$

where K is a constant. Put

$$T_k=-\log W_k$$

so that the asymptotic distribution of nT_k is $\chi^2_{(p-k)(q-k)}$ when H_k is true. The appropriate multiplier of T_k can be obtained by finding its expected value. If we let E_c denote expectation taken with respect to the conditional distribution (60) of $r_{k+1}^2,\dots,r_p^2$, given $r_1^2,\dots,r_k^2$, the following theorem gives the asymptotic distribution of the likelihood ratio statistic and provides additional information about the accuracy of the χ^2 approximation.

THEOREM 11.3.9. When the null hypothesis H_k is true the asymptotic distribution of the statistic

$$L_k=-\left[n-k-\tfrac{1}{2}(p+q+1)+\sum_{i=1}^{k}r_i^{-2}\right]\log W_k$$

is $\chi^2_{(p-k)(q-k)}$, and

$$E_c(L_k)=(p-k)(q-k)+O(n^{-2}).$$

Proof. The conditional distribution (60) is the same as the distribution given by (30) of Section 10.7.4, where we put

$$u_i=r_i^2,\qquad m=p,\qquad n_1=q,\qquad n_2=n-q.$$

The theorem now follows by making these substitutions in Theorem 10.7.5.

It follows from Theorem 11.3.9 that if n is large an approximate test of level α of H_k is to reject H_k if $L_k > c_r(\alpha)$, the upper $100\alpha\%$ point of the χ_r^2 distribution, with $r=(p-k)(q-k)$.

It should be noted that this test, like the test of independence between two sets of variables, is extremely sensitive to departures from normality. If it is believed that the distribution being sampled is elliptical with longer tails than the normal distribution, a much better procedure is to adjust the test statistic for nonzero kurtosis. For work in this direction the interested reader is referred to Muirhead and Waternaux (1980).

PROBLEMS

11.1. Let W be the statistic defined by (8) of Section 11.2 for testing independence between k sets of variables. Show that when $k=3$, $m_2=m_3=1$ the null density function of W can be expressed for all m_1 in the form

$$f_W(x)=\frac{\Gamma(\frac{1}{2}n)^2}{\Gamma[\frac{1}{2}(n-m_1)]\Gamma[\frac{1}{2}(n-m_1-1)]\Gamma(m_1+\frac{1}{2})}$$

$$\cdot x^{(n-m_1-3)/2}(1-x)^{m_1-1/2}$$

$$\cdot {}_2F_1(\tfrac{1}{2}m_1,\tfrac{1}{2}m_1;\; m_1+\tfrac{1}{2};\; 1-x) \qquad (0<x<1).$$

[*Hint*: With the help of the result of Problem 5.11(b) find the hth moment of this density function and show that it agrees with the hth moment of W given in the proof of Theorem 11.2.4 (see Consul, 1967).]

11.2. Let $\mathbf{X}$ be $p\times 1$, $\mathbf{Y}$ be $q\times 1$ $(p\le q)$, and suppose that

$$\operatorname{Cov}\begin{pmatrix}\mathbf{X}\\ \mathbf{Y}\end{pmatrix}=\left[\begin{array}{cccc:cccc} 1 & \alpha & \cdots & \alpha & \beta & \beta & \cdots & \beta\\ \alpha & 1 & \cdots & \alpha & \beta & \beta & \cdots & \beta\\ \vdots & & & & \vdots & & & \\ \alpha & \alpha & \cdots & 1 & \beta & \beta & \cdots & \beta\\ \hdashline \beta & \beta & \cdots & \beta & 1 & \gamma & \cdots & \gamma\\ \beta & \beta & \cdots & \beta & \gamma & 1 & \cdots & \gamma\\ \vdots & & & & \vdots & & & \\ \beta & \beta & \cdots & \beta & \gamma & \gamma & \cdots & 1 \end{array}\right]=\begin{bmatrix}\Sigma_{11} & \Sigma_{12}\\ \Sigma_{21} & \Sigma_{22}\end{bmatrix}\begin{matrix}p\\ q\end{matrix}.$$
$$\qquad\qquad\qquad\qquad\qquad\qquad\qquad\qquad\qquad\qquad\qquad\quad p \qquad q$$

Find both the canonical correlation coefficients between **X** and **Y** and the canonical variables.

11.3. Let $\hat{\rho}_i$ be the maximum marginal likelihood estimates of the ith population canonical correlation coefficient ρ_i, $i=1,\dots,p$, given by (53) of Section 11.3 (assuming $1>\rho_1>\cdots>\rho_p>0$). Putting $\hat{z}_i=\tanh^{-1}\hat{\rho}_i$, show that

$$E(\hat{z}_i)=\xi_i+O(n^{-2}),\qquad \operatorname{Var}(\hat{z}_i)=\frac{1}{n}+O(n^{-2}),$$

where $\xi_i=\tanh^{-1}\rho_i$.

11.4. (a) Let $M=\{\mathbf{x},\mathbf{y};\mathbf{x}\in R^m,\mathbf{y}\in R^m,\mathbf{x}\neq\mathbf{0},\mathbf{y}\neq\mathbf{0},\mathbf{x}'\mathbf{y}=0\}$. If Σ is an $m\times m$ positive definite matrix with latent roots $\lambda_1\geq\cdots\geq\lambda_m>0$ and associated latent vectors $\mathbf{x}_1,\dots,\mathbf{x}_m,\mathbf{x}_i'\mathbf{x}_i=1, i=1,\dots,m,\mathbf{x}_i'\mathbf{x}_j=0\ (i\neq j)$, prove that

$$\sup_{\mathbf{x},\mathbf{y}\in M}\frac{\mathbf{x}'\Sigma\mathbf{y}}{(\mathbf{x}'\Sigma\mathbf{x}\mathbf{y}'\Sigma\mathbf{y})^{1/2}}=\frac{\lambda_1-\lambda_m}{\lambda_1+\lambda_m}$$

and that

$$\frac{\mathbf{x}'\Sigma\mathbf{y}}{(\mathbf{x}'\Sigma\mathbf{x}\mathbf{y}'\Sigma\mathbf{y})^{1/2}}=\frac{\lambda_1-\lambda_m}{\lambda_1+\lambda_m}$$

when $\mathbf{x}=\mathbf{x}_1+\mathbf{x}_m$ and $\mathbf{y}=\mathbf{x}_1-\mathbf{x}_m$.

(b) Suppose Σ is partitioned as

$$\Sigma=\begin{bmatrix}\Sigma_{11} & \Sigma_{12}\\ \Sigma_{21} & \Sigma_{22}\end{bmatrix},$$

where Σ_{11} is $p\times p$ and Σ_{22} is $q\times q$ with $p+q=m$. The largest canonical correlation coefficient is

$$\rho_1=\sup\boldsymbol{\alpha}'\Sigma_{12}\boldsymbol{\beta},$$

where the supremum is taken over all $\boldsymbol{\alpha}\in R^p$, $\boldsymbol{\beta}\in R^q$ with $\boldsymbol{\alpha}'\Sigma_{11}\boldsymbol{\alpha}=1$, $\boldsymbol{\beta}'\Sigma_{22}\boldsymbol{\beta}=1$. Show that

$$\rho_1\leq\frac{\lambda_1-\lambda_m}{\lambda_1+\lambda_m},$$

where $\lambda_1 \geq \cdots \geq \lambda_m > 0$ are the latent roots of Σ.

(c) For the covariance matrix

$$\Sigma = \begin{bmatrix} I_p & \tilde{P} \\ \tilde{P}' & I_q \end{bmatrix}, \qquad \tilde{P} = \left[\begin{array}{ccc:c} \rho_1 & & 0 & \\ & \ddots & & 0 \\ 0 & & \rho_p & \end{array}\right] \qquad (p \times q),$$

where $1 \geq \rho_1 \geq \cdots \geq \rho_p \geq 0$, show that the largest and smallest latent roots are $\lambda_1 = 1 + \rho_1$ and $\lambda_m = 1 - \rho_1$. This shows that the inequality in (b) is sharp (see Eaton, 1976).

11.5. Suppose that $\mathbf{X}$ $(p \times 1)$ and $\mathbf{Y}$ $(q \times 1)$ are jointly distributed with $p \leq q$, and let $\rho_1, \ldots, \rho_p$ be the population canonical correlation coefficients $(\rho_1 \geq \cdots \geq \rho_p)$.

(a) Suppose that r extra variables, given by the components of $\mathbf{Z}(r \times 1)$, are added to the q set, forming the vector $\mathbf{Y}^* = (\mathbf{Y}' : \mathbf{Z}')'$. Let $\sigma_1, \ldots, \sigma_p$ $(\sigma_1 \geq \cdots \geq \sigma_p)$ be the population canonical correlation coefficients between $\mathbf{X}$ and $\mathbf{Y}^*$. Show that

$$\sigma_i \geq \rho_i \qquad (i = 1, \ldots, p).$$

(b) Suppose that the r extra variables in $\mathbf{Z}$ are added to the p set, forming the vector $\mathbf{X}^* = (\mathbf{X}' : \mathbf{Z}')'$. Assume $p + r \leq q$. Let $\delta_1, \ldots, \delta_{p+r}$ be the canonical correlation coefficients between $\mathbf{X}^*$ and $\mathbf{Y}$. Show that

$$\delta_i \geq \rho_i \qquad (i = 1, \ldots, p)$$

(see Chen, 1971). This shows that the addition of extra variables to either set of variables can never decrease any of the canoncial correlation coefficients.

11.6. Obtain Corollary 11.3.8 from Theorem 11.3.6.

11.7. Suppose that $\mathbf{X}_1, \ldots, \mathbf{X}_N$ is a random sample from the $N_m(\mu, \Sigma)$ distribution, where $\Sigma = (\sigma_{ij})$. Let $R = (r_{ij})$ be the sample correlation matrix formed from $\mathbf{X}_1, \ldots, \mathbf{X}_N$. Show that the likelihood ratio statistic for testing the null hypothesis $H: \sigma_{ij} = 0$ for all $i \neq j$ against the alternative hypothesis $K: \sigma_{ij} \neq 0$ for exactly one unspecified pair (i, j), is

$$\Lambda = \left[1 - (\max|r_{ij}|)^2\right]^{-N/2}.$$

Moran (1980).

APPENDIX

Some Matrix Theory

A1. INTRODUCTION

In this appendix we indicate the results in matrix theory that are needed in the rest of the book. Many of the results should be familiar to the reader already; the more basic of these are not proved here. Useful references for matrix theory are Mirsky (1955), Bellman (1970), and Graybill (1969). Most of the references to the appendix earlier in the text concern results involving matrix factorizations; these are proved here.

A2. DEFINITIONS

A $p\times q$ matrix A is a rectangular array of real or complex numbers $a_{11}, a_{12},\dots,a_{pq}$, written as

$$A=\begin{bmatrix} a_{11},\dots,a_{1q} \\ \vdots \\ a_{p1},\dots,a_{pq} \end{bmatrix},$$

so that a_{ij} is the element in the ith row and jth column. Often A is written as $A=(a_{ij})$. We will assume throughout this appendix that the elements of a matrix are real, although many of the results stated hold also for complex matrices. If $p=q$ A is called a *square matrix* of order p. If $q=1$ A is a *column vector*, and if $p=1$ A is a *row vector*. If $a_{ij}=0$ for $i=1,\dots,p$, $j=1,\dots,q$, A is called a *zero matrix*, written $A=0$, and if $p=q$, $a_{ii}=1$ for $i=1,\dots,p$ and $a_{ij}=0$ for $i\neq j$ then A is called the *identity matrix* of order p, written $A=I$ or $A=I_p$. The *diagonal elements* of a $p\times p$ matrix A are $a_{11}, a_{22},\dots,a_{pp}$.

The *transpose* of a $p \times q$ matrix A, denoted by A', is the $q \times p$ matrix obtained by interchanging the rows and columns of A, i.e., if $A = (a_{ij})$ then $A' = (a_{ji})$. If A is a square matrix of order p it is called *symmetric* if $A = A'$ and *skew-symmetric* if $A = -A'$. If A is skew-symmetric then its diagonal elements are zero.

A $p \times p$ matrix A having the form

$$A = \begin{bmatrix} a_{11} & a_{12}, & \dots, & a_{1p} \\ 0 & a_{22}, & \dots, & a_{2p} \\ \vdots & & & \\ 0 & 0, & \dots, & a_{pp} \end{bmatrix},$$

so that all elements below the main diagonal are zero, is called *upper-triangular*. If all elements above the main diagonal are zero it is called *lower-triangular*. Clearly, if A is upper-triangular then A' is lower-triangular. If A has the form

$$A = \begin{bmatrix} a_{11} & 0, & \dots, & 0 \\ 0 & a_{22}, & \dots, & 0 \\ \vdots & & & \\ 0 & 0, & \dots, & a_{pp} \end{bmatrix},$$

so that all elements off the main diagonal are zero, it is called *diagonal*, and is often written as

$$A = \text{diag}(a_{11}, \dots, a_{pp}).$$

The *sum* of two $p \times q$ matrices A and B is defined by

$$A + B = (a_{ij} + b_{ij}).$$

If A is $p \times q$ and B is $q \times r$ (so that the number of columns of A is equal to the number of rows of B) then the *product* of A and B is the $p \times r$ matrix defined by

$$AB = \left(\sum_{k=1}^{q} a_{ik} b_{kj} \right).$$

The product of a matrix A by a scalar α is defined by

$$\alpha A = (\alpha a_{ij}).$$

The following properties are elementary, where, if products are involved, it is assumed that these are defined:

$$
\begin{aligned}
&A+(-1)A=0\\
&(AB)'=B'A'\\
&(A')'=A\\
&(A+B)'=A'+B'\\
&A(BC)=(AB)C\\
&A(B+C)=AB+AC\\
&(A+B)C=AC+BC\\
&AI=A.
\end{aligned}
$$

A $p\times p$ matrix A is called *orthogonal* if $AA'=I_p$ and *idempotent* if $A^2=A$.

If $A=(a_{ij})$ is a $p\times q$ matrix and we write

$$
\begin{aligned}
&A_{11}=(a_{ij}),\ i=1,\ldots,k;\ j=1,\ldots,l\\
&A_{12}=(a_{ij}),\ i=1,\ldots,k,\ j=l+1,\ldots,q\\
&A_{21}=(a_{ij}),\ i=k+1,\ldots,p;\ j=1,\ldots,l\\
&A_{22}=(a_{ij}),\ i=k+1,\ldots,p;\ j=l+1,\ldots,q
\end{aligned}
$$

then A can be expressed as

$$A=\begin{bmatrix} A_{11} & A_{12}\\ A_{21} & A_{22}\end{bmatrix}$$

and is said to be *partitioned* into submatrices A_{11}, A_{12}, A_{21} and A_{22}. Clearly if B is a $p\times q$ matrix partitioned similarly to A as

$$B=\begin{bmatrix} B_{11} & B_{12}\\ B_{21} & B_{22}\end{bmatrix},$$

where B_{11} is $k\times l$, B_{12} is $k\times(q-l)$, B_{21} is $(p-k)\times l$ and B_{22} is $(p-k)\times(q-l)$, then

$$A+B=\begin{bmatrix} A_{11}+B_{11} & A_{12}+B_{12}\\ A_{21}+B_{21} & A_{22}+B_{22}\end{bmatrix}.$$

Also, if C is a $q\times r$ matrix partitioned as

$$C=\begin{bmatrix} C_{11} & C_{12} \\ C_{21} & C_{22} \end{bmatrix},$$

where C_{11} is $l\times m$, C_{12} is $l\times(r-m)$, C_{21} is $(q-l)\times m$, and C_{22} is $(q-l)\times(r-m)$, then it is readily verified that

$$AC=\begin{bmatrix} A_{11} & A_{12} \\ A_{21} & A_{22} \end{bmatrix}\begin{bmatrix} C_{11} & C_{12} \\ C_{21} & C_{22} \end{bmatrix}=\begin{bmatrix} A_{11}C_{11}+A_{12}C_{21} & A_{11}C_{12}+A_{12}C_{22} \\ A_{21}C_{11}+A_{22}C_{21} & A_{21}C_{12}+A_{22}C_{22} \end{bmatrix}.$$

A3. DETERMINANTS

The *determinant* of a square $p\times p$ matrix A, denoted by $\det A$ or $|A|$, is defined by

$$\det A=\sum_{\pi}\varepsilon_{\pi}a_{1j_1},a_{2j_2},\ldots,a_{pj_p}$$

where Σ_{π} denotes the summation over all $p!$ permutations $\pi=(j_1,\ldots,j_p)$ of $(1,\ldots,p)$ and $\varepsilon_{\pi}=+1$ or -1 according as the permutation π is even or odd. The following are elementary properties of determinants which follow readily from the definition:

(i) If every element of a row (or column) of A is zero then $\det A=0$.

(ii) $\det A=\det A'$.

(iii) If all the elements in any row (or column) of A are multiplied by a scalar α the determinant is multiplied by α.

(iv) $\det(\alpha A)=\alpha^{p}\det A$.

(v) If B is the matrix obtained from A by interchanging any two of its rows (or columns), then $\det B=-\det A$.

(vi) If two rows (or columns) of A are identical, then $\det A=0$.

(vii) If

$$A=\begin{bmatrix} b_{11}+c_{11} & b_{12}+c_{12}, & \ldots, & b_{1p}+c_{1p} \\ a_{21} & a_{22}, & \ldots, & a_{2p} \\ \vdots & & & \\ a_{p1} & a_{p2}, & \ldots, & a_{pp} \end{bmatrix},$$

so that every element in the first row of A is a sum of two scalars,

then

$$\det A = \det\begin{bmatrix} b_{11} & b_{12}, & \ldots, & b_{1p} \\ a_{21} & a_{22}, & \ldots, & a_{2p} \\ \vdots & & & \\ a_{p1} & a_{p2}, & \ldots, & a_{pp} \end{bmatrix}$$

$$+\det\begin{bmatrix} c_{11} & c_{12}, & \ldots, & c_{1p} \\ a_{21} & a_{22}, & \ldots, & a_{2p} \\ \vdots & & & \\ a_{p1} & a_{p2}, & \ldots, & a_{pp} \end{bmatrix}.$$

A similar result holds for any row (or column). Hence if every element in ith row (or column) of A is the sum of n terms then $\det A$ can be written as the sum of n determinants.

(viii) If B is the matrix obtained from A by adding to the elements of its ith row (or column) a scalar multiple of the corresponding elements of another row (or column) then $\det B = \det A$.

The result given in the following theorem is extremely useful.

THEOREM A3.1. If A and B are both $p \times p$ matrices then

$$\det(AB) = (\det A)(\det B)$$

Proof. From the definition

$$\begin{aligned} \det(AB) &= \sum_{\pi} \varepsilon_{\pi} \prod_{i=1}^{p} \left(\sum_{k=1}^{p} a_{ik} b_{kj_i} \right) \\ &= \sum_{k_1=1}^{p} \cdots \sum_{k_p=1}^{p} \sum_{\pi} \varepsilon_{\pi} \left(\prod_{i=1}^{p} a_{ik_i} \right) \left(\prod_{i=1}^{p} b_{k_i j_i} \right) \\ &= \sum_{k_1=1}^{p} \cdots \sum_{k_p=1}^{p} \det B(k_1, \ldots, k_p) \left(\prod_{i=1}^{p} a_{ik_i} \right), \end{aligned}$$

where $B(k_1, \ldots, k_p)$ denotes the $p \times p$ matrix whose ith row is the k_ith row of B. By property (vi) $\det B(k_1, \ldots, k_p) = 0$ if any two of the integers

$k_1,\ldots,k_p$ are equal, and hence

$$\det(AB)=\sum_{\substack{k_1=1\\ k_1\neq k_2\neq\cdots\neq k_p}}^{p}\cdots\sum_{k_p=1}^{p}\det B(k_1,\ldots,k_p)\left(\prod_{i=1}^{p}a_{ik_i}\right).$$

By property (v) it follows that

$$\det B(k_1,\ldots,k_p)=\varepsilon_\alpha \det B,$$

where $\varepsilon_\alpha=+1$ or -1 according as the permutation $\alpha=(k_1,\ldots,k_p)$ of $(1,\ldots,p)$ is even or odd. Hence

$$\begin{aligned}\det(AB)&=\sum_{\alpha}\varepsilon_\alpha\left(\prod_{i=1}^{p}a_{ik_i}\right)\cdot\det B\\ &=(\det A)(\det B).\end{aligned}$$

A number of useful results are direct consequences of this theorem.

THEOREM A3.2. If $A_1,\ldots,A_n$ are all $p\times p$ matrices then

$$\det(A_1A_2\ldots A_n)=(\det A_1)(\det A_2)\ldots(\det A_n).$$

This is easily proved by induction on n.

THEOREM A3.3. If A is $p\times p$, $\det(AA')\geq 0$.

This follows from Theorem A3.1 and property (ii).

THEOREM A3.4. If A_{11} is $p\times p$, A_{12} is $p\times q$, A_{21} is $q\times p$, and A_{22} is $q\times q$ then

$$\det\begin{bmatrix}A_{11} & A_{12}\\ 0 & A_{22}\end{bmatrix}=\det\begin{bmatrix}A_{11} & 0\\ A_{21} & A_{22}\end{bmatrix}=(\det A_{11})(\det A_{22}).$$

Proof. It is easily shown that

$$\det\begin{bmatrix}I_p & 0\\ 0 & A_{22}\end{bmatrix}=\det A_{22}$$

and

$$\det\begin{bmatrix}A_{11} & A_{12}\\ 0 & I_q\end{bmatrix}=\det A_{11}.$$

Then from Theorem A3.1,

$$\det\begin{bmatrix} A_{11} & A_{12} \\ 0 & A_{22} \end{bmatrix} = \det\begin{bmatrix} I_p & 0 \\ 0 & A_{22} \end{bmatrix} \det\begin{bmatrix} A_{11} & A_{12} \\ 0 & I_q \end{bmatrix} = (\det A_{11})(\det A_{22}).$$

Similarly

$$\det\begin{bmatrix} A_{11} & 0 \\ A_{21} & A_{22} \end{bmatrix} = \det\begin{bmatrix} A_{11} & 0 \\ 0 & I_q \end{bmatrix} \det\begin{bmatrix} I_p & 0 \\ A_{21} & A_{22} \end{bmatrix} = (\det A_{11})(\det A_{22}).$$

THEOREM A3.5. If A is $p \times q$ and B is $q \times p$ then

$$\det(I_p + AB) = \det(I_q + BA).$$

Proof. We can write

$$\begin{bmatrix} I_p + AB & A \\ 0 & I_q \end{bmatrix} = \begin{bmatrix} I_p & A \\ -B & I_q \end{bmatrix} \begin{bmatrix} I_p & 0 \\ B & I_q \end{bmatrix}$$

so that

(1) $$\det(I_p + AB) = \det\begin{bmatrix} I_p & A \\ -B & I_q \end{bmatrix}.$$

Similarly

$$\begin{bmatrix} I_p & A \\ 0 & I_q + BA \end{bmatrix} = \begin{bmatrix} I_p & 0 \\ B & I_q \end{bmatrix} \begin{bmatrix} I_p & A \\ -B & I_q \end{bmatrix},$$

so that

(2) $$\det(I_q + BA) = \det\begin{bmatrix} I_p & A \\ -B & I_q \end{bmatrix}.$$

Equating (1) and (2) gives the desired result.

Two additional results about determinants are used often.

(ix) If T is $m \times m$ triangular (upper or lower) then $\det T = \prod_{i=1}^{m} t_{ii}$.

(x) If H is an orthogonal matrix then $\det H = \pm 1$.

A4. MINORS AND COFACTORS

If $A=(a_{ij})$ is a $p\times p$ matrix the *minor* of the element a_{ij} is the determinant of the matrix M_{ij} obtained from A by removing the ith row and jth column. The *cofactor* of a_{ij}, denoted by α_{ij}, is

$$\alpha_{ij}=(-1)^{i+j}\det M_{ij}.$$

It is proved in many matrix theory texts that det A is equal to the sum of the products obtained by multiplying each element of a row (or column) by its cofactor, i.e.,

$$\det A=\sum_{j=1}^{p} a_{ij}\alpha_{ij} \qquad (i=1,\dots,p)$$

$$=\sum_{i=1}^{p} a_{ij}\alpha_{ij} \qquad (j=1,\dots,p).$$

A *principal minor* of A is the determinant of a matrix obtained from A by removing certain rows and the same numbered columns of A. In general, if A is a $p\times q$ matrix an *r-square minor* of A is a determinant of an $r\times r$ matrix obtained from A by removing $p-r$ rows and $q-r$ columns.

A5. INVERSE OF A MATRIX

If $A=(a_{ij})$ is $p\times p$, with $\det A\neq 0$, A is called a *nonsingular* matrix. In this case there is a unique matrix B such that $AB=I_p$. The $i-j$th element of B is given by

$$b_{ij}=\frac{\alpha_{ji}}{\det A},$$

where α_{ji} is the cofactor of a_{ji}. The matrix B is called the *inverse* of A and is denoted by A^{-1}. The following basic results hold:

(i) $AA^{-1}=A^{-1}A=I.$

(ii) $(A^{-1})'=(A')^{-1}.$

(iii) If A and C are nonsingular $p\times p$ matrices then $(AC)^{-1}=C^{-1}A^{-1}$.

(iv) $\det(A^{-1})=(\det A)^{-1}$.

(v) If A is an orthogonal matrix, $A^{-1}=A'$.

(vi) If $A=\operatorname{diag}(a_{11},\dots,a_{pp})$ with $a_{ii}\neq 0$ $(i=1,\dots,p)$ then $A^{-1}=\operatorname{diag}(a_{11}^{-1},\dots,a_{pp}^{-1})$.

(vii) If T is an $m\times m$ upper-triangular nonsingular matrix then T^{-1} is upper-triangular and its diagonal elements are t_{ii}^{-1}, $i=1,\dots,m$.

The following result is occasionally useful.

THEOREM A5.1. Let A and B be nonsingular $p\times p$ and $q\times q$ matrices, respectively, and let C be $p\times q$ and D be $q\times p$. Put $P=A+CBD$. Then

(1) $$P^{-1}=A^{-1}-A^{-1}CB(B+BDA^{-1}CB)^{-1}BDA^{-1}.$$

Proof. Premultiplying the right side of (1) by P gives

$$\begin{aligned}
&(A+CBD)\left[A^{-1}-A^{-1}CB(B+BDA^{-1}CB)^{-1}BDA^{-1}\right]\\
&=I-CB(B+BDA^{-1}CB)^{-1}BDA^{-1}+CBDA^{-1}\\
&\quad -CBDA^{-1}CB(B+BDA^{-1}CB)^{-1}BDA^{-1}\\
&=I+CB\left[B^{-1}-(I+DA^{-1}CB)(B+BDA^{-1}CB)^{-1}\right]BDA^{-1}\\
&=I+CB\left[B^{-1}-B^{-1}(B+BDA^{-1}CB)(B+BDA^{-1}CB)^{-1}\right]BDA^{-1}\\
&=I,
\end{aligned}$$

completing the proof.

The next theorem gives the elements of the inverse of a partitioned matrix A in terms of the submatrices of A.

THEOREM A5.2. Let A be a $p\times p$ nonsingular matrix, and let $B=A^{-1}$. Partition A and B as

(2) $$A=\begin{bmatrix}A_{11} & A_{12}\\ A_{21} & A_{22}\end{bmatrix},\qquad B=\begin{bmatrix}B_{11} & B_{12}\\ B_{21} & B_{22}\end{bmatrix},$$

where A_{11} and B_{11} are $k\times k$, A_{12} and B_{12} are $k\times(p-k)$, A_{21} and B_{21} are $(p-k)\times k$ and A_{22} is $(p-k)\times(p-k)$; assume that A_{11} and A_{22} are nonsingular. Put

(3) $$A_{11\cdot 2}=A_{11}-A_{12}A_{22}^{-1}A_{21},\qquad A_{22\cdot 1}=A_{22}-A_{21}A_{11}^{-1}A_{12}.$$

Then

$$B_{11}=A_{11\cdot 2}^{-1},\qquad B_{22}=A_{22\cdot 1}^{-1},\qquad B_{12}=-A_{11}^{-1}A_{12}A_{22\cdot 1}^{-1},$$

$$B_{21}=-A_{22}^{-1}A_{21}A_{11\cdot 2}^{-1}.$$

Proof. The equation $AB=I$ leads to the following equations:

$$(4)\qquad A_{11}B_{11}+A_{12}B_{21}=I$$

$$(5)\qquad A_{11}B_{12}+A_{12}B_{22}=0$$

$$(6)\qquad A_{21}B_{11}+A_{22}B_{21}=0$$

$$(7)\qquad A_{21}B_{12}+A_{22}B_{22}=I.$$

From (6) we have $B_{21}=-A_{22}^{-1}A_{21}B_{11}$ and substituting this in (4) gives $A_{11}B_{11}-A_{12}A_{22}^{-1}A_{21}B_{11}=I$ so that $B_{11}=A_{11\cdot 2}^{-1}$. From (5) we have $B_{12}=-A_{11}^{-1}A_{12}B_{22}$, which when substituted in (7) gives $A_{22}B_{22}-A_{21}A_{11}^{-1}A_{12}B_{22}=I$ so that $B_{22}=A_{22\cdot 1}^{-1}$.

The determinant of a partitioned matrix is given in the following theorem.

THEOREM A5.3. Let A be partitioned as in (1) and let $A_{11\cdot 2}$ and $A_{22\cdot 1}$ be given by (3).

(a) If A_{22} is nonsingular then

$$\det A=\det A_{22}\det A_{11\cdot 2}.$$

(b) If A_{11} is nonsingular then

$$\det A=\det A_{11}\det A_{22\cdot 1}.$$

Proof. To prove (a) note that if

$$C=\begin{bmatrix} I_k & -A_{12}A_{22}^{-1} \\ 0 & I_{m-k}\end{bmatrix}$$

then

$$CAC'=\begin{bmatrix} A_{11\cdot 2} & 0 \\ 0 & A_{22}\end{bmatrix}.$$

(This was demonstrated in Theorem 1.2.10.) Hence

$$\det(CAC')=(\det C)(\det A)(\det C')=\det A=\det A_{11\cdot 2}\det A_{22},$$

where we have used Theorems A3.2 and A3.4. The proof of (6) is similar.

A6. RANK OF A MATRIX

If A is a nonzero $p\times q$ matrix it is said to have *rank* r, written $\operatorname{rank}(A)=r$, if at least one of its r-square minors is different from zero while every $(r+1)$-square minor (if any) is zero. If $A=0$ it is said to have rank 0. Clearly if A is a nonsingular $p\times p$ matrix, $\operatorname{rank}(A)=p$. The following properties can be readily established:

(i) $\operatorname{rank}(A)=\operatorname{rank}(A')$.

(ii) If A is $p\times q$, $\operatorname{rank}(A)\le\min(p,q)$.

(iii) If A is $p\times q$, B is $q\times r$, then

$$\operatorname{rank}(AB)\le\min[\operatorname{rank}(A),\operatorname{rank}(B)].$$

(iv) If A and B are $p\times q$, then

$$\operatorname{rank}(A+B)\le\operatorname{rank}(A)+\operatorname{rank}(B).$$

(v) If A is $p\times p$, B is $p\times q$, C is $q\times q$, and A and C are nonsingular, then

$$\operatorname{rank}(ABC)=\operatorname{rank}(B).$$

(vi) If A is $p\times q$ and B is $q\times r$ such that $AB=0$, then

$$\operatorname{rank}(B)\le q-\operatorname{rank}(A).$$

A7. LATENT ROOTS AND LATENT VECTORS

For a $p\times p$ matrix A the *characteristic equation* of A is given by

$$\det(A-\lambda I_p)=0. \tag{1}$$

The left side of (1) is a polynomial of degree p in λ so that this equation has exactly p roots, called the *latent roots* (or *characteristic roots* or *eigenvalues*) of A. These roots are not necessarily distinct and may be real, or complex, or both. If λ_i is a latent root of A then

$$\det(A-\lambda_i I)=0$$

so that $A-\lambda_i I$ is singular. Hence there is a nonzero vector $\mathbf{x}_i$ such that $(A-\lambda_i I)\mathbf{x}_i=\mathbf{0}$, called a *latent vector* (or *characteristic vector* or *eigenvector*) of A corresponding to λ_i. The following three theorems summarize some very basic results about latent roots and vectors.

THEOREM A7.1. If $B=CAC^{-1}$, where A, B and C are all $p\times p$, then A and B have the same latent roots.

Proof. Since

$$B-\lambda I=CAC^{-1}-\lambda I=C(A-\lambda I)C^{-1},$$

we have

$$\det(B-\lambda I)=\det C\det(A-\lambda I)\det C^{-1}=\det(A-\lambda I)$$

so that A and B have the same characteristic equation.

THEOREM A7.2. If A is a real symmetric matrix then its latent roots are all real.

Proof. Suppose that $\alpha+i\beta$ is a complex latent root of A, and put

$$B=[(\alpha+i\beta)I-A][(\alpha-i\beta)I-A]=(\alpha I-A)^2+\beta^2 I.$$

B is real, and singular because $(\alpha+i\beta)I-A$ is singular. Hence there is a nonzero real vector $\mathbf{x}$ such that $B\mathbf{x}=\mathbf{0}$ and consequently

$$\begin{aligned}0&=\mathbf{x}'B\mathbf{x}=\mathbf{x}'(\alpha I-A)^2\mathbf{x}+\beta^2\mathbf{x}'\mathbf{x}\\&=\mathbf{x}'(\alpha I-A)'(\alpha I-A)\mathbf{x}+\beta^2\mathbf{x}'\mathbf{x}.\end{aligned}$$

Since $\mathbf{x}'(\alpha I-A)'(\alpha I-A)\mathbf{x}\geq 0$ and $\mathbf{x}'\mathbf{x}>0$ we must have $\beta=0$, which means that no latent roots of A are complex.

THEOREM A7.3. If A is a real symmetric matrix and λ_i and λ_j are two distinct latent roots of A then the corresponding latent vectors $\mathbf{x}_i$ and $\mathbf{x}_j$ are orthogonal.

Proof. Since

$$A\mathbf{x}_i=\lambda_i\mathbf{x}_i, \qquad A\mathbf{x}_j=\lambda_j\mathbf{x}_j,$$

it follows that

$$\mathbf{x}_j'A\mathbf{x}_i=\lambda_i\mathbf{x}_j'\mathbf{x}_i, \qquad \mathbf{x}_i'A\mathbf{x}_j=\lambda_j\mathbf{x}_i'\mathbf{x}_j.$$

Hence $(\lambda_i-\lambda_j)\mathbf{x}_i'\mathbf{x}_j=0$, so that $\mathbf{x}_i'\mathbf{x}_j=0$.

Some other properties of latent roots and vectors are now summarized.

(i) The latent roots of A and A' are the same.

(ii) If A has latent roots $\lambda_1,\ldots,\lambda_p$ then $A-kI$ has latent roots $\lambda_1-k,\ldots,\lambda_p-k$ and kA has latent roots $k\lambda_1,\ldots,k\lambda_p$.

(iii) If $A=\text{diag}(a_1,\ldots,a_p)$ then $a_1,\ldots,a_p$ are the latent roots of A and the vectors $(1,0,\ldots,0)$, $(0,1,\ldots,0),\ldots,(0,0,\ldots,1)$ are associated latent vectors.

(iv) If A and B are $p\times p$ and A is nonsingular then the latent roots of AB and BA are the same.

(v) If $\lambda_1,\ldots,\lambda_p$ are the latent roots of the nonsingular matrix A then $\lambda_1^{-1},\ldots,\lambda_p^{-1}$ are the latent roots of A^{-1}.

(vi) If A is an orthogonal matrix $(AA'=I)$ then all its latent roots have absolute value 1.

(vii) If A is symmetric it is idempotent $(A^2=A)$ if and only if its latent roots are 0's and 1's.

(viii) If A is $p\times q$ the nonzero latent roots of AA' and $A'A$ are the same.

(ix) If T is triangular (upper or lower) then the latent roots of T are the diagonal elements.

(x) If A has a latent root λ of multiplicity r there exist r orthogonal latent vectors corresponding to λ. The set of linear combinations of these vectors is called the *latent space* corresponding to λ. If λ_i and λ_j are two different latent roots their corresponding latent spaces are orthogonal.

An expression for the *characteristic polynomial* $p(\lambda)=\det(A-\lambda I_p)$ can be obtained in terms of the principal minors of A. Let $A_{i_1,i_2,\ldots,i_k}$ be the $k\times k$ matrix formed from A by deleting all but rows and columns numbered $i_1,\ldots,i_k$, and define the kth *trace* of A as

$$\text{tr}_k(A)=\Sigma_{1\le i_1<i_2\ldots<i_k\le p}\det A_{i_1,i_2,\ldots,i_p}.$$

The first trace ($k=1$) is called the *trace*, denoted by $\text{tr}(A)$, so that $\text{tr}(A)=\sum_{i=1}^{p} a_{ii}$. This function has the elementary properties that $\text{tr}(A)=\text{tr}(A')$ and if C is $p\times q$, D is $q\times p$ then $\text{tr}(CD)=\text{tr}(DC)$. Note also that $\text{tr}_m(A)=\det(A)$. Using basic properties of determinants it can be readily established that:

(xi) $p(\lambda)=\det(A-\lambda I_p)=\sum_{k=0}^{p}(-\lambda)^k\,\text{tr}_{p-k}(A)$ $\quad[\text{tr}_0(A)\equiv 1]$.
Let A have latent roots $\lambda_1,\dots,\lambda_p$ so that

$$p(\lambda)=(-1)^p\sum_{i=1}^{p}(\lambda-\lambda_i).$$

Expanding this product gives

(xii) $p(\lambda)=\sum_{k=0}^{p}(-\lambda)^k r_{p-k}(\lambda_1,\dots,\lambda_p)$,
where $r_j(\lambda_1,\dots,\lambda_p)$ denotes the jth *elementary symmetric function* of $\lambda_1,\dots,\lambda_p$, given by

$$r_j(\lambda_1,\dots,\lambda_p)=\sum_{1\le i_1<i_2<\cdots<i_j\le p}\lambda_{i_1}\lambda_{i_2}\dots\lambda_{i_j}.$$

Equating coefficients of λ^k in (xi) and (xii) shows that

(xiii) $r_k(\lambda_1,\dots,\lambda_p)=\text{tr}_k(A)$.
It is worth noting that $p(\lambda)$ can also be written as

$$\begin{aligned} p(\lambda)&=(-\lambda)^p\det A\det(A^{-1}-\lambda^{-1}I)\\ &=(-\lambda)^p\det A\sum_{k=0}^{p}(-\lambda^{-1})^k\,\text{tr}_{p-k}(A^{-1})\end{aligned}$$

and equating coefficients of λ^k here and in (xii) gives

(xiv) $\text{tr}_k(A^{-1})=\det A^{-1}\,\text{tr}_{p-k}(A)$.

A8. POSITIVE DEFINITE MATRICES

A $p\times p$ symmetric matrix A is called *positive* (*negative*) *definite* if $\mathbf{x}'A\mathbf{x}>0$ (<0) for all vectors $\mathbf{x}\neq\mathbf{0}$; this is commonly expressed as $A>0$ ($A<0$). It is called *positive* (*negative*) *semidefinite* if $\mathbf{x}'A\mathbf{x}\ge 0$ (≤ 0) for all $\mathbf{x}\neq\mathbf{0}$, written as $A\ge 0$ (≤ 0). It is called *non-negative definite* if $A>0$ or $A\ge 0$, i.e., if $\mathbf{x}'A\mathbf{x}\ge 0$ for all $\mathbf{x}$, and *non-positive definite* if $A<0$ or $A\le 0$.

We now summarize some well-known properties about positive definite matrices.

(i) A is positive definite if and only if $\det A_{1,\ldots,i} > 0$ for $i=1,\ldots,p$, where $A_{1,\ldots,i}$ is the $i \times i$ matrix consisting of the first i rows and columns of A.

(ii) If $A>0$ then $A^{-1}>0$.

(iii) A symmetric matrix is positive definite (non-negative definite) if and only if all of its latent roots are positive (non-negative).

(iv) For any matrix B, $BB' \geq 0$.

(v) If A is non-negative definite then A is nonsingular if and only if $A>0$.

(vi) If $A>0$ is $p \times p$ and B is $q \times p$ $(q \leq p)$ of rank r then $BAB'>0$ if $r=q$ and $BAB' \geq 0$ if $r<q$.

(vii) If $A>0$, $B>0$, $A-B>0$ then $B^{-1}-A^{-1}>0$ and $\det A > \det B$.

(viii) If $A>0$ and $B>0$ then $\det(A+B) \geq \det A + \det B$.

(ix) If $A>0$ and

$$A=\begin{bmatrix} A_{11} & A_{12} \\ A_{21} & A_{22} \end{bmatrix},$$

where A_{11} is a square matrix, then $A_{11}>0$ and $A_{11}-A_{12}A_{22}^{-1}A_{21}>0$.

A9. SOME MATRIX FACTORIZATIONS

Before looking at matrix factorizations we recall the Gram-Schmidt orthogonalization process which enables us to construct an orthonormal basis of R^m given any other basis $\mathbf{x}_1, \mathbf{x}_2, \ldots, \mathbf{x}_m$ of R^m. We define

$$\mathbf{y}_1 = \mathbf{x}_1$$

$$\mathbf{y}_2 = \mathbf{x}_2 - \frac{\mathbf{x}_1'\mathbf{x}_2}{\mathbf{y}_1'\mathbf{y}_1}\mathbf{y}_1$$

$$\mathbf{y}_3 = \mathbf{x}_3 - \frac{\mathbf{y}_2'\mathbf{x}_3}{\mathbf{y}_2'\mathbf{y}_2}\mathbf{y}_2 - \frac{\mathbf{y}_1'\mathbf{x}_3}{\mathbf{y}_1'\mathbf{y}_1}\mathbf{y}_1$$

$$\cdots\cdots\cdots\cdots\cdots\cdots\cdots\cdots$$

$$\mathbf{y}_m = \mathbf{x}_m - \sum_{i=1}^{m-1} \frac{\mathbf{y}_i'\mathbf{x}_m}{\mathbf{y}_i'\mathbf{y}_i}\mathbf{y}_i,$$

and put $\mathbf{z}_i = [1/(\mathbf{y}_i'\mathbf{y}_i)^{1/2}]\mathbf{y}_i$, with $i=1,\ldots,m$. Then $\mathbf{z}_1,\ldots,\mathbf{z}_m$ form an orthonormal basis for R^m. Our first matrix factorization utilizes this process.

THEOREM A9.1. If A is a real $m \times m$ matrix with real latent roots then there exists an orthogonal matrix H such that $H'AH$ is an upper-triangular matrix whose diagonal elements are the latent roots of A.

Proof. Let $\lambda_1,\ldots,\lambda_m$ be the latent roots of A and let $\mathbf{x}_1$ be a latent vector of A corresponding to λ_1. This is real since the latent roots are real. Let $\mathbf{x}_2,\ldots,\mathbf{x}_m$ be any other vectors such that $\mathbf{x}_1,\mathbf{x}_2,\ldots,\mathbf{x}_m$ form a basis for R^m. Using the Gram-Schmidt orthogonalization process, construct from $\mathbf{x}_1,\ldots,\mathbf{x}_m$ an orthonormal basis given as the columns of the orthogonal matrix H_1, where the first column $\mathbf{h}_1$ is proportional to $\mathbf{x}_1$, so that $\mathbf{h}_1$ is also a latent vector of A corresponding to λ_1. Then the first column of AH_1 is $A\mathbf{h}_1=\lambda_1\mathbf{h}_1$, and hence the first column of $H_1'AH_1$ is $\lambda_1 H_1'\mathbf{h}_1$. Since this is the first column of $\lambda_1 H_1'H_1=\lambda_1 I_m$, it is $(\lambda_1,0,\ldots,0)'$. Hence

$$H_1'AH_1=\begin{bmatrix}\lambda_1 & B_1\\ \mathbf{0} & A_2\end{bmatrix},$$

where A_2 is $(m-1)\times(m-1)$. Since

$$\det(A-\lambda I)=(\lambda_1-\lambda)\det(A_2-\lambda I)$$

and A and $H_1'AH_1$ have the same latent roots, the latent roots of A_2 are $\lambda_2,\ldots,\lambda_m$.

Now, using a construction similar to that above, find an orthogonal $(m-1)\times(m-1)$ matrix H_2 whose first column is a latent vector of A_2 corresponding to λ_2. Then

$$H_2'A_2H_2=\begin{bmatrix}\lambda_2 & B_2\\ \mathbf{0} & A_3\end{bmatrix},$$

where A_3 is $(m-2)\times(m-2)$ with latent roots $\lambda_3,\ldots,\lambda_m$.

Repeating this procedure an additional $m-3$ times we now define the orthogonal matrix

$$H=H_1\begin{bmatrix}1 & \mathbf{0}'\\ \mathbf{0} & H_2\end{bmatrix}\begin{bmatrix}I_2 & 0\\ 0 & H_3\end{bmatrix}\cdots\begin{bmatrix}I_{m-2} & 0\\ 0 & H_{m-1}\end{bmatrix} \tag{1}$$

and note that $H'AH$ is upper-triangular with diagonal elements equal to $\lambda_1,\ldots,\lambda_m$.

An immediate consequence of this theorem is given next.

THEOREM A9.2. If A is a real symmetric $m\times m$ matrix with latent roots $\lambda_1,\dots,\lambda_m$ there exists an orthogonal $m\times m$ matrix H such that

$$H'AH=D\equiv\operatorname{diag}(\lambda_1,\dots,\lambda_m). \tag{2}$$

If $H=[\mathbf{h}_1,\dots,\mathbf{h}_m]$ then $\mathbf{h}_i$ is a latent vector of A corresponding to the latent root λ_i. Moreover, if $\lambda_1,\dots,\lambda_m$ are all distinct the representation (2) is unique up to sign changes in the first row of H.

Proof. As in the proof of Theorem A9.1 there exists an orthogonal $m\times m$ matrix H_1 such that

$$H_1'AH_1=\begin{bmatrix}\lambda_1 & B_1\\ \mathbf{0} & A_2\end{bmatrix},$$

where $\lambda_2,\dots,\lambda_m$ are the latent roots of A_2. Since $H_1'AH_1$ is symmetric it follows that $B_1=0$. Similarly each B_i in the proof of Theorem A9.1 is zero $(i=1,\dots,m-1)$, and hence the matrix H given by (1) satisfies $H'AH=\operatorname{diag}(\lambda_1,\dots,\lambda_m)$. Consequently, $A\mathbf{h}_i=\lambda_i\mathbf{h}_i$ so that $\mathbf{h}_i$ is a latent vector of A corresponding to the latent root λ_i. Now suppose that we also have $Q'AQ=D$ for a orthogonal matrix Q. Then $PD=DP$ with $P=Q'H$. If $P=(p_{ij})$ it follows that $p_{ij}\lambda_j=p_{ij}\lambda_i$ and, since $\lambda_i\neq\lambda_j$, $p_{ij}=0$ for $i\neq j$. Since P is orthogonal it must then have the form $\tilde{P}=\operatorname{diag}(\pm1,\pm1,\dots,\pm1)$, and $H=Q\tilde{P}$.

THEOREM A9.3. If A is a non-negative definite $m\times m$ matrix then there exists a non-negative definite $m\times m$ matrix, written as $A^{1/2}$, such that $A=A^{1/2}A^{1/2}$.

Proof. Let H be an orthogonal matrix such that $H'AH=D$, where $D=\operatorname{diag}(\lambda_1,\dots,\lambda_m)$ with $\lambda_1,\dots,\lambda_m$ being the latent roots of A. Since A is non-negative definite, $\lambda_i\geq0$ for $i=1,\dots,m$. Putting $D^{1/2}=\operatorname{diag}(\lambda_1^{1/2},\dots,\lambda_m^{1/2})$, we have $D^{1/2}D^{1/2}=D$. Now define the matrix $A^{1/2}$ by $A^{1/2}=HD^{1/2}H'$. Then $A^{1/2}$ is non-negative definite and

$$A^{1/2}A^{1/2}=HD^{1/2}H'HD^{1/2}H'=HD^{1/2}D^{1/2}H'=HDH'=A.$$

The term $A^{1/2}$ in Theorem A9.3 is called a non-negative definite square root of A. If A is positive definite $A^{1/2}$ is positive definite and is called the positive definite square root of A.

THEOREM A9.4. If A is an $m\times m$ non-negative definite matrix of rank r then:

(i) There exists an $m\times r$ matrix B of rank r such that $A=BB'$.

(ii) There exists an $m\times m$ nonsingular matrix C such that

$$A=C\begin{bmatrix} I_r & 0\\ 0 & 0\end{bmatrix}C'.$$

Proof. As for statement (i), let $D_1=\text{diag}(\lambda_1,\dots,\lambda_r)$ where $\lambda_1,\dots,\lambda_r$ are the nonzero latent roots of A, and let H be an $m\times m$ orthogonal matrix such that $H'AH=\text{diag}(\lambda_1,\dots,\lambda_r,0,\dots,0)$. Partition H as $H=[H_1:H_2]$, where H_1 is $m\times r$ and H_2 is $m\times(m-r)$; then

$$A=H\begin{bmatrix} D_1 & 0\\ 0 & 0\end{bmatrix}H'=[H_1:H_2]\begin{bmatrix} D_1 & 0\\ 0 & 0\end{bmatrix}\begin{bmatrix} H_1'\\ H_2'\end{bmatrix}=H_1D_1H_1'$$

Putting $D_1^{1/2}=\text{diag}(\lambda_1^{1/2},\dots,\lambda_r^{1/2})$, we then have

$$A=H_1D_1^{1/2}D_1^{1/2}H_1'=BB',$$

where $B=H_1D_1^{1/2}$ is $m\times r$ of rank r.

As for statement (ii), let C be an $m\times m$ nonsingular matrix whose first r columns are the columns of the matrix B in (i). Then

$$A=C\begin{bmatrix} I_r & 0\\ 0 & 0\end{bmatrix}C'.$$

The following theorem, from Vinograd (1950), is used often in the text.

THEOREM A9.5. Suppose that A and B are real matrices, where A is $k\times m$ and B is $k\times n$, with $m\le n$. Then $AA'=BB'$ if and only if there exists an $m\times n$ matrix H with $HH'=I_m$ such that $AH=B$.

Proof. First suppose there exists an $m\times n$ matrix H with $HH'=I_m$ such that $AH=B$. Then $BB'=AHH'A'=AA'$.

Now suppose that $AA'=BB'$. Let C be a $k\times k$ nonsingular matrix such that

$$AA'=BB'=C\begin{bmatrix} I_r & 0\\ 0 & 0\end{bmatrix}C'$$

(Theorem A9.4), where rank $(AA')=r$. Now put $D=C^{-1}A$, $E=C^{-1}B$ and partition these as

$$D=\begin{bmatrix} D_1 \\ D_2 \end{bmatrix}, \qquad E=\begin{bmatrix} E_1 \\ E_2 \end{bmatrix},$$

where D_1 is $r\times m$, D_2 is $(k-r)\times m$, E_1 is $r\times n$, and E_2 is $(k-r)\times n$. Then

$$EE'=\begin{bmatrix} E_1E_1' & E_1E_2' \\ E_2E_1' & E_2E_2' \end{bmatrix}=C^{-1}BB'C^{-1\prime}=\begin{bmatrix} I_r & 0 \\ 0 & 0 \end{bmatrix}$$

and

$$DD'=\begin{bmatrix} D_1D_1' & D_1D_2' \\ D_2D_1' & D_2D_2' \end{bmatrix}=C^{-1}AA'C^{-1\prime}=\begin{bmatrix} I_r & 0 \\ 0 & 0 \end{bmatrix}$$

which imply that $E_1E_1'=D_1D_1'=I_r$ and $D_2=0$, $E_2=0$, so that

$$D=\begin{bmatrix} D_1 \\ 0 \end{bmatrix}, \qquad E=\begin{bmatrix} E_1 \\ 0 \end{bmatrix}.$$

Now let $\tilde{E}_2$ be an $(n-r)\times n$ matrix such that

$$\tilde{E}=\begin{bmatrix} E_1 \\ \tilde{E}_2 \end{bmatrix}$$

is an $n\times n$ orthogonal matrix, and choose an $(n-r)\times m$ matrix $\tilde{D}_2$ and an $(n-r)\times(n-m)$ matrix $\tilde{D}_3$ such that

$$\tilde{D}=\begin{bmatrix} D_1 & 0 \\ \tilde{D}_2 & \tilde{D}_3 \end{bmatrix}$$

is an $n\times n$ orthogonal matrix. Then

$$E=\begin{bmatrix} I_r & 0 \\ 0 & 0 \end{bmatrix}\tilde{E}$$

and

$$[D:0]=\begin{bmatrix} I_r & 0 \\ 0 & 0 \end{bmatrix}\tilde{D},$$

and hence

$$E=[D:0]\tilde{D}'\tilde{E}=[D:0]Q,$$

where $Q=\tilde{D}'\tilde{E}$ is $n\times n$ orthogonal. Partitioning Q as

$$Q=\begin{bmatrix} H \\ \cdots \\ P \end{bmatrix},$$

where H is $m\times n$ and P is $(n-m)\times n$, we then have $HH'=I_m$ and

$$C^{-1}B=E=DH=C^{-1}AH$$

so that $B=AH$, completing the proof.

The next result is an immediate consequence of Theorem A9.5.

THEOREM A9.6. Let A be an $n\times m$ real matrix of rank m $(n\geq m)$. Then:

(i) A can be written as $A=H_1B$, where H_1 is $n\times m$ with $H_1'H_1=I_m$ and B is $m\times m$ positive definite.

(ii) A can be written as

$$A=H\begin{bmatrix} I_m \\ \cdots \\ 0 \end{bmatrix}B,$$

where H is $n\times n$ orthogonal and B is $m\times m$ positive definite.

Proof. As for statement (i), let B be the positive definite square root of the positive definite matrix $A'A$ (see Theorem A9.3), so that

$$A'A=B^2=B'B.$$

By Theorem A9.5 A can be written as $A=H_1B$, where H_1 is $n\times m$ with $H_1'H_1=I_m$.

As for statement (ii), let H_1 be the matrix in (i) such that $A=H_1B$ and choose an $n\times(n-m)$ matrix H_2 so that $H=[H_1:H_2]$ is $n\times n$ orthogonal. Then

$$A=H_1B=H\begin{bmatrix} I_m \\ \cdots \\ 0 \end{bmatrix}B.$$

We now turn to decompositions of positive definite matrices in terms of triangular matrices.

THEOREM A9.7. If A is an $m\times m$ positive definite matrix then there exists a unique $m\times m$ upper-triangular matrix T with positive diagonal elements such that $A=T'T$.

Proof. An induction proof can easily be constructed. The stated result holds trivially for $m=1$. Suppose the result holds for positive definite matrices of size $m-1$. Partition the $m\times m$ matrix A as

$$A=\begin{bmatrix} A_{11} & \mathbf{a}_{12} \\ \mathbf{a}'_{12} & a_{22} \end{bmatrix},$$

where A_{11} is $(m-1)\times(m-1)$. By the induction hypothesis there exists a unique $(m-1)\times(m-1)$ upper-triangular matrix T_{11} with positive diagonal elements such that $A_{11}=T'_{11}T_{11}$. Now suppose that

$$A=\begin{bmatrix} A_{11} & \mathbf{a}_{12} \\ \mathbf{a}'_{12} & a_{22} \end{bmatrix}=\begin{bmatrix} T'_{11} & \mathbf{0} \\ \mathbf{x}' & y \end{bmatrix}\begin{bmatrix} T_{11} & \mathbf{x} \\ \mathbf{0}' & y \end{bmatrix}=\begin{bmatrix} T'_{11}T_{11} & T'_{11}\mathbf{x} \\ \mathbf{x}'T_{11} & \mathbf{x}'\mathbf{x}+y^2 \end{bmatrix},$$

where $\mathbf{x}$ is $(m-1)\times 1$ and $y\in R^1$. For this to hold we must have $\mathbf{x}=(T'_{11})^{-1}\mathbf{a}_{12}$, and then

$$y^2=a_{22}-\mathbf{x}'\mathbf{x}=a_{22}-\mathbf{a}'_{12}T_{11}^{-1}(T'_{11})^{-1}\mathbf{a}_{12}=a_{22}-\mathbf{a}'_{12}A_{11}^{-1}\mathbf{a}_{12}.$$

Note that this is positive by (ix) of Section A8, and the unique $y>0$ satisfying this is $y=(a_{22}-\mathbf{a}'_{12}A_{11}^{-1}\mathbf{a}_{12})^{1/2}$.

THEOREM A9.8. If A is an $n\times m$ real matrix of rank m $(n\geq m)$ then A can be uniquely written as $A=H_1T$, where H_1 is $n\times m$ with $H_1'H_1=I_m$ and T is $m\times m$ upper-triangular with positive diagonal elements.

Proof. Since $A'A$ is $m\times m$ positive definite it follows from Theorem A9.7 that there exists a unique $m\times m$ upper-triangular matrix with positive diagonal elements such that $A'A=T'T$. By Theorem A9.5 there exists an $n\times m$ matrix H_1 with $H_1'H_1=I_m$ such that $A=H_1T$. Note that H_1 is unique because T is unique and $\operatorname{rank}(T)=m$.

THEOREM A9.9. If A is an $m\times m$ positive definite matrix and B is an $m\times m$ symmetric matrix there exists an $m\times m$ nonsingular matrix L such that $A=LL'$ and $B=LDL'$, where $D=\operatorname{diag}(d_1,\dots,d_m)$, with $d_1,\dots,d_m$ being the latent roots of $A^{-1}B$. If B is positive definite and $d_1,\dots,d_m$ are all distinct, L is unique up to sign changes in the first row of L.

Proof. Let $A^{1/2}$ be the positive definite square root of A (see Theorem A9.3), so that $A=A^{1/2}A^{1/2}$. By Theorem A9.2 there exists an $m\times m$

orthogonal matrix H such that $A^{-1/2}BA^{-1/2}=HDH'$, where $D=\text{diag}(d_1,\dots,d_m)$. Putting $L=A^{1/2}H$, we now have $LL'=A$ and $B=LDL'$. Note that $d_1,\dots,d_m$ are the latent roots of $A^{-1}B$.

Now suppose that B is positive definite and the d_i are all distinct. Assume that as well as $A=LL'$ and $B=LDL'$ we also have $A=MM'$ and $B=MDM'$, where M is $m\times m$ nonsingular. Then $(M^{-1}L)(M^{-1}L)'=M^{-1}LL'M^{-1\prime}=M^{-1}AM^{-1\prime}=M^{-1}MM'M'^{-1}=I_m$ so that the matrix $Q=M^{-1}L$ is orthogonal and $QD=DQ$. If $Q=(q_{ij})$ we then have $q_{ij}d_i=q_{ij}d_j$ so that $q_{ij}=0$ for $i\neq j$. Since Q is orthogonal it must then have the form $\tilde{Q}=\text{diag}(\pm1,\pm1,\dots,\pm1)$, and $L=M\tilde{Q}$.

THEOREM A9.10. If A is an $m\times n$ real matrix ($m\leq n$) there exist an $m\times m$ orthogonal matrix H and an $n\times n$ orthogonal matrix Q such that

$$HAQ'=\left[\begin{array}{ccc:c} d_1 & & 0 & \\ & \ddots & & 0 \\ 0 & & d_m & \end{array}\right],$$

where $d_i\geq0$ for $i=1,\dots,m$ and $d_1^2,\dots,d_m^2$ are the latent roots of AA'.

Proof. Let H be an orthogonal $m\times m$ matrix such that $AA'=H'D^2H$, where $D^2=\text{diag}(d_1^2,\dots,d_m^2)$, with $d_i^2\geq0$ for $i=1,\dots,m$ because AA' is non-negative definite. Let $D=\text{diag}(d_1,\dots,d_m)$ with $d_i\geq0$ for $i=1,\dots,m$; then $AA'=(H'D)(H'D)'$, and by Theorem A9.5 there exists an $m\times n$ matrix Q_1 with $Q_1Q_1'=I_m$ such that $A=H'DQ_1$. Choose an $(n-m)\times n$ matrix Q_2 so that the $n\times n$ matrix

$$Q=\left[\begin{array}{c} Q_1 \\ \hdashline Q_2 \end{array}\right]$$

is orthogonal; we now have

$$A=H'DQ_1=H'[D:0]Q$$

so that $HAQ'=[D:0]$, and the proof is complete.

The final result given here is not a factorization theorem but gives a representation for a proper orthogonal matrix H (i.e., det $H=1$) in terms of a skew-symmetric matrix. The result is used in Theorem 9.5.2.

THEOREM A9.11. If H is a proper $m\times m$ orthogonal matrix (det $H=1$) then there exists an $m\times m$ skew-symmetric U such that

$$H=\exp(U)\equiv I+U+\frac{1}{2!}U^2+\frac{1}{3!}U^3+\cdots.$$

Proof. Suppose H is a proper orthogonal matrix of odd size, say, $m=2k+1$; it can then be expressed as

$$H=Q'\begin{bmatrix} \cos\theta_1 & -\sin\theta_1 & & & & & & 0 \\ \sin\theta_1 & \cos\theta_1 & & 0 & & & & 0 \\ & & \cos\theta_2 & -\sin\theta_2 & & 0 & & 0 \\ & 0 & \sin\theta_2 & \cos\theta_2 & & & & 0 \\ & & & & \ddots & & & \vdots \\ & 0 & & & & \cos\theta_k & -\sin\theta_k & 0 \\ & & & & & \sin\theta_k & \cos\theta_k & 0 \\ 0 & 0 & 0 & 0 & \cdots & 0 & 0 & 1 \end{bmatrix}Q,$$

where Q is $m\times m$ orthogonal and $-\pi<\theta_i\leq\pi$, with $i=1,\ldots,k$ (see, e.g., Bellman, 1970, p.65). (If $m=2k$, the last row and column are deleted.) Putting

$$\Theta=\begin{bmatrix} 0 & -\theta_1 & & & & & & 0 \\ \theta_1 & 0 & & & 0 & & & 0 \\ & & 0 & -\theta_2 & & & & 0 \\ & & \theta_2 & 0 & & & & 0 \\ & 0 & & & \ddots & & & \vdots \\ & & & & & 0 & -\theta_k & 0 \\ & & & & & \theta_k & 0 & 0 \\ 0 & 0 & 0 & 0 & \cdots & 0 & 0 & 0 \end{bmatrix},$$

we then have $\Theta=\exp(Q'HQ)=\exp(U)$, where $U=Q'HQ$ is skew-symmetric.

Table 9. χ^2 **adjustments to Wilks likelihood ratio statistic** W**: factor** c **for upper percentiles of** $-N \log W$ **(see Section 10.5.3)**[a]

	m = 3									
	r = 3					r = 4				
M \ α	0.100	0.050	0.025	0.010	0.005	0.100	0.050	0.025	0.010	0.005
1	1.322	1.359	1.394	1.437	1.468	1.379	1.422	1.463	1.514	1.550
2	1.127	1.140	1.153	1.168	1.179	1.159	1.174	1.188	1.207	1.220
3	1.071	1.077	1.084	1.092	1.098	1.091	1.099	1.107	1.116	1.123
4	1.045	1.049	1.053	1.058	1.062	1.060	1.065	1.070	1.076	1.080
5	1.032	1.035	1.037	1.041	1.043	1.043	1.046	1.050	1.054	1.057
6	1.023	1.026	1.028	1.030	1.032	1.032	1.035	1.037	1.040	1.042
7	1.018	1.020	1.021	1.023	1.025	1.025	1.027	1.029	1.031	1.033
8	1.014	1.016	1.017	1.018	1.019	1.020	1.022	1.023	1.025	1.026
9	1.012	1.013	1.014	1.015	1.016	1.017	1.018	1.019	1.021	1.022
10	1.010	1.011	1.011	1.012	1.013	1.014	1.015	1.016	1.017	1.018
12	1.007	1.008	1.008	1.009	1.009	1.010	1.011	1.012	1.012	1.013
14	1.005	1.006	1.006	1.007	1.007	1.008	1.008	1.009	1.010	1.010
16	1.004	1.005	1.005	1.005	1.006	1.006	1.007	1.007	1.007	1.008
18	1.003	1.004	1.004	1.004	1.005	1.005	1.005	1.006	1.006	1.006
20	1.003	1.003	1.003	1.004	1.004	1.004	1.004	1.005	1.005	1.005
24	1.002	1.002	1.002	1.002	1.003	1.003	1.003	1.003	1.004	1.004
30	1.001	1.001	1.001	1.002	1.002	1.002	1.002	1.002	1.002	1.002
40	1.001	1.001	1.001	1.001	1.001	1.001	1.001	1.001	1.001	1.001
60	1.000	1.000	1.000	1.000	1.000	1.000	1.001	1.001	1.001	1.001
120	1.000	1.000	1.000	1.000	1.000	1.000	1.000	1.000	1.000	1.000
∞	1.000	1.000	1.000	1.000	1.000	1.000	1.000	1.000	1.000	1.000
χ^2_{rm}	14.6837	16.9190	19.0228	21.6660	23.5894	18.5494	21.0261	23.3367	26.2170	28.2995

[a]Here, M = number of variables; r = hypothesis degrees of freedom; $n-p$ = error degrees of freedom; $M = n - p - m + 1$; and

$$C = \frac{\text{level for } -\left[n-p-\frac{1}{2}(m-r+1)\right]\log W}{\text{level for } \chi^2 \text{ on } mr \text{ degrees of freedom}}$$

Source: Adapted from Schatzoff (1966a), Pillai and Gupta (1969), Lee (1972), and Davis (1979), with the kind permission of the Biometrika Trustees.

Table 9 (*Continued*)

	$m=3$									
	$r=5$					$r=6$				
M \ α	0.100	0.050	0.025	0.010	0.005	0.100	0.050	0.025	0.010	0.005
1	1.433	1.481	1.527	1.584	1.625	1.482	1.535	1.586	1.649	1.694
2	1.191	1.208	1.224	1.245	1.260	1.222	1.241	1.259	1.282	1.298
3	1.113	1.122	1.131	1.142	1.150	1.135	1.145	1.155	1.167	1.176
4	1.076	1.082	1.087	1.094	1.099	1.092	1.099	1.105	1.113	1.119
5	1.055	1.059	1.063	1.068	1.071	1.068	1.072	1.077	1.082	1.086
6	1.042	1.045	1.048	1.051	1.054	1.052	1.056	1.059	1.063	1.066
7	1.033	1.035	1.038	1.040	1.042	1.041	1.044	1.047	1.050	1.052
8	1.027	1.029	1.030	1.033	1.034	1.034	1.036	1.038	1.041	1.042
9	1.022	1.024	1.025	1.027	1.028	1.028	1.030	1.032	1.034	1.035
10	1.019	1.020	1.021	1.023	1.024	1.024	1.025	1.027	1.028	1.030
12	1.014	1.015	1.015	1.017	1.017	1.018	1.019	1.020	1.021	1.022
14	1.010	1.011	1.012	1.013	1.013	1.014	1.014	1.015	1.016	1.017
16	1.008	1.009	1.009	1.010	1.011	1.011	1.012	1.012	1.013	1.014
18	1.007	1.007	1.008	1.008	1.009	1.009	1.009	1.010	1.011	1.011
20	1.006	1.006	1.006	1.007	1.007	1.007	1.008	1.008	1.009	1.009
24	1.004	1.004	1.005	1.005	1.005	1.005	1.006	1.006	1.006	1.007
30	1.003	1.003	1.003	1.003	1.003	1.004	1.004	1.004	1.004	1.004
40	1.002	1.002	1.002	1.002	1.002	1.002	1.002	1.002	1.002	1.003
60	1.001	1.001	1.001	1.001	1.001	1.001	1.001	1.001	1.001	1.001
120	1.000	1.000	1.000	1.000	1.000	1.000	1.000	1.000	1.000	1.000
∞	1.000	1.000	1.000	1.000	1.000	1.000	1.000	1.000	1.000	1.000
χ^2_{rm}	22.3071	24.9958	27.4884	30.5779	32.8013	25.9894	28.8693	31.5264	34.8053	37.1564

	$r=7$					$r=8$				
M \ α	0.100	0.050	0.025	0.010	0.005	0.100	0.050	0.025	0.010	0.005
1	1.529	1.585	1.640	1.708	1.758	1.572	1.632	1.690	1.763	1.816
2	1.251	1.272	1.292	1.317	1.335	1.280	1.302	1.324	1.350	1.370
3	1.156	1.168	1.178	1.192	1.202	1.177	1.190	1.201	1.216	1.227
4	1.109	1.116	1.123	1.132	1.138	1.125	1.133	1.141	1.150	1.157
5	1.081	1.086	1.091	1.097	1.102	1.094	1.100	1.105	1.112	1.117
6	1.063	1.067	1.070	1.075	1.078	1.073	1.078	1.082	1.087	1.091
7	1.050	1.053	1.056	1.060	1.062	1.059	1.063	1.066	1.070	1.073
8	1.041	1.044	1.046	1.049	1.051	1.049	1.052	1.054	1.058	1.060
9	1.034	1.037	1.038	1.041	1.043	1.041	1.043	1.046	1.048	1.050
10	1.029	1.031	1.033	1.035	1.036	1.035	1.037	1.039	1.041	1.043
12	1.022	1.023	1.024	1.026	1.027	1.026	1.028	1.029	1.031	1.032
14	1.017	1.018	1.019	1.020	1.021	1.021	1.022	1.023	1.024	1.025
16	1.014	1.014	1.015	1.016	1.017	1.017	1.018	1.018	1.019	1.020
18	1.011	1.012	1.012	1.013	1.014	1.014	1.014	1.015	1.016	1.017
20	1.009	1.010	1.010	1.011	1.011	1.011	1.012	1.013	1.013	1.014
24	1.007	1.007	1.008	1.008	1.008	1.008	1.009	1.009	1.010	1.01
30	1.005	1.005	1.005	1.005	1.006	1.006	1.006	1.006	1.007	1.007
40	1.003	1.003	1.003	1.003	1.003	1.003	1.004	!.004	1.004	1.004
60	1.001	1.001	1.001	1.001	1.002	1.002	1.002	1.002	1.002	1.002
120	1.000	1.000	1.000	1.000	1.000	1.000	1.000	1.000	1.000	1.001
∞	1.000	1.000	1.000	1.000	1.000	1.000	1.000	1.000	1.000	1.000
χ^2_{rm}	29.6151	32.6706	35.4789	38.9322	41.4011	33.1963	36.4151	39.3641	42.9798	45.5585

Table 9 (*Continued*)

	$m=3$									
	$r=9$					$r=10$				
M \ α	0.100	0.050	0.025	0.010	0.005	0.100	0.050	0.025	0.010	0.005
1	1.612	1.676	1.737	1.814	1.871	1.650	1.716	1.781	1.862	1.921
2	1.307	1.331	1.354	1.382	1.403	1.333	1.359	1.383	1.413	1.435
3	1.198	1.211	1.224	1.240	1.251	1.218	1.232	1.245	1.262	1.274
4	1.141	1.150	1.158	1.169	1.176	1.157	1.167	1.175	1.187	1.195
5	1.107	1.113	1.119	1.127	1.132	1.120	1.127	1.133	1.141	1.147
6	1.084	1.089	1.094	1.099	1.103	1.095	1.101	1.106	1.112	1.116
7	1.068	1.072	1.076	1.080	1.084	1.078	1.082	1.086	1.091	1.094
8	1.057	1.060	1.063	1.067	1.069	1.065	1.068	1.072	1.075	1.078
9	1.048	1.051	1.053	1.056	1.058	1.055	1.058	1.061	1.064	1.066
10	1.041	1.043	1.045	1.048	1.050	1.047	1.050	1.052	1.055	1.057
12	1.031	1.033	1.034	1.036	1.038	1.036	1.038	1.040	1.042	1.043
14	1.025	1.026	1.027	1.028	1.030	1.029	1.030	1.031	1.033	1.034
16	1.020	1.021	1.022	1.023	1.024	1.023	1.024	1.025	1.027	1.028
18	1.016	1.017	1.018	1.019	1.020	1.019	1.020	1.021	1.022	1.023
20	1.014	1.014	1.015	1.016	1.016	1.016	1.017	1.018	1.019	1.019
24	1.010	1.011	1.011	1.012	1.012	1.012	1.012	1.013	1.014	1.014
30	1.007	1.007	1.008	1.008	1.008	1.008	1.009	1.009	1.009	1.010
40	1.004	1.004	1.004	1.005	1.005	1.005	1.005	1.005	1.006	1.006
60	1.002	1.002	1.002	1.002	1.002	1.002	1.002	1.003	1.003	1.003
120	1.001	1.001	1.001	1.001	1.001	1.001	1.001	1.001	1.001	1.001
∞	1.000	1.000	1.000	1.000	1.000	1.000	1.000	1.000	1.000	1.000
χ^2_{rm}	36.7412	40.1133	43.1945	46.9629	49.6449	40.2560	43.7730	46.9792	50.8922	53.6720

	$r=11$					$r=12$				
M \ α	0.100	0.050	0.025	0.010	0.005	0.100	0.050	0.025	0.010	0.005
1	1.685	1.754	1.821	1.907	1.969	1.718	1.791	1.860	1.949	2.013
2	1.358	1.385	1.410	1.442	1.466	1.382	1.410	1.437	1.470	1.495
3	1.237	1.252	1.266	1.284	1.297	1.256	1.272	1.287	1.306	1.319
4	1.173	1.183	1.192	1.204	1.213	1.188	1.199	1.209	1.221	1.230
5	1.133	1.140	1.147	1.156	1.162	1.146	1.154	1.161	1.170	1.176
6	1.106	1.112	1.117	1.124	1.128	1.117	1.123	1.129	1.136	1.141
7	1.087	1.092	1.096	1.101	1.105	1.097	1.101	1.106	1.111	1.115
8	1.073	1.077	1.080	1.084	1.087	1.081	1.085	1.089	1.093	1.097
9	1.062	1.065	1.068	1.072	1.074	1.069	1.073	1.076	1.080	1.082
10	1.054	1.056	1.059	1.062	1.064	1.060	1.063	1.066	1.069	1.071
12	1.041	1.043	1.045	1.047	1.049	1.046	1.048	1.050	1.053	1.054
14	1.033	1.034	1.036	1.037	1.039	1.037	1.039	1.040	1.042	1.043
16	1.027	1.028	1.029	1.030	1.031	1.030	1.032	1.033	1.034	1.035
18	1.022	1.023	1.024	1.025	1.026	1.025	1.026	1.027	1.029	1.029
20	1.019	1.020	1.020	1.021	1.022	1.021	1.022	1.023	1.024	1.025
24	1.014	1.014	1.015	1.016	1.016	1.016	1.017	1.017	1.018	1.019
30	1.009	1.010	1.010	1.011	1.011	1.011	1.011	1.012	1.012	1.013
40	1.006	1.006	1.006	1.007	1.007	1.007	1.007	1.007	1.008	1.008
60	1.003	1.003	1.003	1.003	1.003	1.003	1.003	1.004	1.004	1.004
120	1.001	1.001	1.001	1.001	1.001	1.001	1.001	1.001	1.001	1.001
∞	1.000	1.000	1.000	1.000	1.000	1.000	1.000	1.000	1.000	1.000
χ^2_{rm}	43.745	47.400	50.725	54.776	57.648	47.2122	50.9985	54.4373	58.6192	61.5812

Table 9 (*Continued*)

	$m=3$									
	$r=13$					$r=14$				
M \ α	0.100	0.050	0.025	0.010	0.005	0.100	0.050	0.025	0.010	0.005
1	1.750	1.824	1.896	1.988	2.055	1.780	1.857	1.931	2.026	2.095
2	1.405	1.434	1.462	1.497	1.522	1.427	1.458	1.486	1.523	1.549
3	1.274	1.291	1.306	1.326	1.340	1.292	1.309	1.326	1.346	1.361
4	1.203	1.214	1.225	1.238	1.247	1.217	1.229	1.240	1.254	1.264
5	1.158	1.167	1.174	1.184	1.191	1.171	1.179	1.188	1.198	1.205
6	1.128	1.134	1.140	1.148	1.153	1.138	1.145	1.152	1.159	1.165
7	1.106	1.111	1.116	1.122	1.126	1.115	1.121	1.126	1.132	1.136
8	1.089	1.094	1.098	1.102	1.106	1.097	1.102	1.106	1.111	1.115
9	1.076	1.080	1.083	1.088	1.090	1.084	1.088	1.091	1.095	1.099
10	1.066	1.069	1.072	1.076	1.078	1.073	1.076	1.079	1.082	1.085
12	1.052	1.054	1.056	1.059	1.061	1.057	1.059	1.061	1.064	1.066
14	1.041	1.043	1.045	1.047	1.048	1.046	1.048	1.049	1.052	1.053
16	1.034	1.035	1.037	1.038	1.040	1.037	1.039	1.041	1.042	1.044
18	1.028	1.029	1.031	1.032	1.033	1.031	1.033	1.034	1.035	1.036
20	1.024	1.025	1.026	1.027	1.028	1.027	1.028	1.029	1.030	1.031
24	1.018	1.019	1.019	1.020	1.021	1.020	1.021	1.022	1.023	1.023
30	1.012	1.013	1.013	1.014	1.014	1.014	1.015	1.015	1.016	1.016
40	1.008	1.008	1.008	1.009	1.009	1.009	1.009	1.009	1.010	1.010
60	1.004	1.004	1.004	1.004	1.004	1.004	1.004	1.005	1.005	1.005
120	1.001	1.001	1.001	1.001	1.001	1.001	1.001	1.001	1.001	1.001
∞	1.000	1.000	1.000	1.000	1.000	1.000	1.000	1.000	1.000	1.000
χ^2_{rm}	50.660	54.572	58.120	62.428	65.476	54.0902	58.1240	61.7768	66.2062	69.3360

	$r=15$					$r=16$				
M \ α	0.100	0.050	0.050	0.010	0.005	0.100	0.050	0.025	0.010	0.005
1	1.808	1.887	1.964	2.061	2.133	1.335	1.916	1.995	2.095	2.196
2	1.449	1.480	1.510	1.547	1.575	1.469	1.501	1.532	1.571	1.599
3	1.309	1.327	1.344	1.365	1.381	1.325	1.344	1.362	1.384	1.400
4	1.232	1.244	1.256	1.270	1.280	1.245	1.258	1.271	1.285	1.296
5	1.183	1.192	1.200	1.211	1.218	1.195	1.204	1.213	1.224	1.232
6	1.149	1.156	1.163	1.171	1.177	1.159	1.167	1.174	1.182	1.188
7	1.124	1.130	1.135	1.142	1.147	1.133	1.139	1.145	1.152	1.157
8	1.105	1.110	1.115	1.120	1.124	1.114	1.119	1.123	1.129	1.133
9	1.091	1.095	1.099	1.103	1.107	1.098	1.102	1.106	1.111	1.115
10	1.079	1.083	1.086	1.090	1.093	1.085	1.089	1.092	1.097	1.099
12	1.062	1.065	1.067	1.070	1.072	1.067	1.070	1.073	1.076	1.078
14	1.050	1.052	1.054	1.056	1.058	1.054	1.057	1.059	1.061	1.063
16	1.041	1.043	1.045	1.047	1.048	1.045	1.047	1.049	1.051	1.052
18	1.035	1.036	1.037	1.039	1.040	1.038	1.039	1.041	1.043	1.044
20	1.030	1.031	1.032	1.033	1.034	1.032	1.034	1.035	1.036	1.037
24	1.022	1.023	1.024	1.025	1.026	1.025	1.026	1.026	1.027	1.028
30	1.016	1.016	1.017	1.017	1.018	1.017	1.018	1.018	1.019	1.020
40	1.010	1.010	1.010	1.011	1.011	1.011	1.011	1.011	1.012	1.012
60	1.005	1.005	1.005	1.005	1.005	1.005	1.006	1.006	1.006	1.006
120	1.001	1.001	1.001	1.001	1.002	1.002	1.002	1.002	1.002	1.002
∞	1.000	1.000	1.000	1.000	1.000	1.000	1.000	1.000	1.000	1.000
χ^2_{rm}	57.505	61.656	65.410	69.957	73.166	60.9066	65.1708	69.0226	73.6826	76.9688

Table 9 (*Continued*)

	$m=3$									
	$r=17$					$r=18$				
M \ α	0.100	0.050	0.025	0.010	0.005	0.100	0.050	0.025	0.010	0.005
1	1.861	1.944	2.025	2.127	2.203	1.886	1.971	2.053	2.158	2.235
2	1.489	1.522	1.554	1.594	1.623	1.508	1.542	1.575	1.616	1.646
3	1.341	1.361	1.379	1.402	1.419	1.357	1.377	1.396	1.420	1.437
4	1.259	1.273	1.285	1.300	1.312	1.272	1.286	1.299	1.315	1.327
5	1.206	1.216	1.225	1.237	1.245	1.218	1.228	1.238	1.249	1.258
6	1.169	1.177	1.184	1.193	1.200	1.179	1.188	1.195	1.204	1.211
7	1.142	1.149	1.154	1.162	1.167	1.151	1.158	1.164	1.171	1.177
8	1.122	1.127	1.132	1.138	1.142	1.129	1.135	1.140	1.146	1.151
9	1.105	1.110	1.114	1.119	1.123	1.112	1.117	1.121	1.127	1.130
10	1.092	1.096	1.100	1.104	1.107	1.099	1.103	1.107	1.111	1.114
12	1.073	1.076	1.079	1.082	1.084	1.078	1.081	1.084	1.087	1.090
14	1.059	1.061	1.064	1.066	1.068	1.064	1.066	1.068	1.071	1.073
16	1.049	1.051	1.053	1.055	1.056	1.053	1.055	1.057	1.059	1.061
18	1.041	1.043	1.044	1.046	1.047	1.045	1.046	1.048	1.050	1.051
20	1.035	1.037	1.038	1.040	1.041	1.038	1.040	1.041	1.043	1.044
24	1.027	1.028	1.029	1.030	1.031	1.029	1.030	1.031	1.032	1.033
30	1.019	1.020	1.020	1.021	1.022	1.021	1.021	1.022	1.023	1.023
40	1.012	1.012	1.013	1.013	1.013	1.013	1.013	1.014	1.014	1.015
60	1.006	1.006	1.006	1.006	1.007	1.006	1.007	1.007	1.007	1.007
120	1.002	1.002	1.002	1.002	1.002	1.002	1.002	1.002	1.002	1.002
∞	1.000	1.000	1.000	1.000	1.000	1.000	1.000	1.000	1.000	1.000
χ^2_{rm}	64.295	68.669	72.616	77.386	80.747	67.6728	72.1532	76.1920	81.0688	84.5019

	r = 19					r = 20				
$M \backslash \alpha$	0.100	0.050	0.025	0.010	0.005	0.100	0.050	0.025	0.010	0.005
1	1.909	1.996	2.080	2.188	2.267	1.932	2.021	2.106	2.216	2.297
2	1.526	1.561	1.595	1.637	1.668	1.544	1.580	1.614	1.657	1.689
3	1.372	1.393	1.412	1.437	1.454	1.387	1.408	1.428	1.453	1.472
4	1.285	1.300	1.313	1.330	1.341	1.298	1.313	1.327	1.344	1.356
5	1.229	1.240	1.250	1.262	1.271	1.240	1.251	1.261	1.274	1.283
6	1.189	1.198	1.205	1.215	1.222	1.199	1.208	1.216	1.226	1.233
7	1.160	1.167	1.173	1.181	1.186	1.168	1.176	1.182	1.190	1.196
8	1.137	1.143	1.148	1.155	1.159	1.145	1.151	1.157	1.163	1.168
9	1.119	1.124	1.129	1.134	1.138	1.127	1.132	1.136	1.142	1.146
10	1.105	1.109	1.113	1.118	1.121	1.112	1.116	1.120	1.125	1.128
12	1.084	1.087	1.090	1.093	1.096	1.089	1.092	1.095	1.099	1.102
14	1.068	1.071	1.073	1.076	1.078	1.073	1.075	1.078	1.081	1.083
16	1.057	1.059	1.061	1.063	1.065	1.061	1.063	1.065	1.067	1.069
18	1.048	1.050	1.052	1.054	1.055	1.052	1.053	1.055	1.057	1.059
20	1.041	1.043	1.044	1.046	1.047	1.044	1.046	1.048	1.049	1.050
24	1.032	1.033	1.034	1.035	1.036	1.034	1.035	1.036	1.038	1.039
30	1.022	1.023	1.024	1.025	1.025	1.024	1.025	1.026	1.027	1.027
40	1.014	1.015	1.015	1.016	1.016	1.015	1.016	1.016	1.017	1.017
60	1.007	1.007	1.008	1.008	1.008	1.008	1.008	1.008	1.009	1.009
120	1.002	1.002	1.002	1.002	1.002	1.002	1.002	1.002	1.002	1.003
∞	1.000	1.000	1.000	1.000	1.000	1.000	1.000	1.000	1.000	1.000
χ^2_{rm}	71.040	75.624	79.752	84.733	88.236	74.3970	79.0819	83.2976	88.3794	91.9517

Table 9 (*Continued*)

	$m=3$									
	$r=21$					$r=22$				
M \ α	0.100	0.050	0.025	0.010	0.005	0.100	0.050	0.025	0.010	0.005
1	1.954	2.044	2.131	2.243	2.325	1.975	2.067	2.156	2.269	2.353
2	1.561	1.598	1.633	1.677	1.709	1.578	1.616	1.651	1.696	1.729
3	1.401	1.423	1.444	1.470	1.488	1.415	1.438	1.459	1.485	1.514
4	1.310	1.325	1.340	1.357	1.370	1.322	1.338	1.353	1.371	1.384
5	1.250	1.262	1.273	1.286	1.295	1.261	1.273	1.284	1.297	1.307
6	1.208	1.217	1.226	1.236	1.243	1.218	1.227	1.236	1.246	1.254
7	1.177	1.184	1.191	1.200	1.205	1.185	1.193	1.200	1.209	1.213
8	1.153	1.159	1.165	1.172	1.176	1.160	1.167	1.173	1.180	1.183
9	1.133	1.139	1.144	1.150	1.154	1.142	1.147	1.151	1.157	1.161
10	1.118	1.122	1.127	1.132	1.135	1.124	1.129	1.133	1.139	1.141
12	1.094	1.098	1.101	1.105	1.108	1.099	1.103	1.106	1.110	1.115
14	1.077	1.080	1.083	1.086	1.088	1.082	1.085	1.087	1.091	1.093
16	1.065	1.067	1.069	1.072	1.074	1.069	1.071	1.073	1.076	1.078
18	1.055	1.057	1.059	1.061	1.063	1.059	1.061	1.063	1.065	1.065
20	1.048	1.049	1.051	1.053	1.054	1.051	1.052	1.054	1.056	1.057
24	1.036	1.038	1.039	1.040	1.041	1.039	1.040	1.041	1.043	1.044
30	1.026	1.027	1.028	1.029	1.029	1.028	1.029	1.030	1.031	1.031
40	1.016	1.017	1.018	1.018	1.019	1.018	1.018	1.019	1.020	1.020
60	1.008	1.009	1.009	1.009	1.009	1.009	1.009	1.010	1.010	1.010
120	1.002	1.002	1.003	1.003	1.003	1.003	1.003	1.003	1.003	1.000
∞	1.000	1.000	1.000	1.000	1.000	1.000	1.000	1.000	1.000	1.000
χ^2_{rm}	77.745	82.529	86.830	92.010	95.649	81.0855	85.9649	90.3489	95.6257	99.336

	$m=4$									
			$r=4$					$r=5$		
M \ α	0.100	0.050	0.025	0.010	0.005	0.100	0.050	0.025	0.010	0.005
1	1.405	1.451	1.494	1.550	1.589	1.435	1.483	1.530	1.589	1.632
2	1.178	1.194	1.209	1.229	1.243	1.199	1.216	1.233	1.253	1.269
3	1.105	1.114	1.122	1.132	1.139	1.121	1.130	1.139	1.150	1.158
4	1.071	1.076	1.081	1.088	1.092	1.083	1.089	1.094	1.101	1.106
5	1.051	1.055	1.058	1.063	1.066	1.061	1.065	1.069	1.074	1.077
6	1.039	1.042	1.044	1.048	1.050	1.047	1.050	1.053	1.056	1.059
7	1.031	1.033	1.035	1.037	1.039	1.037	1.040	1.042	1.044	1.046
8	1.025	1.027	1.028	1.030	1.032	1.030	1.032	1.034	1.036	1.038
9	1.020	1.022	1.023	1.025	1.026	1.025	1.027	1.028	1.030	1.031
10	1.017	1.018	1.019	1.021	1.022	1.021	1.023	1.024	1.025	1.026
12	1.013	1.014	1.014	1.015	1.016	1.016	1.017	1.018	1.019	1.020
14	1.010	1.010	1.011	1.012	1.012	1.012	1.013	1.014	1.014	1.015
16	1.008	1.008	1.009	1.009	1.010	1.010	1.010	1.011	1.012	1.012
18	1.006	1.007	1.007	1.008	1.008	1.008	1.008	1.009	1.009	1.010
20	1.005	1.006	1.006	1.006	1.007	1.007	1.007	1.007	1.008	1.008
24	1.004	1.004	1.004	1.004	1.005	1.005	1.005	1.005	1.006	1.006
30	1.002	1.003	1.003	1.003	1.003	1.003	1.003	1.004	1.004	1.004
40	1.001	1.002	1.002	1.002	1.002	1.002	1.002	1.002	1.002	1.002
60	1.001	1.001	1.001	1.001	1.001	1.001	1.001	1.001	1.001	1.001
120	1.000	1.000	1.000	1.000	1.000	1.000	1.000	1.000	1.000	1.000
∞	1.000	1.000	1.000	1.000	1.000	1.000	1.000	1.000	1.000	1.000
χ^2_{rm}	23.5418	26.2962	28.8454	31.9999	34.2672	28.4120	31.4104	34.1696	37.5662	29.9968

Table 9 (*Continued*)

	$m=4$									
			$r=6$					$r=7$		
M \ α	0.100	0.050	0.025	0.010	0.005	0.100	0.050	0.025	0.010	0.005
1	1.466	1.517	1.566	1.628	1.674	1.497	1.550	1.601	1.667	1.715
2	1.222	1.240	1.257	1.279	1.295	1.244	1.263	1.281	1.305	1.322
3	1.138	1.148	1.157	1.168	1.177	1.155	1.165	1.175	1.188	1.197
4	1.096	1.102	1.108	1.115	1.121	1.109	1.116	1.122	1.130	1.136
5	1.071	1.076	1.080	1.085	1.089	1.082	1.087	1.092	1.097	1.101
6	1.055	1.059	1.062	1.066	1.068	1.064	1.068	1.071	1.076	1.079
7	1.044	1.047	1.049	1.052	1.055	1.052	1.055	1.057	1.061	1.063
8	1.036	1.038	1.040	1.043	1.045	1.043	1.045	1.047	1.050	1.052
9	1.030	1.032	1.034	1.036	1.037	1.036	1.038	1.040	1.042	1.044
10	1.026	1.027	1.029	1.030	1.032	1.031	1.032	1.034	1.036	1.037
12	1.019	1.020	1.021	1.023	1.024	1.023	1.024	1.026	1.027	1.028
14	1.015	1.016	1.017	1.018	1.018	1.018	1.019	1.020	1.021	1.022
16	1.012	1.013	1.013	1.014	1.015	1.015	1.015	1.016	1.017	1.017
18	1.010	1.010	1.011	1.012	1.012	1.012	1.013	1.013	1.014	1.014
20	1.008	1.009	1.009	1.010	1.010	1.010	1.011	1.011	1.012	1.012
24	1.006	1.006	1.007	1.007	1.007	1.007	1.008	1.008	1.008	1.009
30	1.004	1.004	1.004	1.005	1.005	1.005	1.005	1.005	1.006	1.006
40	1.002	1.002	1.003	1.003	1.003	1.003	1.003	1.003	1.003	1.004
60	1.001	1.001	1.001	1.001	1.001	1.001	1.001	1.002	1.002	1.002
120	1.000	1.000	1.000	1.000	1.000	1.000	1.000	1.000	1.000	1.000
∞	1.000	1.000	1.000	1.000	1.000	1.000	1.000	1.000	1.000	1.000
χ^2_{rm}	33.1963	36.4151	39.3641	42.9798	45.5585	37.9159	41.3372	44.4607	48.2782	50.9933

	$r=8$					$r=9$				
M \ α	0.100	1.050	1.025	0.010	0.005	0.100	0.050	0.025	0.010	0.005
1	1.528	1.583	1.636	1.704	1.754	1.557	1.614	1.669	1.740	1.792
2	1.266	1.286	1.305	1.330	1.348	1.288	1.309	1.329	1.355	1.373
3	1.172	1.183	1.193	1.207	1.216	1.189	1.201	1.212	1.226	1.236
4	1.123	1.130	1.137	1.146	1.152	1.137	1.144	1.152	1.161	1.167
5	1.093	1.099	1.103	1.109	1.114	1.105	1.110	1.115	1.122	1.127
6	1.074	1.078	1.081	1.086	1.089	1.083	1.088	1.091	1.096	1.100
7	1.060	1.063	1.066	1.070	1.072	1.068	1.071	1.075	1.078	1.081
8	1.050	1.052	1.055	1.058	1.060	1.057	1.060	1.062	1.065	1.068
9	1.042	1.044	1.046	1.048	1.050	1.048	1.050	1.053	1.055	1.057
10	1.036	1.038	1.039	1.041	1.043	1.041	.043	1.045	1.047	1.049
12	1.027	1.029	1.030	1.031	1.033	1.032	1.033	1.034	1.036	1.037
14	1.021	1.023	1.023	1.025	1.026	1.025	1.026	1.027	1.029	1.029
16	1.107	1.108	1.109	1.020	1.021	1.020	1.021	1.022	1.023	1.024
18	1.014	1.015	1.016	1.016	1.017	1.017	1.018	1.018	1.019	1.020
20	1.012	1.013	1.013	1.014	1.014	1.014	1.015	1.015	1.016	1.017
24	1.009	1.009	1.010	1.010	1.010	1.010	1.011	1.011	1.012	1.012
30	1.006	1.006	1.007	1.007	1.007	1.007	1.007	1.008	1.008	1.008
40	1.003	1.004	1.004	1.004	1.004	1.004	1.004	1.005	1.005	1.005
60	1.002	1.002	1.002	1.002	1.002	1.002	1.002	1.002	1.002	1.002
120	1.000	1.000	1.000	1.001	1.001	1.001	1.001	1.001	1.001	1.001
∞	1.000	1.000	1.000	1.000	1.000	1.000	1.000	1.000	1.000	1.000
χ^2_{rm}	42.5847	46.1943	49.4804	53.4858	56.3281	47.2122	50.9985	54.4373	58.6192	61.5812

Table 9 (*Continued*)

	$m=4$									
	$r=10$					$r=11$				
α / M	0.100	0.050	0.025	0.010	0.005	0.100	0.050	0.025	0.010	0.005
1	1.585	1.644	1.701	1.774	1.828	—	—	—	—	—
2	1.309	1.331	1.352	1.379	1.398	1.330	1.352	1.374	1.402	1.422
3	1.206	1.218	1.230	1.244	1.255	1.222	1.235	1.247	1.262	1.274
4	1.150	1.159	1.166	1.176	1.183	1.164	1.173	1.181	1.191	1.198
5	1.116	1.122	1.128	1.134	1.139	1.127	1.134	1.140	1.147	1.152
6	1.093	1.097	1.102	1.107	1.111	1.103	1.107	1.112	1.118	1.122
7	1.076	1.080	1.083	1.088	1.090	1.085	1.089	1.092	1.097	1.100
8	1.064	1.067	1.070	1.073	1.076	1.071	1.075	1.077	1.081	1.084
9	1.054	1.057	1.059	1.062	1.064	1.061	1.064	1.066	1.069	1.071
10	1.047	1.049	1.051	1.054	1.055	1.053	1.055	1.057	1.060	1.062
12	1.036	1.038	1.039	1.041	1.042	1.041	1.043	1.044	1.046	1.047
14	1.029	1.030	1.031	1.033	1.034	1.033	1.034	1.035	1.037	1.038
16	1.023	1.024	1.025	1.026	1.027	1.027	1.028	1.029	1.030	1.031
18	1.019	1.020	1.021	1.022	1.023	1.022	1.023	1.024	1.025	1.026
20	1.016	1.017	1.018	1.019	1.019	1.019	1.020	1.020	1.021	1.022
24	1.012	1.013	1.013	1.014	1.014	1.014	1.015	1.015	1.016	1.016
30	1.008	1.009	1.009	1.009	1.010	1.010	1.010	1.010	1.011	1.011
40	1.005	1.005	1.005	1.006	1.006	1.006	1.006	1.006	1.007	1.007
60	1.002	1.003	1.003	1.003	1.003	1.003	1.003	1.003	1.003	1.003
120	1.001	1.001	1.001	1.001	1.001	1.001	1.001	1.001	1.001	1.001
∞	1.000	1.000	1.000	1.000	1.000	1.000	1.000	1.000	1.000	1.000
χ^2_{rm}	51.8050	55.7585	59.3417	63.6907	66.7659	56.369	60.481	64.201	68.710	71.893

M \ α	$r=12$ 0.100	0.050	0.025	0.010	0.005	$r=13$ 0.100	0.050	0.025	0.010	0.005
1	1.638	1.700	1.760	1.838	1.895	—	—	—	—	—
2	1.350	1.373	1.396	1.424	1.446	1.369	1.393	1.417	1.446	1.468
3	1.238	1.252	1.264	1.280	1.292	1.254	1.268	1.281	1.298	1.310
4	1.177	1.186	1.195	1.205	1.213	1.190	1.200	1.209	1.220	1.228
5	1.139	1.145	1.152	1.159	1.165	1.150	1.157	1.163	1.171	1.177
6	1.112	1.118	1.122	1.128	1.132	1.122	1.127	1.132	1.139	1.143
7	1.093	1.097	1.101	1.106	1.109	1.102	1.106	1.110	1.115	1.118
8	1.079	1.082	1.085	1.089	1.092	1.086	1.090	1.093	1.097	1.100
9	1.068	1.070	1.073	1.076	1.079	1.074	1.077	1.080	1.083	1.086
10	1.059	1.061	1.063	1.066	1.068	1.065	1.067	1.070	1.073	1.075
12	1.046	1.047	1.049	1.051	1.053	1.050	1.052	1.054	1.056	1.058
14	1.037	1.038	1.039	1.041	1.042	1.041	1.042	1.044	1.045	1.047
16	1.030	1.031	1.032	1.033	1.034	1.033	1.035	1.036	1.037	1.038
18	1.025	1.026	1.027	1.028	1.029	1.028	1.029	1.030	1.031	1.032
20	1.021	1.022	1.023	1.024	1.024	1.024	1.025	1.026	1.027	1.027
24	1.016	1.017	1.017	1.018	1.018	1.018	1.019	1.019	1.020	1.020
30	1.011	1.011	1.012	1.012	1.013	1.012	1.013	1.013	1.014	1.014
40	1.007	1.007	1.007	1.008	1.008	1.008	1.008	1.008	1.008	1.009
60	1.003	1.003	1.004	1.004	1.004	1.004	1.004	1.004	1.004	1.004
120	1.001	1.001	1.001	1.001	1.001	1.001	1.001	1.001	1.001	1.001
∞	1.000	1.000	1.000	1.000	1.000	1.000	1.000	1.000	1.000	1.000
χ^2_{rm}	60.9066	65.1708	69.0226	73.6826	76.9688	65.422	69.832	73.810	78.616	82.001

Table 9 (*Continued*)

	$m=4$									
	$r=14$					$r=15$				
M \ α	0.100	0.050	0.025	0.010	0.005	0.100	0.050	0.025	0.010	0.005
1	1.686	1.751	1.814	1.896	1.956	—	—	—	—	—
2	1.388	1.413	1.436	1.467	1.489	1.406	1.432	1.456	1.488	1.511
3	1.269	1.284	1.297	1.314	1.327	1.284	1.299	1.313	1.331	1.344
4	1.203	1.213	1.222	1.234	1.242	1.216	1.226	1.236	1.248	1.256
5	1.161	1.168	1.175	1.183	1.189	1.172	1.179	1.187	1.195	1.202
6	1.131	1.137	1.142	1.149	1.154	1.141	1.147	1.153	1.159	1.164
7	1.110	1.115	1.119	1.124	1.128	1.118	1.123	1.128	1.133	1.137
8	1.094	1.097	1.101	1.105	1.109	1.101	1.105	1.109	1.113	1.116
9	1.081	1.084	1.087	1.091	1.093	1.087	1.091	1.094	1.098	1.101
10	1.071	1.073	1.076	1.079	1.081	1.077	1.080	1.082	1.085	1.088
12	1.055	1.058	1.059	1.062	1.064	1.060	1.063	1.065	1.067	1.069
14	1.045	1.046	1.048	1.050	1.051	1.049	1.051	1.052	1.054	1.056
16	1.037	1.038	1.039	1.041	1.042	1.040	1.042	1.043	1.045	1.046
18	1.031	1.032	1.033	1.034	1.035	1.034	1.035	1.036	1.038	1.039
20	1.026	1.027	1.028	1.029	1.030	1.029	1.030	1.031	1.032	1.033
24	1.020	1.021	1.021	1.022	1.023	1.022	1.023	1.023	1.024	1.025
30	1.014	1.014	1.015	1.015	1.016	1.015	1.016	1.016	1.017	1.017
40	1.009	1.009	1.009	1.009	1.010	1.010	1.010	1.010	1.011	1.011
60	1.004	1.004	1.005	1.005	1.005	1.005	1.005	1.005	1.005	1.005
120	1.001	1.001	1.001	1.001	1.001	1.001	1.001	1.001	1.001	1.001
∞	1.000	1.000	1.000	1.000	1.000	1.000	1.000	1.000	1.000	1.000
χ^2_{rm}	69.9185	74.4683	78.5671	83.5134	86.9937	74.397	79.082	83.298	88.379	91.952

M \ α	$r=16$ 0.100	0.050	0.025	0.010	0.005	$r=17$ 0.100	0.050	0.025	0.010	0.005
1	1.731	1.799	1.864	1.949	2.012	—	—	—	—	—
2	1.423	1.450	1.475	1.507	1.531	1.440	1.468	1.494	1.527	1.551
3	1.299	1.314	1.329	1.347	1.360	1.313	1.329	1.344	1.363	1.377
4	1.223	1.239	1.249	1.261	1.270	1.240	1.252	1.262	1.275	1.284
5	1.182	1.190	1.198	1.207	1.213	1.193	1.201	1.209	1.218	1.225
6	1.150	1.157	1.163	1.169	1.174	1.160	1.166	1.172	1.180	1.185
7	1.127	1.132	1.136	1.142	1.146	1.135	1.140	1.145	1.151	1.155
8	1.108	1.113	1.117	1.121	1.125	1.116	1.120	1.124	1.129	1.133
9	1.094	1.098	1.101	1.105	1.108	1.101	1.105	1.108	1.112	1.115
10	1.083	1.086	1.089	1.092	1.094	1.089	1.092	1.095	1.098	1.101
12	1.065	1.068	1.070	1.073	1.074	1.070	1.073	1.075	1.078	1.080
14	1.053	1.055	1.056	1.058	1.060	1.057	1.059	1.061	1.063	1.065
16	1.044	1.045	1.047	1.049	1.050	1.048	1.049	1.051	1.053	1.054
18	1.037	1.038	1.040	1.041	1.042	1.040	1.042	1.043	1.045	1.046
20	1.032	1.033	1.034	1.035	1.036	1.035	1.036	1.037	1.038	1.039
24	1.024	1.025	1.026	1.027	1.027	1.026	1.027	1.028	1.029	1.030
30	1.017	1.018	1.018	1.019	1.019	1.019	1.019	1.020	1.020	1.021
40	1.011	1.011	1.011	1.012	1.012	1.012	1.012	1.012	1.013	1.013
60	1.005	1.005	1.006	1.006	1.006	1.006	1.006	1.006	1.006	1.007
120	1.001	1.002	1.002	1.002	1.002	1.002	1.002	1.002	1.002	1.002
∞	1.000	1.000	1.000	1.000	1.000	1.000	1.000	1.000	1.000	1.000
χ^2_{rm}	78.8597	83.6753	88.0040	93.2168	96.8781	83.308	88.250	92.689	98.028	101.776

Table 9 (*Continued*)

	$m=4$									
	$r=18$					$r=19$				
M \ α	0.100	0.050	0.025	0.010	0.005	0.100	0.050	0.025	0.010	0.005
1	1.773	1.843	1.911	1.999	1.065	—	—	—	—	—
2	1.457	1.485	1.511	1.545	1.570	1.473	1.502	1.529	1.563	1.588
3	1.327	1.343	1.359	1.378	1.392	1.340	1.357	1.373	1.393	1.408
4	1.252	1.264	1.274	1.287	1.297	1.264	1.276	1.287	1.300	1.310
5	1.203	1.212	1.220	1.230	1.237	1.214	1.223	1.231	1.241	1.248
6	1.169	1.176	1.182	1.189	1.195	1.178	1.185	1.191	1.199	1.205
7	1.143	1.149	1.154	1.160	1.164	1.151	1.157	1.162	1.169	1.173
8	1.123	1.128	1.132	1.137	1.141	1.130	1.135	1.140	1.145	1.149
9	1.107	1.111	1.115	1.119	1.122	1.114	1.118	1.122	1.126	1.130
10	1.095	1.098	1.101	1.105	1.108	1.101	1.104	1.107	1.111	1.114
12	1.075	1.078	1.080	1.083	1.085	1.080	1.083	1.086	1.089	1.091
14	1.061	1.063	1.065	1.068	1.069	1.066	1.068	1.070	1.073	1.074
16	1.051	1.053	1.054	1.056	1.058	1.055	1.057	1.059	1.061	1.062
18	1.044	1.045	1.046	1.048	1.049	1.047	1.048	1.050	1.051	1.053
20	1.037	1.039	1.040	1.041	1.042	1.040	1.042	1.043	1.044	1.045
24	1.029	1.030	1.030	1.031	1.032	1.031	1.032	1.033	1.034	1.035
30	1.020	1.021	1.022	1.022	1.023	1.022	1.023	1.023	1.024	1.025
40	1.013	1.013	1.014	1.014	1.014	1.014	1.014	1.015	1.015	1.015
60	1.006	1.007	1.007	1.007	1.007	1.007	1.007	1.007	1.008	1.008
120	1.002	1.002	1.002	1.002	1.002	1.002	1.002	1.002	1.002	1.002
∞	1.000	1.000	1.000	1.000	1.000	1.000	1.000	1.000	1.000	1.000
χ^2_{rm}	87.7431	92.8083	97.3531	102.816	106.648	92.166	97.351	101.999	107.583	111.495

	r = 20					r = 21				
M \ α	0.100	0.050	0.025	0.010	0.005	0.100	0.050	0.025	0.010	0.005
1	1.812	1.884	1.954	2.045	2.113	—	—	—	—	—
2	1.488	1.518	1.545	1.580	1.606	1.504	1.533	1.562	1.598	1.624
3	1.353	1.371	1.387	1.408	1.422	1.367	1.384	1.401	1.422	1.437
4	1.275	1.288	1.299	1.313	1.323	1.287	1.299	1.311	1.325	1.335
5	1.224	1.233	1.241	1.252	1.259	1.234	1.243	1.252	1.262	1.270
6	1.187	1.194	1.201	1.208	1.215	1.196	1.203	1.210	1.218	1.224
7	1.159	1.165	1.170	1.177	1.182	1.167	1.173	1.179	1.186	1.190
8	1.138	1.143	1.147	1.153	1.157	1.145	1.150	1.155	1.160	1.164
9	1.121	1.125	1.129	1.133	1.137	1.127	1.132	1.136	1.140	1.144
10	1.107	1.110	1.114	1.118	1.121	1.113	1.116	1.120	1.124	1.127
12	1.086	1.088	1.091	1.094	1.096	1.091	1.094	1.096	1.099	1.102
14	1.070	1.072	1.074	1.077	1.078	1.075	1.077	1.079	1.082	1.084
16	1.059	1.061	1.062	1.064	1.066	1.063	1.065	1.066	1.069	1.070
18	1.050	1.052	1.053	1.055	1.056	1.054	1.055	1.057	1.059	1.060
20	1.043	1.045	1.046	1.047	1.048	1.046	1.048	1.049	1.051	1.052
24	1.033	1.034	1.035	1.036	1.037	1.036	1.037	1.038	1.039	1.040
30	1.024	1.024	1.025	1.026	1.026	1.025	1.026	1.027	1.028	1.028
40	1.015	1.016	1.016	1.016	1.017	1.016	1.017	1.017	1.018	1.018
60	1.008	1.008	1.008	1.008	1.008	1.008	1.008	1.009	1.009	1.009
120	1.002	1.002	1.002	1.002	1.002	1.002	1.002	1.003	1.003	1.003
∞	1.000	1.000	1.000	1.000	1.000	1.000	1.000	1.000	1.000	1.000
χ^2_{rm}	96.5782	101.879	106.629	112.329	116.321	100.980	106.395	111.242	117.057	121.126

Table 9 (*Continued*)

			$m=4$		
			$r=22$		
M \ α	0.100	0.050	0.025	0.010	0.005
1	1.848	1.922	1.994	2.088	2.158
2	1.518	1.549	1.577	1.614	1.641
3	1.379	1.397	1.414	1.436	1.451
4	1.298	1.310	1.322	1.337	1.347
5	1.243	1.253	1.262	1.273	1.281
6	1.204	1.212	1.219	1.228	1.234
7	1.175	1.181	1.187	1.194	1.199
8	1.152	1.157	1.162	1.168	1.172
9	1.134	1.138	1.142	1.147	1.151
10	1.119	1.123	1.126	1.130	1.134
12	1.095	1.098	1.101	1.104	1.107
14	1.079	1.081	1.083	1.086	1.088
16	1.066	1.068	1.070	1.072	1.074
18	1.057	1.058	1.060	1.062	1.063
20	1.049	1.051	1.052	1.053	1.055
24	1.038	1.039	1.040	1.041	1.042
30	1.027	1.028	1.029	1.030	1.030
40	1.017	1.018	1.018	1.019	1.019
60	1.009	1.009	1.009	1.010	1.010
120	1.003	1.003	1.003	1.003	1.003
∞	1.000	1.000	1.000	1.000	1.000
χ^2_{rm}	105.372	110.898	115.841	121.767	125.913

$m = 5$

α \ M	$r=5$ 0.100	0.050	0.025	0.010	0.005	$r=6$ 0.100	0.050	0.025	0.010	0.005
1	1.448	1.496	1.544	1.604	1.649	1.465	1.514	1.563	1.625	1.671
2	1.212	1.230	1.246	1.267	1.283	1.228	1.245	1.262	1.284	1.300
3	1.132	1.141	1.150	1.161	1.169	1.144	1.154	1.163	1.175	1.183
4	1.092	1.098	1.103	1.110	1.116	1.102	1.108	1.114	1.121	1.127
5	1.068	1.072	1.076	1.081	1.085	1.077	1.081	1.085	1.090	1.094
6	1.053	1.056	1.059	1.063	1.065	1.060	1.063	1.066	1.070	1.073
7	1.042	1.045	1.047	1.050	1.052	1.048	1.051	1.053	1.056	1.059
8	1.035	1.037	1.039	1.041	1.043	1.040	1.042	1.044	1.046	1.048
9	1.029	1.031	1.032	1.034	1.035	1.034	1.035	1.037	1.039	1.040
10	1.025	1.026	1.027	1.029	1.030	1.029	1.030	1.031	1.033	1.034
12	1.018	1.020	1.020	1.022	1.022	1.022	1.023	1.024	1.025	1.026
14	1.014	1.015	1.016	1.017	1.017	1.017	1.018	1.019	1.019	1.020
16	1.011	1.012	1.013	1.013	1.014	1.014	1.014	1.015	1.016	1.016
18	1.009	1.010	1.010	1.011	1.011	1.011	1.012	1.012	1.013	1.013
20	1.008	1.008	1.009	1.009	1.009	1.009	1.010	1.010	1.011	1.011
24	1.006	1.006	1.006	1.007	1.007	1.007	1.007	1.007	1.008	1.008
30	1.004	1.004	1.004	1.004	1.005	1.005	1.005	1.005	1.005	1.005
40	1.002	1.002	1.002	1.003	1.003	1.003	1.003	1.003	1.003	1.003
60	1.001	1.001	1.001	1.001	1.001	1.001	1.001	1.001	1.001	1.002
120	1.000	1.000	1.000	1.000	1.000	1.000	1.000	1.000	1.000	1.000
∞	1.000	1.000	1.000	1.000	1.000	1.000	1.000	1.000	1.000	1.000
χ^2_{rm}	34.3816	37.6525	40.6465	44.3141	46.9279	40.2560	43.7730	46.9792	50.8922	53.6720

Table 9 (*Continued*)

	$m=5$									
	$r=7$					$r=8$				
M \ α	0.100	0.050	0.025	0.010	0.005	0.100	0.050	0.025	0.010	0.005
1	1.484	1.535	1.584	1.648	1.695	1.505	1.556	1.607	1.672	1.721
2	1.244	1.262	1.280	1.302	1.319	1.261	1.280	1.298	1.321	1.338
3	1.158	1.168	1.177	1.189	1.198	1.171	1.182	1.192	1.204	1.213
4	1.113	1.119	1.125	1.133	1.139	1.124	1.131	1.137	1.145	1.151
5	1.086	1.090	1.095	1.100	1.104	1.095	1.100	1.105	1.110	1.114
6	1.068	1.071	1.074	1.078	1.081	1.076	1.079	1.083	1.087	1.090
7	1.055	1.058	1.060	1.063	1.066	1.062	1.065	1.068	1.071	1.073
8	1.046	1.048	1.050	1.052	1.054	1.052	1.054	1.056	1.059	1.061
9	1.038	1.040	1.042	1.044	1.046	1.044	1.046	1.048	1.050	1.052
10	1.033	1.035	1.036	1.038	1.039	1.038	1.039	1.041	1.043	1.044
12	1.025	1.026	1.027	1.029	1.030	1.029	1.030	1.031	1.033	1.034
14	1.020	1.021	1.021	1.022	1.023	1.023	1.024	1.025	1.026	1.027
16	1.016	1.017	1.017	1.018	1.019	1.018	1.019	1.020	1.021	1.022
18	1.013	1.014	1.014	1.015	1.015	1.015	1.016	1.017	1.017	1.018
20	1.011	1.012	1.012	1.013	1.013	1.013	1.013	1.014	1.015	1.015
24	1.008	1.008	1.009	1.009	1.010	1.009	1.010	1.010	1.011	1.011
30	1.005	1.006	1.006	1.006	1.006	1.006	1.007	1.007	1.007	1.008
40	1.003	1.003	1.004	1.004	1.004	1.004	1.004	1.004	1.004	1.005
60	1.002	1.002	1.002	1.002	1.002	1.002	1.002	1.002	1.002	1.002
120	1.000	1.000	1.000	1.000	1.000	1.001	1.001	1.001	1.001	1.001
∞	1.000	1.000	1.000	1.000	1.000	1.000	1.000	1.000	1.000	1.000
χ^2_{rm}	16.0588	49.8019	53.2033	57.3421	60.2748	51.8050	55.7585	59.3417	63.6907	66.7659

M \ α	$r=9$ 0.100	0.050	0.025	0.010	0.005	$r=10$ 0.100	0.050	0.025	0.010	0.005
1	1.526	1.578	1.630	1.697	1.746	1.547	1.600	1.653	1.721	1.772
2	1.278	1.298	1.316	1.340	1.358	1.295	1.315	1.334	1.359	1.377
3	1.185	1.196	1.206	1.219	1.229	1.199	1.211	1.221	1.235	1.244
4	1.136	1.143	1.150	1.158	1.164	1.147	1.155	1.162	1.171	1.177
5	1.105	1.110	1.115	1.121	1.125	1.115	1.120	1.125	1.131	1.136
6	1.084	1.088	1.092	1.096	1.099	1.092	1.097	1.101	1.105	1.109
7	1.069	1.072	1.075	1.079	1.081	1.076	1.080	1.083	1.087	1.089
8	1.058	1.060	1.063	1.066	1.068	1.064	1.067	1.070	1.073	1.075
9	1.049	1.051	1.053	1.056	1.058	1.055	1.057	1.059	1.062	1.064
10	1.043	1.044	1.046	1.048	1.050	1.048	1.050	1.051	1.054	1.055
12	1.033	1.034	1.035	1.037	1.038	1.037	1.038	1.040	1.041	1.043
14	1.026	1.027	1.028	1.029	1.030	1.029	1.031	1.032	1.033	1.034
16	1.021	1.022	1.023	1.024	1.024	1.024	1.025	1.026	1.027	1.028
18	1.018	1.018	1.109	1.020	1.020	1.020	1.021	1.022	1.022	1.023
20	1.015	1.015	1.016	1.017	1.017	1.017	1.018	1.018	1.019	1.019
24	1.011	1.011	1.012	1.012	1.013	1.013	1.013	1.014	1.014	1.014
30	1.008	1.008	1.008	1.008	1.009	1.009	1.009	1.009	1.010	1.010
40	1.005	1.005	1.005	1.005	1.005	1.005	1.005	1.006	1.006	1.006
60	1.002	1.002	1.002	1.002	1.003	1.003	1.003	1.003	1.003	1.003
120	1.001	1.001	1.001	1.001	1.001	1.001	1.001	1.001	1.002	1.002
∞	1.000	1.000	1.000	1.000	1.000	1.000	1.000	1.000	1.000	1.000
χ^2_{rm}	57.5053	61.6562	65.4102	69.9568	73.1661	63.1671	67.5048	71.4202	76.1539	79.4900

Table 9 (*Continued*)

	$m=5$									
	$r=11$					$r=12$				
M \ α	0.100	0.050	0.025	0.010	0.005	0.100	0.050	0.025	0.010	0.005
1	—	—	—	—	—	1.587	1.643	1.697	1.768	1.821
2	1.312	1.333	1.352	1.378	1.396	1.329	1.350	1.370	1.396	1.415
3	1.213	1.225	1.236	1.250	1.260	1.227	1.239	1.251	1.265	1.275
4	1.159	1.167	1.174	1.183	1.190	1.171	1.179	1.186	1.196	1.203
5	1.125	1.130	1.136	1.142	1.147	1.135	1.141	1.146	1.153	1.158
6	1.101	1.105	1.110	1.115	1.118	1.110	1.114	1.119	1.124	1.128
7	1.084	1.087	1.091	1.095	1.098	1.092	1.095	1.099	1.103	1.106
8	1.071	1.074	1.077	1.080	1.082	1.078	1.081	1.084	1.087	1.089
9	1.061	1.063	1.066	1.068	1.070	1.067	1.070	1.072	1.075	1.077
10	1.053	1.055	1.057	1.059	1.061	1.058	1.061	1.063	1.065	1.067
12	1.041	1.043	1.044	1.046	1.047	1.046	1.047	1.049	1.051	1.052
14	1.033	1.034	1.035	1.037	1.038	1.037	1.038	1.039	1.041	1.042
16	1.027	1.028	1.029	1.030	1.031	1.030	1.031	1.032	1.033	1.034
18	1.023	1.023	1.024	1.025	1.026	1.025	1.026	1.027	1.028	1.029
20	1.019	1.020	1.021	1.021	1.022	1.022	1.022	1.023	1.024	1.024
24	1.014	1.015	1.015	1.016	1.016	1.016	1.017	1.017	1.018	1.018
30	1.010	1.010	1.011	1.011	1.011	1.011	1.012	1.012	1.012	1.013
40	1.006	1.006	1.006	1.007	1.007	1.007	1.007	1.007	1.008	1.008
60	1.003	1.003	1.003	1.003	1.003	1.003	1.003	1.004	1.004	1.004
120	1.001	1.001	1.001	1.001	1.001	1.001	1.001	1.001	1.001	1.001
∞	1.000	1.000	1.000	1.000	1.000	1.000	1.000	1.000	1.000	1.000
χ^2_{rm}	68.796	73.311	77.380	82.292	85.749	74.3970	79.0819	83.2977	88.3794	91.9517

M \ α	$r=13$ 0.100	0.050	0.025	0.010	0.005	$r=14$ 0.100	0.050	0.025	0.010	0.005
1	—	—	—	—	—	1.626	1.683	1.740	1.813	1.867
2	1.345	1.367	1.387	1.414	1.433	1.361	1.383	1.404	1.431	1.451
3	1.241	1.253	1.265	1.280	1.290	1.254	1.267	1.279	1.294	1.305
4	1.182	1.191	1.199	1.208	1.215	1.194	1.203	1.211	1.221	1.228
5	1.145	1.151	1.157	1.164	1.169	1.155	1.161	1.167	1.174	1.180
6	1.118	1.123	1.128	1.133	1.137	1.127	1.132	1.137	1.143	1.147
7	1.099	1.103	1.107	1.111	1.114	1.107	1.111	1.115	1.119	1.123
8	1.084	1.088	1.091	1.094	1.097	1.091	1.095	1.098	1.102	1.104
9	1.073	1.076	1.078	1.081	1.083	1.079	1.082	1.085	1.088	1.090
10	1.064	1.066	1.068	1.071	1.073	1.069	1.072	1.074	1.077	1.079
12	1.050	1.052	1.054	1.055	1.057	1.055	1.057	1.058	1.060	1.062
14	1.040	1.042	1.043	1.045	1.046	1.044	1.046	1.047	1.049	1.050
16	1.033	1.035	1.036	1.037	1.038	1.037	1.038	1.039	1.040	1.041
18	1.028	1.029	1.030	1.031	1.032	1.031	1.032	1.033	1.034	1.035
20	1.024	1.025	1.026	1.026	1.027	1.026	1.027	1.028	1.029	1.030
24	1.018	1.019	1.019	1.020	1.020	1.020	1.021	1.021	1.022	1.022
30	1.013	1.013	1.013	1.014	1.014	1.014	1.014	1.015	1.015	1.016
40	1.008	1.008	1.008	1.009	1.009	1.009	1.009	1.009	1.010	1.010
60	1.004	1.004	1.004	1.004	1.004	1.004	1.004	1.005	1.005	1.005
120	1.001	1.001	1.001	1.001	1.001	1.001	1.001	1.001	1.001	1.001
∞	1.000	1.000	1.000	1.000	1.000	1.000	1.000	1.000	1.000	1.000
χ^2_{rm}	79.973	84.821	89.177	94.422	98.105	85.5270	90.5312	95.0232	100.4252	104.2149

Table 9 (*Continued*)

	$m=5$									
			$r=15$					$r=16$		
M \ α	0.100	0.050	0.025	0.010	0.005	0.0100	0.050	0.010	0.005	
1	—	—	—	—	—	1.663	1.722	1.780	1.855	1.911
2	1.377	1.399	1.421	1.449	1.469	1.392	1.415	1.437	1.465	1.486
3	1.267	1.281	1.293	1.309	1.320	1.280	1.294	1.307	1.323	1.334
4	1.205	1.214	1.223	1.233	1.240	1.216	1.226	1.234	1.245	1.253
5	1.164	1.171	1.177	1.185	1.190	1.174	1.181	1.188	1.195	1.201
6	1.136	1.141	1.146	1.152	1.156	1.144	1.150	1.155	1.161	1.165
7	1.115	1.119	1.123	1.127	1.131	1.122	1.127	1.131	1.136	1.139
8	1.098	1.102	1.105	1.109	1.112	1.105	1.109	1.112	1.116	1.119
9	1.085	1.088	1.091	1.094	1.097	1.091	1.095	1.098	1.101	1.104
10	1.075	1.078	1.080	1.083	1.085	1.081	1.083	1.086	1.089	1.091
12	1.059	1.061	1.063	1.065	1.067	1.064	1.066	1.068	1.070	1.072
14	1.048	1.050	1.051	1.053	1.054	1.052	1.054	1.055	1.057	1.059
16	1.040	1.041	1.043	1.044	1.045	1.043	1.045	1.046	1.048	1.049
18	1.034	1.035	1.036	1.037	1.038	1.037	1.038	1.039	1.040	1.041
20	1.029	1.030	1.031	1.032	1.033	1.032	1.033	1.033	1.035	1.035
24	1.022	1.023	1.023	1.024	1.025	1.024	1.025	1.025	1.026	1.027
30	1.015	1.016	1.016	1.017	1.017	1.017	1.018	1.018	1.019	1.019
40	1.010	1.010	1.010	1.011	1.011	1.011	1.011	1.011	1.012	1.012
60	1.005	1.005	1.005	1.005	1.005	1.005	1.005	1.006	1.006	1.006
120	1.001	1.001	1.001	1.001	1.002	1.002	1.002	1.002	1.002	1.002
∞	1.000	1.000	1.000	1.000	1.000	1.000	1.000	1.000	1.000	1.000
χ^2_{rm}	91.061	96.217	100.839	106.393	110.286	96.5782	101.8795	106.6286	112.3288	116.3211

M \ α	$r=17$ 0.100	0.050	0.025	0.010	0.005	$r=18$ 0.100	0.050	0.025	0.010	0.005
1	—	—	—	—	—	1.698	1.758	1.818	1.895	1.953
2	1.407	1.431	1.453	1.482	1.503	1.421	1.445	1.468	1.498	1.519
3	1.293	1.307	1.320	1.336	1.348	1.305	1.320	1.333	1.350	1.362
4	1.227	1.237	1.246	1.257	1.265	1.238	1.248	1.257	1.268	1.277
5	1.184	1.191	1.198	1.206	1.212	1.193	1.201	1.208	1.216	1.222
6	1.153	1.159	1.164	1.170	1.175	1.161	1.167	1.173	1.179	1.184
7	1.130	1.134	1.139	1.144	1.147	1.137	1.142	1.147	1.152	1.156
8	1.112	1.116	1.119	1.124	1.127	1.119	1.123	1.126	1.131	1.134
9	1.098	1.101	1.104	1.108	1.110	1.104	1.107	1.110	1.114	1.117
10	1.086	1.089	1.092	1.095	1.097	1.092	1.095	1.098	1.101	1.103
12	1.069	1.071	1.073	1.075	1.077	1.073	1.076	1.078	1.080	1.082
14	1.056	1.058	1.060	1.062	1.063	1.060	1.062	1.064	1.066	1.067
16	1.047	1.048	1.050	1.051	1.052	1.050	1.052	1.053	1.055	1.056
18	1.040	1.041	1.042	1.044	1.044	1.043	1.044	1.045	1.047	1.048
20	1.034	1.035	1.036	1.037	1.038	1.037	1.038	1.039	1.040	1.041
24	1.026	1.027	1.028	1.029	1.029	1.028	1.029	1.030	1.031	1.031
30	1.019	1.019	1.020	1.020	1.021	1.020	1.021	1.021	1.022	1.022
40	1.012	1.012	1.012	1.013	1.013	1.013	1.013	1.013	1.014	1.014
60	1.006	1.006	1.006	1.006	1.006	1.006	1.007	1.007	1.007	1.007
120	1.002	1.002	1.002	1.002	1.002	1.002	1.002	1.002	1.002	1.002
∞	1.000	1.000	1.000	1.000	1.000	1.000	1.000	1.000	1.000	1.000
χ^2_{rm}	102.079	107.522	112.393	118.236	122.325	107.5650	113.1453	118.1359	124.1163	128.2989

Table 9 (*Continued*)

	$m=5$									
	$r=19$					$r=20$				
M \ α	0.100	0.050	0.025	0.010	0.005	0.100	0.050	0.025	0.010	0.005
1	—	—	—	—	—	1.731	1.793	1.853	1.933	1.992
2	1.436	1.460	1.483	1.513	1.535	1.449	1.474	1.498	1.528	1.551
3	1.318	1.332	1.346	1.363	1.375	1.330	1.345	1.358	1.376	1.388
4	1.249	1.259	1.268	1.280	1.288	1.259	1.270	1.279	1.291	1.300
5	1.203	1.210	1.217	1.226	1.232	1.212	1.220	1.227	1.236	1.242
6	1.170	1.176	1.181	1.188	1.193	1.178	1.184	1.190	1.197	1.202
7	1.145	1.150	1.154	1.160	1.164	1.152	1.157	1.162	1.168	1.172
8	1.126	1.130	1.134	1.138	1.141	1.132	1.137	1.141	1.145	1.149
9	1.110	1.114	1.117	1.121	1.124	1.116	1.120	1.123	1.127	1.130
10	1.097	1.101	1.103	1.107	1.109	1.103	1.106	1.109	1.113	1.115
12	1.078	1.081	1.083	1.085	1.087	1.083	1.086	1.088	1.091	1.092
14	1.064	1.066	1.068	1.070	1.072	1.069	1.071	1.072	1.075	1.076
16	1.054	1.056	1.057	1.049	1.060	1.058	1.059	1.061	1.063	1.064
18	1.046	1.047	1.049	1.050	1.051	1.049	1.051	1.052	1.053	1.054
20	1.040	1.041	1.042	1.043	1.044	1.043	1.044	1.045	1.046	1.047
24	1.031	1.031	1.032	1.033	1.034	1.033	1.034	1.035	1.036	1.036
30	1.022	1.022	1.023	1.024	1.024	1.024	1.024	1.025	1.025	1.026
40	1.014	1.014	1.015	1.015	1.015	1.015	1.015	1.016	1.016	1.016
60	1.007	1.007	1.007	1.008	1.008	1.008	1.008	1.008	1.008	1.008
120	1.002	1.002	1.002	1.002	1.002	1.002	1.002	1.002	1.002	1.002
∞	1.000	1.000	1.000	1.000	1.000	1.000	1.000	1.000	1.000	1.000
χ^2_{rm}	113.038	118.752	123.858	129.973	134.247	118.4980	124.3421	129.5612	135.8067	140.1695

$m=6$

M \ α	$r=6$ 0.100	0.050	0.025	0.010	0.005	$r=7$ 0.100	0.050	0.025	0.010	0.005
1	1.471	1.520	1.568	1.631	1.677	1.481	1.530	1.579	1.642	1.688
2	1.237	1.255	1.272	1.294	1.310	1.249	1.266	1.284	1.306	1.322
3	1.153	1.163	1.172	1.183	1.192	1.163	1.173	1.182	1.194	1.203
4	1.109	1.116	1.122	1.129	1.134	1.118	1.124	1.131	1.138	1.144
5	1.083	1.088	1.092	1.097	1.101	1.090	1.095	1.099	1.105	1.109
6	1.066	1.069	1.072	1.076	1.079	1.072	1.075	1.079	1.083	1.086
7	1.053	1.056	1.058	1.061	1.064	1.059	1.062	1.064	1.067	1.070
8	1.044	1.046	1.048	1.051	1.053	1.049	1.051	1.053	1.056	1.058
9	1.037	1.039	1.041	1.043	1.044	1.042	1.043	1.045	1.047	1.049
10	1.032	1.034	1.035	1.037	1.038	1.036	1.037	1.039	1.041	1.042
12	1.024	1.025	1.026	1.028	1.029	1.027	1.029	1.030	1.031	1.032
14	1.019	1.020	1.021	1.022	1.022	1.022	1.023	1.023	1.024	1.025
16	1.015	1.016	1.017	1.018	1.018	1.108	1.018	1.019	1.020	1.020
18	1.013	1.013	1.014	1.014	1.015	1.014	1.015	1.016	1.016	1.017
20	1.011	1.011	1.012	1.012	1.013	1.012	1.013	1.013	1.014	1.014
24	1.008	1.008	1.009	1.009	1.009	1.009	1.009	1.010	1.010	1.010
30	1.005	1.006	1.006	1.006	1.006	1.006	1.006	1.007	1.007	1.007
40	1.003	1.003	1.003	1.004	1.004	1.004	1.004	1.004	1.004	1.004
60	1.002	1.002	1.002	1.002	1.002	1.002	1.002	1.002	1.002	1.002
120	1.000	1.000	1.000	1.000	1.000	1.000	1.000	1.001	1.001	1.001
∞	1.000	1.000	1.000	1.000	1.000	1.000	1.000	1.000	1.000	1.000
χ^2_{rm}	47.2122	50.9985	54.4373	58.6192	61.5812	54.0902	58.1240	61.7768	66.2062	69.3360

Table 9 (*Continued*)

	$m=6$									
	$r=8$					$r=9$				
M \ α	0.100	0.050	0.025	0.010	0.005	0.100	0.050	0.025	0.010	0.005
1	1.494	1.543	1.592	1.656	1.703	1.508	1.558	1.607	1.671	1.719
2	1.261	1.279	1.297	1.319	1.336	1.275	1.293	1.311	1.333	1.350
3	1.174	1.184	1.194	1.205	1.214	1.185	1.196	1.205	1.218	1.227
4	1.127	1.134	1.140	1.148	1.153	1.137	1.144	1.150	1.158	1.164
5	1.098	1.103	1.108	1.113	1.117	1.107	1.112	1.116	1.122	1.126
6	1.079	1.082	1.086	1.090	1.093	1.086	1.090	1.093	1.098	1.101
7	1.065	1.068	1.070	1.074	1.076	1.071	1.074	1.077	1.080	1.083
8	1.054	1.057	1.059	1.062	1.063	1.060	1.062	1.065	1.067	1.069
9	1.046	1.048	1.050	1.052	1.054	1.051	1.053	1.055	1.058	1.059
10	1.040	1.042	1.043	1.045	1.046	1.044	1.046	1.048	1.050	1.051
12	1.031	1.032	1.033	1.035	1.036	1.034	1.035	1.036	1.037	1.037
14	1.024	1.025	1.026	1.027	1.028	1.027	1.028	1.209	1.031	1.031
16	1.020	1.021	1.021	1.022	1.022	1.022	1.023	1.024	1.024	1.025
18	1.016	1.017	1.018	1.108	1.019	1.019	1.019	1.020	1.020	1.021
20	1.014	1.014	1.015	1.015	1.016	1.016	1.016	1.017	1.017	1.018
24	1.010	1.011	1.011	1.011	1.012	1.012	1.012	1.013	1.013	1.013
30	1.007	1.007	1.008	1.008	1.008	1.008	1.008	1.009	1.009	1.009
40	1.004	1.004	1.005	1.005	1.005	1.005	1.005	1.005	1.005	1.006
60	1.002	1.002	1.002	1.002	1.002	1.002	1.002	1.003	1.003	1.003
120	1.001	1.001	1.001	1.001	1.001	1.001	1.001	1.001	1.001	1.002
∞	1.000	1.000	1.000	1.000	1.000	1.000	1.000	1.000	1.000	1.000
χ^2_{rm}	60.9066	65.1708	69.0226	73.6826	76.9688	67.6728	72.1532	76.1921	81.0688	84.5016

M \ α	$r=10$ 0.100	0.050	0.025	0.010	0.005	$r=12$ 0.100	0.050	0.025	0.010	0.005
1	1.523	1.573	1.623	1.687	1.736	1.554	1.605	1.655	1.722	1.771
2	1.288	1.307	1.325	1.348	1.365	1.316	1.335	1.354	1.378	1.395
3	1.197	1.208	1.218	1.230	1.239	1.221	1.232	1.242	1.255	1.265
4	1.147	1.154	1.161	1.169	1.175	1.167	1.175	1.182	1.190	1.197
5	1.115	1.120	1.125	1.131	1.135	1.133	1.138	1.144	1.150	1.154
6	1.093	1.097	1.101	1.106	1.109	1.109	1.113	1.117	1.122	1.125
7	1.078	1.081	1.084	1.087	1.090	1.091	1.095	1.098	1.102	1.104
8	1.066	1.068	1.071	1.074	1.076	1.078	1.081	1.083	1.086	1.089
9	1.056	1.059	1.061	1.063	1.065	1.067	1.070	1.072	1.074	1.076
10	1.049	1.051	1.053	1.055	1.056	1.059	1.061	1.063	1.065	1.067
12	1.038	1.040	1.041	1.042	1.043	1.046	1.048	1.049	1.051	1.052
14	1.031	1.032	1.033	1.034	1.034	1.037	1.039	1.040	1.041	1.042
16	1.025	1.026	1.026	1.027	1.028	1.031	1.032	1.033	1.034	1.035
18	1.021	1.021	1.022	1.023	1.023	1.026	1.027	1.027	1.028	1.029
20	1.018	1.018	1.019	1.019	1.020	1.022	1.023	1.023	1.024	1.025
24	1.013	1.014	1.014	1.015	1.015	1.017	1.017	1.018	1.018	1.019
30	1.009	1.010	1.010	1.010	1.010	1.012	1.012	1.012	1.013	1.013
40	1.006	1.006	1.006	1.006	1.006	1.007	1.007	1.008	1.008	1.008
60	1.003	1.003	1.003	1.003	1.003	1.004	1.004	1.004	1.004	1.004
120	1.001	1.001	1.001	1.001	1.001	1.001	1.001	1.001	1.001	1.001
∞	1.000	1.000	1.000	1.000	1.000	1.000	1.000	1.000	1.000	1.000
χ^2_{rm}	74.3970	79.0819	83.2976	88.3794	91.9517	87.7430	92.8083	97.3531	102.8163	106.6476

Table 9 (*Continued*)

	$m=6$									
	$r=14$					$r=15$				
M \ α	0.100	0.050	0.025	0.010	0.005	0.100	0.050	0.025	0.010	0.005
1	1.585	1.637	1.688	1.756	1.806	—	—	—	—	—
2	1.343	1.363	1.383	1.407	1.425	1.357	1.377	1.397	1.422	1.440
3	1.244	1.256	1.267	1.281	1.291	1.256	1.268	1.279	1.293	1.303
4	1.188	1.196	1.203	1.212	1.219	1.198	1.206	1.214	1.223	1.230
5	1.151	1.157	1.162	1.169	1.174	1.160	1.166	1.171	1.178	1.183
6	1.125	1.129	1.133	1.139	1.142	1.132	1.137	1.142	1.147	1.151
7	1.105	1.109	1.112	1.117	1.119	1.112	1.116	1.120	1.124	1.127
8	1.090	1.093	1.096	1.100	1.102	1.097	1.100	1.103	1.106	1.109
9	1.078	1.081	1.083	1.086	1.088	1.084	1.087	1.089	1.092	1.095
10	1.069	1.071	1.073	1.076	1.078	1.074	1.076	1.079	1.081	1.083
12	1.055	1.056	1.058	1.060	1.061	1.059	1.061	1.062	1.064	1.066
14	1.044	1.046	1.047	1.049	1.050	1.048	1.050	1.051	1.053	1.054
16	1.037	1.038	1.039	1.040	1.041	1.040	1.041	1.042	1.044	1.045
18	1.031	1.032	1.033	1.034	1.035	1.034	1.035	1.036	1.037	1.038
20	1.027	1.028	1.028	1.029	1.030	1.029	1.030	1.031	1.032	1.032
24	1.020	1.021	1.021	1.022	1.023	1.022	1.023	1.023	1.024	1.025
30	1.014	1.015	1.015	1.016	1.016	1.016	1.016	1.017	1.017	1.017
40	1.009	1.009	1.009	1.010	1.010	1.010	1.010	1.010	1.011	1.011
60	1.004	1.005	1.005	1.005	1.005	1.005	1.005	1.005	1.005	1.005
120	1.001	1.001	1.001	1.001	1.001	1.001	1.001	1.001	1.002	1.002
∞	1.000	1.000	1.000	1.000	1.000	1.000	1.000	1.000	1.000	1.000
χ^2_{rm}	100.9800	106.3948	111.2423	117.0565	121.1263	107.565	113.145	118.136	124.116	128.299

			$r=16$					$r=17$		
M \ α	0.100	0.050	0.025	0.010	0.005	0.100	0.050	0.025	0.010	0.005
1	1.615	1.668	1.721	1.789	1.841	—	—	—	—	—
2	1.370	1.391	1.411	1.436	1.544	1.383	1.404	1.424	1.450	1.469
3	1.267	1.280	1.291	1.305	1.316	1.279	1.291	1.303	1.317	1.328
4	1.208	1.216	1.224	1.234	1.241	1.218	1.226	1.234	1.244	1.251
5	1.168	1.175	1.181	1.188	1.193	1.177	1.184	1.190	1.197	1.202
6	1.140	1.145	1.150	1.155	1.159	1.148	1.153	1.158	1.164	1.168
7	1.119	1.123	1.127	1.131	1.135	1.126	1.130	1.134	1.139	1.142
8	1.103	1.106	1.109	1.113	1.116	1.109	1.113	1.116	1.120	1.123
9	1.090	1.093	1.095	1.099	1.101	1.096	1.099	1.101	1.105	1.107
10	1.079	1.082	1.084	1.087	1.089	1.085	1.087	1.090	1.092	1.094
12	1.063	1.065	1.067	1.069	1.071	1.068	1.070	1.072	1.074	1.075
14	1.052	1.053	1.055	1.056	1.058	1.056	1.057	1.059	1.061	1.062
16	1.043	1.045	1.046	1.047	1.048	1.047	1.048	1.049	1.051	1.052
18	1.037	1.038	1.039	1.040	1.041	1.040	1.041	1.042	1.043	1.044
20	1.032	1.033	1.033	1.034	1.035	1.034	1.035	1.036	1.037	1.038
24	1.024	1.025	1.025	1.026	1.027	1.026	1.027	1.028	1.028	1.029
30	1.017	1.018	1.018	1.019	1.019	1.019	1.019	1.020	1.020	1.021
40	1.011	1.011	1.011	1.012	1.012	1.012	1.012	1.012	1.013	1.013
60	1.005	1.006	1.006	1.006	1.006	1.006	1.006	1.006	1.006	1.007
120	1.002	1.002	1.002	1.002	1.002	1.002	1.002	1.002	1.002	1.002
∞	1.000	1.000	1.000	1.000	1.000	1.000	1.000	1.000	1.000	1.000
χ^2_{rm}	114.1307	119.8709	125.0001	131.1412	135.4330	120.679	126.574	131.838	138.134	142.532

Table 9 (*Continued*)

	m = 6									
	r = 18					r = 19				
α / M	0.100	0.050	0.025	0.010	0.005	0.100	0.050	0.025	0.010	0.005
1	1.644	1.698	1.752	1.822	1.874	—	—	—	—	—
2	1.396	1.417	1.438	1.464	1.483	1.408	1.430	1.451	1.477	1.497
3	1.290	1.303	1.315	1.329	1.340	1.301	1.314	1.326	1.341	1.352
4	1.228	1.237	1.245	1.255	1.262	1.237	1.246	1.255	1.265	1.273
5	1.186	1.193	1.199	1.206	1.212	1.195	1.201	1.208	1.215	1.221
6	1.156	1.161	1.166	1.172	1.176	1.164	1.169	1.174	1.180	1.184
7	1.133	1.138	1.142	1.146	1.150	1.140	1.145	1.149	1.154	1.157
8	1.116	1.119	1.123	1.127	1.129	1.122	1.126	1.129	1.133	1.136
9	1.101	1.105	1.107	1.111	1.113	1.107	1.110	1.113	1.117	1.119
10	1.090	1.093	1.095	1.098	1.100	1.095	1.098	1.101	1.104	1.106
12	1.072	1.074	1.076	1.079	1.080	1.077	1.079	1.081	1.083	1.085
14	1.060	1.061	1.063	1.065	1.066	1.063	1.065	1.067	1.069	1.070
16	1.050	1.051	1.053	1.054	1.055	1.053	1.055	1.056	1.058	1.059
18	1.043	1.044	1.045	1.046	1.047	1.046	1.047	1.048	1.049	1.050
20	1.037	1.038	1.039	1.040	1.041	1.039	1.041	1.041	1.043	1.043
24	1.028	1.029	1.030	1.031	1.031	1.030	1.031	1.032	1.033	1.033
30	1.020	1.021	1.021	1.022	1.022	1.022	1.022	1.023	1.023	1.024
40	1.013	1.013	1.013	1.014	1.014	1.014	1.014	1.015	1.015	1.015
60	1.006	1.007	1.007	1.007	1.007	1.007	1.007	1.007	1.008	1.008
120	1.002	1.002	1.002	1.002	1.002	1.002	1.002	1.002	1.002	1.002
∞	1.000	1.000	1.000	1.000	1.000	1.000	1.000	1.000	1.000	1.000
χ^2_{rm}	127.2111	133.2569	138.6506	145.0988	149.5994	133.729	139.921	145.441	152.037	156.637

$r = 20$

M \ α	0.100	0.050	0.025	0.010	0.005
1	1.672	1.727	1.781	1.853	1.906
2	1.420	1.443	1.464	1.490	1.510
3	1.312	1.325	1.337	1.353	1.364
4	1.247	1.256	1.265	1.275	1.283
5	1.203	1.210	1.217	1.224	1.230
6	1.171	1.177	1.182	1.188	1.193
7	1.147	1.152	1.156	1.161	1.165
8	1.128	1.132	1.136	1.140	1.143
9	1.113	1.116	1.119	1.123	1.126
10	1.101	1.103	1.106	1.109	1.111
12	1.081	1.084	1.086	1.088	1.090
14	1.067	1.069	1.071	1.073	1.074
16	1.057	1.058	1.060	1.061	1.062
18	1.049	1.050	1.051	1.052	1.053
20	1.042	1.043	1.044	1.045	1.046
24	1.033	1.033	1.034	1.035	1.036
30	1.023	1.024	1.025	1.025	1.026
40	1.015	1.015	1.016	1.016	1.016
60	1.008	1.008	1.008	1.008	1.008
120	1.002	1.002	1.002	1.002	1.002
∞	1.000	1.000	1.000	1.000	1.000
χ^2_{rm}	140.2326	146.5674	152.2114	158.9502	163.6482

Table 9 (*Continued*)

	$m=7$									
	$r=7$					$r=8$				
M \ α	0.100	0.050	0.025	0.010	0.005	0.100	0.050	0.025	0.010	0.005
1	1.484	1.532	1.580	1.643	1.689	1.490	1.538	1.586	1.648	1.694
2	1.256	1.273	1.290	1.312	1.329	1.265	1.282	1.299	1.321	1.337
3	1.170	1.180	1.189	1.201	1.210	1.179	1.189	1.198	1.210	1.218
4	1.125	1.131	1.137	1.145	1.150	1.132	1.139	1.145	1.152	1.158
5	1.096	1.101	1.105	1.111	1.114	1.103	1.108	1.112	1.117	1.121
6	1.077	1.081	1.084	1.088	1.091	1.083	1.086	1.090	1.094	1.097
7	1.063	1.066	1.069	1.072	1.074	1.068	1.071	1.074	1.077	1.080
8	1.053	1.055	1.058	1.060	1.062	1.058	1.060	1.062	1.065	1.067
9	1.045	1.047	1.049	1.051	1.053	1.049	1.051	1.053	1.055	1.057
10	1.039	1.041	1.042	1.044	1.045	1.043	1.044	1.046	1.048	1.049
12	1.030	1.031	1.032	1.034	1.035	1.033	1.034	1.035	1.037	1.038
14	1.024	1.025	1.026	1.027	1.028	1.026	1.027	1.028	1.028	1.029
16	1.019	1.020	1.021	1.022	1.022	1.021	1.022	1.023	1.023	1.024
18	1.016	1.017	1.017	1.018	1.019	1.018	1.018	1.019	1.020	1.020
20	1.014	1.014	1.015	1.015	1.016	1.015	1.016	1.016	1.017	1.017
24	1.010	1.010	1.011	1.011	1.012	1.011	1.012	1.012	1.012	1.013
30	1.007	1.007	1.007	1.008	1.008	1.008	1.008	1.008	1.009	1.009
40	1.004	1.004	1.004	1.005	1.005	1.005	1.005	1.005	1.005	1.005
60	1.002	1.002	1.002	1.002	1.002	1.002	1.002	1.002	1.003	1.003
120	1.001	1.001	1.001	1.001	1.001	1.001	1.001	1.001	1.001	1.001
∞	1.000	1.000	1.000	1.000	1.000	1.000	1.000	1.000	1.000	1.000
χ^2_{rm}	62.0375	66.3386	70.2224	74.9195	78.2307	69.9185	74.4683	78.5672	83.5134	86.9938

	r=9					r=10				
M \ α	0.100	0.050	0.025	0.010	0.005	0.100	0.050	0.025	0.010	0.005
1	1.499	1.547	1.594	1.656	1.703	1.509	1.557	1.604	1.666	1.713
2	1.275	1.292	1.309	1.331	1.348	1.285	1.303	1.320	1.342	1.359
3	1.188	1.198	1.207	1.219	1.228	1.197	1.208	1.217	1.229	1.238
4	1.140	1.147	1.153	1.161	1.166	1.148	1.155	1.162	1.169	1.175
5	1.110	1.115	1.119	1.125	1.129	1.117	1.122	1.127	1.132	1.136
6	1.089	1.093	1.096	1.100	1.103	1.096	1.099	1.103	1.107	1.110
7	1.074	1.077	1.080	1.083	1.085	1.080	1.083	1.086	1.089	1.091
8	1.063	1.065	1.067	1.070	1.072	1.068	1.070	1.073	1.075	1.077
9	1.054	1.056	1.058	1.060	1.062	1.058	1.060	1.062	1.065	1.066
10	1.047	1.048	1.050	1.052	1.053	1.051	1.053	1.054	1.056	1.058
12	1.036	1.038	1.039	1.040	1.041	1.040	1.042	1.042	1.044	1.045
14	1.029	1.030	1.031	1.032	1.033	1.032	1.034	1.034	1.036	1.036
16	1.024	1.025	1.025	1.026	1.027	1.026	1.028	1.028	1.029	1.029
18	1.020	1.021	1.021	1.022	1.023	1.022	1.023	1.023	1.024	1.024
20	1.017	1.018	1.018	1.019	1.019	1.019	1.019	1.020	1.020	1.021
24	1.013	1.013	1.013	1.014	1.014	1.014	1.014	1.015	1.015	1.016
30	1.009	1.009	1.009	1.010	1.010	1.010	1.010	1.010	1.011	1.011
40	1.005	1.006	1.006	1.006	1.006	1.006	1.006	1.006	1.007	1.007
60	1.003	1.003	1.003	1.003	1.003	1.003	1.003	1.003	1.003	1.003
120	1.001	1.001	1.001	1.001	1.001	1.001	1.001	1.001	1.001	1.001
∞	1.000	1.000	1.000	1.000	1.000	1.000	1.000	1.000	1.000	1.000
χ^2_{rm}	77.7454	82.5287	86.8296	92.0100	95.6493	85.5271	90.5312	95.0231	100.4250	104.2150

Table 9 (*Continued*)

M \ α	0.100	0.050	$r=11$ 0.025	0.010	0.005	0.100	0.050	$r=12$ 0.025	0.100	0.005
	$m=7$									
1	—	—	—	—	—	1.531	1.579	1.627	1.690	1.737
2	1.297	1.315	1.332	1.354	1.371	1.308	1.326	1.344	1.366	1.383
3	1.207	1.218	1.227	1.239	1.248	1.217	1.228	1.238	1.250	1.259
4	1.157	1.164	1.171	1.179	1.184	1.166	1.173	1.180	1.188	1.194
5	1.125	1.130	1.135	1.140	1.145	1.133	1.138	1.143	1.149	1.153
6	1.102	1.106	1.110	1.114	1.118	1.109	1.113	1.117	1.122	1.125
7	1.086	1.089	1.092	1.095	1.098	1.092	1.095	1.098	1.102	1.104
8	1.073	1.076	1.078	1.081	1.083	1.079	1.081	1.084	1.087	1.089
9	1.063	1.065	1.067	1.070	1.072	1.068	1.070	1.073	1.075	1.077
10	1.055	1.057	1.059	1.061	1.062	1.060	1.062	1.064	1.066	1.067
12	1.043	1.045	1.046	1.048	1.049	1.047	1.049	1.050	1.052	1.053
14	1.035	1.036	1.037	1.038	1.039	1.038	1.039	1.041	1.042	1.043
16	1.029	1.030	1.031	1.032	1.032	1.032	1.033	1.034	1.035	1.035
18	1.024	1.025	1.026	1.027	1.027	1.027	1.028	1.028	1.029	1.030
20	1.021	1.021	1.022	1.023	1.023	1.023	1.024	1.024	1.025	1.025
24	1.016	1.016	1.017	1.017	1.017	1.017	1.018	1.018	1.019	1.019
30	1.011	1.011	1.012	1.012	1.012	1.012	1.012	1.013	1.013	1.013
40	1.007	1.007	1.007	1.007	1.008	1.008	1.008	1.008	1.008	1.009
60	1.003	1.003	1.004	1.004	1.004	1.004	1.004	1.004	1.004	1.004
120	1.001	1.001	1.001	1.001	1.001	1.001	1.001	1.001	1.001	1.001
∞	1.000	1.000	1.000	1.000	1.000	1.000	1.000	1.000	1.000	1.000
χ^2_{rm}	93.270	98.484	103.158	108.771	112.704	100.9800	106.3948	111.2423	117.0565	121.1263

M \ α	r = 13 0.100	0.050	0.025	0.010	0.005	r = 14 0.100	0.050	0.025	0.010	0.005
1	—	—	—	—	—	1.556	1.604	1.652	1.715	1.763
2	1.320	1.338	1.356	1.378	1.395	1.331	1.350	1.368	1.391	1.408
3	1.228	1.238	1.248	1.261	1.270	1.238	1.249	1.259	1.272	1.281
4	1.175	1.182	1.189	1.197	1.203	1.184	1.192	1.198	1.207	1.213
5	1.141	1.146	1.151	1.157	1.161	1.149	1.154	1.159	1.165	1.170
6	1.116	1.120	1.124	1.129	1.132	1.123	1.128	1.132	1.137	1.140
7	1.098	1.102	1.105	1.108	1.111	1.105	1.108	1.111	1.115	1.118
8	1.084	1.087	1.090	1.093	1.095	1.090	1.093	1.096	1.099	1.101
9	1.073	1.076	1.078	1.080	1.082	1.078	1.081	1.083	1.086	1.088
10	1.064	1.066	1.068	1.071	1.072	1.069	1.071	1.073	1.076	1.077
12	1.051	1.053	1.054	1.056	1.057	1.055	1.057	1.058	1.060	1.061
14	1.042	1.043	1.044	1.045	1.046	1.045	1.046	1.047	1.049	1.050
16	1.035	1.036	1.036	1.038	1.038	1.037	1.039	1.039	1.041	1.041
18	1.029	1.030	1.031	1.032	1.032	1.032	1.033	1.033	1.034	1.035
20	1.025	1.026	1.026	1.027	1.028	1.027	1.028	1.029	1.030	1.030
24	1.019	1.020	1.020	1.021	1.021	1.021	1.021	1.022	1.022	1.023
30	1.013	1.014	1.014	1.014	1.015	1.015	1.015	1.015	1.016	1.016
40	1.008	1.009	1.009	1.009	1.009	1.009	1.009	1.010	1.010	1.010
60	1.004	1.004	1.004	1.004	1.005	1.005	1.005	1.005	1.005	1.005
120	1.001	1.001	1.001	1.001	1.001	1.001	1.001	1.001	1.001	1.001
∞	1.000	1.000	1.000	1.000	1.000	1.000	1.000	1.000	1.000	1.000
χ^2_{rm}	108.661	114.268	119.282	125.289	129.491	116.3153	122.1077	127.2821	133.4757	137.8032

Table 9 (*Continued*)

	$m=7$									
	$r=15$					$r=16$				
M \ α	0.100	0.050	0.025	0.010	0.005	0.100	0.050	0.025	0.010	0.005
1	—	—	—	—	—	1.580	1.629	1.678	1.742	1.790
2	1.343	1.362	1.380	1.403	1.420	1.354	1.373	1.392	1.415	1.432
3	1.248	1.259	1.270	1.283	1.292	1.258	1.270	1.280	1.293	1.303
4	1.193	1.201	1.208	1.216	1.223	1.202	1.210	1.217	1.226	1.232
5	1.157	1.162	1.168	1.174	1.179	1.165	1.171	1.176	1.182	1.187
6	1.131	1.135	1.139	1.144	1.148	1.138	1.142	1.147	1.152	1.155
7	1.111	1.115	1.118	1.122	1.125	1.118	1.121	1.125	1.129	1.132
8	1.096	1.099	1.102	1.105	1.107	1.102	1.105	1.108	1.111	1.114
9	1.084	1.086	1.089	1.092	1.094	1.089	1.092	1.094	1.097	1.099
10	1.074	1.076	1.078	1.081	1.082	1.079	1.081	1.083	1.086	1.088
12	1.059	1.061	1.062	1.064	1.066	1.063	1.065	1.067	1.069	1.070
14	1.048	1.050	1.051	1.053	1.054	1.052	1.053	1.055	1.056	1.057
16	1.040	1.042	1.043	1.044	1.045	1.044	1.045	1.046	1.047	1.048
18	1.034	1.035	1.036	1.037	1.038	1.037	1.038	1.039	1.040	1.041
20	1.030	1.030	1.031	1.032	1.033	1.032	1.033	1.034	1.034	1.035
24	1.023	1.023	1.024	1.024	1.025	1.024	1.025	1.026	1.026	1.027
30	1.016	1.016	1.017	1.017	1.018	1.017	1.018	1.018	1.019	1.019
40	1.010	1.010	1.011	1.011	1.011	1.011	1.011	1.012	1.012	1.012
60	1.005	1.005	1.005	1.005	1.006	1.006	1.006	1.006	1.006	1.006
120	1.001	1.001	1.002	1.002	1.002	1.002	1.002	1.002	1.002	1.002
∞	1.000	1.000	1.000	1.000	1.000	1.000	1.000	1.000	1.000	1.000
χ^2_{rm}	123.947	129.918	135.247	141.620	146.070	131.5576	137.7015	143.1801	149.7269	154.2944

M \ α	r = 17 0.100	0.050	0.025	0.010	0.005	r = 18 0.100	0.050	0.025	0.010	0.005
1	—	—	—	—	—	1.605	1.654	1.703	1.768	1.816
2	1.365	1.385	1.403	1.427	1.445	1.377	1.396	1.415	1.439	1.457
3	1.268	1.280	1.291	1.304	1.314	1.278	1.290	1.301	1.315	1.324
4	1.211	1.219	1.226	1.235	1.242	1.220	1.228	1.236	1.245	1.252
5	1.173	1.179	1.184	1.191	1.196	1.181	1.187	1.192	1.199	1.204
6	1.145	1.150	1.154	1.159	1.163	1.152	1.157	1.161	1.167	1.171
7	1.124	1.128	1.131	1.136	1.139	1.131	1.134	1.138	1.142	1.146
8	1.108	1.111	1.114	1.117	1.120	1.114	1.117	1.120	1.124	1.126
9	1.095	1.097	1.100	1.103	1.105	1.100	1.103	1.105	1.108	1.111
10	1.084	1.086	1.088	1.091	1.093	1.089	1.091	1.093	1.096	1.098
12	1.067	1.069	1.071	1.073	1.074	1.072	1.074	1.075	1.077	1.079
14	1.056	1.057	1.058	1.060	1.061	1.059	1.061	1.062	1.054	1.065
16	1.047	1.048	1.049	1.050	1.051	1.050	1.051	1.052	1.064	1.055
18	1.040	1.041	1.042	1.043	1.044	1.043	1.044	1.045	1.046	1.047
20	1.034	1.035	1.036	1.037	1.038	1.037	1.038	1.039	1.040	1.040
24	1.026	1.027	1.028	1.028	1.029	1.028	1.029	1.030	1.031	1.031
30	1.019	1.019	1.020	1.020	1.021	1.020	1.021	1.021	1.022	1.022
40	1.012	1.012	1.013	1.013	1.013	1.013	1.013	1.014	1.014	1.014
60	1.006	1.006	1.006	1.007	1.007	1.007	1.007	1.007	1.007	1.007
120	1.002	1.002	1.002	1.002	1.002	1.002	1.002	1.002	1.002	1.002
∞	1.000	1.000	1.000	1.000	1.000	1.000	1.000	1.000	1.000	1.000
χ^2_{rm}	139.149	145.461	151.084	157.800	162.481	146.7241	153.1979	158.9264	165.8410	170.6341

Table 9 (*Continued*)

	m = 7									
			r = 19					*r* = 20		
$M \backslash \alpha$	0.100	0.050	0.025	0.010	0.005	0.100	0.050	0.025	0.010	0.005
1	—	—	—	—	—	1.629	1.679	1.728	1.793	1.843
2	1.388	1.408	1.427	1.451	1.469	1.399	1.419	1.438	1.462	1.480
3	1.288	1.300	1.311	1.325	1.335	1.298	1.310	1.321	1.335	1.346
4	1.229	1.237	1.245	1.254	1.261	1.238	1.246	1.254	1.263	1.270
5	1.189	1.195	1.201	1.208	1.213	1.196	1.203	1.209	1.216	1.221
6	1.159	1.164	1.169	1.174	1.178	1.166	1.172	1.176	1.182	1.186
7	1.137	1.141	1.145	1.149	1.152	1.144	1.148	1.151	1.156	1.159
8	1.119	1.123	1.126	1.130	1.132	1.125	1.129	1.132	1.136	1.139
9	1.105	1.108	1.111	1.114	1.116	1.111	1.114	1.117	1.120	1.122
10	1.094	1.096	1.099	1.101	1.103	1.099	1.101	1.104	1.107	1.109
12	1.076	1.078	1.080	1.082	1.083	1.080	1.082	1.084	1.086	1.088
14	1.063	1.065	1.066	1.068	1.069	1.067	1.068	1.070	1.072	1.073
16	1.053	1.054	1.056	1.057	1.058	1.056	1.058	1.059	1.060	1.062
18	1.045	1.047	1.048	1.049	1.050	1.048	1.050	1.051	1.052	1.053
20	1.039	1.040	1.041	1.042	1.043	1.042	1.043	1.044	1.045	1.046
24	1.030	1.031	1.032	1.033	1.033	1.033	1.033	1.034	1.035	1.035
30	1.022	1.022	1.023	1.023	1.024	1.024	1.024	1.025	1.025	1.026
40	1.014	1.014	1.015	1.015	1.015	1.015	1.015	1.016	1.016	1.016
60	1.007	1.007	1.007	1.008	1.008	1.008	1.008	1.008	1.008	1.008
120	1.002	1.002	1.002	1.002	1.002	1.002	1.002	1.002	1.002	1.002
∞	1.000	1.000	1.000	1.000	1.000	1.000	1.000	1.000	1.000	1.000
χ^2_{rm}	154.283	160.915	166.816	173.854	178.755	161.8270	168.6130	174.6478	181.8403	186.8468

$m=8$

$r=8$

M \ α	0.100	0.050	0.025	0.010	0.005
1	1.491	1.538	1.585	1.646	1.692
2	1.270	1.288	1.305	1.326	1.342
3	1.185	1.195	1.204	1.215	1.224
4	1.138	1.144	1.150	1.158	1.163
5	1.108	1.113	1.117	1.123	1.126
6	1.088	1.091	1.095	1.099	1.102
7	1.073	1.076	1.078	1.082	1.084
8	1.061	1.064	1.066	1.069	1.071
9	1.053	1.055	1.057	1.059	1.060
10	1.046	1.048	1.049	1.051	1.052
12	1.036	1.038	1.039	1.040	1.041
14	1.028	1.030	1.031	1.031	1.032
16	1.023	1.025	1.026	1.026	1.027
18	1.020	1.021	1.022	1.022	1.023
20	1.017	1.017	1.018	1.018	1.018
24	1.012	1.013	1.013	1.014	1.014
30	1.009	1.009	1.009	1.009	1.010
40	1.005	1.005	1.006	1.006	1.006
60	1.003	1.003	1.003	1.003	1.003
120	1.001	1.001	1.001	1.001	1.001
∞	1.000	1.000	1.000	1.000	1.000
χ^2_{rm}	78.8596	83.6753	88.0041	93.2169	96.8781

Table 9 (*Continued*)

	$m=8$									
	$r=9$					$r=10$				
M \ α	0.100	0.050	0.025	0.010	0.005	0.100	0.050	0.025	0.010	0.005
1	1.495	1.541	1.587	1.648	1.694	1.501	1.547	1.593	1.653	1.698
2	1.277	1.295	1.311	1.333	1.349	1.286	1.303	1.319	1.341	1.357
3	1.192	1.202	1.211	1.222	1.231	1.200	1.209	1.219	1.230	1.239
4	1.144	1.151	1.157	1.165	1.170	1.151	1.158	1.164	1.172	1.177
5	1.114	1.119	1.123	1.129	1.132	1.120	1.125	1.130	1.135	1.139
6	1.093	1.097	1.100	1.104	1.107	1.098	1.102	1.106	1.110	1.113
7	1.077	1.080	1.083	1.086	1.089	1.082	1.086	1.088	1.092	1.094
8	1.066	1.068	1.070	1.073	1.075	1.070	1.073	1.075	1.078	1.080
9	1.057	1.059	1.061	1.063	1.064	1.061	1.063	1.065	1.067	1.069
10	1.049	1.051	1.053	1.055	1.056	1.053	1.055	1.057	1.059	1.060
12	1.039	1.040	1.041	1.043	1.044	1.042	1.043	1.044	1.046	1.047
14	1.031	1.032	1.033	1.034	1.035	1.034	1.035	1.036	1.037	1.038
16	1.026	1.026	1.027	1.028	1.029	1.028	1.029	1.030	1.030	1.031
18	1.021	1.022	1.023	1.024	1.024	1.023	1.024	1.025	1.026	1.026
20	1.018	1.019	1.019	1.020	1.020	1.020	1.021	1.021	1.022	1.022
24	1.014	1.014	1.015	1.015	1.015	1.015	1.106	1.106	1.016	1.017
30	1.010	1.010	1.010	1.010	1.011	1.011	1.011	1.011	1.011	1.012
40	1.006	1.006	1.006	1.006	1.007	1.006	1.007	1.007	1.007	1.007
60	1.003	1.003	1.003	1.003	1.003	1.003	1.003	1.003	1.003	1.004
120	1.001	1.001	1.001	1.001	1.001	1.001	1.001	1.001	1.001	1.001
∞	1.000	1.000	1.000	1.000	1.000	1.000	1.000	1.000	1.000	1.000
χ^2_{rm}	87.7430	92.8083	97.3531	102.8163	106.6476	96.5782	101.8795	106.6286	112.3288	116.3211

	$r=11$					$r=12$				
M \ α	0.100	0.050	0.025	0.010	0.005	0.100	0.050	0.025	0.010	0.005
1	—	—	—	—	—	1.516	1.562	1.608	1.667	1.713
2	1.294	1.312	1.328	1.350	1.366	1.304	1.321	1.338	1.359	1.375
3	1.208	1.218	1.227	1.239	1.247	1.216	1.226	1.236	1.248	1.256
4	1.159	1.165	1.172	1.179	1.185	1.166	1.173	1.180	1.187	1.193
5	1.127	1.132	1.136	1.142	1.146	1.134	1.139	1.143	1.149	1.153
6	1.104	1.108	1.112	1.116	1.119	1.111	1.114	1.118	1.122	1.126
7	1.088	1.091	1.094	1.097	1.100	1.093	1.097	1.099	1.103	1.105
8	1.075	1.078	1.080	1.083	1.085	1.080	1.083	1.085	1.088	1.090
9	1.065	1.067	1.069	1.072	1.073	1.070	1.072	1.074	1.076	1.078
10	1.057	1.059	1.061	1.063	1.064	1.061	1.063	1.065	1.067	1.068
12	1.045	1.046	1.048	1.049	1.050	1.049	1.050	1.051	1.053	1.054
14	1.037	1.038	1.039	1.040	1.041	1.039	1.041	1.042	1.043	1.044
16	1.030	1.031	1.032	1.033	1.034	1.033	1.034	1.035	1.036	1.036
18	1.026	1.026	1.027	1.028	1.028	1.028	1.029	1.029	1.030	1.031
20	1.022	1.022	1.023	1.024	1.024	1.024	1.024	1.025	1.026	1.026
24	1.017	1.017	1.017	1.018	1.018	1.018	1.019	1.019	1.020	1.020
30	1.012	1.012	1.012	1.013	1.013	1.013	1.013	1.013	1.014	1.014
40	1.007	1.007	1.008	1.008	1.008	1.008	1.008	1.008	1.009	1.009
60	1.004	1.004	1.004	1.004	1.004	1.004	1.004	1.004	1.004	1.004
120	1.001	1.001	1.001	1.001	1.001	1.001	1.001	1.001	1.001	1.001
∞	1.000	1.000	1.000	1.000	1.000	1.000	1.000	1.000	1.000	1.000
χ^2_{rm}	105.372	110.898	115.841	121.767	125.913	114.1307	119.8709	125.0001	131.1412	135.4330

Table 9 (*Continued*)

	$m=8$									
	$r=13$					$r=14$				
M \ α	0.100	0.050	0.025	0.010	0.005	0.100	0.050	0.025	0.010	0.005
1	—	—	—	—	—	1.535	1.581	1.626	1.686	1.731
2	1.313	1.331	1.347	1.369	1.385	1.323	1.341	1.357	1.379	1.395
3	1.225	1.235	1.245	1.257	1.266	1.234	1.244	1.254	1.266	1.275
4	1.174	1.181	1.188	1.196	1.201	1.182	1.189	1.196	1.204	1.210
5	1.141	1.146	1.151	1.156	1.161	1.148	1.153	1.158	1.164	1.168
6	1.117	1.121	1.125	1.129	1.132	1.123	1.127	1.131	1.136	1.139
7	1.099	1.102	1.105	1.109	1.111	1.105	1.108	1.111	1.115	1.118
8	1.085	1.088	1.090	1.093	1.096	1.091	1.093	1.096	1.099	1.101
9	1.074	1.077	1.079	1.081	1.083	1.079	1.082	1.084	1.086	1.088
10	1.066	1.067	1.069	1.071	1.073	1.070	1.072	1.074	1.076	1.078
12	1.052	1.054	1.055	1.057	1.058	1.056	1.057	1.059	1.061	1.062
14	1.043	1.044	1.045	1.046	1.047	1.046	1.047	1.048	1.050	1.051
16	1.035	1.036	1.037	1.038	1.039	1.038	1.039	1.040	1.041	1.042
18	1.030	1.031	1.032	1.033	1.033	1.032	1.033	1.034	1.035	1.036
20	1.026	1.027	1.027	1.028	1.028	1.028	1.029	1.029	1.030	1.031
24	1.020	1.020	1.021	1.021	1.022	1.021	1.022	1.022	1.023	1.023
30	1.014	1.014	1.015	1.015	1.015	1.015	1.016	1.016	1.016	1.017
40	1.009	1.009	1.009	1.009	1.010	1.010	1.010	1.010	1.010	1.010
60	1.004	1.004	1.005	1.005	1.005	1.005	1.005	1.005	1.005	1.005
120	1.001	1.001	1.001	1.001	1.001	1.001	1.001	1.001	1.001	1.001
∞	1.000	1.000	1.000	1.000	1.000	1.000	1.000	1.000	1.000	1.000
χ^2_{rm}	122.858	128.804	134.111	140.459	144.891	131.5576	137.7015	143.1801	149.7269	154.2944

M \ α	$r=15$ 0.100	0.050	0.025	0.010	0.005	$r=16$ 0.100	0.050	0.025	0.010	0.005
1	—	—	—	—	—	1.555	1.601	1.646	1.706	1.751
2	1.333	1.351	1.368	1.389	1.406	1.343	1.361	1.378	1.400	1.416
3	1.243	1.253	1.263	1.275	1.284	1.252	1.263	1.272	1.285	1.294
4	1.190	1.198	1.204	1.212	1.218	1.198	1.206	1.212	1.221	1.227
5	1.155	1.160	1.165	1.171	1.176	1.162	1.168	1.173	1.179	1.183
6	1.130	1.134	1.138	1.143	1.146	1.136	1.141	1.145	1.149	1.153
7	1.111	1.114	1.117	1.121	1.124	1.117	1.120	1.123	1.127	1.130
8	1.096	1.099	1.101	1.105	1.107	1.101	1.104	1.107	1.110	1.113
9	1.084	1.087	1.089	1.091	1.093	1.089	1.092	1.094	1.097	1.099
10	1.074	1.076	1.078	1.081	1.082	1.079	1.081	1.083	1.085	1.087
12	1.060	1.061	1.063	1.065	1.066	1.064	1.065	1.067	1.069	1.070
14	1.049	1.050	1.052	1.053	1.054	1.052	1.054	1.055	1.056	1.057
16	1.041	1.042	1.043	1.044	1.045	1.044	1.045	1.046	1.047	1.048
18	1.035	1.036	1.037	1.038	1.038	1.038	1.038	1.039	1.040	1.041
20	1.030	1.031	1.032	1.032	1.033	1.032	1.033	1.034	1.035	1.035
24	1.023	1.024	1.024	1.025	1.025	1.025	1.026	1.026	1.027	1.027
30	1.016	1.017	1.017	1.018	1.018	1.018	1.018	1.019	1.019	1.019
40	1.010	1.011	1.011	1.011	1.011	1.011	1.012	1.012	1.012	1.012
60	1.005	1.005	1.005	1.006	1.006	1.006	1.006	1.006	1.006	1.006
120	1.002	1.002	1.002	1.002	1.002	1.002	1.002	1.002	1.002	1.002
∞	1.000	1.000	1.000	1.000	1.000	1.000	1.000	1.000	1.000	1.000
χ^2_{rm}	140.233	146.567	152.211	158.950	163.648	148.8853	155.4047	161.2087	168.1332	172.9575

Table 9 (*Continued*)

	$m=8$									
	$r=17$					$r=18$				
M \ α	0.100	0.050	0.025	0.010	0.005	0.100	0.050	0.025	0.010	0.005
1	—	—	—	—	—	1.575	1.621	1.667	1.727	1.773
2	1.353	1.371	1.388	1.410	—	1.363	1.381	1.398	1.420	1.437
3	1.261	1.272	1.282	1.294	1.303	1.270	1.281	1.291	1.304	1.313
4	1.207	1.214	1.221	1.229	1.235	1.215	1.222	1.229	1.238	1.244
5	1.170	1.175	1.180	1.187	1.191	1.177	1.183	1.188	1.194	1.199
6	1.143	1.147	1.151	1.156	1.160	1.150	1.154	1.158	1.163	1.157
7	1.123	1.126	1.130	1.134	1.136	1.129	1.133	1.136	1.140	1.143
8	1.107	1.110	1.113	1.116	1.118	1.112	1.115	1.118	1.122	1.124
9	1.094	1.097	1.099	1.102	1.104	1.099	1.102	1.104	1.107	1.109
10	1.084	1.086	1.088	1.090	1.092	1.088	1.091	1.093	1.095	1.097
12	1.067	1.069	1.071	1.073	1.074	1.071	1.073	1.075	1.077	1.078
14	1.056	1.057	1.058	1.060	1.061	1.059	1.061	1.062	1.064	1.065
16	1.047	1.048	1.049	1.050	1.051	1.050	1.051	1.052	1.054	1.055
18	1.040	1.041	1.042	1.043	1.044	1.043	1.044	1.045	1.046	1.047
20	1.035	1.036	1.036	1.037	1.038	1.037	1.038	1.039	1.040	1.040
24	1.027	1.027	1.028	1.029	1.029	1.029	1.029	1.030	1.031	1.031
30	1.019	1.020	1.020	1.021	1.021	1.021	1.021	1.022	1.022	1.022
40	1.012	1.013	1.013	1.013	1.013	1.013	1.014	1.014	1.014	1.014
60	1.006	1.006	1.007	1.007	1.007	1.007	1.007	1.007	1.007	1.007
120	1.002	1.002	1.002	1.002	1.002	1.002	1.002	1.002	1.002	1.002
∞	1.000	1.000	1.000	1.000	1.000	1.000	1.000	1.000	1.000	1.000
χ^2_{rm}	157.518	164.216	170.175	177.280	182.226	166.1318	173.0041	179.1137	186.3930	191.4585

$m = 9$

	$r=9$					$r=10$				
M \ α	0.100	0.050	0.025	0.010	0.005	0.100	0.050	0.025	0.010	0.005
1	1.495	1.540	1.585	1.645	1.690	1.497	1.542	1.586	1.645	1.690
2	1.282	1.299	1.315	1.337	1.353	1.288	1.305	1.321	1.342	1.357
3	1.197	1.207	1.216	1.227	1.236	1.203	1.213	1.222	1.233	1.242
4	1.149	1.156	1.162	1.169	1.175	1.155	1.162	1.168	1.175	1.181
5	1.119	1.123	1.128	1.133	1.137	1.124	1.129	1.133	1.139	1.143
6	1.097	1.101	1.104	1.108	1.111	1.102	1.106	1.109	1.113	1.116
7	1.081	1.084	1.087	1.090	1.093	1.086	1.089	1.092	1.095	1.097
8	1.069	1.072	1.074	1.077	1.079	1.073	1.076	1.078	1.081	1.083
9	1.060	1.062	1.064	1.066	1.068	1.064	1.066	1.068	1.070	1.072
10	1.052	1.054	1.056	1.058	1.059	1.056	1.058	1.059	1.061	1.063
12	1.041	1.043	1.044	1.045	1.046	1.044	1.045	1.047	1.048	1.049
14	1.033	1.034	1.035	1.036	1.037	1.036	1.037	1.038	1.039	1.040
16	1.027	1.028	1.029	1.030	1.031	1.030	1.030	1.031	1.032	1.033
18	1.023	1.024	1.024	1.025	1.026	1.025	1.026	1.026	1.027	1.028
20	1.020	1.020	1.021	1.022	1.022	1.021	1.022	1.023	1.023	1.024
24	1.015	1.015	1.016	1.016	1.017	1.016	1.017	1.017	1.018	1.018
30	1.010	1.011	1.011	1.011	1.012	1.011	1.012	1.012	1.012	1.013
40	1.006	1.007	1.007	1.007	1.007	1.007	1.007	1.007	1.008	1.008
60	1.003	1.003	1.003	1.003	1.003	1.003	1.004	1.004	1.004	1.004
120	1.001	1.001	1.001	1.001	1.001	1.001	1.001	1.001	1.001	1.001
∞	1.000	1.000	1.000	1.000	1.000	1.000	1.000	1.000	1.000	1.000
χ^2_{rm}	97.6796	103.0095	107.7834	113.5124	117.5242	107.5650	113.1453	118.1359	124.1163	128.2989

Table 9 (*Continued*)

	$m=9$									
	$r=11$					$r=12$				
M \ α	0.100	0.050	0.025	0.010	0.005	0.100	0.050	0.025	0.010	0.005
1	—	—	—	—	—	1.506	1.550	1.594	1.652	1.696
2	1.294	1.311	1.327	1.348	1.364	1.302	1.319	1.335	1.355	1.371
3	1.210	1.219	1.229	1.240	1.248	1.217	1.227	1.236	1.247	1.256
4	1.161	1.168	1.174	1.182	1.187	1.168	1.175	1.181	1.188	1.194
5	1.130	1.134	1.139	1.144	1.148	1.136	1.141	1.145	1.151	1.155
6	1.107	1.111	1.114	1.119	1.122	1.113	1.116	1.120	1.124	1.127
7	1.091	1.094	1.096	1.100	1.102	1.095	1.099	1.101	1.105	1.107
8	1.078	1.080	1.083	1.085	1.087	1.082	1.085	1.087	1.090	1.092
9	1.068	1.070	1.072	1.074	1.076	1.072	1.074	1.076	1.078	1.080
10	1.059	1.061	1.063	1.065	1.066	1.063	1.065	1.067	1.069	1.070
12	1.047	1.048	1.050	1.051	1.052	1.050	1.052	1.053	1.055	1.056
14	1.038	1.039	1.040	1.042	1.043	1.041	1.042	1.043	1.044	1.045
16	1.032	1.033	1.034	1.035	1.035	1.034	1.035	1.036	1.037	1.038
18	1.027	1.028	1.028	1.029	1.030	1.029	1.030	1.030	1.031	1.032
20	1.023	1.024	1.024	1.025	1.025	1.025	1.026	1.026	1.027	1.027
24	1.018	1.018	1.018	1.019	1.019	1.019	1.019	1.020	1.020	1.021
30	1.012	1.013	1.013	1.013	1.014	1.013	1.014	1.014	1.014	1.015
40	1.008	1.008	1.008	1.008	1.008	1.008	1.009	1.009	1.009	1.009
60	1.004	1.004	1.004	1.004	1.004	1.004	1.004	1.004	1.005	1.005
120	1.001	1.001	1.001	1.001	1.001	1.001	1.001	1.001	1.001	1.001
∞	1.000	1.000	1.000	1.000	1.000	1.000	1.000	1.000	1.000	1.000
χ^2_{rm}	117.407	123.225	128.422	134.642	138.987	127.2111	133.2569	138.6506	145.0988	149.5994

α / M	r = 13 0.100	0.050	0.025	0.010	0.005	r = 14 0.100	0.050	0.025	0.010	0.005
1	—	—	—	—	—	1.520	1.563	1.607	1.664	1.708
2	1.310	1.326	1.343	1.363	1.379	1.318	1.335	1.351	1.371	1.387
3	1.224	1.234	1.243	1.255	1.263	1.232	1.242	1.251	1.263	1.271
4	1.175	1.181	1.188	1.195	1.201	1.182	1.189	1.195	1.203	1.208
5	1.142	1.147	1.151	1.157	1.161	1.148	1.153	1.158	1.164	1.168
6	1.118	1.122	1.126	1.130	1.133	1.124	1.128	1.132	1.136	1.139
7	1.101	1.104	1.107	1.110	1.113	1.106	1.109	1.112	1.116	1.118
8	1.087	1.089	1.092	1.095	1.097	1.092	1.094	1.097	1.100	1.102
9	1.076	1.078	1.080	1.083	1.084	1.080	1.083	1.085	1.087	1.089
10	1.067	1.069	1.071	1.073	1.074	1.071	1.073	1.075	1.077	1.079
12	1.054	1.055	1.056	1.058	1.059	1.057	1.059	1.060	1.062	1.063
14	1.044	1.045	1.046	1.047	1.048	1.047	1.048	1.049	1.051	1.051
16	1.037	1.038	1.039	1.040	1.040	1.039	1.040	1.041	1.042	1.043
18	1.031	1.032	1.033	1.034	1.034	1.033	1.034	1.035	1.036	1.037
20	1.027	1.028	1.028	1.029	1.029	1.029	1.030	1.030	1.031	1.032
24	1.020	1.021	1.021	1.022	1.022	1.022	1.023	1.023	1.024	1.024
30	1.015	1.015	1.015	1.016	1.016	1.016	1.016	1.016	1.017	1.017
40	1.009	1.009	1.010	1.010	1.010	1.010	1.010	1.010	1.011	1.011
60	1.005	1.005	1.005	1.005	1.005	1.005	1.005	1.005	1.005	1.005
120	1.001	1.001	1.001	1.001	1.001	1.001	1.001	1.002	1.002	1.002
∞	1.000	1.000	1.000	1.000	1.000	1.000	1.000	1.000	1.000	1.000
χ^2_{rm}	136.982	143.246	148.829	155.496	160.146	146.7241	153.1979	158.9624	165.8410	170.6341

Table 9 (*Continued*)

$m = 9$

M \ α	$r=15$ 0.100	0.050	0.025	0.010	0.005	$r=16$ 0.100	0.050	0.025	0.010	0.005
1	—	—	—	—	—	1.536	1.579	1.622	1.679	1.722
2	1.326	1.343	1.359	—	—	1.335	1.352	1.368	1.389	1.404
3	1.240	1.250	1.259	1.271	1.279	1.248	1.258	1.267	1.279	1.288
4	1.189	1.196	1.202	1.210	1.216	1.196	1.203	1.210	1.218	1.223
5	1.155	1.160	1.165	1.170	1.174	1.161	1.166	1.171	1.177	1.181
6	1.130	1.134	1.138	1.142	1.145	1.136	1.140	1.144	1.148	1.152
7	1.111	1.115	1.118	1.121	1.124	1.117	1.120	1.123	1.127	1.130
8	1.097	1.099	1.102	1.105	1.107	1.102	1.104	1.107	1.110	1.112
9	1.085	1.087	1.089	1.092	1.094	1.089	1.092	1.094	1.097	1.099
10	1.075	1.077	1.079	1.081	1.083	1.079	1.082	1.083	1.086	1.087
12	1.061	1.062	1.064	1.065	1.066	1.064	1.066	1.067	1.069	1.070
14	1.050	1.051	1.052	1.054	1.055	1.053	1.054	1.056	1.057	1.058
16	1.042	1.043	1.044	1.045	1.046	1.045	1.046	1.047	1.048	1.049
18	1.036	1.037	1.037	1.038	1.039	1.038	1.039	1.040	1.041	1.042
20	1.031	1.032	1.032	1.033	1.034	1.033	1.034	1.035	1.035	1.036
24	1.024	1.024	1.025	1.025	1.026	1.026	1.026	1.027	1.027	1.028
30	1.017	1.017	1.018	1.018	1.018	1.018	1.019	1.019	1.020	1.020
40	1.011	1.011	1.011	1.012	1.012	1.012	1.012	1.012	1.012	1.013
60	1.005	1.006	1.006	1.006	1.006	1.006	1.006	1.006	1.006	1.006
120	1.002	1.002	1.002	1.002	1.002	1.002	1.002	1.002	1.002	1.002
∞	1.000	1.000	1.000	1.000	1.000	1.000	1.000	1.000	1.000	1.000
χ^2_{rm}	156.440	163.116	169.056	176.138	181.070	166.1318	173.0041	179.1137	186.3930	191.4585

			$m = 10$							
			$r = 10$					$r = 11$		
M \ α	0.100	0.050	0.025	0.010	0.005	0.100	0.050	0.025	0.010	0.005
1	1.496	1.540	1.584	1.641	1.586	—	—	—	—	—
2	1.291	1.308	1.324	1.345	1.360	1.296	1.313	1.329	1.349	—
3	1.208	1.217	1.226	1.238	1.246	1.213	1.222	1.231	1.243	1.251
4	1.160	1.166	1.172	1.180	1.185	1.165	1.171	1.177	1.185	1.190
5	1.128	1.133	1.137	1.143	1.147	1.133	1.138	1.142	1.148	1.152
6	1.106	1.110	1.113	1.117	1.120	1.111	1.114	1.118	1.122	1.125
7	1.090	1.093	1.095	1.099	1.101	1.094	1.097	1.099	1.103	1.105
8	1.077	1.079	1.082	1.084	1.086	1.081	1.083	1.085	1.088	1.090
9	1.067	1.069	1.071	1.073	1.075	1.070	1.072	1.074	1.077	1.078
10	1.059	1.061	1.062	1.064	1.066	1.062	1.064	1.065	1.067	1.069
12	1.047	1.048	1.049	1.051	1.052	1.049	1.051	1.052	1.054	1.055
14	1.038	1.039	1.040	1.041	1.042	1.040	1.041	1.042	1.044	1.044
16	1.031	1.032	1.033	1.034	1.035	1.034	1.034	1.035	1.036	1.037
18	1.027	1.027	1.028	1.029	1.029	1.028	1.029	1.030	1.031	1.031
20	1.023	1.023	1.024	1.025	1.025	1.024	1.025	1.026	1.026	1.027
24	1.017	1.018	1.018	1.019	1.019	1.019	1.019	1.020	1.020	1.020
30	1.012	1.013	1.013	1.013	1.013	1.013	1.013	1.014	1.014	1.014
40	1.008	1.008	1.008	1.008	1.008	1.008	1.008	1.009	1.009	1.009
60	1.004	1.004	1.004	1.004	1.004	1.004	1.004	1.004	1.004	1.005
120	1.001	1.001	1.001	1.001	1.001	1.001	1.001	1.001	1.001	1.001
∞	1.000	1.000	1.000	1.000	1.000	1.000	1.000	1.000	1.000	1.000
χ^2_{rm}	18.4980	124.3421	129.5612	135.8067	140.1695	129.385	135.480	140.917	147.414	151.948

Table 9 (*Continued*)

	$m=10$									
	$r=12$					$r=14$				
M \ α	0.100	0.050	0.025	0.010	0.005	0.100	0.050	0.025	0.010	0.005
1	1.500	1.543	1.585	1.641	1.684	1.509	1.551	1.593	1.648	1.690
2	1.302	1.318	1.334	1.354	1.369	1.315	1.331	1.347	1.367	1.382
3	1.219	1.228	1.237	1.248	1.257	1.232	1.241	1.250	1.261	1.269
4	1.170	1.177	1.183	1.190	1.196	1.182	1.189	1.195	1.203	1.208
5	1.138	1.143	1.148	1.153	1.157	1.149	1.154	1.159	1.164	1.168
6	1.115	1.119	1.123	1.127	1.130	1.126	1.129	1.133	1.137	1.141
7	1.098	1.101	1.104	1.107	1.110	1.107	1.111	1.113	1.117	1.119
8	1.085	1.087	1.090	1.092	1.094	1.093	1.096	1.098	1.101	1.103
9	1.074	1.076	1.078	1.081	1.082	1.082	1.084	1.086	1.089	1.090
10	1.065	1.067	1.069	1.071	1.072	1.073	1.075	1.076	1.078	1.080
12	1.052	1.054	1.055	1.057	1.058	1.058	1.060	1.061	1.063	1.064
14	1.043	1.044	1.045	1.046	1.047	1.048	1.049	1.051	1.052	1.053
16	1.036	1.037	1.038	1.039	1.039	1.040	1.042	1.042	1.043	1.044
18	1.030	1.031	1.032	1.033	1.033	1.035	1.035	1.036	1.037	1.038
20	1.026	1.027	1.027	1.028	1.029	1.030	1.031	1.031	1.032	1.033
24	1.020	1.020	1.021	1.021	1.022	1.023	1.024	1.024	1.025	1.025
30	1.014	1.015	1.015	1.015	1.015	1.016	1.017	1.017	1.018	1.018
40	1.009	1.009	1.009	1.010	1.010	1.010	1.011	1.011	1.011	1.011
60	1.004	1.005	1.005	1.005	1.005	1.005	1.005	1.006	1.006	1.006
120	1.001	1.001	1.001	1.001	1.001	1.002	1.002	1.002	1.002	1.002
∞	1.000	1.000	1.000	1.000	1.000	1.000	1.000	1.000	1.000	1.000
χ^2_{rm}	140.2326	146.5674	152.2114	158.9502	163.6482	161.8270	168.6130	174.6478	181.8403	186.8468

$m = 12$

$r = 12$

M \ α	0.100	0.050	0.025	0.010	0.005
1	1.495	1.536	1.576	1.630	1.671
2	1.306	1.322	1.337	1.356	1.371
3	1.225	1.234	1.243	1.254	1.262
4	1.177	1.184	1.190	1.197	1.202
5	1.145	1.150	1.154	1.160	1.163
6	1.122	1.126	1.129	1.133	1.136
7	1.104	1.107	1.110	1.114	1.116
8	1.091	1.093	1.095	1.098	1.100
9	1.080	1.082	1.084	1.086	1.088
10	1.071	1.072	1.074	1.076	1.078
12	1.057	1.058	1.060	1.061	1.062
14	1.047	1.048	1.049	1.050	1.051
16	1.039	1.040	1.041	1.042	1.043
18	1.034	1.034	1.035	1.036	1.037
20	1.029	1.030	1.030	1.031	1.032
24	1.022	1.023	1.023	1.024	1.024
30	1.016	1.016	1.017	1.017	1.017
40	1.010	1.010	1.011	1.011	1.011
60	1.005	1.005	1.005	1.005	1.006
120	1.001	1.002	1.002	1.002	1.002
∞	1.000	1.000	1.000	1.000	1.000
χ^2_{rm}	166.1318	173.0041	179.1137	186.3930	191.4585

Bibliography

Anderson, G. A. (1965). An asymptotic expansion for the distribution of the latent roots of the estimated covariance matrix. *Ann. Math. Statist.*, **36**, 1153–1173.

Anderson, T. W. (1946). The noncentral Wishart distribution and certain problems of multivariate statistics. *Ann. Math. Statist.*, **17**, 409–431.

Anderson, T. W. (1951). Estimating linear restrictions on regression coefficients for multivariate normal distributions. *Ann. Math. Statist.*, **22**, 327–351.

Anderson, T. W. (1958). *An Introduction to Multivariate Statistical Analysis.* John Wiley & Sons, New York.

Anderson, T. W. (1963). Asymptotic theory for principal component analysis. *Ann. Math. Statist.*, **34**, 122–148.

Anderson, T. W., and Das Gupta, S. (1964). Monotonicity of the power functions and some tests of independence between two sets of variates. *Ann. Math. Statist.*, **35**, 206–208.

Baranchik, A. J. (1973). Inadmissibility of maximum likelihood estimators in some multiple regression problems with three or more independent variables. *Ann. Statist.*, **1**, 312–321.

Bartlett, M. S. (1933). On the theory of statistical regression. *Proc. R. Soc. Edinb.*, **53**, 260–283.

Bartlett, M. S. (1937). Properties of sufficiency and statistical tests. *Proc. R. Soc. Lond. A*, **160**, 268–282.

Bartlett, M. S. (1938). Further aspects of the theory of multiple regression *Proc Camb. Philos. Soc*, **34**, 33–40.

Bartlett, M. S. (1947). Multivariate analysis. *J. R. Statist. Soc.* (Suppl.), **9**, 176–190.

Bartlett, M. S. (1954). A note on multiplying factors for various χ^2 approximations. *J. R. Statist. Soc. Ser. B*, **16**, 296–298.

Bellman, R. (1970). *Introduction to Matrix Analysis*, 2nd ed. McGraw-Hill, New York.

Berger, J. (1980a). A robust generalized Bayes estimator and confidence region for a multivariate normal mean. *Ann. Statist.*, **8**, 716–761

Berger, J. (1980b). Improving on inadmissible estimators in continuous exponential families with applications to simultaneous estimation of gamma scale parameters. *Ann. Statist.*, **8**, 545–571.

Berger, J., Bock, M. E., Brown, L. D. Casella, G., and Gleser, L. (1977). Minimax estimation of a normal mean vector for arbitrary quadratic loss and unknown covariance matrix. *Ann. Statist.*, **5**, 763–771.

Bickel, P. J., and Doksum, K. A. (1977). *Mathematical Statistics: Basic Ideas and Selected Topics.* Holden-Day, San Francisco

Bishop, Y. M., Fienberg, S. E., and Holland, P. W. (1975). *Discrete Multivariate Analysis: Theory and Practice.* M.I.T. Press, Cambridge, Mass.

Box, G. E. P. (1949) A general distribution theory for a class of likelihood criteria *Biometrika*, **36**, 317–346.

Brandwein, A. R. C., and Strawderman, W E. (1978). Minimax estimation of location parameters for spherically symmetric unimodal distributions under quadratic loss. *Ann. Statist.*, **6**, 377–416.

Brandwein, A R C., and Strawderman, W. E. (1980). Minimax estimation of location parameters for spherically symmetric distributions with concave loss. *Ann. Statist.*, **8**, 279–284.

Brown, G. W. (1939). On the power of the L_1 test for equality of several variances. *Ann. Math. Statist.*, **10**, 119–128.

Brown, L. D. (1966). On the admissibility of invariant estimators of one or more location parameters. *Ann. Math. Statist.*, **37**, 1087–1135.

Brown, L D. (1980). Examples of Berger's phenomenon in the estimation of independent normal means. *Ann. Statist.*, **8**, 572–585.

Cartan, E (1922) *Leçons sur les invariants intégraux.* Hermann, Paris.

Cartan, H. (1967). *Formes différentielles.* Hermann, Paris.

Carter, E. M., and Srivastava, M. S. (1977). Monotonicity of the power functions of the modified likelihood ratio criterion for the homogeneity of variances and of the sphericity test *J. Multivariate Anal.* **7**, 229–233.

Chang, T. C., Krishnaiah, P. R., and Lee, J. C. (1977). Approximations to the distributions of the likelihood ratio statistics for testing the hypotheses on covariance matrices and mean vectors simultaneously. In *Applications of Statistics* (P. R. Krishnaiah, ed), 97–108. North-Holland Pub., Amsterdam.

Chen, C. W. (1971). On some problems in canonical correlation analysis. *Biometrika*, **58**, 399–400.

Chmielewski, M. A. (1981). Elliptically symmetric distributions: A review and bibliography. *Int. Statist. Rev.*

Chou, R., and Muirhead, R. J. (1979). On some distribution problems in MANOVA and discriminant analysis. *J. Multivariate Anal.*, **9**, 410–419.

Clem, D. S., Krishnaiah, P. R., and Waikar, V. B. (1973). Tables for the extreme roots of the Wishart matrix. *J. Statist. Comp. Simul.*, **2**, 65–92.

Constantine, A. G. (1963). Some noncentral distribution problems in multivariate analysis. *Ann. Math. Statist.*, **34**, 1270–1285.

Constantine, A G. (1966). The distribution of Hotelling's generalized T_0^2. *Ann. Math. Statist.*, **37**, 215–225.

Constantine, A. G., and Muirhead, R. J. (1972). Partial differential equations for hypergeometric functions of two argument matrices. *J. Multivariate Anal.*, **3**, 332—338

Constantine, A G., and Muirhead, R. J. (1976). Asymptotic expansions for distributions of latent roots in multivariate analysis. *J. Multivariate Anal.*, **6**, 369–391.

Consul, P. C. (1967). On the exact distributions of likelihood ratio criteria for testing independence of sets of variates under the null hypothesis. *Ann. Math. Statist.*, **38**, 1160–1169.

Cook, M. B. (1951). Bivariate k-statistics and cumulants of their joint sampling distribution. *Biometrika*, **38**, 179–195.

Cramér, H. (1937). *Random Variables and Probability Distributions.* Cambridge Tracts, No. 36. Cambridge University Press, London and New York.

Cramér, H. (1946). *Mathematical Methods of Statistics.* Princeton University Press, Princeton, N.J.

Das Gupta, S. (1969). Properties of power functions of some tests concerning dispersion matrices of multivariate normal distributions. *Ann. Math. Statist.*, **40**, 697–701

Das Gupta, S. (1971). Non-singularity of the sample covariance matrix. *Sankhya A*, **33**, 475–478.

Das Gupta, S., Anderson, T. W., and Mudholkar, G. S. (1964). Monotonicity of the power functions of some tests of the multivariate linear hypothesis. *Ann. Math. Statist.*, **35**, 200–205.

Das Gupta, S., and Giri, N. (1973). Properties of tests concerning covariance matrices of normal distributions. *Ann. Statist.*, **1**, 1222–1224.

David, F. N. (1938). *Tables of the Correlation Coefficient.* Cambridge University Press, London and New York.

Davis, A. W. (1968). A system of linear differential equations for the distribution of Hotelling's generalized T_0^2. *Ann. Math. Statist.* 39, 815–832.

Davis, A. W. (1970). Exact distributions of Hotelling's generalized T_0^2. *Biometrika*, **57**, 187–191.

Davis, A. W. (1971). Percentile approximations for a class of likelihood ratio criteria. *Biometrika*, **58**, 349–356.

Davis, A. W. (1977). Asymptotic theory for principal component analysis: Non-normal case. *Austral. J. Statist.*, **19**, 206–212.

Davis, A. W. (1979). On the differential equation for Meijer's $G_{p,p}^{p,0}$ function, and further tables of Wilks's likelihood ratio criterion. *Biometrika*, **66**, 519–531.

Davis, A W. (1980). Further tabulation of Hotelling's generalized T_0^2. *Commun. Statist.-Simula. Computa.*, **B9**, 321–336.

Davis, A. W., and Field, J. B. F. (1971). Tables of some multivariate test criteria. Tech. Rept. No. 32, Division of Mathematical Statistics, C.S.I.R.O., Canberra, Australia.

Deemer, W. L., and Olkin, I. (1951). The Jacobians of certain matrix transformations useful in multivariate analysis. *Biometrika*, **38**, 345–367.

Dempster, A. P. (1969). *Elements of Continuous Multivariate Analysis.* Addison-Wesley, Reading, Mass.

Devlin, S. J., Gnanadesikan, R., and Kettenring, J. R. (1976). Some multivariate applications of elliptical distributions. In *Essays in Probability and Statistics* (S. Ikeda, ed.), pp. 365–395. Shinko Tsusho, Tokyo.

Dykstra, R. L. (1970). Establishing the positive definiteness of the sample covariance matrix. *Ann. Math. Statist.*, **41**, 2153–2154.

Eaton, M. L. (1972). *Multivariate Statistical Analysis* Institute of Mathematical Statistics, University of Copenhagen.

Eaton, M. L. (1976). A maximization problem and its application to canonical correlation. *J. Multivariate Anal.*, **6**, 422–425

Eaton, M. L. (1977) *N*-Dimensional versions of some symmetric univariate distributions. Tech. Rept. No 288, University of Minnesota.

Eaton, M. L. (1981). On the projections of isotropic distributions. *Ann. Statist.*, **9**, 391–400.

Eaton, M. L., and Perlman, M. D. (1973). The non-singularity of generalized sample covariance matrices. *Ann. Statist.*, **1**, 710–717

Efron, B. (1969). Student's *t*-test under symmetry conditions. *J. Am. Statist. Assoc.*, **64**, 1278–1302.

Efron, B, and Morris, C. (1973a). Stein's estimation rule and its competitors: An empirical Bayes approach. *J Am. Statist. Assoc.*, **68**, 117–130.

Efron, B., and Morris, C. (1973b). Combining possibly related estimation problems. *J. R. Statist. Soc. B*, **35**, 379–421.

Efron, B., and Morris, C (1975). Data analysis using Stein's estimator and its generalizations. *J. Am. Statist. Assoc.*, **70**, 311–319.

Efron, B., and Morris, C. (1976). Multivariate empirical Bayes and estimation of covariance matrices. *Ann. Statist.*, **4**, 22–32.

Efron, B., and Morris, C. (1977). Stein's paradox in statistics. *Sci. Am. 237*, pp. 119–127.

Erdélyi, A., Magnus, W., Oberhettinger, F., and Tricomi, F. G. (1953a). *Higher Transcendental Functions*, Vol. I McGraw-Hill, New York.

Erdélyi, A., Magnus, W., Obergettinger, F., and Tricomi, F. G. (1953b). *Higher Transcendental Functions*, Vol. II. McGraw-Hill, New York.

Erdélyi, A., Magnus, W., Oberhettinger, F., and Tricomi, F. G. (1954). *Tables of Integral Transforms*, Vol, I McGraw-Hill, New York.

Farrell, R. H. (1976). *Techniques of Multivariate Calculation*. Springer, New York.

Feller, W. (1971) *An Introduction to Probability Theory and Its Applications*, 2nd ed., Vol. II. John Wiley & Sons, New York.

Ferguson, T. S. (1967). *Mathematical Statistics: A Decision Theoretic Approach* Academic Press, New York

Fisher, R A (1915). Frequency distribution of the values of the correlation coefficient in samples from an indefinitely large population. *Biometrika*, **10**, 507–521.

Fisher, R. A. (1921). On the probable error of a coefficient of correlation deduced from a small sample. *Metron*, **1**, 3–32.

Fisher, R. A (1928). The general sampling distribution of the multiple correlation coefficient. *Proc. R. Soc. Lond. A*, **121**, 654–673.

Fisher, R. A. (1936). The use of multiple measurement in taxonomic problems. *Ann. Eugen.* **7**, 179–188.

Fisher, R A. (1939). The sampling distribution of some statistics obtained from non-linear equations *Ann. Eugen.*, **9**, 238–249.

Flanders, H. (1963). *Differential Forms with Applications to the Physical Sciences*. Academic Press, New York

Fujikoshi, Y. (1968). Asymptotic expansion of the distribution of the generalized variance in the non-central case. *J. Sci. Hiroshima Univ. Ser. A-1*, **32**, 293–299.

Fujikoshi, Y. (1970). Asymptotic expansions of the distributions of test statistics in multivariate analysis. *J. Sci. Hiroshima Univ. Ser. A-1*, **34**, 73–144.

Fujikoshi, Y (1973). Asymptotic formulas for the distributions of three statistics for multivariate linear hypothesis. *Ann. Inst. Statist. Math.*, **25**, 423–437

Fujikoshi, Y. (1974a). The likelihood ratio tests for the dimensionality of regression coefficients *J Multivariate Anal.* **4**, 327–340.

Fujikoshi, Y. (1974b). On the asymptotic non-null distributions of the LR criterion in a general MANOVA. *Canadian J. Statist.*, **2**, 1–12.

Gajjar, A. V. (1967). Limiting distributions of certain transformations of multiple correlation coefficient. *Metron*, **26**, 189–193.

Gayen, A. K (1951). The frequency distribution of the product-moment correlation coefficient in random samples of any size drawn from non-normal universes. *Biometrika*, **38**, 219–247.

Ghosh, B. K. (1966). Asymptotic expansions for the moments of the distribution of correlation coefficient. *Biometrika*, **53**, 258.

Giri, N. C. (1977). *Multivariate Statistical Inference.* Academic Press, New York.

Girshick, M. A. (1939). On the sampling theory of roots of determinantal equations. *Ann. Math. Statist.*, **10**, 203–224.

Gleser, L. J. (1966). A note on the sphericity test. *Ann. Math. Statist.*, **37**, 464–467.

Gleser, L. J., and Olkin, I. (1970). Linear models in multivariate analysis. In *Essays in Probability and Statistics* (R. C. Bose, ed.), pp. 267–292. University of North Carolina Press, Chapel Hill.

Glynn, W. J (1977). Asymptotic distributions of latent roots in canonical correlation analysis and in discriminant analysis with applications to testing and estimation. Ph.D. Thesis, Yale University, New Haven, Conn.

Glynn, W. J. (1980). Asymptotic representations of the densities of canonical correlations and latent roots in MANOVA when the population parameters have arbitrary multiplicity. *Ann. Statist.*, **8**, 958–976.

Glynn, W J., and Muirhead, R. J. (1978). Inference in canonical correlation analysis. *J. Multivariate Anal.*, **8**, 468–478.

Gnanadesikan, R. (1977). *Statistical Data Analysis of Multivariate Observations.* John Wiley & Sons, New York.

Graybill, F. A. (1961). *An Introduction to Linear Statistical Models*, Vol. I. McGraw-Hill, New York.

Graybill, F. A. (1969). *Introduction To Matrices With Applications In Statistics.* Wadsworth, Belmont, CA.

Gurland, J. (1968). A relatively simple form of the distribution of the multiple correlation coefficient *J. R. Statist. Soc. B*, **30**, 276–283.

Haar, A. (1933). Der Massbegriff in der Theorie der kontinuierlichen Gruppen. *Ann. Math.*, **34**, 147–169.

Haff, L. R. (1977). Minimax estimators for a multinormal precision matrix. *J. Multivariate Anal*, **7**, 374–385.

Haff, L. R. (1979). Estimation of the inverse covariance matrix: Random mixtures of the inverse Wishart matrix and the identity. *Ann. Statist.*, **7**, 1264–1276.

Haff, L. R. (1980). Empirical Bayes estimation of the multivariate normal covariance matrix *Ann. Statist.*, **8**, 586–597.

Halmos, P. R. (1950). *Measure Theory.* Van Nostrand Reinhold, New York.

Hanumara, R. C., and Thompson, W. A. (1968). Percentage points of the extreme roots of a Wishart matrix. *Biometrika*, **55**, 505–512.

Heck, D. L. (1960). Charts of some upper percentage points of the distribution of the largest characteristic root. *Ann. Math. Statist.*, **31**, 625–642.

Herz, C. S. (1955). Bessel functions of matrix argument. *Ann. Math.*, **61**, 474–523.

Hotelling, H. (1931). The generalization of Student's ratio. *Ann. Math. Statist.*, **2**, 360–378.

Hotelling, H. (1933). Analysis of a complex of statistical variables into principal components. *J. Educ. Psychol.*, **24**, 417–441, 498–520.

Hotelling, H. (1936). Relations between two sets of variates. *Biometrika*, **28**, 321–377.

Hotelling, H. (1947). Multivariate quality control, illustrated by the air testing of sample bombsights. *Techniques of Statistical Analysis*, pp. 111–184. McGraw-Hill, New York.

Hotelling, H. (1953). New light on the correlation coefficient and its transforms. *J. R. Stat. Soc. B.*, **15**, 193–225.

Hsu, L. C. (1948). A theorem on the asymptotic behavior of a multiple integral. *Duke Math. J.*, **15**, 623–632.

Hsu, P. L. (1939). On the distribution of the roots of certain determinantal equations. *Ann. Eugen.*, **9**, 250–258.

Hsu, P. L. (1941a). On the limiting distribution of roots of a determinantal equation. *J. Lond. Math. Soc.*, **16**, 183–194.

Hsu, P. L. (1941b). On the limiting distribution of the canonical correlations. *Biometrika*, **32**, 38–45.

Hughes, D T, and Saw, J. G. (1972). Approximating the percentage points of Hotelling's generalized T_0^2 statistic. *Biometrika*, **59**, 224–226.

Ingham, A. E. (1933). An integral which occurs in statistics. *Proc. Cambr. Philos. Soc.*, **29**, 271–276.

Ito, K. (1956). Asymptotic formulae for the distribution of Hotelling's generalized T_0^2 statistic. *Ann. Math. Statist.*, **27**, 1091–1105.

Ito, K. (1960). Asymptotic formulae for the distribution of Hotelling's generalized T_0^2 statistic. II. *Ann. Math. Statist.*, **31**, 1148–1153.

James, A. T. (1954). Normal multivariate analysis and the orthogonal group. *Ann. Math. Statist.*, **25**, 40–75.

James, A. T. (1960). The distribution of the latent roots of the covariance matrix. *Ann. Math. Statist.*, **31**, 151–158.

James, A. T. (1961a). The distribution of noncentral means with known covariance. *Ann. Math. Statist.*, **32**, 874–882.

James, A. T. (1961b). Zonal polynomials of the real positive definite symmetric matrices. *Ann. Math.*, **74**, 456–469.

James, A. T. (1964). Distributions of matrix variates and latent roots derived from normal samples. *Ann. Math. Statist.*, **35**, 475–501.

James, A. T. (1968). Calculation of zonal polynomial coefficients by use of the Laplace–Beltrami operator. *Ann. Math. Statist.*, **39**, 1711–1718.

James, A. T. (1969). Test of equality of the latent roots of the covariance matrix. In *Multivariate Analysis* (P. R. Krishnaiah, ed.), Vol. II, pp. 205–218. Academic Press, New York.

James, A. T. (1973). The variance information manifold and the functions on it. In *Multivariate Analysis* (P. R. Krishnaiah, ed.), Vol. III, pp. 157–169. Academic Press, New York.

James, A. T. (1976). Special functions of matrix and single argument in statistics. In *Theory and*

Applications of Special Functions (R. A. Askey, ed.), pp. 497–520. Academic Press, New York.

James, W., and Stein, C. (1961). Estimation with quadratic loss. *Proc. Fourth Berkeley Symp. Math. Statist. Prob.*, Vol. 1, pp. 361–379.

John, S (1971). Some optimal multivariate tests. *Biometrika*, **38**, 123–127.

John, S (1972). The distribution of a statistic used for testing sphericity of normal distributions. *Biometrika*, **39**, 169–174.

Johnson, N. L., and Kotz, S. (1970). *Continuous Univariate Distributions*, Vol. 2. John Wiley & Sons, New York.

Kagan, A., Linnik, Y. V., and Rao, C. R. (1972). *Characterization Problems of Mathematical Statistics.* John Wiley & Sons, New York.

Kariya, T (1978). The general MANOVA problem. *Ann. Statist.*, **6**, 200–214.

Kariya, T. (1981). A robustness property of Hotelling's T^2-test. *Ann. Statist.*, **9**, 210–213.

Kariya, T., and Eaton, M. L. (1977). Robust tests for spherical symmetry. *Ann. Statist.*, **5**, 206–215.

Kates, L. K. (1980). Zonal polynomials. Ph.D. Thesis, Princeton University.

Kelker, D. (1970). Distribution theory of spherical distributions and a location–scale parameter generalization. *Sankhya A*, **32**, 419–430.

Kendall, M. G., and Stuart, A. (1969). *The Advanced Theory of Statistics*, Vol. 1. Macmillan (Hafner Press), New York.

Khatri, C. G. (1959). On the mutual independence of certain statistics. *Ann. Math. Statist.*, **30**, 1258–1262.

Khatri, C. G. (1967). Some distribution problems associated with the characteristic roots of $S_1S_2^{-1}$. *Ann. Math. Statist.*, **38**, 944–948.

Khatri, C. G. (1972). On the exact finite series distribution of the smallest or the largest root of matrices in three situations. *J. Multivariate Anal.* **2**, 201–207.

Khatri, C. G., and Pillai, K. C. S. (1968). On the noncentral distributions of two test criteria in multivariate analysis of variance. *Ann. Math. Statist.*, **39**, 215–226.

Khatri, C. G., and Srivastava, M. S. (1971). On exact non-null distributions of likelihood ratio criteria for sphericity test and equality of two covariance matrices. *Sankhya A*, **33**, 201–206.

Khatri, C. G., and Srivastava, M S. (1974). Asymptotic expansions of the non-null distributions of likelihood ratio criteria for covariance matrices. *Ann. Statist.*, **2**, 109–117.

Kiefer, J., and Schwartz, R. (1965). Admissible Bayes character of T^2—and R^2—and other fully invariant tests for classical normal problems. *Ann. Math. Statist.*, **36**, 747–760.

King, M L (1980). Robust tests for spherical symmetry and their application to least squares regression. *Ann. Statist.*, **8**, 1265–1272.

Korin, B. P (1968). On the distribution of a statistic used for testing a covariance matrix. *Biometrika*, **55**, 171–178

Krishnaiah, P. R., and Lee, J. C. (1979). Likelihood ratio tests for mean vectors and covariance matrices. Tech. Rept. No. 79-4; Dept. of Mathematics and Statistics, University of Pittsburgh.

Krishnaiah, P. R., and Schuurmann, F. J. (1974). On the evaluation of some distributions that arise in simultaneous tests for the equality of the latent roots of the covariance matrix. *J. Multivariate Anal.*, **4**, 265–282.

Kshirsagar, A. M. (1961). The noncentral multivariate beta distribution. *Ann. Math. Statist*, **32**, 104–111.

Kshirsagar, A. M. (1972). *Multivariate Analysis*. Dekker, New York.

Lawley, D. N. (1938). A generalization of Fisher's z test. *Biometrika*, **30**, 180–187.

Lawley, D. N. (1956). Test of significance for the latent roots of covariance and correlation matrices. *Biometrika*, **43**, 128–136.

Lawley, D. N (1959). Tests of significance in canonical analysis. *Biometrika*, **46**, 59–66

Lee, J. C., Chang, T. C., and Krishnaiah, P. R. (1977). Approximations to the distributions of the likelihood ratio statistics for testing certain structures on the covariance matrices of real multivariate normal populations. In *Multivariate Analysis*, (P. R. Krishnaiah, Ed.), Vol. IV, pp. 105–118. North-Holland, Publ., Amsterdam.

Lee, Y. S. (1971a). Distribution of the canonical correlations and asymptotic expansions for distributions of certain independence test statistics. *Ann. Math. Statist.*, **42**, 526–537.

Lee, Y. S. (1971b). Asymptotic formulae for the distribution of a multivariate test statistic: power comparisons of certain multivariate tests. *Biometrika*, **58**, 647–651.

Lee, Y. S. (1972). Some results on the distribution of Wilks's likelihood ratio criterion. *Biometrika*, **59**, 649–664

Lehmann, E. L. (1959). *Testing Statistical Hypotheses*. John Wiley & Sons, New York.

MacDuffee, C. C. (1943). *Vectors and Matrices*. Mathematical Association of America, Menasha, Wisconsin.

Magnus, J. R., and Neudecker, H. (1979). The commutation matrix: Some properties and applications. *Ann. Statist.*, **7**, 381–394.

Mahalanobis, P. C. (1930). On the generalized distance in statistics. *Proc. Natl. Inst. Soc. India*, **12**, 49–55.

Mathai, A. M., and Saxena, R. K. (1978). *The H-Function with Applications in Statistics and Other Disciplines*. John Wiley & Sons, New York.

Mauchly, J. W. (1940). Significance test for sphericity of a normal n-variate distribution. *Ann. Math. Statist.*, **11**, 204–209.

McLaren, M L. (1976). Coefficients of the zonal polynomials. *Appl. Statist.*, **25**, 82–87.

Mikhail, N. N. (1965). A comparison of tests of the Wilks–Lawley hypothesis in multivariate analysis *Biometrika*, **52**, 149–156.

Mirsky, L. (1955). *Introduction to Linear Algebra*. Oxford University Press, London and New York.

Mood, A. M. (1951). On the distribution of the characteristic roots of normal second-moment matrices. *Ann Math. Statist.*, **22**, 266–273.

Moran, P. A. P. (1980). Testing the largest of a set of correlation coefficients. *Austral. J. Statist.*, **22**, 289–297.

Muirhead, R. J. (1970a). Partial differential equations for hypergeometric functions of matrix argument. *Ann. Math. Statist.*, **41**, 991–1001.

Muirhead, R. J. (1970b). Asymptotic distributions of some multivariate tests. *Ann. Math. Statist.*, **41**, 1002–1010.

Muirhead, R J. (1972a). On the test of independence between two sets of variates. *Ann. Math. Statist.*, **43**, 1491–1497.

Muirhead, R J. (1972b). The asymptotic noncentral distribution of Hotelling's generalized T_0^2. *Ann. Math. Statist*, **43**, 1671–1677.

Muirhead, R. J. (1974). Powers of the largest latent root test of $\Sigma = I$. *Comm. Statist*, **3**, 513–524.

Muirhead, R. J. (1978). Latent roots and matrix variates: A review of some asymptotic results. *Ann. Statist.*, **6**, 5–33.

Muirhead, R. J., and Chikuse, Y. (1975a) Asymptotic expansions for the joint and marginal distributions of the latent roots of the covariance matrix. *Ann. Statist.*, **3**, 1011–1017.

Muirhead, R. J. and Chikuse, Y. (1975b). Approximations for the distributions of the extreme latent roots of three matrices. *Ann. Inst. Statist. Math.*, **27**, 473–478.

Muirhead, R. J., and Waternaux, C. M. (1980). Asymptotic distributions in canonical correlation analysis and other multivariate procedures for nonnormal populations. *Biometrika*, **67**, 31–43.

Nachbin, L. (1965). *The Haar Integral.* Van Nostrand–Reinhold, New York.

Nagao, H (1967). Monotonicity of the modified likelihood ratio test for a covariance matrix. *J. Sci. Hiroshima Univ. Ser. A-I*, **31**, 147–150.

Nagao, H. (1970). Asymptotic expansions of some test criteria for homogeneity of variances and covariance matrices from normal populations. *J. Sci. Hiroshima Univ. Ser. A-I*, **34**, 153–247.

Nagao, H. (1972). Non-null distributions of the likelihood ratio criteria for independence and equality of mean vectors and covariance matrices. *Ann. Inst. Statist. Math.*, **24**, 67–79.

Nagao, H (1973a). On some test criteria for covariance matrix. *Ann. Statist.* **1**, 700–709.

Nagao, H. (1973b). Asymptotic expansions of the distributions of Bartlett's test and sphericity test under the local alternatives. *Ann. Inst. Statist. Math.*, **25**, 407–422.

Nagao, H. (1974). Asymptotic non-null distributions of two test criteria for equality of covariance matrices under local alternatives. *Ann. Inst. Statist. Math.*, **26**, 395–402.

Nagarsenker, B. N. and Pillai, K. C. S. (1973a). The distribution of the sphericity test criterion. *J. Multivariate Anal.* **3**, 226–235.

Nagarsenker, B. N., and Pillai, K. C. S. (1973b). Distribution of the likelihood ratio criterion for testing a hypothesis specifying a covariance matrix. *Biometrika*, **60**, 359–394.

Nagarsenker, B. N., and Pillai, K. C. S. (1974). Distribution of the likelihood ratio criterion for testing $\Sigma=\Sigma_0$, $\mu=\mu_0$. *J. Multi. Analysis*, **4**, 114–122.

Narain, R. D. (1950). On the completely unbiased character of tests of independence in multivariate normal systems. *Ann. Math. Statist.*, **21**, 293–298.

Neudecker, H. (1969). Some theorems on matrix differentiation with special reference to Kronecker matrix products. *J. Am. Statist. Assoc.*, **64**, 953–963.

Ogasawara, T., and Takahashi, M. (1951). Independence of quadratic forms in normal system. *J. Sci. Hiroshima University*, **15**, 1–9.

Olkin, I. (1953). Note on the Jacobians of certain matrix transformations useful in multivariate analysis. *Biometrika*, **40**, 43–46.

Olkin, I., and Pratt, J. W. (1958). Unbiased estimation of certain correlation coefficients. *Ann. Math. Statist.*, **29**, 201–211.

Olkin, I, and Roy S. N. (1954). On multivariate distribution theory. *Ann. Math. Statist.*, **25**, 329–339.

Olkin, I., and Rubin, H. (1964). Multivariate beta distributions and independence properties of the Wishart distribution. *Ann. Math. Statist.*, **35**, 261–269.

Olkin, I., and Selliah, J. B. (1977). Estimating covariances in a multivariate normal distribution. In *Statistical Decision Theory and Related Topics* (S. S. Gupta and D. S. Moore, eds.) Vol. II, pp. 313–326. Academic Press, New York.

Parkhurst, A. M., and James, A. T. (1974). Zonal polynomials of order 1 through 12. In *Selected Tables in Mathematical Statistics* (H. L. Harter and D. B Owen, eds.), pp. 199–388. American Mathematical Society, Providence, R.I.

Perlman, M. D. (1980). Unbiasedness of the likelihood ratio tests for equality of several covariance matrices and equality of several multivariate normal populations. *Ann. Statist.*, **8**, 247–263.

Perlman, M. D., and Olkin, I. (1980). Unbiasedness of invariant tests for MANOVA and other multivariate problems. *Ann. Statist.*, **8**, 1326–1341.

Pillai, K C. S. (1955). Some new test criteria in multivariate analysis. *Ann. Math. Statist*, **26**, 117–121.

Pillai, K. C. S. (1956). Some results useful in multivariate analysis. *Ann. Math. Statist.*, **27**, 1106–1114.

Pillai, K. C. S. (1964). On the distribution of the largest of seven roots of a matrix in multivariate analysis. *Biometrika*, **51**, 270–275.

Pillai, K. C. S. (1965). On the distribution of the largest characteristic root of a matrix in multivariate analysis. *Biometrika*, **52**, 405–414.

Pillai, K. C. S. (1967). On the distribution of the largest root of a matrix in multivariate analysis *Ann. Math Statist.*, **38**, 616–617.

Pillai, K. C. S. (1976). Distribution of characteristic roots in multivariate analysis. Part 1: Null distributions. *Can. J Statist.*, **4**, 157–184.

Pillai, K. C. S. (1977). Distributions of characteristic roots in multivariate analysis. Part 2: Non-null distributions. *Can. J. Statist.*, **5**, 1–62.

Pillai, K. C. S., and Bantegui, C. G. (1959). On the distribution of the largest of six roots of a matrix in multivariate analysis. *Biometrika*, **46**, 237–240.

Pillai, K. C. S., and Gupta, A. K. (1969). On the exact distribution of Wilks's criterion. *Biometrika*, **56**, 109–118.

Pillai, K. C. S., and Jayachandran, K. (1967). Power comparisons of tests of two multivariate hypotheses based on four criteria. *Biometrika*, **54**, 195–210.

Pillai, K. C. S., and Jayachandran, K. (1968). Power comparisons of tests of equality of two covariance matrices based on four criteria. *Biometrika*, **55**, 335–342.

Pillai, K. C. S., and Jouris, G. M. (1969). On the moments of elementary symmetric functions of the roots of two matrices. *Ann. Inst. Statist. Math.*, **21**, 309–320.

Pillai, K. C. S., and Nagarsenker, B. N. (1972). On the distributions of a class of statistics in multivariate analysis. *J. Multivariate Anal.*, **2**, 96–114.

Pillai, K. C. S., and Sampson, P. (1959). On Hotelling's generalization of T^2. *Biometrika*, **46**, 160–168.

Pitman, E. J. G. (1937). Significance tests which may be applied to samples from any population. II. The correlation coefficient test. *J. Roy. Statist. Soc.* (*Suppl.*), **4**, 225.

Pitman, E. J. G. (1939). Tests of hypotheses concerning location and scale parameters. *Biometrika*, **31**, 200–215.

Potthoff, R. F., and Roy, S. N. (1964). A generalized multivariate analysis of variance model useful especially for growth curve problems. *Biometrika*, **51**, 313–326.

Rainville, E. D. (1960). *Special Functions*. Macmillan, New York.

Rao, C. R. (1951). An asymptotic expansion of the distribution of Wilks' Λ-criterion. *Bull. Inst. Int. Statist.* **33**, Pt. II, 177–180.

Rao, C. R. (1973). *Linear Statistical Inference and Its Applications*, 2nd ed. John Wiley & Sons, New York.

Roussas, G. G (1973). *A First Course in Mathematical Statistics*. Addison-Wesley, Reading, Mass

Roy, S. N. (1939). *p*-Statistics or some generalization in analysis of variance appropriate to multivariate problems. *Sankhya*, **4**, 381–396.

Roy, S. N. (1953). On a heuristic method of test construction and its use in multivariate analysis. *Ann. Math. Statist.*, **24**, 220–238.

Roy, S. N (1957). *Some Aspects of Multivariate Analysis*. John Wiley & Sons, New York.

Saw, J. G (1977). Zonal polynomials: An alternative approach. *J. Multivariate Anal.*, **7**, 461–467.

Schatzoff, M (1966a). Exact distributions of Wilks' likelihood ratio criterion. *Biometrika*, **53**, 347–358.

Schatzoff, M. (1966b). Sensitivity comparisons among tests of the general linear hypothesis. *J. Am. Statist. Assoc.*, **61**, 415–435.

Searle, S. R. (1971). *Linear Models* John Wiley & Sons, New York.

Seber, G. A F (1977). *Linear Regression Analysis*. John Wiley & Sons, New York.

Siegel, C. L (1935). Über die analytische theorie der quadratischen formen. *Ann. Math.*, **36**, 527–606.

Simaika, J. B. (1941) On an optimum property of two important statistical tests. *Biometrika*, **32**, 70–80.

Siotani, M. (1957). Note on the utilization of the generalized student ratio in the analysis of variance or dispersion. *Ann. Inst. Stat. Math.*, **9**, 157–171.

Siotani, M. (1971). An asymptotic expansion of the non-null distribution of Hotelling's generalized T_0^2-statistics. *Ann. Math. Statist.*, **42**, 560–571.

Srivastava, M. S., and Khatri, C G. (1979) *An Introduction to Multivariate Statistics*. North-Holland Publ, Amsterdam.

Stein, C (1956a). Inadmissibility of the usual estimator for the mean of a multivariate normal distribution. In *Proceedings of the Third Berkeley Symposium on Mathematical Statistics and Probability*, Vol. 1, pp. 197–206. University of California Press, Berkeley.

Stein, C. (1956b). The admissibility of Hotelling's T^2-test. *Ann. Math. Statist.*, **27**, 616–623.

Stein, C (1969). Multivariate analysis. I. (Notes prepared by M. L. Eaton.) Stanford University Tech. Rept. No. 42, Stanford, CA.

Sternberg, S. (1964). *Lectures on Differential Geometry*. Prentice-Hall, Englewood Cliffs, N.J.

Styan, G. P. H (1970). Notes on the distribution of quadratic forms in singular normal variables. *Biometrika*, **57**, 567–572.

Subrahmaniam, K. (1975). On the asymptotic distributions of some statistics used for testing $\Sigma_1 = \Sigma_2$ *Ann. Statist.*, **3**, 916–925.

Sugiura, N. (1969a). Asymptotic non-null distributions of the likelihood ratio criteria for covariance matrix under local alternatives. Mimeo Ser. No. 609, Institute of Statistics, University of North Carolina, Chapel Hill.

Sugiura, N (1969b). Asymptotic expansions of the distributions of the likelihood ratio criteria for covariance matrix. *Ann. Math. Statist.*, **40**, 2051–2063.

Sugiura, N. (1972a). Asymptotic solutions of the hypergeometric function ${}_1F_1$ of matrix argument, useful in multivariate analysis *Ann Inst. Statist. Math.*, **24**, 517–524.

Sugiura, N. (1972b). Locally best invariant test for sphericity and the limiting distributions. *Ann. Math. Statist.*, **43**, 1312–1316.

Sugiura, N. (1973a). Further asymptotic formulas for the non-null distributions of three statistics for multivariate linear hypothesis. *Ann. Inst. Statist. Math.*, **25**, 153–163.

Sugiura, N. (1973b). Derivatives of the characteristic root of a symmetric or Hermitian matrix with two applications in multivariate analysis. *Comm. Statist.*, **1**, 393–417.

Sugiura, N. (1974). Asymptotic formulas for hypergeometric function ${}_2F_1$ of matrix argument, useful in multivariate analysis. *Ann. Inst. Statist. Math.*, **26**, 117–125.

Sugiura, N., and Fujikoshi, Y. (1969). Asymptotic expansions of the non-null distributions of the likelihood ratio criteria for multivariate linear hypothesis and independence. *Ann. Math. Statist.*, **40**, 942–952.

Sugiura, N., and Nagao, H. (1968). Unbiasedness of some test criteria for the equality of one or two covariance matrices. *Ann. Math. Statist.*, **39**, 1686–1692.

Sugiura, N., and Nagao, H. (1972). Asymptotic expansion of the distribution of the generalized variance for noncentral Wishart matrix, when $\Omega = O(n)$. *Ann. Inst. Statist. Math.*, **23**, 469–475.

Sugiyama, T. (1970). Joint distribution of the extreme roots of a covariance matrix. *Ann. Math. Statist.*, **41**, 655–657.

Sugiyama, T (1972). Percentile points of the largest latent root of a matrix and power calculations for testing $\Sigma = I$. *J. Japan. Statist. Soc.* **3**, 1–8.

Thompson, W. A. (1962). Estimation of dispersion parameters. *J. Res. Natl. Bur Standards Sec. B*, **66**, 161–164

Vinograd, B (1950). Canonical positive definite matrices under internal linear transformation. *Proc. Am Math. Soc.*, **1**, 159–161.

Waikar, V. B., and Schuurmann, F. J. (1973). Exact joint density of the largest and smallest roots of the Wishart and MANOVA matrices. *Utilitas Math.*, **4**, 253–260.

Waternaux, C. M. (1976). Asymptotic distribution of the sample roots for a nonnormal population. *Biometrika*, **63**, 639–645.

Watson, G. S. (1964). A note on maximum likelihood. *Sankhya*, **26A**, 303–304.

Weibull, M. (1953). The distribution of t- and F-statistics and of correlation and regression coefficients in stratified samples from normal populations with different means. *Skand. Aktuarietidskr.* (*Suppl.*), **36**, 1–106.

Wijsman, R. A. (1957). Random orthogonal transformations and their use in some classical distribution problems in multivariate analysis. *Ann. Math. Statist.*, **28**, 415–423

Wilks, S. S. (1932) Certain generalizations in the analysis of variance. *Biometrika*, **24**, 471–494.

Wilks, S. S. (1935). On the independence of k sets of normally distributed statistical variables. *Econometrika*, **3**, 309–326.

Wishart, J. (1928). The generalized product moment distribution in samples from a normal multivariate population. *Biometrika*, **20A**, 32–43.

Index

WILEY SERIES IN PROBABILITY AND STATISTICS

ESTABLISHED BY WALTER A. SHEWHART AND SAMUEL S. WILKS

Editors: *David J. Balding, Noel A. C. Cressie, Nicholas I. Fisher, Iain M. Johnstone, J. B. Kadane, Geert Molenberghs. Louise M. Ryan, David W. Scott, Adrian F. M. Smith, Jozef L. Teugels*

Editors Emeriti: *Vic Barnett, J. Stuart Hunter, David G. Kendall*

The ***Wiley Series in Probability and Statistics*** is well established and authoritative. It covers many topics of current research interest in both pure and applied statistics and probability theory. Written by leading statisticians and institutions, the titles span both state-of-the-art developments in the field and classical methods.

Reflecting the wide range of current research in statistics, the series encompasses applied, methodological and theoretical statistics, ranging from applications and new techniques made possible by advances in computerized practice to rigorous treatment of theoretical approaches.

This series provides essential and invaluable reading for all statisticians, whether in academia, industry, government, or research.

† ABRAHAM and LEDOLTER · Statistical Methods for Forecasting
AGRESTI · Analysis of Ordinal Categorical Data
AGRESTI · An Introduction to Categorical Data Analysis
AGRESTI · Categorical Data Analysis, *Second Edition*
ALTMAN, GILL, and McDONALD · Numerical Issues in Statistical Computing for the Social Scientist
AMARATUNGA and CABRERA · Exploration and Analysis of DNA Microarray and Protein Array Data
ANDĚL · Mathematics of Chance
ANDERSON · An Introduction to Multivariate Statistical Analysis, *Third Edition*
* ANDERSON · The Statistical Analysis of Time Series
ANDERSON, AUQUIER, HAUCK, OAKES, VANDAELE, and WEISBERG · Statistical Methods for Comparative Studies
ANDERSON and LOYNES · The Teaching of Practical Statistics
ARMITAGE and DAVID (editors) · Advances in Biometry
ARNOLD, BALAKRISHNAN, and NAGARAJA · Records
* ARTHANARI and DODGE · Mathematical Programming in Statistics
* BAILEY · The Elements of Stochastic Processes with Applications to the Natural Sciences
BALAKRISHNAN and KOUTRAS · Runs and Scans with Applications
BARNETT · Comparative Statistical Inference, *Third Edition*
BARNETT and LEWIS · Outliers in Statistical Data, *Third Edition*
BARTOSZYNSKI and NIEWIADOMSKA-BUGAJ · Probability and Statistical Inference
BASILEVSKY · Statistical Factor Analysis and Related Methods: Theory and Applications
BASU and RIGDON · Statistical Methods for the Reliability of Repairable Systems
BATES and WATTS · Nonlinear Regression Analysis and Its Applications
BECHHOFER, SANTNER, and GOLDSMAN · Design and Analysis of Experiments for Statistical Selection, Screening, and Multiple Comparisons
BELSLEY · Conditioning Diagnostics: Collinearity and Weak Data in Regression

*Now available in a lower priced paperback edition in the Wiley Classics Library.
†Now available in a lower priced paperback edition in the Wiley–Interscience Paperback Series.

† BELSLEY, KUH, and WELSCH · Regression Diagnostics: Identifying Influential Data and Sources of Collinearity
BENDAT and PIERSOL · Random Data: Analysis and Measurement Procedures, *Third Edition*
BERRY, CHALONER, and GEWEKE · Bayesian Analysis in Statistics and Econometrics: Essays in Honor of Arnold Zellner
BERNARDO and SMITH · Bayesian Theory
BHAT and MILLER · Elements of Applied Stochastic Processes, *Third Edition*
BHATTACHARYA and WAYMIRE · Stochastic Processes with Applications
† BIEMER, GROVES, LYBERG, MATHIOWETZ, and SUDMAN · Measurement Errors in Surveys
BILLINGSLEY · Convergence of Probability Measures, *Second Edition*
BILLINGSLEY · Probability and Measure, *Third Edition*
BIRKES and DODGE · Alternative Methods of Regression
BLISCHKE AND MURTHY (editors) · Case Studies in Reliability and Maintenance
BLISCHKE AND MURTHY · Reliability: Modeling, Prediction, and Optimization
BLOOMFIELD · Fourier Analysis of Time Series: An Introduction, *Second Edition*
BOLLEN · Structural Equations with Latent Variables
BOLLEN and CURRAN · Latent Curve Models: A Structural Equation Perspective
BOROVKOV · Ergodicity and Stability of Stochastic Processes
BOULEAU · Numerical Methods for Stochastic Processes
BOX · Bayesian Inference in Statistical Analysis
BOX · R. A. Fisher, the Life of a Scientist
BOX and DRAPER · Empirical Model-Building and Response Surfaces
* BOX and DRAPER · Evolutionary Operation: A Statistical Method for Process Improvement
BOX, HUNTER, and HUNTER · Statistics for Experimenters: Design, Innovation, and Discovery, *Second Editon*
BOX and LUCEÑO · Statistical Control by Monitoring and Feedback Adjustment
BRANDIMARTE · Numerical Methods in Finance: A MATLAB-Based Introduction
BROWN and HOLLANDER · Statistics: A Biomedical Introduction
BRUNNER, DOMHOF, and LANGER · Nonparametric Analysis of Longitudinal Data in Factorial Experiments
BUCKLEW · Large Deviation Techniques in Decision, Simulation, and Estimation
CAIROLI and DALANG · Sequential Stochastic Optimization
CASTILLO, HADI, BALAKRISHNAN, and SARABIA · Extreme Value and Related Models with Applications in Engineering and Science
CHAN · Time Series: Applications to Finance
CHARALAMBIDES · Combinatorial Methods in Discrete Distributions
CHATTERJEE and HADI · Sensitivity Analysis in Linear Regression
CHATTERJEE and PRICE · Regression Analysis by Example, *Third Edition*
CHERNICK · Bootstrap Methods: A Practitioner's Guide
CHERNICK and FRIIS · Introductory Biostatistics for the Health Sciences
CHILÈS and DELFINER · Geostatistics: Modeling Spatial Uncertainty
CHOW and LIU · Design and Analysis of Clinical Trials: Concepts and Methodologies, *Second Edition*
CLARKE and DISNEY · Probability and Random Processes: A First Course with Applications, *Second Edition*
* COCHRAN and COX · Experimental Designs, *Second Edition*
CONGDON · Applied Bayesian Modelling
CONGDON · Bayesian Statistical Modelling
CONOVER · Practical Nonparametric Statistics, *Third Edition*
COOK · Regression Graphics

*Now available in a lower priced paperback edition in the Wiley Classics Library.
†Now available in a lower priced paperback edition in the Wiley–Interscience Paperback Series.

COOK and WEISBERG · Applied Regression Including Computing and Graphics
COOK and WEISBERG · An Introduction to Regression Graphics
CORNELL · Experiments with Mixtures, Designs, Models, and the Analysis of Mixture Data, *Third Edition*
COVER and THOMAS · Elements of Information Theory
COX · A Handbook of Introductory Statistical Methods
* COX · Planning of Experiments
CRESSIE · Statistics for Spatial Data, *Revised Edition*
CSÖRGŐ and HORVÁTH · Limit Theorems in Change Point Analysis
DANIEL · Applications of Statistics to Industrial Experimentation
DANIEL · Biostatistics: A Foundation for Analysis in the Health Sciences, *Eighth Edition*
* DANIEL · Fitting Equations to Data: Computer Analysis of Multifactor Data, *Second Edition*
DASU and JOHNSON · Exploratory Data Mining and Data Cleaning
DAVID and NAGARAJA · Order Statistics, *Third Edition*
* DEGROOT, FIENBERG, and KADANE · Statistics and the Law
DEL CASTILLO · Statistical Process Adjustment for Quality Control
DEMARIS · Regression with Social Data: Modeling Continuous and Limited Response Variables
DEMIDENKO · Mixed Models: Theory and Applications
DENISON, HOLMES, MALLICK and SMITH · Bayesian Methods for Nonlinear Classification and Regression
DETTE and STUDDEN · The Theory of Canonical Moments with Applications in Statistics, Probability, and Analysis
DEY and MUKERJEE · Fractional Factorial Plans
DILLON and GOLDSTEIN · Multivariate Analysis: Methods and Applications
DODGE · Alternative Methods of Regression
* DODGE and ROMIG · Sampling Inspection Tables, *Second Edition*
* DOOB · Stochastic Processes
DOWDY, WEARDEN, and CHILKO · Statistics for Research, *Third Edition*
DRAPER and SMITH · Applied Regression Analysis, *Third Edition*
DRYDEN and MARDIA · Statistical Shape Analysis
DUDEWICZ and MISHRA · Modern Mathematical Statistics
DUNN and CLARK · Basic Statistics: A Primer for the Biomedical Sciences, *Third Edition*
DUPUIS and ELLIS · A Weak Convergence Approach to the Theory of Large Deviations
* ELANDT-JOHNSON and JOHNSON · Survival Models and Data Analysis
ENDERS · Applied Econometric Time Series
† ETHIER and KURTZ · Markov Processes: Characterization and Convergence
EVANS, HASTINGS, and PEACOCK · Statistical Distributions, *Third Edition*
FELLER · An Introduction to Probability Theory and Its Applications, Volume I, *Third Edition,* Revised; Volume II, *Second Edition*
FISHER and VAN BELLE · Biostatistics: A Methodology for the Health Sciences
FITZMAURICE, LAIRD, and WARE · Applied Longitudinal Analysis
* FLEISS · The Design and Analysis of Clinical Experiments
FLEISS · Statistical Methods for Rates and Proportions, *Third Edition*
† FLEMING and HARRINGTON · Counting Processes and Survival Analysis
FULLER · Introduction to Statistical Time Series, *Second Edition*
FULLER · Measurement Error Models
GALLANT · Nonlinear Statistical Models
GEISSER · Modes of Parametric Statistical Inference
GEWEKE · Contemporary Bayesian Econometrics and Statistics

*Now available in a lower priced paperback edition in the Wiley Classics Library.
†Now available in a lower priced paperback edition in the Wiley–Interscience Paperback Series.

GHOSH, MUKHOPADHYAY, and SEN · Sequential Estimation
GIESBRECHT and GUMPERTZ · Planning, Construction, and Statistical Analysis of Comparative Experiments
GIFI · Nonlinear Multivariate Analysis
GIVENS and HOETING · Computational Statistics
GLASSERMAN and YAO · Monotone Structure in Discrete-Event Systems
GNANADESIKAN · Methods for Statistical Data Analysis of Multivariate Observations, *Second Edition*
GOLDSTEIN and LEWIS · Assessment: Problems, Development, and Statistical Issues
GREENWOOD and NIKULIN · A Guide to Chi-Squared Testing
GROSS and HARRIS · Fundamentals of Queueing Theory, *Third Edition*
† GROVES · Survey Errors and Survey Costs
* HAHN and SHAPIRO · Statistical Models in Engineering
HAHN and MEEKER · Statistical Intervals: A Guide for Practitioners
HALD · A History of Probability and Statistics and their Applications Before 1750
HALD · A History of Mathematical Statistics from 1750 to 1930
† HAMPEL · Robust Statistics: The Approach Based on Influence Functions
HANNAN and DEISTLER · The Statistical Theory of Linear Systems
HEIBERGER · Computation for the Analysis of Designed Experiments
HEDAYAT and SINHA · Design and Inference in Finite Population Sampling
HELLER · MACSYMA for Statisticians
HINKELMANN and KEMPTHORNE · Design and Analysis of Experiments, Volume 1: Introduction to Experimental Design
HINKELMANN and KEMPTHORNE · Design and Analysis of Experiments, Volume 2: Advanced Experimental Design
HOAGLIN, MOSTELLER, and TUKEY · Exploratory Approach to Analysis of Variance
HOAGLIN, MOSTELLER, and TUKEY · Exploring Data Tables, Trends and Shapes
* HOAGLIN, MOSTELLER, and TUKEY · Understanding Robust and Exploratory Data Analysis
HOCHBERG and TAMHANE · Multiple Comparison Procedures
HOCKING · Methods and Applications of Linear Models: Regression and the Analysis of Variance, *Second Edition*
HOEL · Introduction to Mathematical Statistics, *Fifth Edition*
HOGG and KLUGMAN · Loss Distributions
HOLLANDER and WOLFE · Nonparametric Statistical Methods, *Second Edition*
HOSMER and LEMESHOW · Applied Logistic Regression, *Second Edition*
HOSMER and LEMESHOW · Applied Survival Analysis: Regression Modeling of Time to Event Data
† HUBER · Robust Statistics
HUBERTY · Applied Discriminant Analysis
HUNT and KENNEDY · Financial Derivatives in Theory and Practice
HUSKOVA, BERAN, and DUPAC · Collected Works of Jaroslav Hajek—with Commentary
HUZURBAZAR · Flowgraph Models for Multistate Time-to-Event Data
IMAN and CONOVER · A Modern Approach to Statistics
† JACKSON · A User's Guide to Principle Components
JOHN · Statistical Methods in Engineering and Quality Assurance
JOHNSON · Multivariate Statistical Simulation
JOHNSON and BALAKRISHNAN · Advances in the Theory and Practice of Statistics: A Volume in Honor of Samuel Kotz
JOHNSON and BHATTACHARYYA · Statistics: Principles and Methods, *Fifth Edition*

*Now available in a lower priced paperback edition in the Wiley Classics Library.
†Now available in a lower priced paperback edition in the Wiley–Interscience Paperback Series.

JOHNSON and KOTZ · Distributions in Statistics
JOHNSON and KOTZ (editors) · Leading Personalities in Statistical Sciences: From the Seventeenth Century to the Present
JOHNSON, KOTZ, and BALAKRISHNAN · Continuous Univariate Distributions, Volume 1, *Second Edition*
JOHNSON, KOTZ, and BALAKRISHNAN · Continuous Univariate Distributions, Volume 2, *Second Edition*
JOHNSON, KOTZ, and BALAKRISHNAN · Discrete Multivariate Distributions
JOHNSON, KOTZ, and KEMP · Univariate Discrete Distributions, *Third Edition*
JUDGE, GRIFFITHS, HILL, LÜTKEPOHL, and LEE · The Theory and Practice of Econometrics, *Second Edition*
JUREČKOVÁ and SEN · Robust Statistical Procedures: Aymptotics and Interrelations
JUREK and MASON · Operator-Limit Distributions in Probability Theory
KADANE · Bayesian Methods and Ethics in a Clinical Trial Design
KADANE AND SCHUM · A Probabilistic Analysis of the Sacco and Vanzetti Evidence
KALBFLEISCH and PRENTICE · The Statistical Analysis of Failure Time Data, *Second Edition*
KASS and VOS · Geometrical Foundations of Asymptotic Inference
† KAUFMAN and ROUSSEEUW · Finding Groups in Data: An Introduction to Cluster Analysis
KEDEM and FOKIANOS · Regression Models for Time Series Analysis
KENDALL, BARDEN, CARNE, and LE · Shape and Shape Theory
KHURI · Advanced Calculus with Applications in Statistics, *Second Edition*
KHURI, MATHEW, and SINHA · Statistical Tests for Mixed Linear Models
* KISH · Statistical Design for Research
KLEIBER and KOTZ · Statistical Size Distributions in Economics and Actuarial Sciences
KLUGMAN, PANJER, and WILLMOT · Loss Models: From Data to Decisions, *Second Edition*
KLUGMAN, PANJER, and WILLMOT · Solutions Manual to Accompany Loss Models: From Data to Decisions, *Second Edition*
KOTZ, BALAKRISHNAN, and JOHNSON · Continuous Multivariate Distributions, Volume 1, *Second Edition*
KOTZ and JOHNSON (editors) · Encyclopedia of Statistical Sciences: Volumes 1 to 9 with Index
KOTZ and JOHNSON (editors) · Encyclopedia of Statistical Sciences: Supplement Volume
KOTZ, READ, and BANKS (editors) · Encyclopedia of Statistical Sciences: Update Volume 1
KOTZ, READ, and BANKS (editors) · Encyclopedia of Statistical Sciences: Update Volume 2
KOVALENKO, KUZNETZOV, and PEGG · Mathematical Theory of Reliability of Time-Dependent Systems with Practical Applications
LACHIN · Biostatistical Methods: The Assessment of Relative Risks
LAD · Operational Subjective Statistical Methods: A Mathematical, Philosophical, and Historical Introduction
LAMPERTI · Probability: A Survey of the Mathematical Theory, *Second Edition*
LANGE, RYAN, BILLARD, BRILLINGER, CONQUEST, and GREENHOUSE · Case Studies in Biometry
LARSON · Introduction to Probability Theory and Statistical Inference, *Third Edition*
LAWLESS · Statistical Models and Methods for Lifetime Data, *Second Edition*
LAWSON · Statistical Methods in Spatial Epidemiology
LE · Applied Categorical Data Analysis
LE · Applied Survival Analysis

*Now available in a lower priced paperback edition in the Wiley Classics Library.
†Now available in a lower priced paperback edition in the Wiley–Interscience Paperback Series.

LEE and WANG · Statistical Methods for Survival Data Analysis, *Third Edition*
LEPAGE and BILLARD · Exploring the Limits of Bootstrap
LEYLAND and GOLDSTEIN (editors) · Multilevel Modelling of Health Statistics
LIAO · Statistical Group Comparison
LINDVALL · Lectures on the Coupling Method
LINHART and ZUCCHINI · Model Selection
LITTLE and RUBIN · Statistical Analysis with Missing Data, *Second Edition*
LLOYD · The Statistical Analysis of Categorical Data
LOWEN and TEICH · Fractal-Based Point Processes
MAGNUS and NEUDECKER · Matrix Differential Calculus with Applications in Statistics and Econometrics, *Revised Edition*
MALLER and ZHOU · Survival Analysis with Long Term Survivors
MALLOWS · Design, Data, and Analysis by Some Friends of Cuthbert Daniel
MANN, SCHAFER, and SINGPURWALLA · Methods for Statistical Analysis of Reliability and Life Data
MANTON, WOODBURY, and TOLLEY · Statistical Applications Using Fuzzy Sets
MARCHETTE · Random Graphs for Statistical Pattern Recognition
MARDIA and JUPP · Directional Statistics
MASON, GUNST, and HESS · Statistical Design and Analysis of Experiments with Applications to Engineering and Science, *Second Edition*
McCULLOCH and SEARLE · Generalized, Linear, and Mixed Models
McFADDEN · Management of Data in Clinical Trials
* McLACHLAN · Discriminant Analysis and Statistical Pattern Recognition
McLACHLAN, DO, and AMBROISE · Analyzing Microarray Gene Expression Data
McLACHLAN and KRISHNAN · The EM Algorithm and Extensions
McLACHLAN and PEEL · Finite Mixture Models
McNEIL · Epidemiological Research Methods
MEEKER and ESCOBAR · Statistical Methods for Reliability Data
MEERSCHAERT and SCHEFFLER · Limit Distributions for Sums of Independent Random Vectors: Heavy Tails in Theory and Practice
MICKEY, DUNN, and CLARK · Applied Statistics: Analysis of Variance and Regression, *Third Edition*
* MILLER · Survival Analysis, *Second Edition*
MONTGOMERY, PECK, and VINING · Introduction to Linear Regression Analysis, *Third Edition*
MORGENTHALER and TUKEY · Configural Polysampling: A Route to Practical Robustness
MUIRHEAD · Aspects of Multivariate Statistical Theory
MULLER and STOYAN · Comparison Methods for Stochastic Models and Risks
MURRAY · X-STAT 2.0 Statistical Experimentation, Design Data Analysis, and Nonlinear Optimization
MURTHY, XIE, and JIANG · Weibull Models
MYERS and MONTGOMERY · Response Surface Methodology: Process and Product Optimization Using Designed Experiments, *Second Edition*
MYERS, MONTGOMERY, and VINING · Generalized Linear Models. With Applications in Engineering and the Sciences
† NELSON · Accelerated Testing, Statistical Models, Test Plans, and Data Analyses
† NELSON · Applied Life Data Analysis
NEWMAN · Biostatistical Methods in Epidemiology
OCHI · Applied Probability and Stochastic Processes in Engineering and Physical Sciences
OKABE, BOOTS, SUGIHARA, and CHIU · Spatial Tesselations: Concepts and Applications of Voronoi Diagrams, *Second Edition*
OLIVER and SMITH · Influence Diagrams, Belief Nets and Decision Analysis

*Now available in a lower priced paperback edition in the Wiley Classics Library.
†Now available in a lower priced paperback edition in the Wiley–Interscience Paperback Series.

PALTA · Quantitative Methods in Population Health: Extensions of Ordinary Regressions
PANKRATZ · Forecasting with Dynamic Regression Models
PANKRATZ · Forecasting with Univariate Box-Jenkins Models: Concepts and Cases
* PARZEN · Modern Probability Theory and Its Applications
PEÑA, TIAO, and TSAY · A Course in Time Series Analysis
PIANTADOSI · Clinical Trials: A Methodologic Perspective
PORT · Theoretical Probability for Applications
POURAHMADI · Foundations of Time Series Analysis and Prediction Theory
PRESS · Bayesian Statistics: Principles, Models, and Applications
PRESS · Subjective and Objective Bayesian Statistics, *Second Edition*
PRESS and TANUR · The Subjectivity of Scientists and the Bayesian Approach
PUKELSHEIM · Optimal Experimental Design
PURI, VILAPLANA, and WERTZ · New Perspectives in Theoretical and Applied Statistics
† PUTERMAN · Markov Decision Processes: Discrete Stochastic Dynamic Programming
QIU · Image Processing and Jump Regression Analysis
* RAO · Linear Statistical Inference and Its Applications, *Second Edition*
RAUSAND and HØYLAND · System Reliability Theory: Models, Statistical Methods, and Applications, *Second Edition*
RENCHER · Linear Models in Statistics
RENCHER · Methods of Multivariate Analysis, *Second Edition*
RENCHER · Multivariate Statistical Inference with Applications
* RIPLEY · Spatial Statistics
RIPLEY · Stochastic Simulation
ROBINSON · Practical Strategies for Experimenting
ROHATGI and SALEH · An Introduction to Probability and Statistics, *Second Edition*
ROLSKI, SCHMIDLI, SCHMIDT, and TEUGELS · Stochastic Processes for Insurance and Finance
ROSENBERGER and LACHIN · Randomization in Clinical Trials: Theory and Practice
ROSS · Introduction to Probability and Statistics for Engineers and Scientists
† ROUSSEEUW and LEROY · Robust Regression and Outlier Detection
* RUBIN · Multiple Imputation for Nonresponse in Surveys
RUBINSTEIN · Simulation and the Monte Carlo Method
RUBINSTEIN and MELAMED · Modern Simulation and Modeling
RYAN · Modern Regression Methods
RYAN · Statistical Methods for Quality Improvement, *Second Edition*
SALTELLI, CHAN, and SCOTT (editors) · Sensitivity Analysis
* SCHEFFE · The Analysis of Variance
SCHIMEK · Smoothing and Regression: Approaches, Computation, and Application
SCHOTT · Matrix Analysis for Statistics, *Second Edition*
SCHOUTENS · Levy Processes in Finance: Pricing Financial Derivatives
SCHUSS · Theory and Applications of Stochastic Differential Equations
SCOTT · Multivariate Density Estimation: Theory, Practice, and Visualization
* SEARLE · Linear Models
SEARLE · Linear Models for Unbalanced Data
SEARLE · Matrix Algebra Useful for Statistics
SEARLE, CASELLA, and McCULLOCH · Variance Components
SEARLE and WILLETT · Matrix Algebra for Applied Economics
SEBER and LEE · Linear Regression Analysis, *Second Edition*
† SEBER · Multivariate Observations
† SEBER and WILD · Nonlinear Regression
SENNOTT · Stochastic Dynamic Programming and the Control of Queueing Systems
* SERFLING · Approximation Theorems of Mathematical Statistics

*Now available in a lower priced paperback edition in the Wiley Classics Library.
†Now available in a lower priced paperback edition in the Wiley–Interscience Paperback Series.

SHAFER and VOVK · Probability and Finance: It's Only a Game!
SILVAPULLE and SEN · Constrained Statistical Inference: Inequality, Order, and Shape Restrictions
SMALL and McLEISH · Hilbert Space Methods in Probability and Statistical Inference
SRIVASTAVA · Methods of Multivariate Statistics
STAPLETON · Linear Statistical Models
STAUDTE and SHEATHER · Robust Estimation and Testing
STOYAN, KENDALL, and MECKE · Stochastic Geometry and Its Applications, *Second Edition*
STOYAN and STOYAN · Fractals, Random Shapes and Point Fields: Methods of Geometrical Statistics
STYAN · The Collected Papers of T. W. Anderson: 1943–1985
SUTTON, ABRAMS, JONES, SHELDON, and SONG · Methods for Meta-Analysis in Medical Research
TANAKA · Time Series Analysis: Nonstationary and Noninvertible Distribution Theory
THOMPSON · Empirical Model Building
THOMPSON · Sampling, *Second Edition*
THOMPSON · Simulation: A Modeler's Approach
THOMPSON and SEBER · Adaptive Sampling
THOMPSON, WILLIAMS, and FINDLAY · Models for Investors in Real World Markets
TIAO, BISGAARD, HILL, PEÑA, and STIGLER (editors) · Box on Quality and Discovery: with Design, Control, and Robustness
TIERNEY · LISP-STAT: An Object-Oriented Environment for Statistical Computing and Dynamic Graphics
TSAY · Analysis of Financial Time Series, *Second Edition*
UPTON and FINGLETON · Spatial Data Analysis by Example, Volume II: Categorical and Directional Data
VAN BELLE · Statistical Rules of Thumb
VAN BELLE, FISHER, HEAGERTY, and LUMLEY · Biostatistics: A Methodology for the Health Sciences, *Second Edition*
VESTRUP · The Theory of Measures and Integration
VIDAKOVIC · Statistical Modeling by Wavelets
VINOD and REAGLE · Preparing for the Worst: Incorporating Downside Risk in Stock Market Investments
WALLER and GOTWAY · Applied Spatial Statistics for Public Health Data
WEERAHANDI · Generalized Inference in Repeated Measures: Exact Methods in MANOVA and Mixed Models
WEISBERG · Applied Linear Regression, *Third Edition*
WELSH · Aspects of Statistical Inference
WESTFALL and YOUNG · Resampling-Based Multiple Testing: Examples and Methods for *p*-Value Adjustment
WHITTAKER · Graphical Models in Applied Multivariate Statistics
WINKER · Optimization Heuristics in Economics: Applications of Threshold Accepting
WONNACOTT and WONNACOTT · Econometrics, *Second Edition*
WOODING · Planning Pharmaceutical Clinical Trials: Basic Statistical Principles
WOODWORTH · Biostatistics: A Bayesian Introduction
WOOLSON and CLARKE · Statistical Methods for the Analysis of Biomedical Data, *Second Edition*
WU and HAMADA · Experiments: Planning, Analysis, and Parameter Design Optimization
YANG · The Construction Theory of Denumerable Markov Processes
* ZELLNER · An Introduction to Bayesian Inference in Econometrics
ZHOU, OBUCHOWSKI, and McCLISH · Statistical Methods in Diagnostic Medicine

*Now available in a lower priced paperback edition in the Wiley Classics Library.
†Now available in a lower priced paperback edition in the Wiley–Interscience Paperback Series.

9 780471 769859